THE OXFORD HANDBOOK OF

# PHILOSOPHY
# OF TIME

# THE OXFORD HANDBOOK OF

# PHILOSOPHY OF TIME

*Edited by*
CRAIG CALLENDER

OXFORD
UNIVERSITY PRESS

# OXFORD

UNIVERSITY PRESS

Great Clarendon Street, Oxford OX2 6DP

Oxford University Press is a department of the University of Oxford.
It furthers the University's objective of excellence in research, scholarship,
and education by publishing worldwide in

Oxford  New York

Auckland  Cape Town  Dar es Salaam  Hong Kong  Karachi
Kuala Lumpur  Madrid  Melbourne  Mexico City  Nairobi
New Delhi  Shanghai  Taipei  Toronto

With offices in

Argentina  Austria  Brazil  Chile  Czech Republic  France  Greece
Guatemala  Hungary  Italy  Japan  Poland  Portugal  Singapore
South Korea  Switzerland  Thailand  Turkey  Ukraine  Vietnam

Oxford is a registered trade mark of Oxford University Press
in the UK and in certain other countries

Published in the United States
by Oxford University Press Inc., New York

British Library Cataloguing in Publication Data

Data available

Library of Congress Cataloging in Publication Data

Data available

Typeset by SPI Publisher Services, Pondicherry, India
Printed in Great Britain
on acid-free paper by
MPG Books Group, Bodmin and King's Lynn

ISBN  978–0–19–929820–4

1 3 5 7 9 10 8 6 4 2

# CONTENTS

## PART I  TIME AND METAPHYSICS

## PART II  THE DIRECTION OF TIME

# PART V  TIME IN A QUANTUM WORLD

# LIST OF FIGURES

# LIST OF CONTRIBUTORS

**Frank Arntzenius** is Professor of Philosophy at Oxford University and is the Sir Peter Strawson Fellow at University College, Oxford. His research, so far, has mostly been in philosophy of physics and in decision theory. He just finished writing a book on the structure of space and time, which will come out with Oxford University Press. He is now switching his research area to political philosophy, and is particularly interested in peace studies and peace activism.

**David Atkinson** studied at Cambridge University; in 1972 he became Professor of Theoretical Physics at the University of Groningen, the Netherlands. He has written four text books on quantum mechanics and quantum field theory (Rinton Press, 2001–2004), and published many articles on time, thought experiments and confirmation. His recent interests centre on infinity in physics and the philosophy of probability.

**Yuri Balashov** is Professor of Philosophy at the University of Georgia, US. His interests are in analytic metaphysics and philosophy of science. He is the author of *Persistence and Spacetime* (Oxford, 2010), a co-editor of *Einstein Studies in Russia* (Birkhäuser, 2002) and *Philosophy of Science: Contemporary Readings* (Routledge, 2002), and has published in major philosophy journals, such as *Noûs*, *Philosophical Studies*, *American Philosophical Quarterly*, *Australasian Journal of Philosophy*, *Philosophical Quarterly*, *Monist*, and *British Journal for the Philosophy of Science*.

**Craig Bourne** is Senior Lecturer in Philosophy at the University of Hertfordshire. His research interests centre on metaphysics, logic, and the philosophical implications of science. In the philosophy of time, he has published a number of articles and a book *A Future for Presentism* (Oxford University Press 2006). He is currently working on the representation of time in logic and in fiction.

**David O. Brink** is Professor of Philosophy at the University of California, San Diego and Director of the Institute for Law and Philosophy at the University of San Diego School of Law. His research interests are in ethical theory, history of ethics, and jurisprudence. He is the author of *Moral Realism and the Foundations of Ethics* (Cambridge, 1989) and *Perfectionism and the Common Good: Themes in the Philosophy of T.H. Green* (Oxford, 2003).

**Barry Dainton** is a Professor of Philosophy at the University of Liverpool. His research is mainly in metaphysics and the philosophy of mind. He is the author of *The*

*Phenomenal Self* (OUP, 2008), *Stream of Consciousness* (Routledge, 2006 [2000]), and *Time and Space* (Acumen, 2001).

**John Earman** is Distinguished University Professor Emeritus of History and Philosophy of Science at the University of Pittsburgh. His research interests center on the history, methodology, and foundations of modern physics. He is the author of *A Primer on Determinism*, *World Enough and Space-Time: Absolute vs. Relational Theories of Space and Time*, and *Bangs, Crunches, Whimpers, and Shrieks: Singularities and Acausalities in Relativistic Spacetimes*. He is co-editor with Jeremy Butterfield of *Handbook of the Philosophy of Science. Philosophy of Physics*.

**Shaun Gallagher** is Professor of Philosophy and Cognitive Sciences at The University of Central Florida. He is Editor of *Phenomenology and the Cognitive Sciences*, an interdisciplinary journal published by Springer. His research interests include phenomenology and the philosophy of mind, philosophical psychology, embodiment, intersubjectivity, hermeneutics, and the philosophy of time.

**Jan Hilgevoord** studied theoretical physics at the University of Amsterdam. From 1967 to 1987 he was Professor of Theoretical Physics in Amsterdam, and from 1987 until his retirement in 1992 Professor in the Philosophy of the Exact Sciences at the University of Utrecht. He edited *Physics and Our View of the World* (Cambridge University Press, 1994) and wrote many articles on the foundations of quantum mechanics.

**Carl Hoefer** is currently an ICREA Research Professor at the Autonomous University of Barcelona (UAB), and has also taught at the University of California (Riverside) and the London School of Economics. His primary interest is in the metaphysics of nature, and what we can learn about it from physics.

**Christoph Hoerl** is in the Department of Philosophy at the University of Warwick. His research interests include philosophy of mind, epistemology, memory, time, and causal understanding.

**Jenann Ismael** is Queen Elizabeth II Research Fellow at the Centre for Time at the University of Sydney and Associate Professor of Philosophy at the University of Arizona. Her work focuses on philosophy of physics with related interests in questions about mind and the nature of perspective. She has published two books and numerous articles.

**Claus Kiefer** is a Professor of Theoretical Physics at the University of Cologne, Germany. He earned his PhD from Heidelberg University in 1988. He has held positions at the Universities of Heidelberg, Zurich, and Freiburg, and was an invited visitor to the Universities of Alberta, Bern, Cambridge, Montpellier, and others. His main interests include quantum gravity, cosmology, black holes, and the foundations of quantum theory, and he has authored several books including the monograph *Quantum Gravity* (second edition: Oxford 2007).

**Douglas Kutach** is Assistant Professor of Philosophy at Brown University. His research ranges over topics in the philosophy of physics and metaphysics. He is an advocate for Empirical Fundamentalism, a naturalistic program based on exploiting a distinction between fundamental and derivative reality to solve a broad array of philosophical problems. He is the author of *Causation and Its Basis in Fundamental Physics* as well as articles on causal asymmetry, reduction, and philosophical methodology.

**Jean-Pierre Luminet** is an astrophysicist at the Paris-Meudon Observatory in France and a leading expert on black holes and cosmology. He has published numerous articles in the most prestigious journals and reviews in these areas. He was awarded many prizes for his work in pure science and in science communication. Luminet has produced more than twenty books, including *Black Holes* (Cambridge University Press, 1992) and *The Wraparound Universe* (AK Peters 2005), as well as historical novels, poetry, and TV documentaries.

**Teresa McCormack** is Professor of Developmental Psychology at Queen's University, Belfast. Her research focuses on temporal and causal cognition and its development in children. She is the co-editor of two interdisciplinary volumes, *Time and Memory: Issues in Philosophy and Psychology* (Oxford University Press, 2001), and *Joint Attention: Issues in Philosophy and Psychology* (Oxford University Press, 2005). A further volume, *Understanding Counterfactuals, Understanding Causation: Issues in Philosophy and Psychology*, will be published by Oxford University Press in 2011. She has also published numerous empirical papers on children's cognitive development.

**Ulrich Meyer** is Associate Professor of Philosophy at Colgate University. His research interests include metaphysics, logic, and the philosophy of science, with a special focus on the philosophy of time. His first book, *The Nature of Time*, is being published by Clarendon Press, Oxford.

**M. Joshua Mozersky** is Associate Professor of Philosophy at Queen's University in Kingston. He works in metaphysics, the philosophy of language and the philosophy of science. The philosophy of time, which is found at the intersection of these three areas, is a long-standing and continuing interest, though current projects also include: truthmaking and modality, the metaphysics of objects, the nature of properties and the question of just what metaphysics is, or could be, after all.

**Jill North** is an Assistant Professor in the Philosophy and physics departments at Yale, where she also studied physics and philosophy as an undergraduate. Her research interests include philosophy of physics, metaphysics, and philosophy of science.

**Huw Price** is ARC Federation Fellow, Challis Professor of Philosophy, and Director of the Centre for Time at the University of Sydney. His publications include *Facts and the Function of Truth* (Blackwell, 1988), *Time's Arrow and Archimedes' Point* (OUP, 1996), and *Naturalism Without Mirrors* (OUP, 2010), a recent collection of his essays on pragmatism and naturalism. He is also co-editor (with Richard Corry) of *Causation, Physics, and the Constitution of Reality: Russell's Republic Revisited* (OUP, 2007).

**Steven Savitt** is a Professor of Philosophy at the University of British Columbia in Vancouver, Canada. He received his PhD from Brandeis University under the supervision of Jean van Heijenoort and began his career as a philosopher of logic. He was subsequently captured by the philosophy of time and has been twisting in its coils ever since. Currently he is collaborating on a monograph, *The Now in Physics*, with Richard Arthur and Dennis Dieks.

**Lawrence Sklar** is the Carl G. Hempel and William K. Frankena Distinguished University Professor of Philosophy at the University of Michigan. He has published work in the philosophy of space and time, the foundations of statistical mechanics, and the methodology of physical science. Some of his published books are *Space, Time and Spacetime*, *Physics and Chance*, *Theory and Truth*, *Philosophy of Physics*, and *Philosophy and Spacetime Physics*.

**Christopher Smeenk** is an Assistant Professor in Philosophy at the University of Western Ontario. He received a B.A. degree in Physics and Philosophy from Yale University in 1995, and pursued graduate studies at the University of Pittsburgh leading to a PhD in History and Philosophy of Science in 2003. Prior to arriving at UWO, he held a post-doctoral fellowship at the Dibner Institute for History of Science and Technology (MIT) and was an assistant professor in the Department of Philosophy at UCLA (2003–2007). His main research interests are history and philosophy of physics, general issues in philosophy of science, and seventeenth-century natural philosophy.

**Jean Paul Van Bendegem** is Professor at the Vrije Universiteit Brussel (Free University of Brussels) where he teaches courses in logic and philosophy of science. His research focuses on two themes: the philosophy of strict finitism and the development of a comprehensive theory of mathematical practice. Among recent publications are *Perspectives on Mathematical Practices* (co-editor Bart Van Kerkhove, Springer, 2006) and *Mathematical Arguments in Context* (co-author Bart Van Kerkhove, Foundations of Science, 2009).

**Christian Wüthrich** is an Assistant Professor of Philosophy and Science Studies at the University of California, San Diego. He received his PhD in History and Philosophy of Science from the University of Pittsburgh after having read physics, mathematics, philosophy, and history and philosophy of science at Bern, Cambridge, and Pittsburgh. He works in philosophy and history of physics, philosophy of science and the metaphysics of science. He has published in various journals in philosophy and in physics.

**Dean Zimmerman** is Professor of Philosophy at Rutgers University. His research interests include the nature of time and persistence, and God's relation to temporal things. He is editor or co-editor of several books in metaphysics and philosophy of religion, including: *The Oxford Handbook of Metaphysics* (Oxford, 2003), *Persons: Human and Divine* (Oxford, 2007), *Contemporary Debates in Metaphysics* (Blackwell, 2007), and the series *Oxford Studies in Metaphysics*.

# INTRODUCTION

TIME is one of the last great mysteries. Like space, it is part of the fundamental stage upon which the events of the world play out. We know it only indirectly through the movement of the actors upon it. For these reasons time seems deep and remote, just as space does. However, we also find in experience a number of important features that distinguish time from space and that are crucial to the way we live our lives. We remember the past but not the future, believe that our actions cause effects in the future but not the past, and prefer that our pains be past and our pleasures future. Indeed, life is ticking by, as our bodies are clocks set to expire in about eighty years whereas we have no such spatial "clocks" As a result of these features, time is often said to be flowing, the present special, the future open, and the past fixed. Whether any of this talk is correct is of course controversial, but the fact remains that it's almost irresistible to think this way about time. While time as a parameter of the arena of the world seems remote, time is also associated with many features that shape the way we live our lives. It is this juxtaposition between its remoteness and familiarity that makes time one of the great mysteries—and the study of time especially captivating.

Today there is little need to sell the subject of time. Topics connected to time seem more popular than ever, not just in rarefied academic circles but also in the popular imagination. Popular science magazines, television documentaries, bookshelves, and blogs are filled with discussion of the possible flow of time, whether time travel is possible, whether time has a beginning or ending, the perception of time, the direction of time, and even whether time fundamentally exists.

None of this interest should be surprising, for this reporting simply reflects what's going on in science and the humanities. The fields of time perception in cognitive science, time awareness in psychology, the direction of time in statistical physics and cosmology, the nature of time in quantum gravity, chronobiology in the life sciences, and time-keeping in society, in addition to many other areas, are fields that have exploded in popularity in recent times.

Future sociologists can perhaps explain this surge of interest in time. I'll hazard no guesses here. However, no doubt as part cause and part effect of this surge, philosophy of time has also experienced an increase of activity and attention. Whereas it was once a respectable but small niche in metaphysics, it has of late grown almost into a major area in its own right. More and more articles, books, and dissertations are written on the subject. What's striking about this work is that while the central issues in metaphysics have probably seen the greatest surge in activity,[1] analytic philosophers have also expanded their repertoire. Two of the more obvious areas of growth are in the phenomenology of time—how we experience time, how it's connected to the self, and more—and in the study of the science of time and its perception, from cognitive neuroscience to physics.

In philosophy, time has always been an especially challenging topic. At root, the problem is that quintessential difficulty that so often motivates philosophical discussion: the problem of disentangling the nature of the entity itself from features that we happen to attribute to it. Better put, it's hard to tease apart our egocentric representation of time from a more "objective" representation of time. Is time itself responsible for the causal asymmetry, the specialness of the present, and so on? This question, in many guises, has been considered for millennia, from the ancient pre-Socratics and Aristotle to the medieval Augustine to contemporary scientists and philosophers. Making progress on this topic therefore requires work in philosophy and the sciences on both the egocentric side of things as well as the objective side of things. This *Handbook* makes substantial progress on both fronts.

In editing this collection, I sought as best as I could to give the *Handbook* a "prospective" look at the field rather than a "retrospective" look at the field's past. Thus, while mindful of and desiring to represent the good work done in central traditional areas, I also tried to accentuate a lot of the new work and topics being investigated in philosophy of time. Thus I sought out chapters on time and ethics, time and probability, the fascinating CPT theorem, time and action, the possibility of time being discrete, and many others. Even where this *Handbook* covers traditional topics, often they are done with some novel slant. One chapter on the direction of time focuses on the electromagnetic arrow of time, a topic little discussed or understood. One chapter on presentism develops a new view of its relationship with relativity. The chapter on the twin paradox develops it within topology, promising something new to even those familiar with the paradox. The one on time travel focuses more on time machines than the typical "grandfather paradoxes." And so on. Part of what helped me achieve this range of topics was the fact that I wanted to avoid redundancy with the chapters on time and metaphysics in the *Handbook of Contemporary Philosophy* and *Handbook of Metaphysics*, plus the one on time and tense in the *Handbook of Philosophy of Language*. In any case, only the future will tell whether some of my guesses about it are correct.

Let me now describe what is to come.

---

[1] A JSTOR search for the keyword 'presentism' across 63 philosophy journals yields 20 matches for 1980–1990 and 146 for 2000–2010.

# TIME AND METAPHYSICS

Is there more to the world than the present moment? This question is one of the most basic one can ask about time. Some metaphysicians ("eternalists") believe that the past, present, and future are all real, others ("possibilists"), believe the past and present are real but the future is not, and yet a third group ("presentists") hold that only the present is real. This dispute runs throughout many of the chapters in this *Handbook* (as well as the *Handbook on Metaphysics*), but the two chapters devoted specifically to this topic are ones by Dean Zimmerman and by Joshua Mozersky. Zimmerman seeks to defend a version of presentism. Among the challenges that presentism faces is a prima facie conflict with relativity theory in physics. Relativity theory affirms the relativity of simultaneity, a thesis that immediately threatens presentism. (See Savitt's and Luminet's contributions on this topic, discussed later.) Zimmerman develops a theory that he hopes will escape this difficulty. Mozersky reviews a quite different attack on presentism. Called the "truth-maker" or "grounding" objection, the idea is that the presentist's temporally impoverished resources are insufficient to allow him or her to say that statements about the past or future are truth-evaluable. The claim that "there were dinosaurs in the Mesozoic Age" is not made true or false by anything on the present time slice, or at least, not obviously so. So the presentist faces a quandary: the claim is true, yet there is nothing that makes it so. Mozersky, in contrast to Zimmerman, does not see a viable way out of this problem for the presentist.

The above problem arose from asserting that the past and future are unreal, but holding them to be real also presents puzzles. Focusing specifically on whether the future is real, a pair of natural questions arise. If the future is real, and our choices are "already" made, in what sense can we do otherwise? If I was "always" a philosopher in 2010, did I really have a choice about professions in 1990? Similarly, if the future is real, and the outcomes of our chancy processes are "already" occurrent, then in what sense is the chancy process genuinely chancy? If the coin "always" landed heads at a specific toss, in what sense did that roll have a 50 per cent chance of landing tails? The first topic is the notorious fatalism issue. Discussed by Aristotle more than two millennia ago, the question is whether various logical principles, when applied to propositions about future events, imply that the future is in some sense fixed—so that you couldn't have done otherwise. If the future is like the past, and the past is fixed, then are future events fixed in the same sense? Craig Bourne gives the latest on this traditional topic, carefully surveying various works and trying a new tack that steers us away from fatalism. The second question is not as venerable as the first but is just as interesting. Advocates of the propensity interpretation conceive of probabilities as dispositions or tendencies inherent in certain physical situations that cause a particular pattern of outcomes. In his contribution, Carl Hoefer argues that the propensity interpretation of probability is implicitly connected to the assumption of an unreal future. Thinking that an unreal future is untenable in light of relativity theory, Hoefer argues that the intuitive "boost" propensities get by this connection is illegitimate. Once sanitized of the connection, Hoefer claims that propensities lose much of their appeal.

Regardless of whether the future or past are real, there is also the question of what objects (like you) are. What is it for an object to persist in time? At one time you are short and at another time tall. Are you a four-dimensional object with different temporal parts—a short part and a tall part? Or are you the wholly present three-dimensional entity who changes properties? Or are there other options? In his essay, Yuri Balashov gives a clear and admirably rich survey of this problem, updating three positions on this problem to the relativistic context, and providing the reader with a solid base from which to later evaluate the positions.

On the eternalist view of time, other non-present times are like other spatial locations that aren't here. Times are like locations. But a very different picture emerges if one thinks instead of other times as analogous to other possible worlds in modal logic. Modal logic is the logic of possibility and necessity. Even if I don't help you move out of your house, you might still insist that it was *possible* for me to have done so. In other words, a possible world exists wherein I do help you move. Modal logic is the set of rules that seem to govern our inferences about such possibilities. From it we can also construct a tense logic, whereby "possibly" is interpreted as "sometimes", "necessary" as "always", and so on. This logic can be developed in various ways and put to various uses. It might be thought to provide a linguistic theory of time—describing the way we ordinarily talk about time—or underpin a new metaphysical picture of times as like possible worlds. In his chapter, Ulrich Meyer argues against the first use and develops the second.

Time is usually assumed to be continuous. No doubt the reason for this is that science generally takes time to be continuous. But the reasons it does so are not directly empirical. That is, no experiment has been done that directly tests whether time is continuous, dense, or discrete. Rather, time is considered continuous because our best theories of the universe say it is, but they do so mostly because it is much easier to write a theory using notions from calculus this way. Could time instead be discrete? What reasons might there be for thinking so, and what puzzles appear when studying this possiblility? In a very original essay, van Bendegem tackles these questions and others.

# THE DIRECTION OF TIME

The future seems very different from the past. These differences are manifested in various ways.

One way is that we, in some sense, know more about the past than the future. With the simple act of looking in yesterday's paper I can now discover yesterday's stock report. But short of a crystal ball, there is no way I can today determine tomorrow's stock reports. Today's newspaper may contain some information about the future— tomorrow's tides, a street fair next Saturday—but this information is very limited. By contrast, going in the past direction, I have access to a much richer kind of information, such as the nearly exact tide levels, how many people attended the street fair, and so on.

Another temporal asymmetry is what we might call the causation or mutability asymmetry. Actions right now can cause me later to eat in a different restaurant than I otherwise would have; but nothing I do now can cause me to have eaten at a different restaurant than I did last week. I can change where I will die, but I can't change where I was born. Causes typically have their effects in the future, not the past. Another way to think about this is via counterfactual dependence. The future depends counterfactually on the present in a way the past seems not to. This asymmetry leads us to believe that the future is open in a way the past is not.

Our valuations of events and objects also manifest a temporal asymmetry. Given a choice in the matter, I would rather *have had* a bad headache than be *about to have* a bad headache. I want my pains in the past and pleasures in the future.

There are other temporal asymmetries, but the above list is sufficient to motivate a set of questions. What is the nature of these asymmetries? How are these asymmetries related, if at all? How are they related to various temporal asymmetries found in physics? Are these asymmetries in time reflecting some kind of asymmetry of time?

Douglas Kutach, in his contribution, considers the nature of the causal asymmetry, or even more generally, the asymmetry of influence. Putting aside explanations that would appeal to an asymmetry in time as explaining this asymmetry, Kutach hopes to show, using current physical theory and no ad hoc time asymmetric assumptions, why it is that future-directed influence sometimes advances our goals, but backward-directed influence does not. The article claims that our agency is crucial to the explanation of the influence asymmetry.

Might the explanation of some temporal asymmetries simply be that time itself is asymmetric? Some people believe that time flows, and others that it is intrinsically directed. But what do such claims mean, precisely? In his chapter, Huw Price considers three ways of understanding flow—through a distinguished present, an objective temporal direction, and a flux-like character—and finds them all wanting. He spends considerable time evaluating, in particular, the idea that the world possesses a time orientation, critically scrutinizing the ideas of John Earman and Tim Mauldin on temporal orientation and time's arrow.

It's often claimed, or hoped, that some of the above temporal asymmetries are explained by the thermodynamic asymmetry in time. Thermodynamics, the macroscopic physics of pressure, temperature, volume, and so on, describes many temporally asymmetric processes. Heat flows spontaneously from hot objects to cold objects (in closed systems), never the reverse. More generally, systems spontaneously move from non-equilibrium states to equilibrium states, never the reverse. What explains the thermodynamic temporal asymmetries? To some it will come as a surprise that this is itself highly controversial. Delving into the foundations of statistical mechanics, Jill North reviews the many open questions in that field as they relate to temporal asymmetry. Taking a stand on many of them, she tackles questions about the nature of probabilities, the role of boundary conditions, and even the nature and scope of statistical mechanics.

# TIME, ETHICS, AND EXPERIENCE

Whatever the true metaphysics of time is, it's clear that time plays a complicated, fascinating role in the way we experience the world and in the way we live our lives. Our temporal experience is richly structured psychologically. Not only do we have memories of the past and anticipations of the future, but it also seems to be the case that the contents of our experiences at a given time represent temporally extended events. Moreover, the timing of events matters to us ethically, although whether it should is another question. The chapters in this Part focus on these aspects of temporal experience, among others.

Four chapters focus directly on our experience of time, and three bounce off from William James' so-called "specious present." The specious present is the claimed temporal breadth in the content of an experience at a particular time. The content of my experience—right now—of a musical note has temporal breadth. And the same for the one "after" that, and so on. Our experience has a stream-like aspect to it.

But how can this be, asks Barry Dainton, if the world itself does not pass? That is, why does my experience have this stream-like quality to it when (as he assumes) time is itself not flowing? While making the eternalist world safe for the specious present, Dainton carves out and evaluates two contrasting understandings of the specious present.

Jenann Ismael begins her analysis with a careful look at the specious present, but this is only the beginning of her survey of many of the psychological temporal structures that arise in creatures like us. She also examines memory, anticipation, and the building up of our experience through time, focusing especially on the contrast between time from an "embedded" perspective and time from an external perspective. She ends with some exciting suggestions for how this work may link to our conception of the self and also the metaphysics of time. In particular, she claims that the apparent fixity of the past emerges from the adoption of the "embedded" perspective she describes.

Shaughn Gallagher's work links phenomenology with cognitive science. But here he is concerned with what he calls the "intrinsic temporality" in both bodily movement and action, some of which is experienced, but some of which happens at the subpersonal levels of analysis. Like Dainton, Gallagher begins with Husserl's dynamic model of retention and protention; but unlike Dainton, he extends this model to unconscious motor processes too. Bringing empirical studies to support his claims throughout, he then focuses on various timescales in an effort to show how the concept of free will becomes important.

The above papers assume that we already possess a rich set of temporal concepts because we are talking about normal adult human beings and their experiences. But what about young children? What temporal concepts do they have, and when do they acquire them? There are many interesting questions here, and a growing body of empirical work upon which to search for answers. When, for instance, do children master the use of past, present, and future? Are kids presentists? Hoerl and McCormack, in their chapter, pay special attention to the acquisition of a concept of a linear time series. Are children fitting experiences into a linear time series, or are they just doing particular actions that happen to make up such a series? When do children grasp the concepts of 'before' and 'after'? Can work on children and the causal priority principle (that causes precede effects) shed light on this question? Hoerl and McCormack defend the idea that children first conceive of events without genuinely employing tenses.

Finally, the temporal locations of benefits and harms matter to us. We prefer past pain to future pain, even when this choice includes more total pain. We regret our death but not our pre-natal non-existence. But *should* the location of benefits and harms matter to us, all else being equal? This question is an ethical one and the subject of David Brink's chapter. Brink is concerned with defending temporal neutrality, the thesis that agents should attach no normative significance to the temporal location of benefits and harms, all else being equal. A powerful argument for temporal neutrality comes from prudence. However, prudence also assigns normative significance only to benefits and harms that occur to you, not other agents. Given the symmetries of the case, can this hybrid position be defended? Brink thinks so, arguing, *contra* Parfit, that the fact that *you* are later compensated for present sacrifice is crucial to assigning equal importance to all parts of an agent's life, but not equally to all agents.

# TIME IN CLASSICAL AND RELATIVISTIC PHYSICS

Two parts of this *Handbook* are devoted to time in the physical sciences. Why so much physics, and not time in other sciences such as economics, biology, or chemistry? Although one can certainly imagine and even find very good work on the intersection of time and these sciences, the answer to this question is, I think, very simple. Apart from the sociological explanation—physics is where many philosophers interested in time work, and hence is where a lot of good work exists—there is also an intellectual answer: physics, and physics alone, is the science that actually considers time itself to be a target of study. Economics, biology, and chemistry of course implicitly model time, and as such, implicitly hold that time possesses various features, for example discreteness. Although confirmation theory is tricky, very few people would hold that the success of an economic model supports the particular conception of time adopted in the model. Time itself isn't a target of what economists, etc, study, either in theory or experiment. By contrast, time itself is very important to what physicists study. Experiments in physics define the second, which in turn defines the meter and many other quantities. Experiments in general relativity also reveal that it's crucially important that spacetime is curved in the so-called timelike directions of spacetime. The violation of time reversal invariance, something central to particle physics, is also tested experimentally. And, of course, in theory many specific features attributed to time are of paramount importance, for example, in the "signature" of the spacetime metric.

Given its centrality in the subject, one might be forgiven for thinking that everything to do with time in non-quantum physics has been worked out already. Fortunately for philosophers, that is hardly the case.

Time in even classical mechanics has yet to be fully appreciated by philosophers. Lawrence Sklar's chapter begins with time as it's presented to us in Newton's famous *Scholium*. He shows how and why Newton developed a notion that has various specific features, namely, those needed for time to play the role it does in classical dynamics. Along the way, several useful distinctions are made that will help the reader in later chapters.

Time in electromagnetism shares many features with time in other physical theories. But there is one aspect of electromagnetism's relationship with time that has always been controversial, yet hasn't always attracted the limelight it deserves, the electromagnetic arrow of time. While philosophers and physicists have expended much effort on the thermodynamic arrow, the electromagnetic arrow has by comparison been relegated to cameo role. This neglect is especially odd because a good argument can be made that the arrow of radiation is at least as important to the direction of time as the thermodynamic arrow. Beginning his chapter with a re-analysis of a famous argument between Ritz and Einstein over the origins of the radiation arrow, John Earman frames the debate between modern Einsteinians and neo-Ritzians. Earman tries to find a clean statement of what the arrow is—a surprisingly difficult problem—and then explains how it relates to the cosmological and thermodynamic arrows. This chapter represents the most developed and sophisticated attack yet, in either the physics or philosophy literature, on the electromagnetic arrow of time.

No theory has offered more shocking lessons with respect to time than relativity theory. Three chapters in this Part consider various ways in which relativity impacts our understanding of time.

One striking feature of relativity is the fact that clocks moving with respect to you "tick" slower. Combined with the relativity of motion, one arrives at the notorious twin "paradox." Your twin leaves the Earth in a rocket and then returns much later. To her you are moving, and hence you should age more slowly; to you she is moving and hence she should age more slowly. However, when she returns she is younger than you—why? Jean-Paul Luminet is an observational cosmotopologist, a physicist who searches the skies for evidence that spacetime has non-trivial topology, and he discusses this paradox in his chapter. Of course, the fact that this paradox is not really paradoxical is a well-worn fact. Yet I daresay that even the most sophisticated reader will learn something new about the twin paradox from this chapter. Looking at the above scenario played out on a hypersphere and on multiply connected finite spaces,

Luminet shows how this complicates the paradox. More importantly, he shows how its resolution in those cases yields new insights into the nature of spacetime and the equivalence between inertial reference frames.

Restricted to special relativity, the most significant change in our concept of time is certainly the relativity of simultaneity. What events are simultaneous with some event for one observer are different from those that are simultaneous with respect to an object travelling in a different inertial ("force-free") frame. If you and I are both inertial observers moving apart, then the events that comprise my "now" are not the ones that comprise your "now." Many believe that this relativity can play a role in an argument for eternalism. In his chapter Steven Savitt critically surveys these arguments before developing his own take on the implications of relativity for the metaphysics of time. First, however, he tackles another topic related to simultaneity, namely the conventionality of simultaneity. Many philosophers of science, especially during the early days of relativity, felt that simultaneity is not only relative but also conventional—that is, that there is a crucial element of choice in deciding what events are simultaneous for any other in a given inertial reference frame, that there is no fact of the matter about what is simultaneous. Savitt gives us the latest arguments on this debate, too.

When we turn from special to general relativity, many features commonly attributed to time once again change. To name a few: typically there are no global inertial frames, so clocks once synchronized and then freely falling will become unsynchronized; the timelike aspects of spacetime enjoy significant curvature; some general relativistic worlds do not even permit a global carving up of the world into global moments of time.

Of all of these features, perhaps the oddest is the possibility in the theory of non-trivial time travel. This is the subject of the chapter by Christopher Smeenk and Christian Wüthrich. After dispelling the logical and metaphysical arguments against the possibility of time travel, they turn to time travel in general relativity. They pay special attention to the possibility of being able to build a time machine, a device that would create a path suitable for time travel where none otherwise would have existed. Then they end their analysis by briefly reviewing if the quantum nature of matter alters the sense in which time travel is possible.

This last question reminds us that the world is governed by quantum theory, too, so an investigation into time should also turn to the strange world of the quantum.

# TIME IN A QUANTUM WORLD

Quantum mechanics presents us with many mysteries. Most readers will have stumbled across all kinds of perplexing topics associated with quantum mechanics, from Schrödinger's cat, to quantum non-locality, to many worlds theory. But quantum mechanics and time? Sometimes one hears that time in quantum mechanics is more or less the same as time in classical mechanics. While there is some truth to this, this claim obscures the fact that quantum mechanics does present those interested in time with several new and distinctively quantum topics. We finish the volume with three different ways in which quantum mechanics bears on time.

First, quantum mechanics *seems* to care about the direction of time, unlike classical mechanics. In 1964 K-long mesons were shown to violate time reversal invariance. But this violation was indirect, inferred from a remarkable theorem known as the CPT theorem. Derived by Gerhart Lüders and Wolfgang Pauli, and independently by John Bell, the CPT theorem states that local quantum field theories that are Lorentz invariant (i.e. relativistic) must also be invariant under the combined operations of charge reversal (replacing matter with anti-matter), parity (replacing "right-handedness" with "left-handedness"), and time reversal (changing the sign of the momenta). Since CPT holds, the 1964 violation of CP meant that T was violated.

The CPT theorem is quite strange. Why should a quantum field theory be invariant under the combination of two spatiotemporal discrete transformations, and then a quite different type of transformation (matter-anti-matter transformation)? What do these transformations really mean, and what does CPT symmetry imply? In one of the first attacks on these and related questions by a philosopher, Frank Arntzenius argues that CPT symmetry is better understood as PT symmetry. If he is right, CPT symmetry is really saying that quantum field theory doesn't care about temporal orientation or spatial handedness.

Second, unlike classical mechanics, quantum mechanics assumes the famous Heisenberg uncertainty relations. One of these concerns time: the energy-time uncertainty relation. However, there is something fishy about this uncertainty relation. Unlike the canonical position-momentum uncertainty relation, the energy-time relation is not reflected in the operator formalism of quantum theory. Indeed, it's often said and taken as problematic that there isn't a so-called "time operator" in quantum theory. Physicists Jan Hilgevoord and David Atkinson shed much-needed light on these questions and others, including the absorbing matter of whether quantum mechanics allows for the existence of ideal clocks (yes, they conclude).

Finally, science does not end with quantum theory and relativity. Quantum theory and relativity conflict in various ways, and a new theory—dubbed "quantum gravity"— is needed. It is appropriate to end the volume with a chapter reminding us of the challenges the study of time will face in the future. Just as reconceiving our classical notions of time was key for Einstein, in his discovery of special relativity, so too many believe that time will again hold the clue for theoretical advancement, but this time with quantum gravity. Claus Kiefer, a noted expert in the subject, details the challenge of reconciling quantum theory with relativity, concentrating especially on why time in particular causes trouble. He describes a result in canonical quantum gravity that is possibly of signal importance, namely, that fundamentally there is no time at all. In the timeless world he describes he shows us one way that time may emerge in particular physical regimes. This breath-taking possibility is, I hope, an excellent way to leave the reader wanting to study time further.

# PART I

# TIME AND METAPHYSICS

CHAPTER 1

....................................................................................................................

# PERSISTENCE

....................................................................................................................

## YURI BALASHOV

# 1. INTRODUCTION: SOME PICTURES

....................................................................................................................

ATOMS and molecules, desks and computers, dogs, butterflies, and persons persist through time and survive change. This much is obvious. But *how* do material objects manage to do it? What are the underlying facts of persistence? This is currently a matter of intense debate. And it is a relatively recent debate; three or four decades ago most philosophers would not have recognized the question of persistence as deserving more than casual attention. But if they had, the issue would quickly have been chalked up to a combination of older themes. Here is my dog Zarbazan, and there she is again. She changed in between, from being calm to being angry. But what is the big deal? Things change all the time but do not thereby become distinct from themselves (provided they do not lose any of their essential properties, some would add). Is there anything more to persistence? Have we made any progress since Aristotle?

We have. Today we recognize the problem of persistence as being primarily about *parthood* and *location*.[1] These two topics continue to drive the debate. What parts do persisting things have? Most people are happy to concede that a table, say, has four legs and a top; each of these, in turn, has smaller parts. A dog is similarly composed of a myriad of cells. These *spatial* parts compose the selfsame entity at any moment of its existence.[2] This does not rule out mereological change. The dog, in particular, retains its identity through continuous replacement of its cells. This is, by and large, the ordinary picture of persistence.

---

[1] For a systematic unified treatment of these two issues, see Casati and Varzi (1999).

[2] Such parts are spatial in the obvious sense that they are distributed in the three dimensions of space. It is true that salient spatial parts of ordinary objects are usually identified, not by their occupying particular spatial regions, but by other features, e.g. their functional role or relations to other such parts and to the whole. This does not detract from the spatial nature of such parts, although it does make the designation 'spatial' sound a bit unnatural, in some cases. Be that as it may, the designation is widely used in the literature and we shall adopt it below.

There are dissenters anxious to revise this picture. They note that most physical objects are spatially composite, but insist that composition takes place along the temporal dimension too. Just as the dog has distinct spatial parts, it also has different temporal parts at different moments of its life. Moreover, the dog's spatial parts also have temporal parts, and vice versa. One can slice up an object first in a spatial direction and then in the temporal direction, or one can do it in a reverse order, and end up with the same—much as you can cut a cake first lengthwise and then across or vice versa. Thus today's temporal parts of the head, the tail, the legs, and the torso compose today's temporal part of Zarbazan the dog. None of these objects, big or small, existed yesterday, and none will exist tomorrow. Yesterday's temporal part of Zarbazan was made of numerically distinct parts of her spatial constituents. The dog persists by being composed of non-identical temporal dog-parts, much as a road persists through space. And just as the road changes from being narrow in the country to being wide in town, the dog changes from being calm to being angry by having parts—temporal parts—that are calm and those that are angry. Change is qualitative variation in space for the road, and in time for the dog. Ditto for other physical objects.

The views sketched above are variously known as *three-dimensionalism* (3Dism), or *endurantism*, and *four-dimensionalism* (4Dism), or *perdurantism*. They disagree about the manner in which physical objects persist over time and through change. Some of the disagreement boils down to the question of what parts persisting objects have. Perdurantists typically insist, while endurantists typically deny, that objects have temporal as well as spatial parts.

In addition, these theories typically disagree about where and how objects are located in spacetime. The locus of a perduring object is a four-dimensional (4D) region of spacetime which, intuitively, incorporates the object's entire career. A 4D perduring object exactly fits in its spacetime career path and is only partially located at what we normally take to be a region of space at a certain moment of time, a three-dimensional (3D) slice of the object's path in spacetime. A 3D enduring object, on the contrary, occupies its spacetime path in virtue of being wholly present (i.e. present in its entirety, with no part being absent) at its multiple instantaneous slices. The key idea here is multilocation: one object—many locations.

This leaves room for another view known as *stage theory*, or *exdurantism*, which seeks to combine certain features of endurantism and perdurantism. Like perdurantism, stage theory endorses the existence of temporal parts, or stages. But like endurantism (and contrary to perdurantism), stage theory typically identifies ordinary objects with 3D entities that are wholly located at momentary regions that lack temporal extension. Despite their temporal shortness, such entities—object stages—persist. They manage to do so by *exduring*—by standing in temporal *counterpart* relations to later and earlier object stages. (The reader will note a close analogy with modal realism.)

These pictures are rough and ready. The real situation is more complex. To begin with, there is no general consensus about how to *state* the rival views of persistence and what exactly is at stake in the debate.[3] This has resulted in crisscrossing taxonomies and continual redrawing of the boundaries that were previously regarded as fixed. Furthermore, the persistence debate is closely entangled with a number of other philosophical disputes, old and new, which are equally complex: the nature and ontology of time, parts and wholes, material constitution, personal identity, modality, causation and properties, reference, and vagueness (the list is hardly complete). Equally important, considerations from physics have an important bearing on the issue of persistence.

Insofar as the rival views on persistence can be stated with sufficient clarity to the satisfaction of the warring parties, endurantism tends to enjoy the default advantage of a "common sense view," while perdurantism and exdurantism have a characteristically "revisionist" flavor and, as a consequence, are expected to do more work to motivate and defend themselves. This may or may not be fair, but the situation is not uncommon: in many other metaphysical disputes, there are "default" positions (e.g. presentism in the philosophy of time, dualism in the philosophy of mind, Platonism in the philosophy of mathematics) and their theoretical rivals (respectively, eternalism, materialism, and nominalism), whose appreciation requires serious ontological commitment, skepticism about the scope of traditional conceptual analysis, and sometimes a deferential attitude towards scientific evidence.

Be that as it may, there has been no shortage of arguments in favor of 4Dism.[4] Many of them are driven by philosophical reflection on the problems of change, intrinsic properties, temporal predication, material constitution, and vagueness.[5] There has been no shortage of 3Dist responses to those arguments.[6] The ensuing discussions continue to benefit not only direct participants, but various neighboring areas mentioned above. Several excellent review articles cover these grounds in detail.[7] But persistence is a dynamic and rapidly evolving topic, and the debate is changing every day. Recent developments have tended to focus on a set of related issues having to do with parthood and location.[8] It is in this context that broadly empirical considerations

---

[3]  For a particularly incisive recent analysis of these issues, see Hawthorne (2006).

[4]  See, in particular, Armstrong (1980); Balashov (1999), (2000abc), (2002), (2005b), (2009), (2010); Hawley (2001); Heller (1990); Hudson (1999); Lewis (1983), (1986: 202–204), (1988); Sider (2001), (2008).

[5]  Other arguments include those based on space-time analogies, supersubstantivalism about spacetime, and time travel. See Sider (2001: Ch. 4). However, Gilmore (2007) has argued that a certain type of time travel poses a problem for perdurantism.

[6]  See, in particular, Baker (2000); Haslanger (1989), (2003); Johnston (1987); Lowe (1988); McGrath (2007a); Merricks (1994), (1999); Rea (1998); Sattig (2006); Thomson (1983); van Inwagen (1990a); Zimmerman (1996), (1998).

[7]  See Haslanger (2003); Hawley (2004); McGrath (2007b); Sider (2008). Haslanger and Kurtz (2006) is a useful anthology of recent work on persistence over time.

[8]  See, in particular, Bittner and Donnelly (2004); Crisp (2005); Hudson (2006); Gilmore (2006), (2007), (2008), (2009); Sattig (2006); Parsons (2007); Balashov (2008), (2010); Donnelly (2009); Eagle (2009); Saucedo (forthcoming).

are increasingly brought to bear on the discussion.[9] The bulk of the present chapter (Sections 3 and 5) is devoted to this decidedly positive tendency. However, I start with a brief review of some older topics (Sections 2 and 4).

## 2. PERSISTENCE AND PHILOSOPHY OF TIME

The current mature stage in the controversy about persistence was preceded by a period when the topic was not clearly disentangled from related but distinct issues in the philosophy of time.[10] The former is about the ontology of physical objects, while the latter are primarily about the nature of time.

Let us begin with the latter. Much of contemporary philosophy of time is an ongoing battle between two rival ontologies of time, *eternalism* and *presentism*. On a popular characterization of eternalism, this doctrine holds that all moments of time and their contents enjoy the same ontological status. Past and future moments, events, and objects are just as real as the present ones; they are just not "temporally here." (Compare: the planet Neptune is not spatially here, but it does not, for that reason, fail to exist.) Eternalism contrasts with presentism, the view that only the present entities are real. Eternalism and presentism are widely regarded as the best incarnations of the older competing views of time known as A- and B-theories, or tensed and tenseless theories.[11]

Along with the ontological claim that all moments of time and their contents are equally real, eternalism incorporates the linguistic thesis that tensed locutions, such as 'Sarah will fail', have tenseless truth conditions expressed by metalinguistic token-reflexive clauses such as 'Sarah (tenselessly) fails later than this utterance of "Sarah will fail"' or by sentences incorporating dates into their content: 'Sarah (tenselessly) fails on July 12, 2009'. Presentism, on the other hand, must locate the truth conditions of tensed sentences in tensed facts about the present state of the world. It is not immediately clear what these facts could be, and the issue figures prominently in the dispute between presentism and eternalism.[12] In addition, many eternalists are ontological realists about the spacetime manifold. They view space and time as inseparable aspects

---

[9]  See Rea (1998); Balashov (1999), (2000abc), 2002, (2003ab), (2005c), (2008), (2010); Sider (2001: §4.4); Gilmore (2002), (2006), (2007), and (2008); Hales and Johnson (2003); Miller (2004); Gibson and Pooley (2006); Eagle (2009).

[10]  A vestige of such entanglement is found in the ambiguity of the term 'four-dimensionalism' used to denote both a certain ontology of persistence and a certain ontology of time, or spacetime.

[11]  The labels 'tensed/tenseless theories' and 'A/B-theories' are still used today to pick out certain combinations of linguistic and ontological doctrines. They are considered as interchangeable by some authors, but not by all. There does not seem to be general agreement on the usage of this terminology in the literature. For a very useful survey of the different versions of A- and B-theories, see Zimmerman, this volume.

[12]  For recent contributions see Bigelow (1996); Markosian (2003); Crisp (2005), (2007); Bourne (2006); Zimmerman (2008 and this volume); and Mozersky (this volume).

or dimensions of a single four-dimensional framework. Presentists typically resist this approach.

It is important to note that adopting eternalism or presentism does not settle the issue of persistence. Indeed, the two issues are arguably independent.[13] To be sure, some combinations seem more natural than others. Thus presentism provides a friendly habitat for endurance,[14] and eternalism for perdurance and exdurance. But there is nothing problematic in combining endurance with eternalism.[15] Indeed, this is a combination that many endurantists recommend.[16] They are led to reject presentism by various philosophical reasons and, increasingly, by scientific considerations as well. Many physical theories, such as classical mechanics and special relativity, are based on distinctive claims about the intrinsic geometry of the spacetime manifold. Since the geometry of relativistic spacetime does not support a frame-invariant notion of simultaneity, it does not allow one to define the concept of *the* present. And without such a concept, presentism cannot get off the ground.[17] To put the point vividly, the presentist is committed to the following: when I click my finger on Betelgeuse, Boris Yeltsin is either alive or dead. But according to special relativity, there is simply no fact of the matter. There is no global present moment cutting across the entire universe that has more than a frame-relative significance.

From now on we shall set presentism aside and embrace spacetime realism. While this may close some doors, other important doors—those having to do with concerns central to this essay—will remain open.

## 3. ENDURING, PERDURING, AND EXDURING OBJECTS IN CLASSICAL SPACETIME

Now that we have settled on a common spacetime framework, we need to sharpen the contrasts among the three rival modes of persistence—endurance, perdurance, and exdurance. For the purpose of discussion in Sections 3 and 4 we shall assume that spacetime is classical and can be foliated into a family of flat hypersurfaces of

[13] This is the majority view. Those who have argued that the issues are *not* independent include Carter and Hestevold (1994) and Merricks (1995).
[14] See, in this connection, Merricks (1994), (1999); and Hinchliff (1996). For a very useful survey of the different versions of A- and B-theories, see Gilmore 2009.
[15] And in combining perdurantism and presentism, although this package is surely the most exotic one. See Lombard (1999); Sider (2001: 68–73); but cf. Sattig (2006: 62–65).
[16] See, in particular, Johnston (1987); Haslanger (1989), (2003); van Inwagen (1990a); Rea (1998).
[17] This seems obvious to many, but it may require support, for one might argue that the doctrine of presentism could be made compatible with the lack of absolute simultaneity by modifying its letter without abandoning its spirit (see, e.g. some contributions to Craig and Smith 2008 and Zimmerman, this volume). Various ways in which this could be done are discussed and shown to be untenable in Savitt (2000); Callender (2000); Sider (2001: §2.4); Saunders (2002); and Balashov and Janssen (2003). The implications of special relativity for the philosophy of time are further explored in Savitt's contribution to this volume.

simultaneity representing global moments of time. Consider a material object, such as a proverbial poker, that persists over time and changes from being hot at $t_1$ to being cold at $t_2$. As already noted, the parties to the debate agree that the poker has a spacetime *career* represented by a 4D *path*—the shaded region in Figures 1.1a–c.[18] But they disagree about the manner of the poker's location at its path. The endurantist will say that the poker is *multilocated* at all 3D *slices* of its path corresponding to different times; call them '*t*-slices'. To pull off this trick, the poker must fit, in its entirety, into every such slice and must, therefore, be a 3D object. The perdurantist, on the other hand, will say that the poker is *singly located* at its path and is, therefore, a 4D object.

In the present context, location means *exact location*. The idea is that a spacetime region at which an object is exactly located is a region into which the object exactly fits and which has exactly the same size, shape, and dimensionality as the object itself. Exact location has a number of notable *negative* properties. If object *o* is located at each of a number of distinct spacetime regions, it does *not* follow that *o* is located at their union. If *o* is located at a proper subregion R of a larger region R′, it does *not* follow that *o* is located at R′. And conversely, if *o* is located at R, it does *not* follow that *o* is located at any proper subregion of R.[19]

Some of these features of the relation of exact location are crucial to understanding the difference between endurance and perdurance. If the poker endures it is (exactly) located at every *t*-slice of its path but not at any other region, including the path in its entirety. On the other hand, if the poker perdures it is exactly located at its entire path but not at any other region, including various *t*-slices of its path. What is located at such slices are distinct temporal parts (hereafter, '*t*-parts') of the poker. Endurance and perdurance are easy to describe and visualize. What about exdurance?

If the poker exdures it is exactly located at some *t*-slice of its path. More carefully, what is located at this slice is the poker *t*-stage, which *is* a poker in its entirety: no parts of the poker are missing from it. In other words, an exduring object is wholly present at exactly one moment of time. In this respect, exdurance is analogous to endurance. One may even be inclined to depict exdurance in a way similar to Figure 1.1a. But that would be misleading. Figure 1.1a represents the relation of location which holds between a single enduring object and many instantaneous regions of spacetime at which it is exactly located (i.e. multiple *t*-slices of the object's path). But an exduring object cannot be a relatum in this sort of relation, because it is not capable of multi-location. The poker *t*-stage is located *only* at the corresponding *t*-slice of the poker's path. What is located at another *t*-slice is a numerically distinct poker stage. All such

[18] It is difficult to draw full-blown 4D diagrams. Accordingly, we shall follow others in suppressing one or two dimensions of space, as in Figure 1.1. Of course, one should never suppress the temporal dimension. Figures 1.1a and 1.1b are inspired by Gilmore (2006: 205).

[19] For a more detailed discussion of the properties of exact location, see Gilmore (2006: 200–204). The above concept of exact location and its analogs are increasingly used in the literature on persistence. See, in particular, Bittner and Donnelly (2004), Hudson (2001) and (2006), Balashov (2008) and (2010: §2.4), and Donnelly (2009). It should be noted, however, that other ways of understanding the basic concept of location are possible. See, in this connection, Parsons (2007), Gilmore (2008), and Saucedo (forthcoming).

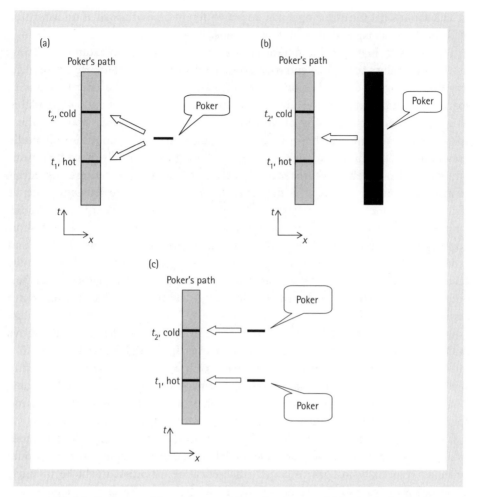

FIGURE 1.1 Endurance (a), perdurance (b), and exdurance (c) in classical spacetime

stages bear to each other a temporal counterpart relation characteristic of pokers. This relation is extensionally equivalent to one that would hold between the corresponding temporal parts of the poker, if the latter perdured. But since it does not—since, by assumption, the poker exdures rather than perdures—then what occupies each $t$-slice of the poker's path is a poker, not a poker part. The distinctive features of exdurance are somewhat elusive, and it is not so clear how to represent them in a diagram. Figure 1.1c may suggest helpful directions, especially when it is viewed against Figures 1.1a and 1.1b.

But there is also a useful non-pictorial analogy. In David Lewis's ontology of concrete possibilia, each object has modal counterparts in many distinct worlds, and each counterpart represents a way an actual object might have been. Our talk of various counterfactual situations involving, say, Hubert Humphrey, is underwritten by Humphrey's distinct modal counterparts located in separate worlds. Similarly, when we talk about what the exduring poker does at different times, this talk is underwritten by distinct poker stages wholly confined to those times.

This helps to highlight the difference between endurance and exdurance. Exdurantists generally accept, but endurantists generally deny, the existence of momentary object stages—entities *exclusively* confined to $t$-slices of the objects' paths in spacetime. They will, however, agree that such a 3D entity *represents* a persisting object in its entirety—directly or vicariously, as the case may be. This sets them apart from perdurantists, who insist that the relevant persisting object is only partly present at a $t$-slice of its path, while being much longer temporally. Note that exdurantists need not deny the existence of such longer entities, which are aggregates of distinct $t$-stages. That is why the title '4Dism' is appropriate for exdurance, as well as perdurance. What the exdurantist must resist is the identification of the long entity with an ordinary persisting object. She will say that a dog is a short dog-stage, not a long spacetime dog-worm. The difference between perdurance and exdurance thus appears to be semantical not ontological, as both parties typically accept instantaneous material stages, as well as their cross-temporal aggregates, in their ontologies. As we shall see shortly, the "merely semantical" difference is difference enough.

But as a preliminary step, we need to correct an unfortunate feature of the above description that makes exdurance needlessly handicapped right from the start. The offending feature has to do with the failure of exduring objects to be multilocated in spacetime. But on a common understanding of persistence, something persists only if it exists at more than one moment,[20] and an instantaneous object stage, strictly speaking, does not. One could, of course, simply accept this consequence and conclude that exduring objects do not persist. Indeed, such considerations have been adduced to disqualify exdurance from being a legitimate mode of persistence. But this would be unfair, especially in light of certain theoretical benefits that might be uniquely associated with exdurance.[21] At the definitional stage, "vicarious persistence" via counterparts should be as respectable as "direct persistence."[22]

What we need is a generalization of the notion of exact location, on which both endurance and exdurance would exemplify *some* mode of multilocation. The sense in which an exduring object accomplishes this feat is similar to the sense in which a worldbound individual of the Lewisian pluriverse, such as Humphrey, can nonetheless be said to exist at (or, as Lewis put it, "according to") multiple worlds. Let us start with (non-modal) counterparthood and stipulate that every object (enduring, perduring, or exduring) is an (improper) counterpart of itself. An object can then be said to be

---

[20]   The locus classicus is probably Lewis (1986: 202): "Something *persists* iff, somehow or other, it exists at various times."

[21]   Such benefits are thoroughly discussed in Sider (2001: 188–208) and Hawley (2001: Chs. 2 and 6). Balashov (2005b) defends exdurance as the best way to account for certain features of the experience of time. In this connection, see also Ismael's contribution to the present volume.

[22]   The situation is very similar to that surrounding Kripke's "Humphrey objection" to modal realism.

*q-located* (*quasi-located*) at a region R just in case one of the object's counterparts is strictly located there:

> (QL)   Object $o$ is (exactly) *q-located* at spacetime region R $=_{df}$ one of $o$'s (non-modal) counterparts is (exactly) located at R.

Note that if $o$ is located at R then it is also q-located there, but not vice versa.

We shall also need a notion of parthood relativized to a temporally unextended, or *achronal*, region of spacetime. We shall take such a three-place relation, expressed by '$p$ is a part of $o$ at achronal region R', as a primitive. It may be interesting to explore the properties of such a relation at a sufficiently general level, where no further restrictions are imposed on R.[23] For our more limited purposes, however, the regions of interest are $t$-slices of objects' paths, which, in the classical context, can be indexed by a single parameter, and we shall restrict further consideration to such slices.[24] This will also help to bring relativized parthood closer to the intuitive sense in which some cells can be part of me at one time but not at another. Where $p$, $o$ and a $t$-slice of $o$'s path, $\mathbb{O}_{\perp t}$, stand in such a relativized parthood relation, we shall say that $p$ is a *spatial part* (*s-part*) of $o$ at $\mathbb{O}_{\perp t}$:

> (SP$^G$)   $p_\perp$ is a *spatial part* (*s-part*) of $o$ at $\mathbb{O}_{\perp t}$ $=_{df}$ $p_\perp$ is a part of $o$ at $\mathbb{O}_{\perp t}$.

Temporal parthood could then be defined as follows:[25]

> (TP$^G$)   $p_{||}$ is a *temporal part* (*t-part*) of $o$ at $\mathbb{O}_{\perp t}$ $=_{df}$ (i) $p_{||}$ is located at $\mathbb{O}_{\perp t}$ but only at $\mathbb{O}_{\perp t}$, (ii) $p_{||}$ is a part of $o$ at $\mathbb{O}_{\perp t}$, and (iii) $p_{||}$ overlaps at $\mathbb{O}_{\perp t}$ everything that is a part of $o$ at $\mathbb{O}_{\perp t}$.

The superscript 'G' reminds us that we are working in the classical framework of Galilean spacetime. The subscripts '$\perp$' and '$||$' indicate that the relevant dimension is, respectively, "orthogonal" or "parallel" to the dimension of time. Although $s$- and $t$-parthood are, strictly speaking, relativized to regions of spacetime (i.e. to $t$-slices of objects' paths), notation can be simplified. In what follows I shall use expressions such as '$p$ is a spatial (temporal) part of $o$ at $t$' in place of '$p$ is a spatial (temporal) part of $o$ at $\mathbb{O}_{\perp t}$', where context makes it clear that '$t$' refers, not to an entire hyperplane of absolute simultaneity, but to a rather small subregion of it: $\mathbb{O}_{\perp t}$.

Notice that on (SP$^G$) and (TP$^G$), spatial and temporal parthood are not mutually exclusive. Indeed, temporal parthood is just a special case of spatial parthood. Hence a temporal part of a perduring or exduring object at $t$ (i.e., at $\mathbb{O}_{\perp t}$) is also its spatial part at $t$. Thus my $t$-part at the current $t$-slice of my path is also my (improper) $s$-part at that slice. On the other hand, *proper* spatial and temporal parthood *may* be exclusive. This depends on how relativized proper parthood is defined. Generally speaking, there are two options here. Relativized proper parthood may be understood as *asymmetrical* relativized parthood:

---

[23]   For details, see Balashov (2010, §2.4.4).
[24]   When we turn to the relativistic context in Section 5 things will become a bit more complex.
[25]   (TP$^G$) follows the general pattern of Sider's definition (2001: 59).

(PP$_1$)    $p_\perp$ ($p_\parallel$) is a *proper spatial (temporal) part of o at* $\circ_{\perp t}$ =$_{df}$ (i) $p_\perp$ ($p_\parallel$) is a spatial (temporal) part of $o$ at $\circ_{\perp t}$; (ii) $o$ is not a spatial (temporal) part of $p_\perp$ ($p_\parallel$) at $\circ_{\perp t}$.

Then if $p_\perp$ is a proper spatial part of $o$ at some $t$-slice of its path $\circ_{\perp t}$ then, intuitively, $p_\perp$ is smaller than $o$ at $\circ_{\perp t}$ and, hence, does not qualify as a temporal part of $o$, at that slice.[26] Thus my hand is a proper $s$-part of me, but not a $t$-part of me, at the current $t$-slice of my path. And if $p_\parallel$ is a temporal part of $o$ at $\circ_{\perp t}$, proper or not, then, intuitively, $p_\parallel$ is "as large as" $o$ at $\circ_{\perp t}$ and, hence, cannot be a proper spatial part of $o$ at $\circ_{\perp t}$.

Relativized proper parthood may, however, be defined differently:

(PP$_2$)    $p_\perp$ ($p_\parallel$) is a *proper spatial (temporal) part of o at* $\circ_{\perp t}$ =$_{df}$ (i) $p_\perp$ ($p_\parallel$) is a spatial (temporal) part of $o$ at $\circ_{\perp t}$; (ii) $p_\perp$ ($p_\parallel$) $\neq o$.

In that case one object could be both a proper spatial and a proper temporal part of another at some $t$-slice of the latter's path. Consider a perduring or exduring statue and the piece of clay of which it is composed. Some would argue that the statue (and hence, its $t$-part) is not identical with the piece of clay (and its corresponding $t$-part). If so then the statue and the piece of clay are both proper spatial and proper temporal parts of each other at the $t$-slice of the path of both objects.[27] (PP$_1$) and (PP$_2$) raise an interesting question of how to develop region-relativized mereology, but we must leave the matter here.[28]

The resources we have developed are sufficient to express the important distinctions among the three modes of persistence in classical (Galilean) spacetime (ST$^G$) a bit more formally.

(END$^G$)    *o endures in* ST$^G$ =$_{df}$ (i) $o$'s path includes at least two points from different moments of (absolute) time, (ii) $o$ is located at every $t$-slice of its path, (iii) $o$ is q-located only at $t$-slices of its path.

(i) ensures that $o$ persists; (ii) says that an enduring object is multilocated in its entirety (or, in the more familiar parlance, is "wholly present") at all moments of classical time at which it exists; (iii) precludes $o$ or its temporal counterparts (if any) from being extended in time.

(PER$^G$)    *o perdures in* ST$^G$ =$_{df}$ (i) $o$'s path includes at least two points from different moments of (absolute) time, (ii) $o$ is q-located only at its path, (iii) the object located at any $t$-slice of $o$'s path is a proper $t$-part of $o$ at that slice.

[26] Here and below one should presuppose an R-relativized analog of Strong Supplementation:

(SS$^R$)    If $x$ is not a part of $y$ at R then there is a part of $x$ at R that does not overlap $y$ at R,

to rule out spurious asymmetry between $x$ and $y$, not grounded in any difference in their parts at R, that would lead to a vacuous satisfaction of the definiens of (PP$_1$). I thank Cody Gilmore for this observation.

[27] We will return briefly to the "problem of co-location" in Section 4.
[28] For a systematic analysis of R-relativized mereology, see Donnelly (2009), which is highly recommended.

(ii) indicates that $o$ is temporally extended and is as long as its path, while (iii) guarantees that $o$ has a distinct proper temporal part at each moment of its career.

(EXD$^G$)    $o$ *exdures* in ST$^G$ =$_{df}$ (i) $o$'s path includes at least two points from different moments of (absolute) time, (ii) $o$ is located at exactly one region, which is a $t$-slice of its path, (iii) $o$ is q-located at every $t$-slice of its path.

(ii) captures the already-noted similarity as well as difference between endurance and exdurance: (a) endurers and exdurers are temporally unextended and "wholly present" at single moments of time, but (b) unlike endurers, exdurers do not literally survive beyond a single moment. However, (iii) states that exdurers survive vicariously, via their temporal counterparts, and that this is as good as it can get.

While the above definitions are true to the spirit of the intuitive pictures of the three modes of persistence, the definitions are far from being watertight and are not intended to satisfy everyone.[29] A few brief comments are therefore in order.

(1) We take the notion of exact location as primitive. It aims to be neutral among the different modes of persistence (thus the sense in which a perduring object is exactly located at a 4D region of spacetime is the same as that in which an enduring object is exactly located at a 3D region) and to do justice to the idea of being "wholly present" at a moment of time, which has figured prominently in earlier discussions of endurance. Could the application of the relevant concept *being wholly present at R* to spatially composite objects be reduced to more basic mereological notions? The question has been thoroughly debated, but that mini-debate is rather peripheral to our concerns here.[30] One reason to be skeptical about the prospect of such reduction is the dilemma of triviality and falsity often associated with it. If '$o$ is wholly present at R' means that no part of $o$ is absent from R, then one wants to know more about the notion of parthood at work. If parthood is taken to be a basic two-place relation (as it is, in classical mereology) then the requirement that no part of $o$ be absent from any region at which $o$ is wholly present is incompatible with persistence through mereological change. On the other hand, if parthood is relativized to time (or more generally, to instantaneous subregions of spacetime, as in this paper) then the requirement becomes trivial and thus empty.[31]

(2) Should parthood be relativized? *Could* it be relativized? One could perhaps argue that the basic two-place relation of parthood is as constitutive of mereology as the corresponding set membership relation is constitutive of set theory and must, therefore, be protected. But many writers tend to think that the theoretical and heuristic benefits of relativized parthood outweigh its costs and that the notion is indispensable.[32] Indeed, much of "folk mereology" is temporally modified. My computer has

---

[29]  It is unclear that such desiderata are met by any set of definitions offered in the literature on persistence.

[30]  The interested reader is advised to consult Rea (1998), Sider (2001: 63–68), McKinnon (2002), Crisp and Smith (2005), Parsons (2007), and references therein.

[31]  For an attempt to get around this problem, see Crisp and Smith (2005).

[32]  "Relativizers" include Hudson (1999), Sider (2001), Bittner and Donnelly (2004), Crisp and Smith (2005), Balashov (2008), (2010), Donnelly (2009), and Gilmore (2009).

just acquired a new video card, a part that it did not have before. To be sure, temporal modification of parthood is more familiar and transparent than its relativization to a region of spacetime. But theories need to turn abstract at some point. Furthermore, as we shall see shortly, a similar sort of relativization is needed by most accounts of persistence to explain change over time, so treating parthood in the same way appears to be natural.[33]

(3) The approach outlined above assumes "achronal universalism," the idea that each enduring object is located at *every* $t$-slice of its path, each perduring object has a $t$-part at *every* $t$-slice of its path, and each exduring object is q-located at *every* $t$-slice of its path. The idea may seem uncontroversial: how could a persisting object, or its $t$-part, *fail* to be located (or q-located) at a $t$-slice of *its path* in spacetime? But putting the matter this way reveals a potentially problematic spot in it, which is the notion of *path* itself. It could be taken as a useful starting point (as is done above), or one could try to build it from more elementary blocks, for example define it as a union of regions at which a persisting object is located (or q-located) (see Gilmore (2006: 204) and Balashov (2008: §2.3)). In the classical context, this automatically gives rise to achronal universalism. Not so in the relativistic context. Indeed, the question of what subregions of $o$'s path contain $o$, its $t$-parts, or counterparts is non-trivial.[34] We shall return to this issue in Section 5.

(4) The three-fold classification of the views of persistence suggested above is somewhat biased (every classification is biased!) in that it abstracts from certain exotic possibilities. Persisting by being singly or multiply located (or q-located) in spacetime and persisting by having or lacking temporal parts are, arguably, two distinct issues. The distinction is made clear by the conceptual possibility of temporally extended simples (Parsons 2000) and instantaneous statues (Sider 2001: 64–65). Taking such cases seriously could motivate a different taxonomy.[35]

(5) Allowing exotica also shows that the above definitions of the three modes of persistence are not watertight. Consider an enduring lump of clay that becomes a statue for only an instant (Sider 2001: 64–65). On (TP$^G$), the statue is a temporal part of the lump at that instant. But endurance is widely regarded as being incompatible with the existence of temporal parts. Also consider an organism composed of perduring

---

[33] Another intriguing question is the *adicity* of the relativized parthood relation. The present paper assumes that this relation is fundamentally *three*-place: $x$ is a part of $y$ at $z$, where $z$ is a place holder for an achronal region of spacetime. This assumption looks natural and appears to accommodate the relevant intuitions about multilocation and other important notions that figure in the debate about persistence. But one may have doubts as to whether a *single* regional modifier can successfully relativize the instantiation of a parthood relation. Gilmore (2009) has recently argued that three-place parthood confronts a number of problems, and that the best way to think of relativized parthood is in terms of a *four*-place relation: $x$ at $w$ is a part of $y$ at $z$. The argument is extended and requires detailed consideration, which cannot be afforded here.

[34] See, in this connection, Gilmore (2006) and (2008), and Gibson and Pooley (2006).

[35] See Gilmore (2006) and (2008) for a detailed discussion of the two issues and the correspondingly more complex classifications of the views of persistence.

cells and stipulate that the cells and their temporal parts are the *only* proper parts of the organism (Merricks 1999: 431). By clause (iii) of (PER$^G$), the organism itself does not perdure, an unwelcome result.

(6) According to (SP$^G$) and (TP$^G$), spatial and temporal parts are achronal, that is, temporally unextended. In this, we deviate from authors who explicitly allow temporally extended temporal parts and make them do some useful work.[36] Such entities also open up another set of exotic possibilities. One could imagine an object satisfying (END$^G$) (hence, a bona fide endurer) but having a set of finitely extended temporal parts. Relatedly, there could be an object satisfying clauses (i) and (ii) of (PER$^G$) but having only finitely extended proper temporal parts. On (PER$^G$), such an object does not perdure, another intuitively wrong result.

As already noted, it is unclear that any set of definitions offered in the literature can handle all exotic cases of persistence. And in any event, cases of that sort are too remote to bear on the agenda of this paper, so we can safely ignore them. For our purposes, (END$^G$), (PER$^G$) and (EXD$^G$) provide good accounts of the three modes of persistence. Even more importantly, these accounts can easily be extended to the special relativistic framework of Minkowski spacetime, as we shall see in Section 5.

But other issues must be addressed first.

# 4. SOME TRADITIONAL ARGUMENTS FOR AND AGAINST 3DISM AND 4DISM

## 4.1 The problem of change and temporary intrinsics

Persisting objects change their properties with time. Many of these properties—shape, temperature, texture and so forth—are "temporary intrinsics": they characterize ways objects are at certain moments of time "in and of themselves," not in relation to anything else. But some temporary intrinsic properties—those that belong to the same family of *determinables*, such as *being hot* and *being cold*—are incompatible. How can the selfsame object have incompatible properties? This is often referred to as the problem of temporary intrinsics or the problem of change. In earlier discussions of persistence these problems were often invoked to favor perdurance over endurance (see, in particular, Lewis 1986: 202–204). To say what properties an object has at *t*, the endurantist who subscribes to spacetime realism must relativize possession of

---

[36] Butterfield (2006), for example, employs non-achronal temporal parts to rebut, on behalf of the perdurantist, Kripke's "rotating disk" argument. This argument is thoroughly debated in Armstrong (1980); Zimmerman (1998), (1999); Lewis (1999); Callender (2000b); Sider (2001: §6.5); Hawley (2001: Ch. 3). For earlier discussions of finitely extended temporal parts, see Heller (1990) and Zimmerman (1996).

temporary properties to time, just as she must relativize parthood to time.[37] She cannot say that a certain poker is hot and stop here, because the selfsame poker is also cold, when it is wholly present at a different time. Several semantically and metaphysically distinct ways of temporal modification have been offered on behalf of the endurance theory.[38] The perdurantist, on the other hand, analyzes possession of temporary properties by temporally extended objects in terms of exemplification of such properties *simpliciter* by the objects' temporal parts.

But what exactly is wrong with relativization of properties to times (or *t*-slices of objects' paths)? Some philosophers have rejected this strategy on the ground that paradigm temporary properties, such as shapes, are not relations. Lewis, in particular, urged that relativization eliminates the having of familiar properties *simpliciter* in favor of having them always in temporally modified ways (Lewis 1986: 204 and 1988: 65) and is, therefore, objectionable. But many others have responded by arguing that relations to times are different in nature from relations to other things and, therefore, relativization to time does not deprive temporary properties of their intrinsicality.[39] This response has merit. Furthermore, *pace* Lewis, it is unclear that perdurantism escapes the need for relativization either. Perduring objects do not have temporary properties simpliciter, but only in virtue of a parthood *relation* to their temporal parts that do.

It begins to look as though exdurance emerges as a winner here: in the ontology of stages, momentary properties are exemplified by persisting objects directly, for such objects are instantaneous stages, not aggregates thereof.[40] This advantage, however, is offset by the need to bring in counterpart relations to neighboring objects stages. Such relations are required to account for exemplification by instantaneous stages of temporal and "lingering" properties (Hawley (2001: 54), such as *will be cold, speeding, traveling across the tennis court*, and *getting wet*.[41] My current stage can *be* wet, but it does not initially seem capable of *getting* wet, for that requires being first dry and then wet, and no instantaneous entity can be both. This apparent problem is resolved by noting that my current stage can be getting wet by being covered with water

---

[37] The presentist endurantist escapes the need for relativization because "a different time" does not exist and the only properties the object has are those it has at the present moment. See, in this connection, Merricks (1994) and Hinchliff (1996). Some authors have argued that this mode of exemplification of temporary properties *simpliciter* favors the combination of presentism and endurance over any other ontology of time and persistence. Those who espouse this approach sometimes go further and argue that endurance *entails* presentism. See, in particular, Merricks (1995), (1999). In light of the problems befalling presentism this entailment may be looked upon as needlessly burdensome. Indeed, most endurantists are eternalists. See note 16.

[38] The general strategy of relativizing temporary properties to times was sketched by Lewis (1986: 202–204). It was then developed by others in a great number of works and in many different forms. For recent contributions and references see MacBride (2001) and Haslanger (2003). But see Sattig (2006) for objections to such relativization strategies.

[39] For recent discussions see MacBride (2001: 84); Hawley (2001: §1.4); Haslanger (2003).

[40] The supremacy of stage theory vis-à-vis the problem of temporary intrinsics is discussed in Sider (2000).

[41] For details, see Hawley (2001: 53–57); Sider (2001: 197–198); and Balashov (2007).

and being counterpart-related to earlier stages that are dry and to later stages that are covered with more water. Single object-stages can have lingering properties— those that take time to be instantiated—by standing in appropriate relations to surrounding stages. There is nothing shocking here. In fact, the situation is familiar: possession of many instantaneous physical properties, such as velocity or acceleration, is partly a matter of what goes on at other instants. But this analysis shows that temporal and lingering properties are highly relational, that their exemplification by object stages is not just a matter of having a certain quality *simpliciter* and *nothing else*.

Thus it appears that everyone must relativize one way or another; hence, the need to do so does not obviously put any theory of persistence at a disadvantage.

## 4.2  Co-location

Consider a piece of clay, C, at $t_1$ that is shaped by a potter into a vase, V, at $t_2$. The endurantist may be hard pressed to say that there are two distinct objects at $t_2$, C and V, which occupy exactly the same region of space and are composed of exactly the same matter. The pressure comes from the observation that C and V differ at $t_2$ in that the former, but not the latter, has the historical property *having existed at $t_1$*. But exactly co-located entities may be regarded as problematic: they gratuitously overcrowd space. The endurantist has several options to avoid co-location, but all require extra commitments that may be deemed undesirable, if not unacceptable.[42]

It has been argued that the perdurantist and, especially, the exdurantist can do a lot better. The former can assimilate temporary co-location of two (or more) persisting objects to the case of harmless *overlap* of two temporally extended perduring entities. The exdurantist can simply deny that C and V are two distinct objects at $t_2$. There is only one object (i.e. an object stage) standing in two different counterpart relations to earlier stages of C and V. Does *that* object exist at $t_1$? Yes and no: it does, *qua piece of clay*, but does not, *qua vase*. But this is no more problematic than saying that someone is rich for a philosopher, but poor for an accountant.

It has also been argued that the stage theorist is uniquely positioned to offer a unified treatment of all versions of co-location puzzles (including their modal versions) and of Parfit-style fission scenarios threatening the notion of personal identity.[43] Are these advantages decisive? The answer depends on other assumptions that go into generating the puzzles of co-location in the first place. These include taking entities such as

---

[42]  Mereological essentialism, a decidedly non-egalitarian treatment of sortal terms, mereological nihilism, and the constitution view are the options usually discussed in this connection. For details and references to earlier work, see Rea (1997); Sider (2001: Ch. 5) and (2008: 247–257). See also McGrath (2007ab) for critical discussion of co-location-inspired arguments for 4Dism.

[43]  See, in this connection, Sider (2001: §5.8).

vases seriously, taking the danger of "overcrowding" seriously, successfully resisting the "constitution" move, and others.[44]

## 4.3 The argument from vagueness

This argument, developed most fully in Sider (2001: §4.9), holds a special place in the debate about persistence. Indeed some authors take it to be central to the case for 4Dism. Its roots are traced to the older argument for *unrestricted composition at a time*, or *synchronic universalism* (SU), the thesis that any collection of objects composes a further object, due to David Lewis (1986: 212f). (SU) can be characterized a bit more formally as follows:

(SU)    Any class of objects existing at $t$ has a fusion at $t$.[45]

Suppose this is not the case and composition at a time is restricted. Then there must be a pair of putative cases of composition connected by a Sorites series such that in one, composition occurs, but in the other, composition does not occur. As a fancy example, consider an orange gradually expanding until its material is scattered across the entire universe. It is easy to get everyone to agree that the bits of the orange material compose something (viz. an orange) at the beginning of the process, but not at the end. But then, either a certain point in the process must feature a sharp boundary between a case of composition and a case of composition failure, or there must be a vague stretch of borderline cases, in which composition neither definitely occurs nor definitely fails to occur. Borderline cases abound in other Sorites series involving vague predicates, such as 'bald'. However, Lewis and others have urged that, unlike 'baldness' and other paradigmatically vague terms, composition does not admit of borderline cases, on pain of implying that existence (viz. the existence of a putative composite object) is itself vague, which would border on the unintelligible. But positing a sharp cut-off in whether composition occurs somewhere in the Sorites series of the states of an expanding orange is extremely implausible too: the cases can be made as close to each other as one wants. According to Lewis, the only way to get around this dilemma is to say that composition *always* occurs, even in the case of the orange molecules scattered throughout the entire universe. Such molecules compose a highly scattered object.

Sider's argument from vagueness to 4D employs similar considerations to establish the thesis of *diachronic universalism* (DU)—the view that, for any interval of time and objects existing at various moments in it, there is something they compose over the

---

[44]   The controversy about co-location has generated extensive literature. The sources mentioned in the previous two notes and references therein provide useful entry points to this literature.

[45]   In general, $y$ is a fusion of the $xs$ iff $y$ contains each of the $xs$ as a part and every part of $y$ overlaps some of the $xs$. The temporally relativized notion of fusion at work in (SU) is easily obtained by appropriately relativizing the parthood relation.

interval. This result is then used to show that temporal parts (i.e. entities satisfying definition (TP$^G$) of Section 3) exist and, therefore, 4Dism (in the form of perdurance or exdurance theory) is true.

To state the argument more precisely, a bit more machinery is needed.[46] An *assignment* $f$ is a function from times to non-empty classes of objects existing at those times. A *diachronic fusion* of assignment $f$ (a *D-fusion* of $f$) is an object $x$ that is a fusion-at-$t$ of $f(t)$ for every $t$ in $f$'s domain. A *minimal D-fusion* of $f$ is a D-fusion of $f$ that exists only at times in $f$'s domain. This can then be put to work by analogy with Lewis's reasoning in support of (SU):

P1:   If not every assignment has a minimal D-fusion, then there must be a pair of cases connected by a continuous series such that in one, minimal D-fusion occurs, but in the other, minimal D-fusion does not occur.

P2:   In no continuous series is there a sharp cutoff in whether minimal D-fusion occurs.

P3:   In any case of minimal D-fusion, either minimal D-fusion definitely occurs, or minimal D-fusion definitely does not occur.

P1, P2, and P3 imply:

(U)   Every assignment has a minimal D-fusion.

Indeed, suppose they do not. Then there is a continuous series connecting a case where minimal D-fusion occurs to a case where minimal D-fusion does not occur (P1). By P3, there must be a sharp cutoff in this series, which, however, is ruled out by P2. But (U) gives the 4Dist what she needs—temporal parts. Consider the assignment $f^* = \langle t, \{x\} \rangle$, where $x$ is an arbitrary object and $t$ a time at which it exists. By (U), $f^*$ has a minimal D-fusion, $z$. On (TP$^G$) (see Section 3), and using the following plausible mereological principle:

(SUPPL*)   If $x$ and $y$ exist at $t$ and $x$ is not part of $y$ at $t$, then $x$ has a part at $t$ that does not overlap $y$ at $t$

it is easy to show that $z$ is a temporal part of $x$ at $t$. QED.

The argument from vagueness has provoked diverse responses questioning virtually every assumption that goes into its construction.[47] Let us take a closer look at the general dialectical situation surrounding it. The argument bases the doctrine of temporal parts on a prior endorsement of unrestricted composition, along the spatial as well as temporal dimensions. Not only is there a momentary object consisting of my nose tip and the Eiffel Tower, but there is also a temporally extended object fusing earlier stages of Napoleon's favorite dog with later stages of the Golden Gate Bridge. To get any mileage from the argument from vagueness, the 4Dist must accept such monsters

---

[46]   The outline of the argument below closely follows Sider (2001: 134–139).
[47]   See, e.g. Koslicki (2003); Markosian (2004); Balashov (2005a); Nolan (2006).

into her ontology. But isn't this price too high? As a matter of sociology, many 4Dists accept the bargain. The question is whether they have to.

They do not. More precisely, the 4Dist is free to treat synchronic and diachronic composition differently. The latter may be causally grounded in a way the former is not. Even in the absence of any principled criterion of restricted composition at a time, the 4Dist can draw a distinction between series of object stages cemented by a broadly causal relation[48] and those that are not. This immediately deflates the argument from vagueness, for premise P2 must be rejected. Contrary to what it asserts, one can provide an example of a continuous series of cases of minimal D-fusion featuring a *motivated* sharp cut-off as regards composition *across* time.[49] Consider an isolated object $o$ and the following two assignments: $f_1 = \{\langle t, \{o\}\rangle, t \in T\}$, where $T$ is $o$'s total lifetime, and $f_2 = \{\langle t, \{o\}\rangle, t \in T'\}$, where $T' = T - \{t^*\}$. These assignments are as close to each other as is possible, but one can reasonably insist that the first has a minimal diachronic fusion (viz. $o$ itself, throughout its life career), whereas the latter does not. Indeed the existence of the object, which would be the minimal D-fusion of $f_2$, would violate a fundamental physical law of conservation of matter, as this fusion would feature an object going out of existence at $t^*$ and popping back into existence *ex nihilo*. As a result, some later phases of such an object would not be connected by immanent causation to its earlier phases (no known dynamical laws act across temporal gaps). One need not accept the existence of such an object. In fact, one has every reason to ban it from one's ontology.

The example thus illustrates an important difference between synchronic and diachronic fusions. "Unnatural" synchronic fusions may be offending, but one cannot rule them out on the ground of their inconsistency with the laws of nature. Indeed, instantaneous composition across space does not appear to be cemented by any causal or nomic glue, so the existence or non-existence of unnatural synchronic aggregates of smaller objects becomes up for metaphysical grabs.[50] In contrast, the alleged existence of certain minimal D-fusions is strictly *incompatible* with some causal laws, and that may be reason enough to reject such "diachronic monsters."[51] The argument from vagueness to unrestricted composition across time is thereby blocked. This, in turn, undermines the case for 4D based on vagueness.

Let's take stock. We have looked at some traditional attempts to defend 4Dism against the default position backed by ordinary intuitions about persistence and identity (i.e. endurantism). The results of such attempts are hardly conclusive. We have not yet fully tapped into the resources of contemporary spacetime theories in physics, which appear to be highly relevant to the issue. It is time to do so.

---

[48] Known in the literature as *immanent causation*; see Zimmerman (1997).

[49] The example is taken, with some modifications and simplifications, from Balashov (2005a).

[50] And thus prevents one from giving an easy answer to the famous "special composition question" (van Inwagen (1990b)).

[51] For related arguments against the marriage of 4Dism and mereological universalism, see Balashov (2003ab), (2007).

# 5. ENDURING, PERDURING, AND EXDURING OBJECTS IN MINKOWSKI SPACETIME

Up until the last decade the debate about persistence was conducted, for the most part, in complete abstraction from physics.[52] The situation has now changed, with the growing number of works exploring various implications of relativity theory for the ontology of persistence.[53] There are two allegedly separate tasks here: (a) to state the rival views in the relativistic context; (b) to investigate whether such statements privilege a particular view over its rivals. There is considerable disagreement as regards both (a) and (b). Indeed some authors have contended that certain views of persistence cannot even be stated in the relativistic framework. While certain considerations to this effect can be shown to arise from confusion about what relativistic persistence properly amounts to,[54] others are highly sophisticated[55] and require detailed commentary which cannot be provided here. In this section I sketch one feasible approach to defining different modes of persistence in Minkowski spacetime,[56] noting, along the way, certain assumptions underlying this approach, which have been perceived by others as controversial. In Section 6, I raise some open questions about the prospects of the persistence debate in physical contexts that go beyond special relativity.

In classical spacetime, locations (and q-locations) of persisting objects, their parts and counterparts, as well as temporary properties were indexed by moments of absolute time—or, more precisely, by $t$-slices of the objects' paths (see Section 3). A natural adaptation of such indexing to the special relativistic framework requires further relativization to inertial frames of reference. This suggests a straightforward way of extending the definitions of Section 3 to Minkowski spacetime ($ST^M$) by replacing the classical '$t$' with a two-parameter index '$t^F$' tracking moments of time in a particular coordinate system adapted to a given inertial frame F. As before, we start with a three-place relation '$p$ is a part of $o$ at achronal region R'. Achronal regions of interest are now intersections of time hyperplanes with the objects' paths in Minkowski spacetime. We shall refer to such regions as '$t^F$-slices' of the objects' paths. Where $p$, $o$

---

[52] With the notable exceptions of Quine (1960: 171–172, 1987) and Smart (1972).

[53] See Rea (1998); Balashov (1999), (2000abc), (2002), (2003ab), (2005c), (2008), (2010); Sider (2001: §4.4); Gilmore (2002), (2006), (2007), (2008); Hales and Johnson (2003); Miller (2004); Hudson (2006: Ch. 5); Gibson and Pooley (2006); and Eagle, (2009).

[54] Such confusions are found, for example, in Hales and Johnson (2003), and are exposed in Miller (2004) and Gibson and Pooley (2006).

[55] See, in particular, Gilmore (2006), (2007), (2008).

[56] I thus set myself in opposition to those who object to the very possibility of relativistic formulations of the major views of persistence. My strategy in this section builds on the classical ideas developed in section 3. For a more systematic exposition see Balashov (2008), (2010, Ch. 5) For related approaches. see Rea (1998); Sider (2001: §4.4); and Sattig (2006: §5.4). For a critique of such approaches, see Gibson and Pooley (2006).

and a $t^F$-slice $\mathbb{O}_{\perp t^F}$ of $o$'s path $o$ stand in such a relation, we shall say that $p$ is a *spatial part* ($s^F$-*part*) *of o at* $\mathbb{O}_{\perp t^F}$:

(SP$^M$)    $p_\perp$ is a *spatial part* ($s^F$-*part*) of $o$ at $\mathbb{O}_{\perp t^F}$ =$_{df}$ $p_\perp$ is a part of $o$ at $\mathbb{O}_{\perp t^F}$.

And we explicate the notion of temporal parthood as follows:

(TP$^M$)    $p_{||}$ is a *temporal part* ($t^F$-*part*) of $o$ at $\mathbb{O}_{\perp t^F}$ =$_{df}$ (i) $p_{||}$ is located at $\mathbb{O}_{\perp t^F}$ but only at $\mathbb{O}_{\perp t^F}$, (ii) $p_{||}$ is a part of $o$ at $\mathbb{O}_{\perp t^F}$, and (iii) $p_{||}$ overlaps at $\mathbb{O}_{\perp t^F}$ everything that is a part of $o$ at $\mathbb{O}_{\perp t^F}$.

Spatial and temporal parthood relations in Minkowski spacetime governed by (SP$^M$) and (TP$^M$) are natural generalizations of their classical counterparts (SP$^G$) and (TP$^G$) from Section 3.

These relations can be employed to define the three modes of persistence in Minkowski spacetime.

(END$^M$)    $o$ *endures* in ST$^M$ =$_{df}$ (i) $o$'s path includes at least two points from different moments of time in a single (inertial) reference frame, (ii) $o$ is located at every $t^F$-slice of its path, (iii) $o$ is q-located only at $t^F$-slices of its path.

(PER$^M$)    $o$ *perdures* in ST$^M$ =$_{df}$ (i) $o$'s path includes at least two points from different moments of time in a single (inertial) reference frame, (ii) $o$ is q-located only at its path, (iii) the object located at any $t^F$-slice of $o$'s path is a proper $t^F$-part of $o$ at that slice.

(EXD$^M$)    $o$ *exdures* in ST$^M$ =$_{df}$ (i) $o$'s path includes at least two points from different moments of time in a single (inertial) reference frame, (ii) $o$ is located at exactly one region, which is a $t^F$-slice of its path, (iii) $o$ is q-located at every $t^F$-slice of its path.

Most of the comments made in connection with the classical versions of these definitions in Section 3 apply in the relativistic context too. Clause (i) of all three statements, in particular, supplies a relativistic analog of the requirement that a corresponding object persist—that its path contain two diachronically (i.e. non-spacelike) separated points.

The above definitions need to be supplemented with an account of the relativization of temporary properties of persisting objects to their locations (in the case of endurance), q-locations (in the case of exdurance), or the locations of their $t^F$-parts (in the case of perdurance). Such (q-)locations are, of course, $t^F$-slices of the objects' path, which can be usefully labeled with the same two-parameter index that figures in the above definitions. Thus, while in the Galilean framework, objects have properties at absolute moments of time (more precisely, at absolute time slices of the objects' paths; see Section 3), in the Minkowskian framework the possession of temporary properties is relativized, in effect, to times-in-frames. This brings some novel features. Consider, for example, a meter stick, whose path is a shaded region in Figures 1.2 abc (with two dimensions of space suppressed). Even if the stick does not change its *proper length* (i.e. the length it has in its rest frame), it exemplifies different length at time slices of its path drawn at different reference frames, such as $Slice_1$ and $Slice_2$. The endurantist will say that the stick is located at both slices and bears the *1 meter-at* relation to $Slice_1$ and *0.5 meters-at* relation to $Slice_2$.[57] The perdurantist will say that the stick is located at its path and has two distinct $t^F$-parts at $Slice_1$ and $Slice_2$ of different corresponding length. The exdurantist will say that the stick is q-located at $Slice_1$ and $Slice_2$ and has simple lengths *1 meter* and *0.5 meters* there, thanks to its counterparts. See Figures 1.2a,b,c.

As they stand, $(END^M)$, $(PER^M)$ and $(EXD^M)$ assume the relativistic version of achronal universalism, the thesis that each enduring object is located at *every* $t^F$-slice of its path, each perduring object has a $t^F$-part at *every* $t^F$-slice of its path, and each exduring object is q-located at *every* $t^F$-slice of its path. In classical spacetime, achronal universalism looked entirely unproblematic, especially when supplemented with the notion that the path of a persisting object is a union of regions at which the object is located or q-located. But things are less straightforward in Minkowski spacetime. Imagine Unicolor, a persisting object, one of whose essential properties is to be *uniformly colored*. Suppose further that Unicolor uniformly changes its color with time in a certain inertial reference frame F. Consider a $t^{F^*}$-slice of Unicolor's path that crisscrosses hyperplanes of simultaneity in F. Whatever (if anything) is located (or q-located) at such a slice is not uniformly colored and, hence, must be distinct from Unicolor, even though it is filled with the (differently colored) material components of Unicolor.[58]

As another example (due to Gilmore 2006: 212–213), consider a complete path of an enduring or exduring object $o$ composed of four atoms $o_1$, $o_2$, $o_3$ and $o_4$ in Minkowski spacetime. See Figure 1.3.

$o_1$, $o_2$, $o_3$ and $o_4$ pop into existence at $t_1^F$ and go out of existence at $t_2^F$. Both $t_1^F$-and $t_2^F$-slices of $o$'s path $o$ are good candidates for $o$'s location (q-location), and so are all the $t^F$-slices for any $t^F \in (t_1^F; t_2^F)$. Consider, however, the $t^{F^*}$-slice in frame $F^*$ distinct from F. According to Achronal Universalism, $o$ must be located in it. But this is problematic, for the $t^{F^*}$-slice of $o$ is a "corner slice" that contains a single atom $o_1$ and can hardly qualify for being a suitable location (or q-location) for the entire object $o$. Recall that on our understanding of the basic notion of (q-)location, a region at which an object is exactly (q-)located, or "wholly present," is the region into which the object exactly fits and which has exactly the same size, shape, and position as the object itself (see Section 3). But the $t^{F^*}$-slice of $o$ is shaped like a single atom and, hence, not shaped like $o$. An object that is, for most of its career, composed of four atoms cannot "fit into" a region shaped like one atom.

---

[57] Alternatively, the endurantist can say (i) that the stick atemporally or tenselessly exemplifies two *timeslice-indexed* lengths, *1 meter-at-$Slice_1$* and *0.5 meters-at-$Slice_2$*; or (ii) that the stick exemplifies two simple length properties, *1 meter* and *0.5 meters*, in two different *timeslice-modified ways*. (i) and (ii) are natural relativistic adaptations of the semantical variants of the endurantist relativization strategies known as "indexicalism" and "adverbialism." For a recent review of such strategies, see Haslanger (2003).

[58] For another illustration of the same point, see Gilmore (2006: 210–211).

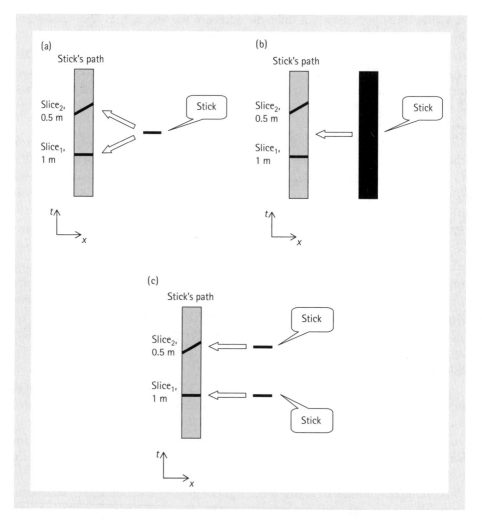

FIGURE 1.2 Endurance (a), perdurance (b), and exdurance (c) in Minkowski spacetime

These examples illustrate some curious properties of the location relation in Minkowski spacetime. Do they go as far as to refute the Minkowskian version of achronal universalism? The case of Unicolor is easy to brush aside as being too metaphysically recherché. More needs to be said about Gilmore's "corner slice" case. Consider the evolution of $o$ in $F^*$. From the physical point of view, $F^*$ is a legitimate frame of reference, which describes $o$ as progressively losing parts, one by one. How many atomic parts could $o$ lose without ceasing to exist? Maybe just a few, or maybe the majority of them. Perhaps there is no general answer and it all depends on the nature of the object in question. Perhaps the answer will remain vague even then. But when the evolution of $o$ is viewed from a perspective in which it looks gradual, it becomes clear that (i) questions of this sort must indeed be settled *before* one

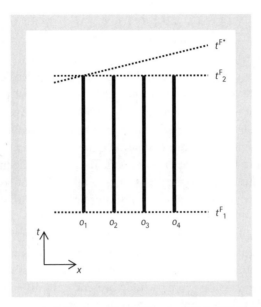

FIGURE 1.3 "Corner slice" (Gilmore 2006: 212–213)

attempts to draw the exact boundaries of $o$'s path,[59] and (ii) *exactly the same* questions would arise if spacetime were classical and $t^{F*}$-planes represented the absolute time planes. But in difference from the classical situation, where settling such questions *ipso facto* determined the boundaries of the object's path in spacetime, thus entailing the classical version of achronal universalism, the situation in Minkowski spacetime is more complicated and shows the need to refine $(\text{END}^M)$, $(\text{PER}^M)$ and $(\text{EXD}^M)$. Simply assuming a relativistic analog of achronal universalism seems to put the cart before the horse.

A separate issue is whether (q-)locations of persisting objects and their parts should be restricted to *flat* slices of their paths, as is done above. Shouldn't *curved* spacelike slices be allowed as well? Limitations of space do not allow us to discuss these interesting possibilities.[60] However, the above brief considerations suffice to show that no view of persistence is an obvious non-starter in Minkowski spacetime. All three views can be adequately stated in that framework in a way that respects their distinctive features. It is a different question whether, *when* so stated, the views fare equally well.[61] On this question, the jury is still out, and the reader is invited to turn to the growing literature on the topic. But we have to leave the topic here.

[59] Cf. Gibson and Pooley (2006: 186–187), who make a very similar suggestion.
[60] For arguments against privileging flat slices, even in the context of special relativity, see Gibson and Pooley (2006, 159–165). For an extended reply to such arguments, see Balashov (2008: §5).
[61] For arguments that they do not, see Quine (1960: 171–172, 1987); Smart (1972); Balashov (1999), (2000abc), (2002); and Gilmore (2006), (2007). For criticisms of all these arguments, see Gilmore (2002), (2008); Miller (2004); Sider (2001: §4.4); Sattig (2006: 182–183); and Gibson and Pooley (2006). For some replies, see Balashov (2005c), (2009) and (2010: Chs. 6–8).

# 6. Beyond Special Relativity?

Insofar as physical considerations are brought to bear on the debate about persistence, discussions in the literature tend to be restricted, for the most part, to the classical (non-quantum) framework of special relativity. This has the advantage of making the discussion manageable. But it certainly brings with it serious limitations. Although Minkowski spacetime is a good approximation of most of the spacetime of our world, it is, in the end, just that: an approximation. It is not immediately clear how to extrapolate the notions central to the debate to general relativistic spacetime, which has no place for global moments of time and intuitive analogs of "momentary" locations (or q-locations).

More importantly, it is even less clear how to think about persistence in the context of (non-relativistic) quantum mechanics, and it is entirely unclear how to begin thinking about it in the context of quantum field theory. Indeed, it is unclear how to apply endurantist concepts (such as 'being wholly present at multiple spacetime regions') even to classical fields. This is not to suggest that questions of this sort should not be raised. But their serious discussion requires extensive preliminary work. We hope that the foregoing considerations go some way towards fulfilling the prerequisites for more ambitious projects.[62]

## References

Armstrong, David (1980), 'Identity Through Time', in P. van Inwagen (ed.), *Time and Cause: Essays Presented to Richard Taylor* (Dordrecht: D. Reidel), 67–78.

Baker, Lynne Rudder (2000), *Persons and Bodies* (Cambridge: Cambridge University Press).

Balashov, Yuri (1999), 'Relativistic Objects', *Noûs* 33: 644–662.

—— (2000a), 'Enduring and Perduring Objects in Minkowski Space-Time', *Philosophical Studies* 99: 129–166.

—— (2000b), 'Relativity and Persistence', *Philosophy of Science* 67 (Proceedings): S549–S562.

—— (2000c), 'Persistence and Space-Time: Philosophical Lessons of the Pole and Barn', *The Monist* 83: 321–340. Reprinted in Haslanger and Kurtz 2006.

—— (2002), 'On Stages, Worms, and Relativity', in C. Callender (ed.), *Time, Reality, and Experience* (Cambridge: Cambridge University Press), 223–252.

—— (2003a), 'Temporal Parts and Superluminal Motion', *Philosophical Papers* 32: 1–13.

—— (2003b), 'Restricted Diachronic Composition, Immanent Causality, and Objecthood: A Reply to Hudson', *Philosophical Papers* 32: 23–30.

—— (2005a), 'On Vagueness, 4D and Diachronic Universalism', *The Australasian Journal of Philosophy* 83: 523–531.

[62] I am indebted to Maureen Donnelly and Cody Gilmore for very helpful discussions of the ideas of this chapter and for extensive comments on earlier drafts. Thanks also to Chuck Cross. Work on this chapter was supported by a Senior Research Fellowship from the Willson Center for Humanities and Arts at the University of Georgia.

—— (2005b), 'Times of Our Lives: Negotiating the Presence of Experience', *American Philosophical Quarterly* 42: 295–309. Reprinted in L. N. Oaklander (ed.), *Philosophy of Time: Critical Concepts in Philosophy* (London: Routledge, 2008), Vol. 3.

—— (2005c), 'Special Relativity, Coexistence and Temporal Parts: A Reply to Gilmore', *Philosophical Studies* 124: 1–40.

—— (2007), 'About Stage Universalism', *Philosophical Quarterly* 57: 21–39.

—— (2008), 'Persistence and Multilocation in Spacetime', in D. Dieks (ed.), *The Ontology of Spacetime*, Vol. 2 (Amsterdam: Elsevier), 59–81.

—— (2009), 'Pegs, Boards and Relativistic Perdurance', *Pacific Philosophical Quarterly* 90: 167–175.

—— (2010), *Persistence and Spacetime* (Oxford: Oxford University Press),

——, and Janssen, Michel (2003), 'Presentism and Relativity: A Critical Notice', *British Journal for the Philosophy of Science* 54: 327–346. Reprinted in L. N. Oaklander (ed.), *Philosophy of Time: Critical Concepts in Philosophy* (London: Routledge, 2008), Volume 3.

Bigelow, John (1996), 'Presentism and Properties', in J. Tomberlin (ed.), *Philosophical Perspectives 10, Metaphysics* (Cambridge, MA: Blackwell), 35–52.

Bittner, Thomas, and Donnelly, Maureen (2004), 'The Mereology of Stages and Persistent Entities', in R. Lopez de Mantaras and L. Saitta (eds.), *Proceedings of the European Conference of Artificial Intelligence* (IOS Press), 283–287.

Bourne, Craig (2006), *A Future for Presentism* (Oxford University Press).

Butterfield, Jeremy (2006), 'The Rotating Discs Argument Defeated', *British Journal for the Philosophy of Science* 57: 1–45.

Callender, Craig (2000a), 'Shedding Light on Time', *Philosophy of Science* 67 (Supplement): S587–S599.

—— (2000b), 'Humean Supervenience and Rotating Homogeneous Matter', *Mind* 110: 25–43.

Carter, William and Hestevold, H. Scott (1994), 'On Passage and Persistence', *American Philosophical Quarterly* 31: 269–283.

Casati, Roberto, and Varzi, Achille (1999), *Parts and Places: The Structures of Spatial Representation* (Cambridge, MA: The MIT Press).

Craig, William Lane, and Smith, Quentin (eds.) (2008), *Einstein, Relativity and Absolute Simultaneity* (London: Routledge).

Crisp, Thomas (2005), 'Presentism and Cross-Time Relations', *American Philosophical Quarterly* 42: 5–17.

—— (2007), 'Presentism and Grounding', *Noûs* 41: 90–109.

——, and Smith, Donald (2005), ' "Wholly Present" Defined', *Philosophy and Phenomenological Research* 71: 318–344.

Donnelly, Maureen (2009), 'Parthood and Multi-Location', in D. W. Zimmerman (ed.), *Oxford Studies in Metaphysics*, Vol. 5 (Oxford: Oxford University Press), 203–243.

Eagle, Antony (2009), 'Location and Perdurance', in D. W. Zimmerman (ed.), *Oxford Studies in Metaphysics*, Vol. 5 (Oxford: Oxford University Press), 53–94.

Gibson, Ian, and Pooley, Oliver (2006), 'Relativistic Persistence', in J. Hawthorne (ed.), *Philosophical Perspectives*, Vol. 20, *Metaphysics* (Oxford: Blackwell), 157–198.

Gilmore, Cody (2002), 'Balashov on Special Relativity, Coexistence, and Temporal Parts', *Philosophical Studies* 109: 241–263.

Gilmore, Cody (2006), 'Where in the Relativistic World Are We?', in J. Hawthorne (ed.), *Philosophical Perspectives*, Vol. 20, *Metaphysics* (Oxford: Blackwell), 199–236.

—— (2007), 'Time Travel, Coinciding Objects, and Persistence', in D. W. Zimmerman (ed.), *Oxford Studies in Metaphysics*, Vol. 3 (Oxford: Clarendon Press), 177–198.

—— (2008), 'Persistence and Location in Relativistic Spacetime', *Philosophy Compass* 3/6: 1224–1254.

—— (2009), 'Why Parthood Might Be a Four-Place Relation, and How It Behaves If It Is', in Ludger Honnefelder, Edmund Runggaldier, and Benedikt Schick (eds.), *Unity and Time in Metaphysics* (Berlin: de Gruyter), 83–133,

Hales, Steven, and Johnson, T. (2003), 'Endurantism, Perdurantism, and Special Relativity', *Philosophical Quarterly* 213: 524–539.

Haslanger, Sally (1989), 'Endurance and Temporary Intrinsics', *Analysis* 49: 119–125.

—— (2003), 'Persistence Through Time', in M. J. Loux and D. W. Zimmerman (eds.), *The Oxford Handbook of Metaphysics* (Oxford: Oxford University Press), 315–354.

——, and Kurtz, Roxanne (2006), *Persistence: Contemporary Readings* (Cambridge, MA: The MIT Press).

Hawley, Katherine (2001), *How Things Persist* (Oxford: Oxford University Press).

—— (2004), 'Temporal Parts', E.N. Zalta (ed.), *The Stanford Encyclopedia of Philosophy*. URL = <http://plato.stanford.edu/entries/temporal-parts>

Hawthorne, John (2006), 'Three-Dimensionalism', in J. Hawthorne, *Metaphysical Essays* (Oxford: Clarendon Press), 85–109.

Heller, Mark (1990), *The Ontology of Physical Objects: Four Dimensional Hunks of Matter* (Cambridge: Cambridge University Press).

Hinchliff, M. (1996), 'The Puzzle of Change', in J. E. Tomberlin (ed.), *Philosophical Perspectives* 10 (Oxford: Basil Blackwell), 119–136.

Hudson, Hud (2001), *A Materialist Metaphysics of the Human Person* (Ithaca: Cornell University Press).

—— (2006), *The Metaphysics of Hyperspace* (Oxford: Oxford University Press).

Johnston, Mark (1987), 'Is There a Problem about Persistence?', *Aristotelian Society* (Supplement) 61: 107–135.

Koslicki, Kathrin (2003), 'The Crooked Path from Vagueness to Four-Dimensionalism', *Philosophical Studies* 114: 107–134.

Lewis, David (1983), 'Survival and Identity', in D. Lewis, *Philosophical Papers, Vol. I* (Oxford: Oxford University Press), 55–77.

—— (1986), *On the Plurality of Worlds* (Oxford: Basil Blackwell).

—— (1988), 'Rearrangement of Particles: Reply to Lowe', *Analysis* 48: 65–72.

—— (1999), 'Zimmerman and the Spinning Sphere', *Australasian Journal of Philosophy* 77: 209–212.

Lombard, Lawrence (1999), 'On the Alleged Incompatibility of Presentism and Temporal Parts', *Philosophia* 27: 253–260.

Lowe, E. J. (1988), 'The Problems of Intrinsic Change: Rejoinder to Lewis', *Analysis* 48: 72–77.

MacBride, Fraser (2001), 'Four New Ways to Change Your Shape', *The Australasian Journal of Philosophy* 79: 81–89.

McGrath, Matthew (2007a), 'Four-Dimensionalism and the Puzzles of Coincidence', in D. W. Zimmernam (ed.), *Oxford Studies in Metaphysics*, Vol. 3 (Oxford: Oxford University Press), 143–176.

—— (2007b), 'Temporal Parts', *Philosophy Compass* 2 (5): 730–748.

McKinnon, Neil (2002), 'The Endurance/Perdurance Distinction', *Australasian Journal of Philosophy* 80: 288–306.

Markosian, Ned (2003), 'A Defense of Presentism', in D. Zimmerman (ed.), *Oxford Studies in Metaphysics,* Vol. 1 (Oxford: Oxford University Press), 47–82.

—— (2004), 'Two Arguments from Sider's *Four-Dimensionalism*', *Philosophy and Phenomenological Research* 68: 665–673.

Merricks, Trenton (1994), 'Endurance and Indiscernibility', *Journal of Philosophy* 91: 165–184.

—— (1995), 'On the Incompatibility of Enduring and Perduring Entities', *Mind* 104: 523–531.

—— (1999), 'Persistence, Parts and Presentism', *Noûs* 33: 421–438.

Miller, Kristie (2004), 'Enduring Special Relativity', *Southern Journal of Philosophy* 42: 349–370.

Mozersky, Joshua, 'Presentism', in Craig Callender (ed.), *Oxford Handbook of Time* (Oxford: Oxford University Press), this volume.

Nolan, Daniel (2006), 'Vagueness, Multiplicity and Parts', *Noûs* 40: 716–737.

Parsons, Josh (2000), 'Must a Four-Dimensionalist Believe in Temporal Parts?', *The Monist* 83: 399–418.

—— (2007), 'Theories of Location', in D. W. Zimmerman (ed.), *Oxford Studies in Metaphysics*, Vol. 3 (Oxford: Oxford University Press), 201–232.

Quine, W. V. (1960), *Word and Object* (Cambridge, Mass.: The MIT Press).

—— (1987), 'Space-Time', in W. V. Quine, *Quiddities* (Cambridge, MA: Harvard University Press), 196–199.

Rea, Michael (1998), 'Temporal Parts Unmotivated', *Philosophical Review* 107: 225–260.

—— (ed.) (1997), *Material Constitution* (Lanham, Maryland: Rowman and Littlefield Publishers).

Sattig, Thomas (2006), *The Language and Reality of Time* (Oxford: Clarendon Press).

Saucedo, Raul (forthcoming), 'Parthood and Location', forthcoming in D. W. Zimmerman (ed.), *Oxford Studies in Metaphysics*, Vol. 6 (Oxford: Oxford Univeresity Press).

Saunders, Simon (2002), 'How Relativity Contradicts Presentism', in C. Callender (ed.), *Time, Reality, and Experience* (Cambridge: Cambridge University Press), 277–292.

Savitt, Steven (2000), 'There's No Time Like the Present (in Minkowski Spacetime)', *Philosophy of Science* 67 (Supplement): S563–S574.

Sider, Theodore (2000a), 'The Stage View and Temporary Intrinsics', *Analysis* 60: 84–88.

—— (2001), *Four-Dimensionalism: An Ontology of Persistence and Time* (Oxford: Clarendon Press).

—— (2008), 'Temporal Parts', in T. Sider, J. Hawthorne, and D. W. Zimmerman (eds.), *Contemporary Debates in Metaphysics* (Oxford: Blackwell), 241–262.

Smart, J. J. C. (1972), 'Space-Time and Individuals', in R. Rudner and I. Scheffler (eds.), *Logic and Art: Essays in Honor of Nelson Goodman* (New York: Macmillan), 3–20.

Thomson, J. J. (1983), 'Parthood and Identity Across Time', *Journal of Philosophy* 80: 201–220.

van Inwagen, Peter (1990a), 'Four-Dimensional Objects', *Noûs* 24: 245–255.

—— (1990b), *Material Beings* (Ithaca: Cornell University Press).

Zimmerman, Dean (1996), 'Persistence and Presentism', *Philosophical Papers* 25: 115–126.

Zimmerman, Dean (1997), 'Immanent Causation', in J. Tomberlin (ed.), *Philosophical Perspectives*, Vol. 11 (Oxford: Blackwell), 433–471.

——(1998), 'Temporal Parts and Supervenient Causation: The Incompatibility of Two Humean Doctrines', *Australasian Journal of Philosophy* 76: 265–288.

——(1999), 'One Really Big Liquid Sphere: Reply to Lewis', *Australasian Journal of Philosophy* 77: 213–215.

——(2008), 'The Privileged Present: Defending an "A-theory" of Time', in T. Sider, J. Hawthorne, and D. W. Zimmerman (eds.), *Contemporary Debates in Metaphysics* (Oxford: Blackwell), 211–225.

——, 'Presentism and the Space-Time Manifold', in Craig Callender (ed.), *Oxford Handbook of Time* (Oxford: Oxford University Press), this volume.

# CHAPTER 2

..........................................................................................

# FATALISM AND THE FUTURE

..........................................................................................

CRAIG BOURNE

# 1. INTRODUCTION

..........................................................................................

STRANGE goings on! Donald did it slowly, deliberately, in the bathroom, with a knife, at midnight. What he didn't do was butter a piece of toast. What he did do was commit murder. The strangeness of the murder did not lie in the act itself, but how it came about. For that Donald would murder someone was foretold by his lodger Potcha, a somewhat sweaty and annoying little man who had a knack with the tea leaves and who relished knowing more about some people than the nosiest of neighbours. Donald was tormented for months, knowing what he would do, and he hated Potcha for telling him. So at midnight, one evening, with knife in one hand (and some toast in the other), Donald cornered Potcha in the bathroom to let the inevitable take place . . . Carefully avoiding the blood as he licked a stray globule of Marmite from his finger, Donald felt a tinge of remorse—after all, despite his insufferable personality, Potcha was guilty only of telling the truth concerning Donald's future midnight murdering escapades. But, Donald reasoned, feeling remorseful was rather pointless—he could hardly undo the deadly deed now—and, in any case, it was nothing compared with the relief he felt in having shed the burden that had been hanging over him for the past few months. He now felt free to get on with the rest of his life.[1]

Despite its literary shortcomings, this tale at least has the virtue of raising many of the interesting questions surrounding fatalism—the view that if something is going to happen, it's going to happen, and there isn't anything anyone could have done to make it otherwise. Certainly, much fatalist reasoning is rather uninteresting since it can swiftly be dismissed as trivial or obviously fallacious. Yet there is real interest

---

[1] My short tale is inspired by Oscar Wilde's (1887) but some details have been changed to bring out the central philosophical issues. Davidson (1967) should also be held responsible.

that lies in uncovering those genuinely perplexing aspects of time that, in the wrong head, give rise to the fatalist confusion. For it arises from the core question of how our world is composed and how we, as free agents, fit into the picture. Central to our experience as agents is the idea that the past is 'fixed'—in some sense *necessary*—but the future is 'open'—in some sense *possible*. It is the interrelation between these temporal and modal notions that gives rise to fatalist thinking. The common sense conception of us as agents is that we are clearly freer to influence the future than the past. But some argue that if the future is not to be treated differently from the past in important respects, it must inherit the necessity—stemming from the idea of fixity—associated with the past. It is understandable, then, why those who feel the pull of fatalist reasoning often concern themselves with trying to break the symmetry between the past and the future. Let us consider one such approach.

## 2. THE BASIC ARISTOTELIAN ARGUMENT

Potcha told the truth about what Donald was to do. If it was true that he would murder Potcha, Donald couldn't make it untrue that he would, any more than he could make it untrue after he had murdered Potcha. So, just as Donald isn't free not to have murdered Potcha after he had done it, he wasn't free not to murder Potcha before he did it. If it was already true, when Potcha spoke, that the murder would take place, then it couldn't fail to take place.

Fatalist reasoning of this sort can be found in Aristotle's *De Interpretatione*, ch. 9.[2] Being no fatalist, however, Aristotle uses the argument to draw a different conclusion: since Donald *was* free not to bring it about that he murdered Potcha, it can't be that it was *true*, when Potcha spoke, that he would. But neither is this to say that it was false. For then Donald would not have been free *to* murder Potcha, which he clearly would have been. Thus, since Donald was free either to murder or not, what Potcha said must have been *neither true nor false*.[3]

What should we make of the fatalist argument on which this conclusion rests? Seductive as the argument is (and not just to first-year undergraduates), there is one way of presenting it that clearly involves fallacious reasoning. We can accept that it must be true that if Donald will murder Potcha, then Donald will murder Potcha. But it doesn't follow from this that, because the murder *will* take place, it *had* to; that it *would*

---

[2]  It is contentious exactly what Aristotle is trying to do throughout *De Interpretatione* Ch. 9 and it is not possible in a short chapter even to survey all the interpretations on offer, let alone offer criticism. In what follows, I shall concern myself mainly with the standard interpretation of his position, whether or not he actually held it.

[3]  For our purposes, I shall not distinguish between two ways of taking this: as a *lack* of one of the two determinate truth-values, or as a *third*, indeterminate truth-value. See Haack (1974) for discussion. Some (from Ammonius and Boethius onwards, it seems) would interpret Aristotle as taking what Potcha said to be *either-true-or-false* but not *definitely* true/false. Whether this is any different from the third, indeterminate truth-value cannot be discussed here (see Kretzmann (1998) for some discussion).

still have occurred *whatever* else happened. Some notation will help us to see this, if it isn't already clear. Let $p$ be a proposition about the present, such as the proposition that *Donald is murdering Potcha*. Let **P** and **F** be past- and future-tense operators on propositions, read 'It was the case that …' and 'It will be the case that …', respectively, and let □ be the necessity operator on propositions, read 'It is necessary that …'. Then, where $p$ is the proposition that *Donald is murdering Potcha*, we can notate the fact that Donald murdered Potcha by '$\mathbf{P}p$', the fact that Donald will murder Potcha by '$\mathbf{F}p$', and the fact that Donald has to murder Potcha by '□$p$'. The 'seductive argument' can then be stated as follows:

> Premise:    □($\mathbf{F}p \supset \mathbf{F}p$) [Necessarily, if Donald will murder Potcha, then Donald will murder Potcha.]
>
> Conclusion:    ($\mathbf{F}p \supset$ □$\mathbf{F}p$) [If Donald will murder Potcha, then necessarily Donald will murder Potcha.]

Because both premise and conclusion are naturally expressed in the same way in English—'If it's true that he did it, then he must have done it!'—it is understandable why so many slide from one to the other. But once the scope of the necessity is brought out, the argument is shown to be invalid. One can and should think that the premise— a harmless, trivial truth—is true, without thinking the conclusion—a more substantial claim—is.

Taken this way, Aristotle's argument for future contingents being neither true nor false crumbles because the fatalist foundations are rotten—the trivial truth alone is not harmful enough to motivate the idea that future contingents are indeterminate. However, although many do commit this seductive fallacy, it is not clear that Aristotle does. For all that has been argued so far is that it is not possible to derive ($\mathbf{F}p \supset$ □$\mathbf{F}p$) from □($\mathbf{F}p \supset \mathbf{F}p$). But Aristotle seems fully aware of this fallacy: '… there is a difference between saying that which is, when it is, must needs be, and simply saying that all that is must needs be …' (*DI* 9, 19$^a$ 25–26). Still, Haack (1974: 79) interprets this in a way that she argues is consistent with Aristotle committing the fallacy: '[it says] that *once an event has happened*, it is necessary, i.e. irrevocable, although not all events are, before they happen, necessary, i.e. inevitable.' This may well be an accurate rendering of what Aristotle meant. But whether or not it is, to accuse anyone of fallacious reasoning on this basis is too quick. Showing that seductive inference is invalid is one thing; showing that we cannot take something of the form (A $\supset$ □A) to feature as a *premise* of the argument for indeterminate future contingents is another. For it hasn't been shown that there are no interpretations under which something of the form (A $\supset$ □A) itself is valid. Since this is something that Aristotle endorses when applied to propositions about the past—they *are* necessary, if true—the crucial thing to establish in assessing whether the seductive fallacy has been committed is whether the *sole* basis for holding something of the form (A $\supset$ □A) is □(A $\supset$ A).

A charge of fallacious reasoning, then, is at this stage premature. The question of whether there are any better reasons for accepting instances of (A $\supset$ □A) is to be addressed in later sections. But whatever the reason, we can present what I shall call the basic Aristotelian argument as follows. Let '$t(\ldots)$' mean that the proposition within the brackets is true and '$f(\ldots)$' mean that the proposition within the brackets is false. Then: $t(\mathbf{F}p) \supset \Box \mathbf{F}p$, $\neg\Box\mathbf{F}p \therefore \neg t(\mathbf{F}p)$. Similarly, $f(\mathbf{F}p) \supset \Box\neg\mathbf{F}p$, $\neg\Box\neg\mathbf{F}p \therefore \neg f(\mathbf{F}p)$. Thus, $\neg(t(\mathbf{F}p) \vee f(\mathbf{F}p))$. This argument is formally valid.[4]

Thus if we insist that what Potcha said when he said it was neither true nor false, Donald did not have to murder him. Fatalism is avoided.[5]

# 3. ŁUKASIEWICZ'S LOGIC

Jan Łukasiewicz in the 1920s and '30s formulated a three-valued logic in an attempt to formalize Aristotle's position. He writes:

> 'I can assume without contradiction that my presence in Warsaw at a certain moment of next year, e.g., at noon on 21 December, is at the present time determined neither positively nor negatively. Hence it is possible, but not necessary, that I shall be present in Warsaw at the given time. On this assumption the proposition "I shall be in Warsaw at noon on 21 December of next year", can at the present time be neither true nor false. For if it were true now, my future presence in Warsaw would have to be necessary, which is contradictory to the assumption. If it were false now, on the other hand, my future presence in Warsaw would be impossible, which is also contradictory to the assumption. Therefore the proposition considered is at the moment *neither true nor false* and must possess a third value, different from "0" or falsity and "1" or truth. This value we can designate by "1/2". It represents "the possible" and joins "the true" and "the false" as a third value.
>
> The three-valued system of propositional logic owes its origin to this line of thought.'                                                       (Łukasiewicz (1930: 53))

It is easy to see why Łukasiewicz has been subject to the criticism that his argument commits the seductive fallacy. But, as with Aristotle, whether this assessment is warranted rests on the reasons for holding (A $\supset$ $\Box$A). If there is an interpretation of $\Box$ under which this is acceptable, we should take Łukasiewicz's reasoning to be a legitimate motivation for his three-valued logic.

In his system, contingent propositions about the future are given the value = $\frac{1}{2}$, true propositions are given the value = 1 and false propositions are given the value = 0. The logical connectives are defined as follows:

We can see that the purely determinate entries match the tables of the classical two-valued system. What needs justification are the other entries. Let us take negation to

---

[4] This is acknowledged by Haack (1974: Ch. 4) but, for the reason quoted above, she argues that Aristotle is ultimately seduced by the fallacy. She doesn't, however, consider the interpretations of the argument I suggest below.

[5] It might be asked why this does not leave us with saying that if $\mathbf{F}p$ is neither true nor false, it is necessarily neither true nor false. I leave this issue for the reader to resolve.

FIGURE 2.1  Łukasiewicz negation

|   | ¬ |
|---|---|
| 1 | 0 |
| ½ | ½ |
| 0 | 1 |

FIGURE 2.2  Łukasiewicz conjunction

| & | 1 | ½ | 0 |
|---|---|---|---|
| 1 | 1 | ½ | 0 |
| ½ | ½ | ½ | 0 |
| 0 | 0 | 0 | 0 |

FIGURE 2.3  Łukasiewicz disjunction

| ∨ | 1 | ½ | 0 |
|---|---|---|---|
| 1 | 1 | 1 | 1 |
| ½ | 1 | ½ | ½ |
| 0 | 1 | ½ | 0 |

illustrate. $Fp$ is indeterminate if $p$ is a proposition about a contingent matter, because there are no actual facts to determine what the truth-value of $p$ is at any future time. The truth-value of the negation of an indeterminate proposition is itself indeterminate, since if $p$'s truth-value could go either way, so must its negation.[6] This reasoning similarly justifies the '½' entries in the other tables.

Łukasiewicz's system works smoothly for most cases of future contingent propositions. Consider, for instance, something of the form (A ∨ B) where each disjunct concerns the future:

(1)    Donald will murder Potcha or Donald will butter some toast

We would intuitively think that if both of the disjuncts are indeterminate, then the whole disjunction must be indeterminate—he might do either action; he might do neither. This is precisely the answer given by Łukasiewicz's truth-tables.

However, the trouble begins when we consider cases where one disjunct is the negation of the other, i.e. something of the form (A ∨ ¬A). In classical two-valued logic, this is a logical truth—it is true no matter what the truth-value of A—and is known

---

[6]  Following, it seems, Aristotle (*DI*9, 19ᵃ 19–20).

FIGURE 2.4 What seems to follow for the material conditional
from Łukasiewicz negation, conjunction, and disjunction

| ⊃ | 1 | ½ | 0 |
|---|---|---|---|
| 1 | 1 | ½ | 0 |
| ½ | 1 | ½ | ½ |
| 0 | 1 | 1 | 1 |

as the *Law of Excluded Middle*. This should be distinguished from the *Principle of Bivalence*, which states that every proposition takes only one of two possible truth-values: *true* or *false*. When a disjunct and its negation concern the past or the present, Łukasiewicz's system agrees with classical two-valued logic: (A ∨ ¬A) comes out true. This is because the Principle of Bivalence is taken to hold for propositions about the past and present.[7] This is a good result because it is hard to see how it could be anything but true. But where the disjuncts concern the future, complications arise. Consider:

(2)    Either Donald will love the Marmite or Donald will not love the Marmite.

Because there seems to be no middle ground to be had—Donald will either love or not love the Marmite[8]—(2) as a whole should come out determinately true. The Law of Excluded Middle should remain as a logical truth. But since each disjunct is indeterminate, according to Łukasiewicz's truth-tables, the whole disjunction must be indeterminate. The Law of Excluded Middle is not a logical truth in this system. Łukasiewicz gives us the wrong answer.

Similarly, in two-valued logic, the Law of Non-Contradiction (¬(A & ¬A)) is a logical truth. But, for Łukasiewicz, something of this form is only true for propositions about the past and present. When applied to propositions about the future, it too takes the value = ½, and again supplies the wrong result.

Finally, it is, for me at least, entirely compelling to think that (A ⊃ B) is equivalent to (¬(A & ¬B)),[9] and so Łukasiewicz should define '⊃' as:

But in the special case of (F*p* ⊃ F)—'what will be, will be'—Łukasiewicz's system should then render it indeterminate. It would be confused to think of this as a virtue— say, as a way of avoiding fatalism—since we have seen from the very beginning that this truth is benign. It not being necessary is no virtue. Curiously, however, Łukasiewicz agrees, for he defines '⊃' as:

---

[7] Following Aristotle (*DI* 9, 18ᵃ 27–8).

[8] Note that this claim is different from that which the Marmite manufacturers make: you'll either love it or hate it.

[9] Some think the same about (¬A ∨ B). However, (¬(A & ¬B)) ≢ (¬A ∨ B) in the system I propose below (consider: A = 1 and B = ½). Since I have a better grasp of conjunction than disjunction, I prefer to think of conditionals in this way. Others might think the opposite and define the conditional using disjunction. For the purposes of the following argument and the other laws at issue in this chapter, either way will do.

FIGURE 2.5  Łukasiewicz conditional

| ⊃ | 1 | ½ | 0 |
|---|---|---|---|
| 1 | 1 | ½ | 0 |
| ½ | 1 | 1 | ½ |
| 0 | 1 | 1 | 1 |

But given that Figure 2.4 is a natural consequence of Łukasiewicz's truth-tables for '¬', '&' and '∨', changing the awkward middle entry to give Figure 2.5 looks like a plain fudge.[10]

Aristotle's position, however, does not fall with Łukasiewicz's, since Łukasiewicz didn't adequately formalize Aristotle's position. Although Aristotle rejects the unrestricted form of the Principle of Bivalence because it doesn't apply to future contingents, he endorses the unrestricted forms of the Laws of Excluded Middle and Non-Contradiction.[11] But this raises the question of whether there can be a satisfactory formalization of Aristotle's position. Is it possible to have a non-bivalent logic and yet hold on to the Laws of Excluded Middle and Non-Contradiction? I shall call this the Problem of Future Contingents.

## 4. NON-TRUTH-FUNCTIONAL SOLUTIONS TO THE PROBLEM OF FUTURE CONTINGENTS

Tooley (1997) adopts Łukasiewicz's system, but his way round the cracks is to drop the assumption that the connectives in three-valued logic are truth-functional—that is, that the truth-value of the whole formula is determined solely by the truth-values of its component parts. For Tooley, a disjunction of the form (A ∨ B), where each disjunct is indeterminate, is indeterminate, whereas those of the form (A ∨ ¬A) are determinately true. So the truth-value of the whole sentence in three-valued logic is not solely a function of its component parts but also a matter of its form. This is a natural reaction: some sentences, we may think, just are different from others because they are true simply in virtue of their form (what Tooley calls 'logical truths' (p. 139)), whereas others require truthmakers external to the proposition to make them true (what Tooley calls 'factual truths' (p. 139)).

But although this solution might initially appeal, it is not satisfactory. For we are left wondering *why* (A ∨ ¬A) and (¬(A & ¬A)) have a privileged status in

---

[10]  This is what I should have said in Bourne (2004: 127) and (2006: 94).

[11]  He explicitly endorses Excluded Middle for propositions about the future (*DI*9, 19$^a$ 27–33) and implicitly endorses unrestricted Non-Contradiction throughout Chapter 9 but in any case explicitly endorses it and Excluded Middle in *Metaphysics*, B, 2(996$^b$ 26–30); Γ, 3(1005$^b$ 19–23); Γ, 7(1011$^b$ 23–4).

three-valued logic. What is so special about these formulae that Tooley feels warranted in holding them to be determinately true, in order to draw the conclusion that the connectives in three-valued logic must therefore be non-truth-functional? Certainly, $(A \lor \neg A)$ and $(\neg(A \,\&\, \neg A))$ are logical truths in two-valued logic: they are true under all possible assignments of truth-values to the component parts and this is what justifies privileging them. But so what, if we think the world is represented by a three-valued logic? Given Łukasiewicz's truth-tables, $(A \lor \neg A)$ and $(\neg(A \,\&\, \neg A))$ are not true under all possible assignments of truth-values to the component parts: they are not 'true in virtue of their form'. In what sense, and why, should they still be considered logical truths? It is hard to see any reason for thinking they are if they do not fall out as a natural consequence of the way the connectives are defined by the truth-tables. Tooley's solution is unsatisfactory.

A related proposal that differs in the details but pushes the non-truth-functional approach is van Fraassen's (1966) method of supervaluations. The method is essentially a way of filling in truth-value gaps. Although well-suited to cases where there is a failure of reference—e.g. 'The present king of France is bald'—its application has been extended to the case of future contingents in resolving indeterminate truth-values one way or the other. It runs as follows. Let $V = \{1, \tfrac{1}{2}, 0\}$ be the set of possible truth-values. A valuation $v$ maps a formula to $V$, i.e. gives the formula one of the three possible truth-values. Call $v^r$ the *resolved valuation* of $v$, where if $v(A) = 1$, $v^r(A) = 1$ and if $v(A) = 0$, $v^r(A) = 0$, but if $v(A) = \tfrac{1}{2}$, $v^r(A) = 1$ or 0. A supervaluation assigns to the formula the value that *all* resolved valuations would assign to it, if there is one; otherwise, it assigns it no value. In the case of the formula $(A \,\&\, B)$, for instance, the resolved valuation that assigns 1 to each of A and B, assigns the value 1 to $(A \,\&\, B)$, whereas all other resolved valuations assign 0 to $(A \,\&\, B)$. Thus, since there is no unique value that all resolved valuations assign to $(A \,\&\, B)$, the supervaluation does not assign a value to it. However, no matter what the resolved valuation assigns to A, in the case of $(\neg(A \,\&\, \neg A))$ it is assigned the value 1; thus, the supervaluation assigns it the value 1. It is easy to see that all classically valid formulae come out true using supervaluations, even when the components are given indeterminate valuations. Advocates of three-valued logic can then argue that it is not all that important whether formulae that are classically valid are currently true or not so long as they *would* be true once the valuations have been resolved (regardless of how they are eventually resolved).

At first sight, this is a rather attractive way of proceeding and it looks to be close to Aristotle's position. However, although the method of supervaluations preserves the classical laws, it fails to preserve the validity of some intuitively valid arguments. To discuss this issue properly, however, would require us to discuss what it is that validity preserves—is it merely truth, or is it everything but falsity, and so on—and how far the notion of supervalidity (where the premises and conclusion involve supervaluations and not just mere valuations) can take us. Interesting as these formal technicalities are, I shall sidestep them to focus on a more fundamental difficulty I have: making *metaphysical* sense of this approach as applied to future-tense contingents.

The method of supervaluations, as applied to future-tense contingents, relies on the idea that an indeterminate truth-value is something to be resolved. But this makes no sense. According to models of time that take $\mathbf{F}p$ to be indeterminate, it *never* will become true or false that $\mathbf{F}p$ – it is *always* indeterminate: *it* never gets resolved.[12] Rather, what happens according to these models of time is that *present*-tense propositions are now either true or false and the determinate truth-value *they* have changes as the relevant actual facts change. *There is a sea-battle occurring* might well come to be true (after being false) but this is not to say that *There will be a sea-battle occurring* changes from being indeterminate to being true. Thus, anyone who thinks that future-tense propositions are indeterminate cannot appeal to this application of the method of supervaluations because its results rely on a supposition that is impossible to satisfy.[13]

But despite the fact that $v^r(\mathbf{F}p) = 1$ or $0$ is impossible on these models of time, perhaps the real thought behind it should be taken to be that at any time we suppose to be the present time, whatever the valuations of present-tense propositions at that time, the classical laws will hold. But this doesn't address our problem. The issue is not whether present-tense propositions will conform to the classical laws. Even if one holds that future contingents are indeterminate, everyone should think that $\mathbf{F}(p \vee \neg p)$ and $\mathbf{F}(\neg(p \& \neg p))$ are true since, at any time, $(p \vee \neg p)$ and $(\neg(p \& \neg p))$ are necessarily true. But this does not touch the issue of what we should take to be the truth-value of $(\mathbf{F}p \vee \neg\mathbf{F}p)$ at any given time. Thus the problem we are really interested in, and which remains, is reconciling the actual valuations of those formulae involving future-tensed propositions at a given time with our intuitions about which formulae should remain true.

## 5. A TRUTH-FUNCTIONAL SOLUTION TO THE PROBLEM OF FUTURE CONTINGENTS

My following suggestion improves on the previous proposals. It allows us to keep the connectives truth-functional; allows us to keep the Laws of Excluded Middle and Non-Contradiction as logical truths and does so by appealing solely to actual valuations and not incoherent supervaluations of future-tense propositions; doesn't introduce

[12] A possible exception, though, where $\mathbf{F}p$ itself might get resolved is if we tie its truth to evidence that we have available for asserting it. It might then change in truth-value over time as evidence changes. Dummett's view (below) is the closest we have to an endorsement of this view, but even he rejects it because it leads to a 'repugnant' conclusion when applied to the past.

[13] Note, however, that this argument does not apply to tenseless propositions involving dates which currently lie in the future. It may well make sense to talk of the truth-values of these very propositions being resolved at some time in the future. See discussion of truth-bearers below.

a distinction between logical and factual truths, and allows us to keep the notion of logical truth as truth under all interpretations, both for two- and three-valued logic.

The solution rests on the following observation: it is the definition of '¬' that causes the trouble. Thus we should stop trying to patch up the cracks in Łukasiewicz's system (as Tooley does) and deal with the foundations directly. For not only does Łukasiewicz's definition of '¬' create the difficulty, I see no reason to think that it is correct, and thus altering it is not fudging it. I claim that the following truth-table is more suitable:

FIGURE 2.6  A more suitable truth table for negation

|  | ¬ |
| --- | --- |
| 1 | 0 |
| ½ | 1 |
| 0 | 1 |

The justification for the ¬(½)=1 entry is as follows. There are two ways in which something can fail to be the case: when it is false and when it is indeterminate. Since being indeterminate is a way of not being the case, then if A is indeterminate, A is not the case; thus, '¬A' must be true. There is, then, no compelling reason for holding that the negation of a proposition can be true if, and only if, that proposition is false, as Aristotle and Łukasiewicz think. If we combine this new definition of '¬' with Łukasiewicz's truth-tables for '&' and '∨' (Figs. 2.2 and 2.3), then the truth-values of all molecular propositions remain intuitive, whilst the Laws of Non-Contradiction and Excluded Middle are retained as logical truths. Moreover, the laws fall out as natural consequences of the truth-tables, which are themselves based on intuitive independent reasoning, without having to stipulate that they are in some way privileged, as Tooley does. Further, whether we take (A ⊃ B) to be equivalent either to (¬(A&¬B)) or to (¬A ∨ B), we get the result that (F$p$ ⊃ F$p$) is necessarily true. This is the desired result that Łukasiewicz's system could deliver only by brute force. Thus, we can keep hold of a plausible non-bivalent logic for future contingents without having to abandon those logical laws.

Some comment, however, is in order. Consider the following

(3)    Donald will murder Potcha
(4)    Donald will not murder Potcha.

Because (3) is a future contingent proposition, it takes the value = ½. According to my truth-tables, (4) should then be assigned the value = 1, if it is the negation of (3). But that seems to be the wrong answer. After all, Donald did end up murdering Potcha! Something appears to have gone wrong. But it hasn't. To see why, we need to understand correctly the propositions involved.

It is obvious that (3) should be analysed as follows:

(3*)    F(Donald murders Potcha)

Care must be taken, however, when analysing (4). It is ambiguous what truth-value it should be assigned. On one reading, the future-tensed operator has wide scope over the negated present-tensed proposition:

(4×)    F¬(Donald murders Potcha).

On this reading, (4) should be assigned the value = $\frac{1}{2}$. To have the value = 1, (4×) would have to be the genuine negation of (3*), which it isn't. The propositions that fall within the scope of the future-tensed operator—*Donald murders Potcha* and *Donald does not murder Potcha*—each have a determinate truth-value, one being the negation of the other; but that doesn't make (4×) the negation of (3*). The correct negation of (3*) is rather where the negation has wide scope over (3*), namely:

(4*)    ¬F(Donald murders Potcha)

It is on this reading that we should take (4) to be true. At the time when Potcha speaks, *Donald will murder Potcha* is indeterminate, so *Donald will murder Potcha* is not the case. Given it is not the case, (4*) is clearly true. But, of course, to say (4*) is true is *not* to say that Donald *won't* murder Potcha. That would be to confuse (4*) with (4×). Thus even if it turns out that Donald murders Potcha, we should still be happy to assign truth to (4*) when it is indeterminate (and so not the case) that Donald will murder Potcha.

This is why assigning the value = $\frac{1}{2}$ to (3*) and (4×) does not entail a failure of the Law of Excluded Middle. Disjoining (3*) and (4×) gives $v(\mathbf{F}p \lor \mathbf{F}\neg p) = \frac{1}{2}$, whereas Excluded Middle requires $(\mathbf{F}p \lor \neg \mathbf{F}p)$, i.e. a disjunction of (3*) and (4*), which, according to my truth-tables, takes the value = 1, as desired.[14]

One final comment is in order concerning the Law of Non-Contradiction. One objection raised against ÂŻukasiewicz taking the value = $\frac{1}{2}$ to represent 'the possible' is that $v(A \,\&\, \neg A) = \frac{1}{2}$ when $v(A) = \frac{1}{2}$, suggesting that (A & ¬A) is possible when each conjunct is. This is unacceptable. But there is a related objection that appears also to affect my solution. Although $v(\neg(A \,\&\, \neg A)) = 1$, for any value of A, as desired, $v(A \,\&\, \neg A) = \frac{1}{2}$, when A is a future contingent proposition. Isn't this as unwelcome a result as any that have plagued Łukasiewicz's system? I say no: when we understand what is being said, this feature is palatable. Unlike in Łukasiewicz's system, my solution does not entail that contradictions are possible, but rather has the welcome result that a proposition and its negation fails to be true. Although this isn't falsity, it still fails to be the case, which is the side of the coin it should be on, and thus cannot be deemed to be on a par with (¬(A & ¬A)) failing to be true. One way to accept $(\mathbf{F}p \,\&\, \neg \mathbf{F}p)$ as indeterminate rather than false is to note that, as with any conjunction with a true

---

[14]  For other features of this system, see Bourne (2006: 94).

conjunct, the burden, so to speak, for determining the value of the conjunction rests with the other conjunct. In this special case, $\neg \mathbf{F} p$ remains true regardless of how $p$ turns out. So, shifting the burden onto the other conjunct, given that it is not fixed whether $p$ will or will not become true, the entire conjunction must be indeterminate.

It is, then, possible to have a plausible non-bivalent logic for future contingents where the classical laws remain intact.[15]

# 6. TRUTH AND TRUTH-BEARERS

There has been much discussion in the literature surrounding future contingents (and on the philosophy of time in general) on which of the possible candidates—sentences, statements, beliefs, judgements, propositions—we should take to be truth-bearers. W. and M. Kneale (1962: 48–52) argue that Aristotle's confusion over truth-bearers (and their supposed changing truth-values over time) is the basis for (what they see as) his mistake in taking future contingents to be indeterminate.[16] They argue as follows. Aristotle takes tensed sentences to be truth-bearers. But had he taken the truth-bearers to be not sentences but the propositions they are used to express, he would have avoided the trouble with indeterminate truth-values. For the proposition expressed today by the use of the sentence 'Tomorrow, Donald will murder Potcha', the proposition expressed tomorrow by the use of the sentence 'Today, Donald murders Potcha', and the proposition expressed in two days by the use of the sentence 'Yesterday, Donald murdered Potcha' are one and the same proposition—presumably, the proposition that *Donald murders Potcha at t*, where *t* is tomorrow's date. If all of these tensed sentences are used at the times they are to express the same proposition, then whether or not that proposition is true or false depends on whether Donald

---

[15]  Note that this solution to the problem of future contingents is not to vindicate Aristotle. Cicero (*De Fato* XVI 37) and Quine (1953) ridicule what they take to be Aristotle's position and, although Hintikka (1973: 163) does not think Aristotle holds the position they think he does, he agrees that if he did, the ridicule would be justified. But Quine et al. take Aristotle's 'fantasy' to be not that non-bivalence and excluded middle can be held together, but rather that 'It is true that $p$ or $q$' is an insufficient condition for 'It is true that $p$ or it is true that $q$'. Whether or not this is a fantasy, it does not affect my solution: in my system, a true disjunction requires a true disjunct. W. and M. Kneale (1962: 46–7) state similar objections to Aristotle, particularly in light of his account of truth and falsity in *Metaphysics*, $\Gamma$, 7(1011$^b$ 26–7), where 'It is true that $p$' is equivalent to '$p$' and 'It is false that $p$' is equivalent to 'not $p$'. I agree that a deflationary approach to truth has difficulty accommodating indeterminate truth-values, and thus that Aristotle is in an uncomfortable position. But *I* take this to be a reason for rejecting such an account of truth, not for thinking that the Principle of Bivalence and Law of Excluded Middle do not peel apart. And the final reason for thinking that my solution cannot lend support to Aristotle's actual position is that it relies on negation having wide scope over a proposition. As Anscombe (1956) and Geach (1956) note, Aristotle was more interested in predicate-negation and didn't have a clear conception of propositional negation.

[16]  See Haack (1974) for a general discussion of truth-bearers, and Anscombe (1956), Ayer ((1956); (1963)), Bradley (1959), and Oaklander (1998) for objections to the idea of truth at a time. I trust it is obvious from the following how I would respond to these papers.

murders Potcha on the day specified by those tenses. The proposition is true if Donald does murder Potcha on that day and false if he doesn't. It isn't true/false *at a time* but true/false *simpliciter*. If true/false, it is true/false for all time. There is no need to countenance indeterminate truth-values.

Hintikka (1973: ch. 8), however, argues that, despite the Kneales focusing their attention on truth-bearing, they have missed what Aristotle is really wrestling with in Chapter 9 of *De Interpretatione*. Rather than being confused over truth-bearing, Aristotle saw to some extent its implications for his theory of modality. For Aristotle equates the *possibility* of *p* with *p* being *sometimes* true, and the *necessity* of *p* with *p* being *always* true.[17] So if there are truths that are true/false for all time, it follows, by this definition, that they are necessary. This gives us one way of understanding (A ⊃ □A). If tenseless propositions are the truth-bearers and are always true/false, they count as being necessarily true/false on this conception, something that provokes Prior (1996: 48) to remark that tenseless statements do not allow for 'real freedom'.[18] But for those whose conception of necessity extends beyond how the actual world is patterned, allowing for the idea that despite what is always true, things could have been different, it will be difficult to see what the fuss is about in taking *p* to be always true.[19] Thus, we still need to address the Kneales' reservations.

It is not feasible here to respond to all of the literature surrounding the debate over truth-bearers, but I can sketch how I think we should understand the situation. First, we cannot assume without argument that propositions are *the* truth-bearers— for there might be good arguments to show that they aren't, or at least are not the *only* truth-bearers (whether or not the other candidates are truth-bearers in some derivative kind of way). So let TB be a place-holder for whatever truth-bearer we may choose—sentences, statements, propositions. Taking token occurrences of TB (if there are any) to be indeterminate is unproblematic. At the time of the occurrence of TB, TB is indeterminate if there is nothing to make TB determinately true or false. Note that this does not require whatever it is that makes TB true to have constituents *located* at the time that TB is located, but simply that whatever it is that makes it true has constituents that *exist*. Neither does this presuppose any particular theory of truth. The demand that at least contingent truths require something to make them true is independent of what truth *is*.[20] Note also that depending on how the truth-conditions for the token TB are specified, it might or might not retain its truth-value over time: it

---

[17]  e.g. *Metaphysics*, Θ, 3(1047ᵃ 10–14).

[18]  Aristotle does not mention tenseless truths in Chapter 9 of *De Interpretatione*, but does say that if we say of anything that takes place that it was always true that it would be, then of necessity it takes place (*DI*9(18ᵇ 10–15)), although there are readings of this that don't require us to take this as an equivalence between omnitemporal truth and necessary truth.

[19]  It may well be that it was Aristotle's intention to show just this; to accept arguments concerning the necessity of our actions but to defuse it by showing it to be a rather innocuous kind of necessity. At least that's how I take Hintikka (1973: 162).

[20]  See Beebee and Dodd (eds. 2005) for some recent work on truthmakers. On the point that truthmaking arises as much for deflationary theories of truth as it does for the correspondence theory, see Lewis (2001) and Smith (2003).

would retain it if given non-token-reflexive truth-conditions, e.g. Lowe (1998: 45); it would change it if given token-reflexive truth-conditions, e.g. Mellor (1981: 101).[21] Either way, the token TB can be said to be indeterminate at a time.

Further, even when the metaphysics allows for the existence of future-located objects to be constituents of truthmakers, whether certain statements can have determinate truth-values depends on how reference works. It is perfectly possible to make determinately true/false singular statements about what presently existing objects *will* do (as Ayer (1963: 248) notes). All that is required is the existence of the relevant facts concerning what will happen to that object (or rather an object causally connected to it in the right way) at the time the statement is made. But it is questionable whether any use of the tenseless sentence 'Aristotle's birth is later than the birth of Socrates', when tokened before Aristotle's conception, could be true, because it is not clear that it is possible to refer to objects wholly located at times later than the location of the use of the sentence. If we think that reference requires some causal connection, then, arguably, the use of the name 'Aristotle' at those times before his conception will fail to refer to him because, whether or not Aristotle exists at later locations, the causal connection between him (or at least the event of his being named) and the earlier usage of 'Aristotle' would have to operate backwards.[22] Thus some tenseless singular statements about the future may well be considered indeterminate in truth-value for this reason. All hangs on our account of reference. Addressing this issue properly, however, would take us too far from our main concern in this paper.[23]

[21]  For a good introductory overview of these and other options, see Le Poidevin (1998).

[22]  With some ingenuity, no doubt cases could be constructed where the use of the name is part of the causal mechanism that brings about the object named. Parents may plan to have a child named 'Aristotle'. But care would have to be taken to secure genuine reference if it is not just to be the general claim that *whoever* is the first son will be called 'Aristotle'. See Kripke (1972) and McCulloch (1989) for a good discussion of the theory of names.

[23]  But here are a few remarks. We might, following Strawson (1950), say that a failure of reference means that no statement has been made (rather than that a statement has been made that is neither true nor false), in which case 'Aristotle's birth is later than the birth of Socrates' is just hot air when uttered before Aristotle is conceived, and is not true nor does it become true just because there are times when another token of that sentence type is used when the name clearly refers (*pace* Ayer (1963: 250)). Our case is slightly different to Strawson's. For him, no statement has been made because there is a false presupposition that the subject exists. However, in the case we are considering, the name fails to refer because of the causal mechanism required for reference, and not because Aristotle does not exist. Either way, the result is the same. I find the quantificational treatment of names (such as Russell's (1911); (1918–19)) the most natural way of dealing with names of wholly future individuals. (See Prior (1968a: 19), who endorses this treatment for the case of past individuals, and Bourne (2006: 99–108) for alternatives.) If we treat names in the quantificational way, then statements about the future are determinately either true or false depending on whether there is a unique object that satisfies the description. Ryle (1954: 27) argues that both singular and general statements can be made about the past but only general statements can be made about the future. Ayer (1963) argues against Ryle that just because reference can often fail because we happen not to know as much about the future as we do the present and past, we shouldn't think genuine reference is not possible. But Ryle's position is based on the future being indeterminate and not on our lack of knowledge. And, so far as I can see, Ayer's examples are really of the quantificational kind, despite their surface resemblance to genuine singular cases.

What of propositions as truth-bearers? Unlike, say, the token sentences used to express propositions, propositions themselves do not have a location in time.[24] Taking tensed propositions to be truth-bearers might be thought to be problematic because it is not clear whether it makes sense to ask whether *There is a sea battle* is true, false, or neither without qualification. For just as *The book is blue* cannot be evaluated for truth-value without a specified context—for it cannot be anchored to a context of evaluation in the way, say, a sentence used to express the proposition can—propositions must contain a date. *There is a sea battle* looks to be incomplete. But once completed, as in *There is a sea battle in 340bce*, depending on what happens in 340bce, the proposition is either eternally true or eternally false. There seems to be no room for indeterminate truth-values.

To think the proposition *There is a sea battle* is incomplete is to think that the world does not have the resources to single out a time of evaluation. (Remember we are not talking about the tokens that are used to express this proposition, but the proposition itself.) But this is simply to beg the question against those who do think that one time—the present—is singled out as ontologically privileged and that which facts are present changes over time. For, given this metaphysical picture, propositions need not be located in time in order to make sense of them being true *now*. The truth of a proposition (as with any truth-bearer) depends on whether there exists something that makes it true. However we spell out this dependence relation, all that a change in truth-value requires is for there to be a change in the truthmaking end of this relation; the proposition need not be located in time.

Further, neither need tenseless propositions be true for all time when true. For even if we say that the proposition expressed today using the sentence 'Tomorrow, Donald will murder Potcha' is the very same proposition expressed tomorrow using the sentence 'Today, Donald murders Potcha', that proposition can be true only when something exists to make it true and, according to those who think the facts change as time moves on, the facts that exist today are not enough to make that proposition determinately true or false, whereas the facts that exist tomorrow are. The very same proposition that is true tomorrow need not be considered true today, even when that proposition is tenseless.

In sum, whether we take propositions or token sentences (etc.), tensed or tenseless, to be truth-bearers, it makes sense to talk of their having indeterminate truth-values. It all hangs on the underlying metaphysical situation.[25]

---

[24]  But see Smith (1998) for a contrary view.

[25]  As Tooley (1997: 132) notes, tenseless propositions can change their truth-value, but only once: they can be indeterminate before they are true/false, but once true/false, they retain that truth-value (see also Broad (1923: 73)). In this respect, tenseless propositions differ from tensed propositions, which can change their truth-values in a number of ways (left for the reader to consider). Tooley's observation about tenseless propositions, however, depends on whether we hold that facts that make determinate propositions concerning earlier times remain in existence. Łukasiewicz (1970: 127–8), for instance, infamously welcomes the idea that propositions could become indeterminate once the time they are about drifts into the past.

## 7. THE METAPHYSICS OF THE FUTURE

There are two broad categories into which metaphysical accounts of time divide: tenseless theories and tensed theories. 'Tense' is meant here to indicate a feature of the world and not to mark a difference concerning language. The important difference between them is that tensed theories claim that there is something ontologically special about the present moment, with the different tensed theories being defined by how they spell out the way in which it is privileged. Tenseless theories claim that there is nothing ontologically special about the present—past and future times are no less real than it, and time's flow should not be understood as a genuine change in what time is ontologically privileged.

The mainstream tenseless theories of time hold that propositions about the future are determinately either true or false.[26] However, the tenseless position, as such, does not entail that propositions about the future are determinately either true or false. For it could be formulated in justificationist terms, as explored by Dummett (2003).[27,28] His views on the truth-values of propositions about the past and future stem from his views on the theory of meaning. The truth of a proposition is traditionally taken to be independent of our means of establishing its truth: it is true or false whether or not we could ever know it. However, the truth of a proposition, according to justificationist theories of meaning, is grounded by the notion of verification and is not independent of our means of establishing it. On a justificationist theory, the Principle of Bivalence need not hold since there might be some things that cannot be established as true or false.

Whether or not this position is tenable,[29] the question we're interested in is whether this version of the tenseless theory bypasses the association of determinate truth-values with fatalism. It doesn't. A proposition about the future is determinately true, on Dummett's account, if it could be verified (such as if someone were there to do

---

[26]    For early versions of the tenseless view, see, for instance, Broad (1921); Goodman (1951); Grünbaum (1967); Quine (1960); Russell (1915); Smart ((1949); (1963)); and Williams (1951). For later, improved, versions, see, for instance, Le Poidevin (1991); Mellor ((1981); (1998)); Oaklander (1984); and Smart (1980).

[27]    See also his (1963); (1969); (1982); (1993a).

[28]    Another possible counterexample where bivalence is not asserted is a branching tenseless model. At any node, on any branch, it is determinate what has happened in its past, but towards the future many possibilities are laid out. Perhaps it is better here to relativize determinate truth-values to a branch, but even so, future-tensed propositions would not be determinately true or false without qualification, as they are according to the mainstream linear model. However, since all branches are actual, perhaps this is a case where it is better to say that propositions about the future *and* their negations are both determinately true. (This is the basis for Lewis's (1986: 207–8) worry about branching accounts of the future. See Bourne (2006: 61) for criticism.) It isn't clear whether or not these suggestions help with the fatalist worry. In any case, apart from contentious speculation over the interpretation of quantum mechanics, I see no reason to adopt this model.

[29]    See Bourne (2007) for reasons to think it isn't.

so). By these standards, the proposition concerning Donald's future murderous act is determinately true.

The picture of times as already laid out before us has led many to think that the tenseless account of the future does not allow for freedom. But, if so, there are those on the tensed side of the fence that are in the same situation. Ruling out obvious non-starters leaves the following tensed positions: the past, present, and future all exist; the past and the present, but not the future exist; the present and the future but not the past exist; the present but not the past or the future exists. The link between each of these options and whether propositions about the future have determinate truth-values is straightforward enough to state without much explanation. The *linear* version of the view that past, present, and future all exist holds that propositions about the future are determinately true or false.[30] The *branching* version holds that time's flow amounts to the dropping out of existence of the possibilities, as one branch becomes the way things actually are (see McCall (1994)). According to this view, propositions about the future should now be taken to be indeterminate.[31] The alternative asymmetric view, associated with Broad (1923) and Tooley (1997), where only the past and present exist and time flows by facts coming into existence, holds that future contingents are indeterminate in truth-value; there is nothing in existence to make such a proposition determinately true/false.[32] And presentism—the view that neither the past nor the future exists—leaves it open, as we shall see, whether the future is determinate or not.

In short, there is determinateness to be found on both sides of the tense-tenseless divide. If determinateness is thought to be the source of our lack of freedom, the mere fact that time flies isn't enough to give us wings, despite what is commonly thought. Indeed, there is an argument, to which we shall now turn our attention, that attempts to show that determinateness when combined with the tenseless theory is compatible with freedom, but is incompatible with freedom when combined with presentism.

## 8. PRESENTISM, DETERMINATENESS, AND FATALISM

It is true that *Donald murders Potcha at t*. So it was true at *t**(1000 years earlier than *t*) that *Donald will murder Potcha in 1000 years*. But Donald has no say in what was true

---

[30] Including its variants, e.g. Smith's (2002) 'degree presentism'.
[31] Despite McCall (1994: 14–5) saying that such propositions are determinately true/false depending on what eventually happens. But because the facts are determinate only at some times, it should be obvious from the previous sections where I think he goes wrong.
[32] This underpins Tooley's views on future contingents discussed above. Broad (1923: 73) differs in not being concerned about the Law of Excluded Middle as applied to the future because he does not take judgements about the future to be genuine judgements.

1,000 years ago. Since what was true 1,000 years ago was that *Donald will murder at t*, Donald has no say in whether he will murder at *t*. Donald is not free.

In response to this kind of argument, Lewis (1976: 151) writes:

> Fatalists—the best of them—are philosophers who take facts we count as irrelevant in saying what someone can do, disguise them somehow as facts of a different sort that we count as relevant, and thereby argue that we can do less than we think—indeed, that there is nothing at all that we don't do but can. [Donald will not let Potcha live. The fatalist argues that Donald not only won't but can't let Potcha live; for his allowing Potcha to live is not compossible with the fact that it was true already 1000 years ago that Donald was going to murder Potcha 1000 years later.] My rejoinder is that this is a fact, sure enough; however it is an irrelevant fact about the future masquerading as a relevant fact about the past, and so should be left out of account in saying what, in any ordinary sense, I can do.

The relevant facts are those that ground the truth-value links. To determine which facts these are, we need to consider what grounds the truth of the present-tense proposition embedded within the tense operators. If *Donald will kill Potcha* is now true, this is not because *it was true 1,000 years ago that Donald will kill Potcha* but rather because the truth of the present-tense proposition that *Donald kills Potcha* is grounded by what facts obtain in the future.

So far so good. But, as Rea (2006) points out, this response seems not to be available to presentism. Only the present exists; the future is that which *will* come to exist, and the past is that which *did* exist.[33] There are no past- or future-located objects. Thus, when *t\** is present there is nothing at *t* to ground the truth at *t\** of *Donald will murder Potcha in 1,000 years*. Thus, the facts that make *Donald will murder Potcha in 1,000 years* true at *t\** must, for the presentist, be grounded at *t\**. Since Donald has no say about what facts exist 1,000 years ago, he cannot influence the facts grounded by what obtains at *t\**; thus, whether or not he will murder Potcha is not something under his influence. Rea concludes that presentists cannot hold that the Principle of Bivalence applies to propositions about the future if humans are to remain free; *Donald will murder Potcha in 1,000 years* must at *t\** be neither true nor false.

Rea's argument imposes constraints on any presentist position that takes the truth-makers for propositions about the future to be grounded in the present. Assuming that allowing for freedom is imperative, the Principle of Bivalence has to go. My response to Rea's argument has two parts. First, bivalence has to be given up anyway by any form of presentism that grounds all truthmakers in the present—but this is due to a fundamental flaw that renders the position untenable in any case, and not primarily because of a worry about freedom. Second, there is a version of presentism not subject to Rea's argument.

There are two ways to develop a presentist position where the truthmakers for future truths are grounded in the present. First, we might hold that what makes propositions

---

[33] Rea focuses on presentism but his argument applies to any model of time that denies the existence of the future.

about the past and future true is due to the causal links between the way the present state of the world is and the way it has been and will be.[34] The pivotal issue here is whether determinate truth-values are 'guaranteed by present fact' (Le Poidevin (1991: 38)). But it is hard to see how present fact can guarantee this, since it would have to ground much more than it can, such as what the laws of nature are, something that is underdetermined by present fact alone. Consequently, because these are the only facts that this version of presentism allows, what the laws of nature are is not just epistemically underdetermined, but ontologically indeterminate. Thus, propositions about the past and future, if their truth-values are grounded in this way, are also indeterminate in truth-value.[35]

Second, it might be thought that determinate truth-values could be had if we held the kind of presentism according to which it is simply *true* now that *p will* come true and now true that *q was* true (where *p* and *q* are present-tensed propositions). So it is now true that *Donald murders Potcha* was true and now true (let's say) that *Donald sleeps soundly* will be true. For such presentists, that's all that needs to be said concerning the truth-values of past- and future-tensed propositions and their determinateness—it's just a brute present fact that such past- and future-tensed propositions are true. (This is the kind of presentism on which Rea focuses.)

The fatal flaw in this version of presentism, however, is that it cannot guarantee the truth-value links between truths holding at different times. It may well be presently true that *Donald murdered Potcha* and that *Donald will sleep soundly*. But there is nothing in this view to guarantee that when a different set of present facts comes into existence it will link in the right way with the set that has gone before. This is absurd. What the problem shows is that there must be something that transcends the present in order for past-tense propositions that are true at the present time to link with present-tense propositions that were true at past times, and future-tense propositions that are true at the present time to link with present-tense propositions that will be true at future times. Without a bridge that extends beyond the present to other times, this view is reduced to claiming that there are determinate present truths about the past and future which do not entail anything about what is true at other times. We need not agonize over whether this then counts as propositions about the future being ultimately indeterminate: it does not matter what we decide here because the fundamental flaw in the view renders its details not worth pursuing.[36] That this version of presentism is incompatible with freedom is the least of its worries. It doesn't even help this position to take Rea's way out and give up on bivalence: we would still be left with a position that does not allow for any real freedom. For how can we have any meaningful influence over our lives when none of the links assumed to hold from time to time can be guaranteed?

[34]  See, e.g. Ayer (1936: 102–3); Lewis (1929: 150–53); and Ludlow (1999) for endorsements of similar views, and Ayer's ((1950); (1956: 153–64)) reconsideration of his earlier position.
[35]  See Bourne (2006: 47–51) for more on this argument.
[36]  See Bourne (2006: 41–6) for further criticism of the position.

If presentism is to be at all palatable, it must accord with our saying that at least some propositions about the past are true, including those that involve non-existents, and it must guarantee that the truth-value links hold across different times. A way of doing this, whilst preserving the idea that there is something ontologically privileged about the present, is to hold that there *are* other times, past and future, at which certain present-tense truths hold, but that these times are not the sort of things where the flesh and blood Socrates and my flesh and blood great-great-grandchildren can exist. Rather they are (and contain) abstract objects that *represent* what Socrates did and what my great-great-grandchildren (if I have any) shall do. They are surrogates for the concrete times of the tenseless theory.[37] Although I think there are good reasons for taking the future to branch on this view, arguably, the view leaves it open whether the future branches or is linear. If linear, and unrestricted bivalence holds, this version of presentism is no more subject to Rea's argument than is the tenseless theory: *Donald will kill Potcha* is now true because there is a future time that represents the fact that Donald kills Potcha, such that when what that time represents becomes present, the present-tense proposition *Donald kills Potcha* will be made true by what Donald freely does *then*.

In sum, the indeterminateness of the future avoids fatalism. But fatalism does not follow from determinateness either. So long as the truths we are interested in are grounded in the right way by matters under our influence, we can avoid any fatalist worries. Our final task is to establish the extent of the influence we have.

## 9. FIXED PAST AND OPEN FUTURE

Lucas (1973: 262) remarks that '[it] is natural to try and view time as the passage from the possible to the necessary'. But how should we interpret this thought? One way is to think of the past as *irrevocable*. Aristotle writes elsewhere:

> It is to be noted that nothing that is past is an object of choice, e.g., no one chooses to have sacked Troy; for no one *deliberates* about the past, but about what is future and capable of being otherwise, while the past is not capable of not having taken place; hence Agathon is right in saying
> For this alone is lacking even to God, To make undone things that have once been done.                                            (*Nicomachean Ethics* VI, 2, 1139$^b$ 7–9)

It is because we cannot undo what has been done that we do not deliberate about the past, according to this view. This is one way in which our influence is constrained and a way in which the past is necessary. But what irrevocability explains is limited, since it is nothing more than an assertion of $\Box(A \supset A)$, something that applies whether the

---

[37] See Bourne (2006) for a development of this view.

proposition is about the past, present, or future, and thus something that cannot be the basis for any difference between them.[38]

Nevertheless, it might be thought that invoking the irrevocability of the past as the basis for thinking the past is necessary works under the assumption that the past is determinate and the future isn't. But the worry here is that the reason given for the future being indeterminate was its not being necessary, leaving us in a tight circle. One way out would be to understand the change in modal status from the possible future to the necessary past as having an independent ontological underpinning, such as we get with the ontologically asymmetric accounts of time. But, apart from some serious difficulties with such views,[39] it doesn't supply a reason for thinking the modal status has changed from possible to *necessary*—which would rule out our having some influence—rather than simply possible to *actual*—which doesn't in itself rule out our having some influence.

A better way to proceed, then, is to invoke causation to ground the modal status of the past and future: one cannot now influence whether or not a proposition about the past is true, whereas one can influence what will be true.[40]

We might question whether, in articulating this sense of fixity, a genuine species of necessity has been identified and thus whether it is legitimate to interpret in this way. But we need only note that it is common and natural to think of necessary truths as those that hold independent of how we act. Mathematical and logical truths fit this idea, as do, arguably, the laws of nature. Under this interpretation of $\Box$, anyone, including Aristotle, would be right to hold that $(\mathbf{P}p \supset \Box\mathbf{P}p)$. Similarly, if there is no simultaneous causation to allow us to influence what is presently true, we should hold what is presently true to be necessary, in which case $(p \supset \Box p)$.[41] But since we *can* do something to bring about whether $\mathbf{F}p$ or $\neg\mathbf{F}p$, in the way we cannot do for propositions about the present and past, neither $\Box\mathbf{F}p$ nor $\Box\neg\mathbf{F}p$ holds.

Thus, we can hold that there are true instances of $(A \supset \Box A)$ without thinking (fallaciously) that they follow from $\Box(A \supset A)$. However, this does not vindicate the Aristotelian position, since this way of taking the necessity involved does not require us to give up the Principle of Bivalence for future contingents, it being determinately true that $\mathbf{F}p$ is perfectly compatible with our now having some influence over its obtaining; that $\mathbf{F}p$ is true is not to say that it would still have obtained whatever we did. Thus, it seems, the basic Aristotelian argument is flawed: either it relies on the seductive fallacy at some point, or it is valid but unsound since the premises $(t(\mathbf{F}p) \supset \Box\mathbf{F}p)$ and $(f(\mathbf{F}p) \supset \Box\neg\mathbf{F}p)$ are false.

But there's more to be got from considering the fatalist position. For the sting lies in pointing out that the arguments put forward for thinking that we are unable to affect

---

[38] This passage supports Haack's interpretation of Aristotle above.
[39] See e.g. Bourne (2002); (2006: Ch. 1).
[40] This also seems to be endorsed by Aristotle (*DI9*, 19ᵃ 6–10).
[41] It isn't clear whether Aristotle thought that what is presently true is necessary. See, e.g. Hintikka (1973), for a discussion of the somewhat inconclusive evidence.

the past seem to apply equally to the case of the future. Conversely, it seems that any argument given for how we can affect the future can be applied to show how we can affect the past. This leaves us with a dilemma, each horn of which is something that requires us to revise our common sense conception of the world: either fatalism is true or backwards causation is possible.[42]

Consider the following. Nobody thinks they have a say in what happened last week, or even a minute ago, let alone 1,000 years ago. Suppose that Potcha's son, Pitcha, has not heard from his dad for a while and fears the worst. Is it worth his while praying so that Potcha is still alive? It seems not. Either Potcha has been murdered or he has not. If he has, then nothing can be done about it. If he hasn't, then there is nothing to do about it. Either way, nothing can be done to affect whether or not he has died. This argument looks good, yet we can see how it mirrors the fatalist argument perfectly simply by swapping the tenses. Either Potcha will be murdered or he will not. If he will, then there is nothing to be done about it. If he will not, then there is nothing to do about it. Either way, nothing can be done to affect whether or not he will be murdered. Yet nobody sensible believes we cannot affect the future, so the challenge is to say something sensible about our influence over the past: either show that it is possible to affect it, or show what grounds the difference in the direction of our causal influence.[43]

There was an explosion of interest in backwards causation in the 1950s and 60s.[44] For instance, Dummett (1964) argues that Pitcha's prayer for Potcha not to have died is worthwhile. For given that Pitcha does not know that Potcha has been murdered, Pitcha can hope that God, who knows everything and thus knows that Pitcha will pray for the survival of Potcha, will answer the later prayer by having made it the case earlier that Potcha had not died. So nobody can do anything now to undo what has happened (nor can anyone change the future in this way), but that's not to say that it isn't possible to do something now in order to bring it about that something happened; something that wouldn't have happened had they not done it. Affecting the past looks perfectly coherent.

As Dummett notes, this example is special because it features God's foreknowledge. For humans, he argues, the difference between the past and future is that it is possible to know whether or not any given past event has happened independently of our present intentions, but for many future events we cannot have such knowledge independently of our present intentions. I suspect that it is this that gives rise to the

---

[42]  See Dummett (1964) and Taylor (1974).

[43]  It is a rather curious feature of this debate that determinate truths are taken to entail a passive approach to life. Why isn't it just as natural a response to think: if it's true that next week this paper will be submitted, I'd better get a move on and do it! Of course, this is a joke position since it ignores our influence over what happens, but it is no more confused than the traditional passive fatalist reaction to determinate truths about the future.

[44]  Beginning with a debate between Dummett (1954) and Flew (1954) and then a flood of papers to the journal *Analysis* in particular. See also Ayer (1956: 170–5). An excellent, relatively recent discussion is Price (1996).

phenomenology of freedom. The notions of agency and deliberation are intimately linked to what we know and intend. It is natural to think that deliberation is futile when the agent already knows what has been or will be the case. Not knowing what he has in store, we can understand Donald feeling freer to get on with the rest of his life after the murder than before, when he knew what was to happen. But, natural as this thought is, thinking deliberation is futile is only valid when what is known about is independent of the agent's action: Donald's mistake was to ignore that what was in store for him was not independent of what he chose to do. Indeed, it is an intriguing trait of many who consider these matters to ignore this dependence, resulting in mistakes being made concerning what foreknowledge entails. Although we cannot delve too deep here, two points are worth making.[45] First, foreknowledge does not entail fatalism: God can know as much as he likes about what I freely choose to do. And from Potcha and Donald knowing that the murder will take place, it does not follow that Donald was not free not to murder. Knowing he did it entails that he did it. But he didn't have to do it; and had he not done it, he wouldn't have known it. Certainly, our lives would be very different if we had foreknowledge, but it wouldn't lead to us being any less free, and shouldn't lead to us feeling that we were any less free; if anything, it would make us more aware of just how effective our free choices are.

This leads to the second point. The foreknowledge that Potcha and Donald have seems problematic. It was Potcha's giving Donald that information that caused him to murder Potcha: a condition that was required for Donald to have the knowledge in the first place.[46] Causal loops like this are familiar from time travel stories and they give rise to an explanatory mystery: how the loop itself got up and running. For some, this very feature makes the loop impossible. For others (me included), this just makes the situation odd rather than incoherent—after all, it doesn't stop each event *in* the loop having an explanation.[47]

In short, it is notoriously difficult finding ways to rule out causal loops (if we wish to) and giving an account of the direction of causation. The accounts that have been offered are too numerous to discuss here. Yet one that Dummett dismisses in his

---

[45]  See Anscombe (1956); Ayer ((1956); (1963)); Horwich (1987: 201–2) for discussion of knowledge asymmetry. See Price (1996) and Ramsey (1929) for the 'agency' theory of causation, which is based around our deliberative practice. Dummett argues that it is not possible to combine the beliefs: (1) that there is a positive correlation between doing action A and the occurrence of event B; (2) that action A is in my power to do or not do, and (3) that I can know whether event B has taken place independently of my intention to perform action A. Mellor ((1981: 160–87); (1998: 125–35)) uses the idea that causation requires the logical independence of cause and effect to argue against the possibility of causal loops and to explain the direction of causation. See Bourne (2006: 131–4) for criticism.

[46]  This also raises an interesting question concerning who should be held responsible. Was it right to say that Potcha was guilty only of telling the truth? Feinberg (1992: 144), for instance, argues that 'where one person causes another to act voluntarily either by giving him advice or information or by otherwise capitalizing on his carefully studied dispositions and policies, there is no reason why *both* persons should not be held responsible for the act.'

[47]  For the best discussion of time travel, see Lewis (1976). See also Mellor in previous footnote.

discussion, but which is pertinent to our interests in this paper, exploits those models of time that take the future to be indeterminate. We might argue as follows: if it is indeterminate which facts will obtain, then nothing determinate can be said to be a later cause of a previous fact's obtaining; thus, backwards causation is not possible. This, of course, is only the beginnings of a theory. Nevertheless, if it could be made good, it would draw all the sting from the fatalist's argument, and Aristotle would have been right, albeit for different reasons, that the proper response to the fatalist is to deny the determinateness of the future.[48]

## References

Ammonius, *On Aristotle's On Interpretation 9*, trans. D. Blank and N. Kretzmann *Ammonius On Aristotle's On Interpretation 9 with Boethius On Aristotle's On Interpretation 9* (Ithaca, New York: Cornell University Press, 1998).

Anscombe, G. E. M. (1956), 'Aristotle and the Sea-Battle', *Mind* 65: 1–15.

Aristotle, *De interpretatione*, trans. E. M. Edghill from W. D. Ross (ed.) *The Works of Aristotle* (Chicago: Encyclopædia Britannica, Inc., 1952).

—— *Metaphysics*, trans. W. D. Ross from W. D. Ross (ed.) *The Works of Aristotle* (Chicago: Encyclopædia Britannica, Inc., 1952).

—— *Nicomachean Ethics*, trans. W. D. Ross from W. D. Ross (ed.) *The Works of Aristotle* (Chicago: Encyclopædia Britannica, Inc., 1952).

Ayer, A. J. (1936), *Language, Truth and Logic* (London: Penguin).

—— (1950), 'Statements about the Past', *Proceedings of the Aristotelian Society*, 1950–1. Also in his *Philosophical Essays* (London: Macmillan, 1954), 167–90.

—— (1956), *The Problem of Knowledge* (London: Peguin).

—— (1963), 'Fatalism', in his *The Concept of a Person and Other Essays* (London: Macmillan, 1963), 235–68.

Beebee, H., and Dodd, J. (eds.) (2005), *Truthmakers* (Oxford: Clarendon Press).

Boethius, *On Aristotle's On Interpretation 9*, trans. D. Blank and N. Kretzmann *Ammonius On Aristotle's On Interpretation 9 with Boethius On Aristotle's On Interpretation 9* (Ithaca, New York: Cornell University Press, 1998).

Bourne, C. P. (2002), 'When am I? A Tense Time for some Tense Theorists?', *Australasian Journal of Philosophy* 80: 359–71.

—— (2004), 'Future Contingents, Non-Contradiction and the Law of Excluded Middle Muddle', *Analysis* 64: 122–8.

—— (2006), *A Future for Presentism* (Oxford: Clarendon Press).

—— (2007), 'Review of Michael Dummett *Truth and the Past*', *Mind* 116: 1110–4.

Bradley, R. D. (1959), 'Must the Future Be What it is Going to Be?', *Mind* 68: 193–208.

Broad, C. D. (1921), 'Time', in J. Hastings (ed.), *Encyclopaedia of Religion and Ethics* (New York: Scribner), 334–45.

—— (1923), *Scientific Thought* (London: Routledge and Kegan Paul).

Cicero, M. T. *De Fato*, in *Cicero in Twenty-Eight Volumes, Vol.4* (London: Loeb Classical Library, 1942).

---

[48] Many thanks to Emily Caddick for comments.

Davidson, D. (1967), 'The Logical Form of Action Sentences', in N. Rescher (ed.) *The Logic of Decision and Action* (University of Pittsburgh Press). Also in his *Essays on Actions and Events* (Oxford: Oxford University Press, 1980), 105–48.

Dummett, M. (1954), 'Can an Effect Precede its Cause?' *Aristotelian Society Supplementary Volume* 28; repr. in Dummett (1978: 319–32).

——(1963), 'Realism', in Dummett (1978: 145–65).

——(1964), 'Bringing About the Past', *Philosophical Review* 73: 338–59; repr. in Dummett (1978: 333–50).

——(1969), 'The Reality of the Past', *Philosophical Review* 78: 239–58; repr. in Dummett (1978: 358–74).

——(1978), *Truth and Other Enigmas* (London: Duckworth).

——(1982), 'Realism', *Synthese* 52: 55–112; repr. in Dummett (1993b: 230–76).

——(1993a), 'Realism and Anti-Realism', in Dummett (1993b: 462–78).

——(1993b), *The Seas of Language* (Oxford: Oxford University Press, 1993).

——(2003), *Truth and the Past* (Columbia University Press).

Feinberg, J. (1992), 'Limits to the Free Expression of Opinion', in his *Freedom and Fulfillment* (New Jersey: Princeton University Press, 1992), 124–51.

Flew, A. (1954), 'Can an Effect Precede its Cause?' *Aristotelian Society Supplementary Volume* 28.

Geach, P. T. (1956), 'The Law of Excluded Middle', *Aristotelian Society Supplementary Volume* 30; also in his *Logic Matters* (Oxford: Blackwell, 1972), §2.5.

Goodman, N. (1951), *The Structure of Appearance* (Indianapolis, Ind.: Bobbs Merrill).

Grünbaum, A. (1967), 'The Status of Temporal Becoming', in his *Modern Science and Zeno's Paradoxes* (Middleton, Conn.: Wesleyan University Press), 7–36.

Haack, S. (1974), *Deviant Logic* (Cambridge: Cambridge University Press).

Hintikka, J. (1973), *Time and Necessity* (Oxford: Clarendon Press).

Horwich, P. (1987), *Asymmetries in Time* (Cambridge, Mass.: MIT Press).

Kneale, W., and M. (1962), *The Development of Logic* (Oxford: Clarendon Press).

Kretzmann, N. (1998), 'Boethius and the Truth about Tomorrow's Sea Battle', in D. Blank and N. Kretzmann trans. *Ammonius On Aristotle's On Interpretation 9 with Boethius On Aristotle's On Interpretation 9* (Ithaca, New York: Cornell University Press, 1998).

Kripke, S. (1972), 'Naming and Necessity', in D. Davidson and G. Harman (eds.), *Semantics of Natural Languages* (Dordrecht: D. Reidel), 253–355. Also as his *Naming and Necessity* (Oxford: Blackwell, 1980).

Le Poidevin, R. (1991), *Change, Cause and Contradiction* (Basingstoke: Macmillan).

——(1998), 'The Past, Present, and Future of the Debate about Tense' in R. Le Poidevin (ed.) *Questions of Time and Tense* (Oxford: Oxford University Press, 1998), 13–42.

Lewis, C. I. (1929), *Mind and the World Order* (New York: Scribner's Sons).

Lewis, D. K. (1976), 'The Paradoxes of Time Travel', *American Philosophical Quarterly* 13: 145–52. Also in his *Philosophical Papers Volume II* (Oxford: Oxford University Press, 1986), 67–80.

——(2001), 'Forget about the "Correspondence Theory of Truth"', *Analysis* 61: 275–80.

Lucas, J. R. (1973), *A Treatise on Time and Space* (London: Methuen & Co.).

Ludlow, P. (1999), *Semantics, Tense, and Time* (Cambridge, Mass.: MIT Press).

Łukasiewicz, J. (1920), 'On Three-Valued Logic', in S. McCall (ed.), *Polish Logic 1920–1939* (Oxford: Clarendon Press, 1967), 16–18. Also in his (1970: 87–8).

Łukasiewicz, J. (1930), 'Philosophical Remarks on Many-Valued Systems of Propositional Logic', in S. McCall (ed.), *Polish Logic 1920–1939* (Oxford: Clarendon Press, 1967), 40–65.

—— (1970), *Selected Works*, edited by L. Borkowski (Amsterdam: North Holland).

McCall, S. (1994), *A Model of the Universe* (Oxford: Clarendon Press).

McCulloch, G. (1989), *The Game of the Name* (Oxford: Clarendon Press).

McTaggart, J. McT. E. (1908), 'The Unreality of Time', *Mind* 17: 457–84.

—— (1927), *The Nature of Existence, Volume II*, edited by C. D. Broad (Cambridge: Cambridge University Press).

Mellor, D. H. (1981), *Real Time* (Cambridge: Cambridge University Press, 1998) *Real Time II* (London: Routledge).

Oaklander, L. N. (1984), *Temporal Relations and Temporal Becoming* (Lanham: University Press of America).

—— (1998), 'Freedom and the New Theory of Time', in R. Le Poidevin (ed.) *Questions of Time and Tense* (Oxford: Clarendon Press, 1998), 185–205.

Price, H. (1996), *Time's Arrow and Archimedes' Point* (Oxford: Oxford University Press).

Prior, A. N. (1967), *Past, Present and Future* (Oxford: Oxford University Press).

—— (1968), 'Changes in Events and Changes in Things', in his *Papers on Time and Tense* (new edn., Oxford: Clarendon Press, 2003), 7–19.

—— (1996), 'Some Free Thinking about Time', in J. Copeland (ed.), *Logic and Reality: Essays on the Legacy of Arthur Prior* (Oxford: Oxford University Press), 47–51.

Quine, W. v. O. (1953), 'On a Supposed Antinomy', in his *The Ways of Paradox and Other Essays* (Cambridge, Mass.: Harvard University Press, 1966), 19–21.

—— (1960), *Word and Object* (Cambridge, Mass.: MIT Press).

Ramsey, F. P. (1929), 'General Propositions and Causality', in D. H. Mellor (ed.) *F.P. Ramsey: Philosophical Papers* (Cambridge: Cambridge University Press, 1990).

Rea, M. C. (2006), 'Presentism and Fatalism', *Australasian Journal of Philosophy* 84: 511–24.

Russell, B. (1905), 'On Denoting', *Mind* 14: 479–93.

—— (1911), 'Knowledge by Acquaintance and Knowledge by Description', *Proceedings of the Aristotelian Society* 11: 108–28.

—— (1915), 'On the Experience of Time', in *Monist* 25: 212–33.

—— (1918–19), 'Lectures on the Philosophy of Logical Atomism', in his *Logic and Knowledge*, edited by R. C. Marsh (London: Routledge, 1956), 175–281.

Ryle, G. (1954), 'What Was to Be', in his *Dilemmas* (Cambridge: Cambridge University Press), ch. 2.

Smart, J. J. C. (1949), 'The River of Time', *Mind* 58: 483–94.

—— (1963), *Philosophy and Scientific Realism* (London: Routledge and Kegan Paul).

—— (1980), 'Time and Becoming', in P. van Inwagen (ed.), *Time and Cause: Essays Presented to Richard Taylor* (Dordrecht: D. Reidel), 3–16.

Smith, P. (2003), 'Deflationism: the Facts', in H. Lillehammer and G. Rodriguez-Pereyra (eds.), *Real Metaphysics* (London: Routledge), 43–53.

Smith, Q. (1998), 'Absolute Simultaneity and the Infinity of Time', in R. Le Poidevin (ed.) *Questions of Time and Tense* (Oxford: Clarendon Press, 1998), 135–83.

—— (2002), 'Time and Degrees of Existence: A Theory of "Degree Presentism"', in C. Callender (ed.), *Time, Reality and Experience* (Cambridge: Cambridge University Press), 119–36.

Strawson, P. F. (1950), 'On referring', *Mind* 59: 320–44.

Taylor, R. (1974), *Metaphysics*, second edition (Englewood Cliffs, N.J.: Prentice-Hall).

Tooley, M. (1997), *Time, Tense and Causation* (Oxford: Clarendon Press).

van Fraassen, B. C. (1966), 'Singular Terms, Truth-Value Gaps, and Free Logic', *Journal of Philosophy* 63: 481–95.

Wilde, O. (1887), 'Lord Arthur Savile's Crime: A Study of Duty', *The Court and Society Review*; repr. in O.Wilde, *Lord Arthur Savile's Crime and Other Stories* (1891).

Williams, D. C. (1951), 'The Myth of Passage', *Journal of Philosophy* 48: 457–72.

CHAPTER 3

..................................................................................................................

# TIME AND CHANCE PROPENSITIES[1]

..................................................................................................................

CARL HOEFER

# 1. INTRODUCTION

..................................................................................................................

WHAT is an *objective probability*? Most philosophers of science believe that there are such things, *contra* the claims of de Finetti and some other hard-line subjectivists. Suppose we believe that there is an objective probability (or *chance*, as I will often write) $x$ of some outcome A occurring, under certain conditions. The word 'objective' indicates that the probability ascription is meant to say something about the external world, over and above our own beliefs. Typically we mean to be saying something not just about the external world as a whole, but about the particular conditions (the *chance setup*) that may or may not lead to A. That set-up, we might say, has a tendency or propensity to give rise to A, and the strength of that tendency is $x$. If the set-up conditions are realized many times, we say that it is likely that the frequency with which A results will be something close to $x$, getting closer (usually) as the number of 'trials' is increased. If we know the objective chance and have to make a bet as to whether A will result the next time the set-up conditions occur, we are rational to set our credence in A equal to $x$. And so on.

This is the way we are intuitively inclined to talk, but philosophers are usually (and I believe, rightly) not content to accept the attribution of tendencies, dispositions, capacities, and suchlike things to parts of reality, without some further story about what these things are, and what it is in the 'occurrent' facts that grounds these ascriptions of non-Humean, modally loaded powers. For the most part, however, advocates

[1] This chapter has benefitted from audience and reader feedback over a number of years; I would especially like to thank Steve Savitt, Craig Callender, Mauricio Suárez and Nick Huggett, for discussion on earlier versions. Research for the paper has been generously supported by Spanish Ministry grants HUM2005-07187-C03-02 and FFI2008-06418-C03- 03/FISO.

of propensity accounts of chance reject the need to give any such further story. It is acceptable, they feel, to declare the existence of chance propensities as *primitive* posits of their ontology and metaphysics.[2] "It's not", they will say, "that we can say *nothing* about what these propensities are, after all: they are those features of reality that you described just now, with their relation to frequencies, degrees of belief, etc. But you can't expect us to give an analysis or reductive definition of what a chance propensity is."

For many philosophers—Humean empiricists, for example, but not only them—this philosophical move is illegitimate. As Lewis (1994) eloquently put it: "Be my guest— posit all the primitive unHumean whatnots you like. . . . But play fair in naming your whatnots. Don't call any alleged feature of reality 'chance' unless you've already shown that you have something, knowledge of which could constrain rational credence." These philosophers, who are either reductionists about objective chances, or skeptics, think that the propensity advocates are playing language tricks to make it *seem* that an account or theory of objective probability is being offered, when in fact something rather less is being provided. The associations and connotations those words ('propensity', 'tendency', etc.) evoke make it seem as though the linkage to frequencies, credences, and so forth are justified, when in fact they are not—at least, not without further work and explanation.

I myself am a skeptic/reductionist about objective chance (see Hoefer (2007)). In this paper I want to advance the skeptic/reductionist cause by criticizing the primitive-propensity view along a new front. I will focus on an aspect of the connotations of propensity-talk that I believe has not been clearly noticed by its opponents (though it *has* been noticed by some advocates), an aspect that turns out to say a lot about the notion of primitive chance propensities. That aspect concerns the linkage between the notion of a chance propensity and *time*. I will argue that those who define objective chances in terms of propensities, tendencies, or dispositions implicitly invoke a picture of time as divided into *past, future,* and a *moving "now"* (i.e. the A-series, or "tensed time"). My claim is that we implicitly adopt the A-series perspective when thinking of objective chances in terms of propensities, and that when we force ourselves away from that perspective, the notion of chance propensities loses most, if not all, of its apparent intuitive content. Understanding this connection will help us to see more clearly how

---

[2]   There are exceptions to this, of course, such as Suárez (2007) who advocates that quantum probabilities be understood as propensities, but does not close the door to further analysis of what propensity-claims mean.

Although I sketch here a unified picture of the propensity view, recent commentators such as Hájek and Gillies have distinguished two branches of propensity views: long-run frequency, and single-case. The former puts emphasis on the tendency of a chance set-up to generate, in the "long run", frequencies identical to the probabilities, while the latter emphasizes that each and every instantiation of the set-up has a tendency to produce each outcome, the strength of that tendency being what the objective probability quantifies. The long-run frequency variant strikes me as an unstable mixture, able to avoid the problems of single-case views, if at all, only by collapsing into the hypothetical frequency interpretation. I will not try to argue this point here, however.

difficult it is to pin down the notion of probabilistic propensities, as well as why the difficulty often goes unnoticed.

In the next section, I will review the A-series/B-series distinction, and some of the reasons why our everyday-life A-series view of time is physically and metaphysically problematic. In section 3, I will present some textual evidence for the linkage between propensity-talk and A-series time. In section 4, I show that the propensity theorist cannot claim that the linkages are merely colorful metaphor, unrelated to the content of the view: we will see that any attempt to strip away the A-series linkage leaves the expression of the propensity account altogether *too* bare and primitive to be taken seriously. Finally, in section 5 I will suggest that the problems of A-series time and of propensities can be overcome by the same route: accepting that these views capture not something physical and "out there", but rather something inextricably tied to our perspective as agents in the world.

# 2. A-SERIES TIME: A PROBLEMATIC ONTOLOGY

I am lumping together a variety of theses concerning time, and ontological views of time, under the heading of "A-series time".[3] To participants in the ongoing debates about time's ontology, this lumping may seem outrageous, but for our purposes, I believe it does no harm. To forestall confusion, however, let me first note that while many authors associate the "A-series/B-series" terminology with a language-oriented approach to the philosophy of time, there will be no such connection here. In the language-oriented approach, the A-series or "tensed" predicates 'is past', 'is present', and 'is future' are held to apply, or not apply, to events, or moments in time; are held to be contradictory, or not; and so forth. McTaggart's original argument for the contradictoriness of time as an A-series suggests this approach, though I believe that the argument can be given a less language-based formulation. In any case, my use of "A-series" will be as a catch-all for a variety of theses or views about time itself. What all the views have in common is at least this: that there is something *ontologically* correct in our intuitive division of time into the past, present, and future, the present being not solipsistic and punctual but universal and shared.[4] Recent debates include various forms of *presentism*, which assert that only present events exist, and various forms of *possibilism*, which assert that not only the present but also past events exist. (Both are opponents of *eternalism*, the sort of view I will associate below with the "B-series"

---

[3]  For further discussion and finer distinctions, see Zimmerman (2010), Chapter 7.

[4]  Howard Stein and some others have advocated a punctual (literally having no spatial extension) "now", one for each person or event. The motivation for this drastic idea, as we will see shortly, is the apparent incompatibility of the traditional present with special relativity theory. But we are getting ahead of ourselves.

or the "block universe".) Both presentists and possibilists embrace A-series time as something real, objective, and not merely part of subjective temporal experience.

Nevertheless, A-series time *is* the time of everyday life, experience and human agency: time as divided into *past, present, and future*, with special importance attaching to the present, or "now". The now is the cusp at which the merely possible or potential becomes actual or real, and fixed forever after—the past being, as we all know to our sorrow, fixed and unchangeable (see Figure 3.1). There is also some sort of *movement* involved in A-series time, which we may think of sometimes as a movement of the "now" into the future, and other times as a movement of the future "towards us" (permanent now-dwellers that we are). This movement is sometimes referred to as "absolute Becoming". What is the same in either metaphorical presentation is the relative motion, and its velocity: one second per second, one hour per hour, etc. Even if the future is not "open" in the sense that past facts plus the laws of nature determine

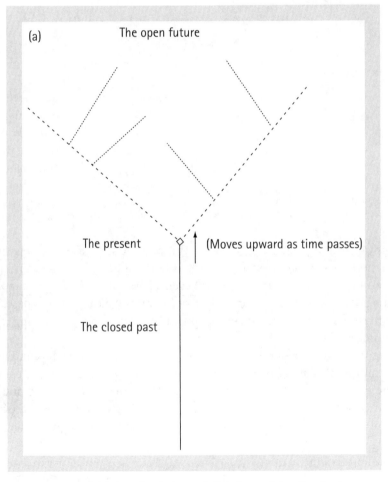

FIGURE 3.1a  A-series, indeterministic "open future" variant

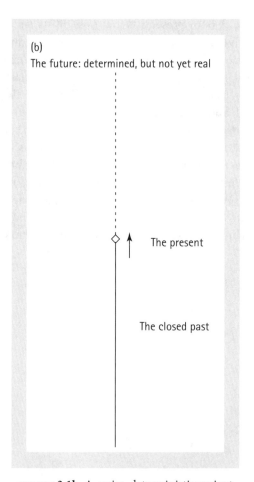

(b)

The future: determined, but not yet real

The present

The closed past

FIGURE 3.1b  A-series, deterministic variant

everything the future will bring (Figure 3.1**b**), most A-series views have it that the future is nevertheless *not yet real*, since it has not "happened" yet, and thus has not the status of *being* that past and/or present events possess. "Unfolding", "happening", "becoming" and "giving rise to" are words and phrases typically used to describe this process of movement in time that seems so familiar to us—though puzzling when we stop to think about it.

McTaggart (1908) famously attacked the A-series as internally contradictory. Though only a minority of philosophers believe McTaggart's argument was success-ful, there are other complaints typically made against the A-series.[5] The notion of a "velocity" of one hour per hour, for example, is troubling. By thinking of the 'now' as in motion up the tree of possibility, as in Figure 3.1**a**, we seem to invoke an *external*

---

[5] Horwich (1987) endorses McTaggart's argument, suitably clarified. Maudlin (2002) defends the notion of "passage of time" against some of the worries mentioned here, and others. It should be noted that Maudlin does not endorse the A-series per s*e*, but rather only the notion of (local) passage of time.

or second-order time—different from time inside the universe, which is after all just the vertical direction in the figure—which looks metaphysically suspect. And while the moving now feels familiar from everyday psychological experience, it seems not to have any place in our best *physical* descriptions of the world. No physical theory invokes or endorses a 'now' in its description of the world.[6] Worse, as we will see, physical theory appears to be incompatible with the A-series, when the latter is understood in its most straightforward sense.

A B-series (often called "tenseless time") is illustrated in Figure 3.2a. The B-series jettisons the categories of past, present ('now'), and future, and simply depicts time as a linear ordering (or partial ordering) of events, according to the relations of before/after (perhaps including a metric determining the temporal "distance" between

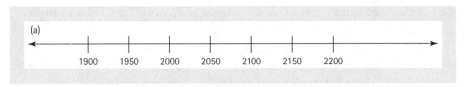

FIGURE 3.2a  A B-series (linear temporal ordering of events)

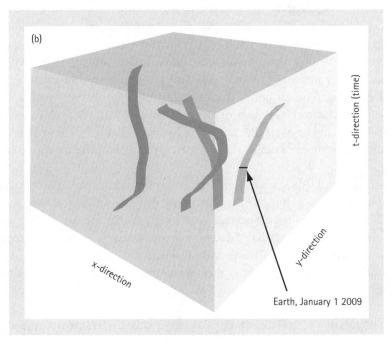

FIGURE 3.2b  Block universe

---

[6] It is true that some philosophers' interpretations of quantum mechanics do invoke the 'now' and give it a physical role. Arguably, as we shall see later, Popper does so; see also McCall (1994), Maxwell (1985).

an earlier and a later event). B-series time, and the related notion of a "block universe" (Figure 3.2**b**) abolish any ontological distinction between past and future, and the notion of a moving 'now'.

Physical theories seem to treat time in B-series or block-universe term. They describe events as ordered linearly in a time-series, but do not invoke the categories of past, present, and future.

Relativistic theories seem not merely to fail to include A-series concepts, but rather to positively rule them out. This was first argued in a forceful and clear way by Putnam (1967). Since special relativity and most models of general relativity do not allow the singling out of a privileged foliation of space-time (the block), they are not compatible with the A-series ontological distinctions.[7] Without a privileged foliation of the block into space-at-a-times, there can be no objective present or 'now' at *any* time—at least, not in the classical sense—nor can the parts of world history that are to count as past and as future be objectively specified.

Relativistic theories are, in fact, not only hostile to the A-series' distinctions, but also the B-series. In such theories, the B-series too is done away with, or at least transformed into something which, if it can be well defined at all, is only well defined with respect to a chosen reference frame or coordinate chart. This is a simple consequence of the relativity of simultaneity. With respect to one reference frame, events A and B, which occur on planets orbiting different stars, may be ordered in a B-series that has A occuring before B, but with respect to a different reference frame, with its own B-series ordering, B occurs before A. This can happen only if events A and B are not connectable by a light-ray or some other, slower-moving signal; events that are connectable by physical rays or signals have an objective, non-frame-dependent temporal ordering. The upshot is that, strictly speaking, relativity makes both the A-series and the classical B-series orderings of time ontologically problematic, and so what we will demand below is that our metaphysics of chance be compatible with the Block Universe, with its partial B-series orderings and complete absence of A-series notions.

The incompatibility of A-series time with current physics is widely accepted among philosophers of physics. But there are certainly dissenters who feel that acceptable senses of "now" and of "time's passage" can be given that are compatible (enough) with special and general relativity, and which adequately fit our classical/common-sense view of time. It will be worthwhile to discuss the ideas of two of these dissenters briefly; the senses in which they are correct (and incorrect) will turn out to be relevant to getting a clear view of the relationships between objective chances and time.

One particularly forceful dissenter is Maudlin (2002, 2007), who argues for the coherence and indeed necessity of accepting time's *passing*. (Maudlin sympathizes with critics of the metaphor of time's *flow*, and urges us to drop the latter metaphor in favor of talk of time's passing; we will come back to the question of whether this

---

[7] Stein tried to rebut Putnam's argument ((1968), (1991); see also Callender and Weingard (2000)).

is also metaphorical below.) Maudlin's defense of temporal passage begins by offering rebuttals to the arguments presented in Price (1996). As part of this he argues, successfully in my opinion, that there is nothing bizarre or incoherent about the *rate* at which time passes: one second per second, one hour per hour, etc. But curiously, Maudlin fails to mention that according to relativity theory, there is a sense in which the rate at which time passes for an observer depends on her state of motion—this being the root of the twin-"paradox" effect, among other things. This frame- or observer-relativity of the rate of time's passing may be seen as undercutting the objectivity of temporal passage, or at least as a feature of relativistic time that fits awkwardly with our traditional view of time. At any given point of space-time, there is no unique answer to the question: "How is time passing, at this point?"[8]

In fact, despite having defended the idea of a measure or rate for time's passage, Maudlin goes on to advocate a non-metrical representation of the passage of time. Maudlin's temporal passage is mathematically represented by a simple distinction, at each point of space-time, between past-directions and future-directions, that is, an orientation of space-time.[9] Maudlin himself recognizes the problem that a mere orientation seems rather poor and barren for the task of representing the alleged passage of time, but he blames the tool rather than questioning the contentfulness of the thing represented.

Setting aside the poverty of this representation of the passage of time for the moment, we should go on to ask: is Maudlin defending, in any sense, the correctness of the A-series view of time? Here things get a bit murky. Maudlin fully accepts the B-series, and accepts "the block universe"—if by this we mean that space-time is a four-dimensional whole with no ontological distinction between parts lying to the future of us vs parts lying to the past. But after defending time's passage in Chapter 4 of (2007), Maudlin returns to the A-series momentarily in his concluding remarks. He points out that rather than there being one A-series, there is an infinity of them, one for each moment that may be chosen as "the present" and thus serve as the divider between past and future. This is simply a terminological point that can be readily conceded. Maudlin claims that by having the B-series, he gets all these A-series "for free". But this begs the question: what sort of "now" or *present* is Maudlin laying claim to, to serve as the divider between past and future? Perhaps his "now" is a mere point in space-time, the intersection point of the past and future light cones? A point is not enough though: it does not divide the rest of the points into the 3 A-series classes of "past", "present", and "future", even if we draw a timelike line through the point and label one direction away from the point "+" or "future".

---

[8] Of course, one can make it look as though there is a unique answer, by noting that in each inertial reference frame, time passes for objects at rest in that frame at 1 second per second etc. But equally, one can say that time passes at $y$ seconds per second for objects moving in that reference frame (where $y$ is some number strictly between 0 and 1, a function of the object's velocity).

[9] Certain space-times allowed by general relativity, for example Gödel's, do not admit a global orientation; Maudlin does not say what we should think about such models, but it seems to me he is committed to saying that they do not represent physically possible worlds at all, since there can be no *time* in a 4-d object accurately represented by such a model.

In order to claim to get the A-series, Maudlin will have to tell us what is the spatial structure that represents the present, or at least serves to separate all those points which are future from those which are past. The problem is that special relativity offers us a variety of options here, even if we insist on using only intrinsic/objective structures built into Minkowski space-time. Maudlin does not discuss what structure should be chosen. Rather, he hints that he might be happy to see the non-relativistically-invariant *classical* A-series structure (involving a privileged foliation of space-time) make a comeback, since it seems necessary for the Bohmian approach to quantum mechanics. But he does not advocate a preferred temporal foliation outright, so we will leave the question hanging.

Steven Savitt, in a recent paper (2009; see also Chapter 18 of this volume), tackles head-on the question of what sort of structure, if any, can be compatible with special relativity theory and yet play something close to the role played by "the present" in our classical/everyday world view. Key to Savitt's proposal is the idea that we should incorporate something like William James' *specious present*. That is, rather than having the present represented by something with no duration, we should instead opt for a "thick" present that accords with the non-sharp nature of the "now" as we experience it.[10] In principle, a number of different structures might be chosen to represent the relativistic present, and we will see some of them below; what all have in common is that their specification is centred on, or relative to, a chosen world line (presumably, the world line of the conscious agent whose "present" we are seeking to model). Savitt chooses a structure he learned from Winnie (1977) called an *Alexandroff interval* or, for Savitt, an *Alexandroff present*. Taking the worldline segment from $e_0$ to $e_1$ to represent the duration of an experienced specious present, the Alexandroff present is defined as the intersection of the forward light cone of $e_0$ with the back light cone of $e_1$. (See Figure 3.3).

Depicted thus in units where the velocity of light $c$ is 1, this looks like a very odd thing to call "the present": it is extremely restricted in space, and also looks rather fat in the time-direction. The appearance is deceptive. Figure 3.4 shows part of an Alexandroff present (it continues off the page quite some distance) in which the vertical units are seconds, and the horizontal units are 100 kilometers. The reader will forgive me for using straight lines to represent the upper and lower boundaries of the Alexandroff present, since the diagram would have to extend for about 500 pages to the right before the upper and lower boundaries meet. Clearly, in everyday units the Alexandroff present looks considerably closer to our classical concept of "now", when the latter is supplemented with a specious-present thickness.

---

[10]  In human experience there is always some range of time within which events may be experienced as simultaneous (though they in fact are not), or even have their real time-order reversed in experience (due to the brain's processing); the "thickness" of the experiential *now* seems to be variable between people, between different times in the same person, and even dependent on what is the focus of attention. So Savitt's decision to look for a relativistically-acceptable structure representing the specious present is certainly not to be understood as an attempt to characterize a stable, intersubjective "now".

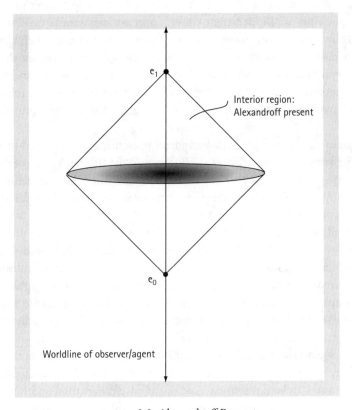

Interior region:
Alexandroff present

Worldline of observer/agent

FIGURE 3.3  Alexandroff Present

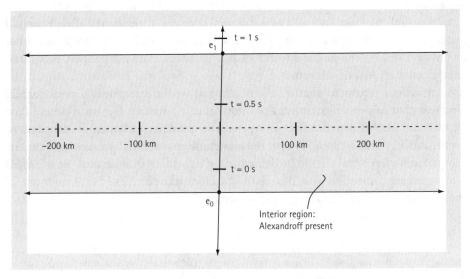

FIGURE 3.4  Alexandroff Present, reasonable units

Savitt's presents are still quite finite in space, so there is no answer to the question "What's happening on Jupiter right now?", nor are any events on Earth ever *co-present* with any events on planets circling other stars. On the other hand, despite their dependence by definition on a selected world line, one can say that different *human* observers share largely overlapping *nows*: we can choose Alexandroff presents for any two persons alive on this planet which will overlap in considerably more than 99% of their volumes, allowing us to speak of their *shared now* without too much distortion.

Savitt makes a nice case for his Alexandroff presents, but we should keep in mind that their thickness is rather arbitrarily chosen, and the thinner we make the specious present "in the middle", the more spatially restricted the whole structure becomes. If the severe spatial restriction seems hard to swallow, we can try the structure of Figure 3.5**a**, or we can reverse the directions of the light cones at $e_0$ and $e_1$, opting for a thickened "absolute elsewhere" (Figure 3.5**b**).

This AE-present will look the same as Savitt's locally; it is as well illustrated by Figure 3.4 as the Alexandroff present was. But it extends spatially to infinity, so there is always an answer to questions like "What is happening now on Alpha Centauri?". Indeed, too much of an answer: eight years' worth of events around Alpha Centauri fall within the scope of one AE-present, and further away still entire alien civilizations may rise and fall within the moment we call "now".[11]

No matter what relativistically acceptable structure one chooses to represent "the present", there will be clashes with our intuitive concept of the present, and I suspect no decisive reasons to be given in favor of one choice rather than another. If we want a series of "nows" (i.e. a collection of A-series) to point to, to represent the passage of time, we are free to choose one of these options and run with it. But we should not lose sight of these two facts: (a) any option we choose will be chosen-worldline-relative, not universal, and (b) the choice will be at bottom arbitrary, dependent on what sorts of features of our classical worldview we most want to preserve, not dictated in any sense by physical reality. The physical world itself, objectively, does not contain, anywhere, anything like an A-series structure.

So now let's return to chance. If the physical world, intrinsically, does contain objective chance propensities, but does not contain anything like an A-series, then we have a right to demand that any proposed account of chance propensities be expressible in terms *compatible* with the block universe and *not presupposing*, in any way, the A-series. What should be demanded of a metaphysical account, in order for it to be judged compatible with the relativistic block universe? The bare minimum, of course, is that it should not presuppose a universal 'now', nor any preferred foliation of space-time into simultaneity slices (the world-at-a-time). But this is not enough. The

---

[11]  Notice that this way of putting things exaggerates the problematicity. Although the rise and fall of said alien civilization may occur within my "now", it is going to be equally true that those same events fall within any "now" we select, say, one year from now on Earth. Far away events are simply neither to the past, nor to the future, of events here—the same thing that we must say if we use Alexandroff presents.

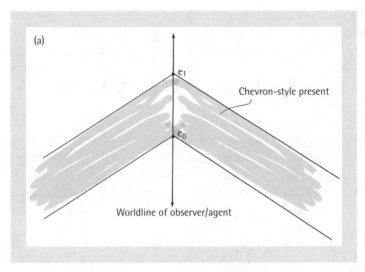

FIGURE 3.5a  Non-spatially-restricted relativistic "present" candidate

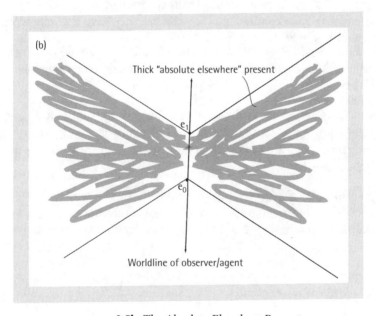

FIGURE 3.5b  The Absolute Elsewhere Present

account also should not rely on any implicit or explicit distinction (of an ontological sort) between the past and the future. And it should be clearly expressible without depending on metaphors that are either directly A-series related ("unfolding", "fixed", etc.), or related to human agency, which is itself something we understand only within an A-series perspective.

The problem I will discuss now is this: primitive propensity accounts of chance appear unable to meet this legitimacy test.

## 3. PROPENSITIES AND TIME

The language with which chance propensities are introduced and characterized is redolent with A-series- or passage-related notions such as *unfolding, happening, becoming, giving rise to, actualizing,* and so on. For a showcase example, I can do no better than the following from Karl Popper, the first post-positivist proponent of propensities. Here is a passage from his 1990 book *A World of Propensities*:

> In all these cases the propensity theory allows us to work with an *objective* theory of probability. Quite apart from the fact that we do not *know* the fututre, the future is *objectively not fixed*. The future is *open: objectively open*. Only the past is fixed; it has been actualized and so it has gone. The present can be described as the continuing process of the actualization of propensities; or, more metaphorically, of the freezing or the crystallization of propensities. While the propensities actualize or *realize* themselves, they are a continuing process. When they have realized themselves, then *they are no longer real processes....*
>
> The world is no longer a causal machine—it can now be seen as a world of propensities, as an unfolding process of realizing possibilities and of unfolding new possibilities.

Although this passage is one of the most striking and overt in its linkage of chance propensities to flowing, A-series time and the tree figure of 3.1a., it is by no means an isolated aberration. Here are some further quotations in which we can see the A-series lurking in the background:

> The problem here is that a realist single-case interpretation of probability is useful only in an indeterministic universe because otherwise the probabilities are all trivial. In such universes the future is "open" with respect to the present and past.
> (Milne 1986: 131)

Discussing probabilities of various outcomes on two successive flips of a coin, subscripted '1' and '2', Milne continues:

> The conditional probability $p(H_2|H_1)$, which is defined as $p(H_1H_2)/[p(H_1T_2) + p(H_1H_2)]$, is susceptible to a realist single-case interpretation. The formally symmetric $p(H_1|H_2)$ has no such interpretation. When $H_2$ is realised the first toss is over and done with, there is no matter of chance, no indeterminacy about the outcome of the first toss, $H_1$ either has or has not been realised....
>
> What makes the probabilities 0 or 1 is that the occurrence or otherwise of the conditioned event is determinate before or concurrently with the occurrence of the conditioning event.

Summing up the dilemma he is posing for realist single-case probability theorists (a class that includes propensity views), Milne says:

> One could ignore conditional probabilities altogether and thus obtain a realist single-case interpretation of normalised measures, not the probability calculus. One could sunder the link between realist single-case interpretations and the indeterminate. Or one could claim that the past and present are as indeterminate as the future.

The ontological distinction between past/present and future is evidently built into these views of chance, for Milne.

These remarks arise in the context of Milne's discussion of an apparent problem for propensities as an interpretation of the probability calculus (for objective probabilities), a problem first articulated by Paul Humphreys.[12] The problem is this: in some circumstances, Bayes' theorem allows us to start from an objective $P(E|C)$, that we have no problem understanding as a propensity, and (with the aid of other objective probabilities in the context) calculate a "backward-looking" probability $P(C|E)$. In the formulation of Salmon, we are able to calculate the probability of a frisbee *having been manufactured by machine 1, given that it is sampled and found defective*. The problem, acknowledged by most philosophers who discuss the case, is that it seems nonsensical to think of this backward-looking probability as a propensity. Humphreys himself (1986, 2004) turns the awkwardness into an outright contradiction, by showing that propensity theorists appear to be committed to two contradictory probability assignments to the "backward-looking" probabilities, at least if they agree that their propensities are subject to the axioms and theorems of conditional probability.

In Miller's (2002) response to Humphreys' paradox we see further evidence of the importance of the fixity of the past in philosophers' understanding of propensities:

> Intuitively we may read the term $P_t(A|C)$ as the propensity at time t for the occurrence A to be realized given that the occurrence C is realized. If C precedes A in time, this presents no extraordinary difficulty. But if C follows A, or is simultaneous (or even identical) with A, then it appears that there is no propensity for A to be realized given that C is realized; for either A has been realized already, or it has not been realized and never will be. (p. 112).

One might try to diagnose the problem as one due to the direction of causation rather than the fixity of the past. Since propensities are often said to be something like non-sure-fire causes, they share the temporal asymmetry we usually ascribe to causality, and the backward-looking probability gets the temporal direction wrong. But I wonder: is that really the correct diagnosis of what seems wrong to us in considering **P***: P(made-by-1| sampled&defective)? Or is what bothers us not rather this: it is crazy to think of a propensity such as this, because *by the time the frisbee is sampled, which machine it came from is already in the past, hence fixed and determined.* This frisbee can have no backward-looking intrinsic, single-case propensities, because everything about its past

---

[12]  See Humphreys (2004) for an extensive and updated discussion.

is done, fixed forever after. This certainly seems to be the way that Milne and Miller express the difficulty. While I don't deny that a worry about backward causation may be part of what ensures our agreement that **P**\* can't be a propensity, it seems to me that our agreement is also—perhaps mainly—rooted in our strong intuitions about the fixity of the past. I therefore read much of the discussion of Humphreys' problem as further evidence for the conceptual linkage of chance propensities with A-series time.[13]

# 4. Stripping away the A-series

So far what we have seen is that the language used to discuss metaphysical, single-case chance propensities is typically laced with A-series linked notions. We have not, however, seen anything like an argument to demonstrate that propensities can only exist in the context of an assumed A-series time. Nor do I think that such an argument or demonstration is feasible, given the vagueness of the metaphorical language with which propensity views are presented. It would seem possible for the advocate of propensities to admit that the language used to present them is infected with A-series connotations, but to insist that these are not essential to the content of the view. Time asymmetry (i.e. *passage* in Maudlin's thin sense) may well be essential, as the issues around Humphreys' paradox show (at a minimum), but perhaps not the A-series itself. In order to assess the adequacy of this response, we need to see what happens if we thoroughly and consciously strip away all such A-series infected metaphor, and try to re-present the propensity view in terms that are compatible with the block universe. What we will see is that the propensity view loses most, if not all, of its intuitive content.

How do we go about presenting propensities in block-universe terms? Let's try to do it first in words, and then using figures. Suppose we are describing a set-up *S* that has propensity of strength 0.6 to give rise to outcome *A*. Instead of saying something about how the set-up conditions have a tendency to *give rise to* or *produce* various outcomes—since these terms are linked to our intuitive A-series view of an unfolding or passing time—we will have to say something like this:

> (1)  "Specific set-up conditions *S* are such that they tend to be followed by outcome *A* with strength 0.6"

But this is meaningless—we have no notion of what it means to say that X tends to be followed by Y "with strength 0.6". In fact, we don't really have any clear, non-

---

[13]  Even if the core of the problem people seem to have with backwards-looking propensities is the intuitive clash with the "fixity of the past", one might argue that the fixity of the past, in turn, is compelling to us only because of the apparent impossibility of backwards causation. This may indeed be the case. But whether propensities have a direct intuitive link with A-series time (as I think), or only a mediated link that arises due to a blanket rejection of backwards causation, either way the propensity theorist needs to be able to divorce her ontology of chance from any connection to A-series time.

anthropomorphic idea of what the "tends to" locution means to begin with, never mind assigning it a numerical "strength". That is, we don't have any idea of what all this could mean, unless perhaps we unpack it as a rough way of speaking about the *frequencies* of *A*-outcomes shortly after *S*-set-ups:[14]

(2)    "The frequency with which *S*-setups are followed by *A*-outcomes is about 0.6"

This is a perfectly good block universe statement, but unfortunately it is explicitly not what the single-case propensity theorist means to say. It fails in two ways. Not only does she want to make an assertion about something that might, in principle, only be instantiated once in world history (hence having frequency 1 or 0); she also wants to leave conceptual room for the possibility that the actual frequencies do not match the propensity-strengths, even if the numbers are large. She *also* means to be saying that there is something causal in the relationship between *S* and *A*, although the causation falls short of being necessitation. So let's try again, now focusing on the causal aspect:

(3)    "Setup *S* is a non-sure-fire cause of various possible later events. Its causal strength for *A* is 0.6 (on a scale from 0 to 1)".

Here we are introducing the notion of 'cause', which itself should perhaps be checked for A-series contamination. For now let's set that aside. Still, even if the notion of cause is unproblematic here, we have just traded the unexplicated, anthropomorphic notion of "tendency" for the unexplicated (and, I suspect, just as anthropomorphic) notion of "causal strength". We can rightly demand an explanation of this notion—and it had better be an explanation that shows us why *causal strength* only ranges from 0 to 1, without ruining the explication by building in a link to probability (thus making the view circular as an analysis of objective probabilities).

So far we are getting nowhere, fast. We have found a B-series compatible translation of the propensity theorist's claim—*if* causation itself is understandable in B-series compatible terms—but due to its reliance on the notion of "strength", it is as much in need of explication as an objective chance statement (the sort of statement it is meant to analyze). But perhaps the seeds of a way out are visible in (3). It describes *S* as a "cause of possible later events".[15] The modality here, possibility, is presumably *physical* possibility. So (3) implicitly adverts to the laws of nature, and the various events to the future of *S* (and suitably connected to it) whose occurrence is compatible

---

[14]  I tend to think that our understanding of the locution "tends to ____" rests on our intuitive understanding of how this applies to ourselves as free agents. "I *tend to* have a beer with lunch on weekends." We all understand this because we are familiar with what it is like to be a free agent faced with choices to make, conflicting desires of varying strengths, and so on, and often forced to choose somewhat arbitrarily, but not "randomly". Coins do not choose how to land, however, nor do radium atoms choose when to decay. I understand how to unpack talk of the tendencies of flipped coins and radium atoms, if I am allowed to translate it into talk either of actual frequencies, or of rational expectations based on past experience. But propensity theorists wish to avoid both of these translations.

[15]  Perhaps this should have been written: "cause of one of a set of possible later events", so that it does not seem we are saying that *S* causes each possible later event. I will not try to clean up this problem.

with natural law. If we take this idea and run with it, we can solve our cause/strength problem quickly:

(4)    "The laws of nature entail that set-up **S** may be followed (after a suitable physical process) by various possible "outcome" events. The laws are at least partially *probabilistic* laws, and they entail that $Pr(A|S) = 0.6$".

This gets rid of the problematic notion of causal strength, and I suspect it is close to what many propensity advocates have in mind. But it relies on a strong reductionism of objective chances to law-derived probabilities that some philosophers would no doubt find uncongenial. Moreover, it simply pushes the bulge under the rug to a different location: now we may justly demand an explication of the notion of "probability" at work in *probabilistic laws*. Elliott Sober (2005) offers us a *no-theory* theory of objective probability that urges us not to ask for such an explication, but rather to take the probability calculus plus the probabilistic laws of the sciences as giving a sufficient implicit definition of objective probability. But if you find this an attractive route to take, you are no longer analyzing objective probabilities in terms of propensities at all.

Finally, we may as well try to express the propensity view without relying on any of these terms that are either suspect (A-series linked, and/or anthropomorphic) or in need of a block universe compatible analysis of their own:

(5)    "There is a metaphysically primitive, temporally asymmetric, numerical relation holding between **S** and its "possible outcomes" such as **A**, such that the sum of these numbers over all the possible outcomes is always equal to 1."

All the content we seem to be left with is this: probabilistic propensities sum to 1. But that is something held in common with all accounts of probability, objective and subjective alike, and is also true of normed measures in general, which need have nothing to do with chance at all.

When I try to re-pose a propensity view of objective chances in terms that are both clear, and block universe compatible, I find that I just cannot do it. But perhaps the fault is with the words being chosen, rather than with the notion of a primitive probability-like relation that is meant to be picked out. Let's see now what happens when we try to illustrate propensities (along the lines of (5) above) in a block universe, using the by now familiar tool of a space-time diagram.

Figure 3.6a shows us a bunch of propensities "at work": the usual philosophers' example of coins being fairly flipped, and having propensity 0.5 of landing heads and 0.5 of landing tails. The lines connecting F's and H's or T's may be thought of as the world line of the spinning coin, or as indicating a primitive relation whose "strength" (numerical value) is given by the number to the side. But there are two problems with this figure, so understood. First, the unexplicated notion of "strength" is not really admissible, as we saw above, nor is it clear how we can give it a genuine explication without departing from the propensity view. Second, the propensities should not really be thought of as a relationship between the set-up events and the *actual outcomes* alone, but rather as a set of relationships between the set-up events and each of the *possible* outcomes (here, H and T—both, in every case of a flip). Perhaps we should

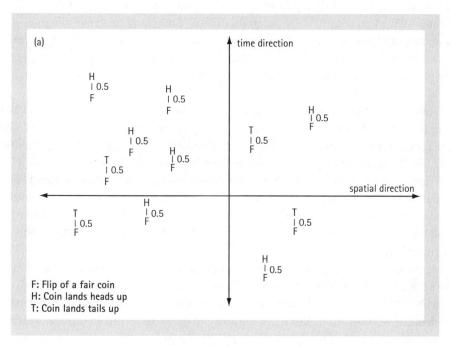

FIGURE 3.6a Propensities in space-time

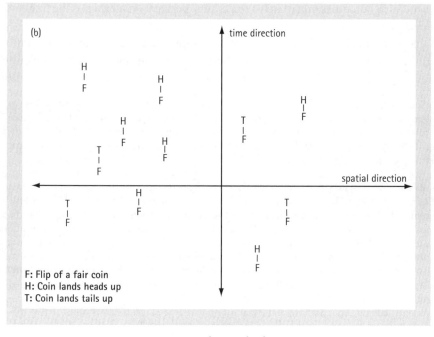

FIGURE 3.6b Just the facts

insert a ghostly letter H on top of every T in the figure (and vice versa) and a similarly ghostly "0.5" to go with it?[16] Or we could draw a "branching universe" diagram such as we saw in Figure 3.1a, but with branches ramifying from the "trunk" (representing what actually happens) at all times, that is, without singling out a "present" moment or any past/future difference other than the causal (and branching) asymmetry, and we could put the numbers on the branches.

However we decide to write in the alleged propensity relations, the point to bear in mind is that all we are doing is taking a diagram that accurately represents what actually exists and occurs, such as Figure 3.6 b, and superimposing on it some numbers (and ideally, branches or ghostly letters for non-actual events).

Aside from their adding up to 1, what reason is there to treat these numbers as probabilities? That is to say, what reason is there for asserting that these numbers *explain* the frequency with which F is followed by T, or make *likely* that about half of all F's are followed by T's, if we look at many F's, or constitute fair betting rates on individual flips, and so on?

My point is this: using a space-time diagram and sticking to neutral talk of "primitive numerical relationships" between F's and (actual and possible) H's and T's, we can perhaps cobble together something meant to represent the propensity account, that is block universe compatible. But stripping away all the A-series linked metaphors leaves us with something that has lost so much of its intuitive content, it no longer clearly has anything to do with what we *mean by* "objective probabilities"—that is, something about the world that deserves to guide expectations for future events, makes certain frequencies in outcomes more likely than others, perhaps even *explains* the frequencies we do see, and so on.[17]

Two points must be discussed before we go on to fix this quandary in section 5. First, I want to note that many philosophers tend to be unmoved by the difficulty of expressing the content of a primitive propensity ascription. The reason for this attitude

[16]  Some propensity advocates would urge, in addition, that the whole representation of these Head-producing and Tail-producing propensities should be superimposed onto the letters F, since they believe that propensities are always, strictly speaking, properties of set-ups at the time that they occur. This might get a bit messy, but could no doubt be adequately represented with the help of a better diagramming program than mine.

[17]  There is one strategy for meeting some of these objections that deserves a brief comment: Ramsification. That is, declare that objective chance propensities are *whatever aspect of the physical world does, in fact, have the right features to let it play the roles demanded by our uses of objective probabilities* (relations to frequency, inductive discoverability, guidance of credence, etc.). This strategy can be seen, for instance, in Mellor (1995). While it may seem to sidestep some of the problems I raise in the text, the cost is excessively high. We leave ourselves unable to say whether or not there are any objective probabilities in the world, and indeed unable to say what future events could begin to mitigate this ignorance. We also leave open the possibility that a reductive analysis of chance is in fact correct (and picked out by our Ramsifying strategy, even though we don't know it!). But if there is in fact nothing in the physical/objective world that can play the chance-role, then the Ramsifier is inadvertently talking about nothing, and if a non-primitive type of fact is in fact suited to play the chance-role, then what we want is to discover what those facts are. The Ramsifying strategy therefore only makes sense if we are confident that chances (a) exist, and (b) are primitive something-or-others out there in the physical world that cannot be further explained, described, or understood. Since I cannot see on what grounds one might have this confidence, I will set the Ramsifying strategy aside.

has, I think, to do with precisely the above-mentioned connotations of objective prob-ability talk, in particular the links to *what we should expect*. The charge that primitive chance-talk lacks content seems obviously mistaken, because to say that the chance of A-type events is 0.7 is to say that we should be confident to degree 0.7 that A will happen, that if the conditions are repeated many times we should expect A to come about approximately 7/10 of the time, that a bet on A's occurrence is fair if placed at 3:7 odds, etc. Is this not contentfulness?

The problem is that all these items have to do with the beliefs of rational agents, and therefore cannot be considered explications of something physical and "out there" in the external world. In fact, these items of content are only obtained by use—explicit or tacit—of the Principal Principle (PP), a principle linking objective probability with rational degree of belief (see Lewis 1986).[18] But why should we think that the Principal Principle is valid for primitive chance propensities? Lewis (1994) famously challenged the advocate of propensities to show why it is rational to apply PP to chance propensi-ties. But clearly, on the basis of the vanishingly slender content we have seen primitive chance propensities to have, no such demonstration is possible.

Second, as a terminological point, I should acknowledge that the word "propensity" plays no real role in the arguments of this section, other than being suggestive of A-series linked notions such as "happening", "producing", and "unfolding"—which is why we decided to see what its content or meaning might be, when such linkages are explicitly avoided. "Propensity" seems to be, in a sense, just a place-holder word, albeit one with strong flavor. Even if we drop the word "propensity", if we retain the idea that objective probabilities are to be understood as *primitive* chances, not further analyz-able or reducible, the lack-of-content problem will remain. What this goes to show is that, if we want to overcome the problem, what we will have to give up is the insistence on unanalyzability/irreducibility. That is to say, objective chances will have to be given some sort of analysis in other terms.[19] Putting together the lessons of sections 3 and 4 will give us hints about how the analysis or reduction might best proceed.

## 5. CHANCE AS A USER-CENTERED CONCEPT

The upshot of our discussion of A-series, flowing time was the following. An extremely thin notion of time's passage may be given that is compatible with relativistic physics: Maudlin's *orientation*, a simple distinguishing of the past-direction from the future-direction at every point in the block universe. But this thin notion does not support

---

[18]  PP says, to a first approximation, this: If all you know concerning whether a possible event A will happen or not is that the objective chance $Ch(A)$ of A is $x$, then if you are reasonable/rational your personal degree of belief in A's occurrence, $Cr(A|K)$, will also be $x$. (Here $K$ represents your background knowledge, and $Cr$ is a reasonable *credence* function, i.e. subjective probability).

[19]  Once such a non-primitivist account is in place, it may then be perfectly appropriate to use the word "propensities" to describe the objective chances.

the traditional connotations of time's *flow*, the *unfolding* of events, the *fixity* of the past, and openness of the future, and so forth. Nor does it give us anything like an A-series, since it does not give us a division of events into past events, present events, and future events. Savitt tries offering a much less thin structure to represent the present, and his proposal captures several of our common-sense notions about time. Yet it fails to capture other notions that are part of the common-sense view of time, and it looks arbitrary in two ways. The first way is that other structures exist that are different, but arguably capture our common-sense notion of "now" just as much as Savitt's Alexandroff presents do. The second way is in my view more important: Savitt's structure, and the others that one might advocate, all are spatially *centered* on a chosen world line. Rather than offering us something that is a correlate of "the present", they offer us the-present-for-Steve, the-present-for-Jerry, and so on. This all suggests that both "passage" and "now" are intrinsically *perspectival* concepts (as the indexical account advocated by most B-theorists has always maintained), *not* elements of physical reality itself. Agents or observers have *nows*; the physical world itself does not.

The upshot of our look at the A-series entanglements of primitive chances was that when we try to express the concept of such chances in explicitly block universe-friendly terms, we find that the content slips through our grasp. Talk of primitive chances seems to need the support of A-series related metaphors in order to seem contentful—that, or some tie to rational credence, that is, to what agents should expect to see (as time unfolds!). In other words, *both* A-series time and objective chance seem to be notions essentially linked to agents, and their perspectives on the world, rather than being part of the physical ontology of the universe itself. For *time* this is no problem: the relativistic understanding of time in the block universe remains, perfectly objective and physically acceptable, even supplemented if we wish by Maudlin's notion of passage. But for *chance,* the situation needs a remedy, if we are to maintain the existence of truly *objective* chances that are, in an appropriate sense, "out there" in the world.

As we saw above, objective chance requires some further, presumably reductive, analysis. But if we are to preserve what seems to be the core of our understanding and uses of chance, the reductive analysis has to be one that lets us see clearly and explicitly how and why chances deserve to guide credence and expectation, that is, why they make the Principal Principle rationally justified. Chance needs to be both fixed by what is "out there" in the physical world (to be objective), yet also clearly useful-to-agents (to play its expectation-guiding role). The analysis of objective chance offered in Hoefer (2007) is one attempt to give an account of objective chance that meets these desiderata.[20]

---

[20]  Lewis' (1994) reductive account of chance is meant to satisfy these goals as well, but he makes no real attempt to show that his account would make PP justified. The account sketched in Hoefer (2007) is developed further in my book manuscript *Chance in the World*. Parts of this article overlap significantly with some sections of Chapter 1 of the book.

## References

Callender, C. (2000), 'Shedding Light on Time', in *Philosophy of Science* (PSA Proceedings) 67: S587–S599.

Gillies, D. (2000), *Philosophical Theories of Probability* (London: Routledge).

Hájek, A. (2003), 'Interpretations of Probability', in *Stanford Encyclopedia of Philosophy* (online), http://plato.stanford.edu.

Hoefer, C. (2007), 'The Third Way on Objective Probability: A Skeptic's Guide to Objective Chance', in *Mind* 116(463): 549–596.

Horwich, P. (1987), *Asymmetries in Time* (Cambridge, MA: MIT Press).

Humphreys, P. (1985), 'Why Propensities Cannot Be Probabilities', in *The Philosophical Review* 94: 557–570.

—— (2004). 'Some Considerations on Conditional Chances', in *British Journal for the Philosophy of Science* 55: 667–680.

Lewis, D. (1986), 'A Subjectivist's Guide to Objective Chance', in *Philosophical Papers, vol. II* (Oxford: Oxford University Press).

—— (1994), 'Humean Supervenience Debugged', in *Mind* 103: 473–490.

Loewer, B. (2004), 'Lewis's Humean Theory of Objective Chance', in *Philosophy of Science* 71: 1115–1125.

McCall, S. (1994), *A Model of the Universe* (New York: Oxford University Press).

McTaggart, J. (1908), 'The Unreality of Time', in *Mind* (New Series) 68: 457–484.

Maudlin, T. (2002), 'Remarks on the Passing of Time', in *Proceedings of the Aristotelian Society* vol. CII (part 3): 237–252.

—— (2007), *The Metaphysics Within Physics* (New York: Oxford University Press).

Maxwell, N. (1985), 'Are Probabilism and Special Relativity Incompatible?', in *Philosophy of Science* 52: 23–44.

Mellor, H. (1995), *The Facts of Causation* (London: Routledge).

Miller, D. (2002), 'Propensities May Satisfy Bayes' Theorem', in *Proceedings of the British Academy* 113: 111–116.

Milne, P. (1986), 'Can There be a Realist Single-Case Interpretation of Probability?', in *Erkenntnis* 25: 129–132.

Popper, K. (1990), *A World of Propensities* (Bristol: Thoemes Press).

Price, H. (1996), *Time's Arrow and Archimedes Point: New Directions for the Physics of Time* (Oxford: Oxford University Press).

Putnam, H. (1967), 'Time and Physical Geometry', in *The Journal of Philosophy* 64: 240–247.

Savitt, S. (2009), 'The Transient *Nows*', in Wayne C. Myrvold and Joy Christian (editors), *Quantum Reality, Relativistic Causality, and Closing the Epistemic Circle: Essays in Honour of Abner Shimony*, (Berlin: Springer).

Stein, H. (1968), 'On Einstein-Minkowski Space-Time', in *The Journal of Philosophy* 65: 5–23.

—— (1991), 'On Relativity Theory and Openness of the Future', in *Philosophy of Science* 58: 147–167.

Suárez, M. (2007), 'Quantum Propensities', in *Studies in the History and Philosophy of Modern Physics* 38: 418–438.

Winnie, J. (1977) 'The Causal Theory of Spacetime', in Earman, J. et al. (eds.) *Foundations of Space-Time Theories* (*Minnesota Studies in the Philosophy of Science vol. 8)* (Minneapolis: University of Minnesota Press), 134–205.

Woodward, J. (2003), *Making Things Happen: A Theory of Causal Explanation.* (New York: Oxford University Press).

# CHAPTER 4

# TIME AND MODALITY

## ULRICH MEYER

## 1. INTRODUCTION

WITH the rigorous development of modal logic in the first half of the twentieth century, it became the custom amongst philosophers to characterize different views about necessity and possibility in terms of rival axiomatic systems for the modal operators '◊' ('possibly') and '□' ('necessarily'). From the late 1950s onwards, Arthur Prior argued that temporal distinctions ought to be given a similar treatment, in terms of axiomatic systems for sentential tense operators like 'P' ('it was the case that') and 'F' ('it will be the case that').[1] My aim here is to give a brief survey of the extent to which time can be treated on the model of modality. I shall not try to address the further question of whether such 'modal' accounts of time are to be preferred over 'spatial' accounts that treat times more like places. (Spatial accounts of time can be distinguished further, depending on which of the two main views about space—substantivalism and relationism—they take as their model.)

Since the pioneer days, much has been done to develop and study various systems of tense logic but it is not always clear what purpose they are supposed to serve. Apart from uses in computer science Pnueli (1977), there are two applications of tense logic that are of philosophical interest. One possibility is to regard tense logic as a linguistic theory of verb tense. This was Prior's view. He believed that tense in English functions in a similar way to the modal adverbs 'necessarily' and 'possibly'. Systems of tense logic might be too simple to provide an exact replica of ordinary temporal discourse, but they are claimed to provide useful toy models in which the underlying linguistic machinery can be studied in abstraction from the misleading embellishments of natural language. Just as standard first-order logic can be used to illuminate the truth-functional and quantificational structure of English, Prior thought that tense logic can help to explain the essential function of verb tense. The second

---

[1]  Prior (1957b, 1967, 1968c). See also Flo (1970) and Ørstrøm and Hasle (1993).

possibility is to regard tense logic as a metaphysical theory of the nature of time. Similar to modal primitivism, which argues that all there is to modality are boxes and diamonds, such a *tense primitivism* would aim to spell out all temporal distinctions in terms of a handful of conceptually primitive tense operators. Unlike spatial views of time, tense primitivism has no ontological commitments of a temporal nature: all its commitments are 'ideological' in the sense of Quine (1951).

Early advocates of tense logic thought that it could do double duty as a linguistic and a metaphysical thesis, but I think the two are best kept separate. The primary function of ordinary language is to facilitate communication about mundane facts, and there is no reason why our best linguistic theory of this role must also serve as our default metaphysics. The theory of time that is fossilized in the structure of verb tense might well be mistaken, or it might operate on a different level of abstraction than the best metaphysical theory of time. This is not to deny that there are some areas where the two projects get entangled. In trying to figure out which arguments involving tense operators are valid, we might be unable to separate linguistic analysis from metaphysics. But the main question about the linguistic theory is whether verb tense functions as a sentential operator in the first place, not what axioms to adopt for our tense operators. This more elementary issue is easily separated from metaphysical questions about the nature of time.

This chapter is organized as follows. After a brief review of modal logic in section 2, the basic results of propositional tense logic are presented in section 3. Section 4 develops a system of quantified tense logic. With these technical preliminaries out of the way, section 5 explains why tense logic ultimately fails as a linguistic theory of verb tense.[2] The remainder of the chapter is devoted to tense logic as a metaphysical theory of time. Section 6 presents the main objection to tense primitivism: that tense logic has insufficient expressive resources to serve as a metaphysical theory of time. Section 7 argues that the tense primitivist can overcome these problems by treating times as maximally consistent sets of sentences. Section 8 discusses a key difference between time and modality: the lack of a temporal analogue of actualism.

## 2. REVIEW OF MODAL LOGIC

As the base case, consider a language of propositional logic with atomic sentences '$A$', '$B$', '$C$', ... and the negation symbol '$\neg$' ('not') and the material conditional '$\rightarrow$' ('if ... then ...') as the only logical constants. The connectives '$\wedge$' ('and'), '$\vee$' ('or'), and '$\leftrightarrow$' ('if and only if') are defined in terms of '$\neg$' and '$\rightarrow$' in the usual way. The properties of these logical constants can be characterized in either of two ways. One method is to present an axiomatic proof system:

---

[2] Once we abandon the linguistic thesis, it would be more appropriate to call our logical system *temporal logic*, as suggested by Rescher and Urquhart (1971), but this proposal has not caught on.

A1   $\varphi \to (\psi \to \varphi)$
A2   $(\neg\varphi \to \neg\psi) \to (\psi \to \varphi)$
A3   $(\varphi \to (\psi \to \omega)) \to ((\varphi \to \psi) \to (\varphi \to \omega))$
MP   If $\varphi \to \psi$ and $\varphi$ then $\psi$

A proof of $\varphi$ from premises $\psi_1, \ldots, \psi_n$ is an ordered sequence of sentences ending with $\varphi$ such that each sentence in the sequence is either one of the premises, one of the axioms A1–A3, or follows from preceding ones by the inference rule of *modus ponens* MP. If there is such a proof then we write $\psi_1, \ldots, \psi_n \vdash \varphi$ and say that $\varphi$ is provable from these premises. A sentence is a theorem of our system, $\vdash \varphi$, if and only if it is provable from zero premises.

The second method characterizes logical properties in terms of a model theory. Let 'T' stand for the truth value *true* and 'F' for *false*. Then a model for propositional logic is a map $m$ from the sentences of the formal language to truth values that satisfies the following two conditions:[3]

$$m(\ulcorner \neg\varphi \urcorner) \quad = T \text{ iff } m(\varphi) = F$$
$$m(\ulcorner \varphi \to \psi \urcorner) = F \text{ iff } m(\varphi) = T \text{ and } m(\psi) = F$$

A sentence $\varphi$ is said to be true in the model $m$ if and only if $m(\varphi) = T$. The sentence is a logical consequence of premises $\psi_1, \ldots, \psi_n$, written $\psi_1, \ldots, \psi_n \vDash \varphi$, just in case $\varphi$ is true in all models in which all of the premises are true. Two sentences $\varphi$ and $\psi$ are logically equivalent, $\varphi \sim \psi$, just in case they have the same truth values in all models. A sentence is a logical truth if and only if it is true in all models. The proof system and the model theory yield the same result. According to the soundness and completeness theorem for propositional logic, a sentence is provable from a set of premises if and only if it is one of their logical consequences:

$$\psi_1, \ldots, \psi_n \vdash \varphi \quad \text{iff} \quad \psi_1, \ldots, \psi_n \vDash \varphi$$

The proof can be found in any introductory logic book. In particular, this means that a sentence is a theorem of our system if and only if it is a logical truth, and that two sentences are logically equivalent just in case each is provable from the other.

If the atomic sentences of our language stand for simple indicative sentences of English, then propositional logic allows us to describe the way things actually are. Modal logic is an extension of propositional logic that also lets us make claims about how things must be, or how they might have been. Suppose we introduce a sentential operator '$\Diamond$' ('possibly') as an additional logical constant and treat '$\Box$' ('necessarily') as an abbreviation for '$\neg\Diamond\neg$'. We then get a system of modal logic by adding the following axioms and inference rules to our propositional logic:[4]

---

[3] The quasi-quotation marks $\ulcorner$ $\urcorner$ were introduced by Quine (1940, §6) to permit the inclusion of variables in quoted material.

[4] Introductions to modal logic can be found in Hughes and Cresswell (1996), Bull and Segerberg (2001), Garson (2006), and Cocchiarella and Freund (2008).

M1   $\Box(\varphi \to \psi) \to (\Box\varphi \to \Box\psi)$
M2   $\Box\varphi \to \varphi$
M3   $\Box\varphi \to \Box\Box\varphi$
M4   $\Diamond\varphi \to \Box\Diamond\varphi$
NEC  If $\vdash \varphi$ then $\vdash \Box\varphi$

These axioms claim that $\psi$ is necessary if both $\ulcorner\varphi \to \psi\urcorner$ and $\varphi$ are necessary (M1); that necessary truths are actually true (M2); that necessary truths are *necessarily* necessary (M3); and that $\varphi$ is necessarily possible if it is possible (M4). The necessitation rule NEC stipulates that all theorems of our logic are necessary.

A model for modal logic is a quadruple $\langle W, R, i, @ \rangle$ that consists of a set $W$ of objects called possible worlds, a binary accessibility relation $R$ on $W$, and an element $i$ of $W$ chosen as actual world. There is also a true-at relation @ between sentences and worlds that needs to satisfy the following conditions:

| | | |
|---|---|---|
| $@(\ulcorner\neg\varphi\urcorner, w)$ | iff | it is not the case that $@(\varphi, w)$ |
| $@(\ulcorner\varphi \to \psi\urcorner, w)$ | iff | either $@(\ulcorner\neg\varphi\urcorner, w)$ or $@(\psi, w)$ |
| $@(\ulcorner\Diamond\varphi\urcorner, w)$ | iff | $@(\varphi, w)$ for some $w' \in W$ such that $Rww'$ |

A sentence $\varphi$ is true in a model if and only if it is true at its actual world, $@(\varphi, i)$. Hence $\ulcorner\Diamond\varphi\urcorner$ is true in a model just in case $\varphi$ is true in some possible world $R$-accessible from $i$, and $\ulcorner\Box\varphi\urcorner$ is true if and only if $\varphi$ is true in all accessible worlds. Logical consequence, logical truth, and logical equivalence are defined as before, in terms of truth in a model. The true-at relation @ is uniquely determined by the truth values it assigns to the atomic sentences in each possible world. Moreover, any world $w$ in a model of modal logic defines a model $m_w$ of ordinary propositional logic via:

$$m_w(\varphi) = \text{T} \quad \text{iff} \quad @(\varphi, w)$$

We can thus think of the models of modal logic as nothing but $R$-ordered sets of models of the underlying propositional logic.

Like the negation symbol, the possibility operator is a unary sentence connective that yields another sentence when prefixed to a sentence. The syntactic properties of '$\neg$' and '$\Diamond$' are thus quite similar, but there are important semantic differences between them. Like the material conditional, negation is a truth-functional sentence connective. The truth value of a molecular sentence constructed with the help of '$\neg$' and '$\to$' is always a function of the truth values of its atomic components. In particular, two negations have the same truth value whenever the negated sentences do. By contrast, $\ulcorner\Diamond\varphi\urcorner$ and $\ulcorner\Diamond\psi\urcorner$ are only guaranteed to have the same truth value if $\varphi$ and $\psi$ are logically equivalent:

If $\varphi \sim \psi$ then $\models \Diamond\varphi \leftrightarrow \Diamond\psi$

Let me call an operator that always takes the same truth value on logically equivalent sentences a *propositional* connective. All truth-functional connectives are propositional, but not vice versa.[5]

Propositional logic has one soundness and completeness theorem that links the model theory and the proof system. The modal case exhibits a more nuanced structure. There is an entire sequence of such theorems that illustrates how successively adding axioms M1 through M4 to our propositional logic imposes more and more stringent restrictions on the accessibility relation:[6]

K   The basic modal system K consists of axioms A1–A3, rules MP and NEC, and the one modal axiom M1. This system requires '$\Diamond$' and '$\Box$' to be propositional operators. Its theorems are those sentences that are true in all models, without any restrictions on $R$.

T   By adding axiom M2 to K, we get the system T. The additional modal axiom ensures that '$\Box$' is factive and only applies to sentences that are actually true. By contraposition, this means that any sentence that is actually true is also possible. The theorems of T are those sentences that are true in all models in which $R$ is reflexive; that is, for which $Rww$ holds for all $w \in W$.

S4  The system S4 is obtained by adding M3 to T, which requires that all true claims of type $\ulcorner \Box\varphi \urcorner$ are themselves necessary. Its theorems are those sentences that are true in all models in which $R$ is reflexive and transitive; that is, for which $Rww''$ holds whenever $Rww'$ and $Rw'w''$. In models with transitive accessibility relation any sentence that is *possibly* possible is itself possible.

S5  The system S5 contains the axiom M4 in addition to everything we had before. The new axiom is a complement to M3 and requires that all true claims of type $\ulcorner \Diamond\varphi \urcorner$ are necessary. To ensure this, the accessibility relation must be symmetric and $Rww'$ hold whenever $Rw'w$. The system S5 is thus sound and complete relative to all models in which $R$ is an equivalence relation (i.e., is reflexive, transitive and symmetric).

We get a similar sequence of soundness and completeness theorems for tense logic, with earlier-than and later-than playing the role of accessibility relations.

# 3. PROPOSITIONAL TENSE LOGIC

The basic idea of tense logic is to treat temporal distinctions in the same way in which modal logic treats modal distinctions, by introducing a family of primitive tense operators as additional logical constants. There are many possible choices of

---

[5] Humberstone (2009) and Williamson (2006) call propositional operators *congruent* connectives. Cresswell (1970) discusses *non*-propositional operators.

[6] Many logic texts give the axioms the same traditional names as the corresponding proof systems. Axiom M1 is often called axiom K, M2 axiom T, M3 axiom 4, and M4 axiom 5.

**Table 4.1 Two Interpretations of S5**

| Formalism | Modal Logic | Tense Logic |
|---|---|---|
| ◊ | possibly | sometimes |
| □ | necessarily | always |
| *W* | possible worlds | times |
| *i* | actual world | the present |
| @ | true in a world | true at a time |

primitives that one could consider and it would go beyond the scope of this article to try to discuss all of them. Let me therefore restrict myself to a few basic systems.[7] As Prior (1957a) notes, one easy way of obtaining a tense logic is by simply reinterpreting our propositional modal logic. Suppose we read '◊' as 'sometimes', '□' as 'always', and reserve sentences without any such operators for making claims about what is presently the case. Interpreted this way, our axiom M1 then claims that $\psi$ is always true if $\ulcorner \varphi \rightarrow \psi \urcorner$ and $\varphi$ are always true; M2 that what is always true is true now; M3 that what is always true is *always* always true; and M4 that what is sometimes true is always sometimes true. NEC becomes a temporal generalization rule that stipulates that theorems are always true. The model theory for modal logic can be recycled in a similar manner. All we need to do is call the elements of *W* 'times' rather than 'possible worlds' and *i* the 'present' rather than 'actual world'. We have one formalism that can serve both as a modal logic and a simple tense logic.

This tense reading of S5 illustrates some of the formal similarities between tense and modal logic, but it also ignores an important difference between the two: the distinction between past and future, which has no modal analogue. To do justice to this additional structure, tense logic typically employs more fine-grained operators. One popular approach uses a past tense operator 'P' ('it was the case that') and a future tense operator 'F' ('it will be the case that'). For expository purposes, it is also convenient to define 'H' ('it always has been the case that') as an abbreviation for '¬P¬', and 'G' ('it will always be the case that') as shorthand for '¬F¬'. We then get a simple system of tense logic Z (*Zeitlogik*) by adding the following axioms and rules of temporal generalization to our propositional logic:

Z1  $H(\varphi \rightarrow \psi) \rightarrow (H\varphi \rightarrow H\psi)$
Z2  $G(\varphi \rightarrow \psi) \rightarrow (G\varphi \rightarrow G\psi)$
TG  If $\vdash \varphi$ then $\vdash H\varphi$ and $\vdash G\varphi$

We can think of this system as consisting of two copies of the modal system K discussed earlier. The tense operators 'P' and 'F' play the role of the possibility operator '◊', and 'H' and 'G' that of the necessity operator '□'.

---

[7] Introductions to tense logic can be found in McArthur (1976), van Benthem (1983), and Burgess (2002). See Gabbay et al. (2002) for a treatment of more advanced topics.

The model theory for Z is a doubled-up version of the model theory for K. A model is a quintuple $\langle T, <, >, @, i \rangle$, where $T$ is a set of objects called 'times' and $i$ an element of $T$ chosen as 'present'. The earlier-than and later-than relations $<$ and $>$ on $T$ can be thought of as accessibility relations for the two tense operators. The true-at relation $@$ between sentences and times is required to satisfy:

| | | |
|---|---|---|
| $@(\ulcorner \neg\varphi \urcorner, t)$ | iff | it is not the case that $@(\varphi, t)$ |
| $@(\ulcorner \varphi \to \psi \urcorner, t)$ | iff | either $@(\ulcorner \neg\varphi \urcorner, t)$ or $@(\psi, t)$ |
| $@(\ulcorner P\varphi \urcorner, t)$ | iff | $@(\varphi, t')$ for some $t' \in T$ such that $t' < t$ |
| $@(\ulcorner F\varphi \urcorner, t)$ | iff | $@(\varphi, t')$ for some $t' \in T$ such that $t' > t$ |

A sentence $\varphi$ is true in a model if and only if it is true at its present time, $@(\varphi, i)$. Logical consequence, logical truth and logical equivalence are defined as before. The true-at relation is completely determined by the truth, values it assigns to the atomic sentences at the various times. Similar to the modal case, each time $t$ in a model of tense logic defines a model $m_t$ of the underlying propositional logic:

$$m_t(\varphi) = \mathrm{T} \quad \text{iff} \quad @(\varphi, t)$$

We can thus think of the models of tense logic as nothing but stacks of models of propositional logic, ordered by the relations $<$ and $>$.

By plagiarizing the proof of the soundness and completeness theorem for K, we can easily show that a sentence is a theorem of Z if and only if it is true in all models. Just as K does not impose any restrictions on the accessibility relation, Z does not impose any conditions on the earlier-than and later-than relation, not even that they are transitive, or that one is the converse of the other. Z is a very weak tense logic that does little more than guarantee that our tense operators are propositional connectives. But there are extensions of Z that impose more stringent restrictions. As in the modal case, we get a sequence of soundness and completeness theorems:[8]

1. By adding the axioms $\ulcorner \varphi \to HF\varphi \urcorner$ and $\ulcorner \varphi \to GP\varphi \urcorner$, we can ensure that the later-than relation is the converse of the earlier-than relation, and that $t' < t$ if and only if $t > t'$ for all $t, t' \in T$.
2. To ensure that the earlier-than and later-than relations are transitive, we could add $\ulcorner PP\varphi \to P\varphi \urcorner$ and $\ulcorner FF\varphi \to F\varphi \urcorner$.
3. Even with these additions, it is still possible for the time series to branch. In the case of a forward branch, the times in the branches are later than the point of divergence, but no time in one branch is later than a time in another one. We can rule out forward branches by adopting $\ulcorner (F\varphi \wedge F\psi) \to (F(\varphi \wedge \psi) \vee F(\varphi \wedge F\psi) \vee F(F\varphi \wedge \psi)) \urcorner$, which requires the later-than relation to be comparable. That is, for all times $t$ and $t'$, we would always have either $t > t'$, $t = t'$, or $t' > t$. Similar remarks apply to backwards branches and the axiom schema $\ulcorner (P\varphi \wedge P\psi) \to (P(\varphi \wedge \psi) \vee P(\varphi \wedge P\psi) \vee P(P\varphi \wedge \psi)) \urcorner$, which requires $<$ to be comparable.

[8] Proofs are in Cocchiarella (1966a,b), Bull (1968), and Gabbay (1975). See also Prior (1966) and Burgess (2002).

4. We can guarantee the existence of a first time by adding $\ulcorner H\bot \vee PH\bot \urcorner$, and the existence of a last time with $\ulcorner G\bot \vee FG\bot \urcorner$. We can rule out such endpoints by adopting $\ulcorner \varphi \rightarrow PF\varphi \urcorner$ and $\ulcorner \varphi \rightarrow FP\varphi \urcorner$. Here '$\bot$' is shorthand for a fixed, but arbitrary contradiction.

5. A dense time series, in which there is always another time between any two given ones, can be enforced with $\ulcorner P\varphi \rightarrow PP\varphi \urcorner$ and $\ulcorner F\varphi \rightarrow FF\varphi \urcorner$. The axioms $\ulcorner (\varphi \wedge H\varphi) \rightarrow FH\varphi \urcorner$ and $\ulcorner (\varphi \wedge G\varphi) \rightarrow PG\varphi \urcorner$ yield a discrete time (without endpoints) by requiring every time to have an immediate predecessor and successor.

6. The time series is complete if and only if any non-empty subset of $T$ with an upper (lower) bound has a least upper (greatest lower) bound. To ensure completeness, we could add $\ulcorner (F\varphi \wedge FG\neg\varphi) \rightarrow F(HF\varphi \wedge G\neg\varphi) \urcorner$ and $\ulcorner (P\varphi \wedge PH\neg\varphi) \rightarrow P(GP\varphi \wedge H\neg\varphi) \urcorner$.[9]

One important property that is absent from this list is anti-symmetry. The earlier-than relation is anti-symmetric if and only if $t \not< t'$ whenever $t' < t$. To see why there is no axiom schema for tense logic that could ensure anti-symmetry, consider two models $M_1$ and $M_2$ of eternal recurrence in which the same episode $H$ keeps repeating itself over and over again. In the model $M_1$, the time series looks like the real numbers, with infinitely many episodes $H$ succeeding one another in a linear fashion. (See Figure 4.1 The model $M_2$ looks like a circle, with only one episode $H$ stretched out along its circumference. If we define transitive earlier-than relations in the two models in the obvious way, then $<$ is anti-symmetric in $M_1$ but not in $M_2$, where every time is both earlier and later than itself.[10] If we also put the present moment $i$ at the same place within an episode $H$ in both cases, then the same sentences of our tense logic are true in both models. Since our tense logic cannot distinguish between $M_1$ and $M_2$, it could only guarantee an anti-symmetric earlier-than relation if it managed to prohibit eternal recurrence altogether. One could easily rule out eternal recurrence by adopting axioms that ensure the existence of a first or last time, but this would also rule out any models in which the time series looks like the real numbers, and most of our physical theories assume that time has that structure. To guarantee non-recurrence in a way that permits such a time series, we need assurance that the future is always different from the past, and that cannot be enforced with an axiom schema. Schemata can do some of the work of universal quantification over sentences, but they cannot be used to formulate the existential claim needed here: that there is a sentence that will be

---

[9] What is at issue here is the completeness of $T$ under $<$ and $>$, and not the completeness of an axiomatic system. Examples of complete sets are the integers and the real numbers, ordered by the less-than relation; the rational numbers are not complete.

[10] The situation is a little bit more complicated than suggested here. If $<$ and $>$ are transitive in $M_2$ then every time is both later and earlier than every other one. Since nothing would distinguish points on one side of $i$ from those on the other, there would be no real substance to the claim that the time series of $M_2$ looks like a *circle*. Van Fraassen (1970, sec. 3.1) and Newton-Smith (1980, sec. 3.2) claim that a cyclical time cannot be described in terms of binary earlier-than and later-than relations, but this is only true if we insist that these relations are transitive. Reynolds (1994) develops a tense logic for cyclical time series that gives up transitivity.

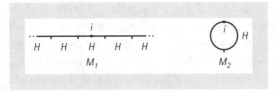

FIGURE 4.1 Two types of eternal recurrence

true relative to the time at issue, but never was true in the past. (I say more about the problem of anti-symmetry in sections 6 and 7.)

## 4. TIME AND EXISTENCE

The final step in the development of tense logic is to add quantifiers and identity. As before, let me start by considering the untensed case. The vocabulary of the language of quantificational logic consists of individual constants '$a$', '$b$', ..., '$s$', variables '$t$', '$u$', ..., '$z$, and predicates '$A$', '$B$', ..., '$Z$'. As logical constants we have the two truth-functional connectives '$\neg$' and '$\rightarrow$', plus the existential quantifier '$\exists$' and the identity symbol '$=$'. Numerals may be used as subscripts to ensure an inexhaustible supply of constants, variables, and predicates. We could also add superscripts to predicates to indicate their adicity, which is the number of empty slots that need to be filled with individual constants to form a sentence. I will follow custom by omitting them. An atomic formula of this language is either an $n$-ary predicate followed by $n$ individual constants or variables, or the identity sign flanked by two constants or variables. Molecular formulae are obtained by combining atomic ones with the help of the logical constants. Both $\ulcorner\neg\varphi\urcorner$ and $\ulcorner(\varphi \rightarrow \psi)\urcorner$ are formulae if $\varphi$ and $\psi$ are, and if $\varphi$ is a formula and $\xi$ a variable then $\ulcorner\forall\xi\,\varphi\urcorner$ is a formula. In such a case, the quantifier is said to bind all occurrences of $\xi$ within $\varphi$. An occurrence of a variable not bound by any quantifier is called free, and a formula without free variables is a sentence. The universal quantifier '$\forall$' is defined in terms of the existential quantifier and negation, by treating $\ulcorner\forall\xi\,\varphi\urcorner$ as an abbreviation for $\ulcorner\neg\exists\xi\,\neg\varphi\urcorner$.

We get an axiomatic system for untensed quantificational logic by adding the following axioms and inference rule of universal generalization to A1–A3 and MP:

Q1  $\forall\xi\,\varphi \rightarrow \varphi[\tau/\xi]$
Q2  $\forall\xi\,(\varphi \rightarrow \psi) \rightarrow (\varphi \rightarrow \forall\xi\,\psi)$   $\xi$ may not be free in $\varphi$
Q3  $\tau = \tau$
Q4  $\tau = \theta \rightarrow (\varphi \rightarrow \varphi[\tau//\theta])$
UG  If $\vdash \varphi$ then $\vdash \forall\xi\,\varphi[\xi/\tau]$

Here $\tau$ and $\theta$ are individual constants. As usual, $\varphi[\tau/\xi]$ denotes the result of substituting all occurrences of $\xi$ in $\varphi$ by $\tau$, and $\varphi[\tau//\xi]$ the result of substituting some (but perhaps all) of these occurrences.

A model for quantificational logic is a triple $M = \langle D, b, e \rangle$, where $D$ is a set of objects chosen as the domain of the model. All non-logical vocabulary of the language gets interpreted within this domain. The baptismal map $b$ assigns elements of $D$ to all singular terms. Some objects in the domain may go unnamed, and some may get more than one name, but no singular term can fail to name something. The extension map $e$ assigns a set of $n$-tuples of elements of $D$ to each $n$-ary predicate. Each unary predicate is assigned a set of objects from the domain, each binary predicate a set of ordered pairs of elements of $D$, and so on. The identity symbol functions like a binary predicate, but it does not get interpreted by the extension map. Identity always gets assigned the diagonal set $\{\langle x, x \rangle : x \in D\}$ as its extension. In these terms, we can now define a map $m$ from sentences to truth values via:

$$
\begin{array}{lll}
m(\ulcorner \Pi a_1 \ldots a_n \urcorner) & = \mathrm{T} \quad \text{iff} & \langle b(a_1), \ldots, b(a_n) \rangle \in e(\Pi) \\
m(\ulcorner a_1 = a_2 \urcorner) & = \mathrm{T} \quad \text{iff} & b(a_1) = b(a_2) \\
m(\ulcorner \neg \varphi \urcorner) & = \mathrm{T} \quad \text{iff} & m(\varphi) = \mathrm{F} \\
m(\ulcorner \varphi \to \psi \urcorner) & = \mathrm{T} \quad \text{iff} & m(\varphi) = \mathrm{F} \text{ or } m(\psi) = \mathrm{T} \\
m(\ulcorner \forall \xi \varphi \urcorner) & = \mathrm{T} \quad \text{iff} & m^*(\varphi[a/\xi]) = \mathrm{T} \text{ for all } a\text{-variants } M^* \text{ of } M.
\end{array}
$$

If $a$ is a singular term that does not occur in $\varphi$ then the model $M^*$ is said to be an $a$-variant of $M$ if their domains and extension maps are the same and their baptismal maps agree everywhere, with the possible exception of what they assign to $a$. If we say that a sentence $\varphi$ is true in the model $M$ if and only if $m(\varphi) = \mathrm{T}$ then our axiomatic system is sound and complete with regard to this model theory.

If we try to add tense operators to this quantificational logic, we need to deal with a number of complications that do not arise when we consider tense operators and quantifiers separately. We saw earlier that models for propositional tense and modal logic are stacks of models of the underlying propositional logic. This result does not extend to the quantificational case. Not every collection of models of ordinary quantificational logic constitutes a model of quantified tense logic.

One complication concerns the interpretation of individual constants. Suppose we accept Q4 in cases where the sentence $\varphi$ may contain tense operators, and let $\alpha$ and $\beta$ be two individual constants. Consider the tense reading of S5 discussed earlier, which takes '$\Box$' to mean 'always' and '$\Diamond$' to mean 'sometimes'. If we take $\ulcorner \Box \beta = \beta \urcorner$ as our choice of $\varphi$ then Q4 delivers:

$$
\alpha = \beta \to (\Box \beta = \beta \to \Box \alpha = \beta)
$$

Since $\ulcorner \Box \beta = \beta \urcorner$ follows by NEC from $\ulcorner \beta = \beta \urcorner$, which is an instance of Q3, this entails $\ulcorner \alpha = \beta \to \Box \alpha = \beta \urcorner$ and thus requires all individual constants to be *temporally rigid*. Unlike definite descriptions, such as 'the Chancellor of Germany', whose referent may change over time, our constants must always denote the same object. If we regard singular terms as the formal analogues of proper names in ordinary language, then this temporal rigidity view is a very natural one. The same kind of evidence that Saul Kripke (1980) adduces in support of the view that proper names in English are modally rigid could be used to support the view that they are temporally rigid as well. But if we treat individual constants in this way then not any arbitrary collection of models of untensed quantificational logic will do as a model for a quantified tense logic; we could only combine those models whose baptismal maps coincide, and which assign the same objects to all singular terms.

A second problem concerns the range of quantifiers. There are two views one could adopt about this, plus a myriad of variations on these two main themes. On a *time-relative* quantifier view, the domain of quantification varies from time to time. At each time $t$, the quantifiers range over the set $D_t$ of all objects that exist then. On an *untensed* quantifier view, the domain comprises all objects that exist at some time or other, which means that the quantifiers range over the set:

$$D = \bigcup_{t \in T} D_t$$

In the modal case, the view that the quantifiers have different domains in different worlds works reasonably well. The same cannot be said about the tense case. David Lewis (2004) gives the following example of a sentence that cannot be rendered with time-relative quantifiers:

There were three kings named Charles.

If we tried to formalize this as $\exists x \exists y \exists z P(\ldots)$ then the sentence would incorrectly assert the present existence of three people named Charles who were kings in the past. The opposite combination does not work, either, because $P \exists x \exists y \exists z (\ldots)$ claims that there was a past time at which there were three kings named Charles simultaneously, and that is false, too. To formalize this sentence, we need an untensed quantifier that ranges over objects that exist at different times:

$$\exists x \, \exists y \, \exists z \, (x \neq y \wedge y \neq z \wedge z \neq x \wedge P(Kx \wedge Gx) \wedge P(Ky \wedge Gy) \wedge P(Kz \wedge Gz))$$

Read the untensed way, the leading quantifiers do not assert the present existence of objects $x$, $y$, and $z$ with the specified properties, but only that there were such objects at some time or other. While such untensed quantifiers can easily be accommodated in our tense logic, this would prevent us from making different choices for the domain of quantification at different times. In constructing a model for quantified tense logic, we could only combine models of ordinary quantificational logic that have the same domain.

If we adopt an untensed view of quantification then we also need an independent *existence predicate* 'E!' to formulate time-relative existence claims. At each time $t$, the extension of this predicate would comprise the set $D_t$ of all objects that exist then. In ordinary quantificational logic, we can define an existence predicate in terms of the existential quantifier via:

$$E!x \leftrightarrow \exists y \, y = x$$

Even though this still works with time-relative quantifiers, this biconditional can fail to be true in a tense logic with untensed quantifiers. At any time $t$, every object in the large domain $D$ would satisfy the right-hand side, but only the objects in the smaller domain $D_t$ would satisfy the left-hand side. If our quantifiers are untensed then 'E!' must be taken as primitive.

There are also complications that arise from the interpretation of predicates. In English, we can make true claims about objects at times at which they do not exist. For example, 'Vincent van Gogh is famous' is true now, even though 'Vincent van Gogh' presently lacks a referent. While he existed, van Gogh was not famous at all. Some predicates are *existence entailing* in that they can only be truly attributed to objects at times at which they exist. Existence-entailing predicates work like the predicates in ordinary quantificational logic, but our quantified tense logic also needs to make room for predicates that are not existence entailing, like the predicate 'is famous' at issue here.[11] This means that we must allow our predicates to take extensions in the larger domain $D$, rather than restrict them to set $D_t$ of objects that exist at the time $t$ at which we are evaluating the predicate.

Putting all of this together, a model for quantified tense logic is a septuple $M = \langle T, <, >, i, D, b, e \rangle$, where $T$, $<$, $>$ and $i$ are as in the case of propositional tense logic. The additional structure comprises a set of objects $D$ as the domain of the model, and two maps. The baptismal map $b$ assigns an element $b(a)$ of $D$ to each singular term $a$. For any time $t \in T$ and $n$-ary predicate $\Pi$ other than identity, the extension map $e$ assigns a set $e(\Pi, t)$ of n-tuples of elements of $D$. In these terms, we can then recursively define a true-at relation:

| | | |
|---|---|---|
| $@(\ulcorner \Pi a_1 \ldots a_n \urcorner, t)$ | iff | $\langle b(a_1), \ldots, b(a_n) \rangle \in e(\Pi, t)$ |
| $@(\ulcorner a_1 = a_2 \urcorner)$ | iff | $b(a_1) = b(a_2)$ |
| $@(\ulcorner \neg \varphi \urcorner, t)$ | iff | it is not the case that $@(\varphi, t)$ |
| $@(\ulcorner \varphi \to \psi \urcorner, t)$ | iff | either $@(\ulcorner \neg \varphi \urcorner, t)$ or $@(\psi, t)$ |
| $@(\ulcorner F\varphi \urcorner, t)$ | iff | $@(t', \varphi)$ for some $t' \in T$ such that $t < t'$ |
| $@(\ulcorner P\varphi \urcorner, t)$ | iff | $@(t', \varphi)$ for some $t' \in T$ such that $t' < t$ |
| $@(\ulcorner \forall \xi \varphi \urcorner, t)$ | iff | $@^*(\varphi[a/\xi], t)$ for all $a$-variants $M^*$ of $M$ |

A sentence is true in a model if and only if it is true at its present time.

Suppose we take such a model of quantified tense logic and restrict its domain to the set $D_t = e(\text{E!}, t)$ of all the objects that exist at a given time $t$. Since not every singular term needs to have a referent within $D_t$, this restriction does not yield a model of ordinary quantification logic, but what is known as an *outer-domain model* of free logic.[12] While any collection of models of propositional logic qualifies as a model of propositional tense logic, in most cases even a collection of outer-domain models fails to constitute a model of quantified tense logic. It only does so if the outer domains and baptismal maps of all these models coincide (but they may disagree on the extensions of predicates other than identity).

---

[11] For more on existence-entailing predicates, see Prior (1967, 161), Woods (1976), Chisholm (1976, 100), and Cocchiarella (1968, 1969a b).

[12] See Bencivenga (2002) for a survey of free logic.

The model theory for quantified tense logic is more complicated than for propositional tense logic, but an axiomatic system can be obtained in a fairly straightforward fashion. We start out with our system of untensed quantificational logic that consists of the truth-functional axioms A1–A3, the rule of modus ponens MP, the quantificational axioms Q1–Q4, and the universal generalization rule UG. To this, we add the tense axioms Z1, Z2, and the temporal generalization rule TG. We also need the *Tensed Barcan Formulae* to account for the interaction between quantifiers and operators:

$$\text{BF} \quad P\exists \xi \varphi \rightarrow \exists \xi P\varphi \qquad F\exists \xi \varphi \rightarrow \exists \xi F\varphi$$

These are tense analogues of the modal principle $\ulcorner \Diamond \exists \xi \varphi \rightarrow \exists \xi \Diamond \varphi \urcorner$ due to Ruth Barcan (1946). On a time-relative reading of the quantifiers, the Tensed Barcan Formulae would falsely entail that everything that did or will exist exists now, but both are logical truths on an untensed reading. If the domain of quantification does not change over time then quantifiers and tense operators commute. The system of quantified tense logic we get in this way consists of two copies of Bernard Linsky and Edward Zalta's 'in Defense of the Simplest Quantified Modal Logic'.[13] By plagiarizing their soundness and completeness proof, we can show that a sentence is a theorem of our quantified tense logic if and only if it is true in all models. This result easily extends to the stronger systems of tense logic discussed n section 3. While quantified tense logic is indeed more complicated than propositional tense logic, it is not *as* complicated as some people think. All the complications can be located in the model theory.[14]

# 5. TENSE LOGIC AS A LINGUISTIC THEORY

Let me now turn to the applications of systems of tense logic. One possibility is to treat tense logic as a linguistic theory of verb tense. After its heyday in the early 1970s, such a modal account of verb tense has largely fallen out of favor, but it is instructive to review the reasons why many authors were initially attracted to it, and why it was subsequently abandoned. A good starting point is Prior's claims that tense functions in an essentially adverbial manner:

> Putting a verb into the past or future tense is exactly the same sort of thing as adding an adverb to the sentence. 'I *was* having my breakfast' is related to 'I am having my breakfast' in exactly the same way as 'I am *allegedly* having my breakfast' is related to it, and it is only an historical accident that we generally form the past tense by modifying the present tense, e.g. by changing 'am' to 'was', rather than tacking on an adverb. In a rationalized language with uniform constructions for similar functions we could form the past tense by prefixing to a given sentence the

---

[13] Linsky and Zalta (1994); Williamson (1998) defends a similar view.
[14] Other views on quantification do not permit us to combine the axioms of quantificational logic with those of tense or modal logic. See Garson (2001) for a survey.

phrase 'It was the case that' … and the future tense by prefixing 'It will be the case that'.[15]

In standard quantificational logic, adverbs tend to get short shrift. Consider the following adverbial constructions, which are due to Donald Davidson (1980):

> Jones buttered the toast.
> Jones buttered the toast slowly.
> Jones buttered the toast slowly, deliberately.

If we formalize these sentences in the way this is done in introductory logic courses then each sentence requires the introduction of a new primitive predicate:

$$Bjt \quad Sjt \quad Djt$$

This might be acceptable for some applications, but it is unsatisfactory as a theory of adverbs. For one, it hides the fact that 'buttered' is an atomic component of the molecular 'buttered slowly' and it fails to illuminate the way in which more complex adverbs can be constructed out of simpler ones. More seriously, such a treatment of adverbs obscures the logical relations between these sentences. Each sentence on our list entails all preceding ones, which does not become clear from this formalization.

There are two main proposals for improving the logical treatment of adverbs. One is due to Davidson (1980), who wants to treat adverbs as properties of events. The last sentence on our list would be regimented as '$\exists e(e$ is performed by Jones $\wedge\, e$ is a buttering of the toast $\wedge\, e$ is performed slowly $\wedge\, e$ is performed deliberately)'. It would be a matter of elementary logic that this entails the two preceding sentences, regimented in a similar fashion. On this view, the molecular structure of adverbial constructions would be spelled out in terms of the conjunction of property attributions to the same event. The second proposal treats adverbs as propositional operators.[16] Suppose we introduce operators 'S' ('slowly') and 'D' ('deliberately'). We could then render our three sentences as:

$$Bjt \quad SBjt \quad DSBjt$$

Since one cannot do something slowly or deliberately without actually doing it, both 'S' and 'D' are factual operators. The necessity of $\ulcorner S\varphi \to \varphi \urcorner$ and $\ulcorner D\varphi \to \varphi \urcorner$ then yields the desired entailment relations between the three sentences. For each new adverb that gets added to a sentence, we would prefix an additional propositional operator. Since any string of propositional operators is again a propositional operator, the molecular structure of adverbial constructions could thus be spelled out in terms of the concatenation of the corresponding operators.

---

[15] Prior (1968c, 7–8); see also Montague (1974) and Dummett (1981, 389).
[16] Apart from Prior, this view is advocated by Clark (1970), Montague (1974) and Parsons (1970).

Those who defend tense logic as a linguistic theory of tense usually do so because they endorse both the adverbial theory of tense and the operator theory of adverbs. The resulting theory of tense might be appealing from a purely theoretical perspective, but it gives a less than adequate account of the way verb tense actually works. Even though one could imagine a language that employs propositional operators to make temporal distinctions, this is not how English and other natural languages deal with this issue.

One problem is that verb tenses do not iterate the way tense operators do. For example, according to our model theory, $\ulcorner PP\varphi \urcorner$ is true at a time $t$ just in case there is an earlier time $t'$ such that $\ulcorner P\varphi \urcorner$ is true then, which in turn requires an even earlier time $t''$, prior to $t'$, such that $\varphi$ is true at $t''$. Each additional past tense operator 'P' that we prefix to the sentence shifts the time at which $\varphi$ gets evaluated further into the past. Mürvet Enç (1987, 635) gives the following example to show that the past tense in English functions differently:

John heard that Mary was pregnant.

On a 'shifted' reading, this sentence asserts that, at some time in the past, John heard that Mary was pregnant at an even earlier time, prior to the hearing. It also has a second, and perhaps more natural reading on which it asserts that John heard that Mary was pregnant at the time of the hearing. On such a 'simultaneous' reading, the sequence of tenses does not effect a successive shift in time, as Prior's account would predict.[17]

Our tense logic is ultimately concerned with the relations between two times: the time of speech $s$ when the sentence is uttered, and the time $x$ at which whatever the sentence is about takes place. The tense operators 'P' and 'F' permit us to make three basic distinctions: that $x$ is before $s$, simultaneous with $s$, or later than $s$. This corresponds to three tense forms in English:

| Present Perfect | Present | Simple Future |
|---|---|---|
| $x \quad s$ | $s, x$ | $s \quad x$ |
| The apple has been red | The apple is red | The apple shall be red |
| $PRa$ | $Ra$ | $FRa$ |

Other tenses are not as simple. Take the past perfect in English, 'The apple had been red'. We still have the time of speech and the time $x$ at which the apple is red, but now there is a third time point $r$, which Hans Reichenbach (1947, §51) calls the *point of reference*. Our sentence asserts that there is a suitable time $r$ in the past of $s$ such that, relative to $r$, the time of apple-redness $x$ is in the past. A number of tenses in English have this kind of structure:

[17]   More examples of this type can be found in Ogihara (1996), Giorgi and Pianesi (1997), Higginbotham (2002), Kuhn (2002), and King (2003). Of related interest are Partee (1973), Parsons (1973), and Stalnaker (1973).

| Past Perfect | Simple Past | Future Perfect |
|---|---|---|
| $x\ r\quad s$ | $x,r\quad s$ | $s\quad x\ r$ |
| The apple had been red | The apple was red | The apple shall have been red |

Indeed, one could say that the reference point is not really absent in the case of simple tenses, either, but that it merely coincides with the time of speech.

What is noteworthy about these cases is not just that there is an extra time at issue, but that the reference point usually gets determined by the *context* of speech. For instance, in a succession of sentences in the simple past, the order in which the sentences are uttered indicates the order in which the events described succeeded each other. Hans Kamp and Christian Rohrer (1983, 256) give the following example, which uses the French plus-que-parfait instead of the English simple past:

> Le téléphone sonna. C'était Mme Dupont à l'appareil. Son mari avait pris deux cachets d'aspirine, il avait avalé sa lotion contre les aigreurs d'estomac, il s'était mis un suppositoire contre la grippe, il avait pris un comprimé à cause de son asthme, il s'était mis des gouttes dans le nez, et puis il avait allumé une cigarette. Et alors il y avait eu une énorme explosion. Le docteur réfléchit un moment; puis il lui conseilla d'appeler les pompiers.

The first sentence uses the passé simple and establishes the past ringing of the telephone as the main reference point. The following sentences in the plus-que-parfait then describe a sequence of events that occurred, in the order described, prior to the main reference point. The last sentence, which is again in the passé simple, describes an event that takes place in the past, but after the reference point. We thus get a description of a sequence of events that took place in the past, and which we might summarize as: aspirin—anti-acid pills—suppository—asthma tablet—nasal drops—cigarette—explosion—telephone call—doctor's advice.

In such cases, tense functions as a vehicle for establishing intersentential relations between the various components of a text.[18] Each successive sentence provides more information about the sequence of events. Its tense indicates where this information is to be inserted in the partial temporal representation constructed from the preceding sentences, which means that the same sentence can make quite different claims when placed within two different contexts. Of course, the sequence of events that befell the unfortunate Monsieur Dupont could also have been described—albeit in a less elegant way—in terms of Prior's tense operators. But the point here is not one of expressive power. By trying to give the truth-conditions of tensed sentences in insolation from their location in a given discourse, Prior ignores a pivotal function of tense in natural languages. Tense logic does a poor job at accounting for the linguistic phenomena. As James Higginbotham sums up the situation, 'the modal theory of tenses is inadequate: there is no basic part of our language for which it is correct' (1999, 199).

---

[18]  Kamp and Reyle (1993, ch. 5) develop a theory of this function of tense.

# 6. DEFINITIONAL INCOMPLETENESS

As noted before, there is no reason why our best theory of verb tense should also serve as our default metaphysics. If that is right, then the demise of tense logic as a linguistic theory tells us little about its use as a theory of the nature of time. According to *tense primitivism*, all temporal distinctions are ultimately to be spelled out in terms of conceptually primitive tense operators. For the sake of concreteness, I shall assume that these tense operators are 'P' and 'F', but other choices of primitives are possible as well.

Even if we do not care about having a theory that closely mirrors our ordinary way of talking about time, it surely ought to have enough expressive resources to at least represent it, and that is where tense primitivism is said to run into trouble. There are a number of otherwise innocuous temporal claims that seem to resist regimentation in terms of the tense operators 'P' and 'F' alone. These problems are familiar from modal logic, where there is said to be more to modality than can be expressed in terms of boxes and diamonds. A standard example in the modal case is the counterfactual conditional $\ulcorner\varphi\square\!\!\rightarrow \psi\urcorner$ ('if $\varphi$ were the case then $\psi$ would obtain'). Nelson Goodman (1983) tried to reduce this binary modal connective to the possibility operator, but nowadays most authors think that this project is bound to fail. According to Lewis (1973), the problem is that '$\lozenge$' only allows us to express plain possibility claims. An analysis of the counterfactual needs a notion of comparative possibility that allows us to say whether one world is more possible than another. Similar issues arise for the binary tense operators 'S' ('since') and 'U' ('until'). For example, $\ulcorner U\psi\varphi\urcorner$ is true just in case $\varphi$ is true at some future time and $\psi$ is true at all times that are *less future* than that time, and similar for 'S':

$$@(\ulcorner S\psi\varphi\urcorner, t) \quad \text{iff} \quad @(\varphi, t') \text{ for some } t' < t \text{ and } @(\psi, t'') \text{ for all } t' < t'' < t.$$
$$@(\ulcorner U\psi\varphi\urcorner, t) \quad \text{iff} \quad @(\varphi, t') \text{ for some } t' > t \text{ and } @(\psi, t'') \text{ for all } t' > t'' > t.$$

Kamp (1968) proves that these operators cannot be expressed in terms of 'P' and 'F', which only care about whether a time is past or future at all, and not whether it is more future or past than some other times.[19]

A second range of problems concerns *two-dimensional* modal operators in the sense of Krister Segerberg (1973). Consider the modal actuality operator 'A', which is governed by the stipulation that $\ulcorner A\varphi\urcorner$ is true at a world $w$ if and only if $\varphi$ is true at the actual world of the model:

$$@(\ulcorner A\varphi\urcorner, w) \quad \text{iff} \quad @(\varphi, i)$$

---

[19] See also Gabbay et al. (2002). Lewis (1973, sec. 5.2) discusses the parallels between these binary tense operators and the counterfactual conditional. Note also that some of the work of binary tense operators can be done by suitably chosen two-dimensional ones. Gabbay (1977) presents a system with monadic two-dimensional tense operators that is equivalent to the S/U-calculus.

This operator allows us to evaluate sentences at the actual world of a model even if they occur within the scope of other operators. Allen Hazen (1978) proves that 'A' is redundant in propositional modal logic, where its role is already performed by the convention that sentences without modal operators get evaluated at the actual world. However, in a quantified modal logic a quantifier can occur between a leading modal operator and a nested occurrence of the actuality operator, and in some such cases 'A' is said to be ineliminable. Here are two standard examples:

> There could be something that does not actually exist.
> Not all objects could have failed to exist.

These sentences are easily rendered with the help of 'A', but they appear to have no counterparts in a quantified modal logic with '◊' as its only modal primitive.[20] The tense-analogue of the actuality operator is the two-dimensional 'now' operator 'N', which always sends us back to the present time of a model:

$$@(\ulcorner N\varphi \urcorner, t) \quad \text{iff} \quad @(\varphi, i)$$

Kamp (1971) proves that this two-dimensional operator is redundant in propositional tense logic, but he also presents the following example to show that the situation is different in quantified tense logic:

> A child was born that will become ruler of the world

Using 'N' and time-relative quantifiers, this can easily be formalized as 'P∃x (Bx ∧ NFRx)'. Kamp claims that it cannot be rendered without 'N'.[21]

One might suggest that these problems merely show that we have made a poor choice of primitives, but it is not clear that adding further tense operators gets rid of the underlying problem. Similar issues keep cropping up. Frank Vlach (1973) argues that the following sentence requires the introduction of yet another primitive, a two-dimensional 'then' operator 'T':

> One day, all persons alive then would be dead

The operator 'T' works in the opposite way as 'N': when placed within the scope of other tense operators, it allows us to regard a different time as present than the one of the model in which we are evaluating the entire sentence. Johan van Benthem (1977) takes it one step further, by presenting a sentence that cannot be formalized in terms of all of the tense operators discussed so far. In ordinary propositional logic, there are finite sets of truth-functional connectives that are *definitionally complete*

---

[20]  See Hazen (1976), Crossley and Humberstone (1977), and Hodes (1984).

[21]  It is not clear that this is really a problem. In a tense logic with untensed quantifiers, like the one developed above, this occurrence of 'N' is easily eliminated in favor of '∃x (PBx ∧ FRx)'. More generally, Meyer (2009a) shows that all occurrences of 'now' and 'then' are eliminable in a tense logic that has sufficient quantificational resources available. For more on 'now' and 'then' see also Prior (1968a) and van Benthem (1977).

in that all other such connectives can be expressed in terms of them. For example, one can show that our choice of primitives $\{\neg, \rightarrow\}$ is a set with this property. The problem facing the tense primitivist is that there is no comparable result in tense logic. There appears to be no finite set of tense operators that is definitionally complete.

Tense logic also fares poorly in two other respects. One problem was already noted in section 3, namely the inability of tense logic to express the anti-symmetry of the earlier-than and later-than relations. The second problem concerns time intervals. The tense operators 'P' and 'F' can do some of the work of quantifying over individual times, but there are a number of applications—notably in accounts of verb aspect[22]— that require the ability to quantify over time periods. This is something that our simple tense logic cannot do, either. Indeed, there is not even a general way of expressing the claim that a sentence $\varphi$ is true at *all* times. One might propose to formalize this as $\ulcorner H\varphi \wedge \varphi \wedge G\varphi \urcorner$, but in a branching time it does not follow from $\varphi$'s being true at all times that are past, present, or future relative to the present, that it is true at all times. The sentence could still be false in some other branch of time. Unless there is a limit on the number of branches, $\varphi$ is true at all times if and only if $\ulcorner \Theta\varphi \urcorner$ holds for *all* sequences $\Theta$ of the tense operators 'H' and 'G', and this is not something we can express in a finite language.[23]

Tense logic appears to suffer from a systemic shortcoming in expressive capacity. The problem is not that it says anything *false* about time, but that it says *too little*. There is a wide range of temporal claims that it cannot pass judgment on because it cannot even express them. Many authors take this as a reason for abandoning intensional theories of time in favor of extensional accounts such as temporal substantivalism, which postulates a one-dimensional manifold of metaphysically basic time points that exist independently of what is happening within time.[24]

## 7. TIMES AS ABSTRACTIONS

To rebut these objections, tense primitivism would have to acquire the ability to talk about *times* as well. Without abandoning the central tenet that all temporal distinctions be spelled out in terms of tense operators, times cannot be regarded as metaphysically basic entities. Instead, they must be taken as certain abstractions. In the modal case,

---

[22]  See Bennett (1977), Kamp (1979) and Kuhn (2002). Logic for time intervals are developed by Humberstone (1978), van Benthem (1980, 1983, 1984), and Burgess (1982).

[23]  If we adopt the comparability axioms from section 3 then we can embed the 'always' interpretation of S5 into the P/F-calculus as a proper fragment, by defining $\ulcorner \Box\varphi \urcorner$ as an abbreviation for $\ulcorner H\varphi \wedge \varphi \wedge G\varphi \urcorner$. See Hughes and Cresswell (1975) for details. Thomason (1974, 1975) considers the converse question of whether the P/F tense logic can be embedded into S5.

[24]  Similar arguments are used against the modal primitivism, which aims to spell out all modal notions in terms of a conceptually primitive possibility operator. See Lewis (1986) and Cresswell (1990).

a popular view about possible worlds is what Lewis (1986) dismissively calls *linguistic ersatzism*. As spelled out by Andrew Roper (1982), this proposal takes possible worlds to be maximally consistent sets of sentences of modal logic. A set of sentences $s$ is consistent if and only if there are no $\varphi_1, \ldots, \varphi_k \in s$ such that $\vdash \neg(\varphi_1 \wedge \ldots \wedge \varphi_k)$, where '$\vdash$' denotes derivability in the appropriate system of modal logic. Such a set is maximal if and only if, for every sentence $\varphi$ of the formal language, either $\varphi \in s$ or $\ulcorner \neg\varphi \urcorner \in s$. Maximally consistent sets of sentences are closed under derivability, and we can say that a sentence is true in such a set if and only if it is an element.

A similar strategy suggests itself in the temporal case, where a tense primitivist could propose to treat times as *possible presents*, which are maximally consistent sets of sentences of tense logic. The technical details of this abstraction are largely the same as for linguistic ersatzism about possible worlds, but there is one crucial difference. While every maximally consistent set of sentences of modal logic qualifies as a possible world, not every maximally consistent set of sentences of tense logic counts as a time. At best, times are those possible presents that did, do, or will happen, and some of them never do. The abstraction of times therefore requires one more step than the abstraction of possible worlds. After we have abstracted all the possible presents, we still need to figure out which of them are times. Here is how this can be done. First, we define earlier-than and later-than relations on arbitrary sets of sentences:

$$t' < t \text{ iff } \ulcorner P\varphi \urcorner \in t \text{ for all } \varphi \in t' \qquad t' > t \text{ iff } \ulcorner F\varphi \urcorner \in t \text{ for all } \varphi \in t'$$

Now let $p$ be the maximally consistent set containing all the sentences of our tense logic that are presently true. This set correctly describes how things presently are, but it also contains an implicit characterization of all other times that is hidden within the scope of the tense operators occurring in its elements. We can then define the time series derived from $p$ as the smallest set of possible presents that contains $p$ and is closed under $<$ and $>$.[25]

This account of times has two noteworthy features. First, it is a contingent matter which possible presents count as times. Any given possible present is a time on some choices of $p$, but not on others. This is in marked contrast to linguistic ersatzism about possible worlds, according to which the space of all possible worlds does not depend on what the facts are. Second, what times we end up with depends on the expressive resources of our language. By choosing a formal language that possesses more fine-grained predicates, and thus allows us to describe the world in more detail, we could get more times than we might obtain otherwise.

This much was of course to be expected from a view that takes times to be sets of sentences. For the tense primitivist, there is no language-independent question of getting the 'right' number of times. But it is important is that the number of times is not unduly limited by their construction. In the modal case, W. V. Quine objects

---

[25] The technical details of this construction can be found in Meyer (2009b), which also shows that the time series obtained in this way have all the structural properties that they ought to have.

that linguistic ersatzism cannot deliver enough possible worlds. If the sentences of our language are finite sequences of symbols from a countable vocabulary, then there are at most continuum many sets of such sentences, but there are more possibilities. Consider a continuum of spatial points. All of these points could be either occupied by matter or not, which gives us as many possible distributions of matter in a continuous space as there are subsets of the continuum. We know from Cantor's theorem that the cardinality of the set of all subsets of the continuum is strictly larger than that of the continuum itself, which means that there are fewer maximally consistent sets of sentences than there are possibilities.[26] Since nobody believes that there are more than continuum many times, this problem does not arise in the tense case. In this respect, the account of times as sets of sentences is in much better shape than linguistic ersatzism about possible worlds.[27]

By treating times as abstractions in this way, tense primitivists could then rebut the objection that their account lacks the expressive resources to serve as a theory of time. Whenever their opponents use quantification over metaphysically basic time points to explain some temporal notion, tense primitivists would do exactly the same, just that their temporal variables range over 'ersatz' times. Since the expressive limitations of the underlying tense logic would be compensated by adding some set theory to the meta-language, this does not yield a sentence-by-sentence translation of every temporal claim into the original tense logic. But this is not something that tense primitivists need, anyway. What is important is only that the additional structure does not expand the temporal ideology or ontology of the account. Given that two maximally consistent sets of sentences of tense logic only differ if they disagree about the truth value of some sentence of the underlying tense logic, tense primitivists would still be entitled to their view that 'P' and 'F' capture everything there is to be said about time.

An alternative to constructing times as sets of sentences is to regard them as maximal propositions. Instead of using quantification over sets of sentences in the meta-language, this approach enriches the object language with propositional quantifiers. Using '$\Box$' and '$\Diamond$' as modal operators, a proposition $p$ is maximally consistent just in case it is possible for $p$ to be true and to entail all other truths:

$$\Diamond(p \wedge \forall q(q \to \Box(p \to q)))$$

We can also define earlier-than and later-than relations on propositions via:

$$P < Q \text{ iff } \Box(Q \to Pq) \quad p > q \text{ iff } \Box(q \to Fp)$$

The time series derived from a proposition $p$ could then be defined as the smallest set of maximally consistent propositions that contains $p$ and is closed under these

---

[26]  Quine (1969); see also Lewis (1973, 90, 1986, 143–147).
[27]  Meyer (2009b) explains how to get continuum many times by adopting a countable language that allows us to describe the motion of an object on a continuous spatial manifold.

earlier-than and later-than relations.[28] At first sight, this might look very similar to the view of times as sets of sentences, but there are important differences between the two proposals. Following Mark Richard (1981), say that a proposition is *eternal* if it always has the same truth value and *temporal* if its truth value changes over time. While the times-as-sets view permits tense primitivism to remain agnostic about the controversial question of whether there are any temporal propositions, the times-as-propositions view requires that at least some propositions are temporal. Otherwise, there could never be more than one time. For the view to work as advertised, we also could not restrict our propositional quantifiers to those propositions that are expressible by a sentence of our formal language. In all but the most trivial cases, maximally consistent propositions could only be described by an 'infinitely long' sentence. Temporal discourse would be partly about a language-independent realm of temporal propositions, thus saddling tense primitivism with additional ontological commitment of a decidedly 'temporal' nature.

The times-as-propositions view has the advantage of allowing us to speak about times in the object language itself, but the advocate of the times-as-sets view can easily come up with something similar. Suppose we extend the vocabulary of the language of tense logic by adding singular terms for times (dates) and time quantifiers with associated time variables. Let us also borrow the true-at operator '|' from George Myro (1986a,b). To form a sentence, this operator needs to be complemented with a date $a$ on the left and a sentence $\varphi$ on the right. Expressions of the form $\ulcorner a \mid \varphi \urcorner$ are then read as 'at time $a$, $\varphi$'. In interpreting this extended language, we start out with an interpretation of the underlying tense logic Z. From these ingredients, we then construct a time series of maximally consistent sets of sentences, using the abstraction method discussed earlier. The time variables range over these maximally consistent sets. The theorems of this expanded system include all the theorems of Z and all theorems of standard quantificational logic, adopted to time quantifiers and variables. Since the time variables range over maximally consistent sets, we also get:

$$a \mid (\varphi \to \psi) \to (a \mid \varphi \to a \mid \psi) \quad \neg a \mid \varphi \leftrightarrow a \mid \neg \varphi$$

For any date $a$, $\ulcorner a \mid \quad \urcorner$ is thus a propositional operator that always takes the same truth values on logically equivalent sentences. Moreover, any sentence that is true at the present time $p$ is true simpliciter, and the tense operators behave as expected:

$$p \mid \varphi \leftrightarrow \varphi \quad a \mid P\varphi \leftrightarrow \exists t(t < a \wedge t \mid \varphi) \quad a \mid F\varphi \leftrightarrow \exists t(t > a \wedge t \mid \varphi)$$

Since we are considering Z as the underlying tense logic, there are no valid schemata for $<$ and $>$ alone, but this can easily be extended to stronger systems of tense logic. In this way, also the times-as-sets view can quantify over times in the object language

---

[28] The *locus classicus* for this view about times is Fine (2005). See also Bourne (2006). Zalta (1988, ch. 4) presents a view on which times are uniquely determined by the temporal propositions that are true then, but which does not claim that times are themselves propositions. Propositional quantifiers are discussed in Bull (1969), Fine (1970), and Kaplan (1970).

and there seems to be no reason for abandoning it in favor of the more problematic times-as-propositions view. Once the language of tense logic has been extended in this way, tense primitivists can easily account for the problem cases discussed in section 6. All they need to do is to replicate the model-theoretic definitions of 'S', 'N', etc. in the object language, by using '|' in place of '@'.

## 8. PRESENTISM AND ACTUALISM

The previous section took tense and modal operators as primitive and tried to construct times and possible worlds out of them. In the modal case, many authors favor the opposite approach, which eliminates the modal operators in favor of quantification over possible worlds. Section 3 used 'possible world' as a fancy name for the elements of a model of modal logic. The primary aim of any model theory is to provide a characterization of the theorems of the logical system in question and it usually does not matter what we take as our models as long as there are enough of them to perform this function. Many philosophers think that modal logic is different. They are impressed by the intuitive appeal of the possible-worlds picture and argue that it goes beyond a mere model theory to provide an *analysis* of the modal operators.

Amongst all the models of modal logic, there is one that is special: the *intended model* $\langle \Omega, R, a, @ \rangle$. While we usually speak of what is true in a model, that qualification can be dropped for the intended model. Any statement that is true in this model is true simpliciter:

$\varphi$ is true iff $@(\varphi, a)$

With the last clause in the model theory for modal logic, this yields:

$\Diamond \varphi$ iff $@(\varphi, \omega)$ for some $\omega \in \Omega$ such that $Ra\omega$

If we take S5 as our theory of possibility then what is possible does not depend on what is actual and every possible world is accessible from every other one. In this case, we can just drop all reference to the accessibility relation:

$\Diamond \varphi$ iff $@(\varphi, \omega)$ for some $\omega \in \Omega$

If we reserve the term 'possible world' for the elements of the intended model of modal logic, then this can be put more succinctly as:

$\Diamond \varphi$ iff $\varphi$ is true in some possible world

According to the possible-worlds analysis, this biconditional permits a reductive elimination of '$\Diamond$' in favor of quantification over possible worlds.

What precisely this amounts to depends on what we say about the intended model of modal logic. There are two main views to choose from. According to *modal realists* like Lewis (1986), the intended model consists of a vast number of equally concrete worlds

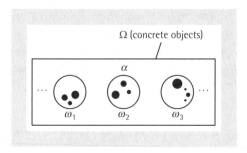

FIGURE 4.2 Modal Realism

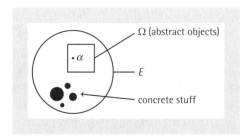

FIGURE 4.3 Actualism

that are spatio-temporally separated from one another, like raisins in a pudding. The actual world $\alpha$ in the intended model consists of whatever is spatio-temporally connected to us. It is the world that we happen to live in, but other possible worlds are as concrete as ours. The rival view of *actualism* (see Figure 4.2) denies the existence of the merely possible objects that make up the possible worlds of the modal realist. According to the actualist, the possible worlds in the intended model are actually existing abstract ways the world might have been. In particular, the actual world $\alpha$ is an abstract object that must be distinguished from the mereological sum-total $E$ ('everything') of all objects that actually exist (see Figure 4.3). The compound $E$ contains the actual world $\alpha$ along with all the other abstract worlds in $\Omega$, but it also contains all concrete objects: sticks, stones, etc. It is the contingent parts of $E$ that determine which element of $\Omega$ correctly describes the way things actually are and thus counts as the actual world $\alpha$.

Section 7 presented linguistic ersatzism as a way of defending modal primitivism, but it is more commonly thought of as a brand of actualism.[29] Rather than bolstering the view that '$\Diamond$' can be taken as our only modal primitive, it is thought to provide a way of eliminating it. The way Roper (1982) presents the view, possible worlds are maximally consistent sets of sentences of *modal* logic. His possible worlds therefore contain sentences that themselves contain modal operators. This simplifies the exposition of the proposal, but it yields larger sets of sentences than one actually needs. It suffices to consider sentences of the underlying extensional logic only, in which case

---

[29] There are other ways of being an actualist; see Stalnaker (1976) and Lewis (1986).

possible worlds would be more like Rudolf Carnap's (1947) state descriptions. Given a sentence $\varphi$ without any modal operators, we could then say that $\ulcorner \lozenge \varphi \urcorner$ is true if and only if $\varphi$ is contained in some maximally consistent sets of sentences of the underlying non-modal language. If our modal logic is S5 then this extends to sentences with nested occurrences of modal operators,[30] thus allowing us to eliminate modal operators in favor of quantification over actually existing sets of sentences.

With the exception of Quine's cardinality worry, all the standard objections to linguistic ersatzism are really directed towards linguistic ersatzism as a way of being an actualist. For instance, Lewis (1986, sec. 3.2) complains that linguistic ersatzism does not eliminate modal notions because it still assumes the notion of consistency. This is especially pressing if we assume that there are necessary truths, such as mathematical truths, that are not logical necessities. In this case, the possible-worlds analysis could only succeed in picking out all necessary truths if it used a more restrictive notion than S5-consistency in characterizing possible worlds. Another complaint, also by Lewis, is that linguistic ersatzism lacks the resources to deal with 'alien' properties, which are properties that are only instantiated in worlds other than the actual one. There is a debate about whether these are serious problems for the linguistic ersatzer,[31] but that need not concern us here.

Whatever we say about the actualist's account of modality, the tense case is different. While we might be able to eliminate modal operators in favor of actually existing abstract possible worlds, we cannot eliminate tense operators in favor of presently existing abstract times. This is perhaps most obvious for the account of times as sets of sentences developed in section 7. On this view, times are presently existing abstract objects but the construction only succeeds as desired because the sentences that make up times contain the very operators that an aspiring presentist would try to eliminate. Every maximal consistent set of sentences of tense logic contains a maximal consistent set of sentences of the underlying propositional logic, but what matters for the construction of the time series are the other sentences, those with occurrences of the tense operators. It is the information contained in these sentences that determines the earlier-than and later-than relations, and thus allows us to construct a time series from a choice of present $p$.

The underlying problem is that it is a mistake to think of past and future times in analogy with merely possible worlds. The temporal analogue of possible worlds are possible presents, and on any view of what possible presents are (set of sentences, maximal propositions, or some other abstracta), not all possible presents are times. The vast majority of all the ways the present might have been never happen. If anything, times are more similar to the *actual world* in modal logic (see Figure 4.4). Just as it is a contingent fact which possible presents correctly describe the actual sequence of events, it is a contingent fact which possible world correctly describes the way things

---

[30] In S5, every sentence with nested modal operators is equivalent to one in which no modal operator occurs within the scope of another. See Hughes and Cresswell (1996, 98).

[31] See, e.g. Roy (1995), Heller (1998), Sider (2002), and Leuenberger (2006).

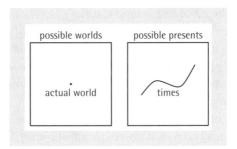

FIGURE 4.4  Possible Worlds and Possible Presents

actually are. On pain of circularity, we cannot eliminate tense operators in favor of quantification over possible presents because we need to appeal to such operators to figure out which of them are times.

In light of the fact that S5 can serve as both a modal logic and a tense logic, these differences might be a little bit surprising, but they are easily explained. The considerations in section 3 show that S5-as-a-modal-logic and S5-as-a-tense-logic have structurally isomorphic theorems. What is at issue in an analysis, though, is the potentially much larger class of all truths involving '$\Diamond$'. Consider the axiom M4, $\ulcorner \Diamond\varphi \rightarrow \Box\Diamond\varphi \urcorner$. Read as a principle of modal logic, this claims that anything that is possible is *necessarily* possible. According to the modal logic S5, there are no contingent truths of type $\ulcorner \Diamond\varphi \urcorner$, and similar remarks apply to M3 and claims of the form $\ulcorner \Box\varphi \urcorner$. But if we read M4 as a principle of tense logic then it only claims that everything that is sometimes true is *always* sometimes true, and not that it is necessarily so. In fact, most tensed truths of type $\ulcorner \Diamond\varphi \urcorner$ are only contingent.

The possible-worlds analysis succeeds because what possible worlds there are does not depend on what the facts are (i.e. which world is actual). The tense case is different. There is *no* model of tense logic that necessarily assigns the correct truth values to all tensed claims. We cannot find a model of tense logic that is made up of abstract objects such that, necessarily, $\ulcorner$sometimes, $\varphi\urcorner$ is true if and only if $\varphi$ at true in some time of *that* model. Depending on what the facts are, different possible presents need to be chosen as times. We can make coherent sense of the actualist thesis that nothing exists that is not actual, but this does not translate into an account of the presentist thesis that nothing exists that is not present.[32]

# 9. CONCLUDING REMARKS

Tense logic might fail as a linguistic theory of tense, but I think it does better as a metaphysical theory of time than is usually supposed. My case for this claim must remain

---

[32]  I say more about the differences between actualism and presentism in Meyer (2006). In Meyer (2005), I argue that presentism is either trivially true or obviously false.

somewhat tentative, though, since there are two questions about tense primitivism that I could not address here. While I discussed time and modality separately, I said very little about the interaction between them. That is a significant omission because time often features prominently in counterfactual reasoning. Consider: 'Had there been no railroad strike, Pierre would have arrived in London sooner'. On a standard possible-worlds analysis of the counterfactual conditional, this sentence is true just in case, in some suitably chosen strike-free possible world, Pierre arrives at an earlier time than the one at which he actually arrives. If times are sets of sentences, and if sets have their elements essentially, then it is impossible for the same time at which Pierre actually arrives to be one at which he does not arrive in some other possible world. Since that other-worldly time would have different elements, it would be distinct from the actual arrival time. To get the truth conditions for such counterfactuals right, one needs a suitable counterpart semantics for reasoning about times and I did not develop one here.[33] The second issue concerns the comparative virtues of tense primitivism. Even if the view fares well on its own terms, this does not yet tell us whether it ought to be preferred over rival accounts of the nature of time, such as temporal substantivalism or relationism about time. Indeed, it is often said that the theory of relativity mandates that we treat space and time alike, which would favor such 'spatial' views of time over 'modal' ones like tense primitivism. This is an objection that tense primitivism must deal with if the view is to be taken seriously.

## References

Barcan, Ruth (1946), 'A Functional Calculus of First Order Based on Strict Implication', *Journal of Symbolic Logic* 11: 1–16.

Bencivenga, Ermanno (2002), Free Logics. in Gabbay and Guenthner (2002), 147–96.

Bennett, Michael (1977), 'A Guide to the Logic of Tense and Aspect in English', *Logique et Analyse* 20: 491–517.

Bourne, Craig (2006), *A Future for Presentism*, Clarendon Press, Oxford.

Bull, R. A. (1968), 'An Algebraic Study of Tense Logics with Linear Time', *Journal of Symbolic Logic* 33: 27–38.

——(1969), 'On Modal Logic with Propositional Quantifiers', *Journal of Symbolic Logic* 34: 257–63.

Bull, Robert and Segerberg, Krister (2001), 'Basic Modal Logic', in Gabbay and Guenthner (2001), 1–81.

Burgess, John P. (1982), 'Axioms for Tense Logic, II: Time Periods', *Notre Dame Journal of Formal Logic* 23: 375–83.

——(2002), 'Basic Tense Logic', In Gabbay and Guenthner (2002), 1–42.

Carnap, Rudolf (1947), *Meaning and Necessity*, University of Chicago Press.

Chisholm, Roderick (1976), *Person and Object*, Open Court, Chicago.

Clark, Romane (1970), 'Concerning the Logic of Predicate Modifiers', *Noûs* 4: 311–35.

---

[33] Another question about the interaction of tense and modality concerns the notion of historical necessity. See Thomason (2002) for a survey.

Cocchiarella, Nino (1966a), 'A Completeness Theorem for Tense Logic', *Journal of Symbolic Logic* 31: 689–91.

—— (1966b), 'Tense Logic: A Study of Temporal Reference', PhD thesis, University of California, Los Angeles.

—— (1968), 'Some Remarks on Second Order Logic with Existence Attributes', *Noûs* 2: 165–75.

—— (1969a), 'Existence Entailing Attributes, Modes of Copulation and Modes of Being in Second Order Logic', *Noûs* 3: 33–48.

—— (1969b), 'A Second Order Logic of Existence', *Journal of Symbolic Logic* 34: 57–69.

Cocchiarella, Nino and Freund, Max (2008), *Modal Logic: An Introduction to its Syntax and Semantics*, Clarendon Press, Oxford.

Cresswell, M. J. (1970), 'Classical Intensional Logic', *Theoria* 36: 347–72.

—— (1990), *Entities and Indices*, Kluwer, Dordrecht.

Crossley, John and Humberstone, Lloyd (1977), 'The Logic of "Actually"'. *Reports on Mathematical Logic* 8: 11–29.

Davidson, Donald (1980), 'The Logical Form of Action Sentences', in *Essays on Actions and Events*, 105–22. Clarendon Press, Oxford.

Dummett, Michael (1981), *Frege: Philosophy of Language*, Harvard University Press, Cambridge, Mass., 2nd edn.

Enç, Mürvet (1987), 'Anchoring Conditions for Tense', *Linguistic Inquiry* 18: 633–57.

Evans, Gareth and McDowell, John, eds. (1976), *Truth and Meaning: Essays in Semantics*, Clarendon Press, Oxford.

Fine, Kit (1970), 'Propositional Quantifiers in Modal Logic', *Theoria* 36: 336–46.

—— (2005), 'Prior on the Construction of Possible Worlds and Instants', in *Modality and Tense*, 133–75, Clarendon Press, Oxford.

Flo, Olav (1970), 'Bibliography of the Philosophical Writings of A. N. Prior', *Theoria* 36: 189–213.

French, Peter, Uehling, Theodore, and Wettstein, Howard (1986), *Studies in Essentialism*, vol. 11 of *Midwest Studies in Philosophy*, University of Minnesota Press, Minneapolis.

Gabbay, Dov (1975), 'Model Theory for Tense Logics', *Annals of Mathematical Logic* 8: 185–236.

—— (1977), 'A Tense System with Split Truth Table', *Logique et Analyse* 20: 359–93.

Gabbay, Dov, Finger, Marcelo, and Reynolds, Mark (2002), 'Advanced Tense Logic', in Gabbay and Guenthner (2002), 43–203.

Gabbay, Dov and Guenthner, Franz, eds. (2001), *Handbook of Philosophical Logic*, vol. 3, Kluwer, Dordrecht, 2nd edn.

—— (2002), *Handbook of Philosophical Logic*, vol. 7, Kluwer, Dordrecht, 2nd edn.

Garson, James (2001), 'Quantification in Modal Logic', in Gabbay and Guenthner (2001), 267–323.

—— (2006), *Modal Logic for Philosophers*, Cambridge University Press.

Giorgi, Alessandra and Pianesi, Fabio (1997), *Tense and Aspect: From Semantics to Morphosyntax*, Oxford University Press, New York.

Goodman, Nelson (1983), *Fact, Fiction, and Forecast*, Harvard University Press, Cambridge, Mass., 4th edn.

Hazen, Allen (1976), 'Expressive Completeness in Modal Languages', *Journal of Philosophical Logic* 5: 25–46.

——(1978), 'The Eliminability of the Actuality Operator in Propositional Modal Logic', *Notre Dame Journal of Formal Logic* 19: 617–22.

Heller, Mark (1998), 'Property Counterparts in Ersatz Worlds', *Journal of Philosophy* 95: 293–316.

Higginbotham, James (1999), 'Tense, Indexicality, and Consequence', in Jeremy Butterfield, ed., *The Arguments of Time*, 197–215. Clarendon Press, Oxford.

——(2002), 'Why is Sequence of Tense Obligatory?', in Gerhard Preyer and Georg Peter, eds., *Logical Form and Language*, 207–27, Clarendon Press, Oxford.

Hodes, Harold (1984), 'Some Theorems on the Expressive Limitations of Modal Languages', *Journal of Philosophical Logic* 13: 13–26.

Hughes, G. E. and Cresswell, M. J. (1975), 'Omnitemporal Logic and Converging Time', *Theoria* 41: 11–34.

——(1996), *A New Introduction to Modal Logic*, Routledge, London.

Humberstone, Lloyd (1978), 'Interval semantics for Tense Logic: Some Remarks', *Journal of Philosophical Logic* 8: 171–96.

——(2009), 'The Connectives', Unpublished manuscript.

Kamp, Hans (1968), 'Tense Logic and the Theory of Linear Orders', PhD thesis, University of California, Los Angeles.

——(1971), 'Formal Properties of "now"', *Theoria* 37: 227–73.

——(1979), 'Events, Instants and Temporal Reference', in Rainer Bäuerle, Urs Egli, and Armin von Stechow, eds., *Semantics from Different Points of View*, 376–417. De Gruyter, Berlin.

——and Reyle, Uwe (1993), *From Discourse to Logic*. Kluwer, Dordrecht.

——and Rohrer, Christian (1983), 'Tense in Texts', in Rainer Bäuerle, Christoph Schwarze, and Armin von Stechow, eds., *Meaning, Use and Interpretation of Language*, 250–69, De Gruyter, Berlin.

Kaplan, David (1970), 'S5 with Quantifiable Propositional Variables', *Journal of Symbolic Logic* 35: 355.

King, Jeffrey (2003), 'Tense, Modality, and Semantic Values', in John Hawthorne and Dean Zimmerman, eds., *Language and Philosophical Linguistics*, vol. 17 of *Philosophical Perspectives*, 195–245, Blackwell, Malden, Mass.

Kripke, Saul (1980), *Naming and Necessity*. Blackwell, Oxford.

Kuhn, Thomas (2002), 'Tense and Time', in Gabbay and Guenthner (2002), 277–346.

Leuenberger, Stephan (2006), 'A New Problem of Descriptive Power', *Journal of Philosophy* 103: 145–62.

Lewis, David (1973), *Counterfactuals*, Harvard University Press, Cambridge, Mass.

——(1986), *On the Plurality of Worlds*, Blackwell, Oxford.

——(2004), 'Tensed Quantifiers', in Dean Zimmerman, ed., *Oxford Studies in Metaphysics*, vol. I, 3–14, Clarendon Press, Oxford.

Linsky, Bernard and Zalta, Edward (1994), 'In Defense of the Simplest Quantified Modal Logic', in James Tomberlin, ed., *Logic and Language*, vol. 8 of *Philosophical Perspectives*, 431–58, Ridgeview, Atascadero, California.

Magalhães, Ernâni and Oaklander, Nathan, eds. (2010), *Presentism: Essential Readings*, Rowman & Littlefield, Lanham, Md.

McArthur, Robert (1976), *Tense Logic*, vol. 111 of *Synthese Library*, Reidel, Dordrecht.

Meyer, Ulrich (2005), 'The Presentist's Dilemma', *Philosophical Studies* 122: 213–25, Reprinted in Magalhães and Oaklander 2010.

Meyer, Ulrich (2006), 'Worlds and Times', *Notre Dame Journal of Formal Logic* 47: 25–37.

—— (2009a), ' "Now" and "then" in Tense Logic', *Journal of Philosophical Logic* 38: 229–47.

—— (2009b), 'Times in Tense Logic', *Notre Dame Journal of Formal Logic* 50: 201–19.

Montague, Richard (1974), *Formal Philosophy*, Yale University Press, New Haven. Edited by Richmond Thomason.

Myro, George (1986a), 'Existence and Time', in Richard Grandy and Richard Warner, eds., *The Philosophical Grounds of Rationality*, 383–409, Clarendon Press, Oxford. Reprinted in Rea 1997.

—— (1986b), 'Time and Essence', in French et al. 1986, 331–41.

Newton-Smith, W. H. (1980), *The Structure of Time*, Routledge, London.

Ogihara, Toshiyuki (1996), *Tense, Attitude, and Scope*, Kluwer, Dordrecht.

Ørstrøm, Peter and Hasle, Per (1993), A. N. Prior's Rediscovery of Tense Logic, *Erkenntnis* 39: 23–50.

Parsons, Terence (1970), 'Some Problems Concerning the Logic of Grammatical Modifiers', *Synthese* 21: 320–34.

—— (1973), 'Tense Operators Versus Quantifiers', *Journal of Philosophy* 70: 609–10.

Partee, Barbara Hall (1973), 'Some Structural Analogies between Tenses and Pronouns in English', *Journal of Philosophy* 70: 601–9.

Pnueli, Amir (1977), 'The Temporal Logic of Programs', in *18th Annual Symposium of Computer Science*, 46–57, Institute of Electrical and Electronics Engineers, New York.

Prior, Arthur (1957a), 'A Tense-Logical Analogue of S5', in Prior 1957b, 18–28.

—— (1957b), *Time and Modality*, Clarendon Press, Oxford.

—— (1966), 'Postulates for Tense Logic', *American Philosophical Quarterly* 3: 153–61.

—— (1967), *Past, Present and Future*, Clarendon Press, Oxford. For errata, see Williams (1969).

—— (1968a), 'Now', *Noûs* 2: 101–19. See also Prior 1968b.

—— (1968b), 'Now' corrected and condensed, *Noûs* 2: 411–12.

—— (1968c), *Papers on Time and Tense*, Clarendon Press, Oxford. For errata, see Williams 1969.

Quine, W. V. (1940), *Mathematical Logic*, Norton, New York.

—— (1951), 'Ontology and Ideology', *Philosophical Studies* 2: 11–15.

—— (1969), 'Propositional Objects', in *Ontological Relativity and Other Essays*, 139–60, Harvard University Press, Cambridge, Mass.

Rea, Michael, ed. (1997), *Material Constitution*, Rowman & Littlefield, Lanham, Md.

Reichenbach, Hans (1947), *Elements of Symbolic Logic*, Macmillan, New York.

Rescher, Nicholas and Urquhart, Alasdair (1971), *Temporal Logic*, Springer-Verlag, New York.

Reynolds, Mark (1994), Axiomatisation and Decidability of *F* and *P* in Cyclical Time, *Journal of Philosophical Logic* 23: 197–224.

Richard, Mark (1981), 'Temporalism and Eternalism', *Philosophical Studies* 39: 1–13.

Roper, Andrew (1982), 'Towards an Eliminative Reduction of Possible Worlds', *Philosophical Quarterly* 32: 45–59.

Roy, Tony (1995), 'In Defense of Linguistic Ersatzism', *Philosophical Studies* 80: 217–42.

Segerberg, Krister (1973), 'Two-Dimensional Modal Logic', *Journal of Philosophical Logic* 2: 77–97.

Sider, Theodore (2002), 'The Ersatz Pluriverse', *Journal of Philosophy* 99: 279–315.

Stalnaker, Robert (1973), 'Tenses and Pronouns', *Journal of Philosophy* 70: 610–12.

—— (1976), 'Possible Worlds', *Noûs* 10: 65–75.

Thomason, Richmond (2002), 'Combinations of Tense and Modality', in Gabbay and Guenthner (2002), 205–34.

Thomason, S. K. (1974), 'Reduction of Tense Logic to Modal Logic, I', *Journal of Symbolic Logic* 39: 549–51.

—— (1975), 'Reduction of Tense Logic to Modal Logic, II', *Theoria* 41: 154–69.

van Benthem, Johan (1977), 'Tense Logic and Standard Logic', *Logique et Analyse* 80: 395–437.

—— (1980), 'Points and Periods', in Christian Rohrer, ed., *Time, Tense and Quantifiers*, 39–57. Niemeyer, Tübingen.

—— (1983), *The Logic of Time*, Reidel, Dordrecht.

—— (1984), 'Tense Logic and Time', *Notre Dame Journal of Formal Logic* 25: 1–16.

van Fraassen, Bas (1970), *An Introduction to the Philosophy of Time and Space*, Random House, New York.

Vlach, Frank (1973), ' "Now" and "Then": A Formal Study in the Logic of Tense Anaphora', PhD thesis, University of California, Los Angeles.

Williams, C. J. F. (1969), 'Prior on Time', *Ratio* 11: 145–58.

Williamson, Timothy (1998), 'Bare Possibilia', *Erkenntnis* 48: 257–73.

—— (2006), 'Indicative Versus Subjunctive Conditionals, Congruential Versus Non-Hyperintensional Contexts', *Philosophical Issues* 16: 310–33.

Woods, Michael (1976), 'Existence and Tense', in Evans McDowel 1976, 248–62.

Zalta, Edward (1988), *Intensional Logic and the Metaphysics of Intentionality*, The MIT Press, Cambridge, Mass.

# CHAPTER 5

·····································································

# PRESENTISM

·····································································

## M. JOSHUA MOZERSKY

## 1. PRESENTISM AND THE TENSELESS COPULA
·····································································

PRESENTISM is the doctrine that everything is present; put another way, it is the claim that that, and only that, which is present exists.[1] Presentists are typically opposed by eternalists, who argue that all times, and their contents, are equally real: the past and future are as much a part of the 'furniture of reality' as is the present.[2]

What does it mean to assert that only that which is present exists? Such a proposition threatens to collapse to either absurdity or triviality (see Meyer 2005). In one sense it is obvious that Napoleon does not exist; since he is dead, he *did* exist. Similarly, the first baby born in the twenty-second century does not exist; he or she *will* exist. This is hardly controversial, and is surely not the core claim of a substantive metaphysical position. What, then, could the presentist have in mind?

Here is how some recent defenders have put their position:

> presentism is the claim that it is always the case that, quantifying unrestrictedly, for every *x*, *x* is present. (Crisp 2007: 107)

> I am using the sentence 'Only presently existing things exist' and its companions, such as 'Only presently red things are red', to distinguish the presentists from the eternalists. To serve this purpose, these "test" sentences must be given a particular interpretation. The quantifiers should be read as unrestricted quantifiers; the tokens of 'exists' should be understood as tokens of a nonindexical 'exists' and as tokens of the same word; and 'presently' should be read as an indexical tense operator. Working under these stipulations, we can see that the sentence 'Newton exists' will express the same proposition at different times, and the sentence 'Newton presently exists' will express different propositions at different times. (Hinchliff 2000: 576–7)

---

[1] See, for example, Bigelow (1996), Chisholm (1990), Craig (1998), Crisp (2005, 2007), Hinchliff (1996, 2000), Markosian (2004); Prior (1962, 1970), Sider (1999); Zimmerman, (1998).

[2] See, for example, Hawley (2001), Quine (1950), Sider (2001), Smart (1980).

These are instructive quotations. What they point to, and what I want to argue, is that presentism requires recognition of both a tensed and tenseless copula, as well as a tensed and tenseless sense of the verb 'to exist'.

The tensed 'is' entails 'is now'; the tensed 'exists' entails 'exists now'. The tenseless versions, on the other hand, have no such entailments: they are neutral with respect to temporal context of utterance. So, letting $is_1$ be the tensed copula ($exists_1$ the tensed verb) and $is_2$ ($exists_2$) the tenseless, the following hold: 'Abraham Lincoln $is_1$ tall', uttered in 2008, expresses a false proposition; 'Abraham Lincoln $is_2$ tall', expresses a true proposition (assuming, as I am, that Lincoln was in fact tall). By the same token, 'Napoleon $exists_1$', uttered today, expresses a falsehood, while 'Napoleon $exists_2$' expresses a truth.

The tenseless copula functions much like predication in standard, first order predicate calculus, where a sentence such as 'Fa' predicates something of an object, a, without entailing anything about when a is F. Similarly, the existential quantifier is, typically, understood tenselessly, that is, such that 'there exists an x such that' is equivalent to 'there $exists_2$ an x such that'.

Suppose, for *reductio*, that there is only a tensed copula and a tensed notion of existence. Accordingly, 'is' means only 'is now', though it retains its usual past and future conjugations, 'was' and 'will be'. Moreover, 'exists' means 'exists now' and will have, as tenses, 'did exist' (or 'existed') and 'will exist'. It follows from this that presentism is either trivially true or false. For under this supposition the claim that *that and only that which is present exists* can mean only one of the following:

(1)    That and only that which is present exists
(2)    That and only that which is present did exist, exists, or will exist
(3)    That and only that which is, was, or will be present exists
(4)    That and only that which is, was, or will be present did exist, exists, or will exist.

Given that 'is present' (in the tensed, temporal sense) is equivalent to 'exists now', (1) is the trivial truth outlined above—that and only that which is now present exists now—which is not under dispute and, so, cannot form the core doctrine of presentism. (2), while not trivially false, is obviously false for it entails that past individuals did not exist and future individuals won't exist, which amounts to a kind of solipsism of the present moment. If (2) is true, then since Abraham Lincoln does not exist, it follows that he did not exist; since Napoleon does not exist, he did not exist; since the first baby born in the twenty-second century does not exist, he or she will not exist; and so on. Though not necessarily false in the sense of being contradictions, the denial of presentism is more plausible than the assertion of such claims. Hence, (2) is a non-starter. (3) is trivially false in much the same way that (1) is trivially true. The stepping of Neil Armstrong onto the surface of the moon, for example, is an event that was present (i.e. is past), but it certainly does not exist (now). Therefore, just because something was present, it simply does not follow that it exists now. Finally, (4) is trivially true. Since to be present at some time or other is to exist at some time or other, obviously only that which was, is, or will be present existed, exists, or will exist (I am ignoring, here, the case of abstract entities, should there be any, for if they exist it is, arguably, not in a tensed sense since they stand in no relations to times).

The way out for the presentist is to suppose that there is a tenseless sense of 'exists' that is temporally unrestricted: it assigns as values to variables any entity that exists at any time. The obvious candidate for such a verb is the existential quantifier; or, put better, the logical role of such a verb is best captured by the existential quantifier. If we combine a tenseless existential quantifier with the supposition that it is a metaphysical fact, and so metaphysically necessary, that this quantifier is only ever able to 'reach' entities that exist now, then presentism becomes the following metaphysical claim:

(5)    Only entities that exist now can satisfy the (temporally unrestricted, i.e. tenseless) existential quantifier.

This has the prospect of a non-trivial yet true interpretation. For consider that presentism is most plausibly combined with the claim that time passes: what was present is no longer present; what is present both will be past and was future; and so on. Hence, while at any given time, the domain over which the existential quantifier can range is metaphysically limited, this domain constantly changes.[3]

Now consider the following characterization of presentism:

> According to Presentism, if we were to make an accurate list of all the things that exist—i.e. a list of all the things that our most unrestricted quantifiers range over—there would be not a single non-present object on the list.    (Markosian 2004: 47)

This indicates that there is supposed to be some kind of metaphysical restriction on quantification that is imposed by time. The quantifiers are, as Markosian suggests, temporally unrestricted, happy to bind variables that take on non-present objects as values. It simply is necessarily, metaphysically the case that the universe will forever disappoint them if they attempt to spread their wings outside of the present moment.

---

[3] Barbour (1999) defends a view of reality that is devoid of temporal passage and consists of unchanging moments (what he calls 'time capsules') that contain only memories, records, or anticipations of anything not part of that time. This opens up the possibility of a universe that consists of just one such time capsule, a single 'frozen' moment of time that seems to be part of a sequence of times but is not. I am not considering such a position here for the following reasons. First, the prominent presentists in the literature generally defend a dynamic world in which what is present changes. Second, Barbour defends his collection of temporally unrelated time capsules as the best way to make sense of quantum indeterminacy. Whether or not this is plausible, note that my concern here is with the *philosophical* understanding of the concept of time, i.e. that which rests on logical, semantic, and metaphysical propositions. Such considerations do not support a version of temporal solipsism. If, on the other hand, scientific evidence does, then we face a difficulty. We must either find a way to reconcile two apparently conflicting but well-grounded conceptions of time, or else we must find reasons for favouring one over the other. Such a project must await another occasion, where scientific and philosophical conceptions of time can be directly compared. In the meantime, it remains worthwhile to pursue a purely philosophical account of time for then, at least, we have a clear understanding of what it is that might conflict with a scientific account of time.

If this is plausible, then the attempt to drive presentism into the dilemma of either obvious falsehood or trivial truth rests, it might be argued, on conflations of two senses of 'to be'. Consider, again, (1) above. It is alleged to be equivalent to:

(1*)    That and only that which exists, exists.

But if one occurrence of 'exists' stands for the temporally unrestricted existential quantifier, the other points to the presentist's metaphysical thesis according to which the world is simply devoid of non-present entities, then (1) is in fact equivalent to:

(1**)    That and only that which $exists_1$, $exists_2$.

This certainly does not have the form of a triviality, so if it is either true of false, substantive argument will be required to make the case.

## 2. WHY THERE ARE TRUTHS ABOUT THE NON-PRESENT

It is, perhaps, rather obvious that propositions that refer to non-present entities, even those that refer exclusively to non-present entities, can be true. Propositions expressed by sentences such as 'World War II ended prior to the first Moon Landing' can be, and in many cases are, true. However, can one provide reasons for this stance other than adverting to its obviousness? I think so.

First, denying that there can be true claims about the past and future reduces presentism to a severely sceptical position, according to which we can have no knowledge of the past or future. This ought, at best, to be a manoeuvre of last resort (though see the section on 'quasi-truth' below).

Second, consider the following argument:

> It will be the case (200 years hence) that the Prime Minister of England is a woman.
> The Prime Minister of England is Tony Blair.
> Therefore, it will be the case (200 years hence) that Tony Blair is a woman.

This argument is clearly invalid. But then it must be possible for all the premises to be true and the conclusion false. Since the two premises refer to entities that are non-contemporaneous, it must be possible for claims that make reference to entities that exist at different times to nevertheless be true.

Third, there are other aspects of temporal reasoning that seem to depend on the possibility of past and future tense claims being true. Consider, for example, the following proposition:

(6)    Bill spoke yesterday.

(6) seems to entail the following:

    (7)   Bill spoke
    (8)   Something occurred yesterday
    (9)   Somebody spoke
    (10)  Somebody spoke the day before today.

Note, however, that if the move from (6) to each of (7) through (10) is one of entailment, then one of two situations must be the case. Either it is impossible for (6) to be true without (7)–(10) being true or it is impossible for (6) to be true, in which case (6) $\supset p$ holds for all $p$. But it seems clear that (6) has restricted entailments so it must be possible for it to be true.

    Such restrictions follow, I presume, from the logical form of the proposition expressed by (6). A natural suggestion for the logical form of (6) might be the following:

    (6**)   $(\exists e)(\exists x)(\exists t)(\exists t^*)[S(e, x, b)\ \&\ \text{Sim}(e, t)\ \&\ (t^* = n)\ \&\ (t = t^* - 1)]$

where:

    $S(x, y, z) = x$ is a speaking of y by z
    $\text{Sim}(x, y) = x$ is at/simultaneous with y
    $n = $ the date of utterance

This appears plausible because the inferences from (6) to (7)–(10) are now a simple matter of basic, truth-functional operations on (6*). For instance, existential generalization and &-elimination give us:

    (7*)   $(\exists e)(\exists x)S(e, x, b)$
    (8*)   $(\exists e)(\exists t)[\text{Sim}(e, t)\ \&\ (t = n\text{-}1)]$
    (9*)   $(\exists e)(\exists x)(\exists p)S(e, x, p)$
    (10*)  $(\exists e)(\exists x)(\exists p)(\exists t)[S(e, x, p)\ \&\ \text{Sim}(e, t)\ \&\ (t = n\text{-}1)]$

which are equivalent to (7)–(10) respectively.

    The problem with this, of course, is that it ontologically commits one to non-present entities, which violates the presentist's metaphysical outlook.[4] As a result, presentists

---

[4] Prior is suspicious of analyses such as this, not only because they are in tension with presentism, but because he rejects the ontological category of events:

> the real point, one might say, is not that events 'are' only momentarily, but that they don't 'be' at all. 'Is present', 'is past', etc., are only quasi-predicates, and events only quasi-subjects. '$X$'s starting to be $Y$ is past' just means 'It has been that $X$ is starting to be $Y$, and the subject here is not '$X$'s starting to be $Y$' but $X$. And in 'It will always be that it has been that $X$ is starting to be', the subject is still only $X$; there is just no need at all to think of *another* subject, $X$'s starting to be $Y$.                   (Prior 1962: 18)

However, a distinct advantage of reifying events is that it helps to explain the obviously valid entailment relations alluded to above. Davidson (1980) notes further that recognizing the ontological category of events makes good sense of causal and intentional talk (see also Katz, 1983).

tend to propose that the logical form of tensed propositions is that of a tense operator attached to a core, present tense proposition. For instance, (6) is typically analysed as follows:

(6**)    **P**(Bill is speaking),

where **P** = 'it was the case that' (similarly **F** = 'it will be the case that). The assumption is that the tense operators form ontologically non-committing contexts.

Fair enough, but at this point I think that a natural question arises for the presentist. If what exists is constantly changing, how is it possible for it to be *true* that there will be or was a different set of entities for us to quantify over if we have, right now, no quantifiers that can 'reach' those future (past) objects? What makes it the case that there was or will be other entities, when we cannot refer to those entities? The presentist owes us, I think, an account of how past and future tense claims, embedded in ontologically non-committing operators, can nevertheless be true. If they cannot be true, then presentism itself cannot be formulated, at least not if presentism includes the claim that what did (or will) exist, doesn't exist. The best we could make of the doctrine would then be that at any time, *t*, one must assert that only the entities that exist at *t* exist at all, and that it is false that anything else did or will exist. So, explaining how tense operators can be both ontologically non-committing yet support true claims about other times seems vital to the presentist cause.

## 3. THE TRUTH-MAKER PRINCIPLE

It is fair to demand that what is true depend on what exists. This is often referred to as the 'truth-maker principle', and can be paraphrased as *truth supervenes on (at least partly) non-propositional being* (see, for example, Sider 2001: 35–42). The point is to rule out true propositions that 'hang free' of reality: bare truths are unacceptable.

Simon Keller usefully points out that there are two ways of understanding the truth-maker principle, one weaker and one stronger (Keller 2004: 85). The stronger version of the truth-maker principle is that there exists a truth-maker for every true proposition. This may be correct, but if one accepts it, one has to account for negative existentials such as 'there are no ten-foot tall philosophers'. This is true, but it is unclear what its truth-maker is. One might suppose there to be negative facts or states of affairs for every negative truth, but that is a controversial stance that one may wish to avoid.

Accordingly, there is a weaker version of the truth-maker principle. As David Lewis notes, there is nothing to say about a negative existential except that there exists nothing of the relevant sort (Lewis 1992: 218). So, in a certain sense the true proposition that there are no ten-foot tall philosophers depends on how the world is; were the world to contain philosophers of such height, the proposition would be true. Similarly, the proposition would be true only if the world were to contain at least one ten-foot tall philosopher. So on the weaker version of the principle, what is true/false depends counterfactually on what exists; any world where the truth predicate is distributed differently among propositions is one which contains different non-propositional entities (see Keller 2004: 86).[5]

In fact, however, neither version of the truth-maker principle, as stated by Keller, seems quite correct. Imagine a world, $w$, in which only one thing exists, namely, the following proposition: *Exactly one proposition exists* (or, alternatively, *I exist*, where 'I' is taken to be a non-personal, self-referential indexical). In such a world, what exists supervenes on being, but the being it supervenes on is entirely propositional. Still, what is true in this world depends on what exists: if $w$ contained no propositions, or two propositions, then what is true in $w$ would differ. Or, consider the following example. A world, $w^*$, in which two propositions exist, each of which is the proposition that exactly one other proposition exists. If we alter the propositional arrangement, we alter the truths in $w^*$.

The modification I recommend we make to the truth-maker principle, in order to retain what is correct in it, is the following.

(TM)    The truth of any proposition that does not refer to a proposition, super-venes on non-propositional existents.

So, if propositions of the form $\mathbf{P}\phi$ and $\mathbf{F}\phi$ can be true, it is fair to ask the presentist to tell us what exists *right now* upon which their truth supervenes. Prior, for one, accepts that claims about the non-present must supervene on what exists in the present. Consider a proposition such as:

(11)    $\mathbf{P}$(John wins the race).

Prior argues that the presentist can coherently hold both that (11) is true and that nothing non-present exists because (11) expresses what he calls 'a general fact', such as 'someone stole my pencil', which is not about any specific person. In other words, Prior suggests that past and future tense truths are not, in fact, truths concerning the non-present:

> the fact that Queen Anne has been dead for some years is not, in the strict sense of "about", a fact about Queen Anne; it is not a fact about anyone or anything—it is a *general* fact. Or if it is about anything, what it is about is not Queen Anne—it is about the earth, maybe, which has rolled around the sun so many times since there was a person who was called "Anne", reigned over England, etc.    (Prior 1962: 13)

So claims about the past and future still depend on what exists, for they are about things such as the sun and Earth, which are present. If this is correct, then the following

---

[5]  Keller points out that it is best to conceive of the truth-maker principle as a statement of explanatory priority, a commitment to appeal to claims about what exists in explaining what is true, rather than vice versa (Keller 2004: 86). It is not, in other words, to be thought of as picking out some kind of natural, mind-independent relation that holds between the propositions and the things in the world. I see no reason to take issue with his stance here.

must hold: if there were different truths concerning the past and future, there would be different truths about the (currently existing) Earth and sun.

So, the question that seems to me to arise is as follows. What generally exists in the present to underwrite the truth of claims about the past and the future, as well as their obvious entailments? It is to this question that I turn next.

## 4. THE GROUNDING PROBLEM FOR PRESENTISM

The objection toward which I have been working is commonly referred to as the 'grounding objection' to presentism, which runs as follows. Putting scepticism to the side, there are determinate truths about the past and future. Truth, however, supervenes on being. Hence, for the presentist, truths about the past must supervene on what is present. The present, however, radically underdetermines the past since what exists is compatible with many different past histories; for example the world *could* be just as it is today whether Caesar stepped into the Rubicon with his left foot rather than his right, or vice versa. This conclusion is, however, absurd, for it entails that there is no determinate fact of the matter as to what has occurred. Hence, presentism must be rejected.

We can view the argument as a *reductio*:

(12)   There exist determinately true and false propositions about the past.
(13)   Truth supervenes on what exists.
(14)   What exists in the present underdetermines what is true in the past.
(15)   All and only that which is present exists.
(16)   Therefore, there are no determinately true or false propositions about the past.

As I mention above, premise (12) can only be denied at the cost of adopting a severe form of scepticism concerning the non-present, one that would entail that there is effectively nothing we can know about the past, for instance. This is difficult to take seriously. Premise (13) seems equally secure: it is hard to conceive of 'ungrounded' truths. (14) also seems impeccable; as Dummett writes: 'the state of the universe at any one instant is *logically* independent of its states at all other instants' (Dummett 2000: 500). We may think that, for example, one couldn't have a stomach ache right now had one not eaten some improperly cooked food an hour ago, so that the present state of things isn't independent of the past. This is, however, to overlook that such dependencies are *causal* rather than logical or metaphysical. It is entirely possible to have a stomach ache at *t* and have eaten nothing an hour before. (15) is just a statement of the presentist doctrine.

The grounding objection casts doubt on presentism because it seems to be the weakest link in the argument outlined above. If we find the conclusion, (16) to be unpalatable, then presentism appears to be the natural premise to jettison. If, on the other hand, we are willing to accept (16), we must face the unhappy situation of answering why a speculative metaphysical theory can justify overturning our commitment to at least one of (12)–(14).

In what follows, I consider some of the most prominent proposals that presentists have wielded in reply to the grounding objection.

## 4.1 Referring to the non-existent

Mark Hinchliff suggests that the presentist could reject (13) by combining the following claims: (i) that nothing non-present exists; and (ii) reference to the non-existent is possible. He writes:

> What seems to keep the presentist from a more natural and appealing solution is the assumption that we cannot refer to what does not exist—most arguments I know in favour of the assumption appear to beg the question. It certainly seems that we can refer to people and things in the past, such as Cicero and Pompeii, even though they no longer exist.                                         (Hinchliff 1996: 124–5)

This is certainly an honest solution to the problem. Hinchliff's presentist admits that past and future entities don't exist, denies that there are surrogate or ersatz entities that stand in for them, and also claims that we can refer to them. This allows for truths about the past/future that don't supervene on anything existent.

I don't think that this is acceptable. If x refers to y, then x and y stand in relation to each other. But two entities can stand in relation to each other if, and only if, they both exist. This is simply a logical truth.

Furthermore, the view seems to rest on a confusion. If we are to claim that there is something that is referred to in an utterance, thought, or proposition, then this is simply contradicted by the claim that what is referred to doesn't exist.

Finally, reference to the non-existent allows for the possibility of successfully referring to fictional entities such as Santa Claus. It then becomes hard to see what the difference is between what has existed/will exist and what is purely fictional. 'Reference' is a success word; to say we can refer to what doesn't exist is akin to saying we can know what is false. Surely, in saying that Santa Claus doesn't exist, one is, among other things, pointing out that 'Santa Claus' doesn't refer to anything.

## 4.2 Tensed Properties

John Bigelow believes that the presentist ought to reject premise (14) of the grounding argument. He writes:

> We do not need to suppose the existence of any past or future things…only the possession by present things of properties and accidents expressed using the past or future tenses.                                         (Bigelow 1996: 46)

So the presentist, Bigelow argues, may deny the reality of the non-present so long as she admits the present existence of past and future tense properties, which are the truth-makers for past and future tense claims. For example:

> The causal relation does not, in fact, ever hold between things that exist at different times. At any given time, the causal relation holds between properties, perhaps between world properties, each of which is present and is presently instantiated.
> (Bigelow 1996: 47)

Of course, there must exist something that can exemplify such properties. Chisholm (1990) opts for eternal entities, such as the property 'blue'. Bigelow remains open-minded:

> What exists in the present is a tract of land which has the accident, of being the earth on which a spark of love kindled a dazzling blaze of pitiless war. Or, what exists in the present is a region of space which has the accident, of being the space within which a Wooden Horse set the towers of Ilium aflame through the midnight issue of Greeks from its womb.
> (Bigelow 1996: 46)

This is fine, but it doesn't go very far. After all, the eternalist may well agree that such-and-such area of land has the property of being where $x$ occurred. The eternalist will, however, parse this out as the proposition that a past occurrence tenselessly exists at the same location as this piece of land, or that an earlier (but real) stage of this land is (tenselessly) spatially coincident with $x$. What is needed is an account of properties such as 'being where $x$ occurred' that distinguishes them from tenseless properties of entities that tenselessly exist at various times.

Not only is an account of such properties needed to distinguish presentism from eternalism, but it is needed because past and future tense claims have logical structure. This can be seen by the fact that logical entailments hold between tensed propositions. For example:

(17)  'Waterloo was a fierce battle' entails 'Waterloo was a battle'

If the proposition expressed by the first sentence in (17) is true because a current parcel of land, Waterloo, has the property 'being where a fierce battle occurred', then the description of this property must have sufficient logical structure to support the entailment. We might suppose that we can embed a description in a past tense operator as follows:

(17*)  W is a parcel of land & $\mathbf{P}[(\exists y)(y$ is a battle & $y$ is at W & $y$ is fierce$)]$

So long as the standard logical rules can be applied within the scope of a tense operator, this will, it seems, entail the right-hand side of (17).

Now we may ask what kind of sense the property presentist can make of a proposition such as:

(18)  $\mathbf{P}[(\exists y)(y$ is a battle & $y$ is at W & $y$ is fierce$)]$.

This cannot be a singular proposition, as concerns the value of y, for y doesn't exist to be a component of any proposition. This is problematic. For consider the past time at which the Battle of Waterloo is occurring. It seems that one could have expressed a singular proposition about the battle, perhaps by use of a demonstrative; for example, 'this is fierce'. Suppose the proposition expressed by S's utterance of 'this is fierce' is:

   (19)   Fierce(b).

Next, suppose that the next day S utters 'the Battle of Waterloo was fierce'. The proposition expressed would have to be non-singular, along the lines of:

   (19a)   $P[(\exists y)(y$ is the referent of 'the Battle of Waterloo' & y is fierce$)]$

On the assumption that a very different (kind of) battle might have been called 'the Battle of Waterloo', it seems possible for (19) and (19a) to diverge in truth value. For instance, if S's utterance of 'this' rigidly designates the battle S witnesses, then there could be a possible world in which b doesn't exist, but there was a fierce battle at Waterloo. But this seems incorrect: if, when looking at the Battle of Waterloo, S utters 'this is fierce', then what S expresses is true if and only if what is expressed tomorrow by 'the Battle of Waterloo was fierce' is true.

The property-presentist might deny the existence of singular propositions (see Markosian 2004: 52–3), but their value in explaining demonstratives renders this move problematic (see Kaplan 1989). Since, moreover, the eternalist has no motivation for such a denial, this in fact should cause the presentist some concern, as the view is inheriting substantive semantic commitments.

There is, however, a more fundamental problem that I see with property presentism. Consider that if we assume the past and future do not exist, then all past and future tense properties must be temporally monadic: they are not relations to past or future times or events. But then any past or future tense property, P, will be intrinsic to the current state of the world, that is its logical form will be Px, rather than, where t is a non-present time, P(x, t). But then the present state of the world cannot depend counterfactually on what occurred or will occur. If x is the kind of entity, the world say, that can have the intrinsic property of 'being where the Battle of Waterloo occurred', then there is a possible world that contains a world-state that is intrinsically indistinguishable for x's, even though there was no battle there.

For example, suppose at t, a battle occurs at W. It seems perfectly possible for it to be the case that by t*, there exist no traces of the battle whatsoever. It doesn't follow that it is no longer true that a battle occurred at W. But if there exist no traces of the battle then there exist no intrinsic features of the world that distinguish it from a world in which no battle occurred at W. That is, there appear to be no features of the world that underwrite the instantiation of the property. Hence, Bigelow's view does in fact run into the grounding objection.

## 4.3  The ersatz B-series

Thomas Crisp (2007) suggests that the presentist view times as abstract entities, in particular sets of maximal, consistent propositions, which he spells out as follows:

> $x$ is a time $=_{df}$ For some class $C$ of propositions such that $C$ is *maximal* and *consistent*, $x = [\forall y(y \in C \supset y$ is true$)]$[6]

Intuitively but loosely put, such maximal propositions constitute an abstract representation of the complete state of the world at a moment (Crisp 2007: 99).

All of these abstract objects are, according to Crisp, ordered by a semantically primitive 'earlier than' relation that forms an 'ersatz B-series' in which non-present times have ersatz existence. Non-present times are temporally ordered entities but, since they are abstract, they are at no temporal distance from the present and can be referred to right now (Crisp 2007: 98–105). Accordingly, properties such as *being an x such that x ate improperly cooked food on hour ago* are properly viewed along the lines of *being an x such that the proposition that x eats improperly cooked food is included in an earlier time* (Crisp 2007: 105).

For Crisp, then, presentism becomes the doctrine that only one time is ever true/concrete—all others are false/abstract—and that which time is true is an ever changing affair. These temporally ordered abstracta serve as grounds for all the properties to which someone such as Bigelow might wish to commit. Hence the past is not, after all, underdetermined by the present, and premise (14) is rejected. In this way the presentist can, Crisp concludes, escape the grounding objection (see also Bourne 2006a, 2006b).

There is, I think, a deep problem with this view. Suppose that Crisp is correct and the world currently contains an ersatz B-series ordering all non-present, but ersatz, times. In a certain sense, then, it is no longer true that the present state of the world underdetermines what has occurred and what will occur. The world could, however, be just the way it is, whether or not the following proposition is true or false:

(20)   Past (future) ersatz times once were (will be) concrete, present times.

The present state of the world contains such properties as 'being the tract of land such that the proposition that dinosaurs roam that tract of land is included in an earlier time' (see Crisp 2007: 98), and surely that couldn't be the case unless there is an earlier time that contains that proposition. Since, however, that earlier time is abstract and consists of abstract entities entirely (propositions), then the world state does not entail that such times once were concrete. This must be either an additional primitive property of the world or else be unexplained. If the former, however, then not only is it a primitive fact that what is real constantly changes, it is also a primitive fact that what is abstract once was concrete; the eternalist requires neither primitive fact. If the latter, then presentism is simply an incomplete view.

---

[6]  Crisp 2007: 99. For Crisp, square brackets indicate a proposition; i.e. [P] = the proposition that P.

To see the problem, consider the proposition that Napoleon lost at Waterloo. If Crisp is right, then the truthmaker for this is the (current) existence of an ersatz entity that includes an abstract general Napoleon and an abstract battle. That Napoleon lost at Waterloo also entails that Napoleon and the Battle of Waterloo once were concrete for what is past once was present. The abstract entity that grounds the truth of this proposition does not, however, entail this proposition. An ersatz Napoleon can be part of an ersatz loss to Wellington at an ersatz time that stands in the primitive earlier than relation to 2008 *without* it ever having been true that that ersatz time was concrete. If the ersatz time suffices for the truth of this proposition then it can be true even if Napoleon and the Battle of Waterloo never were concrete: a renewed version of the truthmaker objection arises in force for there is nothing to ground the truth that what once was present once was concrete.

A second problem with Crisp's view is that it is incompatible with the existence of singular propositions concerning non-present entities. After all, if 1850 exists now as a maximal proposition, then surely anything contained in 1850 is abstract. Accordingly, any reference to, say, Abraham Lincoln can only have an abstract person as a referent rather than a concrete, physical human being. But then, the proposition expressed by 'Abraham Lincoln was tall' cannot be a singular proposition. It must be something more like:

(21)    $P(\exists x)(x =$ the referent of 'Abraham Lincoln' and x is tall).

Now, however, consider what someone would have expressed, in 1850, by the utterance 'Abraham Lincoln is tall'. If we suppose it to have been a singular proposition such as:

(22)    Abraham Lincoln is tall,

then what is expressed in 1850 by 'Abraham Lincoln is tall' cannot be captured by the past tense transform of that sentence, uttered in 2008. Since (21) can be true in a world that never contains Abraham Lincoln—that is, the Abraham Lincoln of our world—but (22) entails that Abraham Lincoln exists, (21) can be true even if (22) is false, and vice versa, so the truth-value links are broken. This leaves presentism in the position of asking us to reject ordinary, seemingly uncontroversial principles of temporal reasoning. Eternalism, by contrast, faces no such difficulty.[7]

A third and final problem that I find Crisp's account to have is that it appears to be formally analogous to an argument that is clearly unsound. Consider a view that I shall call 'hereism', a view that has, so far as I know, no philosophical proponents. The hereist argues that only one spatial position exists, the *here*. We can imagine an objector pressing the point that the here radically underdetermines truths about the there; things could be radically different in spatially distant parts of the world, yet

---

[7] One may wish to reject the view that proper names refer directly (and are rigid designators), but that is a controversial move—one that, again, eternalism need not make.

nothing differ here. Since it is absurd to deny that there are determinate truths about the spatially distant, hereism must be rejected.

We can now imagine a hereist follower of Crisp responding as follows. While there are, of course, determinate truths about the there, the here does not in fact underdetermine what is true concerning the spatially distant. The reason is that places other than what is here are abstract representations of a point (or, perhaps, point state) of the world:

> $x$ is a place $=_{df.}$ For some class $C$ of propositions such that $C$ is *maximal* and *consistent*, $x = [\forall y(y \in C \supset y \text{ is true})]$.

Such propositions would have to make reference to different times, but we shall ignore those complications here, since the hereist needn't be a presentist, for example. Accordingly, every proposition is either true or not of a point at a time.

Now, the hereist continues, these abstract places are ordered by three, primitive relations: 'to the left/right of', 'above/below', 'in front of/behind' (or, possibly, three primitive relations of relative placement on the x, y and z axes). This is all consistent with the non-existence of the non-here, since other points, being abstract objects, have no spatial distance from the here-point (the point that is true).

I think that none of this would convince us to be hereists. Even if we accept that abstract there-points are consistent with singular propositions about such points and, also, are such that they were or will be concrete when they are 'here', the view remains unpersuasive. The reason, I suggest, is the starting point. There is simply no reason to believe that spatially distant points differ in ontological status from the here. The mere fact that we can't directly experience the spatially distant provides insufficient grounds for doubting its concrete reality. Though hereism might be coherent, it is not motivated. I believe, similarly, that presentism is not motivated. Not only is eternalism the natural construal of claims about the non-present, but eternalism has fewer unexplained primitives than presentism and is consistent with singular propositions about the past and future. As it stands, eternalism has more going for it than does Crisp's ersatz B-series.

## 4.4 Haecceities

Another reply to the grounding objection is to suggest that the present includes the haecceities of non-present beings. Let us understand a haecceity to be the property of an entity's being itself: the property that corresponds to the entity's self-identity or 'thisness', as Adams (1979, 1989) puts it. For haecceities to underwrite truths about the past and future, an individual's haecceity must be capable of existing even when the individual does not. In that case, it might be true that, say, Napoleon lost at Waterloo because Napoleon's thisness exists and, presumably, has the property of having lost at Waterloo (Waterloo's haecceity?).

One question that arises for this view concerns the relation between Napoleon's haecceity and Napoleon. If the former can exist without the latter, why can't the former exist even if the latter *never* existed? Couldn't it be true that Napoleon lost at Waterloo, even though the world never contained Napoleon, simply because the haecceity of Napoleon, which exists, has the property of having lost at Waterloo? In other words, a grounding problem arises similar to that which vexes Crisp's ersatz view: if current, abstract entities can make it true that Napoleon lost at Waterloo, then that proposition can be true even if the relevant concrete entities never existed.

Keller (2004) considers the haecceity approach to be promising. He writes:

> When the [eternalist] talks about Ann Boleyn, he is talking about a person—a concrete thing—but when the presentist talks about Anne Boleyn, she is talking about a property—an abstract thing.                      (Keller 2004: 98)

This is troubling for at least two reasons. First, it seems simple to understand what it is to refer to or talk about a person. If, however, I am told that when discussing Anne Boleyn I am really discussing a heacceity-property, I am left feeling quite mystified; this simply does not seem to be what I am discussing. Second, it makes the version of the grounding objection that Crisp's view faces much harder to answer. If the referent of 'Anne Boleyn' is an abstract property and this property, and other abstracta, can make 'Anne Boleyn existed in 1536' true, then what need have we to suppose that a concrete Anne ever existed? The view is consistent with her never having existed concretely.

The only option is to suppose that it is a primitive, metaphysical fact that an object's haecceity exists if and only if the object did, does, or will exist. Such a move, however, is ad hoc. Keller claims that Anne's haecceity and its relations to other haecceities must, as he writes, 'make it the case that on the $19^{th}$ of May, 1536, Anne did exist' (Keller 2004: 99). What this 'making it the case that' comes to remains unexplained.

## 4.5 Evidence

Thus far we have considered the prospects of grounding truths about the past and future in currently existing but abstract entities. The prospects for success along these lines are beginning to look dim. Might there, however, be concrete entities that exist now and can ground claims about the past and future?

Suppose that the presentist were to choose to adopt an epistemic theory of content for past and future tense assertions; verificationism, for example. In other words, tensed assertions are rendered true or false strictly in virtue of currently existing evidence. This evidence can provide reason for one to assert, say, 'John ran', even if John and his run don't exist. To avoid ontologically committing to the non-present, this view must conceive of currently existing evidence as standing in relation to assertions or propositions rather than to putatively past or future entities. That, however, seems plausible enough if one takes the content of tensed claims to be epistemic in character. In short, existing records, documents, artefacts, predictions, and the like, are combined with various principles of inference to render tensed assertions warranted to greater and lesser degrees.

The main problem for such a view concerns the truth-value links, a pair of semantic principles that connect (the content of) sentences that differ only in tense:

(TV1)$\phi \Rightarrow \mathbf{FP}\phi$
(TV2)$\phi \Rightarrow \mathbf{PF}\phi$

As Michael Dummett points out (Dummett 1978), it is particularly difficult for someone whose instincts lean toward epistemic accounts of content to deny these, for we learn the function of tense in our language in part by accepting principles such as these. However, an epistemic account of the content of past and future tense expressions threatens the links. For, after all, there may be evidence today that warrants an assertion even though there will be (was) no evidence that warrants the past (future) tense of that assertion. Dummett puts the objection as follows:

> For you [the epistemic theorist] must surely agree that if, in a year's time, you still maintain the same philosophical views, you will in fact say that, on the supposition we made, namely that all evidence for the truth of the past-tense statement $B$ is then lacking, and will always remain so, $B$ is *not* true (absolutely). And surely also you must maintain that, in saying that, you will be correct. And this establishes the sense in which you are forced to contradict the truth-value link
> (Dummett 1978: 372–3).

It could plainly be that the assertion of $\phi$ is warranted even though it will not be the case that the assertion of $\mathbf{P}\phi$ is warranted, which just is the denial of the truth value links. But this, of course, is a problem:

> The truth-value links are not to be compared to those features of, say, classical logic which undergo modification by the intuitionists. The intuitionistic modifications are deep-reaching; but they are relatively conservative with respect to our ordinary notion of valid inference. Wholesale rejection of the truth-value links, in contrast, would be bound to leave us, it seems, with no clear conception of how tensed language was supposed to work at all                                    (Wright 1993: 179)

Dummett initially attempts to circumvent this problem by arguing that our understanding of the truth value links is itself best accounted for in a present tense, verificationist manner. In other words, in virtue of accepting the links one commits *now* to there being evidence in the future for the assertion of the past tense of current assertions. However, one doesn't commit to being so committed in the future. Since meaning is tied to verification, and what there is evidence for changes in time, what we can express with a given set of word types is not constant:

> even if he [the epistemic theorist/anti-realist] is not convicted of contradicting his earlier contention that the statement $B$, if made in a year's time, will, in virtue of the present truth of $A$, be (absolutely) true, it appears that the realist is justified in maintaining that here at least the anti-realist must diverge from the truth-value link. But the anti-realist replies that he will not in a year's time mean the same

by 'absolutely true' as he now means by it: indeed, he cannot by any means at all now express the meaning which he will attach to the phrase in a year's time... the anti-realist's position is that a statement is true (absolutely) if there is something in virtue of which it is true. He will agree completely with the realist that the truth-value link requires us to recognize that a past-tense statement, made in the future, may be true in virtue of some fact relating to a time before the making of the statement... but he denies that this can legitimately make us conclude that a past-tense statement, made now, can be true in virtue of some past fact, if 'past fact' means something other than that by means of which we can recognise the statement as true... to say that we are in time is to say that the world changes; and, as it changes, so the range of even unrestricted quantifiers changes, so that that over which I quantify now when I say, 'There *is* something in virtue of which ...', is not the same as that over which I shall be quantifying when I sue the same expression in a year's time. The anti-realist need not hang on to the claim that the meaning of the expression alters: he may replace it by the explanation that he cannot now *say* what he will in a year's time be saying when he uses it        (Dummett 1978: 373).

The idea is that at any given time one commits to the truth value links: insofar as there is evidence for $\phi$, there is *then* evidence for **PF**$\phi$ and **FP**$\phi$. However, the verificationist must recognize that as the world changes the expressive power of language changes so that, for example, the future meaning of **P**$\phi$ could differ from the current meaning of the same expression (type) and, crucially, could differ from the current meaning of $\phi$. Hence the future denial of **P**$\phi$ is consistent with the present commitment to both $\phi$ and if $\phi$, then **FP**$\phi$ since the conditional is understood to be based on current evidence.

Wright, ultimately, rejects this solution because it seems to eliminate the possibility of disagreement across time: 'the threatened cost is an inability to explain how there can be such a thing as conflicting views held by protagonists who are sufficiently separated in time' (Wright 1993: 194). Ludlow, on the other hand, comes to Dummett's defence:

one should hold that diachronic disagreements are possible but reject the idea that they can be thought of as conflicts between positions held at $t_1$ and $t_2$. Rather, the idea would be that the relevant historical disagreements are all "in the present." Thus, if I now dispute Plato's doctrine of the forms, it does not mean that the view I hold at $t_2$ conflicts with the content of something that Plato said at $t_1$, but rather that the conflict between our views must take place as they are couched at $t_2$
(Ludlow 1999: 152).

Dummett presents the situation as something of a stalemate between the realist and anti-realist (Dummett 1978: 374; see also Gallois 1997 and Weiss 1996). But, realism vs anti-realism isn't my primary focus here. If, in attempting to construct a semantics for past and future tense claims, one ends up as a general verificationist or anti-realist, then one undercuts the claim that there is an ontological distinction between the present and non-present. If *all* claims are made true by the evidence available at the time of utterance, then the difference between present and past/future tense claims is at best one of degree, not one of kind.

The resultant position is a universal anti-realism that is unable to support any substantial metaphysical difference between the present and non-present—unlikely to upset a philosopher such as Dummett—and so stands in conflict with the presentist's assertion that the present is real in a sense that other times are not. The point of resorting to epistemicism, recall, was to find a semantics that would allow the semantic values of assertions about the past and future to rest solely on present evidence, while denying this for present tense claims. In short, the most plausible epistemicism-presentism combination is one in which it is claimed that assertions about the present can be rendered true or false in virtue of evidence-transcendent entities, but that assertions about the past and future have their content determined entirely by evidence. It must, in short, be a hybrid realist/anti-realist view.

The question, then, is what sense the epistemicist-presentist can make of a *current* assertion of the truth-value links. What, for her, does it mean to say that, for example, $\phi \Rightarrow \textbf{FP}\phi$? If $\phi$ is a present tense, true assertion, then what are we to say, right now, about the future utterance of the past tense of $\phi$? Surely, to remain consistent with presentism, the epistemicist-presentist must admit that in the future, only a particular future time will exist. So, any past tense version of $\phi$, uttered then, will only be warranted if there is evidence then to justify the assertion of $\textbf{P}\phi$. But is that the sense in which she *now* commits to the truth of $\phi$? Clearly not. Currently, $\phi$ is true in virtue of what exists now, and this needn't be evidence. Later, $\textbf{P}\phi$ will be true only in virtue of evidence that obtains then. The verificationist-presentist must, to remain consistent, admit all of this right now. So, right now, she must admit that that in virtue of which $\phi$ is true is not that in virtue of which $\textbf{P}\phi$ will be true; the link between their truth/warrant conditions has been broken.

Let us examine an example. Suppose that today an event, e, say a political rally, occurs and is loud, F. Today, 'Fe' is true. Moreover, 'Fe' is true if and only if Fe. In general, where 'is' is tensed:

(23)    what 'e is F', asserted at t, expresses is true if and only if ($\exists$e)(Fe and e is simultaneous with t)

But, what 'PFe' expresses at t + n will have a different truth condition; in fact it only has a warrant or verification condition. Importantly, this is how the epistemicist-presentist must put things right now: she is committed, now, to there being only evidence in the future for the assertion of 'PFe'. In general, at *t*, the epistemicist-presentist must insist that what 'e was F', said at t + n, will express will be warranted if and only if there will, at t + n, be evidence that justifies the assertion of 'e was F'. In other words,

(24)    what 'PFe', uttered at t + n, will express is true if and only if $\textbf{F}(\exists x)$(x suffices to warrant the assertion, at t + n, of 'PFe').

Notice, again, that this captures the epistemicist-presentist's current commitments; this is intended to be a statement of what she expresses now about what she will express in the future, rather than a statement of what she will be expressing in the future.

Let us, now, ask whether the right-hand sides of (23) and (24) stand, currently, in the appropriate relation. In other words, is it the case that:

(25)    ($\exists$e)(Fe and e is simultaneous with t) only if **F**($\exists$x)(x suffices to warrant the assertion, at t + n, of '**PFe**').

Recall that we are making this assessment at t, which is by hypothesis present. In other words, the epistemicist-presentist must commit now to the future existence of a piece of evidence for an assertion. But the fact that evidence can be lost indicates that (25) cannot be recognized as an entailment. To emphasize, even right now, at *t*, the epistemicist-presentist is forced to admit that her current commitment to the truth-value link commits her to the availability of certain evidence in the future. Since there is no guarantee of such availability, the verificationist-presentist must conclude that $\phi \Rightarrow$ **FP**$\phi$ does not express anything like entailment: it does not even express 'commitment in virtue of the current evidence'.

Now, in discussing the truth-value links, the Dummettian anti-realist typically supposes that anything that warrants the present assertion of 'Fe' warrants right now (but perhaps only right now) the assertion of '**FPFe**'. But notice that the epistemicist-presentist can't avail herself of this way out. This is because, for the epistemicist-presentist 'Fe' might be true without any evidence in its favour. That is, the point of being an epistemicist-*presentist* is the reliance on the claim that truths concerning the present are evidence transcendent, while claims about the past and future are not. Hence it is simply impossible, on this view, to argue that one can commit *right now* to the truth of $\phi \Rightarrow$ **FP**$\phi$ in virtue of the fact that any reason I have *right now* for asserting $\phi$ is reason for asserting right now **FP**$\phi$. The reason it is impossible is that the antecedent of $\phi \Rightarrow$ **FP**$\phi$ does not express an evidence-bound proposition. Hence, *right now* the verificationist-presentist, in asserting that $\phi \Rightarrow$ **FP**$\phi$, cannot help herself to the claim that $\phi \Rightarrow$ there is (now) evidence for $\phi$. Only the general anti-realist can make this move, but general anti-realism, as we saw, cannot support presentism.

Hence, I conclude that the epistemicist-presentist is forced, at any time, to deny the truth-value links: she cannot re-interpret them as Dummett suggests we do in order to resuscitate anti-realism, namely as capturing our current commitments to what will be assertible in the future, while making no claims about what our commitments will be in the future about what is assertible then. When the epistemicist-presentist asserts, right now, her version of the truth-value links, she ends up destroying the links (unless she wishes to assert that evidence is indestructible, a claim that I shall take as simply incredible).

Moreover, the grounding objection re-asserts itself. Since the difference between a correct and incorrect assertion of 'e occurred' cannot be explained by reference to e itself, the difference must be explained by appeal to evidence that exists now. What exists now, however, underdetermines whether or not e occurred: it is logically possible for all the current evidence to exist, even if e never did. The epistemicist-presentist cannot appeal to causal relations between e and the evidence, since such a relation entails the existence of its relata.

I conclude that the combination of presentism with epistemicism is unstable. Universal epistemicism is not a version of presentism, while epistemicism-presentism leads to the denial of the truth-value links. An epistemic account of the content of tensed assertions does not help the presentist combine her metaphysics with a plausible understanding of tensed discourse.

## 4.6 Quasi-truth

Philosophers such as Markosian (2004) and Sider (1999) suggest that the presentist can plausibly deny (10) so long as she can replace the truth of past and future tense claims with something 'close enough' to truth. In particular, the suggestion is that, though propositions about the past and future are not true, they are 'quasi-true'.

According to Markosian, a proposition is quasi-true if and only if it is not literally true but that this is the result of non-empirical, philosophical facts (Markosian 2004: 69). For instance, the presentist, Markosian insists, does not think that the current state of the world differs empirically from what it would be were eternalism true. Still, eternalism is false for philosophical reasons.

Sider's suggestion is to spell out quasi-truth as follows: though the proposition, P, is false, it is quasi true if there is a true proposition, Q, such that were some ontological doctrine, X, true, Q would still be true and would entail P (Sider 1999: 332–3). In particular, the presentist may argue that, say, a singular proposition about Socrates, such as that Socrates was wise, is quasi-true because there is a true, general proposition:

$P[(\exists x)(x = $ the referent of 'Socrates' and x is wise)]

that is: (i) true; (ii) would have been true had eternalism been true; and (iii) would have entailed the truth of the singular proposition that Socrates was wise. I don't think that such an account can render presentism either plausible or preferable to eternalism.

First, I think Crisp (2005) is right to point out that this notion of quasi-truth is subject to a powerful technical criticism. Consider any ontological doctrine you wish, such as 'sockism', the doctrine that *necessarily* everything is composed of tiny, undetectable, grey socks; if true, this proposition is necessarily true. Next, consider a proposition that both is true and would be true were sockism true, for example the proposition that everything is self-identical; this is a necessary truth so it is true and would be true were sockism true. Now consider the proposition that there is at least one tiny, undetectable, grey sock. This proposition comes out as quasi-true because there is a proposition, namely that everything is self-identical, that is: (i) true; (ii) would have been true had sockism been true; (iii) would have entailed the truth of the proposition that at least one tiny, undetectable, grey sock exists. If the proposition that at least one tiny, undetectable, grey sock exists has the same alethic status, quasi-truth, as the proposition that Socrates was wise, then quasi-truth is not near enough to truth to give the presentist any genuine comfort after all.

A second problem with the quasi-truth strategy is that it presupposes an answer to the grounding objection and so can't be used in response to it. The Sider-Markosian, quasi-truth strategy assumes that there are at least general propositions about, for example, past figures such as Socrates. The problem, as I argue above, is that there is no promising strategy for providing grounds for such propositions. Whether the presentist supposes such truths to supervene on currently existing properties, ersatz times, haecceities, or evidence, the view runs into trouble.

Third, even if we ignore the arguments above and suppose that the truth of general propositions such as $\mathbf{P}[(\exists x)(x = \text{the referent of 'Socrates' and } x \text{ is wise})]$ is consistent with presentism, such propositions will not entail the singular propositions that the eternalist believes to exist. Consider two eternalist worlds, $w_1$ and $w_2$. In $w_1$, *the* Socrates, that is, the Socrates of our world, exists and lives in ancient Greece. In $w_2$, a different person exists who is named 'Socrates' who lives in ancient Greece and is wise. In $w_2$, our Socrates does not exist. In $w_2$, the general proposition, $\mathbf{P}[(\exists x)(x = \text{the referent of 'Socrates' and } x \text{ is wise})]$, is true. It does not, however, entail the singular proposition that *Socrates* was wise, since he does not exist in that world. Hence the singular proposition that Socrates was wise is not quasi-true after all.

Finally, there is a general worry about any view that denies that claims about, for example, the past are simply true. There does not seem to be sufficient difference between, for example, my eating cereal this morning and my eating lunch now, such that it is true that the latter is occurring, but not true that the former occurred. It strikes me as incredible to suppose that it is true that I exist but only quasi-true that Socrates did exist. Markosian has three replies. First, he thinks that we often slip into eternalist ways of thinking when we are being careless (Markosian 2004, 68–9). Second, he thinks that quasi-truth explains this intuition, since it asserts that there is no empirical difference between our, presentist world and the eternalist's world (Markosian 2004, 69). Third, he thinks this complaint rests on a misunderstanding of the truth conditions of tensed statements. He insists that, though singular propositions about, say, past individuals are literally false, they are nonetheless meaningful (Markosian 2004, 69–73). We know what it would take for such a proposition to be true (namely for the entities allegedly referred to to exist).

In response to his first reply, I suggest that the analysis above shows that when we are careful, it is presentism that is cast in an unfavourable light and ought to be resisted. In response to his second, I contend that there are too many problems with the notion of quasi-truth to render that proposed explanation satisfying. As for Markosian's third rejoinder, I don't think the concern is how we might be able to understand and believe untrue propositions. At issue is the claim that there is no truth about past individuals, objects, or events. It is this claim that strikes me as incredible, particularly given the other difficulties the presentist faces.[8]

---

[8]  Markosian (2004) suggests we take the analogy between tense and modality seriously, so that being formerly real is analogous to being possibly real. This makes the past seem less like the present

# 5. CONCLUSION

If presentism is to remain plausible, it must be possible to ground all truths about the past and future in that which currently exists. The grounding objection presents an argument designed to show that this leads to absurdity:

(12)    There exist determinately true and false propositions about the past.
(13)    Truth supervenes on what exists.
(14)    What exists in the present underdetermines what is true in the past.
(15)    All and only that which is present exists.
(16)    Therefore, there are no determinately true or false propositions about the past.

As (15) is simply a statement of presentism, the presentist cannot reject it. (13) appears too plausible for anyone to deny; though Hinchliff thinks it is a possibility, I indicate above why I think he is mistaken. (12) is similarly secure, though Sider and Markosian have tried, unsuccessfully I argue, to replace it with something more presentist-friendly.

Most defences of presentism, therefore, rest on attempts to undermine (14); to show that what exists in the present is sufficient to underwrite all the apparent truths concerning the past and future. The strategy is to assert that there exist enough abstract objects—properties, ersatz times, haecceities—to do the grounding work; an alternative is to suggest that there exist enough concrete pieces of evidence to do the job. I have examined the prominent attempts along these lines, arguing that none is satisfactory.

Presentism has been subjected to critiques other than the grounding objection, including the charges that it is incoherent or inexpressible and that it cannot account for diachronic relations (see Crisp 2005, Sider 1999). I have chosen in the foregoing to focus on the grounding objection for I consider it to be the most fundamental problem facing the presentist; if it cannot be rebutted then the view is in serious difficulty independent of other concerns.[9]

## REFERENCES

Adams, R. (1979), 'Primitive Thisness and Primitive Identity', in *Journal of Philosophy* 76: 5–26.

—— (1989), 'Time and Thisness', in J. Almog, J. Perry, and H. Wettstein (eds.), *Themes from Kaplan* (Oxford University Press), 5–26.

Barbour, J. (1999), *The End of Time* (Oxford University Press).

and more like the unreal, which can help weaken the feeling that it is unacceptable to treat the past differently from the present. The time-modality analogy is, however, problematic (see, for example, Meyer (2006)).

[9] I would like to thank James R. Brown, Bernard Katz, L. Nathan Oaklander, V. Alan White, students in my graduate seminars and audiences at Queen's University, Ryerson University, and the 2006 Central Division meeting of the APA for comments, criticism, and encouragement.

Bigelow, J. (1996), 'Presentism and Properties', in J. E. Tomberlin, ed., *Philosophical Perspectives, Volume X: Metaphysics*, (Blackwell), 35–42.

Bourne, C. (2006a), 'A Theory of Presentism', in *Canadian Journal of Philosophy* 36: 1–24.

Bourne, C. (2006b), *A Future for Presentism* (Oxford University Press).

Chisholm, R. (1990), 'Referring to Things that No Longer Exist', in *Philosophical Perspectives* 4: 545–556.

Craig, W. L. (1998), 'McTaggart's Paradox and the Problem of Temporary Intrinsics', in *Analysis* 58: 122–127.

Crisp, T. (2005), 'Presentism and Cross-Time Relations', in *American Philosophical Quarterly* 42: 5–17.

—— (2007), 'Presentism and The Grounding Objection', in *Noûs* 41: 90–109.

Davidson, D. (1980), 'Causal Relations', in *Essays on Actions and Events* (Oxford University Press), 149–162.

Dummett, M. (1978), 'The Reality of the Past', in *Truth and Other Enigmas*, (Harvard University Press), 358–374.

—— (2000), 'Is Time a Continuum of Instants?', in *Philosophy* 75: 489–515.

Gallois, A. (1997), 'Can an Anti-Realist Live with the Past?', in *Australasian Journal of Philosophy* 75: 288–303.

Hawley, K. (2001), *How Things Persist* (Oxford University Press).

Hinchliff, M. (1996), 'The Puzzle of Change', in J. E. Tomberlin (ed.), *Philosophical Perspectives 10: Metaphysics*, (Blackwell), 119–136.

—— (2000), 'A Defense of Presentism in a Relativistic Setting', in *Philosophy of Science* 67 (Proceedings): 575–586.

Kaplan, D. (1989), 'Demonstratives', in J. Almog, J. Perry, and H. Wettstein (eds.), *Themes From Kaplan* (Oxford University Press), 461–563.

Katz, B. D. (1983), 'Perils of an Uneventful World', in *Philosophia* 18: 1–12.

Keller, S. (2004), 'Presentism and Truthmaking', in D. Zimmerman (ed.), *Oxford Studies in Metaphysics: Volume 1* (Oxford University Press), 83–104.

Lewis, D. (1992), 'Critical Notice: Armstrong, D. M., *A Combinatorial Theory of Possibility*', in *Australasian Journal of Philosophy* 70: 211–224.

Ludlow, P. (1999), *Semantics, Tense and Time* (MIT Press).

Markosian, N. (2004), 'A Defence of Presentism', in D. Zimmerman (ed.), *Oxford Studies in Metaphysics: Volume 1* (Oxford University Press), 47–82.

Meyer, U. (2005), 'The Presentist's Dilemma', in *Philosophical Studies* 122: 213–225.

—— (2006), 'Worlds and Times', in *Notre Dame Journal of Formal Logic* 47: 25–37.

Prior, A. N. (1962), *Changes in Events and Changes in Things* (University of Kansas).

—— (1970), 'The Notion of the Present', in *Studium Generale* 23: 245–248.

Quine, W. V. O. (1950), 'Identity, Ostension and Hypostasis', in *Journal of Philosophy* 47: 621–632.

Sider, T. (1999), 'Presentism and Ontological Commitment', in *The Journal of Philosophy* 96: 325–347.

—— (2001), *Four-Dimensionalism: An Ontology of Persistence and Time*, (Clarendon Press).

Smart, J. J. C. (1980), 'Time and Becoming', in P. van Inwagen (ed.), *Time and Cause: Essays Presented to Richard Taylor* (D. Reidel), 3–15.

Weiss, B. (1996), 'Anti-Realism, Truth-Value Links and Tensed Truth Predicates', in *Mind* 105: 577–602.

Wright, C. (1993), *Realism, Meaning and Truth* (Blackwell).

Zimmerman, D. (2008), 'Temporary Intrinsics and Presentism', in P. van Inwagen and D. Zimmerman (eds), *Metaphysics: The Big Questions* (Blackwell), 269–281.

# CHAPTER 6

# THE POSSIBILITY
# OF DISCRETE TIME

JEAN PAUL VAN BENDEGEM

## ABSTRACT

THIS chapter examines the possibility of discrete time. At first sight, the answer seems trivial, but actually it raises a number of interesting questions, both philosophical and scientific. First, I explain what interpretations of discrete time I will not deal with here. Then, two key philosophical problems are addressed: if there are such things as chronons, smallest 'bits' of time, do they have extensions and can a distance function, that is, a (kind of) duration, be defined on them? Second, the relation between discrete time and discrete space is briefly discussed, showing that the former implies the latter. Thus, if we have applications in mind, both time and space are to be seen as discrete. This leads, third, to the hardest problem of all: is discrete time applicable in physical theories? While fully aware of the host of problems needing answers, I will defend a positive answer.

## 1. INTRODUCTION

The problem of whether time could be meaningfully thought of as being discrete in some sense has quite a lot in common with Zeno's paradoxes. I am not referring here to the well-known fact that one of the paradoxes actually deals with the discreteness of space and time (see further discussion in section 2.3), but to the observation that for many philosophers and mathematicians, the paradox of the runner, to take the best known example, has a trivial solution. Although the runner has to cover an infinite number of distances, each distance shortens by half, and today we know that the summation of an infinite number of terms can produce a finite answer, for example in the specific case where the terms form a geometric series. End of story, so it seems. However for perhaps not that many (remaining) philosophers and mathematicians, the problem is not solved at all by this answer. What is still left open, is the question how I am to 'picture' what is happening: how does a runner achieve the formidable task of crossing an infinite number of distances in a finite time, no matter how small these distances become? The first position in the debate is satisfied with a formal or syntactical solution, whereas the second position needs something more, that could be labelled a semantic or an interpretative solution. The possibility of discrete time generates a similar distinction.

When we discuss the possibility of discrete time, we can be satisfied with a syntactical solution and then the problem is nearly trivial. Of course, discrete time must be possible: take any set $T$ of time units $t$, call them *chronons* if you like, and impose a discrete order, $<$, on $T$, such that for every $t$, there is a $t'$, such that, if $t < t'$ then there is no $t''$, such that $t < t'' < t'$. In the (formal) language of first-order predicate logic, this translates neatly into: $(\forall t)(\exists t')(t < t' \supset \sim (\exists t'')(t < t'' < t'))$. Once again, end of story.[1] Unless one demands in a Zeno-like fashion to know what chronons are, how they can 'fit' together to constitute something like 'time' or even suggest such a thing as 'the flow of time'. These questions are not answered by the formal proposal on its own. The underlying assumption in this essay will indeed be that semantical or interpretative answers are needed or, to inverse this statement, if there were to be a fundamental problem with such a semantic account, then this would constitute a severe critique, perhaps leading up to the rejection of the possibility of discrete time. Therefore such problems as whether or not chronons have an extension, in a sense to be specified, will be at the heart of the discussion.

A separate question that will be discussed is not about the possibility of discrete time, but about is applicability. After all, suppose it turns out that discrete time is indeed possible, in the sense that some nice formalism and semantical interpretation can be formulated and justified, but that it cannot be implemented in our descriptions of the world, then surely little has been gained. To narrow down this broad question, the most obvious and important place to investigate is physics, though not excluding, for example, psychology, where the discreteness of time is an important issue (see further down in this section). Nearly all physical theories use the real number line as a model for time, usually quite simply indicated by a parameter $t$ running over the set **R** of reals. This implies that time is not merely dense, that is, that between every pair of different times, there is a third one, but also continuous, that is, informally speaking, that if a sequence of time points tends to a limit, then that limit exists (and is hence a time point as well). It seems hard to imagine two structures more

---

[1]  Although some might remark that a number of questions remain unanswered. Is this order supposed to be total, i.e. for every pair of time units, $t$ and $t'$, does the statement '$t < t'$ or $t = t'$ or $t' < t''$ hold? Need it be transitive? Can it contain loops? Should it be linear? Is it necessarily asymmetric, as suggested here? In contradistinction with the runner paradox, it is not a mere matter of applying a particular mathematical technique.

radically different than discrete and continuous time in this sense. Thus the question is legitimate whether physics can 'survive' such a replacement. A related matter has to do with the discreteness or continuity of space: are these topics separated—can time be discrete without any implications as to the structure of space?—or are they in some way, perhaps necessarily so coupled? If the latter, then unavoidably, all that is being written here is under the assumption that the corresponding space problem has been adequately solved.[2]

A last topic to be briefly mentioned in this introduction is what the paper will not be about. There are too many questions and problems evoked by the matter of the possibility of discrete time, so justified choices have to be made. This chapter will not deal with:

- The existing techniques to transform equations of whatever sort involving continuous time into similar equations involve discrete time. The most typical example is the transformation of differential equations, involving derivates and integrals, into difference equations, as is done in numerical analysis.[3] As an example, take the exponential equation, that is, $dy/dt = y$, or, in words, the change in the variable $y$ in time is equal to that variable itself. The solution is the exponential function $e^t$ (ignoring some constants). If $t$ is replaced by discrete time, then we have a minimum step $t_{i+1}-t_i$, so that the equation becomes $(y_{i+1}-y_i)/(t_{i+1}-t_i) = y_i$. If, for simplicity's sake, the minimum step equals 1, then $y_{i+1} = 2.y_i$, so that the solution is now $y_i = 2^{t_i}$, again an exponential equation.[4] It is clear that we are talking here not so much about discrete time, but about a *sampling* problem: if a finite selection of time points is made, how do computations and calculations on this discrete sample relate to the original continuous problem? Whereas here we are interested in the possibility of discrete time *on its own*.
- The problem whether time is infinite or not 'in the large'. Or, if you like, the question whether time has a beginning and/or an end. Discreteness is in this sense a local property: given a time point, what can I say about neighbouring time points? And as a derived question: is it possible to patch all these local structures together into a global structure? Put differently, both discrete and continuous time can have a beginning and an end (and suffer from the same philosophical problems),[5] thereby implying that the two questions are independent of one another.
- The geological, biological, psychological, sociological, and historical themes related to the possibility of discrete time. The biologist might explain the rich

---

[2]  I will not discuss this matter here any further, but refer the reader to Van Bendegem (2002a).

[3]  In the well-known *Mathematical Subject Classification of 2000* (http://www.ams.org/msc/classification.pdf), this constitutes a separate heading, 65-XX, including, e.g. 65D30, Numerical integration, 65L07, Numerical investigation of stability of solutions, and 65T50, Discrete and fast Fourier transforms.

[4]  This is a very rough approximation only meant as an illustration. Better results can be obtained if the time-interval is not set equal to 1.

[5]  Such as the Kantian-flavoured paradox: if there is a first instant, what was before that (if that question makes sense) and, if not, how did we get here today?

variety of internal biological clocks in our body that all seem to have finite lower and upper thresholds, thus showing that in the biological sense, time is indeed discrete. But then the psychologists might wonder, if biologically time is discrete, how we are to explain our direct experience of continuity as we perceive the world or, in contrast, why we experience, for example, thoughts as discrete. The sociologist might note that, although the exact sciences, such as physics, use continuous time, nevertheless in everyday social life discrete time is what is most often used, with the exception of professional or Olympic sports, down to the level of seconds, and more often minutes. The historian might attract our attention to the fact that continuous time is historically speaking a very recent phenomenon and that discrete time fits in quite well with Christian thinking, where the finitude of man and world is to be contrasted with the infinite capacities and capabilities of God.[6]

All interesting themes, topics, problems, and conundrums, no doubt, but let me start first of all with a discussion of the semantical questions concerning discrete time.

# 2. THE SEMANTICAL PROBLEMS OF DISCRETE TIME

## 2.1 The extension problem

Let $T$ be a non-empty time interval. $T$ has an extension, call it $Ext(T)$. The first thing to note is that $Ext(T) \neq 0$, no matter what mathematical measure is proposed. (The intuition here is that time has duration and that duration presupposes an extension.) The most important property of an extension is that it can be divided; thus, if the time interval is split into two, say $T_1$ and $T_2$, then (a) both $T_1$ and $T_2$ have thereby an extension, thus $Ext(T_1)$ and $Ext(T_2)$ exist, and (b) both are non-zero, $Ext(T_1) \neq 0$ and $Ext(T_2) \neq 0$. If time is discrete, then this process of division can only be executed a finite number of times. Thus there must be a moment where we reach the chronons, call them $c_1, c_2, \ldots, c_n$ (as they need not be equal, if chronon equality makes sense). By the above reasoning, they will all have a non-zero extension, $Ext(c_i) \neq 0$, for all $i$. But, if they have a non-zero extension, the process can be repeated and therefore they are not chronons. The suggestion that, for chronons, the extension is indeed 0, raises the problem of how an addition of a finite number of zero extensions can lead to a non-zero extension for an interval. Apart from the Zenonian echo, it is clear that this is a fundamental problem for any proposal of discrete time (or, by extension, for any discrete concept that has extension as a property).

---

[6] Either a nearly infinite list of references should be produced here, or an easier solution suggests itself. Check out the website of the *International Society for the Study of Time*, founded in 1966 by J. T. Fraser (http://www.studyoftime.org/), where all references can be found to their publication series, *The Study of Time*, now up to volume XII. The focus of *ISST* is precisely on the interdisciplinary study of time.

However, an answer is possible. It suffices to look at the underlying principles used in the above argument. It is then sufficient to deny one of these principles. A very good candidate is that idea that, if $A$ has a certain property $P$, and $B$ is (in some sense) a (proper) part of $A$, then $B$ has the property $P$. A very strong argument to support the rejection of this principle is the observation that there are plenty of examples to be found in favour of the above idea. The classic cases are the wetness of water—water is wet, but an individual water- or $H_2O$-molecule is not—the colour of an object, and so on. A trivial formal example is, of course, to take for $P$, the property of being identical to $A$. As only $A$ is identical to itself, all other $B$s, including parts of $A$, cannot be identical to $A$. So there is at least one formal property that breaks the rule. Take now for $A$ any set of chronons, for $B$ a chronon $c$ of that set, for $P$ the property to have an extension, and for the part-relation simply membership, then it is perfectly compatible to claim that $Ext(A) \neq 0$, $c \in A$ and $Ext(c) = 0$. In other words, the property to have an extension is a *threshold* property: individual chronons do not have the property, but any aggregation of chronons does. Of course, one might justifiedly remark that it is highly artificial to claim that, for example, $c_1$ and $c_2$ have no extension, but the set $\{c_1, c_2\}$ does. Here too an answer is possible, though perhaps less attractive.

When we are talking about concepts or properties such as 'to have an extension', 'to be divisible in parts', 'to add up', ..., it is plausible to assume that not all elements of a given set need have all these properties. Let us make a comparison with heaps. The following statements will surely sound reasonable to everyone and, thereby, mutually compatible:

(a)   A heap $H$ can be split up in two parts $H_1$ and $H_2$,
(b)   There will be cases where the two parts $H_1$ and $H_2$ are themselves again heaps,
(c)   There will also be cases where at least one of the two parts $H_1$ and $H_2$ is itself not a heap.

We easily see examples to illustrate cases (b) and (c). If a sufficiently large heap is split up in two almost equal parts, then these parts will be heaps as well, but if I take away one grain of sand from a heap, then I will be left with a heap and a grain of sand, both parts of the original heap. So, as far as plausibility is concerned, there is a good case to be made. However, it is clear what the real challenge here is, and why this solution is not really attractive. Talking about heaps (it could also have been other predicates such as 'having a beard', 'being small', ...) introduces a philosophically speaking daring topic: *vagueness*.

The core of the problem about vagueness is neatly summarized in the Wang paradox:

Given the premises (where $Heap(n)$ means that $n$ grains form a heap):

(a)   $Heap(n)$,
(b)   For all $n > 0$, if $Heap(n)$ then $Heap(n-1)$,

the conclusion inevitably follows that

(c)   $Heap(0)$.

Under the reasonable assumption that the first premise is acceptable and the conclusion not, it follows that only the second premise can be attacked.[7] However, the problem is that the negation of (b) introduces a curious problem:

not-(b): There is a $n > 0$, such that $Heap(n)$ and not $Heap(n-1)$.

That is to say, removing one grain transforms miraculously a heap into something else. The conclusion must be that classical logic is not the ideal candidate to find a solution (unless, of course, one opts for the view that vague concepts need to be eliminated for precisely this reason). However, alternatives do exist, ranging from the use of fuzzy logic, paraconsistent logic, supervaluations, contextualized logics, and so on.[8] However, at present none of these are really satisfactory, hence the claim that it is a daring topic. In other words, it need not be a major advantage to consider 'being a chronon' as a vague predicate. The other side of the coin is that, as far as we are successful, it will make sense to say that chronons and (perhaps) small sets of chronons have no extension, whereas sufficiently large sets of chronons do.

For the sequel, we will consider the extension problem as in principle solvable, so that we can deal with the next problem, namely, how to measure time.

## 2.2 A distance function for discrete time

Suppose for simplicity's sake that we can indeed model discrete time as a set $T$ of chronons $c_1, c_2, \ldots, c_n$, together with a linear order, $<$, that obeys the basic properties:

(a)   Irreflexivity: for all $c_i$, it is not the case that $c_i < c_i$,
(b)   Antisymmetry: for all pairs $c_i$ and $c_j$, $i \neq j$, if $c_i < c_j$ then it is not the case that $c_j < c_i$,
(c)   Transitivity: for all triples $c_i, c_j, c_k$, if $c_i < c_j$ and $c_j < c_k$, then $c_i < c_k$.
(d)   Linearity: for all pairs $c_i$ and $c_j$, $c_i < c_j$ or $c_i = c_j$ or $c_j < c_i$.

Expressing discreteness requires an additional, definable predicate, namely, the immediate neighbour of a chronon:

$N(c_i, c_j)$ iff there is no $c$, such that either $c_i < c < c_j$ or $c_j < c < c_i$.

Note that it follows that immediate neighbours are unique, that is, if $N(c_i, c_j)$ and $N(c_i, c_k)$, then $c_j = c_k$. This is easy to see: if $c_j \neq c_k$, then according to linearity either $c_j < c_k$ or $c_k < c_j$. In either case one of the chronons ceases to be the immediate neighbour. The same predicate also allows for the definition of an interval $I$. An interval $I$ is a subset $T'$ of $T$ such that for every $c_i$ in $I$, there is a $c_j$ such that $N(c_i, c_j)$.

---

[7]   Assuming, of course, that we accept the repeated application of modus ponens (If A and 'If A, then B', then B). Note that, in this version, mathematical induction is not strictly required since the first premise will refer to a specific finite number and so modus ponens only needs to be applied a finite number of times to reach the conclusion.

[8]   An excellent place to start is Sorensen (2006).

From these definitions it follows that any interval $I$ can only contain a finite number of chronons. Sketch of the proof: suppose it is not finite. Start with the leftmost chronon in the order of $I$, say $c_1$. $c_1$ has an immediate neighbour say, $c_2$. Between $c_1$ and $c_2$ no chronons are to be found, so we are now left with an interval $I'$ starting with $c_2$ as lowest element and still counting an infinite number of chronons. Continuing this process, we get an infinite row of intervals such that between the lowest and the highest element in the interval, there is still a chronon to be found, contradicting the discreteness.

Equipped with these facts, we can plausibly propose the following distance function for any two chronons $c_i$ and $c_j$. First consider the interval $I$ that has $c_i$ and $c_j$ as the boundary elements. Then:

$d(c_i, c_j) = \#(I) - 1$, where $\#(I)$ stands for the number of elements in $I$ or its cardinality.

(Actually, there is a hidden assumption here: it remains to be shown that the interval that has two specific chronons as its boundary elements is *unique*. Again we will see that rather sophisticated proof procedures are required to prove this fact. Although, of course, intuitively speaking, the matter seems clear: if there were two intervals $I$ and $I'$, then, moving from $c_i$ to $c_j$ and, using the idea that neighbours are unique, the result will follow.)

It is easy to see that $d$ is indeed a distance function:

(a)   $d(c_i, c_i) = 0$, for all $c_i$,
(b)   $d(c_i, c_j) = d(c_j, c_i)$, for all pairs $c_i$ and $c_j$,
(c)   $d(c_i, c_j) + d(c_j, c_k) = d(c_i, c_k)$, for all triples $c_i$, $c_j$ and $c_k$.

To illustrate property (c), suppose that $\#([c_i, c_j]) = n$ and $\#([c_j, c_k]) = m$, then it follows, first, that $\#([c_i, c_k]) = n + m - 1$, as $c_j$ is counted twice, and, secondly, that $d(c_i, c_j) = n-1, d(c_i, c_k) = m-1, d(c_j, c_k) = n + m - 2$, so that, indeed, $d(c_i, c_j) + d(c_j, c_k) = d(c_j, c_k)$.

Most authors seem to agree that the existence of this type of distance function is to be counted as a very strong argument in favour of discrete time. Usually the distinction is made between an intrinsic and an extrinsic measure. 'Intrinsic' here means that the distance function is somehow related to the elements that are being measured. Obviously the function proposed here is intrinsic, since we actually count the chronons in the interval. An extrinsic measure is added to the things being measured. Take the continuous case: it does not make sense to count the number of time points in an interval, since any connected interval on the real line still counts a continuum of elements, so they would all have the same distance. Note however that it does not imply that the distance function is *completely* determined by the chronons. Any function $f$, such that $d(c_i, c_j) = f(\#(I))$, satisfying the properties (a), (b) and (c), will do. One might remark that there are not that many choices for $f$, as soon as one takes into account the immediate neighbour relation. Is it plausible to suppose that $d(c_i, c_j) = 1$ if $N(c_i, c_j)$? Plausible yes, necessary no. An additional assumption of homogeneity is required in the sense that chronons are interchangeable. It is therefore a better strategy to claim that intrinsic measures are co-determined by the number of chronons making up the interval one wants to measure.

It is also worth noting that, assuming that time is modelled as a one-dimensional structure, then the famous Weyl objection does not apply (see Weyl 1949: 43). Suppose one has a two-dimensional plane and suppose one wants to define a distance function that corresponds to the classical Euclidean distance function, viz., given two points $p_1(x_1, y_1)$ and $p_2(x_2, y_2)$, then the distance between $p_1$ and $p_2$ is equal to $d(p_1, p_2) = \sqrt{(x_1 - x_2)^2 + (y_1 - y_2)^2}$. But imagine a right-angled triangle such that the two legs are parallel to the $x$- and $y$-axis and that the distance is co-determined by the number of elementary parts that form the leg. Then it is easy to see that the hypothenuse of the triangle must count the same number of parts as one of the legs. So it can never satisfy or even approximate the Euclidean distance function. In one dimension, this distance function reduces to the definition in the section above, but does not cause the same problem to occur, because rather than a triangle *in*equality, we now have a triangle equality.

This raises another interesting problem that needs to be addressed: what is the relation between discrete time and discrete space? If the former implies the latter, then we will have to deal with the Weyl problem for space. So let us have a closer look at this relation via a short detour leading us to Zeno.

## 2.3  Zeno's relevance

The aim in this section is not to fully explore Zeno's paradoxes, and especially not in a historical setting, as an attempt to try to understand what Zeno could have meant with his arguments. What we can and should do is look at the paradoxes from the perspective of the problems presently under consideration. Paradoxes such as the arrow and the race between Achilles and the tortoise do not really represent a problem in terms of the structure of time as they deal with the divisibility of space. True, there is the argument that an infinite number of tasks cannot be executed in a finite time, but that does not (necessarily) involve the discreteness or continuity of time itself. But the paradox of the stadium does seem to be the odd one out and raises a deep and important problem about time: what is its relation with space? Not that this was Zeno's intention (as far as we can reconstruct what it was he was trying to tell us) for his aim was to show that, if space and time are discrete, then the duration of a chronon must be equal to the duration of half a chronon. That is, according to most authors, a problem of relative velocity and is in principle solvable (see, for example, Faris 1996). What is interesting for our theme is that Zeno assumes that discrete time goes together with discrete space. The question quite simply is this: is this necessarily so? Or, in other words, given the continuity or discreteness of space on the one hand and time on the other hand, are there impossible combinations or not?

One thing is obvious: continuous space and continuous time go perfectly together as all our physical theories demonstrate. Most authors that reflect about the discreteness of space and time assume that, if you have the one, then you must have the other. But what about the mixed cases? Can we imagine continuous space together with discrete time, or discrete space together with continuous time?

An argument that carries little weight is to claim that space and time are sufficiently similar to be treated in the same way. As a matter of fact, they are clearly not. Just a few differences: time is usually thought as one-dimensional, space as (at least) three-dimensional (not counting the infinitesimal ones); space has no favoured directions, time seems to have a preferred direction (ignoring here the complex problem of its source); space can be thought of as something tangible, something that contains objects, time is rather imagined as a structure between objects.

A better argument is to see what happens in a mixed case. Let us consider continuous space and discrete time: is this possible? To simplify things somewhat, imagine that, from the continuous point of view, a chronon has a beginning, $t_b$, and an ending time, $t_e$. Imagine a mass $m$ moving along a straight line at a fixed speed $v$. Clearly, between $t_b$ and $t_e$ the mass cannot change position, for, if it did, we would have to indicate when such a change took place and, as a consequence, the chronon would have parts, which is impossible. It follows that the mass will show a 'jerky' motion, making jumps from one chronon to another and remaining in the same position 'during' chronons. That implies, even if the background is supposed to be continuous, nevertheless movements of masses correspond to a discrete geometry in the sense that 'during' one chronon, nothing can change. That position in space corresponds to a position that cannot be further analysed. In analogy with chronons, call such space particles *hodons*. In summary, discrete time does not exclude continuous space, but it reduces continuous space to something that is similar to a discrete space and in that sense, one may conclude that discrete time forces continuous space not so much to become discrete, but at least to appear so.

For completeness' sake: what about discrete space going together with continuous time? Imagine two hodons and a mass oscillating from one hodon to the other. Imagine time to be discrete such that at chronon $c_i$, the mass is at hodon 1 and the next chronon $c_{i+1}$ it is at hodon 2, the next chronon $c_{i+2}$ it is back again at hodon 1, and so on. It is clear that, from the continuous perspective, the 'duration' of the chronon is of no importance whatsoever. It can be as long or as short as one likes. Or, to use a metaphor, suppose one films discrete movement. Then no matter how fast the recording is replayed, there will still be a discrete movement (ignoring physical limitations of course).

In summary, we arrive at the following conclusion: let DS, CS, DT, and CT stand, respectively for discrete space, continuous space, discrete time, and continuous time, then one possibility (CS, DT) seems to reduce to (DS, DT), whereas the three other combinations (CS, CT), (DS, CT), and (DS, DT) are indeed possible and, more importantly, it follows that DT implies DS. This is undoubtedly an additional challenge for the defender of the possibility of discrete time: she will also have to show that discrete space is possible. Fortunately, some work has been done on this topic and the tentative answer at this moment is that indeed discrete space is a genuine possibility.[9]

Probably, at this stage, the critic might remark that it is high time to address the applicability problem. If both time and space have to be thought of as discrete, this really demands a complete rewriting of physics as we know it today. If this task proves to be impossible, what has been gained except for a demonstration that an imaginary structure, discrete time, can be coherently imagined and described?

# 3. THE APPLICABILITY OF PHYSICS PROBLEM

Imagine ourselves in the best of all possible worlds: we have a 'decent' model of discrete space and time and we wish to apply it. What needs to be shown? There are, at least, two answers possible to that question:

(a) Show that, for any classical physical theory $T$, there is an analogue rephrased in terms of discrete time and space. Thus, for example, take classical mechanics and show what the discrete version of this theory should look like.

(b) Show that, for any classical physical theory $T$, there is un underlying theory $T^*$, such that $T^*$ is a discrete theory and $T$ can be derived (in some sense) from $T^*$.

Perhaps this might seem a subtle distinction, but it is nevertheless important. In case (a), what is required is that, for any concept $C$ belonging to $T$, there is a discrete analogue $C^*$, whereas in case (b), $T^*$ can use a completely different set of concepts $C_*$, provided it can be shown that any concept $C$ can be derived from $C_*$.

## 3.1 Discrete analogical theories: the classical case

At first sight this task might seem truly impossible: how does one show that an existing physical theory $T$ has a discrete counterpart? As mentioned in the first section of this paper, it is not sufficient to remark that continuous physics can be treated as discrete, say for computational purposes, for such a treatment does not constitute an analogue or counterpart theory: it is a reduction of a given theory to deal with a specific problem or, phrased differently, the continuous theory remains present in the background. As it happens, although formidable, the task need not be seen as necessarily impossible. As an example I will look at classical mechanics.

The first thing to note is that a lot of work has already been done to find (as) complete (as possible) axiomatizations of existing physical theories (it was actually one of

---

[9]  See the already mentioned Van Bendegem (2002a).

Hilbert's 23 problems).[10] For classical mechanics pioneering work has been delivered in the early fifties and it still being referred to today. What is particularly interesting is that such axiomatizations look for the barest minimum of concepts, probably inspired by the logical idea to ensure independence of the axioms. If it turns out that only a few axioms are needed, it follows that a discrete version will be, hopefully, easier to formulate. McKinsey, Sugar, and Suppes (1953) have shown that a mere six (basic) concepts will do. A model $M$ for classical mechanics consists of six elements,

$$M = < P, T, m, s, g, f >,$$

where:

(a)   $P$ is the set of particles, supposed to be non-empty and finite,

(b)   $T$ is a set of time points, seen as an interval of the real line $\mathbf{R}$,

(c)   $m$ is a function of $P$ to an interval of the real line $\mathbf{R}$, so $m(p) = r$, i.e. $m$ gives the mass of particle $p$,

(d)   $s$ is a function that takes couples of particles and times to a triple of real numbers, so $s(p, t) = < r_1, r_2, r_3 >$, i.e. $s$ states where the particle $p$ is at time $t$,

(e)   $g$ is a function that takes couples of particles and times to a triple of real numbers, so $g(p, t) = < r_1, r_2, r_3 >$, i.e. $g$ is the external force exerted on $p$ at time $t$,

(f)   $f$ is a function that takes a triple, consisting of two particles and a time moment to a triple of real numbers, so $f(p, q, t) = < r_1, r_2, r_3 >$, i.e. $f$ is the internal force exerted by $p$ on $q$ at time $t$.

So far, our task seems to be easy: replace $\mathbf{R}$ by some finite analogue and we have a discrete model. However, there is at least one specific condition that needs to be satisfied, for the simple reason that otherwise Newton's basic law cannot be expressed: if $s(p, t)$ is defined, then $d^2s(p, t)/dt^2$, the second derivative, is also defined. We therefore need an analog of continuous derivatives to satisfy this condition. Of course, there is a straightforward way to achieve this. If time is discrete, we can enumerate the time moments consecutively: $t_0, t_1, t_2, \ldots, t_n, \ldots$ Instead of $dx/dt$ we now write $(x_{i+1} - x_i)/(t_{i+1} - t_i)$ or, if the 'distance' between two consecutive moments is a fixed constant, say $c$, then the derivate simply becomes $(x_{i+1} - x_i)/c$. Although this now seems to be a function of two arguments, it can be easily transformed into a function of one argument, as $x_{i+1} = x(t_{i+1}) = x(t_i + c)$. If we now define $y_i = x(t_i + c) - x(t_i)$, then $y_i$ is a function solely of $t_i$. Similar to the first derivative, using this function, the second order derivative can be defined as well. If $c$ is sufficiently small, then the discrete formulation will approximate the continuous derivative.[11]

---

[10]   As is well known, in 1900 in Paris, David Hilbert held his famous speech where he listed twenty-three problems that would occupy mathematicians for the twentieth century, including the question of the provability of consistency of arithmetic, the continuum hypothesis, and Riemann's hypothesis. The sixth problem, however, in all its simplicity asks for an axiomatization of all of physics.

[11]   I fully realize that this is a very bold statement and it is therefore to be understood with all necessary caution. The claim refers to all 'well-behaved' functions and not necessarily to exceptional cases, where the idea itself of an approximation is not clear at all. I am thinking, e.g., of functions such

One might wonder whether the constant $c$ is not acting like an infinitesimal. And if so, then does not the problem of all infinitesimal reasoning arise, namely, you cannot get rid of them as the following classical argument shows? Take for $x(t)$ the function $t^2$, then the derivative according to the definition given will be

$$((t + c)^2 - t^2)/c = 2t + c.$$

However $2t$ is the desired answer to have a good approximation. But putting $c = 0$ reminds one of the fundamental problem of infinitesimal reasoning: $c$ is supposed to be different and equal to 0, which is obviously contradictory.[12] After all, this was the very reason why infinitesimal reasoning was replaced in mathematics by reasoning in terms of limits. However, one might remark that non-standard analysis, developed by Abraham Robinson, does allow for infinitesimals without any problems (see, for example, Keisler (1976), for an excellent, accessible presentation). This statement being true without any discussion, it does not help us out as far as the possibility of discrete time is concerned. In Robinson's approach one starts with the set of reals $\mathbf{R}$ to which the infinitesimals are added to obtain an extended structure $\mathbf{R}^*$, where an infinitesimal $x$ satisfies the condition that for all $r \neq 0$ in $\mathbf{R}$, $x < r$. From the viewpoint of discreteness, this is a move in the opposite direction as we are adding elements to a structure that is already uncountable. On the positive side, it should be noted that we have reformulated the original problem we wanted to solve. Instead of formulating explicitly what a discrete time could look like, we are now led to the question whether a discrete analog of $\mathbf{R}$ is imaginable.

A possible answer is given in Van Bendegem (2002b). Suppose we have a theory $T_R$ of the real numbers $\mathbf{R}$, and we want to calculate the derivative of a given function, say $t^3$, to use another example. Performing the calculation generates a finite set of sentences $\mathbf{F} = \{F_1, F_2, F_3, F_4, F_5\}$

$F_1$: $((t + c)^3 - t^3)/c = ((t^3 + 3t^2c + 3tc^2 + c^3) - t^3)/c$
$F_2$: $((t^3 + 3t^2c + 3tc^2 + c^3) - t^3)/c = (3t^2c + 3tc^2 + c^3)/c$
$F_3$: $(3t^2c + 3tc^2 + c^3)/c = 3t^2 + 3tc + c^2$
$F_4$: $3t^2 + 3tc + c^2 = 3t^2 + (3t + c).c$
$F_5$: $((t + c)^3 - t^3)/c = 3t^2 + (3t + c).c.$

The variable $t$ is supposed to be a real variable and $c$ an infinitesimal. If, so to speak, we restrict our attention to the set $\mathbf{F}$, then we can always find a model $M$ for this set, such that a value can be chosen for $c$ that is not only distinct from $t$, but from all terms involving $t$. In short, since the set $\mathbf{F}$ is finite, such values can always be chosen.

as $x = \sin(1/t)$, defined over $\mathbf{R}$, that starts to oscillate infinitely fast as $t$ approaches 0. It remains to be seen, of course, how dramatic the situation would be, if it turns out that such functions have no discrete counterpart. On the other hand, Kac and Cheung (2002) give a detailed presentation of discrete calculus that is immediately applicable. Without going into details, let me just remark that they consider two kinds of derivatives: $(x(t + c) - x(t))/c$, the version we have been discussing here, and $(x(qt) - x(t)/(1 - q)t$.

[12] Or, as Bishop Berkeley expressed it so well in *The Analyst*: they behave like the ghosts of departed quantities.

The first objection to make is, of course, that this might well work for finite sets **F**, but not for the full language of $T_R$. This is quite true. The question is how damaging this observation is. Seen from a more abstract viewpoint, one can answer that the damage is (only) marginal. We never use the full language of any theory as such: we always look at finite fragments. So, if every finite fragment has a model, that is what we really need. In logical terms, what we are giving up is *compactness*. This property states that a(n infinite) set (formulated in first-order logic) has a model if and only if all finite subsets have a model. In this proposal every finite subset has a model, but the full set does not. From the metamathematical viewpoint, it is, one must admit, less clear what the loss of compactness entails.[13] The important conclusions to draw here are that, firstly, such a discrete counterpart for the reals is indeed possible, and, secondly, that a discrete counterpart for classical mechanics is at least imaginable. It remains of course to be shown that the same strategy can work for electromagnetism, thermodynamics, statistical dynamics, and relativity, both special[14] and general, to name the most important classical physical theories.

## 3.2  Discrete analogical theories: the quantum case

Quantum mechanics (QM) deserves to be treated separately. In contrast with the classical theories, within the development of QM, there have been on and off voices that suggested that it should be treated as a discrete theory, namely, that time and space should be seen and interpreted as discrete. An excellent survey for the period 1925–1936 is given by Kragh and Carazza (1994). It is extremely helpful to make a number of distinctions and to clear up some confusion:

- It is often stated that QM, as we know it today, is in fact a discrete theory, since all observables are quantized (in some sense or other). This is a serious misunderstanding as it is sufficient to look at any formal presentation of QM—Schrödinger's equations, Hilbert operator formalism, Dirac's bra-ket formalism, Heisenberg's matrix formulation, to name the most important ones—to see immediately that

---

[13]  One important example, relevant for this paper, is precisely the loss of the construction of $\mathbf{R}^*$ in non-standard analysis. Although here too, one might wonder whether this is a loss. It shows that discrete time with infinitesimals excludes the construction of $\mathbf{R}^*$. Thus it might count as an argument in favour of discrete time as an independent notion.

[14]  At least for special relativity, there is the early attempt of Ludwik Silberstein (1936). He reformulates special relativity in terms of difference equations (as one might expect), and manages to derive a discrete analog of the Taylor series. If one were to think that this is a rather straightforward, if not rather boring task, I will mention only the fact that at a certain point in deriving the basic equations that link space-time events in one frame of reference to another, a case of Pell's equation—$x^2 - ny^2 = 1$, where $n$ is a non-square number and solutions are sought within the integers—appears, something that is entirely lacking in the continuous approach (Silberstein, 1936, 31–33). He does remark however on page 41 that general relativity will not be that easy, as he does not see how to formulate a discrete counterpart for tensors.

they all presuppose some form or other of continuous mathematics. The basic case is that, given some particular physical problem, a variable can only take a finite set of distinct values, for example, the discrete spectrum of an atom, but these values can be selected from a continuous domain, such as **R**. At best one can speak of a mixture of discreteness and continuity. This often not recognized subtlety leads immediately to the second confusion.

- It is often assumed that the Heisenberg uncertainty relations prove that time and space are discrete. Again this is a misinterpretation. Given the relation $\Delta x \Delta p \geq \hbar/2$, where $\hbar$ stands for $h/2\pi$, $h$ being Planck's constant, it does not follow that $\Delta x$ and $\Delta p$ separately have to satisfy a lower limit. Either $x$, respectively $p$ can be determined as precisely as one wants; the implication will be that $p$, respectively $x$ will become less and less determined.

- When some authors speak about the discreteness of QM, they often mean this statement in a very restricted sense. As in any other physical theory, QM contains a number of constants: the velocity of light $c$, Planck's constant $h$, the fine-structure constant $a$, the mass $m_e$ of the electron, the charge $e$ of the electron, and so on (see, e.g. Davies 1982). These constants all have, quite unavoidably, discrete values. Meaningful relations are then sought between these constants, in such a way that these relations have explanatory value and thus (could) form the basis of a fundamental theory.[15] Some researchers have sought to define a chronon in these terms, the most well-known case being, of course, Planck time $t_p$. Its value is, approximately, $5.4 \times 10^{-44}$ s, and its formal definition is $\sqrt{\frac{\hbar G}{c^5}}$, where $G$ is the gravitational constant. Such arguments and considerations can support the case for a discrete time, but do not *prove* its possibility.

These observations make immediately clear that the problem for QM is just as acute as the problem was in the previous section for classical theories. To introduce time as a discrete variable, means to rewrite QM along the lines sketched above. The task does not become easier in any way, rather the opposite. To mention just one problem that did not pose itself in the case of classical mechanics: what we will need for a reformulation of QM in discrete terms is a discrete version of probability theory, as the outcomes of measurements in QM are typically probability statements, that is, the chance to find, for example, this or that particle at this or that location.

There is, however, one special feature that needs to be mentioned. Suggestions that the 'size' of chronons can be determined by the other constants, could lead to experimental verification of the discreteness of time (and space). That opens up a different discussion altogether. Whether time (and space) is (or are) discrete is then no longer a matter of what seems plausible, what seems philosophically defensible, or what is easier to calculate. Now it will be a consequence of observational results. The mere fact that such a scenario is imaginable indicates that what is often labelled as quantum

---

[15] The most famous example is of course Eddington (1948). The book actually has precisely that title and still is a source of controversy.

'numerology'—the magic of the universal constants, so to speak—may prove to be more interesting than thought previously. To present at least one illustration, a brief look at the work of G. I. Pokrowski may suffice. If there are chronons and they have a precise numerical value, say $\theta$, the same for all chronons, it follows that all time-periods $v = 1/\theta$ are multiples of this value, and thereby also all wavelengths, $\lambda = c/v = c.\theta$, so any wavelength can be expressed as $\lambda = n.c.\theta$, for some natural number $n$. Thus, any measurement of the spectrum of a given element should show that all the wavelengths turn out to be multiples of the value of a chronon. Note that this type of reasoning has had at least one important (and successful) precursor, namely, in the discussion about atomic weights. However, here, the issue is less clear-cut and not much came of it, according to Kragh and Carazza (1994: 445).

As we move to more fundamental theories, it is clear that the question of discrete time will only gain in importance, as often the very nature of what space and time are, is called in question. This is exactly what happens today in the search for the unification of QM and general relativity.

## 3.3  Underlying discrete theories

This section will necessarily be a rather sketchy presentation as we are now moving into, to a rather large extent, uncharted territory, namely, the reconciliation in whatever form of general relativity on the one hand and QM (or rather, quantum field theory) on the other hand. Or, as I should perhaps write, many have entered this area from quite different directions (and sometimes with different intentions), and it is extremely difficult at the present moment to piece together all these partial explorations to arrive at a somewhat more global view. There are, however, some common elements that return:

- It is often assumed that in such a unification proposal, time and space themselves will appear as derived concepts. In other words, they cease to be primitive concepts and therefore need to be defined on the basis of other, more fundamental concepts.
- The basic concepts are usually quite abstract, as might be expected since any intuitions about time and space have to be left behind. Networks, graphs, or cellular automata are the favourite models. The starting point then becomes a couple $G = < V, E >$, where $V$ is a set of 'points' (no further specification required) and $E$ is a set of 'edges', i.e. a non-empty subset of $V \times V$. Time appears in the basic form of a discrete clock ticking away.
- As we are in an abstract realm, there is a large freedom in defining, e.g. a distance function over $V$. There is no need to limit oneself to definitions that are restricted to the plane. Extreme example: nothing prevents to define a distance $d$ between two vertices $v_1$ and $v_2$ equal to 1 if there is an edge $e$ connecting $v_1$ and $v_2$. In a totally connected network all distances will therefore be equal to 1. It is

important to note that, no matter how the basic distances are determined, it is always possible to define a distance function that satisfies the usual requirements.[16]

- The most important step forward is now to show that the 'classical' concepts can somehow be derived from this abstract basis. Or, if you like, how continuity can be restored from discreteness. What is often proposed is some form of averaging. Rather than speak of individual transitions of this or that vertex or this or that edge, one looks at large sets and defines averages over that set, for example, the average distance between two vertices. This makes possible the transition from natural numbers to fractions, an important transition.[17] Note that the numbers so obtained, need not be interpreted in terms of the fundamental concepts, very much so as it is a wrong question to ask how it is biologically possible for an average family to have, say, 1.8 children.
- Finally, just as in the case of QM, numerology is often present.

I will not attempt to give a full description of all proposals, but refer the reader to the excellent survey paper, Meschini et al. (2005). Let me restrict myself to one example of the numerological kind and not dealt with in the paper just mentioned. In 1981 Parker-Rhodes published an intriguing book, entitled *Theory of Indistinguishables*. This theory uses one basic operation, namely, to make a distinction. You start out with two elements $a$ and $b$, such that $a = a$, $b = b$, and $a$ and $b$ can be distinguished, $a \neq b$ and $b \neq a$. Time, as already mentioned, is once again seen as a universal clock ticking away. With these elements, special sets can be constructed, so-called discriminatory closed sets (dcs). Given the discriminations, it is possible to define an operation $+$ on $a$ and $b$, such that $a + a = b + b = 0$ and $a + b = b + a = 1$. If $a$ and $b$ themselves are equated with 0 and 1, we end up with an operation that corresponds to addition modulo 2. What sets are closed under this operation? Three of them: $\{0\}$, $\{1\}$ and $\{0,1\}$. In the next step, consider all linear transformations between the three elements of this set, such that $\{0\}$, $\{1\}$ and $\{0,1\}$ are mapped on themselves. There are precisely $2^3 - 1 = 7$ such transformations. This brings the total number of dcss up to ten. Repeat the process and the next step produces $2^7 - 1 = 127$ dcss. Add to these the previous 10 and we find 137, a first number pointing in the direction of a meaningful physical interpretation, namely, the fine-structure constant $a$. The next step produces $2^{127} - 1 = (approx.)10^{39}$, a second meaningful number, related to Newton's gravitational constant $G$, and, given this particular construction, it can be shown that a next level cannot exist. This construction is

---

[16] Define $d(e_1, e_2)$ as the shortest path that connects $e_1$ and $e_2$. Then it is obvious that $d(e, e) = 0$ and that $d(e_1, e_2) = d(e_2, e_1)$. That leaves the triangle inequality: $d(e_1, e_2) + d(e_2, e_3) \geq d(e_1, e_3)$, which is obvious for, if the inequality were not the case, i.e., $d(e_1, e_2) + d(e_2, e_3) < d(e_1, e_3)$, then $d(e_1, e_3)$ is not the shortest route, since connecting the routes from $e_1$ to $e_2$ and from $e_2$ to $e_3$, produces a shorter distance.

[17] Suppose you have a set of distances, ranging from 1 to $n$. If all averages can be taken, including the possibility of taking $k$ times the same element, then it is clear that all fractions $a/b$ will be obtained, for $a$ and $b$ ranging between 1 and $n$, so if $n$ is sufficiently large, all fractions will appear. This is an important step towards a set of distances that is, at least locally, dense.

known as the 'combinatorial hierarchy'. A more detailed treatment can be found in Bastin and Kilmister (1995). It is obviously not an easy task to evaluate such proposals, but, unless all such ideas are to be dismissed as ridiculous, they do show that time, if anything, is best seen as discrete, which is after all, the basic issue of this chapter.

# 4. CONCLUSION

In 1936 Lindsay and Margenau wrote that, while discussing the work of Pokrowski, 'one is hardly justified at the present time in taking the chronon hypothesis too seriously' (77) and, on the next page, they added that 'if time itself is assumed discrete, this (continuous, my addition) background is lost and the whole question of the use of time in physical description must be examined anew' (78). Discussing ideas of discrete time and space, Penrose 2004 wrote that 'it would not altogether surprise me to find these notions playing some significant role in superseding conventional spacetime notions in the physics of the 21st century' (960). One cannot escape the impression that, during seventy years, not that much has changed. Of course on the positive side, it is important to note that, although in a rather weak voice and not overly enthusiastic, nevertheless physicists and philosophers remain interested in the possibility of discrete time (and, as they go hand in hand, discrete space). That in itself shows that a positive answer to the question whether discrete time is possible, is reasonable and defensible. It will remain to be seen whether or not the most fundamental physical theory of nature will indeed start with a discrete clock-time. But, independent of that development, it is clear that, first, discrete time need not be a trivial notion, as it generates at least some important philosophical problems concerning extensions and vagueness, and that, second, it is not excluded that it should have physical importance as well. All that being said, the focus in this paper was on the question of possibility. It requires another paper to show that it is a more interesting concept than continuous time, especially if one has certain qualms about the notion of infinity.

## REFERENCES

*Notes*: Perhaps it is somewhat astonishing, but apparently there is not a single book to be found entirely devoted to the possibility of discrete time. So, for some of the book references below, usually a chapter or some sections in a chapter are devoted to the issue.

Bastin, E. W. and Kilmister, C. W. (1995), *Combinatorial Physics* (Singapore: World Scientific).
Faris, J. A. (1996), *The Paradoxes of Zeno* (Aldershot: Avebury).
Davies, P. C. W. (1982), *The Accidental Universe* (Cambridge: Cambridge University Press).
Eddington, A. S. (1948), *Fundamental Theory* (Cambridge: Cambridge University Press).

Kac, V. and Pokman, C. (2001), *Quantum Calculus* (New York: Springer).

Keisler, H. Jerome (1976), *Elementary Calculus* (Boston: Prindle, Weber & Schmidt).

Kragh, Helge and Carazza, Bruno (1994), 'From Time Atoms to Space-Time Quantization: the Idea of Discrete Time, ca 1925–1936', *Studies in History and Philosophy of Modern Physics*, vol. 25, no. 3, 437–462.

Lindsay, R. B. and Margenau, H. (1936), *Foundations of Physics* (New York: Dover, 1957, original edition: 1936).

Meschini, Diego, Lehto, Markku and Piilonen, Johanna (2005), 'Geometry, Pregeometry and Beyond', *Studies in History and Philosophy of Modern Physics*, vol. 36, no. 3, 435–464.

Nerlich, Greaham (1994), *The Shape of Space* (Cambridge: Cambridge University Press).

Mckinsey, J. C. C., Sugar, A. C., and Suppes, Patrick (1953), 'Axiomatic Foundations of Classical Particle Mechanics', *Journal of Rational Mechanics and Analysis*, vol. 2, no. 2, 253–272.

Newton-Smith, W. H. (1980), *The Structure of Time* (London: RKP).

Parker-Rhodes, A. F. (1981), *Theory of Indistinguishables* (Dordrecht: Reidel).

Penrose, Roger (2004), *The Road to Reality. A Complete Guide to the Laws of the Universe* (London: Jonathan Cape).

Silberstein, Ludwik (1936), *Discrete Spacetime. A Course of Five Lectures delivered in the McLennan Laboratory* (Toronto: University of Toronto Press).

Sorensen, Roy (2006), 'Vagueness'. *The Stanford Encyclopedia of Philosophy* (Spring 2009 Edition), Edward N. Zalta (ed.), url = http://plato.stanford.edu/entries/vagueness/, The Metaphysics Research Lab at the Center for the Study of Language and Information, Stanford University, Stanford, CA.

Van Bendegem, Jean Paul (2009), 'Finitism in Geometry'. *The Stanford Encyclopedia of Philosophy* (Spring 2009 Edition), Edward N. Zalta (ed.), url = http://plato.stanford.edu/entries/geometry-finitism/, The Metaphysics Research Lab at the Center for the Study of Language and Information, Stanford University, Stanford, CA, 2002a.

—— (2002b), 'Inconsistencies in the History of Mathematics: The Case of Infinitesimals', in: Joke Meheus (ed.): *Inconsistency in Science* (Dordrecht: Kluwer Academic Publishers), 43–57 (*Origins: Studies in the Sources of Scientific Creativity, volume 2*).

Weyl, Hermann (1949), *Philosophy of Mathematics and Natural Sciences* (Princeton: Princeton University Press).

CHAPTER 7

·····································································································

# PRESENTISM AND THE SPACE-TIME MANIFOLD*

·····································································································

DEAN ZIMMERMAN

## 1. INTRODUCTION: PRESENTISM AMONG THE A-THEORIES

·····································································································

### 1.1 The A-theory of time and the B-theory of time

MᴄTᴀɢɢᴀʀᴛ gave the name "A-series" to "that series of positions which runs from the far past through the near past to the present, and then from the present through the near future to the far future, or conversely"; and the name "B-series" to "[t]he series of positions which runs from earlier to later, or conversely".[1] McTaggart's rather bland labels have stuck, and been put to further use. The "determinations" (his word), or properties, *being past*, *being present*, and *being future* are generally called the "A-properties". The relations of *being earlier than*, *being later than*, and *being simultaneous with*, are the "B-relations". These days, philosophers are said to hold an "A-theory of

* In my work on the topic of this chapter, I have incurred many debts; I feel sure I will fail to acknowledge some who helped me over the years, and I had better apologize in advance. My greatest debt is to Tim Maudlin, whose seminars, papers, and books have taught me whatever I know about relativity. In addition, Tim provided all kinds of useful advice and criticism, helping me to avoid many mistakes (though he was not, in the end, able to save me from the biggest mistake, namely, defending presentism!). I also owe a great deal to conversations or correspondence with: Frank Arntzenius, Yuri Balashov, Craig Callender, Robin Collins, William Lane Craig, Tom Crisp, Shamik Dasgupta, Kit Fine, John Hawthorne, Hud Hudson, Barry Loewer, Peter Ludlow, Ned Markosian, Bradley Monton, Oliver Pooley, Alan Rhoda, Jeff Russell, Steve Savitt, Ted Sider, Zoltan Szabo, and Timothy Williamson. I learned much from questions and conversations after talks at the University of Georgia, NYU, Western Washington University, Oxford, and Rutgers; and in the Rutgers philosophy of religion and metaphysics reading group.

[1] J. McT. E. McTaggart (1927: 10).

time" or a "B-theory of time", depending upon their attitudes to these properties and relations.

Some philosophers suppose that there are objective distinctions between what is present and what is past and what is future. In order to tell the full truth about time, they think, one must advert to the A-properties. Naturally enough, such philosophers are called "A-theorists". Although A-theorists disagree about many details, they agree that the present is distinguished from past and future in a deep and important way. Exactly how to describe this difference is a vexed question, and some philosophers have argued that would-be A-theorists inevitably fail to stake out a coherent position.[2] I shall not attempt a full-scale defense of the coherence of the A-theory here; but hopefully the following characterization will suffice to convey, in a rough-and-ready way, the nature of the A-theorists' convictions about time: The A-theorist grants that every thing in time (setting aside the possibility of a beginning or end of time itself) is "past relative to" some things, "future relative to" others, and "present relative to" itself—just as every place on Earth (setting aside the poles) is south relative to some places, north relative to others, and at the same latitude as itself. But the A-theorist insists that this attractive analogy between spatial and temporal dimensions is misleading, for, of any time or event that is past, present, or future in this merely relative way, one can also ask whether it is, in addition, past, present, or future in a *non*-relative way—past, present, or future *simpliciter*. The A-theorist takes the merely relative A-determinations to be based upon facts concerning which times and events are *really* past, present, or future, not merely relatively so. "B-theorists", by contrast, deny the objectivity of the division of time into past, present, and future; they regard the spatial north-south analogy as deeply revelatory of the purely relative nature of this division (though many B-theorists admit that there is some intrinsic difference between spatial and temporal distances). To arrive at more objective facts about time, one must turn to relations like *being earlier than*, *being later than*, and *being simultaneous with*—the "B-relations".[3]

The A-theory is almost certainly a minority view among contemporary philosophers with an opinion about the metaphysics of time.[4] Nevertheless, it has many

---

[2] Dolev (2007), for example, rejects the current A-theory–B-theory debate altogether; he advocates an anti-metaphysical approach to questions about the past, present, and future, hoping to bypass all the traditional metaphysical issues. For less radical forms of skepticism about the A-theory–B-theory distinction, see Lombard (1999, forthcoming), Williams (1998), and Callender and Weingard (2000).

[3] B-theorists who affirm that Relativity provides the deepest, most objective description of the relations between events will want to fiddle with McTaggart's B-relations: outmoded conceptions of *being earlier than* and *being simultaneous with* must give way to the more fundamental relations of spatio-temporal distance encoded in the metric of the manifolds of Special or General Relativity. Objective (i.e. not-merely-relative) temporal precedence relations remain, though they only hold between events, one of which could reach the other by a flash of light or slower means—and simultaneity is at best a relative affair.

[4] Although it seems that most philosophers who take a position on the matter are B-theorists, nevertheless, A-theorists have made up a significant proportion of the metaphysicians actually working on the A-theory–B-theory debate during the past ten or fifteen years. We A-theorists might be inclined to explain this as a case in which the balance of opinion among the experts diverges from that of the

defenders—Ian Hinckfuss, J. R. Lucas, E. J. Lowe, John Bigelow, Trenton Merricks, Ned Markosian, Thomas Crisp, Quentin Smith, Craig Bourne, Bradley Monton, Ross Cameron, William Lane Craig, Storrs McGall, Peter Ludlow, George Schlesinger, Robert M. Adams, Peter Forrest, and Nicholas Maxwell, to name a few.[5] Several of the most eminent twentieth-century philosophers were A-theorists, notably C. D. Broad, Arthur Prior, Peter Geach, and Roderick Chisholm.[6]

The B-theory can claim support from two of the founders of the analytic movement in philosophy: Gottlob Frege and Bertrand Russell.[7] In the years since, it has achieved broad acceptance. D. C. Williams, W. V. O. Quine, Adolf Grünbaum, J. J. C. Smart, David Lewis, D. H. Mellor, Paul Horwich, Tim Maudlin, Frank Arntzenius, Theodore Sider, Robin Le Poidevin, Nathan Oaklander, Steven Savitt, and Thomas Sattig are just the tip of the B-theorist iceberg.[8]

B-theorists have raised many kinds of objections to the A-theory and to the particular kind of A-theory I find most attractive, namely, *presentism*.[9] What follows is a defense of presentism in the face of just one of these: that the view has been refuted or at least badly undermined by discoveries in physics. The rejection of Newton's substantival space by the natural philosophers and physicists of the nineteenth century had already created an environment somewhat hostile to presentism, as shall emerge in my discussion of one of Theodore Sider's objections to presentism (the objection described near the end of Section 3, based on cross-temporal relations involving motion). Einstein's Special Theory of Relativity (SR) and General Theory of Relativity (GR) seem only to have made things worse. Both imply the "relativity of simultaneity"; and this raises obvious questions for all A-theorists. If, as A-theorists believe, there is

hoi polloi. There is an alternative explanation, however. I have the impression that there is a much larger proportion of incompatibilists (about free will and determinism) among those actually writing on free will than among philosophers more generally. A similar phenomenon may be at work in both cases: The B-theory and compatibilism are regarded as unproblematic, perhaps even obviously true, by a majority of philosophers; they seem hardly worth defending against the retrograde views of A-theorists and incompatibilists. Philosophers sympathetic to A-theories or incompatibilism, on the other hand, are more likely to be goaded into defending their views in print precisely because they feel their cherished doctrines are given short shrift by most philosophers.

[5] See Hinckfuss (1975), Lucas (1989), Lowe (1998: Ch. 4), Bigelow (1996), Merricks (1999), Markosian (2004), Crisp (2004, 2003), Smith (1993), Bourne (2006), Monton (2006), Cameron (2010), Craig (2000), McCall (1994), Ludlow (1999), Schlesinger 1980, 1994), Adams (1986), Forrest (2004, 2006), and Nicholas Maxwell (2006). See also Zimmerman (1996, 1997, 2006a), and Gale (1968) (though Gale has since repudiated the A-theory). Tooley (1997) sounds like an A-theorist, although I am not so sure that, in the end, he is one; for discussion of the question, see (Sider, 2001: 21–5).

[6] See Broad (1923) [an excerpt in which Broad defends an A-theory is reprinted in (van Inwagen and Zimmerman, 2008: 141–9)], Prior (1970, 1996, 2003), Chisholm (1990a, 1990b, 1981), and Geach (1972). Charles Hartshorne is another famous twentieth-century A-theorist (Hartshorne, 1967: 93–6).

[7] See Frege (1984: 370) and Russell (1938: Ch. 54).

[8] See Williams (1951), Quine (1960: §36), Grünbaum (1967: Ch. 1), Smart (1963: Ch. 7; and 1987), Lewis (1976, 1979, 2002, 2004), Mellor (1981a, 1981b, 1998), Horwich (1987), Sider (2001), Le Poidevin (1991), Oaklander (1991), Savitt (2000), and Sattig (2006).

[9] Sider (2001: Ch. 2) gives a litany of serious objections; for a useful strategy presentists might use to respond to certain kinds of objections, see (Sider, 1999).

an objective fact about what is presently happening, there must be an objective fact about which events are simultaneous with one another—in other words, a fact about simultaneity that is not relative to anything, including the frames of reference of SR, or the local frames of GR. But, on the face of it, these scientific theories require that simultaneity be frame-relative.

Before launching into description of the variety of A-theories, and defense of the kind I prefer, I should mention some things that will not be discussed in this chapter. The first is a philosophical debate that is sometimes conflated with the A-theory–B-theory dispute: namely, the question whether time has an intrinsic direction to it; that is, whether the relation *being earlier than* holds between moments or space-time locations independently of the contents of space-time. No one, to my knowledge, has defended the following combination of views: time lacks an intrinsic direction but includes objective distinctions between past, present, and future. The idea is not completely incoherent: one might attempt to reconstruct the difference between distance-into-the-past and distance-into-the-future on the basis of facts about what is present and "direction-neutral" temporal distance facts.[10] But the project will probably seem quixotic to most A-theorists (I know it does to me!). B-theorists, however, are sharply divided on the question whether time has an intrinsic direction. Two temporal directions are distinguishable from the spatial ones in the space-time of SR and in manifolds compatible with GR—at least, in the less bizarrely shaped ones. But some philosophers of physics deny that one direction is more deserving of the label "forward in time", when considered all by itself, in abstraction from space-time's material contents. These deniers of intrinsic direction claim that the actual difference between forward and backward in time does not supervene upon facts about intrinsic structure alone, but only upon such facts *plus* contingent facts about the contents of space-time, such as the distribution of matter. Both A-theorists and the friends of intrinsic temporal direction are often said to believe that "time flows", or to believe in "objective passage"; and their opponents may be said to champion a more "static" conception of time. But the two controversies are obviously quite different. Only the bona fide A-theory–B-theory debate shall figure in this chapter; for an up-to-date discussion of the question whether time has an intrinsic direction, see Huw Price, "The Flow of Time", in this volume.[11]

Many other important issues will be set aside, including the allegation that A-theorists and B-theorists are simply talking past one another. Some philosophers claim not to be able even to understand the A-theory–B-theory debate, and they cannily manage to extend this inability so that it includes all the terms that might be used to explain it to them.[12] Fortunately, a number of effective therapies have been developed

---

[10]  Sider (2005) discusses the prospects for carrying out such a project using binary, undirected tense operators.

[11]  See also Jill North's chapter in this volume for discussion of the closely related topic of "Time in Thermodynamics".

[12]  The strategy can be found in Lombard (1999, forthcoming), Williams (1998), Callender and Weingard (2000), Dorato (2006), and Savitt (2006)—though Savitt tries to salvage some kind of disagreement that could occur in the context of certain scientific theories about space-time.

for those who find themselves losing their grip on the nature of the disagreement between A-theorists and B-theorists, or between presentists and non-presentists. The best treatment for the condition is something I discuss elsewhere.[13]

Other issues to be set aside include the majority of the objections that have been raised against the A-theory in general and presentism in particular. Chief amongst these is the argument that truths about the past need "truthmakers" (where a truth-maker for a proposition is some part of the world in virtue of which the proposition is true), and that the presentist lacks the resources to provide them.[14] In the face of this challenge, some presentists cast about for present states of the world that will ground truths about the past;[15] while others question the legitimacy of the demand for truthmakers.[16] I shall take it for granted that presentists can appeal, unproblematically, to facts about what was the case at each instant in the past—at least, that purely *qualitative* facts about each past moment are well grounded. Even granting so much as that, presentism's critics can raise serious objections based on what physics seems to say about the nature of space-time.

The two problems upon which I focus are these: (1) Determinate facts about the state of the universe at each instant may fail to provide an adequate basis for certain cross-temporal facts that are physically important and objective. Theodore Sider has challenged the presentist to find a basis in reality for these physical facts; and I offer a couple of ways in which a presentist could meet the challenge. (2) The very idea of well-defined instants seems to be inconsistent with SR and GR, but crucial to any version of the A-theory, presentist or not. The most common objection along these lines, made by Putnam, Sider, and others, is that presentism is inconsistent with SR; and therefore false.

The chapter is a long one; so a little guidance may be helpful. I spend the remainder of this first section describing the kind of presentism I prefer, and contrasting it with other A-theories. It turns out that there is considerable pressure on presentists to accept the existence of certain things that, intuitively, are "in the past". (A plausible response to the pressure, introduced in section 1, will prove relevant to the question of section 3: what theory of the manifold should presentists adopt in the face of Theodore Sider's worries about cross-temporal states of motion?) Section 2 introduces SR and GR as theories about the structure of a four-dimensional manifold; I claim it is "safe" to assume substantivalism, and I focus on the Minkowskian manifold structure of SR. In section 3, I argue that a presentist who accepted SR should have to suppose that the present slices the Minkowskian manifold in a certain way, and that its past and future

---

[13]  See Zimmerman (2005, 2006a), and also Sider (1999, 2006), and Crisp (2004). See also Mozersky, "Presentism", in this volume; and Ludlow (2004).

[14]  See Keller (2004), Armstrong (2004: Ch. 2), Lewis (1999a), Sider (2001), Tooley (1997), and Mozersky, this volume.

[15]  Examples include Crisp (2007a), Bourne (2006), Cameron (2010), and Rhoda (forthcoming). Although I have some sympathy with the idea that the demand for truthmakers is illegitimate, I have also argued that defenders of truthmaker arguments must admit that Rhoda's proposed truthmakers are adequate to the task; see Zimmerman (2009).

[16]  For example, Merricks (2007: Ch. 6), and Kierland and Monton (2007).

locations would constitute a foliation of the manifold. I offer presentists two ways of thinking about the metaphysics of this ostensibly four-dimensional entity, including one that manages to reject "past" (i.e. formerly occupied) and "future" (i.e. soon to be occupied) space-time points. I show that the presentist who takes my approach to the manifold can deal with the kinds of fundamental cross-temporal relations needed by post-Newtonian theories of motion and gravitation. In section 4, I argue that the conflict with SR is not as deep as has been suggested. In particular, the oft-heard charge of "inconsistency" is not so straightforward as it is made to seem. Whatever disharmony remains between the A-theory and SR is of dubious significance. SR is of interest mainly as an approximation to GR, and it is even less clear whether presentism is inconsistent with GR. Furthermore, quantum theory may well call for radical changes in our conception of space-time; and some of the proposed changes promise to reinstate a way of slicing the manifold that would *have to* coincide with the A-theorist's division of the manifold into a series of successive presents. It is unclear whether these versions of quantum theory will win out over competitors that leave space-time looking more as Relativity sees it. But SR is false, and GR faces challenges from an even more impressively confirmed physical theory. These facts can hardly be irrelevant to the significance of arguments that assume their truth.

## 1.2  Principal varieties of A-theory: presentism and the growing block

A-theorists disagree among themselves about the exact nature of the distinction between past and present things and events, and also about the distinction between present and future things and events. Presentism is an extreme form of the A-theory, but perhaps also its most popular variant. Analogous to a doctrine in the metaphysics of modality called "actualism", presentism is the view that all of reality (with the possible exception of utterly atemporal things, if such there be) is confined to the present—that past and future things simply do not exist, and that all statements that seem to carry an existential commitment to past or future things are either false or susceptible of paraphrase into statements that avoid the implication.

Some other A-theorists, though not presentists themselves, are like the presentists in distinguishing themselves from B-theorists by the restrictions they place upon what exists. "Growing block" theorists, such as C. D. Broad, regard future events and things as non-existent, and present things as special in virtue of being the latest parts to have been added to a four-dimensional reality. According to the "growing blocker", to become past is to cease to be on the "cutting edge" of a growing four-dimensional manifold of events. For Broad, ceasing to be present, and becoming past, involves no intrinsic change whatsoever: "Nothing has happened to the present by becoming past except that fresh slices of existence have been added to the total history of the world."[17]

---

[17]  Broad (1923: 66).

Contemporary growing blockers, unlike Broad, tend to insist that events and things that become past do more than merely "recede" into the block. The reason they disagree with Broad can be brought out using a famous argument of Arthur Prior's. Prior (a presentist, himself) claimed that, if our past headaches were as painful as present ones, their merely becoming past would hardly be a matter for celebration.[18] Prior was wrong to think that this observation constitutes a knock-down argument for the A-theory.[19] Still, Prior's worry about the status of past thoughts can be turned into a cogent critique of Broad's version of the growing block theory, along the following lines.

A-theorists are trying to develop a metaphysical theory of time that validates a conviction shared by most people in most times and places: namely, that the change from being present to being past is a deep and important one. It would be strange to believe this, while professing utter ignorance about which things have undergone the supposedly radical change. According to Broad, however, any particular judgment I make about which events actually *are* present will be correct only briefly (for brief events, at any rate), and then forever wrong as the event of my judging them to be present recedes into the past, intrinsically unchanged. Equating the present with the edge of Broad's growing block is costly: it leads to absolute skepticism about what time it is![20]

Trenton Merricks points out that Broad could, perhaps, dodge this skeptical bullet by distinguishing between two notions of "being present": an objective one and a subjective one. To say that some event is *subjectively* present is just to say that it is simultaneous with one's location (at the time one makes the judgment) within the four-dimensional growing block; while to speak of the *objective* present is to speak of what is on the block's "cutting edge". "Being past" and "being future" would admit similar disambiguation into subjective meanings ("being earlier than one's location" and "being later than one's location") and objective, more "metaphysical" meanings ("being embedded within the block" and "not existing at all . . . yet!"). According to this modification of Broad's growing block theory, in ordinary thought and speech we use "A-determinations" subjectively, so that we remain forever by-and-large right about what time it is; however, when discussing the metaphysics of time, the same words are to be given the growing blocker's metaphysical interpretation. But Merricks has a challenge for a growing blocker who would make such a distinction: If all our ordinary judgments about past, present, and future are *subjective* ones, what has the growth of the block to do with *time*? The growing blocker's "objective present", "objective past", and "objective future" have become technical terms within an unmotivated metaphysical theory.[21]

---

[18]  Prior (1959, 1996).
[19]  For B-theoretic replies to Prior's argument, see Mellor (1981b), MacBeath (1983), and Hardin (1984).
[20]  See Braddon-Mitchell (2004) and Bourne (2002) for statements of this objection to the growing block.
[21]  Merricks (2006).

The canny growing blocker should part company from Broad, insisting instead that events and objects change radically when they cease to be present. An event is only *really happening* when it is on the cutting edge. Although the growing blocker admits that events continue to *exist* when they are past, she can maintain that they are only *doing something* (e.g. they are only in the process of bringing about other events) when they are present. Something similar should be said of the objects to which events happen; Bucephalus (Alexander the Great's horse) is only stamping his hooves, reflecting light, and rearing on his hind legs when he is *presently* doing these things. On the proposed version of the growing block view, past events and objects more closely resemble *merely possible* ones than presently occurring ones.[22]

Elsewhere in this volume, Barry Dainton calls this sort of metaphysics of time "Growing Block + Glowing Edge", or "Grow-Glow" for short. I prefer to call it a "ghostly growing block" metaphysics. Were I to defend this sort of theory, I should prefer to say that objects and events on the edge are not "aglow" with a strange light; rather, they are *normal*, simply existing or occurring in the fashion to which we are accustomed. It is the objects and events entirely in the past that are strange—that is, strangely intangible.

The ghostly growing blocker can do justice to the feeling that an important change has taken place when an event or object has become past, and she can plausibly explain how we know what time is present. Suppose that Newton once observed a shadow on a sundial and made the following judgment: "At present, it is exactly 3 p.m." Within a ghostly growing block, Newton, the sundial, and the events of observation and judgment have all undergone radical changes since that judgment was made. Newton has no shape or size or location; no brain with which to think or eyes to see. Since neither Newton nor any of his temporal parts (if such there be) can see or think, the events that are the observation and thought, whatever they are now like, can hardly be said still to be *occurring*. And if Newton has neither brain nor eyes, he can hardly be convicted of making an ongoing mistake about what time it is. The ghostliness strategy can be extended to all the interesting properties of events and objects; to be *truly* loud, tall, hungry, etc. is to be *presently* loud, tall, hungry, etc.

Although this view makes sense of our relief when pain is past, and of our knowledge of what time is present, it has less appealing consequences as well. Consider first the objects that linger on, after having undergone processes that would ordinarily be said to have destroyed them utterly. The ghostly growing blocker must say that a horse can exist although it is not actually alive or even spatially located; a hand-grenade can exist, though it has been blown to bits. Indeed, every particle in an object could have been converted into energy within the sun, and the object (and the particles, for that matter) would still exist—though it would then lack all spatial location, and be in some sense "outside of space". A physical object that ceases to have any shape or size or location

[22] As I read Adams (1986), he holds a version of this view; Forrest (2004) develops a related view, according to which the past changes in such a way that consciousness disappears. For more discussion, see Zimmerman (2008: 212–216).

at all is extremely "thin", insubstantial. Recoiling from this result is what drives many of us toward presentism. Consider, secondly, the events that must be supposed to exist after they have ceased to occur, such as a game of horseshoes or an explosion. Now, on one conception of events, their continued existence would not be too surprising. Some "event-talk" seems naturally to be construed as reference to something like Chisholm's "states of affairs", events of a kind that can *exist* though they do not occur—and, in fact, may *never* occur.[23] My playing a game of horseshoes with Bob Dylan is one of the things I hope will happen, but which probably will not occur. If the subject term of the previous sentence—"My playing a game of horseshoes . . ."—refers to anything, it refers to a kind of thing that may or may not occur; and that, even if it does occur, could have existed without occurring. Names for events that are constructed by nominalizing sentences (e.g. "my playing a game of horseshoes with Dylan", "The exploding of the hand grenade") seem more susceptible to this construal than names for events that do not have a "verb alive and kicking" inside them (e.g. "the game of horseshoes between me and Dylan", "the explosion of the hand-grenade").[24] Most of our talk about events does not go by way of nominalizing sentences, and is harder to construe as referring to things that could exist without occurring. The game of horseshoes I played with my uncle, the explosion of the hand-grenade in the Knesset in 1957 . . . it is harder to imagine that these would have existed whether or not we had thrown any horseshoes, and whether or not the grenade had malfunctioned.[25] If there are such things as events or states that could not exist without occurring—particular games, explosions, headaches, and kickoffs that would not have existed if no horseshoes were thrown, nothing ever exploded, no one felt pain, and nothing was kicked—then the presentist is likely to doubt whether such things can continue to exist after they have ceased to occur, but the ghostly growing blocker must suppose that they never go away.[26]

Similar morals apply to a third, less popular style of A-theory, sometimes called the "moving spotlight" view. According to moving spotlighters, reality has always consisted of everything that will ever have existed—in other words, spotlighters resemble B-theorists in their acceptance of "eternalism" about what there is; existence claims are eternally true if they are ever true. Here is a bad version of the spotlight theory, susceptible to the same objections that were lodged against Broad's growing block:[27]

[23] Chisholm (1970, and 1976: Ch. 4).

[24] Bennett (1988) provides a subtle analysis of the systematic differences in these two ways of talking about events.

[25] I should not want to put too much weight on our initial inclination to disbelieve in explosions and games that do not occur, given that contexts can be created in which it seems fine to talk about such things; for example: "In 2005, a grenade was thrown at George Bush in Tbilisi, Georgia; the explosion of the grenade was prevented by a malfunction of the firing mechanism." Perhaps the use of certain kinds of names for events can create a tacit restriction of the domain of quantification to *events that occur.*

[26] Heathwood (2005) makes the point that, if the principle reason for being a growing blocker (rather than a presentist) is to have better truthmakers for statements about the past, this sort of growing blocker is in trouble.

[27] Although this sort of A-theory may never have been held by any historical individual, it is in fact the version of the A-theory McTaggart attacks in his argument against the existence of time; and

All events and objects are spread out in a great four-dimensional block, and the *only* changes that happen to them are changes in their A-properties as they go from being far future to near future, near future to present, present to near past, near past to distant past, etc. Unless some other changes accompany the change from being future to being present, and from being present to being past, there could be no way for us to know what time it is. Nothing in our brains or minds could possibly be sensitive to facts about what is present, if the "presentness" of events behaves like a mere spotlight (one might as well say "shadow"), passing over the block without affecting the things it strikes. The natural way out for the spotlighter is to utilize the ghostliness strategy I urged upon growing blockers: deny that merely future and merely past events are really happening; and strip merely future and merely past individuals of all their interesting, manifest properties.

As it happens, few A-theorists are spotlighters; most of us want to say that the future is "open" in a way an eternalist must deny.[28] So I shall ignore the view, and concentrate on presentism and (the ghostly version of) the growing block.

## 1.3  Pressures to accept some past things

Growing blockers may be saddled with an implausibly ghostly afterlife for everything that ever exists. But presentism's utterly empty past creates problems of its own—for there are many cross-temporal relations for which the presentist seems to lack adequate grounds, and even simple cross-temporal truisms like "England had three kings named 'Charles'" prove difficult for presentists to interpret.[29] Solving these problems may well push the presentist at least some distance toward acceptance of a ghostly growing block. In this section, I describe a couple of stock puzzles about cross-temporal relations, in order to show how the pressures arise.

The A-theory, in every version, carries with it a commitment to *tense logic*. The fundamental truth-bearers (for which I shall use the term "proposition" as a place-holder, leaving their nature up-for-grabs) must be susceptible of change in truth-value. Objective facts about what is present, past, and future require propositions that are flat-out *true* (not merely true relative to one time, but false relative to others) but that will not always be true and that have not always been true.

---

McTaggart has an (unconvincing) argument to the effect that it is the only potentially sustainable version of the A-theory (McTaggart, 1927: 12–18).

[28]  There is a form of the eternalist growing spotlight view that can leave the future radically open. Eternalism requires the existence, already, of every future event and individual that will ever occur or exist, thereby threatening to "close off" other future possibilities. But if every possible future event and individual already exists, though many of them will never occur or be made concrete (as in, for example, Timothy Williamson's densely populated ontology (Williamson, 1999)); then the bare existence of the future history that will, in fact, have occurred could not be thought by anyone to raise the specter of fatalism.

[29]  Lewis (2004).

Typically, the presentist's fundamental machinery for talking about the past and future truth of these temporally variable propositions consists in what Theodore Sider has called "slice-operators": tense operators that (when attached to simple present tense sentences) take us from a sentence expressing a proposition about the present to a sentence expressing a proposition about an *instantaneous* location in the past or future. The simple tense operators "$F$" and "$P$" in Arthur Prior's systems of tense logic are slice-operators.[30] "$Fp$" and "$Pp$" could be given the informal glosses: "It will be true, at some future instant of time, that $p$" and "It was true, at some past instant of time, that $p$". Tenses, construed as sentential operators, allow presentists to make distinctions of scope when tense operators and quantifiers interact. A past tense operator can, for instance, take wide scope or narrow scope in a sentence containing quantification over past things. The distinction can be detected in the pair: "It was true, at some past instant, that some dinosaurs roam the earth" (in which tense has wide scope), and "There are things that used to be dinosaurs roaming the Earth" (in which tense has narrow scope). The presentist supposes that these sentences express distinct propositions; that each is a plausible candidate for the meaning of "Dinosaurs roamed the Earth"; and that the latter does not follow from the former. Presentists can accept the wide scope interpretation of "Dinosaurs roamed the Earth"; they need harbor no skepticism about whether there were dinosaurs, but their presentist scruples direct them to reject the narrow scope interpretation.

Slice operators of this sort "require talk of the past and future to proceed 'one time at a time'".[31] Alternative operators that allow a false proposition to *have been true* in virtue of what went on over a span of time—"span operators", as they're called— are problematic in various ways, at least for presentists.[32] But slice operators make it difficult for the presentist to find adequate grounding for various kinds of cross-temporal relations, and even some simple past-tense claims become hard to translate into a formal slice-operator language.[33] If my great-grandfather and I never co-exist, at no time do I stand in the relation of great-grandson to him. Of course I *am* his great-grandson, and he *was* my great-grandfather; but what does this assertion amount to, if there is no instant at which we are related? Perhaps this assertion is not really as relational as it sounds, or perhaps it is true in virtue of a relation between me and some kind of surrogate for my no-longer-existing grandfather.[34] The growing blocker avoids the problem by affirming the ongoing—albeit ghostly—existence of my grandfather.

The ghostly growing blockers are committed to the automatic, ongoing existence of *everything* that comes into existence; according to their metaphysics, going out of

---

[30]  See Prior (2003) for an informal presentation of his tense logic.

[31]  Sider (2001: 25).

[32]  For objections to use of span operators by presentists, see Lewis (2004) and Sider (2001: 26–27); for attempts to introduce span operators while respecting presentist scruples, see Brogaard (2007) and Bourne (2007).

[33]  See Lewis (2004) and Sider (2001: 25–35).

[34]  For a variety of strategies presentists might use to tackle various problems of cross-temporal relations, see Chisholm (1990a, 1990b), Markosian (2004), Sider (1999), Szabo (2007), Crisp (2005), Bourne (2006: 95–108), Bigelow (1996), and Zimmerman (1997); for more criticism, see Davidson (2003).

existence is an impossibility. Although we would-be presentists shy away from ghostly past things, the puzzles about cross-temporal relations may nevertheless force us to posit the ongoing existence of certain entities that would normally be said to be "in the past". For example, continuous causal processes seem to require fundamental relationships between events or states of affairs that occur at non-overlapping instants or intervals. If some such causal relationships really are fundamental, and really are relations, then they had better hold—at some time or other—between some pairs of entities. One response to the challenge is to construe the causally related events as something like Chisholm's states of affairs. By consigning causal relata to a category of entity that can be expected to exist even if they never occur, the puzzle about cross-temporal causal relations begins to look more tractable.[35]

Of course, if one must keep doing this sort of thing for all manner of entities, including ones that violate our deepest presentist instincts, one should begin to wonder whether the ghostly growing block theory of time is not, after all, the best version of the A-theory. I am hopeful that we A-theorists will not be forced to accept the continued existence of too many of the things we ordinarily say have "ceased to be". To this category, I would consign all spatially located particulars that seem hard to imagine existing while no longer being located anywhere. Events may also belong in this category—but only if events are *not* taken to be Chisholmian, proposition-like states of affairs. Many metaphysicians think that—instead of, or in addition to, states of affairs that can *exist* without *occurring*—there are "concrete" events that must occur in order to exist. Such event-like entities have been called "particularized qualities", "particular characters", "abstract particulars", and "tropes";[36] they belong in the same category as what Aristotle called "individual accidents", and seem similar to what the early moderns called "modes".[37] In our ordinary "event talk" we seem to shift back and forth between these two conceptions.

For present purposes, I shall assume that, in addition to states of affairs that may or may not obtain, there are also events of this more "concrete" sort. The stingy metaphysician in me would like to avoid commitment to both kinds of event. If I accept all of the more abstract states of affairs, so I can have the right kind to serve as causal relata, do I really need the more concrete events as well? Is there anything left for them to do? For now, I shall allow for the existence of both states of affairs and more concrete, trope-like events. Throughout this essay, when I speak of past events ceasing to be, I have in mind events construed in this second, concrete fashion; the existence of the more abstract states of affairs, after they have ceased to occur, strikes me as relatively unproblematic.

Presentists face further pressures to accept a plethora of states and events, besides the need for co-existing causal relata. All manner of states and events may be needed

[35]  This was the strategy I advocated in (Zimmerman, 1997); Bigelow (1996) defends a very similar idea. Of course some will find these entities too "abstract" (a word with no fixed meaning, but often used as a term of abuse) to serve as causal relata, since they are rather like propositions (indeed, Chisholm held that they simply *were* a certain kind of proposition).

[36]  These are terms introduced by Cook Wilson (1926: 713), Stout (1921/22), and Williams (1953).

[37]  For discussion of "tropes throughout history", see Mulligan et al. (1984: 290–93).

in order to respond to (what Zoltan Szabo calls) "semantic arguments" against presentism: arguments to the effect that a certain inference pattern must be respected in
any plausible semantics of, say, English, and that such a semantics validates inferring
"There are $Fs$" from some obviously true claim, though the presentist says "There
are no Fs". David Lewis, for example, argues that the presentist cannot make sense
of simple assertions like "England has had two kings named 'Charles'". It is difficult
to state tense-logical truth-conditions for the sentence that do not imply the truth,
at some past time, of the proposition that there are two men named 'Charles' who
are or were or will be kings of England"—which need not have been true, in order
for the original sentence to be true. But, as Szabo points out, similar problems arise
for sentences like "The election could have had three different outcomes" (where the
outcomes are incompatible), and "Three ghosts are supposed to inhabit the woods"
(where the ghosts are entirely imaginary). A Davidsonian "event" or "state" semantics
is suggested by these examples and by a wide variety of linguistic data. Many philosophers and linguists accept such a semantics for reasons other than its usefulness with
possible events and imaginary entities, but, once it is taken on board, it further earns
its keep by helping to make sense of statements about the elections and the ghosts,
without requiring commitment to the possibility of an election that has an impossible
outcome or the existence of imaginary ghosts. The first example becomes something
like "There are three different states of the-election's-possibly-having-an-outcome";
the second, "There are three states of the-wood's-being-supposed-to-have-a-ghost".

Szabo argues that the postulation of non-obvious quantification over events or states
is independently motivated, and provides the presentist with natural materials for
dealing with Lewis's example (by means of "There are two states of England's having
had a king named 'Charles'") and many others. Of course, there is an ontological cost
for the presentist who makes this move: the continued existence of the two states
in question, long after the kings have ceased to exist. The automatic and ongoing
existence of such "resultant states" generates commitment to a host of entities: for
every past state or event, there is the current state of its having occurred. Szabo argues
that the commitment to resultant states is in harmony with other things presentists
often say, and not so costly:

> They can be seen as shadows of the past, and as such, they are the sort of things
> presentists like to appeal to when they seek truth-makers for past tense sentences.
> But I am not pulling them out of a hat—I claim that a good semantic account of
> simple natural language sentences quantifies over them.[38]

Suppose the presentist accepts Szabo's offer: every event and state that ever occurs
to an entity of any sort leaves behind, as a kind of shadow or echo, a forevermore
existing state that does not depend upon the ongoing existence of the entity to which
the original event or state occurred. These resultant states would comprise a ghostly
image of the past, rather like the ghostly growing blocker's faded objects and events,
but with most of the original objects removed.

---

[38] Szabo (2007: 414). Szabo points out that the term "resultant states" was coined by Terence
Parsons.

The presentist may, then, be forced, by various problems about cross-temporal relations, to recognize certain kinds of more-or-less abstract entities—states that exist without occurring, resultant states that automatically appear and cannot go away. A growing blocker *might* be able to tell a more plausible story than the presentist about the nature of resultant states. A B-theorist *might* be able to do without them altogether—if he can find an alternative to an event or state semantics for non-temporal examples (the election outcomes, the witches) and other data that seem to require a generous attitude toward such states. But, if presentists are forced to accept them, they should not feel too badly about it. A ghostly manifold of resultant states can be independently motivated from within the philosophy of language, and is easier to believe in than a ghostly manifold of horses and hand-grenades.

A very different response to problematic cross-temporal claims is to grant that, strictly speaking, the statements are false—though they come extremely close to being true. If the presentist can find plausible grounding for present truths describing, in general terms, the past history of the universe on the most fundamental level; then she can say that truths about particular non-existent things, and about the enumeration of such things, are at least "quasi-true"—that is, true but for the falsity of a certain metaphysical thesis, namely, eternalism. And that is true enough for ordinary purposes.[39] Later, I consider an argument against presentism based on certain cross-temporal relations that play a crucial role in contemporary physics—a case in which the presentist does not, I believe, have the luxury to allow that, strictly speaking, the relational claims are false-but-quasi-true. The argument may drive the presentist to accept a different kind of more-or-less abstract or ghostly thing: a four-dimensional manifold of empty points, representing places where events were once occurring or where matter was once located. In this context, too, I will explore the possibility that something like resultant states—"trajectories" that happen to a series of points—may provide us with a less objectionable surrogate for a block of past things.

# 2. THE RELATIVISTIC MANIFOLD: ITS ONTOLOGICAL STATUS AND INTRINSIC STRUCTURE

## 2.1 Space-time substantivalism

Precise physical laws are expressible as mathematical relationships. All decent candidates for the laws of motion or electrodynamics or other fundamental physical phenomena occurring in space and time appeal to mathematically describable relationships holding among locations in space and time. At the end of the day, different

---

[39]  Sider (1999) offers presentists a version of this quasi-truth response; Markosian (2004) defends a slightly different one.

metaphysicians want to say different things about the ontological status of "locations in space and time". But we must all somehow make sense of the idea that space-time locations stand in precise distance relations of various kinds, and so constitute a "manifold"—for a manifold is any set of things that are interrelated in such a way that their structure can be described geometrically. A *space-time* manifold is a set of minimal-sized locations in space and time, "points" at which something could happen or be located.

"Substantivalists" and "relationalists" disagree about just how seriously one ought to take the manifold and its points. Substantivalists advocate "admitting space-time into our ontology": it is an extra part of "the furniture of the universe"—a sort of invisible jell-o filling up the otherwise empty spaces between objects (and suffusing their insides as well). Relationalists, on the other hand, hope to be able to treat the web of spatiotemporal relations attributed to the manifold as a sort of imaginary scaffolding; the fundamental spatiotemporal relations hold among the kinds of things we say "fill" locations or "happen at" them—particles, fields, and events, for instance.

I have encountered quite a few philosophers who have no patience for speculation about the metaphysics of worlds that are not governed by the *actual* laws governing physical things in *our* world—whatever those laws might turn out to be. I can think of only a couple of philosophical positions that justify the rejection of all such speculation. Some may be deeply skeptical about our ability to answer metaphysical questions concerning worlds with different laws: we should not pretend to have *any* insight into the realm of the merely possible, because metaphysical possibility and necessity are meaningless or beyond our grasp. Since fewer philosophers nowadays are quite so skeptical about our ability to understand questions about necessity and possibility, I sometimes suspect that impatience with speculation about contra-legal worlds is based upon a tacit metaphysical commitment to the *necessity* of whatever the actual laws turn out to be—a respectable metaphysical position defended by Sydney Shoemaker, among others.[40] But false physical theories about our world, when they do not contain hidden inconsistencies, certainly do not *seem* impossible, so, in the absence of argument to the contrary, I take them to be descriptions of possible space-time worlds. Some of these descriptions seem to require a space-time manifold with built-in structure of its own, and others do not.[41] By my lights, then, the debate between space-time subtantivalists and relationalists should be seen as an argument over whether or not the physics of *our* world requires taking the manifold seriously as a cosmic jell-o. Its existence is a contingent and, broadly speaking, empirical matter.

Since the modern era, four kinds of space-time have proven most appealing to scientists: (1) Newton's space-time consists of a persisting, infinite, three-dimensional Euclidean space, together with the series of times at which it exists. These days, Newtonian space-time is often described, somewhat anachronistically, as

---

[40]  Shoemaker (1998).
[41]  Maudlin (1993) explains why Newton's space-time and that of SR are more promising contexts for relationalism than are Galilean space-time and the manifolds countenanced by GR.

a four-dimensional manifold of points. A Newtonian four-dimensional manifold is a series of distinct, infinite, Euclidean, three-dimensional spaces; each is instantaneous in temporal length, and spread out continuously in a fourth, temporal, dimension. An objective relation of same-place-at-a-different-time holds between the points of different three-dimensional spaces. (2) Galilean space-time is just like Newton's at each moment, but does not include a non-relative same-place-at-a-different-time relation, though it does admit of another kind of fundamental cross-temporal relation, namely, straight lines representing possible inertial paths through space-time. (3) The space-time posited by SR is a manifold exemplifying a less intuitive geometrical structure, to be described below. It is often called "Minkowski space-time", since Minkowski is responsible for formulating Einstein's theory in terms of the geometrical structure of a four-dimensional manifold of points. (4) The various four-dimensional manifolds consistent with GR approximate the structure of Minkowski space-time in arbitrarily small regions around each point, but can have variable curvature on a larger scale. The different kinds of space-time manifold follow from, or fit together with, different physical theories, which require different fundamental measurable distance relations among the parts of the manifold. And there are ongoing arguments among metaphysically-minded philosophers of physics about which of these theories, if any, requires our taking a substantivalist attitude toward the space-time manifold it describes.

The substantivalist-relationalist debates are made possible by the fact that, on the spectrum between observable and theoretical entities, a space-time manifold is far to the theoretical side. A substantival space-time manifold, and the points of which it consists, are things that *seem* to be required by various physical theories, but there is room for doubt about whether this requirement is real—in other words, relationalism does not fly in the face of experience, at least not directly. Like quarks or "dark matter", if we come to believe in the existence of a space-time manifold at all, it will be because physics needs it, and we should therefore let our best physics tell us what it is like. SR is, at best, only approximately true; for a long time now, GR has looked like our best theory of the structure of space-time. Gradually, the difficulties in squaring GR with quantum theory have become clear to more and more people working on foundational physics; it is no longer obvious what space-time will look like, in tomorrow's theory of quantum gravity. In this paper, I will eventually try to evaluate the following very popular style of argument against presentism: Presentism (and other versions of the A-theory) are metaphysical theories that conflict with Relativity; but Relativity is our best physical theory of space-time (or so it has been assumed); since our best physical theories are better grounded than any metaphysical theory could possibly be, presentism should be rejected. Since critics ought to grant that SR is false, "Relativity" in such arguments had better mean GR. The basic question, then, is whether GR is in conflict with presentism. However, SR is a simpler theory, and more familiar to non-specialists. The manifold it describes can seem like a "special case" of GR (its metric looks like that of an infinitely large, flat, empty GR space-time), and the theories posit similar *local* space-time structure. Furthermore, similar larger-scale features generate an apparent conflict between each version of Relativity, on the one hand, and presentism, on the other. Understandably, many B-theorist critics have made use of SR in their arguments against presentism; I will often follow their lead, while keeping in mind that the most important question is how the arguments play in the context of GR. (When I say, in what follows, that "Relativity implies such-and-such", I mean that both theories have this implication.)

In my responses to arguments against presentism, I shall assume substantivalism about the manifold. Here, briefly, is my justification for regarding this as a safe assumption. The serious question on the table is not whether presentism conflicts with SR, but rather whether it conflicts with GR.[42] The fact that GR is probably not the final word on space-time structure will become relevant later. For now, however, I shall be asking what the presentist should say were it to turn out that GR is the best theory of the space-time manifold, and, for much of the time, the issues raised by SR will be similar enough so that Minkowski space-time can go proxy for whatever relativistic manifold we might be thought actually to inhabit. Although relationalism within SR may not be hopeless, attempts to understand GR in a relationalist fashion wind up positing something that, for present purposes, is enough like substantival space-time as to make no difference.

GR puts constraints upon the structure space-time could display, but it is consistent with an infinite variety of differently shaped manifolds. The space-time manifolds of GR can be finite or infinite in size, and their metrical properties generally vary from place to place. By contrast, the space-times posited by Newton and Minkowski are infinitely-extendable, and everywhere-the-same—features that make it easier to use certain relationalist tricks to avoid serious commitment to the manifolds the theories seem to describe.[43] Some philosophers have special reasons to want such tricks to work: Leibniz has his principle of sufficient reason, which would be violated by God's choosing to create in one part of such a manifold rather than another. Some philosophers subscribe to a kind of causal criterion of existence, according to which one should not posit anything that cannot, at least in principle, be affected by something else, and the manifolds described by Newton and SR are unaffected by their contents. Other philosophers (plausibly, to my mind) argue that, although the space-time described by SR would not be changed by the presence of matter, and may not in any straightforward sense *cause* the motions of particles (at least, not in the way *forces* cause motion); nevertheless, a substantival manifold earns its keep by playing an important explanatory role in the theory; namely, that of defining the "default" states of motion, thereby allowing for a sharp distinction between dynamics and kinematics. These are deep waters.[44] But, however the debates about SR and substantivalism turn

---

[42] In Belot's words: "The observation that the structure of Minkowski spacetime is incompatible with the lapse of time and the existence of genuine change would be of limited interest if similar conclusions did not follow in more fundamental contexts", such as "general relativistic cosmology" (Belot, 2005: 263).

[43] Maudlin (1993).

[44] Balashov and Janssen (2003: 340–1) ask whether "the Minkowskian nature of space-time explain[s] why the forces holding a rod together are Lorentz invariant or the other way around". They opt for the first answer, and regard the explanatory power of Minkowski space-time as a reason for

out, there are powerful reasons to think that GR (and, for that matter, Galilean space-time) requires substantivalism. In a broad survey of the substantivalism-relationalism dispute, Tim Maudlin concludes that, given GR, substantivalism—or something near enough to it—is inevitable, since "[t]he set of all spatiotemporal relations between occupied event locations cannot generally provide enough information to uniquely settle the geometry of the embedding spacetime."[45] The relationalist needs a "plenum" of entities—a field of some kind—upon which to hang GR's web of spatiotemporal relations; and the best candidate for this field is very hard to distinguish from the kind of entity substantivalists have always wanted.

It is true that more philosophers of physics are defending something they call "relationalism"; but the kinds of relationalism are, from my point of view, so close to substantivalism as to make no difference. Advocates of the "hole argument" for relationalism about GR's space-time do not, to my knowledge, question the need for a four-dimensional plenum to bear the properties of the metric field.[46] Julian Barbour's recent work, advocating "the disappearance of time" in the context of GR, may require some qualification of Maudlin's arguments. But the qualifications would not prove relevant to my purposes. Barbour articulates a kind of eliminativism about relations between time-like slices of a manifold satisfying GR; but his theory is, and arguably needs to be, substantivalist about the space-time points of the slices themselves.[47]

It is relatively safe, then, to assume substantivalism about GR's manifold. And, since the space-time of SR is mainly of interest for its approximating the structure of GR, and raising the same problems for presentism in a simpler context, it will be safe to treat its manifold in a substantivalist fashion as well.

## 2.2 Relativistic space-time structure

If Relativity is the best physical theory of the space-time manifold, presentists have some difficult questions to answer: How should we think about its intrinsic structure?

realism about its structure and existence. Brown and Pooley (2006) take the second option, and it leads them to declare Minkowski space-time (in the words of their paper's title) "a glorious non-entity". Saunders remarks that, since Brown and Pooley do not follow Lorentz in positing a privileged rest frame, "they suppose that the forces which yield the contraction and dilation effects may be explanatory, even if there is no fact of the matter as to what they really are" (Saunders, 2002: 290, n. 13).

[45]  Maudlin (1993: 199); see also Nerlich (1994).

[46]  See Norton (1989), and Earman (1989). I should also note that the "hole argument" depends upon a number of highly abstract metaphysical theses that have been called into question by its critics. The argument depends upon a quite technical definition of "determinism", and then assumes that any decent theory of space-time has to be consistent with the possibility of determinism, *in this precise sense*. It also presupposes a kind of haecceitism about space-time points that is not beyond question. For criticism of the "hole" argument, see, e.g. Maudlin (1989).

[47]  See Barbour (1999: 165–181). Belot (1999) also champions revival of the relationalism–substantivalism debate in the context of GR; but, again, the relationalism he articulates makes free use of a plenum of point-like entities, and is close enough to substantivalism for my purposes.

Does only one slice of it exist? Does the relativity of simultaneity conflict with the presentist's need for objective facts about what is present? Before tackling these questions in the next two sections, I need to place the bare bones of the theory on the table. The aspects of Relativity that are supposed to raise the most trouble for the presentist can best be described by contrasting Minkowski space-time (and, ultimately, the manifolds of GR) with Newton's theory of space-time, and with Galilean theories of space-time.

Different theories about the nature of space-time say different things about the kind of structure the manifold contains—they describe its parts as interrelated in different ways. A crucial part of the manifold's structure is *metrical*. Metrical structure is what makes the manifold measurable; for present purposes, it can be thought of as the sum total of all the fundamental distance relations holding among the manifold's points. On the Newtonian conception, there are, at any given moment, facts about the spatial distance relations among the points that comprise all of space at that time. There are also facts about the temporal distances between any two temporal locations in the manifold. But Newton posited a further kind of metrical structure: objective relations of spatial distance between points at different times. On his view, there is a single right answer to the question: What is the spatial distance between this point, at this time, and that point, at that other time? The answer might be one mile, or one inch, or 100,000 miles. But the answer might also be zero, in which case the two points represent *the very same location* at different times. (Newton understood absolute sameness of position in the four-dimensional manifold of space-time locations as due to the presence of a three-dimensional Euclidean object—absolute space—that persists through time. For two events to occur at "the same place at different times" is for them to occur in a bit of space that has persisted from one time to another. The Newtonian manifold of distinct possible event-locations is still, in a sense, four-dimensional. For it is one thing for an event to occur in a given region of space at one time, and quite another thing for the same sort of event to occur in that same region, say, five minutes later. So the two possible-event-locations must be regarded as separated in a fourth dimension, the temporal one.)

Newton's notion of absolute sameness of place over time can be contrasted with that of merely relative sameness of place. Suppose I forget my book on a train, and return to find it in the very same place I left it, relative to the parts of the train (it is still there on my seat). If the train has been traveling in the meantime, the book is in a different place relative to the surface of the Earth (it was in New York but is now in New Jersey). In addition to all such merely relative relations of same-place-at-a-different-time, Newton's space-time includes a non-relative, objective relation of same-place-at-a-different-time, a relation built into the structure of space-time itself. The other theories about the metrical structure of space-time mentioned above—the Galilean theory, SR, and GR—deny that the points of space-time stand in such relations. Galilean space-time rejects absolute sameness of place over time; so it rejects Newton's brand of space-time, in which an objective relation of same-place-at-a-different-time is underwritten by the persisting parts of a three-dimensional space. It does recognize absolute simultaneity, however; and, at each instant, there exists a set of possible places at which events could occur at that time, spread out in three spatial dimensions, constituting a Euclidean space. Although absolute sameness of place does not hold between points in different instantaneous spaces, there are cross-temporal relations built into Galilean space-time geometry: Some paths constitute straight lines in a time-like direction. Their straightness consists in the fact that they are the "natural" or "default" paths of particles through the manifold. The time-like straight lines represent possible *inertial states of motion*, motion explicable in purely kinematical terms. The physical significance of their straightness is most naturally described dispositionally: if an object occupies a portion of such a line, and there are no forces at work, it stays on the line.

The intrinsic metrical structure attributed to the manifold by SR and GR is radically different from that of Newtonian space-time, and quite different from that of Galilean space-time as well. The structure of the Newtonian manifold, as I described it, is based upon three fundamental types of relations among points: (i) spatial distance relations within each momentary three-dimensional space, (ii) temporal distance relations between the points in different spaces, and (iii) a "same-place-at-a-different-time" relation between points in different momentary spaces. SR (upon which I will mainly focus) bases the structure of space-time upon a very different relation of "space-time distance". As in Galilean space-time, in SR there are sets of points lying on straight lines in time-like directions; and their straightness represents the fact that they are the inertial paths of (subluminal) particles. Again, as in Galilean space-time, Newton's same-place-at-a-different-time relation is eliminated, at least as a basic metrical feature; instead, only highly derivative, relative notions of same-place-at-a-different-time make sense. But SR goes further down the road of relativization than Galilean space-time. In SR, even separable spatial and temporal distances between points come to seem second-rate, because they, too, are merely relative. What are the truly intrinsic, not-merely-relative metrical features of space-time? Relations of space-time distance among points—or, better, path-dependent distance relations in terms of which distance between points can be defined. Space-time distances in SR come in three quite different flavors: Points can be separated by positive, negative, and null space-time distances. Without plunging into the mathematics of space-time distances, it is not easy to explain what these distance relations really amount to. They are measurable quantities closely tied to the explanations of motion that Relativity affords; and parts of their roles can be described dispositionally, in much the way I explained the role of straight time-like lines in Galilean space-time.

Here are some connections to motion that will hopefully shed a little light on the nature of the fundamental geometrical features in the manifold of SR. SR's relations of space-time distance give a sense to "straight lines" in the manifold—the shortest distance between two points. But what is SR saying about points when it says they lie along a straight line and stand in *positive* distance relations? These straight lines play the same basic function as the straight, time-like paths in Galilean space-time. Such a line is said to have a time-like direction; and it corresponds to the possible path of a particle that is moving at subluminal speeds and neither accelerating nor decelerating—an object in a state of inertial motion. What does it mean, in SR, to say that points within the manifold are at zero space-time distance from one another? Not that they are "the same place" or "the same point". It means that they correspond to points along a path that light would take in a vacuum. What does it mean, in SR, to say that points are on a straight line and standing in *negative* space-time distance relations? In that case, the line is "space-like": it corresponds to a straight line in a certain kind of three-dimensional region of the manifold—a region that, according to at least one inertial frame, has no depth in the time-like direction.

The straight lines of Minkowski space-time may usefully be compared with those of Galilean space-time by means of familiar space-time diagrams. Figure 7.1 depicts Galilean space-time around a point, $x$. The temporal dimension "goes up" (i.e. higher points represent the locations of later events). One spatial dimension is represented by the horizontal lines, another is depicted by imagining the parallelograms as flat squares passing through the paper, and a third spatial dimension is suppressed. Straight lines passing through $x$ in a temporal direction represent space-time paths through $x$ that could be taken by particles moving inertially—that is, undergoing no acceleration or deceleration. A particle that has occupied a series of points on one of these straight lines will be told to "stay on this line, in the future", unless forces come into play.

In a diagram of Galilean space-time, one must ignore the fact that some of these lines are perfectly vertical, and others slanted. The vertical ones do not represent "the same place again", and objects remaining on these lines are not objectively stationary, while objects occupying slanted lines are in absolute motion. In Newtonian space-time, there is such a thing as absolute sameness of place and absolute motion,

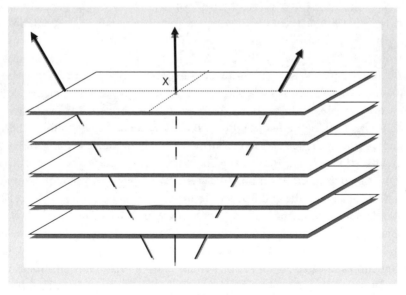

FIGURE 7.1  Galilean Space-Time

but in Galilean space-time, all velocity is relative. Nevertheless, acceleration and deceleration—departures from inertial motion—are *not* merely relative in Galilean space-time. The straightness of temporally oriented lines does indicate something objective about a space-time path—the fact that it is an inertial path—but the angle of such a line does not.

The straightness of lines in the two spatial dimensions of Figure 7.1 is more straightforward: it is just the familiar straightness of spatial lines in a two-dimensional Euclidean plane. Since there is a suppressed third spatial dimension, each plane really represents a three-dimensional Euclidean space—a different, instantaneous, three-dimensional space for each instant in the temporal dimension. Crucially, at each point on a particle's space-time path, there is exactly one of these three-dimensional Euclidean spaces—the space of the entire universe, as it exists simultaneously with the event of the particle's occupying that point. The straight spatial lines through $x$ that are depicted in the diagram constitute exactly one two-dimensional plane; and that plane stands in for exactly one three-dimensional space: the universe at the moment simultaneous with the event-location labeled $x$.

Space-time diagrams of the Minkowskian manifold are similar in many ways. In Figure 7.2, up-down represents time, left-right represents one space-like direction, a second spatial direction is suggested by the imagined "depth" of the cones, and a third spatial dimension is suppressed. The straight lines along the surfaces of the cones that meet at $x^*$ represent an objective feature of the Minkowskian manifold's structure that has no counterpart in Galilean space-time. The points on a straight line passing through $x^*$ and staying on the surfaces of the lower and upper cones are all said to be at zero- or null-distance from $x^*$—though this does not mean the points are at the same place in a way that would make sense in Newtonian space-time, say. The

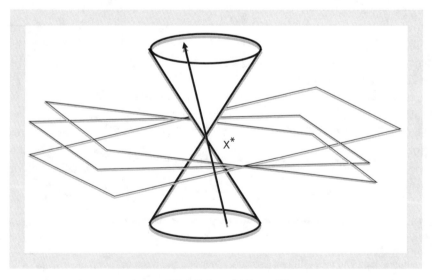

FIGURE 7.2  Minkowski Space-Time

cones are called "light-cones" because a straight line on $x^*$'s lower (backward) cone represents a path along which a flash of light could have reached $x^*$; a straight line on $x^*$'s upper (forward) cone represents a path that could be taken by a flash originating at or passing through $x^*$. Straight lines extending from $x^*$ *within* its forward or backward light-cone contain points in positive distance relations from $x^*$; like the vertical lines in Figure 7.1, they lie along a path in a time-like direction from $x^*$, and represent possible trajectories of objects moving inertially through $x^*$.

The relations of positive and null space-time distance represented by these two kinds of straight line give space-time the structure it needs to tell particles and photons "what to do next". Where will a photon go if it is located at a certain point (and in a vacuum)? The sets of points that are on light-like paths (points each of which is at zero space-time distance from the others) tell a photon that has been moving along one of them to "stay on this line, in future". Where will a particle go if it is located at a certain point and moving inertially? The points constituting straight lines with positive distance relations tell a particle that has been moving along one of them, and that is not acted upon by any forces, to "stay on the same line, in future". One might sum up these relations by saying that a path of points at zero-distance from one another is connected by a relation of "light-like accessibility", and that the straight lines with positive distances among their points are connected by a relation of "inertial accessibility".[48]

The third kind of straight line in Minkowski space-time extends from $x^*$ into the "bow-tie-shaped" region outside the two light-cones. (As I mentioned above, space-time distance relations holding between space-like separated points on a line are represented as negative numbers in typical formulations of SR.) Inspection of Figure 7.2 reveals, however, that straight lines at right angles can be inscribed through $x^*$ in many different ways, determining many different two-dimensional planes passing through $x^*$—planes that cut across one another but share a line that includes $x^*$. In the full, four-dimensional manifold, these planes correspond to various three-dimensional regions, overlapping in two-dimensional planes that include $x^*$. In SR, the space-time distance relations in these regions give each of them the geometrical structure of a three-dimensional Euclidean space. But what are these different spaces like?

Each flat spatial plane extending from $x^*$ represents a slice of the manifold with a special status: it is intimately related to exactly one of the inertial paths passing through $x^*$. So, whatever else these planes are like, they can at least be seen to have a relative kind of privileged status; inertial paths play a special role in the geometry of Minkowski space-time, and each combination of a point, and an inertial path through that point, determines just one of these planes. The respect in which a given plane is privileged,

<hr />

[48] One need not regard the "telling" as anything like a causal process. And one need not regard the dispositional characterizations of these relations as fundamental. Perhaps the distance relations are "categorical", and it is a merely contingent fact that they play these roles with respect to the propagation of light and the motion of particles. Mundy (1986) reconstructs SR on the basis of five physical relations tied closely to the kinds of physical processes Einstein employed in constructing coordinate systems.

relative to a point and an inertial path, is often presented in these terms: It is the plane an observer at that point, on that inertial path, would choose as containing events simultaneous with her, if she accepted a certain operational definition of "distant simultaneity".

The proposed definition of "distant simultaneity" is often motivated by telling this sort of story: When you are trying to find out "what time it is elsewhere", you naturally use whatever signals are most reliably constant in speed. SR gives a special, objective role within its space-time structure to light (the null-paths are specially reserved for flashes of light in a vacuum); what could be more reliably constant than that? And so you might be led by this thought to adopt the "Poincaré-Einstein" or "Radar" method for determining a relation plausibly worthy of the name "distant simultaneity": You send a light signal, noting your local clock time; you ask the distant recipient to note her local arrival time, returning a light signal just as she sees yours; then you note the arrival time of her signal; divide your total time by half; and figure that, whatever was happening to you at the halfway point was simultaneous with the arrival of your flash at her location. Applying this method in all directions to discover "what is going on simultaneously with $x$" at every point in the universe would yield a different two-dimensional plane for different inertial paths through $x$. "Observers" in different states of motion "passing through" one another at $x$ would slice the bow-tie region of space-time in different ways if they rely upon the Radar method for determining distant simultaneity.

However "natural" it may be to use light signals and the Radar method, hoping thereby to discover facts about distant simultaneity, it is not obvious that this procedure will deliver a relation coinciding with the relation of simultaneity between occurrences that we appeal to when we say they are "both happening now". Of course the B-theorist may say, "It is obvious that there is *nothing better*, so we might as well use this one." But one might reasonably wonder whether there is some other means of signaling that gives self-consistent results, but that sometimes conflicts with those delivered by the Radar method—in which case, there would be a competing set of candidate simultaneity relationships among the same events. The Radar method gives us a relation worthy of the name "optical simultaneity"; but one can imagine discovering other methods that yield different results—"telepathic simultaneity", say, if there were such a thing as faster-than-light telepathy; or "quantum measurement simultaneity", a notion needed for some interpretations of quantum theory (below, I say a little more about the reasons some versions of quantum theory introduce a simultaneity-like relation—one that is especially relevant to quantum measurements).

More to the point, the A-theorist is committed to thinking that there are facts about simultaneity that go deeper than any operational definition given in terms of a particular means of signaling—objective facts about which events are presently happening, for example.[49] So the A-theorist is bound to be suspicious of the proposed identification of simultaneity with the deliverances of the Radar method. As many A-theorists have

---

[49]  For similar reactions, see Prior (1996), and Bourne (2006: 172–76).

pointed out, Einstein's claim that, in SR, the relation of distant simultaneity must be defined by means of optical simultaneity, is based upon profoundly verificationist assumptions; those who are not verificationists are free to wonder whether simultaneity might be something deeper and non-relative, as A-theorists believe.[50]

For now, however, I shall set aside worries about the relationship between "*real simultaneity*" and the Radar method's surrogate for distant simultaneity. Instead, I shall ask merely whether the method picks out an interesting structural feature of the manifold posited by SR. And the answer is, clearly, yes, it does.[51] Assuming SR, the Radar method provides a perfectly natural way for an inertially moving observer at the point $x$ to divide up all of space-time into three-dimensional regions. Extend a time-like straight line through $x$ in both directions along your inertial trajectory; use the Radar method at each point on the infinitely long line to pick out a three-dimensional slice of the manifold; and the result will be an exhaustive division of the manifold into non-overlapping, flat surfaces, on the basis of optical simultaneity.

The important fact, for present purposes, is that such a *foliation* of a manifold— an exhaustive division into three-dimensional, non-overlapping regions each of which "slices" the manifold "all the way through"—is the result of applying the Radar method to just one of an infinity of inertial paths through $x$, none intrinsically better than the others, at least so far as the geometry of SR is concerned. If two observers were in different states of motion but passing right through one another at $x$, they would come up with different answers to the question "what is happening right now?" using the Radar method; and, extending their inertial paths into the past and future, they would come up with complete foliations of the manifold that cut across one another.

The metrical structure described by SR does not, then, privilege just one way to "slice" the manifold into non-overlapping, continuous, three-dimensional, space-like regions. According to SR, there are infinitely many ways to exhaustively divide the four-dimensional manifold into a series of slices, each slice corresponding to a three-dimensional space that at no point has any thickness in a time-like direction.

The same can be said in the context of GR, but with important qualifications. One serious issue that arises in GR, but which I shall have to set to one side here, concerns some of the stranger shapes that have been contemplated as possible space-time manifolds. An empty, flat GR manifold has a metric like that of SR around each point, but in a manifold containing matter, the light-cones must be bent toward the location of mass, with greater curvature near larger masses. GR is, in effect, a set of equations that puts constraints on the varieties of possible combinations of manifold-plus-contents. GR manifolds come in all sorts of shapes; but the ones that

[50] Prior (1996), Smith (1993: 225–38), Craig (2001: Ch. 8), and Bourne (2006: 172–176) are examples of A-theorists who charge Einstein with verificationism, and advocate A-theoretically-based notions of simultaneity.

[51] In almost all GR manifolds, there is no slicing of the manifold definable by anything like the Radar method; this might lead one to dismiss it as an unimportant structural feature of the SR manifold as well, if one is thinking of SR as a sort of special case of a GR manifold; see, e.g. Maudlin (2008: 156).

look like they come close to resembling our space-time (black holes and all) include a "global time parameter"—which means they can be exhaustively sliced up into ordered, space-like three-dimensional regions without time-like depth at any point. And, most importantly for present purposes, just as in the case of Minkowski space-time, these manifolds admit foliations that cut across one another.

The equations of GR have solutions that allow the manifold to take on all kinds of bizarre shapes. For example, GR does not rule out the possibility of "closed time-like curves"—paths through space-time that loop back upon themselves. And some of the more oddly-shaped manifolds do not submit to a natural, exhaustive slicing into space-like regions—they are "non-foliable". Such space-times pose difficult questions for the presentist, who expects a manifold with a time-like direction to have a privileged foliation—a division of the manifold into slices each of which contains events that were all happening at once. Physicists sometimes dismiss these non-foliable models of space-time as "not physically real" or "pathological", in much the way retarded solutions to Maxwell's wave equations are thrown out as "unphysical" for implying waves that move backwards in time. I take it that, in both sorts of cases, the physicists who say these things mean more than just that such models do not describe the actual world—after all, lots of solutions to Einstein's or Maxwell's equations are *known* to be failures as descriptions of our universe, but are not deemed "unphysical".[52] I should think the A-theorist may, with a relatively clear conscience, reject non-foliable models as *not genuinely possible ways for time to be*.[53]

One might think that philosophers are overstepping their bounds if they deny, for philosophical reasons, the existence of non-foliable manifolds when, on the face of it, current physics does not rule them out. Perhaps ... but philosophers frequently engage in such forays into physical territory. Physics, by itself, is unlikely to rule out the possibility of lots of bizarre things that philosophers routinely reject as not *genuinely* possible. For example, physics has nothing to say about such controversial mental phenomena as "qualia" or "irreducible intentionality", but, according to physicalist orthodoxy in philosophy of mind, these are impossible phenomena, whether or not they are ruled out by physics. The A-theorist should not lose much sleep over GR's failure, by itself, to imply the impossibility of temporal loops and other surprising space-time shapes that create problems for her view.[54] More worrisome is the prospect of a conflict between presentism and the attempts to describe, at least approximately, the actual structure of our space-time by means of SR and GR.

---

[52]  Confronting certain non-foliable space-times discovered by Gödel (1949), Einstein, for example, asked "whether these are not to be excluded on physical grounds" (Einstein, 1949: 688). I take it he was not merely considering whether to exclude them from the category of likely candidates for our space-time.

[53]  Crisp (2008: 274) takes the same attitude toward non-foliable GR manifolds. Monton (2006: 274–6) points out that GR is false, anyway; and he speculates that non-foliable manifolds may well turn out to be physically impossible according to whatever theory of quantum gravity supercedes GR.

[54]  Gödel's (1949) argument against the reality of time makes use of a particularly oddly shaped GR manifold. For a recent criticism of Gödel's much-discussed argument, see Belot (2005).

# 3. A MANIFOLD FOR PRESENTISTS

## 3.1 The shape of the present

I have advocated the assumption of substantivalism, at least as a theory about our universe's manifold. The following seems to me to be regarded as a good bet, at least among philosophers of physics: Even if GR turns out not to be, strictly, true, nevertheless, a physics that adequately describes the laws of motion in our universe will likely imply the existence of a manifold with similar metrical structure. It would, then, be a bad bet for presentists to hitch our wagons to relationalism about the space-time manifold, given the difficulty of being a relationalist about a manifold with GR-like structure; and a fairly safe bet for us to appeal to space-time points, substantivally construed, in our attempts to develop an adequate presentist metaphysics for a universe like ours. In this section I consider what presentists should say about the nature of this manifold.

In the first half of this section, I consider what the presentist should say about the shape of that part of the manifold that contains currently occurring events and currently existing objects. I attempt to articulate the basic convictions that drive me toward the A-theory, and then, supposing the manifold to have the metric of Minkowski space-time, I try to work out what part of it must be filled with objects and events right now. The conclusion is not surprising: On plausible assumptions about causation in a world correctly described by SR, the presently filled region is a "slice" of the manifold, with no depth in a time-like direction at any point. Here, the potential for conflict with Relativity becomes apparent; it will be addressed in the final section of the chapter.

Having reached a conclusion about how much of the manifold is presently filled with events and objects, I turn in the second half of this section to the question: How much of the manifold *exists*? Just one ultra-thin slice in which present events are happening and at which presently existing objects are located, or also parts of the manifold that merely *were* the locations of events and objects? Theodore Sider has advanced an argument against presentism on the basis of cross-temporal relations needed to distinguish between different states of motion. One way to respond to his challenge is to accept the existence of past (i.e. formerly occupied) points. Although that would not be a terrible result, I will suggest an alternative response as well. In either case, it shall become clear that "space-time" is a bit of a misnomer for the presentist's manifold: it is a substantial, though theoretical, object that persists *through* time—it is not something that could contain time itself as one of its dimensions.

## 3.2 Why I believe in an A-theoretically privileged foliation

I am an A-theorist because I am convinced that there is a big difference between an event that is really happening to me, and one that merely has happened or will happen to me. The ones that are *really* happening are—in some objective, non-relative way—"more real" than the others. They constitute a minuscule proportion of the events that occur over the course of my life. Assuming SR, I inhabit a four-dimensional manifold, and the events in my life occur at different points along a path in one of its four-dimensions, my "world-line". So I affirm:

(1)   There is an objective, important difference between events that are really happening to me, and ones that merely *did* or *will* happen to me; and the events that are really happening to me are confined to a tiny region, $r$, on the world-line I will eventually have traced through the manifold.

Although I think only some of the events on my world-line are special in this way, I should not take myself, or my world-line (including my current state of motion, which is a function of the shape of my world-line near $r$) to be deeply special—I should be surprised to discover that I play a unique role in the physics or metaphysics of space-time.[55] Because there are so many other people, places, and world-lines, and no reason to think I occupy a privileged place relative to all of them, I should adopt the following as an extremely likely working hypothesis:

(2)   I am not metaphysically special, unique among all human beings with respect to some important, objective feature of the manifold; neither is the region $r$, nor is my world-line.

Of course I should not rule out, a priori, the idea that, lo and behold! and despite all odds, I just happen to be standing at "the center of the universe" (if there were such a thing) or at some other spatiotemporally special place. But any discovery along those lines would have to be based on serious study of the actual structure of our manifold, and my place in it. Only the extreme egotist would assume such a thing right off the bat in his theorizing about time, and anyone bold enough to do so is almost certainly mistaken, and surely does not *know* that he occupies a special place.

It is a favorite pastime of philosophers to contemplate a kind of "me-now" solipsism: What if nothing else is going on but what is happening to me? The supposition is useful for shaking loose some kinds of philosophical fruit. But I have no reason to *believe* such a thing, because:

(3)   If the only events in the universe that are really happening are the ones happening to *me* at $r$, then $r$ and I would be very special.

From these three assumptions, it follows that *what is really happening* excludes many events that have already happened to me and many that have not yet happened to me; but also includes many other events that do not happen to me at all.

(4)   Events are really happening to me, at $r$, and to many other objects at points on their paths through the manifold. (From 1, 2, & 3)

---

[55]  The argument leading to (8), below, is similar in some respects to arguments that can be found in Sider (2001: 42–52), Savitt (2000), Saunders (2002), and Callender and Weingard (2000).

The extent of what is really happening must "stick out" into the manifold beyond the brief events happening at $r$ on my world-line. One could call the part of the manifold at which events are really happening, "the present". The question then becomes: What is the shape of the present?

The metrical properties of the SR manifold have been sketched. It is an infinitely large, connected set of points, each of which lies at the intersection of infinitely long straight lines consisting of points at positive, negative, and null distances. These lines inscribe forward and backward light-cones around each point, including points at time-like separation within the cones, and space-like separation within the "bowtie" region outside the cones. SR attributes this much geometrical structure to the manifold, and nothing more. If the present, around $r$, has a shape that is recognized as "natural" by the lights of SR, it must be definable in terms of these fundamental metrical properties of the manifold. The choices are quite limited.

(5)   According to SR, the only geometrically distinguished subsets of points that include $r$, along with many other locations in the manifold, are the following: (a) the points at space-like distances from $r$, that is, the ones filling the "bowtie" region around $r$ in a two-dimensional space-time diagram; (b) the points in or on $r$'s forward light-cone; (c) the points in or on $r$'s backward light-cone; (d) the points on the various planes associated, by the Radar method, with continuous paths passing through $r$; (e) three "hyperboloids of revolution" about $r$; or (f) some set of points definable in terms of these distinctions.

But each of these alternatives has its problems. If the only events really happening were the ones at space-like separation from me (i.e. ones occurring in the "bow-tie"-shaped space-time region around the point at which my real experiences occur), then I would occupy a very special place in the cosmos—the present would "emanate" from me, so to speak. Something similar would be true if the only events really happening were the ones on the surface of my backward or forward light-cone. I might think my research is really "cutting edge", a critic may find my views old-fashioned, retrograde, but neither of us will be inclined to link reality to one of my light-cones, putting me ahead or behind everyone else in my progress through the manifold. Similar criticisms would apply to identification of the present with the entire contents of my backward or forward light-cone, or the combination of the two; associating the present with any of these regions would not only make me special, but would be quite perverse, at least in the latter two cases.

Forget the light-cones, then. Suppose, instead, that I try to make use of the flat planes of simultaneity that include $r$. There are, however, infinitely many, none of which can claim any geometrical distinction. The only hope for selecting one is to choose a particular state of motion through $r$—for instance, my own. So suppose the only events in the universe that are really happening are the ones on the plane that would be picked out by use of the Radar method at $r$ by someone on an inertial world-line having the state of motion I have, at that point. If that were the case, my world-line in the vicinity of $r$ would be very special; use of the Radar method by observers in relative motion would place my current experience in a different plane, one that cuts across mine. I would be able to use light signals and assumptions about the equality of the round-trip speed of light to correctly determine the shape of the present, but those in relative motion would get the wrong results, were they to use the same method and assumptions. Choosing some other observer's state of motion will simply privilege a different path through $r$.

The hyperboloids of revolution are included only for completeness. I will not describe them in detail, except to say that they are surfaces consisting of points at a constant space-time interval from $r$.[56] Some are hyperbolas stacked within my rearward light-cone, others are stacked within my forward light-cone, and another family divides up the bow-tie area around $r$. Choosing the points on one of the surfaces in one or more of these families as "the present for $r$" would be odd for all sorts of reasons. Such a choice would not only make $r$ very special (serving as a sort of generating point for the hyperboloids from which the surface was chosen), it would also leave a space-time gap between $r$ and the other events going on now, since none of the hyperbolas in any of the three families is connected to $r$.

Sets of points distinguished by some combination of these distinctions will remain centered around me and my world-line. For example, one might focus on the points *on or below* my inertial plane at $r$ but above my rearward light-cone; but, again, that would make $r$ and my current state of motion very special.[57]

I hold a privileged place in all the divisions of the manifold that include $r$ and that can be defined in terms of SR's fundamental metrical properties. There simply are no more "objective lines" that SR can discern passing through my current experiences and out into the rest of space-time; no more regions that SR recognizes as natural or objectively special. That is to say:

(6)    If the region in which events are happening were restricted to (a), (b), (c), (d), or (e), I or $r$ or my world-line would be very special.

Which leads to the conclusion:

(7)    If the region in which events are really happening coincided with a set of points including $r$ that are geometrically distinguished, according to SR, then I or $r$ or my world-line would be very special. (From 5 & 6)

---

[56]  For a better description, see Maudlin (2002: 104–108).

[57]  Arthur (2006) introduces an interestingly shaped "present" that he calls "the interactive present"; it is defined in terms of the region around a part of a world-line that contains all the places from which there could be "mutual physical connection" between the place and the world-line. In SR, it turns out to be the intersection of the backward light-cone of the latest point on the world-line and the forward light-cone of the earliest part of the world-line. Arthur does not intend his interactive present to provide anything like the objective "shape of the present" required by an A-theorist. According it an objective status would give $r$ and some world-line segment including $r$ a unique role to play in the evolution of the universe.

I should suppose that it is vanishingly unlikely that the present takes one of these distinguished shapes centered on me, given the infinite number of alternative perspectives that could have been privileged instead of mine. And so, from (2) and (7), together with conclusion (4) (which affirms that there *is* a larger region of the manifold including $r$ in which events are really happening) and (1) (which requires that this region be less than the whole of the manifold), I must conclude that:

(8)   There is a region of the manifold in which events are really happening; it includes $r$ and many other points, and it does not coincide with any region that is geometrically distinguished, according to SR.

In all likelihood, then, the present "lights up" a part of the Minkowskian manifold that is geometrically undistinguished, according to SR.

Can I, on general presentist principles, reach a more precise judgment about the shape of the region in which events are really happening? How much of the manifold, in my vicinity, should I suppose is "lit up"? I shall allow myself what I regard as a quite reasonable assumption about the connection between causation or causal dependence and an event's *really happening*; and I shall suppose that the physics of SR correctly describes the structure of the manifold, and that it implies that causal processes propagate along continuous paths. On these assumptions, it turns out that any physical system that includes my body would have to occupy an exceedingly thin slice through the bow-tie around my current position. Although this is the natural conclusion for an A-theorist to draw about the shape of the present, I will be deriving it *not* from outmoded assumptions about the structure of the manifold—that is, I do not assume that it is Newtonian or Galilean, with a built-in, geometrically privileged foliation. Instead, the conclusion follows from the very general A-theoretic principles (1), (2), and (3), and assumptions about the way causation works in a Minkowskian manifold—assumptions that seem to be close approximations of the way causation typically works in the real world.

Events in my history seem always to have been caused by events that have already happened. (This is a conviction about the direction of causation relative to the A-series, not a view about the direction in which causes tend to produce effects within the Minkowskian manifold.) Generalizing from my own experience, the following, then, seems likely:

(9)   For any events $e^1$ and $e^2$, $e^2$ is causally dependent upon $e^1$ only if, when $e^2$ was happening, $e^1$ had already happened.

SR puts constraints on the propagation of energy, and the paths of particles and light—all of which, I assume, carry causal dependencies. If all interactions are mediated by processes no faster than light—a prohibition that need not be built into SR, but that is generally coupled with it—then causal dependencies within the manifold follow continuous paths of light-like or inertial accessibility.[58] As noted earlier, it is a nice

---

[58]   Even if Relativity were taken to imply that "what has already happened" is relative to a frame of reference, (9) would not imply that there is anything relative about causal facts, so long as light is an

question how a presentist should make sense of causation between non-existent past events and existing present events, and I suggested that the causal relata are not concrete events—the kind of thing that would have to pass away when it ceases to happen—but the more abstract category of states of affairs. However the metaphysics of causation is to be handled, the presentist should agree with SR that causal relations or dependencies of some kind hold along the paths traced by particles and processes.

And so I shall assume:

> (10)   If a particle, photon, or wave occupies a path in the manifold, its occupancy of a point $r$ on that path is causally dependent upon its having occupied the points on that path that stand in light-like or inertial accessibility relations to $r$.

But if occupation of a point $p$ by a particle $x$ is an event that is causally due to the particle's occupation of locations "lower" on its world-line, then (9) implies that $x$'s occupying those inertially accessible points has already happened when $x$ occupies $p$. All the points on $x$'s world-line from which $p$ is inertially accessible represent places $x$ occupied in the past, since $x$'s existence in those places was partially causally responsible for its continued existence at points between them and $p$. More generally:

> (11)   If the way things presently are at a given point $p$ is causally dependent upon a certain event occurring at a point $p^*$ from which $p$ is light-like or inertially accessible, then the occurrence of the event at $p^*$ has already happened.

All of this will be true for each particle and process in a larger physical system, such as my body or, indeed, the universe as a whole. Suppose the full history of such a system winds its way through a four-dimensional, connected region in the imagined Minkowskian manifold, and suppose that the evolving system includes fields and particles everywhere within it, displaying causal dependencies along all the internal lines of inertial and light-like accessibility. (11) leads to a very intuitive picture of the shape of the present state of such a system. Let $x$ be one particle in the system; let $p$ be $x$'s present location; and let $S$ be the set of locations presently occupied by the rest of the particles and processes going on within the system. What is $S$'s shape, and how is it related to $p$? $S$ must extend continuously in space-like directions from $p$, and it must be extremely thin in the fourth dimension, the dimension of inertial accessibility.

Why must $S$ extend in space-like directions from $p$? The events in $S$ that, together with $x$'s state at $p$, have an immediate effect upon $x$'s subsequent states must be arbitrarily close to $p$—i.e., right up against it. But they cannot *now* be happening on or within $p$'s rearward light-cone. To affect $x$'s state immediately after $p$, SR requires that their effects upon $x$ propagate along a light-like or time-like path to reach $x$ when it is at $p$—which is to say, *now*. Since that would violate (11), the nearest, currently occurring events that will affect $x$ in the future must surround $p$ in space-like directions. The same moral applies to events that could influence the events that

upper limit on the propagation of causal dependency. Events on or within the backward light-cone of another event are earlier than it according to every frame.

could influence $x$, and so on; so $S$ is guaranteed to be a continuous region extending into the bow-tie region around $p$ all the way to the boundaries of the system. Given (9) and (10), each spatially unextended subprocess within the system can only occupy, at present, one point on its world-line. So the sum of these points, $S$, will be thin in the fourth dimension; nowhere will it contain a line extended in the direction of light-like or inertial accessibility.

If I were part of an infinitely large, continuous, evolving physical system that filled all of the imagined Minkowskian manifold, then the other locations at which events would presently be happening would include a space-like region extending out from my current location and slicing the entire manifold. If the physical system of which my body is part were smaller than that, the presently occurring events might extend only to its boundaries. So far, so good; the picture is a natural one, in which the currently occurring events of "my universe"—the extended web of causally interrelated fields and particles in which my body is embedded—fill a thin, three-dimensional slice of the manifold.[59]

Nothing in the reasoning that has led to this conclusion has carried the slightest suggestion that the slice in question must be the hyperplane associated with my inertial frame, or with the inertial frame of any other object or system. Nor has there so far emerged any reason to think that the slice takes the form of a perfectly flat hyperplane at all. Were I to believe that the actual manifold was Minkowskian, then, for all I would know at this point, the universe of occupied points around me might have the shape of a "non-standard simultaneity slice"—that is, it may not be one of the planes in the manifold resulting from employment of the Radar method within a single inertial frame. Perhaps that would be the most natural thing for me to think, in these hypothetical circumstances; but nothing about SR or the A-theory implies that the A-theoretic foliation of a Minkowskian manifold *must* take this form.

Eventually, we must set aside the fiction that our world inhabits the infinite flat manifold described by SR. Even if GR is not the final word about space-time structure,

---

[59]  William Lane Craig has brought to my attention a paragraph from Bergson in which he argues for a single plane of simultaneity, the same for all, in a way that is somewhat analogous to the strategy pursued in this section. Bergson's argument is, however, based on our experience of "indivisible duration": "Each of us feels himself endure: this duration is the flowing, continuous and indivisible, of our inner life. But our inner life includes perceptions, and these perceptions seem to us to involve at the same time ourselves and things. We thus extend our duration to our immediate material surroundings. Since, moreover, these surroundings are themselves surrounded, there is no reason, we think, why our duration is not just as well the duration of all things. This is the reasoning that each of us sketches vaguely, I would almost say, unconsciously" (Gunter, 1969: 128–9). I do not know whether the durations of which I am aware are supposed, by Bergson, to be measurable using my clocks. If they are, I would reject this argument. Since my state of motion changes, using the Radar method to measure distant simultaneity will yield inconsistent results, and, assuming Relativity is right about the amount of metrical structure that is built into the manifold, some possible distributions of matter will provide no other physical phenomena to ground objective distance relations between slices in the A-theoretically preferred foliation. In that case, I should want to say that there is no objective fact of the matter how much time has passed between the occurrence of an event in one slice and the occurrence of an event in another. As Crisp (2008: 266–8) points out, one can combine the A-theory with Relativity (Crisp is concerned, in particular, with GR) while denying that the series of co-present slices displays an "intrinsic temporal metric".

at least our manifold comes closer to satisfying the metrical constraints of GR. Gravity is the manifestation of mass as it warps space-time, and SR's metric is not ours. GR predicts that there will not, in worlds like ours, even *be* universe-wide inertial frames. It would be injudicious (to say the least!) for the A-theorist to suppose that being present is essentially tied to a physical phenomenon that *does not actually exist*! This is one of the points at which one must keep in mind that Minkowskian metrical structure is only approximately correct, and that structure peculiar to it will not prove useful to any A-theorist who is trying to find the shape of the present in the real world.

The foregoing arguments are supposed to show that, under plausible assumptions about causation in a Minkowskian manifold, the largest physical system within which I am embedded should, right now, be limited to a single space-like slice. But might there be other physical systems evolving elsewhere in the manifold, *right now*? Might events be occurring in parts of the manifold at a distance from the current location of my universe?

The question presupposes a flat-footed interpretation of the manifold of SR, according to which it is a four-dimensional, eternally existing, geometrically uniform space in which some of the straight lines constitute paths of inertial and light-like accessibility. On this simple conception of the manifold, the possibility arises, in principle at least, of one family of causally interacting particles and fields moving through one part of the manifold, while another family of particles and fields is moving through regions of the manifold formerly occupied by the first family but (at least so far) causally unconnected with them. Shortly, I will consider three different conceptions of the nature of the non-present parts of the manifold. One of them (my favorite) treats presently occupied points as the only real locations, past and future ones being, strictly, nonexistent (though we have means to describe what they were like, and which current locations are related to them). On such conceptions of the Minkowskian manifold, and within the more realistic manifolds consistent with GR, the possibility of non-interacting "parallel universes" within the same manifold will not arise. (In GR, the geometry of the manifold is not independent of its material contents: it makes no sense to imagine its remaining unchanged while matter comes and goes.)

Still, the possibility of parallel universes in the same manifold is not something that must be ruled out at this point, and its conceivability makes vivid the fact that distances along a path of inertial accessibility are not, strictly speaking, temporal distances. If simultaneous events could, at least in principle, occur at points separated in this dimension, then the distance along the shortest path between them could hardly be a distance within *time itself.* The straightness of paths in the so-called "time-like" directions of the manifold is part of the structure of a substantival entity postulated (like "dark matter") to explain certain observable phenomena. It can do this explanatory job even though extension in this direction is not literally temporal.

So much for the current shape and location of the rest of my universe, on the assumption of SR. What about its past shapes and locations? Everything said so far could have been said by me when I occupied points on my world-line that fall within my rearward light-cone, and could be said in the future when I occupy points in my forward light-cone. The causal constraint requires that the present stages of the processes in my bow-tie region move ever forward, and so one can see the beginnings of an argument for an A-theoretically privileged *foliation* of the manifold (or at least an exhaustive slicing of the parts of the manifold through which my universe moves): a series of slices, each member of which is a set of points in the manifold at which events were happening all at once. This sort of foliation of a Minkowskian manifold need not coincide with any foliation that has a simple metrical description using the resources of SR, like a series of hyperplanes associated with an inertial path through the manifold; and, even if it did coincide with such a foliation, many other foliations would be equally metrically "special", by SR's standards. So the presentist has added some distinctions not found in SR's description of the manifold.

The final section will ask: Just how bad would this addition be? Would it generate a theory *inconsistent* with SR? Would adding an A-theoretically privileged foliation call for a revolution in physics? Or could it merely be the addition of something SR does not, by itself, describe—like, for example, the unique center of mass that some finite universes would contain, and the "privileged frame" that would be associated with that center of mass. Anyhow, what does this matter, if SR is not true? But first, I consider what a presentist should say about the existence and nature of not-currently-occupied points within the manifold; and, in doing so, I respond to an objection to presentism due to Theodore Sider.

## 3.3  Points where nothing now happens

So far, I have been describing the SR manifold as though it is an eternally existing, unchanging four-dimensional object. Many presentists, once they are convinced that we must posit a four-dimensional substantival manifold for physical reasons, will suppose that its points come into existence with the events that occur at them, and immediately (and permanently) cease to exist when the instantaneous events in them have occurred. Call these A-theorists "one-slice presentists". As shall appear, presentists may have reasons to believe in the co-existence of points that stand in relations of light-like and inertial accessibility, and, more generally, to believe in the ongoing existence of all locations in which events *were* but *are no longer* occurring. I give the name "growing-manifold presentists" to A-theorists who accept the existence of just points that are presently or were formerly filled, but who deny the existence of future objects and points not yet filled, and who also deny the existence of Bucephalus and other paradigmatic cases of things that have ceased to be. Above, I surveyed some of the pressures that might force presentists to accept the ongoing existence of rather abstract *resultant states* for every event or state that ever occurs or obtains. Such things would constitute a sort of echo of the entire past—a ghostly history with something like a fourth dimension corresponding to the order in which the resultant states came into existence. Manifold substantivalism, coupled with post-Newtonian theories of motion, may require a similar concession: the ongoing existence of an empty (but haunted) "house" in which the ghostly events once occurred.

Sider has given what he takes to be a powerful argument against presentism, based on the fundamental status accorded to certain cross-temporal relations by physics; but his argument may just as easily be reinterpreted as an argument that the presentist should accept the existence of a four-dimensional manifold that includes, in addition to points at which things are presently happening, also at the very least points that *were* once similarly occupied. (I shall generally ignore questions about future things, including not-yet-filled regions; like many presentists, I am perfectly content to let facts about them remain unsettled.) Of course, if acceptance of empty, formerly filled points represents a high cost for the presentist, Sider's argument still packs a punch. I shall argue that (a) the cost is not so high, and (b) the presentist may not even have to pay it.

Sider's starting point is the fact that, since the rejection of Newton's absolute space, the states of motion physics ascribes to objects seem to require cross-temporal relations that cannot be captured with the resources afforded one-slice presentists by slice-operators alone.[60] The sentences expressible by means of "one-time-at-a-time" tense operators provide the presentist (or other A-theorist) with a series of instantaneous "snapshots" of the world. In the absence of Newton's persisting substantival space— for example, in the space-times of the Galilean theory, SR, and GR—the snapshots merely tell us the relative spatial locations of objects at each instant. But, says Sider, "the sentences [expressible by slice-operators] do not specify how the snapshots line up with each other spatially, since such facts are not facts about what things are like at any one time."[61] The one-slice presentist cannot simply let cross-temporal spatial relations slide, making do only with cross-temporal comparisons of position that are relative to objects persisting throughout the times at which the comparisons are made. It turns out that, according to any of these theories, there is a big physical difference between, for instance, a particle that is moving inertially throughout a period, and a particle that is undergoing acceleration during a period. The relative velocities of two particles may be recovered from the snapshots (assuming the snapshots include information about the identities of particles in different slices), but which particle is moving inertially, it would seem, cannot be recovered. Compare a particle $a$, accelerating to catch up to and pass a particle $b$ in inertial motion during a period $T$, with an unaccelerated particle $a^*$ steadily overtaking and passing a particle $b^*$ that is rapidly decelerating during a similar period $T^*$. $a$ and $b$ may stand in the same relative velocities at each instant in $T$ as $a^*$ and $b^*$ in the corresponding instants of $T^*$. So snapshots of the particles at each instant would not seem to be able to distinguish the two cases. Indeed, in the absence of a space-time manifold that continues to exist, it is not clear how to derive even the *continuity* of the paths of particles from the facts about slices alone.[62]

But why do the slice-operators only "take pictures" of the parts of the manifold that are filled at a given instant? Sider assumes that a presentist must be a one-slice

[60] Sider (2001: 27–33).    [61] Sider (2001: 32).    [62] Sider (2001: 32–35).

presentist, rejecting points that merely *were* or *will be* occupied by events and objects. But suppose that at least the formerly filled points do still exist. Their ongoing existence could preserve the distinctions between continuous and discontinuous, inertial and non-inertial paths taken by particles, and they could do so under at least two different assumptions about their behavior once they are empty. (1) The presentist could adopt a sort of "empty box" view of formerly filled points, and could suppose that formerly occupied locations in the manifold continue to exist with their relations of light-like and inertial accessibility intact—for example, points once occupied by a photon moving through empty space remain at null distances from one another, the end-points of a path along which a particle moved inertially remain at positive distances from one another, and the path remains straight. (2) The presentist could instead conceive of formerly filled points as constituting a sort of "ghostly box"; she could treat these regions in the way the ghostly growing blocker treats past individuals: the formerly filled points continue to exist, but they have only *backward-looking* properties and relations, where the empty-boxer sees spatio-temporal geometry still intact.

A simple example illustrates the difference between the two approaches: Consider a continuous series of points $S$, and another point, $x$; and suppose SR's description of the manifold would have us say that $S$ constitutes a segment of a straight line with positive length, with $x$ as its endpoint in the direction of inertial accessibility (i.e. $x$ is accessible to particles moving along $S$). The empty-boxer will take this to mean that, were a particle now to begin occupying the points along $S$, successively, then—in the absence of forces—it would come to occupy $x$ (in the fullness of time—which, for the empty-boxer, is clearly not just another dimension of the substantival manifold). The ghostly-boxer, on the other hand, will take this to mean that, had a particle successively occupied the points along $S$, then—in the absence of forces—it would have come to occupy $x$, but she will deny that these dispositional facts are *still* the case. Points, once occupied, are mere shadows of their former selves, no longer connected to one another by robust accessibility relations. Either they can no longer be occupied by anything—perhaps because they no longer belong to the kind, *locations*—in much the same way that, in the ghostly growing block, past horses are no longer horses. Or the points *could* be occupied, in principle—that is, it is not absolutely impossible. However, if, somehow, something were in one of them, it would no longer be near any other locations. An object at such a point would be at a place that *used to be* on a path to somewhere, but that is now a dead end.

Given either an empty but intact box, or a ghostly one, past-tense slice operators can describe the facts about the points a particle occupied at every instant in its history; and the present truths about the relations between the points in the empty or ghostly box—truths about which ones are or were mutually inertially or light-like accessible—will fully characterize the shape of the trajectories constituted by these points.

Does acceptance of the existence of a four-dimensional manifold, in the form of an empty or ghostly box, constitute a great cost or ontological burden for the growing-manifold presentist? One-slice presentism does not seem to me vastly superior to presentism with a four-dimensional manifold of either empty or ghostly regions, because the reasons for positing the box are simply the largely empirical reasons which, I take it, support substantivalism as a somewhat surprising, contingent thesis, true because of the kind of universe we inhabit.

Why should one-slice presentism be the "default" version? I do not see that the pull I feel toward presentism has much to do one way or the other with ontological commitment to the structured four-dimensional manifold of points described by Relativity. I begin my philosophical reflection convinced that there exist only a relatively few events and objects. I exist, and the sounds I am hearing; but I find it hard to believe that there are any such things as the Peloponnesian War and Alexander-the-Great's horse, Bucephalus; or the first manned Martian landing and my first great-grandchild. At least, if there *are* such things, I need to be argued into believing in them! On the other hand, I do not—or at least *should* not—begin my philosophical reflection with strong convictions about the existence of quarks, or dark matter. The space-time manifolds of SR and GR resemble quarks and dark matter more than they resemble horses and wars, with respect to our reasons for believing in them. They are theoretically posited entities that earn their keep by the crucial roles they play in successful scientific theories. Suppose I come to believe in a four-dimensional manifold with a specified structure because interactions among objects alone are not enough to explain why observable things behave as they do. Should this bother me, *as a presentist*?

Not much, I think. A space-time manifold is a strange beast—at least, when it is construed substantivally, as a sort of four-dimensional, invisible, permeable cosmic jell-o. The manifold of Galilean or Minkowskian space-time, and the manifolds allowed by GR, are not the kinds of thing one should have posited, had they not seemed necessary to play a role in some well-confirmed scientific theory. An A-theorist, like everyone else, should look to science for information about the structure of such things, including their metrical properties and the number of dimensions they have. My convictions about the unreality of past and future objects and events, on the other hand, are convictions about horses and wars and people: they have little to do with questions about what sorts of theoretical entities should be allowed to figure in scientific theories.[63]

Accepting the ongoing existence of formerly occupied parts of the manifold provides one way to ground the cross-temporal relations to which Sider has drawn attention, and the costs to the presentist do not strike me as terribly high. But I see

---

[63] Shamik Dasgupta and Peter Forrest have pointed out to me that, given *supersubstantivalism* (the thesis that material objects are made up out of those points of space-time which we would ordinarily say are occupied by the objects), the independence of our judgments about the existence of ordinary objects and scientific information about the nature of space-time cannot be kept apart so neatly. I must admit that, for someone attracted to supersubstantivalism, these judgments may be less independent than I have portrayed them. But even an empty-boxer supersubstantivalist need not suppose that Bucephalus still exists, just because the space-time points that once made him up continue to exist and bear metrical relations to one another. If they no longer exemplify the material properties they did when they were present, it seems to me that a supersubstantivalist with presentist inclinations should say, *not* that Bucephalus exists but is now an empty region, but rather that the points in this empty region once constituted a horse, but do so no longer.

the makings of a still more excellent way—at least, in a Minkowskian manifold, and probably in foliable GR space-times as well.

If one takes for granted the metric structure of Minkowskian space-time or a not-too-bizarre manifold satisfying GR's constraints, surrogates for past points can easily be constructed out of the points in the present slice. For each past point, there is a region in the presently existing slice of the manifold that contains all and only the points on the slice that were inertially or light-like accessible from the past point; the region in question is the presently existing slice of the point's forward light-cone. In SR and foliable GR space-times, these regions could be used as descriptive names for each formerly filled, now non-existent space-time point—each such point has exactly one point-surrogate in the presently existing slice. If the presentist is allowed to help herself to the facts about which collections of points constitute point-surrogates, the current geometry of the present slice will include enough information to recover all the facts about which past space-time points constituted inertial and light-like paths. For every presently existing point $p$ and every inertial or light-like path a particle could have taken that leads up to $p$, there is a unique set of point-surrogates consisting of all and only the surrogates for points on that path. In Minkowskian space-time, that is all the metrical structure there is. I hope that the general strategy could be extended to GR. I believe that, in foliable GR manifolds, the present slice can be relied upon to include a surrogate for each past point; and I suspect that all the geometrical properties of paths through the manifold could be recovered, given: (i) facts about which sets of past-point-surrogates lie along geodesics ending in presently existing points, plus (ii) facts about which past points constituted a privileged slice, and what its intrinsic curvature was like. But I confess that a proof of the adequacy of this approach is beyond me.

Although, in general, I am setting truthmaker worries to one side, the nature of the current proposal will be made a bit clearer by advancing a possible account of the ontological grounds in the present for the space-time structure of nested light-cones characterizing past points. The fact that a certain region constitutes a point-surrogate (it represents all and only the present points accessible from a single past point), together with the present facts about overlap of point-surrogates, encodes a lot of information about the past. Take a point $p$, and two past-point-surrogates $R1$ and $R2$. Suppose that what needs present grounding is the fact that the shortest path between these three points was a straight time-like line in Minkowski space-time. A one-slice presentist could fall silent, claiming that there is no more to say about the grounds for this fact than that $p$, $R1$, and $R2$ are "co-trajectoried"—a relationship holding among the point and the two regions just in case they are point surrogates for no longer existing points that stood in inertial accessibility relations to one another and to the present point. But it would be nice to be able to say something more: a relation like the proposed *being co-trajectoried*—one that only holds among instantaneous things— seems a funny sort of relation to be at the basis of *cross-temporal* space-time structure. Would it not be better if the straightness of a path throughout a period were based upon features of something that persists throughout the period?

Earlier, I gave reasons why simple claims about the past (e.g. "England has had two kings named 'Charles'") have been thought to force presentists to recognize a host of resultant states (e.g. two states of England's having had a king named 'Charles'). A one-slice presentist who accepts the ongoing existence of resultant states could make use of them here, positing a persisting state for every inertial path that passes through a presently existing point. For each time-like straight line that a B-theorist sees in a Minkowskian manifold, the one-slice presentist will see a possible inertial trajectory, only one point of which actually exists. A one-slice presentist looks at a point, $p$, in the presently existing slice and sees infinitely many different ways in which a point-sized particle in inertial motion could have reached $p$—infinitely many inertial trajectories meeting at the point. For each of these trajectories, she could posit a distinct state consisting of $p$'s *being part of an inertially connected trajectory*, a state that is occurring to $p$ now, and that has occurred to the continuously many other points in the past that would have been occupied by a particle in inertial motion on that trajectory. On this metaphysics of the manifold's structure, the states I am calling "trajectories" outlive the past space-time points to which they occurred; and facts about past space-time points, and about which ones were mutually inertially accessible, are grounded in facts about these current states, and facts about which ones co-occurred—that is, which such states overlapped by happening to the same point in the past. In fact, it is tempting to regard the trajectories as not just the grounds for the metrical relations among points, but the grounds for their very existence. A point in the manifold could be identified with the set of trajectories that uniquely converge upon it. Reducing points to trajectories would mean that the co-occurrence or intersection of trajectories could not be explained in terms of the trajectories "happening to the same point". However, given substantivalism's requirement that we posit, among the most brutal of facts, a manifold with intrinsic Minkowskian structure, I see no objection to construing this brute structure in terms of brute physical facts about which groups of trajectories have and have not co-occurred or converged. There are general truth-maker worries about what grounds truths about the past of a presently existing thing, but these truths about backward-looking properties of trajectories seem little worse than truths about whether I was happy yesterday.

Is it possible, in the context of GR, to deny the existence of past space-time points and to recover past space-time structure by appeal only to persisting trajectories and their past co-occurrence relations? I am not sure. In foliable space-times, the strategy of constructing unique present surrogates for past space-time points will procure a surrogate for every point from which a signal could have been sent. But one would have to tell a slightly different story about what makes for the straightness of a trajectory in a GR manifold; the gloss I gave about the relations among points along inertial paths in Minkowski space-time could not be quite right. The presence of a particle with any mass alters the shape of space-time, in GR, so it is problematic to treat formerly empty straight time-like lines as having the shape they would have had, had there been particles moving along them. (The surrounding space-time would have to display a different metrical structure, raising doubts about the "transworld identity" of the path itself.) As a first pass, one can say at least this much about the physical meaning of straight time-like paths in a GR manifold: they are the possible paths of idealized massless test particles. There may be other obstacles to utilizing the one-slice presentist strategy in the context of GR, in which case the presentist is once again under pressure to admit the ongoing existence of formerly occupied manifold points.[64]

How ad hoc and revisionary are these one-slice, trajectory-based strategies for the present grounding of past space-time structure? The one-slice presentist must, in general, be willing to allow present truths about the past to be true in virtue of backward-looking states and properties of presently existing things. Given the existence of point-sized locations, if they stand in significant relations to no longer existing things, it must be in virtue of backward-looking states. Relativity requires an infinity of different inertial trajectories by means of which a thing could get into a given point. If we accept the need to posit a resultant state for every past event or state—a view that has some independent support from semantics, and seems to be needed by the presentist in order to deal with more mundane past-tense claims—it is but a short step to recognize an infinity of distinct states happening to a point, each representing one inertial trajectory by which a thing could get to that point. Such states do not seem much stranger than other backward-looking states.[65]

This constitutes my response to Sider's objection to presentism based on the need for cross-temporal relations in order to distinguish different states of motion. In the final section, I shall consider some objections that appeal to Relativity. But first, as an aside, I note that the metaphysics of the manifold advocated here affords the presentist a ready reply to one of the stock philosophical objections to the A-theory, based on the alleged need for "hyper-time".

---

[64]  For a better-informed and more detailed discussion, see Crisp (2007). Crisp reconciles presentism and GR in a way that is neutral between two hypotheses: (i) a one-slice metaphysics, according to which space is constituted by a different set of points at each time; and (ii) a persisting space metaphysics, according to which space is always constituted by the same point-sized parts. (The neutrality of his theory is not emphasized in Crisp's exposition, which favors (ii); but it is there, upon examination.) Crisp's approach combined with (i) yields a theory positing less complexity in the structure of the physical world, since it abolishes absolute sameness of place; while (ii) retains absolute location as a physically inert factor. The choice between (i) and (ii) is a delicate one, however, turning on subtle questions about the weightings of different theoretical virtues. (ii) introduces physically idle objective facts about sameness of region, which I should count as a major strike against it, if an alternative is available. On the other hand, if the presentist finds herself forced toward an empty box or ghostly box with four-dimensions, a three-dimensional persisting space might seem preferable. But whatever costs would be incurred under (ii) by positing the formerly filled space-time points, in addition to presently filled ones, it is eliminated if points can be constructed out of trajectories instead of the reverse.

[65]  There is a quite different approach to lining up snapshots due to Barbour. Barbour's "best matching" technique is designed to take a collection of "Nows or 'instants of time'" (Barbour, 1999: 177), three-dimensional slices of a relativistic manifold corresponding to one of its foliations, and to put them back together in the right order and with the cross-temporal geometry intact. His method works particularly well in the context of GR (Barbour, 1999: 167–77). Barbour may think of the procedure as "eliminating" time, but the A-theorist need not.

## 3.4 Yes, but is it space-time?

For the B-theorist, the slices of the four-dimensional, substantival manifolds described by SR and GR can be taken to include within them a genuinely temporal dimension. Not so for the presentist. The empty boxer, the ghostly boxer, and the one slice presentist can all agree about the following: points that were occupied, but are so no longer, are not in any straightforward sense temporally related to presently occupied points. Separation in the direction of inertial accessibility is not literal temporal separation; to call a point "past" is simply to say that events were happening there, but are not now happening there. The empty boxer cannot say that "past points" are earlier than present events, in some absolute sense. In principle, events could happen once again at formerly occupied points, and parallel universes might be moving through the manifold, the one ahead of the other—in which case, the formerly filled points would also be soon-to-be-filled points; such regions would, in a sense, be both part of the past and the future, and not, in themselves, either earlier or later than the events presently going on. Ghostly boxers and one slice presentists will most likely deny the possibility of formerly occupied parts of the manifold coming to be occupied again. The former will probably want to say: once ghostly, always ghostly. And the latter will likely hold that points can never come back into existence once they have ceased to be. But these presentists, too, will deny that the distance relations between points filled successively are straightforward temporal distances, like *being five minutes earlier than*. The points filled five minutes ago are not five minutes earlier than current events: it is the *event of their being filled* that is five minutes earlier than current events.

The fact that the slices of an A-theorist's manifold are not straightforwardly earlier or later than one another helps defuse a common philosophical objection to the A-theory—namely, that the A-theory requires an implausible commitment to at least two temporal dimensions; and, if two, then infinitely many. Suppose, contrary to fact, that the distances between distinct slices in an A-theorist's manifold *were* genuinely temporal distances, like *five minutes earlier than*. In that case, her manifold would have a temporal dimension built right into it, and when she also affirms that parts of it are filled but will no longer be filled, she would have thereby generated a second temporal dimension—a "hyper-time" in which changes occur to a thing that already possesses its own, distinct, intrinsic temporal ordering.[66] But, as I argued, that is not how presentists should think about the manifold; and, indeed, "space-time" is a very misleading label for our manifold (and so it is a label I have tried not to use when describing the A-theorist's version of Minkowski's four-dimensional manifold). Like the B-theorist's space-time, it consists of possible locations in which events can happen, and both presentist and B-theorist manifolds have a metrical structure that satisfies the mathematical description of a Minkowskian four-dimensional space. Nevertheless, the fourth dimension of the presentist's (so-called) "space-time" is not time itself. The

---

[66] Smart (1963: 136) gives an argument against the A-theorist along these lines.

presentist's manifold has no true temporal extent to it (except in the way ordinary objects have temporal extent: namely, by existing for a while, which the empty box and ghostly box views allow the parts of the manifold to do); and the real meaning of its various dimensions is given by the physical theory that demands that we posit such a manifold. And that theory, SR, ties its most fundamental "time-like" distances to states of motion of particles, and the propagation of light; consequently, instead of calling the relations between such points "time-like" and "light-like", I prefer to speak of inertial accessibility and light-like accessibility. The relations in question may or may not be *fundamentally* dispositional—perhaps there are categorical relations among points that underlie and explain why some are mutually inertially accessible, others mutually light-like accessible, and so on. But whatever their ultimate nature, the basic "time-like" distance relations in the presentist's manifold will not be ordinary temporal relations like *being five minutes earlier than*. The wise presentists will insist, with good reason, that these ordinary, truly temporal relations simply do not characterize the substantival manifold posited by physical theories like SR or GR.

The B-theorist is in no position to criticize the presentist for making use of fundamental manifold structure that does not correspond neatly to ordinary spatial or temporal relations. For the B-theorist should agree with at least this much of the presentist's account of space-time geometry: everyday expressions for temporal relations, like "five minutes later than", do not correspond directly to the truly fundamental distance relations in Minkowski space-time. Many ordinary temporal judgments presuppose that spatially distant events can be simultaneous; if such judgments can be true at all, they must be construed as invoking frame-relative temporal notions. *Being five minutes earlier than*, for example, would seem to be infected with such frame-relativity; for I can ask, "What was going on in New York and L.A. five minutes ago?" The B-theorist can provide plenty of frame-relative temporal distance relations that are pretty good candidates for what I meant by this; but none of them would correspond to a fundamental metrical relation in Minkowski space-time.

# 4. PRESENTISM AND RELATIVITY

## 4.1 Sketch of Putnam's argument

Hilary Putnam, Theodore Sider, and others have claimed that presentism is inconsistent with SR, and that this constitutes a conclusive refutation of presentism.[67] On the basis of conflict with SR, Putnam concludes that "the problem of the reality and the determinateness of future events is solved...; by physics and not by philosophy".[68]

---

[67]  See Putnam (1967), Sider (2001: 42–52). Mellor (1998: 55–57) endorses the objection, although he does not regard it as his main argument against the A-theory.

[68]  Putnam (1967: 247); a similar argument for a slightly different conclusion can be found in Rietdijk (1966).

According to Sider, the argument that SR and presentism are inconsistent "is often (justifiably, I think) considered to be the fatal blow to presentism."[69]

Putnam (rightly, by my lights) attributes to "the man on the street" (and, presumably, the women there too) a combination of views that amounts to presentism: "All (and only) things that exist *now* are real"—and he insists that, by "real", we ordinary people-in-the-street do not mean something merely relative, so that what is real-to-me might not be real-to-you: we mean to be talking about a transitive, symmetric, and reflexive equivalence relation, one that holds between events currently happening to us and at least some other events happening elsewhere, to other things—including events happening to things in motion relative to us.[70] He then assumes that this equivalence relation must by "definable in a 'tenseless' way in terms of the fundamental notions of physics". But the metric of a Minkowskian space-time does not include a relation that fits the bill—one that will carve the manifold into equivalence classes of co-present points in a way that does not look "accidental (physically speaking)".[71] "Simultaneity relative to coordinate system $x$", for some arbitrarily chosen inertial frame of reference, will provide an equivalence relation, all right, but there are infinitely many coordinate systems to choose from, and nothing physically special about just one of them. No other relations look any more promising.[72] So Putnam concludes that presentism is inconsistent with SR, and alleges that this inconsistency proves presentism's falsehood.

Distracting elements in Putnam's presentation and a misunderstanding of his intentions have rendered some discussions of his argument otiose. Stein started things off on the wrong foot, raising objections to Putnam's argument that no presentist or other A-theorist could use.[73] If Stein's response were an adequate rebuttal of Putnam's intended conclusion, Putnam could not have been attacking presentism or the A-theory, after all. Although some philosophers of physics have thought Stein's reply to Putnam was a success,[74] it seems obvious to me (and to many others[75]) that, whatever Stein may have taken to be the target of Putnam's argument, Putnam was indeed attacking presentism, and Stein's reply was simply a red herring. After a couple of muddles are cleared up, Putnam's argument turns out to be relatively simple: The presentist supposes that one foliation of Minkowskian space-time is very special; but no foliation is geometrically special; the presentist must, then, deny that SR tells the full story about space-time structure; and that is tantamount to rejecting SR.

---

[69]  Sider (2001: 42).

[70]  Putnam (1967: 240–1); Putnam does not, in his argument, emphasize the need for symmetry in the "real-for" relation; but, as Saunders (2002: 282–3) points out, Putnam pretty clearly does, and should, assume its symmetry.

[71]  Putnam (1967: 241).

[72]  Putnam (1967: 242–3).

[73]  Stein (1968); see also Stein (1991).

[74]  E.g. Dickson (1998), and Clifton and Hogarth (1995).

[75]  Callender and Weingard (2000), Saunders (2002), and Savitt (2000) provide insightful (and, to my mind, decisive) analysis of the Putnam–Stein debate, and its confusing aftermath. See also Peacock (2006: 248–50).

Ted Sider offers an argument that is similar to Putnam's, at least in its overall thrust. After running through all the ways a presentist might try to define the shape of the present in terms of the manifold's Minkowskian geometry, Sider concludes that the presentist has little alternative but to suppose that the present effects a foliation that is "arbitrary"—that is, one not "distinguished by the intrinsic geometry of Minkowski spacetime."[76] But positing such a thing is "scientifically revisionary"; if presentists take this route, "[a] physical theory of time other than special relativity must be constructed".[77]

Many philosophers have endorsed the Putnam-Sider line of reasoning: SR is incompatible with the introduction of an A-theoretically privileged foliation, and this incompatibility is enough to refute presentism, along with any other A-theory that requires a similar addition.[78] Others are more cautious. For instance, Steve Savitt, Simon Saunders, and Bradley Monton agree that the presentist who found herself inhabiting an otherwise Minkowskian manifold would have to posit a privileged foliation; and that doing so would be incompatible with SR.[79] But Savitt, Saunders, and Monton go on to note that the relevance of such incompatibility is not entirely obvious, given the fact that SR is *not* our best theory of the space-time manifold. It is, after all, only approximately true,[80] and GR is, if not the final word, at least a closer approximation. Saunders and Savitt point out that GR may prove to be a more hospitable environment for the presentist than SR.[81] Saunders and Monton note that the difficulties squaring

---

[76] Sider (2001: 47).

[77] Sider (2001: 52).

[78] Here, for example, is the endorsement of Ian Gibson and Oliver Pooley: "We take it that relativity rules decisively against both the non-eternalist and the tenser [i.e., against all versions of the A-theory]. … Both presentism and tensed theories of time need an objectively privileged set of subregions of spacetime, each of which can serve as the present as 'time passes' (however this is to be interpreted!). Relativistic physics simply does not provide such a set" (Gibson and Pooley (2008: 159). (To be fair, it is possible that they mean GR, not SR, by "relativity"; but since the paper concerns persistence in Minkowski space-time, I take them to be agreeing with Putnam and Sider: incompatibility with SR refutes presentism.)

[79] Saunders, after tidying up Putnam's argument, provides a concise summary: "The argument is so simple that it speaks for itself. No technical result is needed: it is of the essence of the theory of special relativity that absolute simultaneity does not exist. Everyone knows there is nothing else to replace it—there is no other non-trivial symmetric and transitive relation intrinsic to Minkowski space" (Saunders, 2002: 279–80). Savitt, too, thinks SR is incompatible with supposing that a geometrically undistinguished foliation is special. To choose one hyperplane of simultaneity as "*the* present" is "a form of inertial chauvinism". Such chauvinism may simply be "a strategy for rejecting special relativity in favor of presentism rather than accommodating presentism in Minkowski spacetime". But Savitt seems to think it would not constitute rejection of SR if the presentist says the A-theoretically privileged foliation is "metaphysically distinguished" but irrelevant for physics. He thinks no presentist should be comfortable holding this sort of view, for reasons discussed below (Savitt, 2000: S570). Monton thinks "presentism is incompatible with special and general relativity" because the manifolds of these theories "do not have a foliation into spacelike hypersurfaces as part of their structure" (Monton, 2006: 267).

[80] Savitt (2000: S572) cites Misner et al. (1973, 187): "the existence of the gravitational redshift shows that a consistent theory of gravity cannot be constructed within the framework of special relativity."

[81] Saunders (2002: 290–1) and Savitt (2000: S572–3); Savitt is more pessimistic than Saunders about the prospects for presentism in GR.

GR with quantum theory throw its status into question as well; and a successor theory uniting gravity and quantum theory holds out even more hope for the presentist who is looking for a physically privileged foliation.[82]

In this section, I shall tease apart two things all these authors might mean by their talk of "inconsistency" or "incompatibility" with SR. One fairly clear criterion for inconsistency with SR has been put to use in debates about the compatibility of quantum theory and SR: a theory is inconsistent with SR if the laws of the theory appeal to "intrinsic structure" pertaining to "space-time itself", structure that goes beyond the Minkowskian metric. Several interpretations of quantum theory require the addition of a foliation that would constitute additional intrinsic structure, in this sense. But it is worth pointing out that not every added foliation qualifies as inconsistent by this standard. A theory can even add a causally-relevant foliation without violating SR, so long as the laws that make use of the foliation appeal not to intrinsic structure but to a foliation privileged by the material contents of the manifold. On a second, much weaker reading of inconsistency with SR, the A-theory will stand guilty; but the seriousness of the charge is far from clear. After disentangling these two forms of potential conflict with SR, I emphasize the strangely hypothetical nature of the supposed conflict: SR is false, and the outcome of extending the criticism to more realistic contexts is uncertain. Finally, I shall respond to Craig Callender's claim that A-theorists cannot expect aid and comfort from quantum theory.

## 4.2  Stronger and weaker forms of "inconsistency with SR"

The grounds adduced for the inconsistency of SR and presentism seem straightforward enough, on first blush. As Saunders puts it: the metric of a Minkowskian manifold does not contain a "non-trivial symmetric and transitive relation" that would divide the manifold into intrinsically distinctive equivalence classes, each of which could serve as a member of the presentist's privileged foliation. Of course the presentist's successively co-filled regions *might* happen to coincide with a series of slices having a relatively short geometrical description (e.g. they might coincide with the simultaneity-slices of one particular inertial frame). But even that much geometrical naturalness is not required by presentism; and, in any case, even a neat series of hyperplanes would have infinitely many rivals, cutting across the first series but satisfying the same geometrical description. Adding an A-theoretically privileged foliation to a metrically Minkowskian manifold inevitably results in a theory according to which there is more to space-time than what is described by SR.

But more needs to be said about the conditions under which an added foliation constitutes additional structure inconsistent with SR. Obviously, not just any addition of contents to the manifold should count as inconsistent with SR. The theory does not pretend to tell the whole truth about the actual physical universe we inhabit; it must be

---

[82]  Saunders (2002: 291), and Monton (2006: 265–66).

supplemented with additional laws about the behavior of specific physical phenomena, and with contingent facts about the material contents of space-time (e.g. "initial conditions"), before it can begin to predict any actual occurrences. Such supplementary physical theories and contingent facts "add structure to the manifold", in the broadest sense, without falling afoul of Relativity. The presentist posits a privileged foliation, all right; but why is this not simply adding some contents to space-time, no worse than a physical theory that attributes a certain finite shape to the material contents of the universe, say, or a theory of electro-magnetism that ascribes field values to every point in the manifold?

Those who accuse the A-theory of inconsistency with SR must be relying upon a criterion according to which the A-theorist's foliation constitutes an impermissible addition to Minkowski space-time, but a big bang theorist's bounded physical universe counts as harmless additional contents. The best place to look for explicit attempts to develop a criterion that would yield such a ruling is the debate over whether quantum theory can be interpreted so as to be consistent with Relativity. Tim Maudlin's work on this topic is particularly subtle and illuminating. Maudlin formulates his proposed criterion in this way: a theory is inconsistent with SR if it attributes more "intrinsic structure" to the manifold than is found in its Minkowskian geometrical properties. More generally, consistency with SR or GR is, he says, "a matter of formulating a theory so that it employs nothing more than the metric when describing space-time itself."[83] B-theorist critics invoke the prestige of physics, alleging that presentism is refuted "by physics itself" or claiming that it is "scientifically revisionary". By doing so, I take them to have incurred an obligation to use a notion of inconsistency with SR that is relevant to conflicts between SR and other scientific theories. I shall assume that, when scientists or philosophers of physics are asking about the inconsistency of one or another theory with Relativity, it is something like Maudlin's "additional intrinsic structure" criterion that is—or at least should be—in play.

Maudlin explores the subtleties of this criterion—when is a feature "intrinsic", belonging to "space-time itself"?—but one thing seems obvious enough: There are possible distributions of matter in a space-time with Minkowskian metrical properties that are consistent with SR, despite the fact that they effectively "privilege a foliation". Here is what should be an uncontroversial example: suppose there were, spread evenly throughout the cosmos, a kind of particle every member of which is moving inertially and at rest relative to every other. This family of fellow-travelers would select an inertial frame; and there would be exactly one foliation of the Minkowskian manifold in which every slice is orthogonal to the path of every one of the special particles.[84] Since the foliation is the result each particle would get by using the Radar method to settle questions of distant simultaneity, the foliation could be called the "optical simultaneity slices" relative to that frame. The family of particles is, by hypothesis, very special; and the frame they pick out is, for that reason, also special. Should we

---

[83] Maudlin (2002: 230–31).

[84] Compare Belot's (2005: 262–3) "augmented Minkowski spacetime", with its field of inertial observers at relative rest.

say that any physical theory that posited such particles would be inconsistent with SR? If, according to the theory, the particles just *happen* to be traveling together in this way, then surely not. So long as the choice of their inertial frame is a contingent matter determined by initial conditions, it should not be attributed to space-time itself, even if they *must* travel on parallel paths. The particles choose a set of parallel inertial paths, and make these paths and the accompanying foliation special, but there need be nothing intrinsically special about the paths in virtue of which the particles *must* take them, rather than those of some other inertial frame.

A less contrived example of a foliation-selecting physical phenomenon in a Minkowskian manifold is the frame of reference associated with the center of mass of the universe. The center of mass determines an inertial frame and a corresponding set of optical simultaneity slices. Now, it turns out that one cannot, in SR, always count on there *being* such a frame; in some cases of matter distributed in a Minkowskian manifold, there will be too many equally good contenders for "the" center-of-mass frame; and in others, none at all.[85] So, like the field of special particles, the presence or absence of a unique center of mass is a contingent matter—though not so surprising a contingency as a uniform distribution of co-moving particles. But again, the point of the example is that no one would suppose this hypothesis—that the universe has a center of mass—is inconsistent with SR, despite the fact that it makes one foliation of the manifold look special. What makes the A-theorist's foliation worse? The critic alleging inconsistency between presentism and SR owes us some explanation of the difference between an A-theorist's additional structure and these examples of harmless-though-foliation-privileging material contents.

One might try to make heavy weather out of the idea that attributing mass and energy to regions of space-time does not add intrinsic features to the manifold—even if it happens to privilege a foliation—since particles and fields are, in some sense, extrinsic to the manifold in which they are located. This seems a poor place to draw a deep distinction, however. Do we really want to say that the presence of matter or energy in a region is an *extrinsic* fact about that region? The physical contents of a region seem tolerably intrinsic to it. Given the representation of matter and energy fields by tensors, the physical features at a point have *implications* for arbitrarily nearby regions, but the field values at a point still seem *relatively* intrinsic to the point, and certainly intrinsic to regions that include a sphere of any size with the point at its center. One might try saying that the A-theorist's foliation depends upon properties exemplified by parts of the manifold itself, while the mass-energy distribution does not. But this, too, is a doubtful move. The value a field has at a point is often said to be a property of the point itself; and "supersubstantivalists" go further, treating particles as parts of the manifold that display the right sorts of properties. No one has ever suggested that either of these doctrines requires the denial of SR. And, in any case, it is not clear that the A-theorist must regard the fact that the points in a slice are all co-present as being due to properties the points themselves exemplify. A-theorists are not obliged to be supersubstantivalists. Why should we not say that the slices in the

---

[85] See Maudlin (2002: 203–4).

A-theoretically privileged foliation are special in virtue of facts about their contents, not in virtue of anything the points in these slices are doing all by themselves? The points in a special slice could be said to be co-present in virtue of the fact that the particles and fields occupying those points are occupying them all at once.

How else might the critic of presentism distinguish between benign extrinsically imposed structure, and SR-violating intrinsic structure—and do so in such a way that foliation-privileging material contents qualify as extrinsic, while the metrical properties of Minkowskian space-time and the presentist's foliation qualify as intrinsic? One might try to forge a connection between extrinsicness and *contingency*: had the material contents of a region failed to be there, the region could still have existed, but had presentness failed to strike a region of space-time, that region could not have existed. One might think the presentist is committed to the latter thesis, because, on most A-theorists' conception of things, time (and so space-time) is impossible without an objective, moving present.

The hypothetical wielders of the proposed contingency-criterion are granting that the distribution of mass and energy in a region can be consistent with SR, yet impart additional intrinsic structure to the region in at least one sense: the mass-energy distribution is entirely a matter of what is happening *within* the boundaries of the filled region. But they are claiming that the region itself could have existed apart from the physical phenomena filling it—either with different fields and particles, or utterly empty of all but metrical properties. And the claim has plausibility, at least while we are pretending that SR describes the true manifold structure. Consider, for example, a finite universe in a Minkowskian manifold. Did the boundaries of the physical universe *have* to be located at precisely these space-time points? Not if this very manifold could have existed without matter, or with matter elsewhere instead. (Could the region occupied by the physical universe have existed with different metrical properties? That seems much less plausible—though I should not know how to begin assessing the pros and cons.) These suppositions suggest a distinction that is in the right neighborhood; one that puts ordinary, physical content on one side and the metric of space-time on the other: A theory posits additional intrinsic structure belonging to space-time itself if it attributes *non-contingent* intrinsic structure that goes beyond the metric, but attributing contingent intrinsic structure is not a problem. But does this sort of criterion yield the desired result: that the presentist's foliation is intrinsic?

If the contingency of the shape of the physical universe in a Minkowskian manifold is sufficient reason to call that feature extrinsic to the structure of space-time itself, then the presentist's privileged foliation could easily qualify as well. After all, the reasons I rehearsed for being an A-theorist did not compel the presentist to identify the privileged foliation with that of an inertial frame, let alone a *particular* inertial frame. It might be a necessary fact, with respect to each point of space-time, that it is included in some A-theoretically privileged slice or another; but the actual "angle" with which the manifold is cut by successive co-present events, and even whether the slices are flat simultaneity slices, can be regarded as deeply contingent. So, if SR were otherwise adequate as a description of the metrical properties of the manifold, and the contingency-criterion were used to judge whether some bit of additional structure fails to violate SR, the A-theorist's privileged foliation could be—and, by my lights, should be—regarded as contingent, and therefore consistent with SR.

One might try imposing an even stronger contingency-criterion for determining whether something is intrinsic structure that does not count as a violation of SR. The presence of a finite physical universe within a Minkowskian manifold, $M$, would be a quite radically contingent feature of $M$, if $M$ could have existed with *no* matter or other physical phenomena inside it. An A-theorist who believes the A-theory is necessarily true—there could be no temporal universe without objective facts about past, present, and future—might seem to be in a bind here. Could the presentist go a comparable distance toward contingency, supposing that the Minkowskian manifold has a privileged foliation but could have existed with no privileged foliation—no series of slices of successively co-present events? If the unproblematic nature of contingent physical "fillings" of space-time requires that one take seriously the idea of an utterly empty manifold, I do not see why the presentist should not be able to do the same; the empty-boxer, at least, would have an easy time of it. Imagine a community of non-spatial Cartesian souls, communicating telepathically (the detail is added so as to make it clear that time is truly passing in the world they inhabit), and imagine further that, co-existing with these souls, there is the empty-box presentist's four-dimensional manifold of points, satisfying the geometrical description of a Minkowskian manifold. Spatially located events *could* happen at its points, although no such events ever happen in the world inhabited by the souls. The result is a Minkowskian manifold that could have contained an A-theoretically privileged foliation, but that does not as a matter of fact do so. It appears that, just as Minkowski space-time can be imagined without material contents, the empty-boxer, at least, can imagine it without a privileged foliation—even if the A-theory is necessarily true. A one-slice or ghostly-box presentist could make this move, so long as she could convince herself that points of Minkowski space-time which in fact never co-exist, or in fact never co-exist while exemplifying their fundamental causal structure, *could* co-exist while devoid of events and objects. If the contingency of physical phenomena in a region can only be secured by the critic's insistence that the points of Minkowski space-time could have existed with nothing in them, it is far from clear that the presentist, of whatever stripe, cannot legitimately claim to be able to imagine similar scenarios—possible circumstances in which the points exist without ever being present.

I do not say that this way of understanding inconsistency with SR has much going for it. For one thing, it would be difficult to extend this contingency criterion to a manifold satisfying GR, and still derive the result that its material contents are extrinsic but its metrical properties are not. Given GR, one cannot simply imagine away material contents without altering the metrical properties of the manifold—unless the contents are replaced by ones that put the very same constraints upon the space-time metric (for example, a proton in a region would have to be replaced by a particle with the exact same mass and any other properties affecting space-time metric—could such a particle be anything but a proton?).

A salient fact about the two examples of frame-privileging, SR-consistent physical phenomena—the inertial frame of the family of co-moving particles, and the center of mass frame—is that the foliation each privileges is causally inefficacious; at least, nothing in my descriptions of these frames gave them any causal job to do. This thought provides a more promising explanation of the innocence of frames distinguished by the material contents of the manifold. Here is the rough idea: It is no crime against Minkowskian space-time to posit contingent phenomena that happen to distinguish one frame from all the others, but what one cannot do, consistent with SR, is to posit *laws* governing some phenomenon (whether the phenomenon be physical or metaphysical) that *directly invoke* a privileged frame. What is inconsistent with merely Minkowskian intrinsic structure is to explain some fact about the contents of space-time as being due to the special nature of one foliation, and then not be able to appeal to any deeper laws that fail to mention that foliation. If the laws of a theory merely pick out the relevant frame of reference in terms of contingent material contents, and the contents merely *happen* to pick out that frame; then it is the material contents that are doing the work. But if a theory's most basic laws (whether they govern physical or metaphysical features of the manifold) must invoke one inertial frame of reference or foliation "by name", as it were; then there is something special about the frame or foliation itself, quite apart from the manifold's content. The law is an indication that the manifold includes built-in "rails", directing things in a certain way; some structure that is part of space-time itself is doing the work.

This is, put very roughly, the option Maudlin takes in his discussion of the distinction between intrinsic space-time structure and mere material contents. Maudlin shows that, although the basic idea is clear enough, it can be a subtle matter whether a particular theory invokes laws that violate a criterion along these lines. When do the laws of a theory *directly* invoke a privileged foliation, frame of reference, or other aspect of space-time and its contents? It is not always obvious. Maudlin articulates criteria that, "applied with good taste, are the best we have to go on when trying to determine the intrinsic structure of space-time." Still, "it is not always clear how one determines when these criteria have been met."[86] I will not try to do full justice to the details of Maudlin's discussion of the criteria. But the difficulties he notes can be brought out, in an informal way, by briefly describing one of his chief examples: the question whether positing a lumeniferous ether constitutes a return to Newtonian absolute space.

The fundamental metrical structure of Galilean space-time does not include a cross-temporal relation of sameness-of-place—unlike Newton's persisting three-dimensional space, which, when combined with time, results in a manifold of place-times that *does* contain such a relation. In the absence of absolute sameness of place, Galilean space-time does not admit absolute velocities, only velocities that are relative to one or another frame of reference—which can be thought of as a set of (real or merely possible) objects in inertial motion that do not move relative to one another. By the nineteenth century, scientists knew that the velocity of light did not depend

[86] Maudlin (2002: 194).

upon the velocity of its source, strongly suggesting that it behaved like a wave in a material medium; and so they set about the task of discovering the frame of reference in which this medium, the ether, was at rest. Although they failed to find it, one wonders: "[W]hat would have happened if the ether frame had been detected, if light propagated in only one inertial frame?"[87] Would such a discovery have justified rejection of Galilean space-time in favor of the Newtonian variety?

If the laws governing light (and other electromagnetic phenomena) contain velocities, they appeal to what would be, in Galilean space-time, a frame-relative property of light. So long as there is a material medium to distinguish the relevant frame of reference, no additional space-time structure is needed; the laws of the theory need not directly appeal to the frame of reference, but can be formulated in terms of velocity relative to the material medium. But is the ether an independent, material medium? Maudlin offers a number of considerations relevant to answering this sort of question, only a couple of which I shall describe here. Does the so-called "ether" admit of different properties in different places? According to "ether drag" theories, for example, the speed of light would have been determined by different frames of reference in different locations. On such a theory, the ether seems much like a universal fluid, and space-time itself remains Galilean. Does the "ether" have to be everywhere? If the ether had turned out not to be universal (if, say, there were "ether vacuums" through which light could not pass), then again the ether would seem more like material content, less like intrinsic space-time structure. But if the so-called "ether" existed everywhere, determining one inertial frame; and if it were said to have no other properties characteristic of material things (e.g. it cannot be "thinned out or compressed"); then, by Maudlin's lights, it would have to be regarded as "sufficiently 'ethereal' to escape classification as a material substance", and "the existence of the naturally preferred frame of reference would have to be taken as evidence that Newton was right and that there is more to space and time than the Galilean structure."[88]

Maudlin's goal is to develop a criterion for determining whether various interpretations of quantum theory are inconsistent with SR in virtue of positing intrinsic space-time structure beyond the Minkowskian metric. Some versions of quantum theory require non-relative facts about which of two spatially separated events is earlier than which. (Two theories of this sort will be described shortly.) These extra facts about quantum-theoretic priority imply that one foliation of the manifold is especially relevant to the results of quantum measurements. One might think that, just in virtue of including laws about a privileged foliation, such theories would have to be inconsistent with SR. But the morals learned from the case of the ether suggest that there may be subtleties here, and that one must proceed with caution. The mere presence of some contingent phenomenon that privileges a foliation does not automatically add intrinsic space-time structure; the quantum theorist's violation of SR cannot consist merely in positing some relation among points that effects a foliation. Suppose the laws of the proposed version of quantum theory allow that the foliation could have cut the manifold in various ways, or that the relation between points that determines quantum

---

[87] Maudlin (2002: 191).    [88] Maudlin (2002: 192).

priority could have failed to generate a complete foliation. In that case, applications of the criteria that should lead us to say that the ether was mere material contents would yield the same result for the added foliation. The real threat to SR is not the addition of some content that enables one to pick out slices as special, but rather the positing of laws about the quantum foliation that appeal, directly, to a particular reference frame.

Must the laws of these varieties of quantum theory appeal to a *built-in foliation*? Not necessarily. As Maudlin and others have pointed out, if the foliation that determines quantum-theoretic priority were lawfully correlated with the optical simultaneity slices of the center-of-mass frame for the universe, the resulting version of quantum theory would not appeal directly to one particular frame and its accompanying foliation. The location of the quantum-theoretic foliation would, on this supposition, be governed by the material contents of space-time; it would divide the manifold in the way that it does *not* because of any "rails" built into the manifold, telling events which slices they should occupy; instead, the "rails" would be laid down by the way matter is distributed.[89]

In the effort to prove the bare consistency of such a version of quantum theory with SR, one might flesh out the picture a bit: Imagine a kind of world in which space-time is otherwise Minkowskian and there occur phenomena at least superficially similar to our quantum events. Suppose that, as a matter of law, quantum measurements can only occur in such worlds if the universe has a well-defined center of mass: no center of mass frame, no quantum measurements. And which quantum measurements occur earlier than which is determined by priority within the series of optical simultaneity slices associated with the center-of-mass frame. Far-fetched? Yes. But that should not be relevant to the mere question of the *consistency* of SR and a theory that makes use of a privileged foliation in its laws. That question is answered by the existence of possible worlds in which quantum phenomena are tied to a special foliation, but the foliation in question does not constitute intrinsic structure beyond the metric of a Minkowskian manifold.

Here is another sort of law that, by Maudlin's standards (as I understand them), really ought to be judged consistent with SR: The foliation relevant to priority among quantum events is determined by the state of motion of just one particle emerging from the big bang; and, for every particle that existed at the origin of the physical universe, there was an equal chance that its initial state of motion would be the one to "choose" the quantum-priority frame of reference.

Neither of these imagined theories about a quantum-priority foliation would be at all plausible, were it being proposed as a theory of how quantum phenomena actually work. Maudlin rejects the center of mass idea immediately as a non-starter,

---

[89] See Maudlin (2002: 202–4); and Maudlin (2002: 222, n.8): "When a unique center of mass frame does exist, though, one can construct a Lorentz invariant theory" in which a single foliation plays the role of simultaneity for quantum-theoretic purposes. Valentini (2008: 150) makes the same point: "In itself, the mere fact of superluminal interaction is not necessarily incompatible with fundamental Lorentz invariance. For example, the interactions might be instantaneous in the centre-of-mass frame (a manifestly Lorentz-invariant statement)."

not for inconsistency with SR, but for its improbability: one should expect the laws actually governing quantum phenomena to generalize to many different possible mass distributions; in particular, they ought to be able to hold in worlds with systems of particles lacking a unique center of mass. And so Maudlin does not pursue the idea that quantum theory and SR might *actually* be rendered consistent by means of a lawful connection between quantum priority and the center of mass of the universe. The stochastic law selecting a single particle to determine the foliation is my own invention, but I do not suppose it is any more plausible. Still, the conceptual possibility of such laws shows that a quantum theory of this general type—one requiring a non-frame-relative priority relation for quantum events, and thereby adding a privileged foliation to the manifold—is not inconsistent with SR, at least not given Maudlin's understanding of what constitutes additional intrinsic space-time structure.

The criterion we are using for determining inconsistency with SR is: postulating laws that directly appeal to intrinsic space-time structure beyond the Minkowskian metric. There are, in principle, ways for a quantum-theoretically privileged foliation to be governed by laws that respect this criterion. But then there must be, at least in principle, ways for the A-theorist to posit a "wave of becoming" that also respects this criterion. The A-theorist's privileged foliation is, in effect, the "mark" left by the wave of co-present events as they pass through the manifold, and, so long as only *consistency* is at issue, one may suppose that its progress is governed by principles ("metaphysical laws") similar to the proposed quantum theoretic laws: the series of co-present points is determined by the center of mass frame, or selected by the initial state of one randomly chosen particle emerging from a Big Bang. In such worlds, facts about the contingent physical contents of the manifold would, in effect, tell the wave which slices to occupy; the "rails" along which the A-theorist's present must move would be laid down by matter, and no additional intrinsic space-time structure need be invoked in the fundamental laws governing the moving present.

Peter Forrest reaches a similar conclusion, in the course of arguing that his growing-block A-theory is not undermined by the (approximate) truth of SR.[90] Forrest asks us to consider the following hypothesis: as a matter of natural law, "Time passes in such a way that *some* system of parallel hyperplanes are successive presents"; but, within a Minkowskian manifold, the laws of temporal evolution do not determine which of the many possible series of hyperplanes it shall be.

> Once a given system of parallel hyperplanes has established itself, then the law tells us that all subsequent presents are also parallel, but how it got established is some-thing to do with initial conditions, or, more accurately something to do with the early stages of the universe when Special Relativity was not a good approximation.[91]

[90] Forrest (2008) is after bigger game than just defending the consistency of his A-theory with SR. He means to show that the inflationary Big Bang explanation of the universe's nearly isotropic expansion can generate a reason to think that the A-theoretic foliation roughly coincides with that of "cosmic time", and thus the A-theorist can acquire empirical evidence for the location of the successive presents in the manifold.

[91] Forrest (2008: 249).

Forrest identifies SR with the following two theses: "It is a law of nature that the electromagnetic constant $c$ has a fixed value in cm per secs", and "All laws of nature are invariant with respect to changes from one frame of reference to another moving relative to the first with some uniform velocity less than $c$."[92] A law to the effect that the wave of becoming takes the form of *some* series of hyperplanes does not, he says, violate SR, so understood. "All [SR] implies [about the laws governing the A-theoretically privileged foliation] is that, whatever the system of successive presents is, any relativistic transformation of this system would also be a (nomologically) possible system of successive presents."[93] By Forrest's lights, then, endorsing an A-theory that included these laws about the wave of becoming would not violate SR—so long as the additional laws (if any) that govern the "establishment" of the actual A-theoretically privileged foliation are also invariant with respect to the relevant changes in reference frame.

Forrest's requirement that, to be consistent with SR, the laws must display a certain kind of invariance is, in effect, a way of requiring that laws not appeal directly to additional space-time structure. On Forrest's supposition, each of the infinitely many time-like foliations of the otherwise Minkowskian manifold could have been the A-theoretically privileged foliation—that is, whatever laws govern the wave of becoming, they do not rule out any of these alternatives. The manifold has no "built-in" grain telling the present where it must go. What *does* select one series of slices is, he supposes, something to do with initial conditions, and if specifying these conditions does not, in itself, violate the criterion, then neither the laws nor the conditions that select a foliation violate SR. (Forrest has his own ideas about how a wave of becoming actually became established in our universe: he believes there are probably laws linking the privileged foliation with a frame of reference in which the expansion of the physical universe is nearly isotropic.)[94] Positing a wave of becoming inevitably privileges a single foliation; nevertheless, if the laws determining its location do not themselves appeal to non-Minkowskian space-time structure, the privileging does not require that the foliation be special in-and-of-itself—in advance of the contingent conditions that choose one foliation to be the lucky winner. A wave of becoming that obeys this law requires no more help from the manifold than the family of particles envisaged earlier: particles that inevitably move inertially and at rest relative to one another, but that could have been introduced into space-time in any frame. I shall call a "Forrest-style law" any law about the A-theoretically privileged foliation that (i) allows nothing but the Minkowskian metric to determine a range of possible foliations, but (ii) does not dictate which of these foliations is actually selected.

What kind of initial conditions could determine the "angle" at which the present slices the manifold, given Forrest's laws about the evolution of becoming within a Minkowskian manifold? Consider the position of a one-slice presentist who believes in a finitely old physical universe, one that begins with a "bang" and is not caused by earlier events in space and time. If this presentist also takes SR to be the best theory of

---

[92] Forrest (2008: 248).    [93] Forrest (2008: 249).    [94] Forrest (2008: 250–2).

the manifold, she should probably deny that SR requires that any point-sized regions existed, prior to the first physical events. Matter, energy, and four-dimensional manifold come into existence together. One can take the earliest trajectories and *imagine* them having existed, occurring at (or, according to the reductionist about points, constituting) points that no longer exist; but, according to the one-slicer, the physical world has just come into existence, and these past points are mere fictions. Now the question that emerged from discussion of Forrest-style laws is this: Could the initial conditions of the material contents of the universe at the "bang" determine where the wave of becoming will lie, and do so in a way that does not depend upon laws appealing to additional intrinsic space-time structure, beyond the Minkowskian metric? Our one-slice presentist can easily formulate hypotheses that would do the trick. She can use either of the mechanisms considered above for determining a quantum-priority foliation. Perhaps the overall distribution of mass at the beginning of the physical universe selected a becoming frame, or some randomly chosen particle selected a frame. Neither supposition requires laws that directly invoke a particular foliation. Who knows how many other SR-consistent proposals the creative A-theorist might cook up—if, unlike us, she happened to live in a world that seemed, ultimately, to be Minkowskian?

What if *nothing* about the physical contents of the manifold determines the becoming frame? Would that be tantamount to adding intrinsic structure to the manifold itself? The following scenario calls for exercise of "good taste" on the part of anyone trying to judge whether it involves laws appealing to an intrinsically special foliation. Suppose one-slice presentism is true, but that the selection of a foliation is not determined, stochastically or otherwise, by the material contents of the universe. However, Forrest's imagined laws governing the wave of becoming are also true, and they determine its future location, given its past locations. Does this combination satisfy Maudlin's criterion for inconsistency with SR? Does it require laws invoking additional intrinsic manifold structure? The answer is, I submit, not a clear and definite "Yes". Granted, according to this combination of the A-theory with laws about the shape and location of the present, there is an additional *brute fact* about space-time that renders one foliation special. But, given the contingency of this fact, and the nature of the law supposed to govern the wave, it is not obvious that the brute fact requires laws appealing *directly* to a particular frame of reference and its associated foliation. Does the manifold itself provide the "rails" along which the wave must move? Or should we rather say that *nothing* provides the "rails"? An A-theory combined with a Forrest-style law governing the evolution of the wave of becoming, if it leaves its actual location completely unexplained, seems to me to be at best a borderline case of inconsistency with SR. The possibility of borderline cases should not come as a surprise. The moral has already been drawn from Maudlin's discussion of the ether: It is sometimes a subtle matter whether to say that a certain theory attributes structure to the manifold itself, or merely adds material content.

I have rejected a number of ways in which one might try to spell out "inconsistency with SR", and focused on the most promising proposal that one finds actually being applied to scientific theories. There may be some alternative interpretations of "incon-sistency with SR", applicable in scientific contexts, with which I have not engaged. But one gets no help finding them when examining the arguments of Putnam, Sider, and others who allege inconsistency between presentism and SR. Putnam and Sider do not discuss the possibility that the material contents of space-time might play a role in determining the location of the A-theoretic foliation in a Minkowskian manifold; presumably, they do not think the idea is relevant. But, given a Maudlin-style criterion of "inconsistency with SR", the possibility is highly relevant.

Unlike Putnam and Sider, Simon Saunders does consider the idea that a presen-tist might want to posit lawful connections between the series of co-present slices and some foliation-privileging physical phenomenon. He seems tacitly to agree with Maudlin's parallel judgment in the case of a quantum-foliation: If it *could* be made to work, it would save presentism from inconsistency with SR. But he rejects the maneuver:

> Of course, making reference to the matter content of space-time as well, there may well be methods for defining a partitioning of spacetime into spaces (for defining global instants, as required by presentism), but none of them are likely to claim any fundamental status. It is unlikely that any can be taken seriously, if we are concerned with the definition of the totality of what is physically real.[95]

Saunders alleges that a physically definable foliation must be "obviously privileged"[96] if the presentist is going to suppose that it coincides, in a non-accidental way, with the wave of becoming.

There can be no doubt that Saunders is onto something here. When the distinguish-ing mark of a physically privileged foliation is not very "deep" or natural, it becomes less plausible for an A-theorist to suppose that the foliation coincides, contingently but lawfully, with the progress of the wave of becoming. There is good reason to grant this: Positing non-accidental correlations between some highly natural kind and a relatively superficial or gerrymandered kind is always less plausible than positing such correlations between two highly natural kinds. The presentist is bound to think that her relation of absolute simultaneity constitutes an important "joint in nature", and so she should be very surprised to find it lawfully linked to some "grue"-some physical feature.

So I accept something that is certainly in the vicinity of what Saunders is claiming: Suppose there were a presentist whose evidence otherwise supported SR, and who could find only gerrymandered or highly contingent physical phenomena to select a foliation; such a presentist should conclude either that something like Forrest's law is correct (the manifold allows for the wave of becoming to pass through at many angles, and it is a brute fact which one is chosen), or that the true principles governing the wave of becoming advert to space-time structure beyond the Minkowskian metric. I have granted, begrudgingly, that the former course might be thought to qualify as a

---

[95] Saunders (2002: 280–81).    [96] Saunders (2002: 281).

borderline case of inconsistency with SR, by a Maudlin-style criterion, though no more than that; but the latter would obviously qualify as inconsistency with SR.

While I do not deny, then, that this sort of presentist, in these circumstances, would be forced in the direction of inconsistency with SR, I would emphasize that this is a highly hypothetical statement about what a presentist would have to believe in certain circumstances—circumstances which, as shall appear, do not apply to today's presentists (as Saunders himself points out; in fact, I suspect that our differences may be primarily a matter of emphasis). The imagined linkages between the wave of becoming and a certain physically privileged foliation are supposed to be contingent. If somewhat implausible laws governing the present are not impossible, but merely unlikely, then there are possible worlds in which the link is a lawful one—even if the inhabitants of such a world ought not to believe in it. When the question is that of the bare *consistency* of presentism with SR, it would be wrong to require that the linkage must seem obvious to us, or be highly natural: it need only be a possibility. From the point of view of establishing consistency, Kent Peacock is right: "the interesting question is not what *metrical* structures can *necessarily* be found in *all* time-oriented spacetimes, while assuming from the outset that there are no spacelike dynamical interactions [e.g. superluminal motion]. . . . The aim is to determine what is possible, not what is necessary."[97]

When Sider, Putnam, and other critics invoke SR as evidence against presentism, they portray the presentist as *rejecting* a scientific theory. Accepting presentism, they say, would be "scientifically revisionary", it would require that one "reject" SR, and so on.[98] To justify the solemn invocation of science by Putnam, Sider, and others, the kind of inconsistency in question would have to be of a sort that holds between, say, Bohmian quantum theory and SR, or SR and GR. Maudlin's examination of inconsistency in such contexts is hard to gainsay, and, by his standards, there are numerous, non-crazy hypotheses according to which presentism and SR would be consistent. I conclude that Putnam and Sider have not made a case for their conclusion that presentism would *require* the revision of a well-established scientific theory, *even if* SR were a well-established scientific theory (which it is not; but more on that issue later . . . ).

There are ways to force the presentist into imagined circumstances that would demand rejection of SR. The presentist can hardly deny that the wave of becoming moves through the manifold in a way that is governed by *some* kinds of laws or principles. For example, assumptions (9) and (10) were found to put serious constraints on the shape of the present, and accessibility relations within the manifold determine the direction in which it moves. Assuming that a wave of becoming moves through a manifold that, A-theory aside, looks Minkowskian; and that no facts about the material contents of the manifold determine the "angle" at which the series of co-present events slices the manifold; and that the laws or principles governing the successive locations of the wave are not Forrest-style laws; then, at last, one has specified a possible world

---

[97]  Peacock (2006: 255–6).        [98]  See, e.g. Sider (2001: 52), and Mellor (1998: 57).

in which some extra intrinsic fact about the manifold itself *must* be what governs the location of the A-theoretically privileged foliation. But one should hardly, at the end of this series of stipulations, proclaim: Therefore presentism is inconsistent with SR; let alone, proclaiming: Therefore presentism is false.

## 4.3  A weaker form of inconsistency

Some critics have argued that Relativity undermines the A-theory on grounds other than inconsistency of the sort examined above. Some have said that the absence of a privileged foliation in SR (and in many models of space-time consistent with GR) shows that, if we believed the physical theory, we should conclude: as far as physics is concerned, the A-theorist's relation of absolute simultaneity is not *needed*. And then these critics go on to claim that, if physics does not need something, it is not there—or at least we have no reason to believe in it. As Adolf Grünbaum put it:

> It seems to me of decisive significance that no cognizance is taken of nowness (in the sense associated with becoming) in any of the extant theories of physics. If nowness were a fundamental property of physical events themselves, then it would be very strange indeed that it could go unrecognized in all extant physical theories *without detriment to their explanatory success.*[99]

This is more or less how Craig Callender portrays the conflict between scientific theories and the A-theory, though his discussion is more nuanced than Grünbaum's. For a presentist to posit "Minkowski space-time with a preferred foliation", under circumstances when SR seemed otherwise adequate, would be to introduce "otherwise unnecessary unobservable structure to the theory".[100] By Callender's lights, reasonable belief in the A-theory would have to be supported by some powerful philosophical argument for an objective past–present–future distinction, if its existence does not fall out of our best physical theory of space-time. Until such argument is given, "merely as a by-product of scientific methodology, physics will not accommodate [the A-theorist's foliation]"; "physics—and science itself—will always be against tenses [i.e. A-theoretic distinctions] because scientific methodology is always against superfluous pomp."[101]

I do not expect many diehard B-theorists to be moved by the brief objections I shall raise to criticisms of this general style: "Your distinction does not appear in physics; therefore you have no reason to believe it marks a real 'joint' in nature". The one thing I *do* hope will be apparent to all parties is this: An objection in this style is a far cry from the claims of Sider and Putnam. In general, a theory that posits something not found in another theory does not automatically lead to *inconsistency*, even when the theories are describing the same objects.

---

[99] Grünbaum (1967: 20).       [100] Callender (2008: 66).       [101] Callender (2008: 67).

Comparison with epiphenomenalism in the philosophy of mind may be useful. Some philosophers of mind take seriously the idea that consciousness might be an epiphenomenal property of brains—something extra, beyond the physical phenomena described by biochemistry, but just as fundamental. They do not have to say that biochemistry gets the workings of the brain *wrong*, only that it does not tell the whole story. One criticism of epiphenomenalism is that the extra mental phenomena are not needed to do any extra work, or that the work they are supposed to do could not be done by them. Continuing to believe in the phenomena, in such circumstances, would be to believe in something that is *dispensable*. Callender, and others, have criticized presentism and the A-theory along similar lines: everything about time that needs explaining is explained by physical theories that do not mention objective past–present–future distinctions; such distinctions are, therefore, dispensable.

It would be a gross overstatement to characterize such arguments as based on inconsistency, rather than dispensability. They are allegations of intellectual profligacy, of positing more distinctions when one could have gotten by with fewer. A criticism of the A-theory built along these lines, and appealing to SR, would go something like this: if the A-theorist's past–present–future distinction does not play a role in SR (nor in SR conjoined with a physics of particles and fields roughly similar to those of the actual world), then it would be irrational to accept SR while nevertheless retaining this extra distinction.

To make such a dispensability argument persuasive, a good deal should have to be said about the myriad distinctions one *could* rationally continue to regard as objective, despite their absence from fundamental physics: What makes *them* okay, and the past–present–future vulnerable? The less "Scientiphical" amongst us (to use Peter Unger's term[102]) will believe in many things that fail to put in an appearance in fundamental physics; it will be harder to convince us that, if physics does not mention the present, then it isn't there.

I will not attempt to address dispensability arguments in great depth, but confine my attention to a couple of recent versions in which some efforts are made to explain what makes the A-theorist's foliation especially vulnerable, worse off than other things that seem important but are unmentioned in physics: namely, the foliation's "elusiveness". The criticism is similar to a familiar objection to epiphenomenalism about certain aspects of consciousness; so I begin with application of a dispensability argument to that case.[103]

I should not want to deny that, in certain circumstances, if some distinction or property fails to show up in a scientific theory of the things that (allegedly) have it, this should count strongly against the objectivity of that distinction or property. Only an extreme form of "physics-ism", however, would insist that, unless a term appears in (the final, true) physics, it cannot be used to accurately describe real things. But there

---

[102]  Unger (2006: 6–9) outlines the elements of what he sees as the dominant "Scientiphical Metaphysic".

[103]  Comparison with epiphenomenalism is explicit in Prosser's (2007) version of a dispensability argument.

are more plausible requirements one could invoke: for example, that all the objective aspects of resemblance among things supervene upon the exemplification of properties mentioned in (the final, true) physics. An epiphenomenalist about consciousness can obey the letter of this supervenience claim by positing laws of final physics which govern the generation of the epiphenomenal properties, thus drawing them into the "supervenience base". Invocation of psycho-physical laws in order to respect supervenience is not the merest cheat it might at first seem. Physics (a fortiori final physics) has high predictive ambitions, and should probably be regarded as truly *final* only when its laws subsume *all* fundamental phenomena—including, if the epiphenomenalists are right, some aspects of conscious experience, namely, their "qualia" or phenomenal properties. Physicalistically-inclined philosophers will not be mollified by the invocation of such laws by epiphenomenalists, however. Given epiphenomenalism, a fundamental physics that failed to mention such properties would explain every physical event that can be explained; what work is left for qualia to do? Why should we expect them to show up as fundamental, at the end of the day?

At this point, the fans of qualia can point out that we—most of us—find ourselves drawn to recognize possibilities that *require* the extra properties; for example, the possibility of qualia inversions or qualia absence in creatures physically indiscernible from ourselves ("philosophical zombies"). But there is a special reason for being suspicious of adding these distinctions among phenomenal states, *given epiphenomenalism*. Positing a truly epiphenomenal property raises serious questions about our ability to know that it is exemplified. If their epiphenomenal nature means that, even if they *did* exist, we could not know anything about them, then it would not matter that qualia are part of our common-sense conception of things: we have no reason to care about them, and should be skeptical whether they even exist.

Critics of the A-theory have marshaled superficially similar arguments against the presentist: a fundamental physics built around SR would not mention the A-theorist's privileged foliation. The reasons A-theorists posit the distinctions that yield the foliation may well be ordinary beliefs that most of us have, but, were SR fully adequate, the A-theoretic distinctions would be elusive and unknowable. If it would be utterly mysterious how we could know about an objective past–present–future distinction, we have no reason to care about it, and should be skeptical about its existence.

Shortly, I shall consider whether the failure of SR and GR to mesh well with quantum theory has cast doubt upon the significance of this argument. First, though, I look at a couple of concrete attempts to support the main claims: that, given SR, we could not know, and should not care about, the presentist's A-theoretically privileged foliation.

Steven Savitt considers the idea that a Minkowskian manifold could simply be augmented with brute facts about the frame picked out by successively co-present events, and he expresses sentiments like those of Callender and Grünbaum. Savitt identifies stronger and weaker forms of "inertial chauvinism"—that is, choosing one frame of reference, and its planes of simultaneity, as the one selected by the wave of becoming. The stronger form "holds the principle of relativity to be false"; but there is a weaker form that does not go so far as that:

> A weaker version of inertial chauvinism agrees that all admissible frames of reference "are completely equivalent for the formulation of the laws of physics" but asserts that one frame is *meta*physically distinguished. This metaphysically distinguished present cannot, according to the relativity principle, be ascertained by any (classical) physical measurement or experiment. If the present is indeed so elusive, I find it difficult to imagine what aid or comfort it could be to a metaphysician.[104]

Epiphenomenalism (about consciousness) could be faulted for introducing something that would have no impact upon us. There seems to be a similar objection behind Savitt's nice turn of phrase in the final sentence. He insists that, for a feature of space-time to play the role assigned to the privileged foliation by presentism, it must mark a deep and important divide: it must be something we presentists can regard as providing "aid" and "comfort", something that satisfies our conviction that the present is more robustly real than past or future. So far, I agree. But he further alleges that, if the location of the divide between present and past or present and future were "elusive"— and, here, the elusiveness can only consist in uncertainty about what is present at relatively large distances, the kind of uncertainty there would be in an otherwise Minkowskian manifold with no faster-than-light processes—then the present could not mark a divide that we would rationally regard as deep and important.

Put thus sparely, the objection does not seem to me to have much force. Why should the inability to *tell* which distant events are in the past make it irrational for us to *care* whether they are in the past? For most of human history, there has been massive, unavoidable uncertainty about the times at which distant events occur. Yet people often wondered, "What is really happening over there now?", even when "there" was a great distance away and the answer was not determinable to within a small margin of error by any known means. Was it irrational for us—we human beings—to take the answer to this question to be an important one, back in the days when we lacked reliable methods for determining precise relations of simultaneity over significant distances on the earth's surface? Insisting upon the following principle would impugn the rationality of too many people: "If a person faces unavoidable uncertainty about whether something is past, then it automatically follows that they ought not to care whether it is past." Granted, *if* the A-theory is false, and *if* we inhabit a Minkowskian manifold, and *if* there are no faster-than-light processes (something that does not fall out of the Minkowskian metric by itself), then we were making a mistake to think that we could be asking about a deep, objective fact with the words: "What is really happening at that distant location now?" So, given all those assumptions, we were making a mistake; but those are assumptions no one could make before the advent of Relativity, and they are assumptions that the presentist still rejects.

Our ancestors should not be convicted of *irrationality* for thinking that the differences between past, present, and future are important ones, simply because they lacked

---

[104]  Savitt (2000: S570).

precise clocks and rapid signaling methods. And if the contemporary presentist were forced, by relativistic physics, to grant that the lack of precise methods is a matter of physical necessity, I do not see that this should make her any more irrational then our ancestors, were she to continue to regard it as an important one. What could be more important than existing or not existing?

Craig Callender also provides a special reason to think that the presentist's distinctions, if they do not coincide with something found in physics, belong on the chopping block. And his suspicions, too, are based on the idea that merely adding a metaphysically privileged foliation to Minkowski space-time would result in a difference about which we could not know anything. He makes roughly the same point as Savitt (though in a slightly different context). He imagines a situation in which the presentist posits a privileged foliation, while admitting that, due to ineliminable restrictions on the precision of measurements, the "angle" at which it cuts the manifold cannot be empirically determined. In that case, "*[y]our intuitions, introspections, etc., all being species of interactions, can be in principle no guide to which foliation is the true foliation or even whether there is one.* If the world becomes or enjoys an objectively privileged present, then it is not something at all connected to experience (assuming physicalism)" [author's italics].[105] Whether or not the presentist accepts "physicalism" or the thesis that "intuitions" are really "interactions" (i.e. physical events in the brain), she should happily grant that experience would be no guide to the precise location of her privileged foliation, given certain packages of hypotheses that recent physics has sometimes seemed to favor—for example, *if* SR were true, *if* all causation were local, *if* superluminal influences were impossible, and *if* there were no reason to think the present is lawfully connected to some foliation privileged by the physical contents of space-time. But the presentist should not, however, sit still when Callender baldly asserts that, in such circumstances, our "intuitions" could be no guide to the question "*whether there is [an A-theoretically privileged foliation]*". The "intuitions" in question are, presumably, simply the widespread A-theoretic "intuitions" to which I have appealed in the argument for a privileged foliation—"intuitions" here being simply another name (and, in the context, a derogatory one) for mental states also known under more familiar names such as "convictions" and "beliefs". Why is it that these beliefs are useless for determining whether there is an objective past–present–future distinction? The only reason given here is: if we take them seriously, and if SR is true, they will force us to believe in something about which we cannot, in principle, know everything we should like to know. I find this no more impressive than Savitt's argument.[106]

---

[105] Callender (2008).

[106] To be fair, Callender has other things to say against the convictions about past, present, and future upon which the A-theorist relies. He claims that these beliefs "arise solely from ordinary language analysis—a mostly bankrupt enterprise" (Callender, 2008: 67). The real locus of our disagreement seems to me to be this: how much credence should we accord to the beliefs with which we find ourselves before we begin our more systematic inquiries? Reasoning that begins from them yields the kinds of arguments Callender calls "ordinary language analysis", though I should resist the label (except, perhaps, when they happen to be beliefs about language). There are many widely-shared,

I am prepared to grant that someone is in trouble if they try to hold a combination of views along these lines: There is a feature that belongs to certain parts of the physical world; this feature does not seem to play a role in the best science of that domain, and every brain or mind would be exactly the same even if *nothing* had had this feature. But the presentist, at least, is *not* in a position analogous to this sort of extreme epiphenomenalism. It is not as though presentism has the result that we can know nothing about which events are present and which are not. The presentist must admit that our knowledge of what is present does not extend *very far*, if superluminal signaling is impossible, space is Minkowskian, and we have no reason to link the present to a foliation that is privileged by physical contents. The more implausible forms of the moving spotlight and growing block A-theory may be open to this sort of criticism; if being lit up by a primitive property of presentness, or being on the cutting edge of the block, are the only factors distinguishing present events and things from past ones, the minds and brains of ourselves and all our ancestors would be exactly the same in every respect but this one; and we would almost always be wrong in our judgments about the location of the present. It is difficult to see what "aid and comfort" the spotlight or the edge could offer, on a version of the A-theory that makes it impossible to know even which events in one's own life are present and which are past. But the more plausible versions of the A-theory do not fall into this trap.

Presentism is a view that, for many of us, has considerable "intuitive" appeal—by which I mean little more than that, upon reflection, many people find themselves believing it. Scientiphicalist enemies of the A-theory who would use dispensability arguments against presentism may be making a flat-footed claim like Grünbaum's. But if they are to do more than that, they need to produce subtler principles relating the content of physics to our ordinary beliefs about the world, including beliefs with metaphysical implications. Perhaps a more compelling argument against the A-theory can be constructed along these lines, but, so far, I have not seen it done.

## 4.4 How bad would an added foliation be?

What problems would face a presentist who adds an A-theoretically privileged foliation to an otherwise Minkowskian world? Perhaps the laws governing the wave of becoming could be made consistent with the letter of SR, in one of the ways I have indicated. Perhaps it would not be so radically epiphenomenal or elusive as to be irrelevant to human concerns. But could it really be added to a Minkowskian manifold without radically altering SR's description of space-time?

Once again, it proves instructive to consider the case of grafting a quantum-mechanically preferred foliation onto an otherwise Minkowskian space-time. How radically would this alter one's physics? Here is Maudlin's assessment: "This would

---

common-sensical convictions about time that I judge "innocent until proven guilty", while Callender gives the opposite verdict.

not demand the elimination of any relativistic structure, but would undercut the relativistic democracy of frames."[107] By not eliminating the manifold's structure—such as the facts about which time-like paths are straight and which are curved—the quantum theorist who takes this route would not be robbed of the explanatory resources of the Minkowskian manifold.

> [N]o positive part of the relativistic account of space-time is being *rejected*: rather, in addition to the Lorentzian metric, a new structure is being *added*. Because of this, there is a straightforward sense in which no successful relativistic account of any physical phenomenon need be lost or revised: if something can be accounted for without the foliation, then one need not mention it. So there is no danger that existing adequate relativistic accounts of phenomena will somehow be lost: *in this sense, the content of relativity is not being rejected at all* [my italics].[108]

Can the presentist claim that, in an otherwise Minkowskian world, her added foliation would not rob her theory of the explanatory resources provided by SR? Those who take the manifold seriously could plausibly make this claim.

I have argued that, in order to make sense of cross-temporal facts about motion in a theory like SR, the presentist should accept the existence of a manifold with built-in metrical structure. She may be able to get away with only allowing one slice of the manifold to exist at present, talking about the formerly filled points indirectly by means of present surrogates in the form of persisting trajectories, or she might be forced to accept the ongoing existence of an empty or ghostly manifold; but, in any case, she must find a way to ascribe, at least to past and present points, the fundamental metrical properties mentioned in our best physical theories. The presentist who has gone so far as this is at a distinct advantage, if she should happen to inhabit a world that otherwise looks Minkowskian. She can accept the existence of a manifold with built-in paths of inertial and light-like accessibility satisfying the Minkowskian metric. She can insist that the kinematical part of explanations of motion should appeal to the fundamental metrical properties of this Minkowskian manifold (as opposed to those of a Newtonian or Galilean manifold). And this will take some of the sting out of the need to add a privileged A-theoretic foliation as a piece of extra space-time structure. There is a sense in which she does not reject relativity, because she can continue to give the same kinds of explanations of the same physical phenomena on the basis of the same metrical facts about the manifold's intrinsic structure.

The sense in which such a presentist accepts SR can be illustrated by considering one of Craig Callender's arguments against the A-theory. Callender has two options to offer the A-theorist, should she happen to discover that she lives in what otherwise appears to be a Minkowskian world: (i) "[O]ne could adopt the empirically adequate Lorentzian interpretation", thereby rejecting SR on metaphysical grounds and returning to a Newtonian manifold that posits absolute space and absolute time. (ii) "Alternatively, we might keep the Minkowski metric but add more structure to spacetime. We might add a foliation, i.e. a preferred stacking of spacelike hypersurfaces

---

[107]  Maudlin (2002: 202).        [108]  Maudlin (2008: 160).

that divides the spacetime manifold. Becoming, then, could occur with respect to this preferred stacking".[109]

Callender would like to saddle the A-theorist with the first option: "by far the best way for the tenser to respond to Putnam *at al.* is to adopt the Lorentz 1915 interpretation of time dilation and Fitzgerald contraction."[110] He does not, however, explain *why* he thinks the A-theorist should take this route, rather than taking his option (ii). And when, later on, he faults the A-theorist for having to retreat to Lorentz,[111] one begins to suspect that Callender's second alternative was the better of the two all along.

The presentist could hardly be *forced* to become a Lorentzian simply by adding a preferred foliation. Lorentz's immobile, universal ether provided an absolute relation of sameness of place over time, effectively turning Galilean space-time into Newtonian space-time again. The ether serves as a privileged inertial frame. For presentism to force a return to Newtonian space-time, it must do the same: the wave of becoming would have to successively occupy the optical simultaneity slices of some inertial frame. But presentism does not, by itself, require this. The argument I gave for the conclusion that the present takes the shape of a thin slice of the manifold, and that the series of co-present slices constitutes a complete foliation, did not entail that the events in a single slice must be optically simultaneous; for all I said there, the foliation could just as well consist of "nonstandard simultaneity slices", hypersurfaces that do not correspond to the planes of simultaneity determined by the Radar method from a particular inertial path through space-time. If the hypersurfaces were sufficiently irregular, the shortest path between two points on a slice might *never* consist entirely of points within the slice. When an object consists of several disconnected particles, a nowhere-flat present of this sort might nevertheless *sometimes* cut each of the particles' paths at a point that falls on a single flat simultaneity slice. But this would be the exception, not the rule. The actual shapes of objects at various times would rarely correspond to the three-dimensional shapes assigned to the same objects by any optical simultaneity slice.

Adding a foliation of this "wobbly" kind, or slicing the manifold in some other non-standard way, would fail to select an inertial frame to play the role of absolute Newtonian space, and so could not possibly constitute a return to Lorentz. A problem with a wobbly present in a Minkowskian manifold is that no Forrest-style law could govern its progress. One might suppose that its shape is somehow determined by the material contents of the manifold—perhaps the presence of matter in the present slice tends to warp the shape of immediately succeeding presents, in something like the way matter was expected to warp a mutable ether, before the negative results of the Michelson-Morley experiments undermined ether-drag theories. Without some such association, an irregular wave of becoming would be a source of massive indeterminism.

Fortunately (for the sake of this very hypothetical presentist, in a world radically unlike ours…), even coincidence with the simultaneity slices of one inertial frame would not automatically constitute a complete retreat to Lorentz. What does Callender think would be so bad about a Lorentzian approach to the physics of space-

[109] Callender (2000: S595–6).    [110] Callender (2008: 52).    [111] Callender (2008: 67).

time, and how could it be avoided? In Lorentz's Newtonian space-time, the Fitzgerald contraction looks like the work of forces shrinking objects in the direction of motion when they move rapidly relative to "the ether"—an entity that, in Lorentz's theory, has become indistinguishable from absolute space. From the point of view of these rapidly moving things, objects at rest in the ether will look to be contracted and clocks to be temporally dilated, although this would be an illusion. As Callender sees it, the Lorentzian "introduces unexplained coincidences: why do those rods and clocks keep contracting and dilating, respectively? As a kinematical effect in Minkowski space-time, Minkowski space-time is a common cause of this behavior, which is otherwise brute in the Lorentzian framework."[112] It is simpler and more elegant to be able to regard the Lorentz invariance of laws governing many different kinds of forces as stemming from the same source, namely the structure of space-time.

But this the presentist can do, so long as she insists upon the fundamentality of the Minkowskian metrical structure of her manifold. Its structural features—the straightness and space-time lengths of light-like and inertial paths, in particular—provide the kinematical background upon which dynamical theories are to be erected. The location of the A-theoretically privileged slicing is supposed to be a contingent matter; where it falls is a further interesting fact beyond the Minkowskian metric, but the latter can play its role in explaining the shapes of the paths taken by particles, no matter where the A-theorist's present may lie.

Choice of a single frame by the wave of becoming does give one set of distance relations among objects a special status, metaphysically, and Callender will no doubt feel that this "introduces otherwise unnecessary unobservable structure to the theory".[113] But that is a different objection from the one to which I am responding here: that the presentist is unable to explain the Fitzgerald contraction as "a kinematical effect in Minkowski space-time", unable to regard a Minkowskian manifold as "common cause" of all frame-relative spatial contractions and time dilations. If the presentist insists that the Minkowskian metrical structure of her manifold is the real subject matter of SR, the presentist is in the same position as Maudlin's quantum theorist with a quantum-theoretically privileged foliation: the "old" Minkowskian explanations of various phenomena do not become inapplicable merely because some additional structure has been posited.

## 4.5 Could SR really be a description of the presentist's manifold?

It might be objected that the metrical properties the presentist ascribes to her four-dimensional manifold simply cannot be the same as those ascribed to Minkowski space-time by relativistic physics. After all, the latter is a theory about space-time; the presentist's manifold may be four-dimensional, but the fourth dimension is not exactly that of *time*. It is distance in the direction of light-like and inertial accessibility, but

[112] Callender (2008: 67).     [113] Callender (2008: 66).

that is not a temporal direction. The presentist cannot give the same kinds of scientific explanations of motion as would be given by a Minkowskian B-theorist, if she is not even talking about the same metrical relations among points.[114]

It must be granted that the presentist has to give the structural properties of her manifold a metaphysical gloss somewhat different from that of the ordinary B-theorist. Still, all three presentists with a manifold—empty box, ghostly box, and one-slice—have the means to talk about every formerly occupied point; all three can ascribe properties to sets of them in virtue of which they satisfy the Minkowskian metrical description, and they can construe this metrical structure as relevant to the motion of particles through the manifold in a way that certainly *sounds* just like the kind of relevance a B-theorist would ascribe to the properties of his Minkowskian manifold. If there is a difference in their explanations of physical phenomena, it is not apparent in the words (and equations) they use.

Whatever differences there are between the way presentist and B-theorist understand the ontology of the manifold and the nature of the relations between points that give it metrical structure, I do not think they should be called *scientific* differences. It would be a stretch to insist that the laws of physics can only be interpreted as laws about the relations within a B-theorist's block, and that the trajectories in the A-theorist's manifold are clearly *not* what relativistic physics is describing.

The unfairness of such an accusation becomes apparent when one considers other cases of metaphysical disagreements about things that are governed by scientific laws. In such circumstances, the laws articulated by the relevant science provide relatively abstract descriptions of the entities about which there is metaphysical disagreement; philosophers haggle over the best metaphysical scheme for classifying these entities, but, at least in many cases, the laws will stand as accurate descriptions of their behavior no matter which party is right about the metaphysics. For example, some metaphysicians believe that at least some objects with mass are "enduring things"— that is, things that last through time but do not need temporal parts in order to do so. Other metaphysicians believe that *everything* that lasts through time is, automatically, a "perduring thing"—something wholly constituted at each time it exists by a different instantaneous stand-in or temporal part. Physical laws about massive objects— for example, Newton's laws of motion—will be given different metaphysical glosses by the two sorts of metaphysicians. The "endurantist" metaphysicians will say that Newtonian laws are really constraints upon the possible histories open to a single thing, provided that the thing has mass; the friends of temporal parts will regard the laws as describing the possible ways in which a series of massive-object-stages, each causally dependent upon earlier ones, can be spatiotemporally arranged. I suppose an endurantist who accepts Newtonian physics could try to refute perdurantists by arguing along these lines: Newton's laws only mention objects with mass; they do not require the existence of instantaneous object-stages; therefore these laws of motion are about enduring massive objects, not about series of object-stages; and so the laws

---

[114] Here, I am responding to an interesting objection raised by William Lane Craig, in correspondence.

of motion would be *overturned* if perdurantism were true—perdurantists are anti-scientific!

Newtonian perdurantists would be unimpressed by such an argument, and rightly so. Granted, if the endurantists are right, the laws about the behavior of massive objects are not laws about self-perpetuating chains of object-stages. But the perdurantists can plausibly turn around and claim that, if *their* metaphysics is correct, the physicists were in the business of giving laws about object-stage propagation *all along*. A modest sort of externalism about the natural kind, *object with mass*, will yield this result. If perdurantists are right about massive objects, the theory of motion would not be falsified, even if all the physicists who developed the theory had believed that objects endure.

The response I have offered the perdurantist can be used by our (hypothetical) presentist who wants to accept SR, and explain everything that it explains by means of Minkowskian metrical structure. If this A-theorist is right about the structure of the manifold, SR would not be false; rather, SR (whenever it was advanced "realistically", intended as an objectively correct description of the physical universe) was always a theory about the metrical properties of the trajectories constituting the A-theorist's manifold, whether or not the proponents of SR were card-carrying A-theorists and would have accepted this description of their theory.

Fortunately, then, the presentist who adds a foliation to space-time but affirms its fundamentally Minkowskian metrical structure retains the right to draw the line between kinematics and dynamics exactly where it belongs on orthodox versions of SR, and this provides a sense in which "the content of Relativity is not being rejected at all"—to borrow Maudlin's description of the parallel case of the added quantum-theoretically privileged foliation.

But who is this presentist, forced to choose between Callender's two options: (i) a Lorentzian return to absolute space, and (ii) a Minkowskian manifold with an added foliation? It turns out that she is an entirely hypothetical philosopher who inhabits a world quite unlike ours—one in which SR seems otherwise adequate. Why should we actual-world A-theorists think that *our* fortunes are in any way tied to *hers*? That is the question to which I now turn.

## 4.6  The plight of some merely possible presentists

Now that the prospects for reconciling presentism and SR have been explored, it is time to consider the extent to which the difficulties that have turned up (such as they are) should be taken to undermine presentism.

As Bradley Monton has emphasized, there is something decidedly odd about arguments like Putnam's and Sider's: the actual falsity of the A-theory is inferred from SR or GR, despite widespread agreement that SR is false, and that GR is inconsistent with quantum theory and therefore likely to undergo serious revision, at the very least. Just what are these authors assuming about SR that justifies accepting its (alleged) implications as true, despite the theory's falsehood?

When philosophers make use of SR in this way, they must be assuming that, although SR has been superseded by GR, the features of SR that conflict with presentism are preserved in GR. But it is not obvious that this is so. Saunders provides one reason to think that the prospects for presentism might look different, in GR:

> Of course general relativity, just like the special theory, is committed to the principle of arbitrariness of foliation. Nevertheless, for an important class of spacetime models—*hyperbolically complete* spacetimes, for which the Cauchy problem is soluble—there is a natural definition of a global foliation, which has a number of desirable, dynamical properties. It is essentially unique: it is what is actually used in numerical calculations in geometrodynamics; it also has links to a number of open theoretical questions, particularly questions concerning the nature of scale in the classical theory.[115]

Saunders is not claiming that, if our world satisfies GR, presentists would inevitably want to choose his example ("York time") as coincident with the wave of becoming. His point is that it is the sort of thing that *could* be thought to coincide with the A-theoretically privileged foliation. And GR has served up other physically interesting candidates. William Lane Craig, Peter Forrest, and J. R. Lucas, for example, suppose that "cosmic time"—"the fundamental frame of the cosmic expansion"[116]—"*contingently* coincides with metaphysical time", i.e. the A-theoretically privileged foliation.[117]

In short, the presentist's situation with respect to GR is much like it was with respect to SR: In some possible worlds, GR and presentism can be true together, because the laws governing the passage of the present make use of a foliation that is distinguished by the manifold and its contents. Indeed, assuming GR, likely shapes for our manifold would allow a couple of ways to link the wave of becoming to a physically unique foliation. However, if one describes a sufficiently hostile (and probably *merely* possible) combination of manifold-plus-material-contents, no such laws can be formulated. A presentist who, unlike us at the present state of knowledge, found herself in such a universe, would face more pressure to add space-time structure beyond the metric of a GR manifold. Here is the similar argument such a presentist would face: According to her, the history of the universe includes facts about which events were truly simultaneous, and these facts select a series of slices that constitute a complete foliation. Is this wave of becoming acting in a regular way, obeying some law? Given a universe chosen for its hostility to the physical privileging of any particular foliation, either the evolution of the wave of becoming is insanely indeterministic, or it is governed by a law that could only link the present to some extra feature of the manifold itself.

---

[115] Saunders (2002: 290). And many A-theorists have taken GR to promise a place for their wave of becoming; Craig, for example, supposes that "cosmic time"—"the fundamental frame of the cosmic expansion" (Craig, 2001: 234)—"*contingently* coincides with metaphysical time", i.e. the A-theoretically privileged foliation (Craig, 2001: 237).

[116] Craig (2001: 234).

[117] Craig (2001: 237). See also Lucas (2008) and Forrest (2008). Swinburne (2008) argues that genuine simultaneity in an expanding universe like ours would correspond to cosmic time.

But there is reason to worry about the relevance of GR to presentism, as well. An argument from GR to the falsehood of presentism would seem to require, not only that our universe be one of those with contents hostile to all physically privileged foliations, but also that GR be *true*. Quantum mechanics, however, is an even more impressively confirmed theory than GR; the two theories appear to be in conflict; and some of the most promising ways to iron out the conflict turn out to be quite friendly to the A-theory, since they reintroduce a privileged foliation of the manifold.[118] According to Monton, presentists should be encouraged by these developments: "general relativity is incompatible with quantum mechanics, so our most fundamental physics can be found in the nascent theories of quantum gravity, which attempt to resolve the incompatibility. It turns out that there are some theories of quantum gravity, which are compatible with presentism. Thus, ... presentism is unrefuted."[119] A successful argument for inconsistency with GR, at this point in time, would only show that the A-theorist *may or may not* have to posit either an additional layer of space-time structure (which could, as in the case of SR, leave the explanatory role of a substantival GR manifold intact), or admit massive indeterminism about the successive locations of the present. Why get worked up about the possibility of having to concede this, when things could look very different once quantum theory and gravity are successfully put together? I am not qualified to have an independent opinion about Monton's claims concerning the current live options for a unified theory of quantum gravity. But I find them confirmed by reliable sources—including the staunchly B-theorist philosophers of physics who serve as my main informants on such matters.[120] It is too early for presentists to begin hand-wringing.

Can one grant the falsehood of SR and the shakiness of GR, but still find inconsistency with SR or GR relevant to the truth or falsehood of presentism? I am at a loss to see how, and I find no suggestions in Putnam, Sider, or other B-theorist critics who emphasize the (alleged) inconsistency of SR and the A-theory. Perhaps inconsistency with SR is thought to show that, whatever the actual world is like, we are *not too far* from worlds in which SR could truly and completely describe all space-time structure—were it not for that pesky wave of becoming, and the laws governing its progress. Is that enough to undermine the A-theory's credibility, even if *our* space-time manifold turns out, in fact, to be friendlier to presentism?

There is something of an opportunity to make A-theorists uncomfortable here, since most of us reject the very possibility of worlds with temporal phenomena but lacking a wave of becoming—we think the A-theory is necessarily true, if true at all. So, if some possible worlds are temporal but have no A-theoretic foliation, this would show that the A-theory is not necessary and therefore not true.

---

[118] As Callender puts it, "there are reasons for thinking general relativity and quantum field theory are mutually incompatible and must themselves give way to quantum gravity"; and there are "sketches of theories of quantum gravity that yield a preferred foliation", as well as "sketches of those that do not" (Callender, 2008: 65).

[119] Monton (2006: 265).

[120] E.g. Callender (2008: 65) and Maudlin (2002: 240–2).

It seems to me to be completely fair for the A-theorists to point out that: (a) all that has been shown is that, in these supposedly "nearby" worlds, more structure would have to be added to space-time than is supplied by a Minkowskian metrical description; and (b) this additional structure would not rob the presentist of the most important explanatory resources of SR, for reasons rehearsed above. We A-theorists should *not* admit that, in these (allegedly) nearby worlds, the A-theory would be false. What's more, the sense of "nearby" seems largely epistemic, and only dubiously relevant. The nearness consists in the fact that, had our evidence been only slightly different, we would have been justified in accepting SR as the final word about space-time structure. But, from a less anthropocentric, more objective point of view, one should say that the actual metrical features of our manifold are radically different than they would be in a Minkowskian universe. Granted, if GR is right, the world looks more Minkowskian as one looks at smaller and smaller patches of it, but a curved space-time with black holes and other radical deformities is very different from the infinite flat manifold of SR, and who knows whether it will look more or less Minkowskian when gravity and quantum theory are united? So the "nearby" Minkowskian worlds may well be far away from us, by objective measures.

I could even grant that my justification for *believing* the A-theory would, in fact, be undermined had I been in a world where, so far as physics is concerned, SR is adequate as a theory of the manifold, while nevertheless affirming that the A-theory is necessarily true in any world with temporal goings-on. Given the sense in which the presentist need not reject the explanatory virtues of SR, I do not think that the A-theorist should grant even this much. But, for the sake of argument, let it be granted. How much should that affect my convictions about the actual truth of presentism? For any interesting fact, including many necessary ones, it is relatively easy to cook up circumstances in which the fact is *true* but the evidence misleadingly points away from it. And, given the actual falsity of SR, the presentist need not admit that the B-theorist has done any more in this case: We have been asked to imagine a world in which we might be misled into thinking there is no A-theoretically privileged foliation.

## 4.7 Would the rejection of relativity for physical reasons "only make things worse"?

In the actual world, quantum-theoretic phenomena raise difficulties for Relativity, and these difficulties may well require the introduction of a preferred foliation of one sort or another. I agree with Monton that A-theorists should be encouraged by this development. One would have to look closely to see what role a given foliation plays in the physics before reaching a judgment, but there will sometimes be reason to think that it coincides with the A-theorist's—as is the case in the examples to be discussed here: Bohmian quantum gravity and GRW.

Some "interpretations" of quantuam mechanics require non-local causal influences; in fact, it looks as though *most* require non-locality, once one sets aside "many worlds" versions of quantum theory.[121] Two much-discussed proposals—Bohm's theory and GRW, a theory with instantaneous collapses of the wave function—posit a foliation of the manifold unknown to Relativity.[122] Their viability might seem to provide aid and comfort to the presentist—indeed, I believe it does. But Callender argues that they would only make things worse!

Following Callender's discussion, I shall focus mainly on Bohm's theory, which implies that the outcome of the measurement of a particle can depend upon whether another particle, arbitrarily far away (anywhere within the first particle's "bow-tie" region), is measured *first*; but I believe similar morals could be drawn in the case of GRW's instantaneous collapses.[123] The natural development of a Bohmian theory in a Minkowskian manifold would simply add a foliation, not found in the metric, that marks the line between "before" and "after" for quantum measurements. If we could determine the precise locations of pairs of particles as they go off into measurement devices, Bohm's theory predicts that we would discover the exact shape of the quantum-theoretically privileged foliation. But the theory also implies that we *cannot* determine precise locations. Does this constitute an implausible "conspiracy" in nature, a fiddling with the laws that feels like it was carefully designed to hide the shape of the series of quantum-theoretically privileged presents? Maudlin thinks not: 'the only reason we can't "see" the foliation is because we can't "see" the local beables [i.e. particle locations] with arbitrary accuracy (without disturbing the wavefunction), and the reason in turn for this is given by the structure of the basic dynamical laws that govern all physical interactions.'[124]

> When trying to formulate a Bohmian theory in a Relativistic domain, there is an evident need for some structure that will give rise to non-locality, and the postulation of a foliation is the simplest, most natural way to be able to write down non-local dynamical laws. *And once the foliation is postulated, no particular effort or adjustment of parameters is made with the purpose of hiding the foliation.* Rather, one writes down the simplest dynamical equations that look like versions of the non-Relativistic equations, and it then *turns out* that foliation will not be empirically accessible. Once the equations are in place, all the rest is just analysis.[125]

Maudlin argues for a similar conclusion with respect to the instantaneous collapses posited by a quite different version of quantum theory, GRW. The theory needs a foliation to be added to the manifold, if it is to be developed in an otherwise Minkowskian setting. Because it makes slightly different empirical predictions than more orthodox versions of quantum theory, GRW's additional foliation would be,

---

[121] For surveys of the range of (mostly) less wild options, see Maudlin (2002, 2008).
[122] Maudlin (2008).
[123] For an accessible presentation of the details of this approach, see Albert (1992).
[124] Maudlin (2008: 163).
[125] Maudlin (2008: 165).

in principle, detectable—but only by means of experiments we lack time or technical ability to carry out.[126]

Should the presentist be at all encouraged by the fact that a few philosophers of physics and theoretical physicists working on the foundations of quantum theory feel the need to add a foliation to an otherwise relativistic manifold? Should we hope that they are simply discovering a use for the foliation the A-theorist has been positing all along (thereby vindicating Arthur Prior's prediction that, eventually, scientific opponents of the A-theory would come slinking back to make use of his tense logic)?[127] Callender says, No. He describes what a Bohmian would say about the case of measurements of a two-particle system made at space-like separation by two characters, A and B. The precise details of the set-up are irrelevant for present purposes; what is important is that the Bohmian will say that it makes a big difference which one of them *measured first*. If A is measured first, then A determines the outcome of B's measurement; if B is measured first, the reverse is the case. To make sense of non-relative facts about relations of temporal priority among events at space-like distances, the Bohmian needs a preferred foliation that settles facts about which event-locations are earlier than which, for quantum-mechanical purposes.

Callender denies that Bohmianism would make a happy home for the presentist:

> There is an in principle irresolvable *coordination problem* between the two preferred foliations, the metaphysically preferred foliation posited by the [A-theorist] and the physically preferred one by Bohmian mechanics. There is simply no reason to think the two are the same. Only blind faith leads one to expect that the two are coordinated. In our above experiment, A might measure first and then B measure second according to the Bohm frame, yet according to the temporal becoming frame B measures first and A second. Assuming the becoming frame is primary, we would say B really happened before A; meanwhile fundamental physics would say that A happened before B. Since it would be a miracle if the two frames coincided exactly, with near certainty this will be the case for *some* pairs of events. *Hence the tenser is committed to asserting that with near certainty fundamental physics gets the order of some events the wrong way round.* Far be it from quantum mechanics saving tenses, the tenser merely trades one conflict with fundamental physics for another.[128]

Is it only "blind faith" that could convince the A-theorist that the two foliations coincide? That depends upon whether it could only be "blind faith" that makes (9), above, or the slightly weaker principle (9*), seem plausible:

(9*)    For any events $e^1$ and $e^2$, $e^2$ is causally dependent upon $e^1$ only if, when $e^2$ was happening, $e^1$ was happening or had already happened.

Given (9*), the Bohmian's privileged foliation could not cut across the A-theorist's. The Bohmian foliation is introduced precisely to answer questions about causal dependence; whether the outcome of one measurement depends upon the outcome of

---

[126]  Maudlin (2008: 166–70).        [127]  Prior (1996: 51).        [128]  Callender (2008: 62–3).

another measurement, or vice versa, is determined by which of the two occurred first, according to the Bohmian ordering. In Callender's example, the fact that B's measurement came out a certain way is causally dependent upon the fact that A's measurement was made first (relative to the Bohmian foliation); but, if the A-theoretically privileged foliation could cut across the Bohmian one, there will be measurement situations like this in which B's measurement is happening, but A's has not yet happened—a combination ruled out by (9*). No doubt Callender noticed that cross-cutting of the foliations would force an A-theorist to accept that an event can be causally dependent upon one that has not yet happened. He must, then, think that a presentist has no reason to believe principles like (9*), and could only accept them on "blind faith".

Suppose the presentist really has no right to appeal to (9*). If even this mild causal assumption is unwarranted, I had no right to use (9) in my argument about the current shape of the present. If there is no presumption that presently occurring events are caused by events that have already happened, then it should be "up for grabs" whether presently occurring events at locations other than my own include ones inside or on the surface of my rearward or forward light-cone. The only reasons I can find to rule them out are the sort I rehearsed above: the rearward ones may already have affected me, and so must already have occurred; and the forward ones could be affected by me, and so cannot yet have occurred. It would be strange to grant the presentist this much use of (9), while denying her the right to use similar reasoning in the case of the Bohmian foliation. If the presentist is justified in supposing that what is happening now does not include a dinosaur's death or the death of the sun, she is also justified in judging that the Bohmian and A-theoretic foliations coincide.

Callender argues that the Bohmian's foliation would be of no use to the A-theorist, but his argument requires that (9) and (9*) be utterly unmotivated for a presentist. The presentist is bound to disagree: she will regard such principles as extensions of something known to be true with respect to causal dependencies among events in her own life. The presentist needn't produce an indubitable a priori proof of these intuitively plausible and inductively confirmed generalizations in order to believe them on the basis of something more than blind faith.

# 5. CONCLUSION

## 5.1 Summing up

The main problems for presentism discussed in this paper were: (1) Sider's argument that presentists lack adequate grounds for physically important cross-temporal relations involving motion; and (2) objections based on inconsistency with Relativity, especially those based on alleged inconsistency with SR. My main conclusions can be summed up as follows:

(1) Sider's problems about cross-temporal relations require that presentists take manifold structure seriously. In a Galilean or Relativistic universe, physically fundamental cross-temporal relations force the presentist to admit the existence of formerly occupied points, or find some kinds of surrogates for them in the present. The simplest strategy would be to adopt a growing-manifold presentism, but I suggested a way to maintain a one-slice presentism with persisting trajectories that passed through no-longer-existing points.

(2) I tried to show that the conflict with Relativity is not so deep as one might think, while also calling into question the relevance of conflict with SR or GR. If we inhabited a manifold that appeared to have the metrical properties of Minkowski space-time or of some foliable manifold satisfying GR, those of us who are presentists would not automatically be forced to reject SR or GR, because we would not automatically have to posit laws involving additional intrinsic space-time structure. The present might march in step with some physical phenomenon, and so obey laws that do not directly appeal to manifold structure going beyond the metric. Even if the presentist were forced to posit such additional structure, she would not be radically scientifically revisionist, so long as she accepts the existence of the manifold and recognizes the fundamentality of its structure in scientific explanations. Furthermore, the relevance of whatever conflict there might be between presentism and either form of Relativity remains uncertain. SR is false, and GR is challenged by quantum theory. Although most physicists who are looking for a theory of quantum gravity are trying to get by without imposing a preferred foliation upon space-time, some think we need to do so. I lack the expertise to hazard an informed guess about the relative chances of these two kinds of theories. But some of the reasons proposed for introducing a physically distinguished foliation would play right into the presentists' hands. So long as we are allowed to appeal to principles like (9) or (9*), we could legitimately claim that physics had come around at last, and found a use for genuine simultaneity after all.

## 5.2  Pushing Back

Metaphysicians—including the numerous theoretical physicists and philosophers of physics who moonlight as metaphysicians—should naturally like to be able to invoke the prestige of physics in settling disputes. After all, scientific questions actually do get settled occasionally, unlike so many of the larger questions of metaphysics; it would be nice if stable results in physics could provide some leverage on the slippery problems of metaphysics. And this metaphysician, at any rate, agrees that physics simply *must* be relevant to many of the metaphysician's central concerns. What part of metaphysics is more exciting than the attempt to locate ourselves—thinking and feeling human agents—in the physical world? Since physics represents our best efforts to describe the fundamental nature of that world, metaphysicians cannot ignore advances in physics if we are serious about this project.

However, when appealing to findings from empirically well-grounded disciplines, philosophers face a strong temptation to overstate their case—especially if their philosophical opponents can be relied on to be relatively innocent of new developments in the relevant science. I fear that some B-theorists have succumbed to the temptation, judging by the relish with which they often pronounce a verdict based on Relativity. They can practically hear the *crunch* of the lowly metaphysician's armor giving way, as they bring the full force of incontrovertible physical fact down upon our A-theoretically-addled heads.[129] But what actually hits us, and how hard is the blow? SR is false; GR's future is highly uncertain; and the presentist's conflict with either version of Relativity is shallow, since the presentist's manifold can satisfy the same geometrical description as a B-theorist's manifold, and afford explanations of all the same phenomena in precisely the same style. In these circumstances, how could appeal to SR or GR justify the frequent announcements that the A-theory–B-theory dispute has been "settled by physics, not philosophy"?

## REFERENCES

Adams, R. (1986), 'Time and Thisness', in P. French, T. Uehling, and H. Wettstein (eds), *Midwest Studies in Philosophy: Volume 11* (Minneapolis: University of Minnesota), 315–329.

Albert, D. (2002), *Quantum Mechanics and Experience* (Cambridge, Mass.: Harvard University Press).

——and Galchen, R. (2009), 'A Quantum Threat to Special Relativity', *Scientific American Magazine* (March).

Armstrong, D. (2004), *Truth and Truthmakers* (Cambridge: Cambridge University Press).

Arthur, R. T. W. (2006), 'Minkowski Spacetime and the Dimensions of the Present', in Dieks, D. *The Ontology of Spacetime* (Elsevier) 129–155).

Barbour, J. (1999), *The End of Time* (Oxford: Oxford University Press).

Belot, G. (1999), 'Rehabilitating Relationalism', *International Studies in the Philosophy of Science* 13: 35–52.

——(2005), 'Dust, Time and Symmetry', *British Journal for the Philosophy of Science* 56: 255–291.

Bennett, J. (1988), *Events and their Names* (Indianapolis: Hackett Publishing Co.).

Bennett, K. and Zimmerman, D. (2010), *Oxford Studies in Metaphysics: Vol. 6* (Oxford: Oxford University Press).

Bigelow, J. (1996), 'Presentism and Properties', in (Tomberlin, 1996: 35–52).

Bourne, C. (2002), 'When am I? A Tense Time for Some Tense Theorists?', *Australasian Journal of Philosophy* 80: 359–371.

——(2006), *A Future for Presentism* (Oxford: Oxford University Press).

——(2007), 'Numerical Quantification and Temporal Intervals', *Logique et Analyse* 50.

Braddon-Mitchell, D. (2004), 'How Do We Know it is Now Now?', *Analysis* 64: 199–203.

Brogaard, B. (2007), 'Span Operators', *Analysis* 67: 72–79.

---

[129] I also suspect that philosophy of physics and metaphysics differ to some degree in the conventions governing writing style and rhetorical pitch.

Broad, C. D. (1923), *Scientific Thought* (London: Routledge & Kegan Paul).

Brown, H. R. and Pooley, O. (2006), 'Minkowski Space-Time: A Glorious Non-Entity', in Dieks, D. *The Ontology of Spacetime* (Elsevier).

Cameron, R. (2010), 'Truthmaking for Presentists', in (Bennett and Zimmerman, 2010).

Callender, C. (2000), 'Shedding Light on Time', *Philosophy of Science* (supplement) 67: S587–S599.

—— (ed.). (2002), *Time, Reality and Experience* (Cambridge: Cambridge University Press).

—— (2008), 'Finding "Real" Time in Quantum Mechanics', in Craig and Smith, 2008: 50–72.

Chisholm, R. (1970), 'Events and Propositions', *Noûs* 4: 15–24.

—— (1976), *Person and Object: A Metaphysical Study* (La Salle, Ill.: Open Court).

—— (1981), 'Time and Temporal Demonstratives', in K. Weinke (ed.), *Logik, Ethik und Sprache* (Vienna and Munich: R. Oldenburg Verlag), 31–36.

—— (1990a), 'Events Without Times: An Essay On Ontology', *Noûs* 24: 413–428.

—— (1990b), 'Referring to Things That No Longer Exist', in J. Tomberlin (ed.), *Philosophical Perspectives: Vol. 4 (Action Theory and Philosophy of Mind)* (Atascadero, CA: Ridgeview: 545–556).

Clifton, R. and Hogarth, M. (1995), 'The Definability of Objective Becoming in Minkowski Spacetime', *Synthese* 103: 355–387.

Cook Wilson, J. (1926), *Statement and Inference*, Vol. II, A. S. L. Farquharson (ed.), (Oxford: Clarendon Press).

Craig, W. (2000), *The Tensed Theory of Time* (Dordrecht: Kluwer).

Craig, W. (2001), *Time and the Metaphysics of Relativity* (Dordrecht: Kluwer).

Craig, W. L. and Smith, Q. (eds). (2008), *Einstein, Relativity, and Absolute Simultaneity* (New York and London: Routledge).

Crisp, T. (2003), 'Presentism', in (Loux and Zimmerman, 2003: 211–245).

—— (2004), 'On Presentism and Triviality', in (Zimmerman, 2004: 15–20).

—— (2005), 'Presentism and Cross-Time Relations', *American Philosophical Quarterly* 42: 5–17.

—— (2007a), 'Presentism and Grounding', *Noûs* 41: 90–109.

—— (2007b), 'Presentism, Eternalism and Relativity Physics,' in (Craig and Smith, 2008: 262–278).

Davidson, M. (2003), 'Presentism and the Non-Present', *Philosophical Studies* 113: 77–92.

de Clercq, R. (2006), 'Presentism and the Problem of Cross-Time Relations', *Philosophy and Phenomenological Research* 72: 386–402.

Dickson, M. (1998), *Quantum Chance and Nonlocality.* (New York: Cambridge University Press).

Dieks, D. (ed.). (2006), *The Ontology of Spacetime* (Amsterdam: Elsevier).

—— (ed.). (2008), *The Ontology of Spacetime II* (Amsterdam: Elsevier).

Dolev, Y. (2007), *Time and Realism* (Cambridge, Mass: M.I.T.).

Dorato, M. (2006), 'The Irrelevance of the Presentist/Eternalist Debate for the Ontology of Minkowski Spacetime', in (Dieks, 2006: 93–109).

Earman, J. (1989), *World Enough and Spacetime* (Cambridge, Mass.: M.I.T.).

Einstein, A. (1949), 'Remarks Concerning the Essays Brought Together in This Co-operative Volume', in (Schilpp, 1949: 665–688).

Fine, A. and Leplin, J. (eds). (1989), *Proceedings of the 1988 Biennial Meeting of the Philosophy of Science Association, Vol. 2* (East Lansing, Mich.: Philosophy of Science Association).

Forrest, P. (2004), 'The Real but Dead Past: A Reply to Braddon-Mitchell', *Analysis*, 64: 358–362.

—— (2006), 'General Facts, Physical Necessity, and the Metaphysics of Time', in (Zimmerman, 2006: 137–152).

—— (2008), 'Relativity, the Passage of Time and the Cosmic Clock', in (Dieks, 2008: 245–53).

Frege, G. 1984, 'Thoughts', in Frege, *Collected Papers on Mathematics, Logic, and Philosophy* (Oxford: Basil Blackwell), 351–372.

Gale, R. 1968, *The Language of Time*, London: Routledge & Kegan Paul.

Geach, P. 1972, 'Some Problems About Time', reprinted in Geach, *Logic Matters* (Berkeley and Los Angeles: University of California Press), 302–318.

Gibson, I., and Pooley, O. (2008), 'Relativistic Persistence', *Philosophical Perspectives, Vol. 20 (Metaphysics)*: 157–198.

Gödel, K. (1949), 'A Remark About the Relationship Between Relativity Theory and the Idealistic Philosophy', in (Schilpp, 1949: 557–562).

Grünbaum, A. (1967), *Modern Science and Zeno's Paradoxes* (Middletown, Conn.: Wesleyan University Press).

Gunter, P.A. Y., ed. (1969), *Bergson and the Evolution of Physics* (Knoxville: University of Tennessee Press).

Hardin, C. (1984), ' "Thank Goodness It's Over There!" ', *Philosophy* 59: 12–17.

Hartshorne, C. (1967), *A Natural Theology for Our Time* (La Salle, Ill.: Open Court).

Heathwood, C. (2005), "The Real Price of the Dead Past: A Reply to Forrest and to Braddon-Mitchell", *Analysis* 65: 249–251.

Hinckfuss, I. (1975), *The Existence of Space and Time* (Oxford: At the Clarendon Press).

Horwich, P. (1987), *Asymmetries in Time* (Cambridge, Mass.: M.I.T.).

Keller, S. (2004), 'Presentism and Truthmaking', in (Zimmerman, 2004: 83–104).

Kierland, B., and Monton, B. (2007), 'Presentism and the Objection from Being Supervenience', *Australasian Journal of Philosophy* 85: 485–497.

Le Poidevin, R. (1991), *Change, Cause, and Contradiction* (London: Macmillan).

Lewis, D. (1976), 'The Paradoxes of Time Travel', *American Philosophical Quarterly* 13: 145–152.

—— (1979), 'Attitudes *De Dicto* and *De Se*', *Philosophical Review* 88: 513–543.

—— (1999), 'Armstrong on Combinatorial Possibility', in Lewis, *Papers on Metaphysics and Epistemology* (Cambridge: Cambridge University Press).

—— (2002), 'Tensing the Copula', *Mind* 111: 1–13.

—— (2004), 'Tensed Quantifiers', in (Zimmerman, 2004: 3–14).

Lombard, L. (1999), 'On the Alleged Incompatibility of Presentism and Temporal Parts', *Philosophia* 27: 253–260.

—— (forthcoming), 'Time for a Change: A Polemic Against the Presentism/Eternalism Debate', in J. K. Campbell, M. O'Rourke, and H. Silverstein (eds), *Time and Identity: Topics in Contemporary Philosophy, Vol. 6* (Cambridge, Mass.: M.I.T.).

Lowe, E. (1998), *The Possibility of Metaphysics* (Oxford: At the Clarendon Press).

Lucas, J. (1989), *The Future* (Oxford: Blackwell).

—— (2008), 'The Special Theory and Absolute Simultaneity', in (Craig and Smith, 2008: 279–290).

Ludlow, P. (1999), *Semantics, Tense, and Time* (Cambridge, Mass.: M.I.T.).

—— (2004), 'Presentism, Triviality, and the Varieties of Tensism', in (Zimmerman, 2004: 21–36).

Loux, M. and Zimmerman, D. (eds). (2003), *The Oxford Handbook of Metaphysics* (Oxford: Oxford University Press).

Markosian, N. (2004), 'A Defense of Presentism', in (Zimmerman, 2004: 47–82).

Maudlin, T. (1989), 'The Essence of Space-Time', in (Fine et al., 1989: 82–91).

—— (1993), 'Buckets of Water and Waves of Space: Why Space-Time is Probably a Substance', *Philosophy of Science* 60: 183–203.

—— (2002), *Quantum Non-Locality and Relativity*, 2nd edn. (Malden, Mass.: Blackwell).

—— (2008), 'Non-Local Correlations in Quantum Theory: How the Trick Might Be Done', in (Craig and Smith, 2008: 156–179).

MacBeath, M. (1983), 'Mellor's Emeritus Headache', *Ratio* 25: 81–88.

McTaggart, J. McT. E. (1927), *The Nature of Existence: Vol. 2* (Cambridge, UK: Cambridge University Press).

McCall, S. (1994), *A Model of the Universe* (Oxford: At the Clarendon Press).

Mellor, D. H. (1981a), *Real Time* (Cambridge: Cambridge University Press).

—— (1981b), ' "Thank Goodness That's Over" ', *Ratio* 23: 20–30.

—— (1998), *Real Time II* (London: Routledge).

Maxwell, N. (2006), 'Special Relativity, Time, Probabilism and Ultimate Reality', in (Dieks, 2006: 229–245).

Merricks, T. (1999), 'Persistence, Parts, and Presentism', *Noûs* 33: 421–438.

—— (2006), 'Goodbye Growing Block', in (Zimmerman, 2006: 103–110).

—— *Truth and Ontology* (Oxford: Oxford University Press).

Misner, C., Thorne, K., and Wheeler, J. (1973) *Gravitation* (San Francisco: W. H. Freeman).

Monton, B. (2006), 'Presentism and Quantum Gravity', in (Dieks, 2006: 263–280).

Mulligan, K., Simons, P., and Smith, B. (1984), 'Truth-Makers', *Philosophy and Phenomenological Research* 44: 287–321.

Mundy, B. (1986), 'The Physical Content of Minkowski Geometry', *British Journal for the Philosophy of Science* 37: 25–54.

Nerlich, G. (1994), *The Shape of Space*, 2nd edn. (Cambridge: Cambridge University Press).

Norton, J. (1989), 'The Hole Argument', in (Fine et al., 1989: 56–64).

Oaklander, N. (1991), 'A Defense of the New Tenseless Theory of Time', *Philosophical Quarterly* 41: 26–38.

Oaklander, N. and Smith, Q. (eds). (1994), *The New Theory of Time* (New Haven: Yale University Press).

Peacock, K. (2006), 'Temporal Presentness and the Dynamics of Spacetime', in (Dieks, 2006: 247–261).

Prior, A. (1959), 'Thank Goodness That's Over', *Philosophy* 34: 12–17.

—— (1970), 'The Notion of the Present', *Studium Generale* 23: 245–248.

—— (1996), 'Some Free Thinking About Time', in J. Copeland (ed.), *Logic and Reality* (Oxford: Clarendon Press), 47–51; reprinted in (van Inwagen and Zimmerman, 2008).

—— (2003), 'Changes in Events and Changes in Things', in Prior, *Papers on Time and Tense*, New Edition, P. Hasle, P. Ohrstrom, T. Braüner, and J. Copeland (eds), (Oxford: Oxford University Press), 7–19.

Prosser, S. (2007), 'Could We Experience the Passage of Time?', *Ratio* 20: 75–90.

Putnam, H. (1967), 'Time and Physical Geometry', *Journal of Philosophy* 64: 240–247.

Quine, W. V. O. (1960), *Word and Object* (Cambridge, Mass.: M.I.T. Press).

Rhoda, A. (forthcoming), 'Presentism, Truthmakers, and God', *American Philosophical Quarterly*.

Rietdijk, C. W. (1966), 'A Rigorous Proof of Determinism Derived from this Special Theory of Relativity', *Philosophy of Science* 33: 341–344.

Russell, B. (1938), *Principles of Mathematics* (New York: W. W. Norton).

Sattig, T. (2006), *The Language and Reality of Time* (Oxford: Clarendon Press).

Saunders, S. (2002), 'How Relativity Contradicts Presentism', in (Callender, 2002: 277–292).

Savitt, S. (2000), 'There's No Time Like the Present (in Minkowski Spacetime)', *Philosophy of Science* 67 (Proceedings): S663–S574.

—— (2002), 'On Absolute Becoming and the Myth of Passage', in (Callender, 2002: 153–167).

—— (2006), 'Presentism and Eternalism in Perspective', in (Dieks, 2006: 111–127).

—— (2009), 'The Transient *Nows*', in *Quantum Reality, Relativistic Causality, and Closing the Epistemic Circle: Essays in Honour of Abner Shimony*, W. C. Myrvold and J. Christian (eds) (New York: Springer), 349–362.

Schilpp, P. A. (ed.). (1949), *Albert Einstein: Philosopher-Scientist* (New York: Tudor).

Schlesinger, G. (1980), *Aspects of Time* (Indianapolis: Hackett).

—— (1994), 'Temporal Becoming', in (Oaklander and Smith, 1994: 214–220).

Shoemaker, S. (1998), 'Causal and Metaphysical Necessity', *Pacific Philosophical Quarterly* 79: 59–77.

Sider, T. (1999), 'Presentism and Ontological Commitment', *Journal of Philosophy* 96: 325–347.

—— *Four-Dimensionalism* (New York: Oxford University Press).

—— (2005), 'Traveling in A- and B-Time', *The Monist* 88: 329–335.

—— (2006), 'Quantifiers and Temporal Ontology', *Mind* 115: 75–97.

—— (2007), 'Ontological Realism', in D. Chalmers, D. Manley and R. Wasserman (eds), *Metametaphysics* (Oxford: Oxford University Press).

Smart, J. J. C. (1963), *Philosophy and Scientific Realism* (London: Routledge and Kegan Paul).

—— (1987), 'Time and Becoming', reprinted in Smart, *Essays Metaphysical and Moral* (Oxford: Basil Blackwell), 78–90.

Smith, Q. (1993), *Language and Time* (New York: Oxford University Press).

Stein, H. (1968), 'On Einstein-Minkowski Space-Time', *Journal of Philosophy* 65: 5–23.

—— (1991), 'On Relativity Theory and Openness of the Future', *Philosophy of Science* 58: 147–167.

Stout, G. F. (1921/22), 'The Nature of Universals and Propositions', *Proceedings of the British Academy* 10: 157–172.

Szabo, Z. (2007), *Philosophical Perspectives: Vol. 20* (Metaphysics): 399–426.

Tooley, M. (1997), *Time, Tense, and Causation* (Oxford: At the Clarendon Press).

Unger, P. (2006), *All the Power in the World* (Oxford: Oxford University Press).

Valentini, A. (2008), 'Hidden Variables and the Large-Scale Structure of Space-Time', in (Craig and Smith, 2008): 125–155.

van Inwagen, P. and Zimmerman, D. (eds). (2008), *Metaphysics: The Big Questions*, 2nd edn. (Malden, Mass.: Blackwell).

Williams, C. (1998), 'A Bergsonian Approach to A- and B-Time', *Philosophy* 73: 379–393.

Williams, D. (1951), 'The Myth of Passage', *Journal of Philosophy* 48: 457–472.

—— (1953), 'On the Elements of Being: I and II', *Review of Metaphysics* 7: 3–18 and 171–192.

Williamson, T. (1999), 'Existence and Contingency', *Proceedings of the Aristotelian Society*, Suppl. Vol. 73: 181–203.

Zimmerman, D. (1996), 'Persistence and Presentism', *Philosophical Papers* 25: 115–126.

Zimmerman, D. (1997), 'Chisholm and the Essences of Events', in Lewis Hahn, ed., *The Philosophy of Roderick M. Chisholm* (Peru, Illinois: Open Court), 73–100.

—— (ed.). (2004), *Oxford Studies in Metaphysics: Volume 1* (Oxford: Oxford University Press).

—— (2005), 'The A-theory of Time, the B-theory of Time, and 'Taking Tense Seriously"', *Dialectica* 59: 401–457.

—— (2006a), 'Temporary Intrinsics and Presentism' (with Postscript: "Can One 'Take Tense Seriously' and Be a B-theorist?"), in *Persistence*, S. Haslanger and R. Fay (eds) (Cambridge, Mass.: M.I.T.), 393–424.

—— (ed.), (2006b), *Oxford Studies in Metaphysics, Vol. 2* (Oxford: Oxford University Press).

—— (2008), "The Privileged Present: Defending an 'A-theory' of Time", in T. Sider, D. Zimmerman, and J. Hawthorne (eds), *Contemporary Debates in Metaphysics* (Malden, Mass.: Blackwell), 211–225.

—— (2009), 'The A-theory of Time, Presentism, and Open Theism', in Melville Stewart (ed.), *Science and Religion in Dialogue*, Vol. 2 (Malden, Mass.: Wiley-Blackwell), 791–809.

—— (2010), 'From Property Dualism to Substance Dualism', *Proceedings of the Aristotelian Society*, Supplementary Vol. 84: 119–150.

# PART II

## THE DIRECTION OF TIME

CHAPTER 8

..............................................................................................................

# THE ASYMMETRY
# OF INFLUENCE

..............................................................................................................

## DOUGLAS KUTACH

## 1. INTRODUCTION

..............................................................................................................

ATTEMPTS to meddle with the past are futile. While this nugget of folk wisdom serves as a respectable guide to action, its utility is standardly conceived as arising from the general inability of anything to influence the past. This explanation, though, oversimplifies the complex architecture of fact and fiction responsible for the reasonableness of not trying to affect the past. It makes little genuine progress in understanding the asymmetry of influence because it fails to distinguish between two significantly different kinds of explanation. The first kind appeals to the fixity of the past as a strict, fundamental, metaphysical or scientific fact that guarantees the inefficacy of all attempts at past-directed influence. The second kind of explanation says that in virtue of the reasonableness of the folk wisdom, we interpret 'influence' in a way that ignores those senses in which the past can be influenced, so that "nothing can affect the past" is rendered true largely by definition.

A fruitful analogy is the concept of solidity. One might say the reason it is useful to treat a boulder as if it were solid is because the boulder really is solid. But this platitude does not distinguish the false explanation that boulders are solid through and through from the correct explanation that, although the boulder mostly consists of empty space between its atoms, its chemical bonds give rise to a cluster of complex macro-properties like rigidity and impenetrability that make it effectively solid for most ordinary practical purposes. Taking 'solid' to denote this imprecise cluster of imprecise properties makes it literally true that the boulder is solid by discounting respects in which the boulder is not solid.

One strategy for explaining the folk wisdom that affecting the past is futile follows the first kind of explanation by arguing that it is in the nature of time, or somehow built into the universe's fundamental structure, that the past is fixed or that any lawful dependence of the past on the present is not genuine influence. The strategy advocated in this chapter elaborates on one version of the second kind of explanation. Rather than the fixity of the past being a fundamental metaphysical fact, there are numerous respectable senses in which the past can be influenced. Yet, because of a complex assortment of conceptual and physical relations, past-directed influence, also known as 'backward influence', turns out to be critically different from future-directed influence in its practical impact. Our ordinary conceptualization of reality incorporates this pragmatic asymmetry by counting only the future-directed influence as genuine influence. Using the second kind of explanation is arguably preferable because it posits less metaphysical baggage and—more important—clarifies intricate connections between influence, causation, time, and chance, better than in accounts of the first kind.

The resulting account contributes to our understanding of time by providing a crucial component of a static theory of time: an explanation of why it is natural to think of the past and future as essentially different even though they are metaphysically on par. Dynamic theories of time virtually always enforce a fundamental difference between past and future, either by postulating a primitive ontological difference, or by positing a metaphysically robust passage of time which, in turn, somehow vindicates the belief that the future is fundamentally different from the past. The growing block model illustrates both theses by holding that physical reality at any moment consists of a spacetime block of present and past facts but no future facts, and that the block grows. The account presented here provides an alternative explanation for our stubborn conviction that the past and future are essentially different. The hard part of the explanation involves demonstrating how known physical structures make it reasonable for us to believe in the folk wisdom even if we routinely influence the past. The easy part, then, attributes our intuition that the past is fixed to an understandable misinterpretation. People know of the occasional practicality of trying to affect future events, and the universal impracticality of trying to affect the past, but they mistake the asymmetry of *useful* influence for an asymmetry of influence simpliciter.

If successful, this account would explain the influence asymmetry using relatively innocuous resources: the fundamental laws and facts about the world's fundamental material layout. Because it dedicates no structures to the implementation of temporal passage, the only sense in which time passes is that there is something about the combination of laws and material facts that makes it reasonable to think of time as passing. This still permits the possibility that passage is fundamental. If the laws turn out to have the right properties, then the aspect of reality most responsible for the utility of conceiving time as passing might turn out to be a feature of the laws alone and not rely on features of material facts. If so, time will pass fundamentally not because there is some flow of time in the sense of dynamic passage, or some fundamental arrow specifically dedicated to playing the passage role, but merely because, out of all the available structures (resulting from the laws plus facts), what ends up best vindicating talk of passage is an asymmetry in the fundamental laws.

If there is nothing solely in the fundamental laws that plays the passage role, time will either not pass at all, or will pass non-fundamentally, i.e. derivatively. Passing not at all means that nothing plays near enough the role of passage, and passing derivatively means there is something in the content of the laws and historical layout of facts that plays the role of passage near enough, but like temperature or solidity, is non-fundamental. Choosing between these two options depends largely on how one wants to construe passage. Much could be said about what in general constitutes bona fide temporal passage, but once one departs from theories with fundamental passage, the elements of the constitutive role of passage become rather unclear. There do seem to be some platitudes that tether the meaning of passage to other concepts: The unidirectionality of passage is tied to the idea that a number of important psychological attitudes are appropriately temporally asymmetric, for example, that the fear of death is more reasonable than the fear of birth, and that feelings of anticipation are appropriate to future but not past events. Passage is also arguably that which people perceive or seem to cognitively latch onto when they feel (or report that they feel) that time is passing. And, passage is linked to the thesis that the present is somehow special. The influence asymmetry too is closely aligned with the direction of passage in the sense that passage is often conceived as a conversion of the not-yet-settled possibilities of the future into settled facts of the past. None of these conceptual links are intended as necessary conditions for genuine passage, and any of them may turn out in the end to express misguided commitments. They are mentioned only because the goal of accounting for a non-fundamental asymmetry is to provide some story about why it is reasonable for people to believe in some package of beliefs that roughly resembles platitudes like these. Once one abandons the picture of passage as a metaphysically preferred presentness sweeping across the whole of history, the best account of passage will likely involve significant theoretical refinement of the gist of the passage-related platitudes. If the conceptual revision is too extensive, the question of whether the completed account possesses genuine temporal passage may well degenerate into a definitional quibble, where there is no substantive dispute about whether time really passes or whether we just tend to think of time that way. The rest of the discussion will focus on the influence asymmetry and disregard other aspects of passage, but the account I will present is definitely intended to be friendly to the conclusion that passage itself, such as it exists, is little more than the asymmetry of useful influence.

There is a tradition of attacking robust temporal passage by demonstrating the coherence and even physical possibility of backward influence, often as it appears in coherent time-travel stories. Models of general relativity with closed time-like curves (CTCs) bolster the relevance of such time-travel scenarios by treating as physically possible, physical processes that locally instantiate the normal future-directed influence, but which reach back around the CTC into the global past. While I believe such arguments help to undermine many dynamic (or 'moving now') conceptions of temporal passage, they fail to confront static theories of fundamental temporal passage, (e.g. Maudlin 2007), which account for passage without requiring more ontological resources than exist in standard spacetime structure plus a temporal orientation representing metaphysically fundamental passage. Furthermore, arguments relying on processes or topological structures that have not been empirically confirmed might turn out to be based on models that are, on deeper investigation, not physically possible. My explanation of the influence asymmetry differs from this tradition by claiming that backward influence may well obtain amid ordinary physical processes of the kind that pervade our local environment. Regardless of whether time travel into the past successfully undermines appeal to metaphysically robust temporal passage, backward influence can exist without such time travel. In order to set aside the complications introduced by time travel, it will be assumed hereafter that the laws and contingent facts do not permit it. For instance, the topology of spacetime will be assumed to be free of CTCs, and any laws of nature governing the world's development will not allow physical states to have causal impact that skips across time. That is, the state $c_1$ can determine (or fix a probability for) a state $c_2$ only if every state temporally between them is determined by (or has some probability fixed for it by) $c_1$, and determines (or fixes some probability for) $c_2$.

The central contention of this chapter—that what we think of as the influence asymmetry is really an asymmetry of useful influence—is suggested by uncontroversial observations about ordinary future-directed influence. We have good reason for believing in influence because we are aware of events at different times where the existence or character of one event, the effect, correlates with the character or existence of the other event, the cause. Some kinds of forward influence, though, are undetectable by observation of the relevant events, for example, the weak gravitational influence a person has on the motion of Jupiter or the butterfly's chaotic influence on the next decade's weather. A good reason for accepting them as cases of influence comes from believing that laws of physics are at least partially responsible for the observable influence, and that these laws remain operative when the physical connection between the events becomes too weak or too chaotic for us to detect. In brief, we naturally (and correctly) accept that the fundamental laws dictate how far our paradigmatic influence extends to other kinds of influence.

Once one allows that influence should be extended to cases where the physical laws establish dependencies between events at different times, it is not outrageous to ask whether influence could be extended further to other cases of nomological dependence. Since paradigm fundamental physical theories virtually all postulate at least some kind of nomological dependence of past states on future states,[1] one should consider whether such backward nomic dependence should also count as influence. There would be no problem in doing so, I think, if it were not for our steadfast commitment to the folk wisdom that meddling with the past is futile, which is seemingly at odds with backward influence. An obvious strategy for resolving the conflict is to clarify how, despite the presence of both forward and backward influence, the folk wisdom continues to serve as a reliable practical guide.

The hypothesis under consideration is that we influence the future more or less as pre-theoretical intuition dictates but that we also influence the past in a number of

---

[1] Even theories with fundamental chanciness, like GRW-style collapse interpretations of quantum mechanics (Ghirardi and Weber, 1986), fix *some* probabilities for past states. There is a determinate probability at $t$ that no collapse will have occurred at $t - \epsilon$ and that is equivalent to the probability that the state at $t - \epsilon$ implied by the deterministic backward evolution of the wave function at $t$ will have occurred.

different senses. Yet, for each way of affecting the past, some excuse is always available for why it is unexploitable in practice.

- Some instances of past-directed influence are too weak.
- Some are too chaotic.
- Some fall under no regularities that are epistemologically accessible.
- Some are re-interpretable as ordinary, unobjectionable future-directed influence.
- Some count as influence due to ad hoc parameter settings or other ways of precisifying the influence notion that beg the question about the direction of influence.
- Some are systematically unusable due to some physical contingency that constrains events in a suitably large region to systematically fail to display the expected consequences of influence.

While these excuses will be clarified in due course, the intended conclusion is that all past-directed influence falls into at least one of the above categories, and is therefore generally useless for advancing any goals one might have. Unfortunately, the argument is complicated by the fact that influence generally, and past-directed influence in particular, is a vague notion admitting diverse precisifications. For the argument to be successful, it must ensure that for any halfway reasonable notion of influence, some excuse is available to ensure that we cannot exploit our influence over the past.

The advancement asymmetry is the claim that there exists at least one reasonable notion of influence such that future-directed influence is sometimes useful for advancing goals and that there is no reasonable notion of influence such that past-directed influence is useful for advancing goals. The purpose of this paper is to give an exhaustive account of the advancement asymmetry that (1) is connected with fundamental physics, influence, causation, counterfactual dependence, and related notions in palatable ways, (2) derives the asymmetry wholly from scientifically acceptable sources, and (3) does not assume the asymmetry by fiat, whether by positing an asymmetry of robust temporal passage, or by smuggling in a temporal asymmetry through the use of an asymmetric notion of chance, or by deriving the influence asymmetry from a primitive causal relation that exists in addition to the complete history of fundamental properties, relations, and laws. It is possible to have fundamental (asymmetric) temporal passage within a fully scientific and static account of the influence asymmetry, but I will set aside even this possibility because the asymmetry is better explained without it.

## 2. INFLUENCE AND COUNTERFACTUAL DEPENDENCE

The guiding idea behind one family of influence-concepts is that of some kind of counterfactual dependence. For one event $C$ to influence an event $E$ is for $E$ to be different from what it would have been if $C$ had not happened. Regimenting this idea to make it both scientifically legitimate and useful for understanding causation requires care because the use of counterfactual conditionals raises worries of arbitrariness and circularity. If we evaluate counterfactuals in some way that introduces facts that go beyond the physical condition of the world, we may beg the question about influence asymmetry or just be talking nonsense. It is therefore important that any arbitrary aspects of how influence-counterfactuals are evaluated are not misrepresented as facts about the actual world and do not rely on counterfacts—facts about what would have happened that go objectionably beyond facts about what actually happens.

Three clarifications are needed before detailing how to evaluate influence. First, one must quarantine the intuitions about influence that are at the heart of traditional struggles over free will. Incompatibilists hold that under determinism, there is no genuine free will. This might be extended into a claim that, under determinism, nothing really influences anything else—free or not—and that at best one has the illusion of influence. Because the kind of influence elaborated below does not take sides on the free-will debate, the incompatibilist might complain that unless the laws are of a special form, one gets only pseudo-influence out of the account. Even if that is correct, the ability of the account to achieve its goal is unaffected because it can still explain why it is reasonable for us to think of ourselves as influencing the future and not the past. One need only add that the reasonability in part arises because insofar as the utility of the folk wisdom is concerned, it is harmless to mistake pseudo-influence for the genuine article.

Second, it is important to set aside commonly accepted counterfactual logics like the kind developed by Stalnaker (1968) and Lewis (1973). Such logical systems are designed to regiment inferences and truths of natural language. To apply such systems to a scientific investigation risks contaminating one's theory of influence with linguistic quirks and artificially limiting the range of possible theories. If by some bizarre coincidence the logic of natural language conditionals shares important similarities with the physical structures that best explain influence, no harm is done, but to presume from the outset that nature must obey the rules of natural language is to restrict oneself by a fantastically implausible constraint.

Third, there is a natural temptation to balk at equating influence with counterfactual dependence because while some kinds of counterfactual dependence are causal in character, others are apparently evidential. For example, the details of a photographic portrait counterfactually vary in accord with the photographed subject. If the woman's photograph had pictured her wearing a hat, then very probably the woman would have been wearing the hat during the photo shoot. One naturally interprets this counterfactual dependence as a consequence of the ordinary future-directed influence of hats on photographs. Yet, the theory being developed here equates influence with some kinds of counterfactual dependence, and is on some precisifications of 'influence' committed to the counterintuitive claim that the photograph backwardly influenced the woman's appearance. This seeming problem is partly terminological. I could align the terminology to more closely match standard usage by claiming that counterfactual dependence goes both forward and backward in time and by using the label 'influence' to mark the dependence associated with the useful-for-advancement direction. Then, the woman would counterfactually depend on the later photograph, but the photograph would not literally influence her. The more conventional usage, though, would not alter the substantive content of the theory: that some forms of backward counterfactual dependence and forward counterfactual dependence express the same kind of physical link. They both express lawful relationships between states at different times. One need not assume one direction as having a metaphysically privileged status associated with causation. I deliberately choose the unintuitive usage where 'influence' is equivalent to (my preferred versions of) counterfactual dependence, regardless of temporal direction, in order to continually reinforce the controversial character of my conclusion that the seeming temporal directionality of influence is pragmatic, and not necessarily fundamental.

The model I propose for evaluating influence-counterfactuals goes as follows. The counterfactual "If $C$ were true, $E$ would be true," is abbreviated $C \:\square\!\!\!\rightarrow E$, and has as its "semantic" value the objective probability of $E$ across the relevant possible worlds where $C$ obtains. What counts for relevance is determined by developing a theory whose goal is to extract the right amount of information from the laws of nature, the actual circumstances, and the antecedent itself, without being artificially restrictive or begging the question about the direction of influence. The approach I follow is to require enough structure so that the laws of physics entail determinate probabilities for $E$, but allow other precisifications of the relevant worlds to be left as free parameters. The value of the counterfactual will then exist only relative to some specific setting of the parameters.

To fix a determinate probability for $E$, the relevant worlds must obey all fundamental laws of physics. The most important kind of laws for evaluating influence are dynamical laws, rules whereby a precise physical state at one time entails an objective probability for physical states at other times. (States are events that instantiate "what happens at a single time.") It is certainly conceivable that the universe has no dynamical laws, but I will assume their existence on the hope that the fundamental laws turn out to be nice enough, like the paradigm fundamental theories of classical mechanics, relativity, and quantum mechanics. It is also possible that the dynamics permits sources of indeterminism that are unconstrained by probabilistic rules, and again I will just assume such sources either do not exist or are rare enough to be ignorable, at least for ordinary causation in mundane environments. Henceforth, only deterministic dynamics and chancy sources of indeterminism will be countenanced. If empirical evidence reveals these assumptions to be ill founded, this account of the asymmetry will presumably fall as well.

An important quantity for evaluating influence is the counterfactual probability of the effect, given the non-occurrence of the cause. To interpret this counterfactual correctly, one must distinguish between fine-grained events and coarse-grained events. A fine-grained event is by definition a perfectly precise microphysical condition inhabiting some specific spacetime region. A coarse-grained event is defined to be a (usually continuous) set of possible fine-grained events, to be thought of as possible inhabitants of the same spacetime region. A coarse-grained event, in effect, abstracts away from the precise microscopic details and thus serves as an event type. We can speak of a coarse-grained event as occurring in virtue of one of its elements occurring. (Some remarks on notation: Capital letters refer to coarse-grained events and lower case letters to fine-grained events. When the same letter appears in capital and lower case form, the designated fine-grained event is one element of the designated coarse-grained event. Strictly speaking, a coarse-grained event can be trivially coarse-grained, meaning that it contains a single fine-grained event as its only element. Finally, a coarse-grained event need not fall under a relatively simple natural language description, though we are typically concerned with such events.)

In judging influence, the effect must not be understood in a fine-grained way because that would make its counterfactual probability zero under almost any realistic physical laws and would render it a useless quantity. By understanding the effect in a coarse-grained way, insignificantly small alterations to the physics instantiating the effect do not necessarily generate a numerically distinct event.

The characterization of the non-existence of the cause is also a coarse-grained event because when one considers what would have happened had $C$ not occurred, one does not necessarily intend a microscopically specific instantiation of $\neg C$. There exists considerable latitude in what counts as an acceptable way for the cause not to occur. If the stick had not been released, was it still being held? Broken? Burned? Because there is no systematic answer as to which states constitute the cause not happening, the non-occurrence of the cause must be left as another free parameter.

In order to get any probabilities for $E$ from the fundamental laws of nature, one must consider states expansive enough for the laws to be informative. For realistic laws of physics, this requires embedding the non-release of the stick in a large enough physical state for the laws to entail a probability for $E$. How one should accomplish this embedding in full generality is not perfectly systematic, but in practice one can take an actual microstate big enough to count as "a state of the world at the time of $c$'s occurrence" and modify it to instantiate the non-releasing of the stick. Because there are many precise ways to instantiate $\neg C$ in the larger state, and many of them fix different probabilities for $E$, it is useful to postulate a probability measure over all the states so that together they determine a unique counterfactual probability, the probability of $E$ weighted over all these states. The *contextualized event*, $\overline{\neg C}$, is a probability distribution over a set of states each realizing $\neg C$ and some specification of the microscopic background environment. The contextualized event can treat the background in a fine-grained way by having all its elements agree on the microscopic details outside the $\neg C$ region, but can also coarse-grain over the background environment. The objective probability of $E$, given these contextualized events is labeled $p_{\overline{\neg C}}(E)$. This quantity serves as a kind of semantic value for the counterfactual conditional $\neg C \,\square\!\!\rightarrow E$ in the sense of being an objective quantity towards which belief in the counterfactual should be directed. Ordinary beliefs are directed towards truth, in that an ideally knowledgeable person will have degree of belief one in what is true and zero in what is false. Beliefs about influence counterfactuals are directed towards this probabilistic value in that an ideally knowledgeable person—one who knows what the laws entail—will have degree of belief $p_{\neg \overline{C}}(E)$ in $\neg C \ \Box\!\!\!\rightarrow E$, taking into account how $\neg \overline{C}$ fleshes out $\neg C$ in more detail. The probability is not intended to be part of a larger logical structure that one could rightly call a semantics because among other things it does not make sense for propositions generally and is not compositional.

# 3. FINE-GRAINED AND COARSE-GRAINED INFLUENCE

The objective probability of $E$ across the relevant $\neg C$-worlds gives us a measure of what would have happened had the cause not occurred. But what if the cause had occurred? Using standard linguistics-based counterfactual semantics, one would apply the inference rule, $C \& E \vdash C \ \Box\!\!\!\rightarrow E$, and infer that $E$ would have happened had $C$ happened. It is trivial using these semantics that any actual event counterfactually implies any other actual event. Thus, one could only judge counterfactual dependence by the counterfactual worlds where the cause doesn't occur. This leads to a conception I dub the inadequate notion of influence, 'influence$_i$'.

Influence$_i$ is defined by setting the degree to which $c$ influences$_i$ $E$ equal to $1 - p_{\neg \overline{C}}(E)$. Influence$_i$ adequately captures the idea that there is significant influence when $E$ would have been made very improbable by $C$'s non-occurrence. When the laws are deterministic, influence$_i$ is semi-adequate for representing influence. For example, when $c$ does not bear on $E$ by way of the fundamental laws because, say, it is outside of $E$'s past light cone,[2] then $c$ does not influence$_i$ $E$. For another example, waving an arm does not significantly influence$_i$ the position of the moon one second later because, given any plausible alternative activity one could engage in, the moon would very probably be located very near its actual position. For another example, suppose a die is shaken ten times, is tossed, and lands on five. Shaking the die ten times versus seven had a significant influence on its landing on five. Without the extra shakes, its probability of landing on five would have been one sixth. Hence the extra shakes influenced$_i$ the outcome to degree five-sixths.

One could quibble that influence$_i$ is artificially precise, for example that our influence over the die outcome is modeled as exactly five-sixths, when arguably nothing in the phenomena of physical influence forces this particular quantity on us. This is no different in principle from the use of the probability calculus to model degrees of belief. In both cases, it is best just to say that we employ the richer-than-needed structure and

---

[2] One must be careful not to misidentify any illicit smuggling of non-local counterfactual alterations as faster-than-light influence. If one constructs $\neg C$ to include alterations to the state of Alpha Centauri, the star will counterfactually depend on the earthly $C$, not because of non-local physical interactions, but because the dependence was inserted by hand.

then ignore any artificial precision at the end of calculation. No important conclusions will hereafter be drawn from the precise value of influence.

What makes influence$_i$ a poor guide to influence is its inability to properly handle stochastic dynamical laws. Suppose a present event $c$, the flapping of a certain butterfly's wings, contributes to a future lightning strike, $e$ coarse-grained as $E$, ten years from now. Because lightning strikes are so rare, $p_{\neg C}(E)$ is very low; hence, the butterfly strongly influences$_i$ the lightning. This verdict accords with our intuition that, had the butterfly done something else, there would likely not have been a lightning strike (at near enough the same time and place). But its influence$_i$ does not reflect that the butterfly was an insignificant contributor to $E$ in the sense that no matter what the butterfly did, lightning would likely not have struck. Assuming $E$'s great sensitivity to the many chance processes during the intervening years, replacing the butterfly's actual state with any other reasonable state will lead to very nearly the same probability of $E$. The many chance outcomes magnified through numerous chaotic microscopic interactions drown out the butterfly's contribution.

It is more useful to assess influence without any contamination from later chance outcomes. For that we can use a new notion, influence$_f$, which represents fine-grained influence. Influence$_f$ is defined by setting the degree to which $c$ influences$_f$ $E$ equal to $p_{\bar{c}}(E) - p_{\neg C}(E)$. Influence$_f$ captures the idea that there is significant influence when $E$ would have had a very different probability had $C$ not occurred than if $c$ had occurred. This is a reasonable measure of influence in the sense of quantifying how much $c$ is a difference-maker to $E$. When positive, it indicates that $c$ promotes $E$; when negative, that $c$ inhibits $E$.

Influence$_f$ reduces to influence$_i$ when the laws are deterministic. This feature is both good and bad, as illustrated by a case of deterministic chaotic causation. Suppose that (the precise future lightning strike) $e$ is nomologically entailed by the present microstate $\bar{c}$ (a small part of which instantiates $c$, the flapping of the butterfly's wings). Because lightning strikes are so rare, $p_{\neg C}(E)$ is very low, so the butterfly does strongly influence$_f$ the lightning. The good news is that this accords with our intuition that the butterfly's precise action has a significant effect on the future weather. Small differences in the butterfly imply big differences later. Its influence$_f$ is not overwhelmed by chance outcomes because there are none. The bad news is that influence$_f$ is overly sensitive to whether the chaotic processes are at bottom deterministic or stochastic. Under determinism, the butterfly strongly influences$_f$ the lightning, and under a stochastic indeterminism it does not significantly influence$_f$ the lightning. If we are interested in a notion of influence connected to deliberate control, manipulation, and the advancement of goals, the desired precisification of influence should not depend on the precise nature of the underlying physics. Because we do not have arbitrarily precise control over the microscopic implementation of our actions, we cannot reliably generate a specific fine-grained event, and because we do not have arbitrarily precise epistemological access to the microscopic implementation of our actions, we cannot reliably become aware of lawful generalities among fine-grained events and the events that follow them. One might say that knowledge of the laws allows calculation in principle of lawful generalities between $c$ and $E$, but for even the simplest of fundamental theories, practical calculation of fine-grained influence is out of the question.

Though influence$_f$ is a legitimate notion of influence, its limitations motivate a precisification of influence that coarse-grains over the cause. Coarse-grained influence, influence$_c$, is defined by setting the degree to which $C$ influences$_c$ $E$ equal to $p_{\overline{C}}(E) - p_{\overline{\neg C}}(E)$. Influence$_c$ accurately captures the idea that $C$ raises or lowers the chance of $E$ relative to $\neg C$. The value $p_{\overline{C}}(E)$ (and by similar reasoning $p_{\overline{\neg C}}(E)$) is insensitive to whether the seeming chanciness in nature is grounded in a stochastic dynamics or just a chaotic determinism because averaging over all the elements of $\overline{C}$ will smooth out differences among the worlds. So long as the dynamics is chaotic enough and the specification of $C$ is not too narrow, the precise boundary and probability measure will not make much difference. With sufficiently chaotic interactions, the butterfly does not significantly influence$_c$ the lightning because whether we average over all the worlds with flapping or average over all the worlds without flapping, the principal effect will be to average over many possible weather states at the time and location of the actual lightning strike. So the probability either way will just be the low probability of a lightning strike in general.

Influence$_c$ captures the idea that $C$ affects the probability of $E$. With many coarse-grained events being epistemologically accessible to us, we can examine statistical information about the frequencies of $C$ and $E$, and thereby make educated guesses about the existence of law-grounded regularities between them, without knowing the details of fundamental physics. Strictly speaking, the probabilistic relations hold only between coarse-grained effects and contextualized causes. However, the import of contextualized causes can often be approximated by segregating each into a salient cause and a background field. The approximation is useful when the coarse-grained influence of the background is sufficiently simple, for example small enough to be ignorable, or chaotic in a way that makes it effectively chancy, or somehow regular enough to be treatable as an external factor like terrestrial gravity. For example, we know a mallet hitting a bell in ordinary conditions greatly raises the probability of the bell's ringing versus not hitting the bell. This relation can be spelled out in terms of the probabilities nomologically fixed by sufficiently large physical states. In evaluating whether a strike of the bell influences$_c$ the bell's ringing, one takes a sufficiently large piece of the full actual state, $\overline{c}$—a sphere of radius ten light-seconds, say—and considers all the coarse-grainings of this state that involve the mallet hitting the bell. In doing the coarse-graining, one can hold fixed the microstate at locations away from the mallet and bell, or one can coarse-grain those as well. If $\overline{C}$ just coarse-grains over the mallet striking the bell and a bit of nearby air, then $p_{\overline{C}}(E) - p_{\overline{\neg C}}(E)$ measures how much striking affects ringing in this particular environment. If $\overline{C}$ and $\neg\overline{C}$ coarse-grain the whole environment, $p_{\overline{C}}(E) - p_{\overline{\neg C}}(E)$ measures how much striking affects ringing in more general environments. One can coarse-grain to different degrees, measuring for example, how much striking raises the probability of ringing in environments with the air at room temperature and standard pressure, or instead across environments with a wide range of temperatures, pressures, humidity, etc. The most useful coarse-grainings are those that retain the kinds of environments one is likely to find repeated in other places and times. So, a $\overline{C}$ useful for causal inferences involves a mallet hitting the bell with typical air around, but allowing the objects in more distant environment to vary, including people, trees, soil, weather. The corresponding $\overline{\neg C}$ needs to coarse-grain the external environment to the same degree as $\overline{C}$, lest one introduce a spurious probabilistic dependence. Because in this particular example $p_{\overline{C}}(E) - p_{\overline{\neg C}}(E)$ is largely insensitive to how the external entities are varied, it serves as a reasonable quantity to represent how much the mallet strike itself raises the probability of the ringing generally. When there is a kind of event $C$ that raises the probability of a kind of effect $E$, we say that $C$ is a promoter of $E$. Promotion strictly speaking only holds relative to a background and choice of coarse-graining, that is, in how $C$ is contextualized as $\overline{C}$, but we can leave the contextualization of $C$ unmentioned when it is obvious or doesn't make much of a difference.

To review, we evaluate coarse-grained influence by comparing the value of $C \boxminus\!\!\!\Rightarrow E$ and $\neg C \boxminus\!\!\!\Rightarrow E$, which tells us to what degree $E$ is made more likely with $C$ rather than $\neg C$. The degree of influence depends on the following free parameters: the coarse-graining of $\overline{C}$, $\overline{\neg C}$, and $E$, with the first two requiring probability distributions over their elements. Whether $C$ influences $E$ depends on the parameterization, but whether $C$ can be used for advancing $E$ generalizes over all the different parameterizations and precisifications of influence in such a way that advancement of the past always fails, though possibly in different ways for different parameter settings.

All the preceding discussion of influence was to establish two notions of influence, influence$_f$ and influence$_c$, that adequately model future-directed influence, and to define them so they can apply to effects in the past without begging any questions about the direction of influence. The difficulty in applying the definitions to past-directed influence is that both notions of influence are based on objective probabilities and it is not obvious how one should model objective probabilities of past events. Under determinism, past-directed probabilities are well defined, but without modification do not play the usual epistemological role of chances. In the stochastic case, a different issue looms: without modification, there are virtually no past-directed probabilities.

# 4. STOCHASTIC INDETERMINISM

Realistic stochastic theories have rules that provide transition probabilities for any future state given any present state, but no rules that provide probabilities for past states. There is nothing in principle preventing a stochastic theory from providing backward transition inference rules, but any such rules that worked like normal transition rules would not play the standard epistemic role of chances due to the actual world's backward evolution being statistically atypical. Restricting consideration to laws that provide only future-directed transition probabilities, only very limited inferences can be made about how the past would have been different under various counterfactual suppositions about the present. An example is the class of spontaneous collapse interpretations of quantum mechanics, especially the original GRW model (Ghirardi and Weber, 1986). These interpretations posit a wave function $\Psi$ representing the complete state of the universe at each time. At random moments, the state undergoes a special collapse transition according to a probabilistic rule that depends on the state's value at the time of the collapse. At other times it evolves deterministically. Under this kind of interpretation, a state at one time does not entail probabilities for past states because the pre-collapse state alone defines the probability distribution for its possible post-collapse states. Many different possible pre-collapse states can jump to a given post-collapse state, and nothing in the character of the post-collapse state gives a probability distribution in the reverse direction. At best, one can infer via the deterministic rule the probability of there having been no collapses from some time $t$ back to some $t - \epsilon$, and what the state would have been then. This allows only minimal inference towards the past, and in realistic situations is practically useless because the $\epsilon$ must typically be nearly zero for there to be any appreciable probability of non-collapse.

The lack of objective probabilities for past states permits a simple characterization of the influence$_c$ asymmetry. Because there are probabilities for how the world would evolve towards the future, but virtually none for how it would evolve towards the past, there are no relations of backward influence$_c$ to exploit for achieving aims.

If there were no way to augment a stochastic theory to make sense of past-directed probabilities, the explanation of the advancement asymmetry in the context of stochastic dynamics would be complete. But the rhetorical situation is more complex. To explain why there can be no advancement of goals towards the past, one must explain the lack of advancement for any reasonable way of defining past-directed probabilities, and there are several ways of defining them that cannot be dismissed as obviously unreasonable. Theories with no dynamical backward transition probabilities can only acquire non-trivial chances for the past by adding non-dynamical constraints, for example assumptions about past states or boundary conditions. Some ways of doing this obviously beg the question about useful past-directed influence. Holding fixed the construction of the Eiffel Tower under counterfactual alteration of the present ensures that we cannot affect whether the tower was constructed, but this is not an interesting lack of influence. For non-triviality, one's standards should be held constant regardless of the specific content of the counterfactual being evaluated, and should be based solely on scientific principles. For instance, a boundary condition might play a privileged role in an account of chance because it contributes to redeeming scientific features of the overall theory, making it more informative, simple, explanatory, etc.

A simple model illustrates one way to get backward chances. Suppose the actual world possesses a simple spacetime structure as in classical physics or special relativity except that it has an initial time-slice, a past boundary with some initial microstate, $a$. Then, we can just hypothesize that the relevant counterfactual worlds for judging influence are all $a$-worlds, that is, worlds that have the same initial condition as the actual world. The objective probability of $E$, whether forward or backward, attributable to the event $C$ would then be $p_{C\&a}(E)$, the probability of $E$ across the possible worlds that start in state $a$ and evolve through $C$. Call such objective probabilities '$a$-chances'. When $E$ is to the future of $C$, the $a$-chance of $E$ matches the ordinary future-directed chance, and it extends the notion to situations where $E$ is to the past of $C$. Chances of past events are typically understood to be either one or zero, but this is not true of $a$-chances.

No doubt, restricting the counterfactually relevant worlds to just those that started the same way as the actual universe inserts an explicitly time-asymmetric assumption into the evaluation of influence. Although this raises an obvious question of circularity, one could defend the practice (Loewer 2007; Kutach 2002, 2007) on the grounds that any asymmetry of influence must come from somewhere in the physics, and the use of $a$-chances results from being frank about the hypothesis that the asymmetry arises from some physical specialness of the early universe, for example that it has a big bang whereas the future has no big crunch, at least in the near term. Such a defense also allows one to extend the definition of $a$-chance naturally to situations where there is strictly speaking no initial state but there is some physical condition that adequately grounds the special treatment of early universe. One could even have different kinds of $a$-chances associated with holding counterfactually fixed the general character of the universe at different temporal stages, and this could be done regardless of whether the universe stretches temporally back further than the big bang. It makes no ultimate difference whether there are multiple types of $a$-chances associated with different past facts or times, or whether $a$-chances beg the question about the direction of influence because there exist multiple arguments demonstrating that $a$-chances cannot be exploited to promote desired past events.

## 5. DETERMINISM

Under determinism, when a microstate entails an objective probability for some effect $E$, it is by definition either one or zero. Contextualized events, when coarse-grained in natural ways, generate non-trivial deterministic chances that play essentially the same role towards the future as chances generated by stochastic laws of evolution. The chances resulting from a contextualized event $\overline{C}$ are not fully objective in the sense that the elements of $\overline{C}$ and the probability distribution over the elements are stipulated. Yet, often in mundane situations the probability of a possible effect $E$ given by $\overline{C}$ is so insensitive to the specific probability distribution on $\overline{C}$, that it can be treated as an effectively objective chance. There is a large technical literature discussing the nature of non-fundamental chanciness, for example in statistical mechanics, which I cannot discuss here except to note briefly that there are facts about fundamental physics, for example symmetries in the fundamental laws and spacetime structure, that make some probability measures much more useful than others for drawing inferences about macroscopic entities, but one does not need a unique objective probability measure to model influence$_c$ adequately. One only needs the actual microconditions to be congenial enough to a bit of arbitrariness in how they are represented as contextualized events for the contextualized events to serve as a useful surrogate for fundamental chanciness.

One can cook up physical theories where states determine everything about the future but nothing about the past. For such theories, the explanation of the advancement asymmetry is trivial. One could have theories that are deterministic in one direction but chancy in another, but by the end of this chapter, it should be easy to see how to handle such cases in light of the ones I discuss. From here on, I will only discuss deterministic theories that are fully bidirectional. Whatever derivative chanciness these theories provide for future-directed influence, they also trivially provide for backward influence. The only possible problem for past-directed deterministic chances consists in their diverging dramatically from the constitutive role of chance as a kind of idealized best guess at outcomes.

A key result from statistical mechanical explanations of irreversible processes is that when the physical states instantiating these processes have deterministic developments towards the past, these developments are highly unstable under the slightest perturbations. Furthermore, the hypothetical instigation of an instability in one patch of matter will overwhelmingly likely trigger (again, backwards in time) the activation of instabilities in other dynamically connected patches, magnifying the extent of the instability. The upshot is that the actual state of the world is past-atypical, meaning that virtually any alteration of the current microstate, if evolved deterministically towards the past, will differ enormously from the actual past history. By contrast, the actual state of the world is future-typical in the sense that the vast majority of alterations to the current microstate, if evolved deterministically towards the future, will exhibit the same kinds of macroscopic regularities that exist in the actual world.

The relevance of statistical mechanics to deterministic influence is that both the space of $\neg C$-worlds and of $C$-worlds are populated overwhelmingly with past-atypical worlds, from which many counterintuitive results follow. Unless the system for evaluating counterfactuals discussed in section 3 is modified, it follows that if virtually anything were different right now, the probability would be very nearly one that the counterfactual present would have evolved (forward in time) by way of a fantastically unlikely anti-thermodynamic fluctuation. That is, the presence of the planet as we know it—with diverse organisms, architecture, manuscripts, fossils, etc.—would have arisen by way of an overwhelmingly improbable fluctuation. And strangely, if the dynamical laws are local, as in relativistic electrodynamics, there would be a light-cone of anti-thermodynamic physical evolution focused on the counterfactually hypothesized alteration surrounded by ordinary pro-thermodynamic evolution everywhere else.

As crazy as it sounds, there are instructive consequences to be drawn from this temporally symmetric way of evaluating counterfactuals. Interestingly, fine-grained influence over the past turns out to be just like chaotic influence$_f$ over the future. Virtually any mundane event $c$ significantly influences$_f$ virtually every actual past event $e$ (coarse-grained as $E$) because $p_{\neg C}(E)$ is very low as a result of the overwhelming majority of such worlds evolving anti-thermodynamically.[3] The recognition that we continually exert enormous influence$_f$ over the past turns conventional wisdom on its head. People naively believe the past is not at all influenceable, but in the symmetric theory, the past is far more influenceable than the future. Indeed, it is *because* the past is so sensitively influenced$_f$ by our actions, that we cannot manipulate what happened, allowing us to get away with thinking of it as not influenced at all. This closely resembles the kind of influence we have over who will be the first human born in the year 3000, assuming the human race lasts that long. In both cases, we exert a powerful influence$_f$, but because the connection is so sensitive to our precise action and precise environment, it is too difficult in practice to control, making it useless for achieving any goals we might have. There are other reasons to reject the utility of deterministic past-directed influence$_f$, but they also apply to coarse-grained influence, which we can now investigate.

Influence$_c$ over the past is just as controversial as influence$_f$ but even more subtle. For counterfactuals where no temporal asymmetry is built into the antecedent, there will be a corresponding past-directed counterfactual with the same semantic value. For example, a blue hydrangea is in alkaline soil. If the soil had been acidic at $t = 0$, the flower would be pink at $t = 1$ day, but also if the soil had been acidic at $t = 0$, the flower would have been pink at $t = -1$ day. The symmetry follows merely from the lack of any temporal asymmetry in the contextualized events or laws.[4] This not only results in seemingly absurd counterfactual statements but also means that we influence$_c$ the past just like we influence$_c$ the future.

One possible response to these counterintuitive results is to count them as counterexamples to the symmetric theory. Because they seemingly grant us a kind of influence over the past we clearly don't have, it must be misrepresenting influence, and because its interpretation of counterfactuals are so counter to what we intend when we consider, "What would have happened if $\neg C$?" it must count as a reductio of the interpretation of counterfactuals. The problem with such a response is that it proves too much. The rejection of the bizarre anti-thermodynamic counterfactual worlds cannot be justified merely by pointing to the fact that it doesn't capture what we intuitively have in mind when considering counterfactual alternatives, for we intuitively have in mind an asymmetric evolution, which—if taken for granted—would beg the question about the influence asymmetry. Accounting for the

---

[3] If $E$ is so extremely coarse-grained that the possibilities it encompasses occupy a sizable fraction of the universe's state space, then this claim may not hold, but that just speaks to the unnaturalness of the coarse-graining and not to anything interesting about influence$_f$.

[4] My discussion here simplifies by ignoring antecedents where there is significant macroscopic motion, but this does not affect the overall argument.

influence asymmetry is not a matter of concocting a story to justify our naive prejudice that the past is immune to influence. We don't intuitively know the right way to evaluate influence towards the past. We don't even know the right way to evaluate influence towards the future: section 3 uncovered two distinct, reasonable versions of influence towards the future. It turned out that influence$_c$ served as a better account of future-directed advancement, but whether that same influence$_c$ notion turns out to give a good account of the apparent lack of any past-directed advancement turns critically on how it is made more precise to illuminate subtleties in how counterfactual dependence may be usefully evaluated. Section 7 will advocate accepting the symmetric influence$_c$ as is, and demonstrate that it cannot be exploited to further goals.

Another possible response (for example, Albert 2000, Kutach 2002, Kutach 2007, Loewer 2007) to the counterintuitive character of the symmetric theory is to restrict the space of counterfactual possibilities to a privileged set: to worlds that started in near enough the same way as the actual world. We have already seen how a defensible treatment of counterfactuals in the context of stochastic dynamical evolution could assign objective probabilities for past events by restricting the relevant counterfactual worlds to those that started out in the same exact state as the actual world. One could replicate this feat for deterministic theories, by extending the definition of an $a$-chance to range over possible worlds that began in nearly the same initial configuration as the actual world. If any conception of deterministic $a$-chances can be faulted for any reason, no harm is done; such failure only simplifies the overall argument that there is no reasonable notion of probability under which making past states more likely, furthers one's goals. If deterministic $a$-chances do make sense, the general arguments to be provided against the exploitability of $a$-chances holds for the deterministic variety as well.

## 6. THE ADVANCEMENT ASYMMETRY

To summarize, we now have three models that have not been decisively convicted of inadmissibly presuming an asymmetry of influence.

1. A stochastic dynamics with virtually no backward transition probabilities.
2. A symmetrically deterministic dynamics.
3. A dynamics that is either stochastic or deterministic and takes the relevant counterfactual worlds to be restricted to worlds that start out in a state $a$ that is exactly (or near enough) the universe's actual (or surrogate) initial state.

These models, it will be shown, all have the feature that advancing one's goals is asymmetric. The first case is easy to handle. Because there are virtually no probabilities for events in the past, there are virtually no ways to make past events more likely than they would have been if the present were different. Thus, there are no promoters of past events, and consequently no backward advancement of goals. End of story. The other two models give probabilities that ground backward influence$_c$ and demand extensive discussion.

For comparative purposes, it is helpful to consider how to model furthering a goal towards the future, spelled out with an example. Instantiated in some agent's brain right now is both a desire for $E_f$, the existence of a loaf of bread one hour from now (towards the future), and the volition to satisfy that desire. The oven contains a pan of suitable dough, and the agent's volition combined with knowledge of how ovens work leads him to turn on the oven at 220 degrees centigrade. The decision to turn on the oven, $C$, is in the circumstances a promoter of the oven being nearly 220 degrees not long after. Likewise, $C$ promotes $E_f$ because the objective probability of the bread being created minutes later is very high and the probability of $E_f$ is very low given the implicit contrast event $\neg C$, the decision to leave the oven dial in the off position.

Now consider how the agent could act to further the goal of having a fresh loaf of bread an hour ago. Somehow instantiated in the brain is a desire for $E_p$, the past existence of a loaf of bread in the oven, and the volition to satisfy that desire. We can consider numerous ways in which $C$ promotes $E_p$. The first two are far and away the most important and hold for any notion of past-directed influence, including those that employ $\alpha$-chances.

## 6.1 Backward promotion by way of the future

First consider attempts to promote $E_p$ by way of creating some intermediary event in the future, $I_f$. This is called backtracking influence because it first goes one direction in time and then backtracks by going in the other temporal direction. The strategy is to set up a situation where the volition, $C$, promotes $I_f$, which in turn promotes $E_p$. The problem with any such attempt is that the dynamical laws of our world apparently operate in a way where states causally contribute to effects by way of continuous transitions that screen off previous states. All paradigm theories of fundamental physics, at least, exhibit this behavior except for the CTCs allowed in general relativity. In Newtonian gravitation, for example, an arbitrary initial state $S_0$ defined by the relative positions and velocities of all the masses determines both the state $S_1$ one second later and the state $S_2$ two seconds later. But $S_1$ determines $S_2$ by itself as well: it encodes all the information in $S_0$ needed to get $S_2$. If the fundamental laws include stochasticity, the condition still applies. Any probability that $S_1$ fixes for $S_2$ is invariant under conditionalization on all states previous to $S_1$.

In a relativistically local theory, illustrated in Figure 8.1, the contextualized event $\overline{C}$ that instantiates $C$ is just barely big enough to fix a probability for $I_f$. Likewise, the contextualized event $\overline{I_f}$ is just barely big enough to fix a probability for the earlier $E_p$. But any contribution $I_f$ makes to $E_p$ goes through the surface where $\overline{C}$ is instantiated. Thus, any aspect of $C$ that promotes $E_p$ through $I_f$ already exists in the direct connection between $\overline{C}$ and $E_p$. One might think that there is still some room for $I_f$

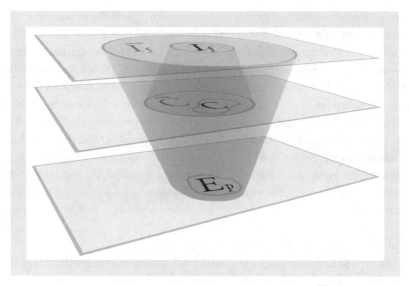

FIGURE 8.1  $I_f$ only contributes to $E_p$ by way of $\overline{C}$

to impact $E_p$ by somehow affecting the details of $\overline{C}$ beyond the occurrence of $C$. But any such possibility would be question begging because, in order to infer anything about the probability of $I_f$ using the fundamental laws, one must first have a fully detailed, complete $\overline{C}$. The fundamental laws do not ground inferences from $C$ itself or an incompletely described $\overline{C}$ to $I_f$. They require fully filled out states at least as large as $\overline{C}$. Thus, any backtracking inferences in filling out how $C$ is translated into a contextualized event does not occur in virtue of any laws, and thus whatever backward influence exists by way of future events is rendered superfluous by influence that goes directly back in time to $E_p$.

Given this redundancy of all backtracking influence, actions to promote a past goal are useless, at least as actions are normally understood. Actions aimed at some goal can be conceived as processes roughly like the following: An action starts with one's attention being focused on some situation, evolves through a deliberative process incorporating desires and beliefs, climaxes in some decision being made, and eventuates in some (typically external) physical effects. When promoting future events, the physical development subsuming the action proceeds through various stages, each of which promotes later stages and finally external future events. But for an action to proceed toward the future through all its stages and then to go back into the past to promote past events constitutes an instance of backtracking promotion, which is guaranteed to be superfluous. Thus, the only non-trivial way an action can promote past events is to directly promote them by way of the reverse sequence of stages, that is, starting with a decision having been made, which promotes (towards the past) the agent's comparing various options, which then promotes the agent's having no idea of what to do, and finally a stage where the agent is not attending to the situation. For such actions, we have a built-in excuse for their lack of utility. Paradigmatic actions and those same actions reinterpreted as evolving towards the past are importantly different in the functional interaction of desires, beliefs, and volition. We naturally reject such reverse-actions as genuine actions directed towards a goal. Hence, we naturally reject the possibility that they could stand as a counterexample to the folk wisdom.

It is possible, given all that has been concluded so far, that an ordinary action to promote a past goal does promote it through this highly unconventional reverse-action process. The rest of this section aims to close loopholes, so that even if one counts such reverse-actions as legitimate, it will still turn out that they cannot regularly be used to promote the goals in the past. But even if such arguments fail, the master argument in this section suffices to vindicate the folk wisdom insofar as its content is restricted to normal actions.

## 6.2  The epistemic inaccessibility of backward promotion

Even if there were some reverse-action influence on the past, we will be unable to become aware of it in the ordinary way we learn of future-directed promotion, that is, by observation, trials, etc. The extent to which our evidence of correlation between $C$'s and $E_p$'s can be properly attributed to genuine causal promotion is the extent to which the probabilistic connection exists after we have controlled for all the evidence for $E_p$ existing in $C$'s larger environment, $\overline{C}$. By definition, $C$ promotes $E_p$ to the extent that the instantiation of $C$ in the volitional component of the agent's brain raises the probability of $E_p$ beyond what probability is fixed by $\neg C$. In order to gather the kind of correlational evidence relevant to establishing promotion between $C$ and $E_p$, we need epistemic access to $C$, which we can easily get just by asking the agent to tell us what she was trying to do, and access to $E_p$ that does not come from $C$'s larger environment. But, our only source of information about $E_p$ that is independent of $C$ itself comes by way of this larger environment, unless the laws of nature permit epistemic access to past facts that skips over intermediate physical states. Presumably, our epistemic access to $E_p$ comes only by way of how reality evolves, which apparently does not involve any state skipping. Thus, any attempt to collect statistics about what $C$ promotes by comparing what happens after $\overline{C}$ occurs and what happens after $\neg C$ will leave us no access (independent of $C$) to previous events, and so we cannot learn by observation or trial the extent to which attempts at past-directed promotion are genuinely successful.

## 6.3  Backward promotion that is interpreted away

One kind of past-directed promotion sure to exist is the connection between the act of will and its immediate precursor, the volitional part of one's brain state microseconds before the act of will. Though technically the act of will influences$_c$ this precursor, such influence is useless for achieving interesting goals, and what impact it has can be reinterpreted as part of the overall volitional process.

A similar kind of past-directed promotion exists, under determinism, in one's ability to promote the state of the world ten minutes ago to be just the right kind of state to lead deterministically towards bread being baked one hour later. This is just an ordinary decision to bake the bread reinterpreted as involving past events. Such promotion is in a sense exploitable, but only to promote the existence of bread in the future. The previous argument (in section 6.1) concluded that a present state cannot influence the past by way of the future more than what one gets by influencing the past directly. Because no asymmetry is implicated in the argument, the same holds in the other temporal direction. One cannot influence the future by way of the past more than what one gets by influencing the future directly. Any influence$_c$ over the past that is defined trivially in terms of what future goals are satisfied does not count as fulfilling goals that are about the past in any interesting sense.

## 6.4 Backward promotion by cheating

One's choice of contextualized events can almost always be rigged to create a contrast that generates spurious influence$_c$ or non-influence$_c$. For comparison, consider cases of spurious future-directed non-influence$_c$. One could countenance a set of $C$-states where one turns on the oven but the energy supply is disconnected, or where a nemesis is lurking nearby with the aim of secretly shutting off the oven. One could also countenance a set of $\neg C$-states where one does not intentionally turn on the oven, but where a defect in the dial is set to malfunction by igniting the oven any way. Any of these would make $C \mathbin{\Box\!\!\rightarrow} E_f$ and $\neg C \mathbin{\Box\!\!\rightarrow} E_f$ roughly equal, entailing that $C$ does not help achieve $E_f$. Made precise in those ways, $C$ does not promote $E_f$, but such cases are clear examples of contriving just the right kind of contrast to ensure that one gets non-influence$_c$. Thus, they do not ground substantive claims of non-influence$_c$.

The same reasoning applies to some purported cases of past-directed influence$_c$. This is clearest under determinism. Let $C$ be the trivial coarse-graining of the precise instantiation of the volition $c$, and let $\overline{C}$ be its extension as far out spatially as is needed. To construct the counterfactual contrast, take the actual state when some chosen past event $E_p$ occurs, replace it with some specific $\neg E_p$ and let that evolve toward the present, labeling that present state $\overline{\neg C}$. By construction, $C \mathbin{\Box\!\!\rightarrow} E_p$ is equal to one and $\neg C \mathbin{\Box\!\!\rightarrow} E_p$ is zero, so that $C$ has influenced$_c$ $E_p$, no matter what that event is. The reason to rule out such influence$_c$ as usable for advancement is that $\overline{\neg C}$ is not plausibly an event representing a natural contrast class. It was selected only because it has coded into it all the microphysics needed to entail $\neg E_p$. Whenever the coarse-graining or contrast is selected in a biased way, that itself counts as a legitimate reason for dismissing the alleged attempt at advancing some goal for the past.

## 7. BACKWARD PROMOTION WITH THE SYMMETRIC THEORY

The arguments in the previous section provide overlapping reasons to reject the utility of backward influence, but one could attack the employed coarse-grained notion of influence. As flagged in section 3, the symmetric theory makes some instances of influence$_c$ temporally symmetric. Since influence$_c$ is a model for the kind of influence supporting the advancement of goals, it seems to imply that we can advance the goal of making bread in the past just by choosing to turn on the oven, which would flatly contradict the folk wisdom.

The solution to this problem is to admit the past-directed influence$_c$ as being legitimate in a broad sense, and to delineate another sense in which that influence$_c$ can never be exploited. What can make the influence$_c$ unusable are historical contingencies that entail a systematic atypicality throughout a large region of spacetime such that the frequency of successful attempts in that region to employ the influence$_c$ are very likely to mismatch what the contextualized event indicates. I will provide two examples of what should be uncontroversial appeals to atypicality to explain away how influence$_c$ can exist even while all attempts to exploit it fail. Because we can legitimately appeal to atypicality to explain these uncontroversial cases, it will be available for rescuing the symmetric theory at no extra cost.

The point of abstracting away from the precise microphysics using contextualized events is to have a theoretical structure that to some extent adequately represents our ability to control things. A suitably constructed pair of contextualized events, $\overline{A}$ and $\overline{\neg A}$, represents the physical difference between our choosing to take action $A$ versus some alternative $\neg A$. (In order to be suitable, the two contextualized events need to agree everywhere except with regard to whatever physical structures instantiate the initiation of action $A$ rather than $\neg A$.) One can think of this difference as representing what we have direct control over. Then, the fundamental laws entail from this pair, probabilities for events at other times, which represents what we have indirect control over, that is, what we influence$_c$.

Taking this fiction too seriously, though, leads to a paradoxical commitment. Under determinism, whenever we take any action of type $A$, the precise state of the world, $a$, at the time of the action, determines the full future historical development, $f$. Suppose that $A$ is an action intended to thwart determinism, to make whatever future is determined, fail to happen. Any representation of $A$ in terms of a (continuous) contextualized event, $\overline{A}$, fixes the probability of $f$ at zero. If we take seriously the idea that we can directly control whether or not $\overline{A}$ occurs, then we can indirectly fix the probability of $f$ at zero, and hence make $f$ fantastically improbable. There is a seeming conflict between (1) the claim that I have direct control over whether or not $\overline{A}$ and thus indirect influence over whether $f$ happens and (2) the claim that no matter how many attempts I make to perform actions of type $A$, $f$ is certain to happen. Obviously, the reason that $f$ occurs despite any attempts to prevent it is just that $f$ was stipulated to be the actual future, but the issue here is whether such cases can be accommodated adequately in how we construe the link between, on the one hand, our notions of control and influence, and on the other, the patterns of outcomes that serve as evidence of our having control and influence.

A poor resolution of this conflict would be to forgo influence$_c$ altogether in favor of influence$_f$. We can say instances of the coarse-grained action do not really fix a probability of zero for $f$ because in every time the coarse-grained action is taken it is instantiated by a fine-grained action that fixes a probability of one for $f$, but using such reasoning to vitiate the notion of a coarse-grained action prevents us from taking advantage of the utility of contextualized events for representing the difference, flagged in section 3, between indirect control and dumb luck.

A better explanation is fortunately available. We do not need to maintain a conception of 'ability' and 'influence' such that the guaranteed repeated failure to prevent $f$ from coming about counts as a strict inability to influence whether $f$. A looser conceptual link allows that whenever we have influence$_c$ over some kinds of future events, that only justifies a defeasible inference to a pattern of future outcomes. Defeasibly, when the frequencies of chance outcomes mismatch what our hypotheses about the operative chances dictate, we have reason to revise our hypotheses, but this default is overridden if special facts about a circumstance imply that our hypotheses should not be expected to hold for the pattern of outcomes. One example of such a situation is when the effect has been rigged using a description that ensures it will happen regardless of the laws or circumstances. Because $f$ denotes the actual future, attempts to prevent $f$ must fail regardless of what probability an action fixes for it. It is fair to apply the term 'unexploitable' for this kind of influence$_c$.

For another example, imagine a universe much like ours except that particles only come into existence through supernova-like explosions. Each current galaxy is composed of matter from a single explosion, with a bit of intergalactic mixing from stray particles or galactic collisions. Suppose the laws of nature are such that whenever any of these explosions occur, the masses of the generated up-quarks all match a single quantity through a fundamentally chancy process. Because the galaxies and the star systems within them are formed from the detritus of these explosions, virtually all the up-quarks have the exact same mass as their neighbors. Yet, throughout the universe as a whole, the masses vary randomly. Suppose that for most macroscopic processes, the variation in mass makes little difference in causal regularities, but there are some rare chemical reactions that sensitively depend on it. For concreteness, let the reaction of white roses to the presence of silver in the soil be one such reaction. Using the statistical distribution of quark masses fixed by the laws, planting a silver coin under the roots of a white rosebush makes highly probable that its blooms turn blue. However, with the very specific improbable value of the quark mass in our galaxy, the silver has no effect on the flower's color. We have both a narrow and broad way of thinking about the influence$_c$ of $C$, placing a silver coin under the rosebush, on $E$, its blooms' being blue one week later. If we coarse-grain narrowly using only states that have the local value of the quark masses, then the counterfactual probability of blue blooms is low regardless of the presence of the coin, implying no influence$_c$. If we coarse-grain broadly to include a range of different quark masses, then the probability of blue blooms is high given the coin and low without the coin, implying influence$_c$ over the color of the flower. Both versions of influence$_c$ are technically legitimate, but the narrow version has a superior connection to actual statistical regularities in the local environment. The broad version gives a good measure of how likely it is that the flower will turn blue in general circumstances across the universe. It is only because we are here on Earth that the probability generated by the broad coarse-graining is less useful than the one that sticks to the local quark mass in the local environment. Because up-quarks on Earth always have the same mass, any testing one does on the effects of silver on roses will reveal only the statistics involving the local masses. The counterfactually different masses are irrelevant to any attempted probing of the influence of earthly silver on earthly roses. This provides a sense in which the precise local value for the mass makes the broad influence$_c$ unexploitable.

A potential worry about breaking the link between "what actions of type $A$ are able to influence" and "what patterns we observe after repeated attempts to $A$" is that in general, it is all too easy to find some fact that entails the widespread failure of an alleged ability to influence. I could claim the ability to turn lead into gold by touching it and explain my poverty by pointing out that every time I employ my virtually foolproof ability, subtle contingencies concerning the particular layout of matter, for example the particular value of some fundamental constant or something about the precise microcondition gets in the way to ensure that the gold does not come into existence. Upon deeper consideration, though, no harm is committed by permitting such specious abilities. There is no need for a precise general rule that separates out frivolous ability claims in a principled way because we can instead just judge each claim's merits individually, based on what kind of fact undermines the default expectation that whenever the laws entail $p_{\overline{A}}(E)$, $E$ will be effected in a fraction $p$ of the cases where action $\overline{A}$ is taken. If the fact is something quite simple like the mass of the up-quark, then we can note the broad ability but not take it seriously on a practical level. If the fact is gerrymandered to justify a desired conclusion about one's abilities, we can reject it as ad hoc.

Given this general argument that attempts at influence$_c$ are unexploitable when actual outcomes are guaranteed to be atypical, one can argue that the past-atypicality of all actual states in our environment makes the symmetric theory's influence$_c$ over the past unexploitable for any purpose, including advancing goals. With regard to future-directed influence, these same natural coarse-grainings lead to values for $p_{\overline{C}}(E)$ that match actual frequencies well enough. With regard to past-directed influence, these same natural coarse-grainings lead to values for $p_{\overline{C}}(E)$ that diverge dramatically from actual patterns in every affected patch of spacetime. In the same sense that we are able to make the flowers turn blue and are able to thwart determinism, we are able to make the previous fourteen billion years of history evolve anti-thermodynamically. But because of a specific, non-gerrymandered, contingent fact about our local environment, any repeated attempts we might make to turn the flowers blue will result in failure. Similarly, because of a specific, non-gerrymandered, contingent fact about our local environment, that is, its past-atypicality, any repeated attempts we make to make a vast swathe of history anti-thermodynamic will result in failure, even though virtually every coarse-grained action we take makes such a bizarre history overwhelmingly probable.

An important point here is that the strategy I have outlined is not some desperate measure to save a baroque theory of backwards influence but is natural given our practice of fuzzing over microscopic details in order to gain a simpler cognitive grip on reality. The only way claims of influence can express something useful is for them to go beyond the occurrent facts, and so the concept of influence must to some degree abstract away from the precise character of actuality. But to the extent influence abstracts from actual conditions, it permits the possibility of mismatch between what the abstraction deems influence and what actually happens when influence claims are tested.

# 8. CONCLUSION

The core phenomena behind the asymmetry of advancement is this:

- There exist uncontroversial, non-ad hoc ways of interpreting the counterfactual dependence implicit in influence such that there exist lots of exploitable regularities of causal promotion going from the present towards the future.
- Though some ways of understanding counterfactual dependence permit causal promotion going from the present towards the past, there are nevertheless no regularities governing such promotion that rightly count as exploitable for achieving aims.

Unless a viable candidate has been overlooked, there is no reasonable notion of influence such that one can advance one's goals in the past. This means it is pointless to try to influence the past, and that therefore in practice it is mostly harmless to think of the past as effectively fixed and independent of one's current action. The asymmetry of influence, charitably interpreted, exists because it is nothing other than the asymmetry of advancement. Alternatively, if the influence asymmetry is construed to require that events predominantly fail to influence the past, it is a bogus but understandable simplification.

The asymmetry of advancement might strike people as too remote from the objective structure of nature to count as the correct explanation of the influence asymmetry. An explanation in terms of how agents are able to promote their goals seems implausible given that processes like stars burning and cliffs eroding possess an apparent causal or influence asymmetry that has nothing to do with goals or agency. Furthermore, some of these temporal asymmetries, for example, the radiation asymmetry and T-violations in weak nuclear decay processes, may turn out to be essential elements of fundamental causation-like relations. If so, a direction of causation would exist at the fundamental level, not—as the analogy to solidity would have it—an asymmetry that arises only macroscopically.

This charge can be rebutted on several points. First, nothing forbids a fundamental asymmetry from grounding the advancement asymmetry. If the laws of nature entail no probabilities for past events and $a$-chances are rejected for any reason, then there is a fundamental asymmetry that directly explains why we cannot make desired past effects more likely. Yet, however comforting such an explanation is to people who want to retain our naive conception of the influence asymmetry, there are two features of this explanation that are somewhat conceptually revisionary. For one thing, such a result would not substantiate the idea that the past is fixed, because past facts would not hold regardless of counterfactual alterations to the present. Rather, the counterfactuals would just be silent about the past, which in a sense makes the past even more open than it would be with determinism or past-directed chanciness, for there would not be any laws constraining what would have happened in the past. For another thing, even if there is a fundamental explanation of the influence asymmetry, it remains true that if the laws were to have bidirectional chanciness or determinism, we would still have an advancement asymmetry. Thus, my arguments would provide a redundancy in the explanation of why it is useful to believe in asymmetric influence. Because advancement is asymmetric regardless of whether the laws themselves are asymmetric, it is not obvious that the postulated fundamental asymmetry alone explains why we believe in the asymmetry of influence. It explains why influence is asymmetric, but if we believe in the asymmetry of influence because of the asymmetry of advancement, both the fundamental and non-fundamental explanations play a role in vindicating our beliefs. The situation is reminiscent of John Locke's (1685, II, 21, §10) character who wakes up inside a room and volunteers to stay because a friend is there, but who could not leave anyway because they are locked inside the room. We can explain why the man does not leave either by citing his volition or the locked door, but either one is redundant given the other. Similarly, we can explain why advancement is asymmetric, and thus why we think of influence the way we do, either by pointing directly to the fundamental asymmetry in the laws or to the more general fact that advancement is asymmetric regardless of any fundamental asymmetry in the laws.

Regarding the seeming oddity of accounting for the advancement asymmetry specifically in terms of achieving goals, it is worth noting that the explanation can be subsumed within a broader account of more general physical processes. The physical basis underlying the advancement asymmetry is the fact that our local environment is composed of spacetime patches with future-typical and past-atypical dynamical development. This typicality asymmetry establishes a local condition where virtually any counterfactual difference leads dynamically to future states falling under familiar rules of thumb about how things evolve. For example, changes to the position of water molecules in the ocean probably make little difference in at least the near-term future. A change to the position of balls in the lottery bin leads to a big difference in who wins (but not to the coarse-grained probability of who wins). Changes to the position of a light switch lead to a significant difference in the amount of light in the room, etc. These rules of thumbs exist because future-typicality ensures the standard connection between the chances and the distribution of chance outcomes. By contrast, past-atypicality ensures that the otherwise-to-be-expected rules of thumb do not apply. This typicality asymmetry holds whether there are agents or not, and subsumes the asymmetry of advancement within an explanation of the broader physical phenomena. The explanation, briefly stated, is that a few general features of the fundamental laws, together with some fact about (what we call) the early universe, makes the typicality asymmetry fantastically likely throughout the known universe. The typicality asymmetry in turn explains the asymmetry of thermodynamic processes and why many macroscopic processes are asymmetric in character and are temporally aligned with each other, for example why the life cycle of ferns is asymmetric with all ferns sprouting, growing, reproducing, and dying in the same temporal direction. The process of agency is just one of many such processes, and its asymmetry is explained in the very same way.

Agency arises as a crucial feature of interest because of its being central to what we naively think of as causation. If we had assembled a bunch of facts about macro-asymmetries—the temporal asymmetries of ferns or volcanos or toasters—then we could point to typicality (and features of fundamental physics) to explain the asymmetry of their functionality. That would be one kind of explanation, but there would still be a further question about what makes those macro-asymmetries *causal*. Our naive notion of cause incorporates a directionality, crudely expressed as the fact that the cause *makes* the effect happen and not the other way around. Merely having asymmetries in ferns and such does not vindicate this conception of causation. One could think of the 20 cm high fern making itself shorter towards what we call the past, and one could think of a charred log as making a fire occur previously. Nothing disastrous would come of such talk, even if for some predictive and explanatory purposes it is inferior. What vindicates talk of 'making happen' is that among the various macro-asymmetries, there is one called 'agency,' whose existence makes causation a concept that is useful, not only for predicting on the basis of the macro-asymmetries or explaining on the basis of the macro-asymmetries, but for manipulation. My observations here closely track what Michael Dummett (1964) observed: an intelligent plant would have reason for an asymmetric notion of cause, but *our* concept of causation is also important to us because of our agency. The intelligent plant would be able to predict and explain macro-asymmetries using the resources of the typicality asymmetry and perhaps other physical asymmetries, but it would have no need for its causal notion to incorporate the idea that present states do not bring about the past and do not influence the past. The implications of the typicality asymmetry for agency is what vindicates our conceiving of causation as essentially involving the inability of the effect to bring about the cause.

The role agency plays in the asymmetry of causation can be illustrated by considering what it implies about the direction of causation in regions where the typicality asymmetry is reversed. Suppose some patch of the universe, $R$, is such that its material content is macroscopically temporally reversed from us and contains sequences that (in reverse) instantiate regularities like those on Earth. (For simplicity, set aside the kind of reasoning that motivated $a$-chances.) Is the arrow of causation reversed in $R$? The answer is ambiguous. If an Earthly agent were hypothetically transported to $R$ in a way that preserved its temporal orientation, then actions that agent took would promote events towards what we call the future. So, the direction of causation would be the same. If we instead consider alterations to the physics in the heads of inhabitants of $R$, those changes will be of the kind that promote *their* goals in the temporal direction we call the past. So, the direction of causation in $R$ would be in that sense reversed. The direction of causation is thus perspectival in the same way as the difference between up and down. On one way of thinking about it, everyone agrees that down is towards the center of the planet. On another way of thinking about it, people on the opposite side of the planet disagree about what direction counts as down.

To summarize the objectivity of the asymmetry of advancement, it is useful to return to the metaphor that initiated this chapter, the solidity of ordinary objects. There is certainly a sense in which a granite block is objectively solid; no fancy jiggering of one's subjective attitudes will negate the damage it inflicts when dropped on a stray finger. Yet, there is also a sense in which objectively the block is mostly empty space; solidity is a conceptual simplification grouping together a cluster of properties for pragmatic purposes. If we were radically different creatures—proton-sized or galaxy-sized or composed of neutrinos—we would not find the distinction between the solidity of the cinder block and the liquidity of oil to be of much significance. In this way, the distinction between solid and liquid is carved in a way that corresponds to the interests of creatures like us. Nevertheless, one must depart rather significantly from the kinds of creatures we are to find beings for whom solidity makes no practical difference. The insensitivity to particular character of humans makes the solid/liquid distinction to that extent objective. Analogously, no adjustment of one's opinions about the direction of influence will negate the fact that people who try to bake a pie for yesterday's dessert are objectively wasting their effort. Yet the asymmetry, as it has been scrutinized and elaborated here, essentially involves relativizing facts about influence to choices about how to conceive of action, and these choices are subjective in the sense that nothing in logic or in the natural world forces one to conceive of actions, for example, as future-directed. This kind of subjectivity is innocuous because the structures of the natural world are such that the coarse-grainings and contrasts that permit useful influence over the past are ones that intuitively count as ad hoc or question-begging. Relativize to uncontrived choices and the laws ensure the unexploitability of past-directed influence. One could imagine hypothetical creatures for whom the contextualized events that ground past-directed influence count as natural ways of coarse-graining, but such creatures would arguably be so bizarre as to be nearly unrecognizable as creatures.

# REFERENCES

Albert, D. (2000), *Time and Chance* (Cambridge: Harvard University Press).

Dummett, M. (1964), 'Bringing About the Past', *The Philosophical Review*, **73** (3), 338–59.

Ghirardi, G. C., Rimini, A., and Weber, T. (1986), 'Unified Dynamics for Microscopic and Macroscopic Systems,' *Phys. Rev. D 34*, 470–91 (doi:10.1103/ PhysRevD.34.470).

Kutach, D. (2002), 'The Entropy Theory of Counterfactuals', *Philosophy of Science*, **69** (1), 82–104.

——(2007), 'The Physical Foundations of Causation', in Price, H. and Corry, R. (eds.), *Causation, Physics, and the Constitution of Reality* (Oxford: Oxford University Press).

Lewis, D. (1973), *Counterfactuals* (Oxford: Blackwell).

Locke, J. (1685), *An Essay Concerning Human Understanding*.

Loewer, B. (2007), 'Counterfactuals and the Second Law', in Price, H. and Corry, R. (eds.), *Causation, Physics, and the Constitution of Reality* (Oxford: Oxford University Press).

Maudlin, T. (2007), 'On the Passing of Time', in *The Metaphysics Within Physics* (Oxford: Oxford University Press).

Price, H. and Corry, R. eds. (2007), *Causation, Physics, and the Constitution of Reality* (Oxford: Oxford University Press).

Stalnaker, R. (1968), 'A Theory of Conditionals', in N. Rescher (ed), *Studies in Logical Theory, American Philosophical Quarterly Monograph Series, No. 2* (Oxford: Basil Blackwell), 98–112, reprinted in E. Sosa (ed), *Causation and Conditional* (Oxford: Oxford University Press), 165–79, and in W. L. Harper, R. Stalnaker, and G. Pearce (eds), *Ifs*, 41–55, and in Jackson, *Conditionals*, 28–45.

CHAPTER 9

........................................................................................................

# THE FLOW OF TIME

........................................................................................................

HUW PRICE

# 1. INTRODUCTION

........................................................................................................

TIME seems to have a transitory character. We have the impression that it *flows*, or *passes*. From here, well-worn paths of philosophical enquiry lead in two directions. One seeks to regard this aspect of the human experience of time as a reflection of the nature of time itself. The other treats it as merely a feature of human experience, requiring a psychological rather than a metaphysical explanation.

In this chapter I want to explore the first path, and to explain why I think it leads to a dead end—or rather to several dead ends, for it turns out that there are several distinct threads bundled together in the intuitive notion of the flow of time, each of which might be held to rescue something of the idea that the passage of time is an objective feature of reality. I want to distinguish these strands, and to argue that they all lead nowhere. I shall say little about the alternative view, that the apparent passage of time should be regarded as a psychological phenomenon, but it will be clear that I regard it as much more promising, if only by default. I close with some remarks about avenues that seem worth exploring in the philosophy of time, when we are done with trying to make sense of passage.

## 1.1 Three paths to passage

What would the world have to be like, for the flow of time to be an objective feature of reality? It seems to me that we can distinguish three possible answers to this question, compatible but largely independent. Hence there are three distinct views, a defence of any one of which would go some way towards vindicating the view that there is something objective about the passage of time. Of course, defending three or two of these claims would be better than defending only one, but one—any one—would rescue something of the intuitive notion:

1. The view that the *present moment* is objectively distinguished.
2. The view that time has an objective *direction;* that it is an objective matter which of two non-simultaneous events is the *earlier* and which the *later.*
3. The view that there is something objectively *dynamic,* flux-like, or "flow-like" about time.

It seems to me that these views have not been sufficiently distinguished, either by defenders or critics of the notion of objective passage—a fact which has allowed the two sides to talk past one another, in various ways. I shall say a little more about why they are distinct in what follows. Mostly, however, my plan is to argue against each view independently. In each case, I shall claim, the view in question is not so much false as doubtfully coherent. On examination, it turns out to be hard to make sense of what the view *could be,* at least if it is to be non-trivial, and of use to a friend of objective passage.

## 2. IS THERE AN OBJECTIVE 'PRESENT MOMENT'?

One major component of the intuitive idea of the passage of time is that it involves a distinguished but continually variable 'present moment'—a single 'box' or 'frame', whose contents are continually changing. One version of this idea is at the heart of *presentism,* a view which holds that the present moment is *all there is*—that the past and future simply don't exist. Of course, presentists normally combine this view with the claim that the present moment (or its contents) *change.* But it is worth noting that this is a separate claim. If presentism itself is coherent, then why not a presentism of a single moment (fictionalist, perhaps, about time and change)? We'll return to this presentism-without-change in a moment.

Presentism is not the only version of the view that there is a distinguished present moment. Another version, more useful for our present purposes, is what has become known as the *moving spotlight view.* As Zimmerman (2008: 213) notes, this image is due to C. D. Broad, who introduces it in this passage:

> We are naturally tempted to regard the history of the world as existing eternally in a certain order of events. Along this, and in a fixed direction, we imagine the characteristic of presentness as moving, somewhat like the spot of light from a policeman's bull's-eye traversing the fronts of the houses in a street. What is illuminated is the present, what has been illuminated is the past, and what has not yet been illuminated is the future.                    (Broad, 1923: 59)

There is something deeply puzzling about Broad's metaphor, however.[1] After all, place yourself at any moment in time, and ask yourself "Is this moment the *present*

---

[1] As Broad himself appreciates—he does not endorse this view.

moment?" The right answer, obviously, is "Yes"—a fact guaranteed by the stipulation that you are to ask the question *of* a moment, *at* that moment. But this means that if the houses in a street are to play the role of moments of time in Broad's analogy, then the answer to the question, "If you place yourself in one of the houses, and open the front door, will the policeman's bull's-eye be shining in your face?" must also be "Yes"—which means that the light must be shining on all the houses, not just on one!

One might reply that this objection ignores the fact the light shines on each house *in succession,* so that we simply need to add a temporal stipulation to the question about what you see when you open the front door, for all to be well. Will you see the light? Perhaps, but it depends on when you open the door.

But this reply overlooks the fact that the series of houses was supposed to be playing the role of the series of temporal moments. To introduce this temporal aspect to the question, we would need a *second* temporal parameter, other than the one assumed to be represented by the series of houses. Some writers have been prepared to bite this bullet, to try to save the notion of a distinguished (but changing) present moment. But the option is neither appealing nor promising. In the present context, the obvious objection is that the issues of making sense of flow will arise all over again, with respect to the second temporal dimension, and hence that we will have made no progress by introducing it.

The source of the difficulty is that the moving spotlight view is trying to combine two elements which pull in opposite directions. On the one hand, it wants to be *exclusive,* saying that one moment is objectively distinguished. On the other hand it wants to be *inclusive,* saying that all moments get their turn—their Warholian instant of fame, when the spotlight turns on them alone. (*Everybody* is a star.)

Imagine an attempt to defend a similar combination in another case: concerning persons, or minds, rather than times. We are familiar with solipsism, which for present purposes we might think of as the view that only our own mind "has the lights on," and that everyone else is a mere zombie. This is an exclusive view, and let us now try to combine it with an inclusive element. Suppose first that "having the lights on" is a matter of being energized by some heavenly bull's-eye—being at the point of focus of the divine gaze, say. And to make the view inclusive, suppose that the focus of the divine gaze varies from person to person. God's eyes follow humanity around the room, as it were. Whoever you happen to be, you'll always find them directed at you. (This is the analogue of the fact that if you place yourself at any moment in time and ask "Is this moment the *present* moment?", the right answer is always "Yes.")

At this point, clearly, we've added inclusivity at the cost of exclusivity. We've lost the idea that one person is objectively distinguished, as *the* object of the divine gaze. We can still make sense of the idea that each person is distinguished *from her own point of view.* But this is too weak to give us what the solipsist wanted: namely, a deep, *non-perspectival* sense in which one person is objectively distinguished. In the temporal case, the same dilemma confronts any attempt to combine an (exclusive) objective "present moment" with the (inclusive) view that all moments get their turn.

The inclusive aspect threatens to overwhelm the exclusive aspect, reducing it to an innocuous and uncontroversial perspectivalism.

For a defender of the objective present, the only option at this point seems to be to try to save exclusivity by back-pedalling on inclusivity. The presentist needs to insist that *this* moment, *now,* has privileged status. There is no legitimate inclusive view, except in the sense that we can make sense of it from this exclusive perspective. In place of the inclusive image of a movie, with a long series of frames, the presentist needs to insist that there is only one frame—one aspect of which is labelled *The Story So Far,* being a representation *within this single frame* of what the exclusive view thinks of as the content of the *previous* frames (and similarly for an aspect labelled *Upcoming Episodes*).

But can this shift to the exclusive view succeed, without fatal damage to the intuitions that inclusivity was intended to respect? Here it is worth noting a difference between presentism and solipsism. For a solipsist, there may be no need for a surrogate for inclusivity.[2] For a presentist, on the other hand, the desire to make sense of passage creates an immediate and pressing need for such a surrogate. After all, the idea that every instant has its own moment in the spotlight was driven by intuitions about passage. But can we make *enough* sense of this idea, for the purposes of a realist view of passage, within an exclusive view of this kind?

It seems to me that there are two obstacles here, which act together to provide a fatal dilemma for any 'distinguished present' version of objective passage.

**1. Frozen-block presentism.**    The first problem is that in defending exclusivity at the cost of inclusivity, the presentist seems to have thrown out not just the baby, but almost the entire bathroom. After all, what did God need to create, in order to create the whole of reality, as our exclusive presentist describes it? Not a long series of world-stages, but just a single moment, complete with its internal representation of a past and future. It is as if we've built just one house in 'Broad Street', relying on stories its occupants tell about imaginary neighbours as surrogates for all the rest. (They tell us about the time the police shine their bull's eye on the ne'er-do-wells at Number 96, but this is just make-believe. There is no such address, and no family who live there.)

For present purposes, what matters most about this is that we seem to have lost the materials for a realist view of passage, change, or temporal transition. All of these notions seem to involve a relation between equals, a passing of the baton between one state of affairs and another. But in this picture, we've lost one party to the transaction. We've lost genuine change, and replaced it, at best, with a kind of fiction about change.

**2. Inclusivity strikes back.**    At this point, our opponent is likely to object that for one reason or another, this conclusion is too strong: the exclusive picture does allow for real transition, change, and passage, in so far as it needs to, to make sense of a objectivist view of these things. It allows that change *has happened* and *will happen,* and that's all the objectivist needs.

---

[2]  Unless perhaps for a kind of fictionalism about other persons, to make some sense of one's own behaviour with respect to the zombies with which one shares the planet.

However, this reply leaves the opponent vulnerable to the second horn of the dilemma. Let's allow that the presentist can make sense of future change which is "real enough" for the purposes of a defence of objective passage. Since, by assumption, this act of making sense is being done from a standpoint that privileges the present moment, it is 'inclusive' with respect to all other moments (in whatever sense it can make of such talk at all), in treating them all on a par. The depiction of change in the Upcoming Episodes part of this picture depicts all future moments in the same way. None of them is privileged *in this representation*, though of course each may be depicted as privileged—harmlessly and trivially, as we noted above—from its own point of view. So the distinguished moment view has now become self-undermining; committed to the view that change can be represented *without* a distinguished moment.

Think of the other houses in Broad Street represented in a picture on the wall of the actual ("present") house. Which, if any, of the other houses should be shown in the picture as illuminated by the spotlight? If *none*, then the aspect of the model that corresponds to change has disappeared completely. If *one*, then which one—what could break the symmetry? And if *all*, then what has happened to the idea that the representation of change requires a distinguished moment?

To summarize, our strategy was this. Faced with an opponent who holds that change requires a distinguished present moment, we granted her view a distinguished moment from which to speak, and argued that the view then falls victim to a dilemma: either it cannot make sense of change at all, from that distinguished standpoint, or it does so without a distinguished standpoint. Either we lose change, because from this distinguished standpoint there is no change in the past and future, or we accept that change can be modelled in an *inclusive* way (i.e. without a distinguished moment), which undermines the motivation for playing the game in this way in the first place.

## 2.1  Comparison with McTaggart

As readers may have noticed, there are similarities between the above discussion and J. M. E. McTaggart's famous (1908) argument for the view that time itself is unreal. McTaggart's argument proceeds in three steps:

1. He argues that there is no time without change.
2. He maintains that real change requires that events change with respect to the properties of *pastness, presentness,* and *futurity*: they begin as *future,* and become *present,* and then *past.* Combined with premise (1), this implies that the reality of time depends on the reality of the properties of pastness, presentness, and futurity.
3. He argues that these properties cannot be real for they are incoherent: each event must have all three, and yet they are contradictory.

Many commentators get off the boat at steps (1) or (2), denying that time requires change, or denying that change needs to be conceived in terms of acquisition and loss of the properties of pastness, presentness, and futurity. But these options hardly reduce the interest of the argument as a whole, which can be thought of, after all, as an argument for the need to jump ship in one or both of these ways, if one wants to be a realist about time and/or change.

Much interest therefore lies in step (3)—especially so, in the present context, since it is easy to reconstrue McTaggart's argument as an argument against the moving spotlight view, using the two properties *illuminated* (i.e. present) and *not illuminated* (not present) in place of the three properties past, present, and future.

At this point, in this slightly modified form, McTaggart's argument claims that each moment must have both properties, *illuminated* and *not illuminated;* and that this is incoherent. My approach has been to take a slightly different tack, emphasizing the pressure on the moving present view to combine an *inclusive* view, in which all moments are on a par with respect to these properties, with an *exclusive* view, in which one moment is distinguished as uniquely illuminated. We saw that the inclusive element threatens to drive out the exclusive element, by requiring us to acknowledge that each moment is equally illuminated, from its own point of view.

I have noted the option of stressing the exclusive view, by insisting that we can speak from nowhere *except* a particular moment, and hence that we can say truthfully that only *that* moment is illuminated. This option is available to opponents of McTaggart's step (3), too. They can simply deny that all events have all three properties, pastness, presentness and futurity, insisting from the standpoint of a particular moment (while denying the availability of any other standpoint) that each event only has one: if it is not present, then it is either past or future (though not both). But for McTaggart's opponents, as for mine, this option leads to a fatal dilemma. Either it discards the baby with the bathwater, leaving them with a frozen presentism, lacking the conceptual resources to make sense of change (by their own lights). Or it sneaks the inclusive view back in, in its own depiction of past and future change.[3]

I conclude that there is no coherent notion of an objectively distinguished present, at least of the sort required to provide an ingredient of an objective flow of time. If objective flow is to be a coherent notion, it will have to be constituted, somehow, from the remaining ingredients.

## 3. AN OBJECTIVE DIRECTION OF TIME?

The second ingredient is the idea that time has an objective *direction*—that there is an objective distinction between past and future, earlier and later. Concerning this

---

[3] Nicholas J.J. Smith (2010) has recently proposed an elegant elaboration of McTaggart's argument, in some ways similar to the one I have offered here.

ingredient, we can get a good sense of what is at issue by comparing some remarks from writers on opposite sides. First, on the "pro" side of the question, consider the following characterization of the passage of time by Tim Maudlin:

> The passage of time is an intrinsic asymmetry in the temporal structure of the world, an asymmetry that has no spatial counterpart. It is the asymmetry that grounds the distinction between sequences which run from past to future and sequences which run from future to past. Consider, for example, the sequence of events that makes up an asteroid traveling from the vicinity of Mars to the vicinity of the Earth, as opposed to the sequence that makes up an asteroid moving from the vicinity of Earth to that of Mars. These sequences might be "matched", in the sense that to every event in the one there corresponds an event in the other which has the same bodies in the same spatial arrangement. The topological structure of the matched states would also be matched: if state B is between states A and C in one sequence, then the corresponding state B* would be between A* and C* in the other. Still, going from Mars to Earth is not the same as going from Earth to Mars. The difference, if you will, is how these sequences of states are oriented with respect to the passage of time. If the asteroid gets closer to Earth as time passes, then the asteroid is going in one direction, if it gets farther it is going in the other direction. So the passage of time provides an innate asymmetry to temporal structure.
>
> (2007: 108)

As Maudlin points out, however, not just any intrinsic asymmetry will do the trick:

> [T]he passage of time connotes more than just an intrinsic asymmetry: not just any asymmetry would produce passing. Space, for example, could contain some sort of intrinsic asymmetry, but that alone would not justify the claim that there is a "passage of space" or that space passes.                           (2007: 109)

So one question we want to keep in mind, in considering the general issue of intrinsic temporal asymmetry and directionality, is whether a particular asymmetry is of the right kind to "produce passing", and what kind this could be.

## 3.1  Boltzmann's Copernican moment

The possibility that there may be no objective direction of time is famously presented in the following remarks by Ludwig Boltzmann, from his *Lectures in Gas Theory* of 1896–1898:[4]

---

[4]  Boltzmann first presents these ideas in an (1895) letter to *Nature,* attributing them to his 'old assistant, Dr Schuetz.' His suggestion about the non-objectivity of the direction of time is offered in the context of an avowedly tentative proposal concerning the origins of the state of thermodynamic disequilibrium in which we now find ourselves. Despite certain striking theoretical advantages—in particular, that it permits the observed thermodynamic time-asymmetry to emerge from a model with no temporal asymmetry at all, at the global level—this proposal about the source of disequilibrium faces seemingly damning objections (see, e.g., Price 2004, for more on these issues). But neither these difficulties, nor the tentative spirit in which Boltzmann presents the proposal, in any way diminish the importance of the recognition that the direction of time might not be objective. In putting that

One can think of the world as a mechanical system of an enormously large number of constituents, and of an immensely long period of time, so that the dimensions of that part containing our own "fixed stars" are minute compared to the extension of the universe; and times that we call eons are likewise minute compared to such a period. Then in the universe, which is in thermal equilibrium throughout and therefore dead, there will occur here and there relatively small regions of the same size as our galaxy (we call them single worlds) which, during the relatively short time of eons, fluctuate noticeably from thermal equilibrium, and indeed the state probability in such cases will be equally likely to increase or decrease. For the universe, the two directions of time are indistinguishable, just as in space there is no up and down. However, just as at a particular place on the earth's surface we call "down" the direction toward the center of the earth, so will a living being in a particular time interval of such a single world distinguish the direction of time toward the less probable state from the opposite direction (the former toward the past, the latter toward the future).                    (1964: 446–447)

Let us compare Boltzmann's viewpoint with Maudlin's. Would Boltzmann deny that there is a difference between asteroid moving from the vicinity of Earth to that of Mars and the same asteroid moving from the vicinity of Mars to that of Earth? I think we can be sure that he would not! Rather, he would maintain that while of course the two cases are *different,* there's no fact of the matter as to which is which. At best, there's a fact of the matter relative to a particular temporal perspective, or choice of coordinate frame. A good analogy is with the issue as to whether, in a Newtonian framework, there is an objective distinction between an asteroid at rest and an asteroid moving at a uniform non-zero velocity. Boltzmann's view compares to the relationist view that there is no such distinction: fix a coordinate frame, and you can certainly distinguish between the two cases, but nothing distinguishes a unique "correct" coordinate frame.[5]

For present purposes, the important issue is this one. If Boltzmann is wrong, what exactly is he wrong *about?* What *extra* feature does the world have, that it does not have in the Boltzmann picture? What could make it the case that there is an objective earlier–later relation?

## 3.2 Orientability: necessary but not sufficient

Our interest in the issue of the direction of time is guided by an attempt to make sense of the notion of the passage or flow of time. If time has a direction in a sense relevant to passage, then presumably it is the same everywhere—passage is supposed to be global, universal, and unidirectional. This implies that a precondition of any relevant notion of the direction of time is that it be possible to *assign* a temporal direction at every

possibility on the table, Boltzmann is, as Hans Reichenbach (1956: 128) puts it, providing "one of the keenest insights into the problem of time."

[5]  Another analogy, to which we'll return, is a reworking of a famous example we have from Kant (opposing Kant's own view about the case): a world with two hands of opposite parity, but no fact of the matter as to which of the two is the left hand and which the right hand.

place and time, in a consistent way. In other words, we want to be able to pick a temporal direction, and label it (say) the positive direction, without our labels suddenly changing, as we move from place to place.

Formalized a little bit, this means that a precondition for the existence of any relevant notion of the direction of time is that the spacetime within which we live be temporally "orientable", in the following sense:

> A relativistic spacetime $(M, g_{ab})$ is *temporally orientable* iff there exists a con-
> tinuous everywhere defined timelike vector field on $M$. If such a field exists,
> reversing the arrows gives another such field. The choice of one of these fields
> as "pointing the way to the future" is what is meant by the assignment of a time
> orientation.                                    (Earman and Wüthrich, 2008, fn. 7)

The important point to note here is that such an "assignment" of a time orientation, while it requires temporal orientability, does not require or imply that there be a unique *correct* assignment, dictated by some objective feature of reality. So, while orientability is necessary for the existence of an objective direction of time, in the intended sense, it is far from sufficient. Orientability ensures that if we decide by convention that one of two asteroids is travelling *from* Mars, *towards* the Earth, then we can extend the convention to the rest of spacetime in a consistent way. It does not ensure that such a choice of convention would be objectively *right* or *wrong*, or even tell us what it would *be* for there to be an objective standard of such a kind.

## 3.3  Earman *v.* Reichenbach

We stressed a moment ago that if there is a direction of time of a kind relevant to the passage of time, it had better be a *global* notion. This point connects closely with an argument John Earman offers against Reichenbach, in a seminal (1974) paper. Reichenbach's posthumous book, *The Direction of Time* (1956), defends a view resembling Boltzmann's, to the effect that the direction of time is potentially a local matter, reducible to the direction of entropy increase in a particular region of spacetime.[6] Earman objects as follows:

> Reichenbach himself grants that spacetime can be assumed to be a manifold with a
> null cone structure, and can be assumed to be temporally orientable. But this seems
> enough to justify the following Principle of Precedence.
>
> PP    Assuming that spacetime is temporally orientable, continuous timelike trans-
>       port takes precedence over any method (based on entropy or the like) of
>       fixing time direction; that is, if the time senses fixed by a given method in two
>       regions of space-time (on whatever interpretation of 'region' you like) disagree
>       when compared by means of transport which is continuous and which keeps
>       timelike vectors timelike, then if one sense is right, the other is wrong.

---

[6]  As I noted above, Reichenbach describes Boltzmann's view on this point as one of the "keenest insights" about time, saying that Boltzmann was the "first to have the courage to draw this conclusion." (1956: 128).

To put the matter in a nutshell, PP says that (assuming temporal orientability) once the time sense is established anywhere in spacetime, the structure of space-time (in particular, the null cone structure and spatio-temporal continuity) serve to fix it everywhere. From PP we can conclude that if there is disagreement, then either (i) neither time sense is right or wrong or else (ii) one is right and the other is wrong, and, therefore, the given method is not generally correct. Using the following fact

F    With Reichenbach's entropy method it is always physically possible and in many cases highly likely (according to statistical mechanics) that there will be disagreement.

we can conclude that it is always physically possible and in many cases highly likely that either (a) there is no right or wrong about time direction—talk about which direction is "really" the future and which is "really" the past, and it is not meaningful anywhere in spacetime—or (b) the entropy method yields the wrong result somewhere in spacetime. Reichenbach can accept neither (a) nor (b) because he claims that the content of the philosophically purified concept of time direction is given by the entropy method.                                    (1974: 22)

However, so long as we recognize that there are two distinct conceptions of the project of giving an account of the direction of time—roughly, Maudlin's, which treats the direction of time as something global and fundamental, and Boltzmann's, which treats it on a par with "up" and "down"—then it is easy to see how Reichenbach might have responded to Earman. He could say that if by "direction of time" we agree to mean Boltzmann's local notion, then he rejects PP, while if we mean Maudlin's global notion, then he can accept Earman's option (a). The trick is simply to notice that a "philosophically purified concept of time direction" requires that we distinguish two notions: about one of them, there may be "no right or wrong"; about the other, the truth may vary from place to place.[7]

I am not sure to what extent this irenic distinction would have appealed to Earman's targets—the proponents of the view that "the" direction of time was reducible to the thermodynamic asymmetry. But for present purposes, since our interest is in the prospects of making sense of a strong, global sense of the direction of time, we can ignore this interpretational issue: Earman seems right that direction in the strong sense cannot be reduced in this way, whether or not Reichenbach and others actually thought that it could be.

## 3.4 Earman's Heresy

Earman took the view that the existence of an objective direction of time turns on issues of time-invariance and reversibility to be so much an orthodoxy that he refers to the alternative as a heresy. As he puts it:

[7] True, it might be argued that the latter is not properly called a notion of the direction *of time*, but at this point the argument is about terminology.

[I]t will be useful for sake of contrast to state explicitly a view which goes directly counter to the reductionist position outlined above. I will refer to this view as *The Time Direction Heresy*. It states first of all that **if it exists,** a temporal orientation is an intrinsic feature of space-time which does not need to be and cannot be reduced to nontemporal features, and secondly that the existence of a temporal orientation does not hinge as crucially on irreversibility as the reductionist would have us believe. I am not at all sure that The Time Direction Heresy is correct, but I am certain that a failure to consider it, if only for purposes of contrast, will only lead to further stagnation.                            (1974: 20, emphasis in bold mine)

As Maudlin (2007: 108) notes, "Earman himself does not unequivocally endorse the Heresy, but does argue that no convincing arguments against it could be found, at that time, in the very extensive literature on the direction of time." "Over three decades later," Maudlin continues, "I think this is still the case, and I want to positively promote the Heresy.... [My] Chapter can be seen as a somewhat more aggressive companion piece to [Earman's]."

Like Maudlin, I am a fan of Earman's Heresy, and I, too, want to support it in blunter form. But unlike Maudlin, I do so by denying its antecedent—that is, by denying that there is any direction of time in the strong sense—not by affirming that there is such a direction (though not one reducible to irreversibility). I think that Earman is right to reject reductionism, and to question the relevance of the issue of time-invariance of the laws of physics to that of the direction of time; but wrong to the extent that he believes that the answer might lie somewhere else. On the contrary, I claim: the right answer is that there is no answer.

## 3.5  Earman and the no direction option

The possibility that there might be no objective direction of time is explicitly on Earman's radar, early in his paper. We have seen that he acknowledges it in the passage quoted above, in which he introduces his Heresy. And a page before that, he sets out some of the issues to be considered, as follows:

P4    Does the world come equipped with a time orientation?
P5    If the answer to P4 is affirmative, where does it come from? If the answer is negative, what explains our psychological feeling of a direction for time?
P6    If the answer to P4 is affirmative, how do we know which of the two possible orientations is the actual one?

P4 and P5 are in rather crude form. One of the purposes of this paper is to sharpen them. If there is a global time order, then P6 amounts to the following: given that either *E(x, y)* or *E(y, x)* [i.e., *x* is earlier than *y*, or *y* is earlier than *x*], how do we know which holds?                            (1974: 19)

However, I think it is fair to say that with one (partial) exception, these questions hardly get addressed in Earman's paper. The main exception is the first part of P5, but even this is addressed only in a conditional and negative fashion. In defending the

Heresy, Earman defends the view that *if* there is a temporal orientation, then it is not reducible to issues of irreversibility and the like. But this is not to argue that there *is* such an orientation; or to explain where it comes from, if so; or to answer P6.

By the end of Earman's paper, however, the possibility of a negative answer to P4 seems to have dropped out of sight:

> In summary, I think that there are grounds for believing that entropy is not as important as some philosophers would have us believe for the temporal asymme-tries discussed above. It also seems to me that these asymmetries are not crucial for some of the aspects of the problem of the direction of time; for on my view, laws and/or boundary conditions might have been such that these asymmetries did not exist and *still it would have made sense to speak of a direction for time,* or else the laws and boundary conditions might have been such that these asymmetries were the reverse of what they are now without it necessarily being the case that *the direction of time is the reverse of what it is now.*    (1974: 45, my emphasis)

But nowhere does Earman venture an opinion about what it would be for the direction of time to be "the reverse of what it is now"—or, what would presumably amount to the same thing, what it is for time to *have* a particular direction.

## 3.6  Great heroics

Earman does consider the opposing view, that there is no direction of time, and to some extent his conception of what it is for there to be a direction of time can be discerned from what he says against the opposing view. But the inference is difficult, because Earman takes his main opponent—the Reichenbach–Gold view, as he calls it—to be arguing that the existence of a direction of time depends on the non-time-invariance of the laws of physics. Against this opponent, he argues (forcefully, in my view) that the considerations which might lead us to deny the existence of a direction of time in the case of a time-invariant physics would apply equally forcefully in the non-time-invariant case. This is a good argument against the Reichenbach–Gold view, but it helps rather than hinders the more radical view that there is no direction of time (whatever the time-invariance properties of the laws of physics). In focusing his attention on the Reichenbach–Gold view, Earman seems to leave his flank open to a Boltzmannian opponent.

Earman introduces his discussion of the Reichenbach–Gold view with these words:

> The radical view that if all the laws of nature were time reversal invariant, then there would be no right or wrong in the matter since there could be no temporal orien-tation, is a view which can also be discussed ['discerned'—HP] in Reichenbach's writings. Since this view need not be based on a reductionistic attitude towards space-time, since it has been held by many others besides Reichenbach, and since it goes to the heart of the obsession with irreversibility, it will be examined in some detail in the following section.    (1974: 23)

Earman describes the Reichenbach–Gold view itself as follows:

> Now imagine that $T$ is a super theory which captures all of the laws of physics and suppose that $T$ is time reversal invariant. In such a case, Reichenbach would maintain that for any $m \epsilon \mathfrak{M}_{TD}$, $m$ and $\mathbf{T}(m)$ are not descriptions of two different physically possible worlds but rather are "equivalent descriptions" of one and the same world and that, therefore, "it would be meaningless to ask which of the two descriptions is true" [Reichenbach, 1956: 31–32]. Reichenbach is not alone in this interpretation. For instance, T. Gold says that in the envisioned situation, $\mathbf{T}(m)$
>
> > is not describing another universe, or how it might be but isn't, but is describing the very same thing [as $m$]. [Gold, 1966: 327]
>
> The upshot is supposed to be that if universal time reversal invariance holds, there is no objective fact of the matter as regards time direction or time order since $m$ and $\mathbf{T}(m)$ involve different time directions and orders.          (Earman, 1974: 23)

Most of Earman's objections to the Reichenbach–Gold view are of the kind mentioned above, turning on the claim that there is a direction of time in the non-time-invariant case. The first, however, is more general:

> Even before going into the details of time reversal invariance, I think the implausibility of the Reichenbach–Gold position should be apparent from several considerations. First, a characterization of invariance under charge conjugation $\mathbf{C}$ and mirror image reflection $\mathbf{P}$ can be given along the same lines as given above for $\mathbf{T}$ invariance. If then the Reichenbach–Gold position is correct for $\mathbf{T}$, why isn't it also correct for $\mathbf{C}$ and $\mathbf{P}$? Why doesn't $\mathbf{C}$ $(\mathbf{P})$ invariance imply that $m$ and $\mathbf{C}(m)$ $(\mathbf{P}(m))$ are not descriptions of two different possible worlds but rather equivalent descriptions of the same world? And would it not then follow that the predicates 'having a positive (negative) charge' and 'having a righthanded (lefthanded) orientation' do not correspond to any objective feature of reality? (If the conditions of the $\mathbf{CPT}$ theorem of quantum field theory apply, then $m$ and $\mathbf{CPT}(m)$ will always be equivalent descriptions.) Either this consequence must be swallowed or else it must be maintained that there is some feature which makes $m$ and $\mathbf{T}(m)$ but not $m$ and $\mathbf{C}(m)$ or $m$ and $\mathbf{P}(m)$ equivalent descriptions. To grasp the second horn of this dilemma would require an explanation of what the relevant feature is. No such explanation seems forthcoming. Grasping the [first] horn would seem to involve a great heroism [Earman notes here that Gold (1967: 229) does grasp this horn]; but *in any case it would be an admission that there is nothing special about time direction per se and that the alleged nonobjectivity is the result of a far flung nonobjectivity.*
>
>                                                                  (1974: 24)

I'll return in a moment to the option Earman takes to "involve a great heroism", which seems to me the right choice here, although now to require no great courage, if it ever did.[8] Earman seems quite right to suggest that the case of $\mathbf{T}$, $\mathbf{C}$, and $\mathbf{P}$ are likely to be

---

[8] From my perspective, the true hero of the contemporary scene is Maudlin, playing a stout Cardinal Bellarmine to Boltzmann's Galileo.

on a par. He returns to the point a little further on in his paper, in connection with the claims of another proponent of the Reichenbach–Gold view, Max Black.

Asking what argument could be given in favour of the Reichenbach–Gold position, Earman (1974: 26) offers the following suggestion:

> One obvious suggestion would be to try to accommodate time reversal to the passive interpretation of symmetry. For on the passive interpretation, a symmetry operation corresponds not to a change from one physical system to another but rather a change, so to speak, of the point of view from which the system is described; if both the original description resulting from the original point of view and the new description resulting from the new point of view are equally legitimate according to $T$, then $T$ is said to be invariant under the transformation which goes from one point of view to the other. Not surprisingly, such an interpretation of time reversal invariance has been offered. Max Black concludes that if all the laws of physics were time reversal invariant, the relation of chronological precedence would be an "incomplete" relation in that it would really be a three-place relation like the relations of being to the left of and being to the right of.

>> We would have to say that the very same series of events $A$, $B$, $C$, might properly be described by one observer as being in temporal arrangement in which $A$ occurred first, while that very same series could properly be described by some other observer as being in the opposite temporal arrangement in which $C$ occured first.
>> [Black, 1962: 192]

To this, Earman responds:

> Black seems correct to the extent that his conclusion would follow if the passive interpretation of **T** were legitimate. But many physicists regard this and similar conclusions about **C** and **P** as a *reductio ad absurdum* of the passive interpretation of the discrete symmetries **C**, **P**, and **T**. The passive interpretation of continuous symmetries, like spatial rotation, is meaningful since one can suppose at least in principle that an idealized observer can rotate himself in space in correspondence with the given spatial rotation.... But how is an observer, even an idealized one, supposed to "rotate himself in time ?"                    (1974: 26–27)

Again, I think Earman is quite correct to treat **C**, **P**, and **T** as on a par here. But the challenge that follows seems unsuccessful. The obvious reply is that the passive interpretation of the symmetries requires not the rotation of the original idealized observer, but only (at most), the idealized postulation of a second observer who stands in the appropriate "rotated" relationship to the first observer. Far from being unimaginable in the temporal case, this is exactly the possibility that Boltzmann imagines—and for real observers, too, not merely idealized ones. (The view Black is describing here is essentially Boltzmann's, of course.)

Earman continues his response to Max Black by pointing out that an appeal to the passive interpretation of the symmetries is much more powerful than Black himself takes it to be:

> *Black's passive interpretation of time reversal is too powerful; for this conclusion*
> *follows whether or not the laws of physics are time reversal invariant.* Thus, it follows
> from the very fact that a passive interpretation of spatial rotation can be given that
> the relations of being to the left of and being to the right of are "incomplete rela-
> tions" irrespective of whether or not the laws of physics are invariant under spatial
> rotations. Similarly, the "incompleteness" of the relation of temporal precedence
> would follow from the existence of a passive interpretation of time reversal irrespec-
> tive of whether or not the laws of physics are time reversal invariant.    (1974: 27)

"As a result," Earman concludes, "the passive interpretation is of no use to someone
who wants to maintain the Reichenbach–Gold position" (1974: 27).

Once again, I think that Earman's objection here is entirely correct, as a point
against the Reichenbach–Gold–Black view (at least to the extent that these authors
really are committed to the claim that if physics is not time-invariant, then there is a
*fundamental* direction of time). But it is no objection at all against what I am treating
as the Boltzmann view, namely, that there is no fundamental direction of time in any
case, whether or not the laws are time-invariant. (Again, Earman has his eye on the
Reichenbach–Gold position, and doesn't watch the threat from the other flank.)


## 3.7 Modal collapse?

Another concern that Earman mentions about the "passive" interpretation of the **C**, **P**,
and **T** symmetries is, if I understand it correctly, that it sets us on (or threatens to set
us on) a slippery slope towards treating *all* differences as "merely notational."

> [I]f on the Reichenbach-Gold position, all possible worlds are not to collapse into
> a single one, there must be some objective feature which separates them and which
> can be ascertained to hold independently of the direction of time.    (1974: 24)

In other words (as I interpret the point), it would clearly be *reductio* of the "passive"
methodology if it required us to treat all differences between theories as mere differences in notation, describing the same possible world. To prevent a slide to this
absurdity, we need some fixed points—roughly, some features which can be identified
independently of alternative theoretical notations on offer. In the temporal case, these
need to be features, as Earman says, "which can be ascertained to hold independently
of the direction of time."

There are some interesting and deep issues here, which deserve much more attention than I can give them in this context. For present purposes, I simply want to point
out that there's a danger in the other direction, too. If the desire for fixed points makes
us *too* unadventurous, we'll decline some of the greatest adventures that science has to
offer—those magnificent Copernican moments, when what we had always assumed to
be part of the structure of reality is revealed as an artefact of our parochial viewpoint.
This is what Boltzmann proposes about the direction of time, and it isn't a good objection to a well-motivated suggestion of this kind—that is, in particular, a suggestion

motivated by our developing sense of the world's basic symmetries—to point out that if we were to follow the same path all the way to the horizon, we would end up as *über*-Kantian idealists, attributing *all* structure to our own viewpoint. After all, there be monsters at the other end of the slope, too—Solipsism and Ignorance, to name but two. So we need to stick out our necks to some extent, and the safest policy seems to be to follow the recommendations of physics, in so far as we can understand what they are. The question is, has Boltzmann got them right, in the case of the direction of time?

## 3.8 Counting worlds

We have been considering such arguments as may be found in Earman's (1974) paper for answering "Yes" to his question:

P4    Does the world come equipped with a time orientation?

There's a prior question, which Earman does not formulate explicitly. Let's call it:

P0    What would it *be* for the world to "come equipped with a time orientation" (in the sense required by a positive answer to P4)?

The answer to P0 implicit in Earman's paper is in terms of the distinctness of worlds. It is the denial of the view he attributes to Reichenbach and Gold:

> [F]or any $m \in \mathfrak{M}_{TD}$, $m$ and $\mathbf{T}(m)$ are not descriptions of two different physically possible worlds but rather are "equivalent descriptions" of one and the same world ....
>
> (1974: 23)

In other words, Earman's answer to P0 seems to be that for the world described by theory $T$ to come equipped with a time orientation is for $m$ and $\mathbf{T}(m)$ to describe different possible worlds.

At first sight, the most helpful part of this suggestion seems to be the idea that a temporal orientation doubles the number of possible worlds. After all, there's going to be an obvious move of claiming that both $m$ and $\mathbf{T}(m)$ can be construed as description of *either one* of the pair of worlds concerned, under an appropriate transformation of "description".[9] The bulwark against the Reichenbach–Gold view is the seemingly objective fact about the number of worlds.

Or is it? *Mere* world counting can't do the trick, presumably. Any additional binary property of world-histories will multiply possible worlds in this way, whether it is temporal or not. World counting alone takes us no closer to an understanding of what it is for two worlds to differ specifically with respect to the direction of time. By analogy, the question of objective parity isn't settled by the issue of what choices God faces when he creates a one-hand world, unless we have already singled-out parity from other properties that might distinguish two hands. (The distinction between red

---

[9]    The move is a baby case of Putnam's model-theoretic argument, and, in virtue of the symmetries, especially hard to resist.

hands and green hands will double God's choices, too, but objective colour has nothing to do with objective parity.)

At this point, I find it very difficult to see, in the abstract, how *anything* could count as a satisfactory answer to P0. Accordingly—conscious of the dialectical deficiencies of stares, whether incredulous, puzzled, or merely blank!—I want to try to bring the issue down to earth. I want to return to questions about possible lines of inference from observed temporal asymmetries to *knowledge* of temporal orientation, in the hope that this will clarify the issue of the *content* of such knowledge, as well as the problems or opportunities for its acquisition.

## 3.9  Three grades of temporal asymmetry

Suppose everything in the universe were to vanish, except a single giant signpost, pointing forlornly to a particular corner of the sky. Or suppose that a universe had always been like this. This spatial asymmetry would not require that *space itself* be anisotropic, presumably, or that the direction in question be distinguished by anything other than the fact that it happened to be the orientation of signpost. Similarly in the case of time. The contents of time—that is, the arrangement of physical stuff—might be temporally asymmetric, without time itself having any asymmetry. Accordingly, we need to be cautious in making inferences from observed temporal asymmetries to the anisotropy of time itself.

This caution-requiring step precedes another, if our interest is the direction of time in Earman's and Maudlin's sense. As we noted earlier, temporal anisotropy is necessary but certainly not sufficient for a direction of time: not all possible anisotropies have the right character to constitute the direction of time. Imagine some simple cases: suppose time is finite in one direction, infinite in the other, or more granular in one direction than the other (in the sense that the gaps in discrete time get progressively smaller). In both cases, it would seem reasonable to say that time itself was anisotropic. But what relevance would factors like this have for the existence or orientation of an objective distinction between earlier and later, or for "passage"? Why shouldn't it remain an open question whether the asymmetrically bounded universe had an objective temporal direction at all, and if so, whether the bounded end was really the past or the future?

In order to exercise due caution at both steps, a defender of the objective direction of time thus needs answers to questions such as these:

1. Is time anisotropic *at all,* and how could we tell if it is? What could constitute good grounds for taking it to be so, and do we have such grounds?
2. What kind of temporal anisotropy would be the right kind, from the point of view of grounding temporal passage? And what kind of grounds do we have for thinking that time is anisotropic *in this sense?*

Can these questions lead us to an answer to P0?

### 3.9.1  *The epistemology of anisotropy*

Taking the easier question first, what would count as evidence of anisotropy? Some writers take the view that time-asymmetry in the physical laws would be evidence of anisotropy.[10] For example, Paul Horwich argues that in this case the anisotropy of time would be an hypothesis with explanatory force:

> Once a genuine instance of nomological irreversibility has been identified, it is not hard to justify the inference that time is anisotropic. Suppose there is a physically possible process ABCD whose temporal inverse is impossible. Let (ABCD) designate the process whose temporal orientation is unspecified—merely that B occurs between A and C, and C between B and D. Then a physically necessary condition for the occurrence of (ABCD) is that A is earlier than B. Thus the relation *earlier than* enters into explanations that are fundamental, for we have no deeper account of that necessary condition. In particular, we cannot suppose that the possibility of (ABCD) will be found to depend on its orientation relative to certain other events: for in that case the reverse of ABCD would not be physically impossible.
>
> (1987: 53)

However, I think this argument overlooks the fact that there will always be a "Machian" reading of the kind of lawlike irreversibility that Horwich has in mind here—simply a law to the effect that all instances of the kind (ABCD) have the same temporal orientation. The Machian law will do the same job of explaining the orientation of any particular (ABCD): the opposite orientation would not match all the other instances. True, it won't explain why it isn't the case that all instances have the opposite orientation. But even if we grant for the moment that this is a distinct possibility, rather than a notational variant (more on this issue below), Horwich's version of the explanation shares an exactly analogous deficit: in his case, there is nothing to explain why we don't find the reverse law, with respect to the temporal anisotropy in question. (Why shouldn't it be *later* rather than *earlier* that does the explanatory job, as it were?)[11]

In the present context, the main relevance of this point is that it suggests that proponents of the view that time has an intrinsic direction are even further away from their goal than might be supposed. Anisotropy isn't sufficient for their case, but it is necessary, and they frequently take heart from the one apparent case of a fundamental lawlike time-asymmetry in contemporary physics, the so-called **T** violation.[12] Thus Maudlin, for example:[13]

---

[10]  Indeed, Horwich (1987: 54) seems to conclude that this is both necessary and sufficient for anisotropy.

[11]  Cf. Pooley (2003) for a convincing presentation of this kind of argument.

[12]  It is controversial whether this should be counted a case of time-asymmetry at all. Arntzenius and Greaves (2007) argue that what it reveals is simply that time-reversal requires CP reversal, as well as T reversal. Under that construal, there is no fundamental irreversibility. But I put that aside.

[13]  In the light of Maudlin's blunt endorsement of Earman's Heresy, it is a little surprising to find him in this camp, taking issues of T invariance to be relevant, one way or the other, to the existence

> The discovery that physical processes are not, in *any* sense, indifferent to the direction of time is important and well known: it is the discovery of the violation of so-called CP invariance, as observed in the decay of the neutral K meson.... [V]iolation of CP implies a violation of T. In short, the fundamental laws of physics, as we have them, *do* require a temporal orientation on the space-time manifold.
>
> (2002: 267)

One way to read this passage—not, as we'll see, the reading that Maudlin intends—would be to interpret "require a temporal orientation on the space-time manifold" as meaning "require that space-time be temporally orient*able*." In that case, the conclusion seems correct. What would it mean to say the laws exhibited a specific temporal asymmetry, unless one could orient a choice of temporal coordinate consistently across space-time, in the manner guaranteed by orientability?

This isn't what Maudlin has in mind, however. Orientability is a much weaker condition than existence of an objective distinction between earlier and later—it doesn't imply even that time is *anisotropic,* much less that it is *objectively directed.* Where Maudlin makes the above remarks, he himself has just distinguished orientability from the existence of a temporal orientation, noting that in the relativistic models with which he is concerned, the former is provided by the light-cone structure: they are "space-times in which the light-cones are divided into two classes, such that any continuous timelike vector field contains only vectors which lie in one of the classes." (2002: 266)

Maudlin then notes what we need to add to such a model, to provide an *orientation,* or distinguished direction:

> [What] we need to do is to identify one of these classes as the *future* light-cones and the other as the *past* light-cones. Once I know which set is which, I can easily distinguish a Mars-to-Earth asteroid from an Earth-to-Mars one.    (2002: 266)

In the passage quoted previously, then, we should read Maudlin as claiming that the T-violation exhibited by the neutral kaon "requires" such an orientation. But as our objection to Horwich's argument makes clear, however, this is simply not true: a lawlike time asymmetry does not even require temporal anisotropy, let alone the true directionality that Maudlin is after.[14] It is true, of course, that the T-asymmetry of the neutral kaon could provide the basis of a universal convention for *labelling* lightcones as "past" and "future". But this no more requires that time even be anisotropic—let alone objectively oriented—than does our universal signpost for space, in the example above.

of an objective orientation of time. However, the passage provides a convenient statement of a common view.

[14] By way of analogy, consider again a modified Kantian example: a universe comprising many hands, in which it is a matter of law that all the hands are congruent. The laws of this world certainly exhibit a strong P violation, but the law as formulated compares hands only to other hands, not to space itself, and requires no fact of the matter about whether they are "really" left hands or right hands. So, despite the lawlike P violation, space itself need not be handed—let alone objectively handed in one sense rather than the other.

### 3.9.2  *An arrow that couldn't point backwards?*

Suppose for the moment that Horwich, Maudlin, and others were right, and that a lawlike T-violation such as that of the neutral kaon did require, or at least provide evidence for, a temporal anisotropy. As we have stressed, such an anisotropy would not necessarily provide an objective direction. It would explain neither what *constitutes* the objective distinction between past and future, if there is such a thing; nor how we could *know* which direction was which. By the lights of a proponent of the view that time has an intrinsic direction, after all, it would still seem to be an *open question* whether this nomologically-characterized arrow really points towards the past or the future. (Couldn't God have made the world with the reverse asymmetry, with respect to the objective earlier–later distinction?)

This challenge gives us a handle on our second question above—on the issue of what kind of temporal anisotropy would be the right kind for a defender of the view that time has an objective direction. It has to be an asymmetry which comes with an answer to the open question objection. This challenge functions much like open question arguments elsewhere in philosophy. It objects to any proposed reduction-base for the fundamental asymmetry that it cannot be what we were after, because, in making sense of the issue of its own temporal orientation, we defeat the proposed identification.

To meet this objection, any proposed fundamental asymmetry needs to "connect" with something already in play—with something that can be argued to be *constitutive* of the distinction between past and future, apparently, since only this will defeat the open question. There seem to be three candidates: the special (low-entropy) *initial conditions* of the universe, *causation,* and *conscious experience.*

### 3.9.3  *Initial conditions*

The case for including initial conditions on this list rests on the argument that the familiar temporal asymmetries of our experience—including the asymmetry of our own memories—turn out to depend on the fact that the universe was in a state of very low entropy, some time in the distant past. Leaving the details entirely to one side, the crucial point, for present purposes, is the suggestion that it is not accidental that the low entropy boundary lies in the *past*. If 'past' means, inter alia, something like *the direction in which we remember things,* then it is not an open question why the low entropy boundary condition lies in the past but not the future.

That's the good news. The bad news comes in three parts:

1. First, and perhaps most obviously, there's the issue put on the table by Boltz-mann. What if the required "initial" conditions are not found uniquely at one temporal extremity of the universe, but can occur in multiple locations, in either temporal direction from our own era?
2. Second, even if we could assume that the required conditions were unique, it is far from clear that a low entropy initial condition need constitute an anisotropy

*of time* at all. On the contrary, it is usually presented as a temporal asymmetry in the physical arrangement of matter *within* space and time—and in the classical case it is hard to see how else one could present it.

3. Third, the point of Boltzmann's speculation—which is the origin of this proposal to tie low entropy boundary conditions with the intuitive idea of the past—is that it leads to a picture in which the direction of time is not fundamental. At best, it is something we have 'locally', in appropriate proximity to non-equilibrium regions. With Boltzmann's picture in play, what reason do we have to think that the direction of time is anything more fundamental, even if the low-entropy boundary condition is unique? If Boltzmann's picture could 'explain the appearances' in the non-unique case, why is something more fundamental needed to do so in the unique case?[15]

In answer to the second point, it might be pointed out that general relativity introduces other possibilities. Roger Penrose (1989, chs. 7–8), for example, proposes that the low entropy boundary condition is a product of a lawlike time-asymmetric constraint on spacetime, namely, that its Weyl curvature be zero (or at least finite) at initial singularities. Here, the source of the low entropy past seems explicitly represented as a feature of spacetime—and in a way which makes it unique, apparently, thus offering a solution to the uniqueness problem, too.

Perhaps the simplest way to see how little this helps with the main problem is to ask what difference it would make if we considered a time-symmetric version of Penrose's proposal, with a lawlike low entropy constraint at both ends of a recollapsing universe (or the middle of a bouncing universe, or simply Boltzmann fluctuations). In that case, for Boltzmann-like reasons, we want to say that there is no global direction of time, but only local 'explanations of the appearances'. But such an addition takes away nothing from the model in the regions in which it coincides with the original model, apparently—there's no objective direction of time which we have to *remove* from the model, in order to make it symmetric. So there's nothing over and above 'explanation of the appearances' in the original case, either.[16]

### 3.9.4 *A causal arrow as the key to the temporal arrow?*

The second possibility is that the causal or 'production' arrow might ground the direction of time. Earman mentions this idea:

> *T* can be time reversal invariant although $T \vdash (E(x, y) \leftrightarrow R(x, y))$ where $[R(x, y)]$ is interpreted to mean that it is physically possible for events at $x$ to cause events at $y$ via causal signals.                                                                              (1974: 28)

---

[15] Maudlin himself wants to challenge the claim that Boltzmann's hypothesis could explain the appearances, in the non-unique case. We'll come to those objections in a moment.

[16] This conclusion is hardly surprising. Once we've seen why *up* and *down* are relative to our standpoint, we see that we wouldn't be tempted to regard *up* as any more objective, if everywhere except Antarctica became uninhabitable.

This is a large topic, but we can deal with it expeditiously. If the causal arrow is to play this role, it needs to be sufficiently objective and global to underpin a universal direction of time, and sufficiently linked to other matters not only to be epistemologically accessible, but also to avoid the open question problem. What could the other matters be? If asymmetric boundary conditions, then we are back to the previous case. If the laws of physics, then there are two problems: first, the whole idea will collapse if the laws are T-symmetric, contrary to Earman's and Maudlin's intention to find a notion of direction of time not tied to failures of T-invariance, and second, lawlike violations of T-symmetry *are* vulnerable to the open question objection, as we've already noted.

The one remaining option seems to be to tie the causal arrow to the temporal perspective of observers and agents. This is a good move, in my view, because, as I've argued elsewhere—for example, Price (1991, 1992a, 1992b, 1996, 2001), Price and Weslake (2010)—nothing else turns out to be capable of explaining the intuitive asymmetry of causation. But in the present context, it takes us immediately to the question as whether, constitutively or at least epistemologically, conscious experience plays a fundamental role in the case for a direction of time.

Conscious experience could only play this role if it itself is unidirectional, but this is precisely the assumption that Boltzmann challenges, in the passage we quoted earlier. Boltzmann offers us a picture of the universe in which not all conscious observers "point the same way" in time—some have the opposite orientation to us—and invites us to find it plausible that there can't be a fact of the matter about who gets it right. So a friend of the view that time has an objective direction—especially one who agrees that conscious experience plays a crucial role in making the case for this view—needs a response to Boltzmann.

## 3.10 Maudlin *v.* Boltzmann on 'backward brains'

Maudlin offers such a response in recent work. Maudlin's immediate target is an argument by D. C. Williams (and a related argument by me). However, Williams's example is simply a philosophically-motivated version of Boltzmann's, so we may take Maudlin to be responding to Boltzmann, too. Here is Maudlin's characterization of the Boltzmann–Williams argument:

> If we accept that the relevant physics is Time Reversal Invariant, then we accept that [our] time-reversed Doppelgänger is physically possible. Let's suppose, then, that such a Doppelgänger exists somewhere in the universe. What should we conclude about its mental life?
>
> The objector, of course, wants to conclude that the mental state of the Doppelgänger is, from a subjective viewpoint, just like ours. So just as we judge the 'direction of the passage to time' to go from our infant stage to our grey-haired, so too with the Doppelgänger. But that direction, for the Doppelgänger, is oppositely oriented to ours. So the Doppelgänger will judge that the temporal direction into the future points opposite to the way we judge it. And if we insist that there is a

direction of time, and we know what it is, then we must say that the Doppelgänger is deceived, and has mistaken the direction of time. But now we become worried: the Doppelgänger seems to have exactly the same *evidence* about the direction of time as we do. So how do we know that (as it were) *we* are not the Doppelgängers, that we are not mistaken about the direction of time? If there *is* a direction of time, it would seem to become epistemically inaccessible. And at this point, it seems best to drop the idea of such a direction altogether. But is this correct?

(2002: 271–272)

Maudlin now introduces a terminological convention, before offering his response to the Doppelgänger argument:

> In order to facilitate the discussion, I will refer to corresponding bits of the Doppelgänger with a simple modification of the terms for parts of the original person. For example, I will speak of the Doppelgänger's neuron*s: these are just the bits of the Doppelgänger that correspond, under the obvious mapping, to the original's neurons....
>
> [G]iven the physical description of the Doppelgänger that we have, what can we conclude about its mental state? The answer, I think, is that we would have no reason whatsoever to believe that the Doppelgänger has a mental state at all. After all, the physical processes going on the Doppelgänger's brain* are quite unlike the processes going on in a normal brain. Nerve impulses* do not travel along dendrites to the cell body, which then fires a pulse out along the axon. Rather, pulses travel up the axon* to the cell body*, which (in a rather unpredictable way) sends pulses out along the dendrite*s. The visual system* of the Doppelgänger is also quite unusual: rather than absorbing light from the environment, the retina*s emit light out into the environment. (The emitted light is correlated with the environment in a way that would seem miraculous if we did not know how the physical state of the Doppelgänger was fixed: by time-reversing a normal person.) There is no reason to belabour the point: in every detail, the physical processes going on in the Doppelgänger are completely unlike any physical processes we have ever encountered or studied in a laboratory, quite unlike any biological processes we have ever met. We have *no reason whatsoever* to suppose that any mental state at all would [be] associated with the physical processes in the Doppelgänger. Given that the Doppelgänger anti-metabolises, etc., it is doubtful that it could even properly be called a living organism (rather than a living* organism*), much less a conscious living organism.

(2002: 272–273)

However, it is easy to imagine an analogous argument against the claim that we might expect to find conscious life, or any sort of life, on distant planets. After all, imagine a Doppelgänger of one of us, on Planet Zogg. Following Maudlin's example, let's use a superscript notation to denote bits of, and processes within, the Doppelgänger that correspond to bits and processes in us, under the obvious Zogg–Earth translation: the Doppelgänger thus has neurons$^z$, for example. The analogous argument now runs like this, with the obvious modifications to Maudlin's text:

> [T]he physical processes going on in the Doppelgänger's brain$^z$ are quite unlike the processes going on in a normal brain. Nerve impulses$^z$ do not travel along dendrites to the cell body, which then fires a pulse out along the axon. Rather,

[impulses$^z$ travel$^z$ along dendrites$^z$ to the cell body$^z$, which then fires$^z$ a pulse$^z$ out along the axon$^z$.] ... There is no reason to belabour the point: in every detail, the physical processes going on in the Doppelgänger are completely unlike any physical processes we have ever encountered or studied in a laboratory, quite unlike any biological processes we have ever met. We have *no reason whatsoever* to suppose that any mental state at all would be associated with the physical processes in the Doppelgänger. Given that the Doppelgänger metabolises$^z$, etc., it is doubtful that it could even properly be called a living organism (rather than a living$^z$ organism$^z$), much less a conscious living organism.

Why is this argument unconvincing? Essentially, because we regard spatial translation as a fundamental physical symmetry, *and therefore expect that it holds in biology and psychology, too.* Far from being "completely unlike any physical processes we have ever encountered," the processes in question are *exactly alike,* by the similarity standards embodied in the fundamental symmetries. It *could be* that these symmetries break down for life, or consciousness. But that would be a huge surprise, surely. And similarly for T or CPT symmetry. So far from having "no reason whatsoever to suppose that any mental state at all would be associated with the physical processes in the Doppelgänger", we have a reason grounded on an excellent general principle: physical symmetries carry over to the levels that supervene on physics.

Maudlin would reply, I think, that this appeal to the symmetries just begs the question against his view. If there is an objective direction of time it is surely part of physics—in which case physics is *not* T-reversal symmetric, and there's no failure of supervenience. The latter claims are correct, as far as they go, as are the analogous claims about spatial translation symmetry. If the position of the Earth is objectively distinguished, in the way imagined, then physics is *not* translation-invariant—a fact evidenced by the zombie-like nature of our Zoggian Doppelgängers. Again, there's no failure of supervenience involved in the huge mental difference between us and them, because it sits on top of a huge physical difference.

But let us be clear about the commitments of this position. Recall Maudlin's central example, that of an asteroid travelling between Mars and Earth. The passage of time is supposed to provide what it takes to make it the case that the asteroid is *actually* moving in one direction rather than the other. If we describe the process without stipulating in which direction time is taken to be passing, we leave something out: our representation is incomplete. A familiar example of an incomplete representation in this context is that of a movie, which can be shown to an audience "forwards" or "backwards"—that is, with either ordering of the frames. Let's call physical processes "time-blind" to the extent that their appearance in such a movie doesn't give the game away. So asteroid motion, in particular, is time-blind.

What else is time-blind? In particular, what about conscious experience? Here the defender of an objective direction of time faces a dilemma: either consciousness is time-blind, too, in which case the internal phenomenology "as of" an orientation in time doesn't *actually* fix the direction of a mental life, or there is a radical discontinuity between consciousness on the one hand, and ordinary physical systems, on the other.

The former option undermines the claim that our conscious experience could be a guide to existence or orientation of a privileged direction of time, while the latter seems contrary to the spirit of physicalism, in the sense that it implies that there is something that can be detected by a conscious instrument that cannot be detected by a physical instrument.

Maudlin is willing to grasp the second horn of this dilemma, and he is not the first to do so. Eddington, too, came this way:

> The view here advocated is tantamount to an admission that consciousness, looking out through a private door, can learn by direct insight an underlying character of the world which physical measurements do not betray.    (Eddington, 1928: 91)

While such views don't violate the letter of physicalism, they are certainly unappealing to physicalist intuitions—spooky both on the side of physics, in requiring that there is an element of the physical world so secretive as to be detectable only by minds, and on the side of the theory of mind, in assuming that minds have the ability to detect a fundamental aspect of reality, detectable in no other way.[17]

True, "unappealing to physicalist intuitions" is very far from "untenable". But Maudlin himself wants to appeal to orthodoxy in philosophy of mind, so it does seem a fair point to use against him. Responding to the challenge in (Price 1996) that the Doppelgänger argument undermines an appeal to conscious experience in support of a flow of time, Maudlin says this:

> [T]he response to Price is even more stark. He imagines a Doppelgänger which is not just reversed in time, but a Doppelgänger in a world *with no objective flow of time at all,* i.e. (according to his opponent) to a world in which there is no time at all, perhaps a purely spatial four-dimensional world. So it not just that the nerve pulse*s of this Doppelgänger go the wrong way (compared to normal nerve pulses), these nerve pulse*s don't go anywhere at all. Nothing happens in this world. True, there is a mapping from bits of this world to bits of our own, but (unless one already has begged the central question) the state of this world is so unlike the physical state of anything in our universe, that to suppose that there are mental states at all is completely unfounded. (Even pure functionalists, who suppose that mental states can supervene on all manner of physical substrate, use temporal notions in defining the relevant functional characterizations. Even pure functionalists would discern no mental states here.)    (2002: 273–274)

This appeal to the authority of functionalists is optimistic, to say the least. Of course, functionalists "use temporal notions." But equally obviously, they don't do so (typically!) under the supposition that Maudlin makes here, that genuine temporality requires flow. If we insist on adding that supposition as a terminological stipulation, a typical functionalist will simply reformulate her view in non-temporal terms, to avoiding signing up for objective passage.

---

[17]  A comparison might be with Eugene Wigner's view that it takes a conscious observer to collapse the wave packet in quantum mechanics. This proposal is unappealing to physicalist intuitions in a similar way.

Indeed, if we want an example of a functionalist who is explicit about *not* signing up for Maudlin's picture—with its lawlike, unidirectional conception of time and causation—we need look no further than the greatest of them all. Here is David Lewis, giving us his view of the character of the causal asymmetry:

> Let me emphasize, once more, that the asymmetry of overdetermination is a contingent, *de facto* matter. Moreover, it may be a local matter, holding near here but not in remote parts of time and space. If so, then all that rests on it—the asymmetries of miracles, of counterfactual dependence, of causation and openness—may likewise be local and subject to exceptions. (1979: 475)

As for Maudlin's suggestion that my argument begs the question against the proponent of objective temporal flow, I think he is mistaken about the dialectic. My argument is a reply to the suggestion that temporal phenomenology provides reason to believe in an objective flow of time. It proceeds by pointing out that to whatever flow-invoking hypothesis is offered in explanation of the agreed temporal phenomenology, there's a parallel hypothesis—generated by an obvious mapping between underlying brain states as described in the flow picture and the corresponding brain states as described in the flowless picture—offering an explanation of the same phenomena *without* invoking flow.

Maudlin objects as if the flow-invoking hypothesis is already confirmed by direct experience, while its rival remains mere speculation. But this is not the relevant dialectical position at all. Rather, we need to suppose ourselves open-minded about whether there is flow, and hence in the business of considering hypotheses that might provide reason for coming down on one side or other. The situation is then as I claimed. The phenomena do not support the existence of flow, because—at least for a physicalist—any flow-invoking explanation of the phenomena is easily matched by a flowless explanation.

## 3.11 Summary

At the end of §3.8, I claimed to be at a loss to find *any* answer, in the abstract, to our question P0: What would it *be* for the world to come equipped with a time orientation? We then set out to investigate the issue from the (epistemological) ground up, by asking what kinds of T-asymmetry might provide evidence either for temporal anisotropy or (more problematically) for an objective temporal orientation. The answer to the latter part of this question has turned out to be Eddington's: if there is evidence for orientation, it lies in the special character of our temporal phenomenology. And in the light of Boltzmann's challenge, it must be held to be evidenced in no other way.

Thus we can say this much in answer to P0. For the world to come equipped with a time orientation *to which we have access* is for there to be some time-asymmetric "underlying character of the world" (as Eddington puts it), on which conscious experience provides "a private door". This helps a little with my puzzlement, in the sense that the box of options no longer seems entirely empty, but it is hardly satisfying. The proposal remains vulnerable to my version of the Doppelgänger objection—namely, that we have been offered no convincing argument that our temporal experience needs such an explanation (the role of the Doppelgänger being to generate "directionless" or "flowless" alternatives to any attempt to show how a direction or flow would explain the phenomenology). And it conflicts with physicalist intuitions, in the sense explained above. Eddington himself nails one aspect of this concern:

> The physicist, whose method of inquiry depends on sharpening up our sense organs by auxiliary apparatus of precision, naturally does not look kindly on private doors, through which all forms of superstitious fancy might enter unchecked.   (1928: 91)

But, apt as it is, this characterization makes the physicist, not the physicalist, the aggrieved party. The physicalist's concern is not that conscious experience tends to be an unreliable guide to nature, but that mind should be thought of, *qua* object rather than observer, as merely a *part* of physical nature. This commitment sits extremely uncomfortably with the view that there is a fundamental feature of the world to which only conscious minds are sensitive.

I conclude that while the proposal that time has an objective orientation is not incoherent, it is both (i) a long way out on a philosophical limb, in virtue of its conflict with physicalism; and (ii) entirely unsupported by its own claimed grounds, in virtue of the ready availability of alternative explanations of the relevant phenomenology. Hence it is hardly an appealing alternative to Boltzmann's view.

# 4. OBJECTIVE FLUX?

The third ingredient of the "passage package" is the idea that time has a transitory, flux-like, or dynamic character, of a kind not captured by the spatialized conception of time that is prevalent in physics (and popular with opponents of objective passage). Usually, of course, this ingredient is bundled with the other two: the transition in question is thought of as that of a distinguished moment, and as possessing a particular orientation. The new ingredient of the bundle—the ingredient I take to be most characteristic of the notion of flux—is that it is something to which a *rate* may sensibly be attached. Time passes at a certain number of seconds per second.

For the purposes of this section, I want to detach this ingredient from the familiar bundle. If we could make sense of this flux-like character of time at all, I think we could make sense of a Boltzmann-friendly version of it, according to which it did not have a preferred direction (and did not require a distinguished present moment). That spare view is my target in this section. (It cannot be too spare, however—we are still looking for something that distinguishes time from space.)

## 4.1  Objections to flow

One objection to the coherency of the notion of flow of time (see, e.g., Price 1996, 13) turns on the fact that it is usually thought to have a preferred *direction*. The objection is then that other flows acquire their objective direction, if any, from that of time itself— think of Maudlin's asteroids. But if the flow of time is itself supposed to be constitutive of the direction of time, it needs to do double duty, so to speak, to provide its own direction.

I think this is a good objection, but in this context I take it to be outranked by the general discussion of objective direction in the previous section. For present purposes, and in keeping with my strategy of distinguishing the three key strands in the usual conception of the passage of time, I want to put the issue of directionality to one side. What is now on the table is a notion of temporal flux that does not have a preferred direction, but merely an (undirected) rate. What can be said about that proposal?

In earlier work I characterized the "stock objection" to the *rate* of flow of time as follows:

> If it made sense to say that time flows then it would make sense to ask how fast it
> flows, which doesn't seem to be a sensible question.                    (1996: 13)

I went on to note that "[s]ome people reply that time flows at one second per second," but say that "even if we could live with the lack of other possibilities," there's a more basic problem. Before criticizing the latter claim,[18] Maudlin offers this response to the original objection:

> What exactly is supposed to be objectionable about this answer? Price says we must
> 'live with the lack of other possibilities', which indeed we must: it is necessary, and,
> I suppose, a priori that if time passes at all, it passes at one second per second.
> But that hardly makes the answer either unintelligible or meaningless. Consider
> the notion of a fair rate of exchange between currencies. If one selects a standard
> set of items to be purchased, and has the costs of the items in various currencies,
> then one may define a fair rate of exchange between the currencies by equality of
> purchasing power: a fair exchange of euros for dollars is however many euros will
> purchase exactly what the given amount of dollars will purchase, and similarly for
> yen and yuan and so on. What, then, is a fair rate of exchange of dollars for dollars?
> Obviously, and necessarily, and a priori, one dollar per dollar. If you think that this
> answer is meaningless, imagine your reaction to an offer of exchange at any other
> rate. We do not need to 'live with' the lack of other possibilities: no objectionable
> concession is required.                                                  (2004: 112)

In reply to Maudlin's suggestion, consider a graph of the amount of money I give you in currency X, against the amount of money you give me in currency Y. On such a graph, there is a straight line marking the fair rate of exchange: as Maudlin says, exchanges taking place at points on that line can be interpreted in terms of equal

---

[18]  Which I won't try to defend here—in this respect, I now think, my bad.

purchasing power. And when X and Y are the same currency, the slope of that line is 1.

So far, so good. But what are the two axes, in the temporal case? One (corresponding to the numerator) measures the amount of time passed between two times, $t_1$ and $t_2$ (cf., for example, the amount of fuel dispensed by a pump between two times, $t_1$ and $t_2$); the other, corresponding to the denominator, measures the amount of time it takes for that amount of time to pass (cf. the amount of time taken for that amount of fuel to be dispensed—the amount of time between $t_1$ and $t_2$).

The problem is not that these amounts of time are necessarily, a priori, of equal length. The problem is that they are *the very same thing.* The claim about the rate is informative to the same degree that the following statement is informative:

> Taking the Hume Highway from Sydney to Melbourne, the traveller completes his journey at a rate of one yard per yard traversed. By the time Melbourne looms on the southern horizon, he has put behind him more than 500 miles, over a distance of the same magnitude.

This tells us that Melbourne is more than 500 miles from Sydney (via the Hume Highway), but the reference to rate is entirely vacuous. We can inform travellers about the number of kangaroos, or fence posts, or public conveniences, they will encounter per mile of their journey. We cannot sensibly inform them of how many miles they will encounter per mile, for here there are not two things—a tally of kangaroos, say, and a tally of miles—but just one (the tally of miles). Maudlin's exchange rate example misses this point, because it provides two things to tally: the dollars you give me, and the dollars I give you.

In defence of Maudlin, one might say that triviality isn't a fault but a feature—isn't that the point of his comparison with the *fair* rate of exchange? Fine, but we've just seen that we can have spatial rates in the same (trivial) sense. What we were after was a notion of flow, or flux, which would capture what's supposed to be special about time. What we have been given is a notion of flow so thin that the only thing that distinguishes time and space is that in one case the progression is at one minute per minute, in the other case at one metre per metre—that is, that in one case it is time, in the other case space!

## 5. PROVING THE PUDDING

I conclude that all three of the paths that seemed variously co-mingled in philosophical accounts of the flow of time—a distinguished present, an objective temporal direction, and a flux-like character, distinctive to time—are theoretical dead ends. In most cases, it is difficult to see what coherent sense can be made of these notions, let alone how they could be supported by evidence or argument.

This is good news for the alternative view of time—for *Boltzmann's Block,* as we might call it, to acknowledge Boltzmann's "keen" insight, that the block need have no preferred temporal direction. Lest we Boltzmannians should become complacent, however, I want to finish by stressing two respects in which the project is very far from complete. One of these tasks is familiar, the other less so. Both are crucial, in my view, and I think that together they should be setting the agenda for future research in philosophy of time.

## 5.1  The flow of time as a secondary quality

The familiar task is that of explaining the temporal character of conscious experience: explaining the phenomenology, in virtue of which the notion of the passage of time has such a powerful grip on us. If consciousness is not, as Eddington suggests, "looking out through a private door", at the "underlying [temporal] character of the world," what gives us the impression that it is doing so?

This is the project of explaining how the flow of time is a secondary quality— "resident solely in the sensitive body", as Galileo puts it,[19] rather than in the objective world. I have nothing to contribute to this project here, but I would like to record a conviction (wholly unoriginal) that at least part of the key lies in the illusion of a persisting self. In a sense, I think, this is a double illusion. First, there's the contribution so nicely nailed by Austin Dobson:

> TIME goes, you say? Ah no!
> Alas, Time stays, we go.[20]

In other words, we (mistakenly) treat ourselves as fixed points, and hence think of time as flowing past us.

But this illusion rests in turn on a deeper one; that of a single persisting self, self-identifying over time. I think that Jenann Ismael is correct about the origin of this deeper illusion, in treating it as what she calls a 'grammatical illusion', resting on an indexical 'abuse of notation':

> When I talk or think about myself, I talk or think about the connected, and more or less continuous, stream of mental life that includes this thought, expressing the tacit confidence that that is a uniquely identifying description (in the same way I might speak confidently of this river or this highway pointing at part of it, expressing the tacit assumption that it doesn't branch or merge), but it need not be. There is no enduring subject, present on every occasion of 'I'-use, encountered in toto in different temporal contexts. The impression of a single thing reencountered across cycles of self-presentation is a grammatical illusion.... (2007: 186)

---

[19]  *Il Saggiotore,* from a passage quoted by Burtt (1954: 85).
[20]  Austin Dobson, 'The Paradox of Time'.

That is, in my view, a key ingredient in an understanding of the *flow of time* as a secondary quality, is an understanding of the *enduring self* as a secondary quality.[21]

## 5.2 Eddington's challenge

For the less familiar task, I return to Eddington. As I noted above, Eddington is well aware of the dangers of private doors:

> The physicist, whose method of inquiry depends on sharpening up our sense organs by auxiliary apparatus of precision, naturally does not look kindly on private doors, through which all forms of superstitious fancy might enter unchecked.   (1928: 91)

But he counters with a challenge which I think his opponents—we friends of Boltzmann's Block—have ignored to our cost. The above passage continues like this:

> But is he [i.e. the physicist who renounces private doors] ready to forgo that knowledge of the going on of time which has reached us through the door, and content himself with the time inferred from sense-impressions which is emaciated of all dynamic quality?
>
> No doubt some will reply that they are content; to these I would say—Then show your good faith by reversing the dynamic quality of time (which you may freely do if it has no importance in Nature), and, just for a change, give us a picture of the universe passing from the more random to the less random state... If you are an astronomer, tell how waves of light hurry in from the depths of space and condense on to stars; how the complex solar system unwinds itself into the evenness of a nebula. Is this the enlightened outlook which you wish to substitute for the first chapter of Genesis? If you genuinely believe that a contra-evolutionary theory is just as true and as significant as an evolutionary theory, surely it is time that a protest should be made against the entirely one-sided version currently taught.                                                        (1928: 91–92)

I want to make two responses to this challenge. The first is to note a respect in which it is a little unfair to Boltzmann's Block, or at least to Boltzmann himself—in one respect, Boltzmann is ahead of Eddington, I think. But the second is to acknowledge that in other respects, Eddington makes a very important point. In general, friends of Boltzmann's Block have not done enough to free themselves from the shackles of the pre-Copernican viewpoint; and in the long run, the best case for Boltzmann's view would flow from the clear advantages of doing so (were such to be found).

Boltzmann is ahead of Eddington in offering us a picture in which the entropy gradient is a local matter in the universe as a whole, entirely absent in most eras and regions, and with no single preferred direction in those rare locations in which it is to be found. Combined with Eddington's own view that the asymmetries he challenges his opponent to consider reversing—asymmetries of inference and explanation, for example—have their origin in the entropy gradient, this means that Boltzmann has an

---

[21] See Ismael (this volume) for more on this project.

immediate answer to the challenge. Of course we can't "[reverse] the dynamic quality of time" *around here,* for we live within the constraints of the entropy gradient in the region in which we are born. But we can tell you, in principle, how to find a region in the picture is properly reversed; and that shows that the fixity of our own perspective does not reflect a fundamental asymmetry in nature. Analogously (Boltzmann might add), the fact that men in Northern Europe cannot live with their feet pointed to the Pole Star does not prove a spatial anisotropy. If you want to live with your feet pointing that way, you simply need to move elsewhere.

Moreover, Eddington associates the entropy gradient directly with the "time of consciousness":

> It seems to me, therefore, that consciousness with its insistence on time's arrow and its rather erratic ideas of time measurement may be guided by entropy-clocks in some portion of the brain.... Entropy-gradient is then the direct equivalent of the time of consciousness in both its aspects.     (1928: 101)

So Boltzmann's hypothesis also threatens the veracity of Eddington's "private door".

In broader terms, however, Eddington's challenge has not been taken up. Most advocates of the 'no flow' view—even those explicit about the possibility that time might have no instrinsic direction—have not explored the question as to what insights might follow from Boltzmann's Copernican shift. I want to conclude with some remarks on this issue. It seems to me that there are at least three domains in which we might hope to vindicate Boltzmann's Copernican viewpoint, by exhibiting the advantages of the atemporal perspective it embodies.[22]

**Cosmology.**     The first domain is that of cosmology. There are two aspects to the relevance of Boltzmann's viewpoint in this context. First, and closest to Boltzmann's own concerns, there is the project of understanding the origin of the entropy gradient, in our region. One of the great advances in physics over recent decades has been the realization that this problem seems to turn on the question as to why gravitational entropy was low, early in the history of the known universe—in particular, why matter was smoothly distributed, to a very high degree, approximately 100,000 years after the Big Bang.[23] As we try to explain this feature of the early stages of the known universe, Boltzmann's hypothesis ought to alert us to the possibility that it is non-unique—ought to open our eyes to a new range of cosmological models, in which there is no single unique entropy gradient.

---

[22]  For the benefit of young scholars reading this chapter in search of a thesis topic, I paraphrase Jehangir's famous tweet from Kashmir: "If time doth conceal a philosophers' prize; here it lies, here it lies, here it lies."

[23]  See Penrose (1989), Price (1996, ch. 2; 2004; 2006), Albert 2001), and Carroll and Chen (2009) for expositions of this story. See Earman (2006) for some interesting criticism—though criticism largely defused, in my view, by the observation (cf. Price 2002, §1.2, 2006, §3.4) that the story in question does not need to be told in *thermodynamic* terms. It can be regarded as an *astrophysical* explanation of the existence of stars and galaxies (themselves by far the most striking manifestations of the entropy gradient, in our region).

There is some work which takes this possibility seriously—see, for example, Carroll and Chen (2004) and Carroll (2009, ch. 15). However, there is much more work in which it is either overlooked, or dismissed for what, with Boltzmann's symmetric viewpoint clearly in mind, can be seen to be fallacious reasons. For example, the possibility that entropy might decrease 'towards the future' is dismissed on statistical grounds, with no attempt to explain why this is a good argument towards the future, despite the fact that (i) it is manifestly a bad argument towards the past, and (ii) that the relevant statistical considerations are time-symmetric. (See Price (1996) for criticism of such temporal 'double standards', and the role of the timeless Copernican viewpoint in avoiding them.)

These considerations point in the direction of the second and broader aspect of the relevance of Boltzmann's Copernican viewpoint in cosmology. It alerts us to the possibility that the usual model of 'explanation-in-terms-of-initial-conditions' might simply be the wrong one to use in the cosmological context, where the features in need of explanation are larger and more inclusive than anything we encounter in the familiar region of our 'home' entropy gradient. Here, the point connects directly with Eddington's challenge, in the way noted above. We can concede our local practices of inference and explanation are properly time-asymmetric, as Eddington observes; while insisting that symmetry might prevail on a larger scale.

**Modal metaphysics.**    Many *modal* properties and relations, such as chances, powers, dispositions, and relations of causal and counterfactual dependence, seem to exhibit a strong 'past-to-future' orientation. Sometimes this passes without comment, but sometimes it is presented as a philosophical puzzle, especially in the light of the apparent temporal symmetry of (most of) fundamental physics. A natural question is what we should say about these modal asymmetries in the context of Boltzmann's globally-symmetric viewpoint. Prima facie, there are several possibilities. We might try to maintain that the modal asymmetries are primitives, not dependent on the local entropy gradient, or the perspective of creatures whose own temporal viewpoint depends on that gradient, in Boltzmann's picture. But this will have the disturbing consequence that some of Boltzmann's intelligent observers will simply be *wrong* about the direction of these modal arrows—and how could we tell that it wasn't us?

For this reason, we might prefer to tie the modal asymmetries either directly to the local entropy gradient, or indirectly to it, by associating it with the temporal perspective or the observers and agents in question. But this has the consequence that—like the direction of time itself—the modal asymmetries are bound to be a lot less fundamental than the pre-Copernican picture assumes, in Boltzmann's model. And this consequence may be of much more than merely philosophical interest, if these asymmetric modal notions are applied unreflectively in science. If their use does reflect a particular, contingent temporal perspective, then some parts of science—and physics, especially—may be less objective than they are usually assumed to be. So I think there is important work to be done on the relation of modal concepts and the temporal contingencies of our physical situation. Once again, the subject as a whole is still in its pre-Copernican phase, and Eddington's challenge goes largely unheeded.

**Microphysics.**   Most interestingly of all, there is the possibility that the pre-Copernican viewpoint might be standing in the way of progress needed in fundamental physics—that is, that there might be explanations to which this viewpoint is at least a major obstacle, if not an impenetrable barrier. Here the most interesting candidate, in my view, is the project of realist interpretations and extensions of quantum mechanics. Discussions of hidden variable models normally take for granted that in any reasonable model, hidden states will be independent of future interactions to which the system in question might be subject. The spin of an electron will not depend on what spin measurements it might be subject to in the future, for example. Obviously, no one expects the same to be true in reverse. On the contrary, we take for granted that the state of the electron may depend on what has happened to it in the past. But how is this asymmetry to be justified, if the gross familiar asymmetries of inference, influence, and explanation are to be associated with the entropy gradient, and this is a local matter? Are electrons subject to different laws in one region of the universe than in another, or "aware" of the prevailing entropy gradient in their region? On the contrary, in Boltzmann's picture: we want microphysics to provide the universal background, on top of which the statistical asymmetries are superimposed.

This topic intersects closely with the last one. For it requires that we be careful about what we mean by the *state* of a physical system—careful that we don't simply take for granted a conception that cements in place the kind of time-asymmetric modal categories just mentioned. As I have noted elsewhere (Price, 1996: 250), we find it very natural to think of the state of a system in terms of its dispositions to respond to the range of circumstances it might encounter in the future. What we use state descriptions *for*, above all else, is predicting such counterfactual, or 'merely possible', responses. However, if we want to allow for the possibility that the present state is affected by future circumstances, this conception of the state will have to go, apparently. After all, if different future circumstances produce different present states, what sense can we make of the idea that the actual present state predicts the system's behaviour in a range of *possible* futures? If the future were different, the *actual* present state wouldn't be here to predict anything.

When we explore these issues, it might turn out that the apparently puzzling assumption that hidden variables cannot depend on future interactions is merely a manifestation of asymmetry of our modal notions—just a kind of perspectival gloss on underlying dynamical principles which are symmetric in themselves. If so, there would be no new *physical* mileage to be gained by adopting the Copernican viewpoint. Certainly, we would understand better what belonged to the physics and what to our viewpoint, but no new physics would be on offer as a result.

However, the more intriguing possibility is that there is a new class of physical models on offer here—models which are being ignored, not for any genuinely good reason, but only because they seem to conflict with our ordinary asymmetric perspective. If that's the case, and if the models presently excluded have the potential they seem to have in accounting for some of the puzzles of quantum mechanics, then Boltzmann's viewpoint will prove to be truly revolutionary, and Eddington's challenge will be well and truly met.[24]

# References

Albert, D. (2001), *Time and Chance* (Cambridge, Mass.: Harvard University Press).

Arntzenius, F. and Greaves, H. (2007), 'Time Reversal in Classical Electromagnetism'. http://philsci-archive.pitt.edu/archive/00003280/

Black, M. (1962), *Models and Metaphors* (Ithaca: Cornell University Press).

Boltzmann, L. (1895), 'On Certain Questions of the Theory of Gases', *Nature*, 51, 413–415.

—— (1964), *Lectures on Gas Theory 1896–1898*, S. Brush (Trans.) (Berkeley: University of California Press).

Broad, C. D. (1923), *Scientific Thought* (London: Routledge & Kegan Paul).

Burtt, E. A. (1954), *The Metaphysical Foundations of Modern Physical Science* (Garden City, NY: Doubleday).

Carroll, S. (2009), *From Eternity to Here: The Quest for the Ultimate Theory of Time* (New York: Dutton).

Carroll, S. and Chen, J. (2009), 'Spontaneous Inflation and the Origin of the Arrow of Time', arXiv:hep-th/0410270v1.

Earman, J. (1974), 'An Attempt to Add a Little Direction to "The Problem of the Direction of Time"', *Philosophy of Science*, 41: 15–47.

—— (2006), 'The "Past Hypothesis", Not Even False', Studies in History and Philosophy of Modern Physics; 37, 3, 399–430.

—— and Wüthrich, C. (2008), 'Time Machines', in Edward N. Zalta (ed.), *The Stanford Encyclopedia of Philosophy (Fall 2008 Edition)*, <http://plato.stanford.edu/archives/fall2008/entries/time-machine/>.

Eddington, A. (1928), *The Nature of the Physical World* (Cambridge: Cambridge University Press).

Gold, T. (1966), 'Cosmic Processes and the Nature of Time', in Colodny, R., ed., *Mind and Cosmos* (Pittsburgh: University of Pittsburgh Press), 311–329.

—— (1967), *The Nature of Time* (Ithaca: Cornell University Press).

Horwich, P. (1987), *Asymmetries in Time* (Cambridge MA: MIT Press).

Ismael, J. (2007), *The Situated Self* (New York: Oxford University Press).

—— This volume. 'Temporal experience'.

Lewis, D. (1979), 'Counterfactual Dependence and Time's Arrow', *Noûs*, 13(4): 455–476.

McTaggart, J. M. E. (1908), 'The Unreality of Time', *Mind*, 17: 456–473.

Maudlin, T. (2002), 'Remarks on the Passing of Time', *Proceedings of the Aristotelian Society*, 102: 237–252.

—— (2007), *The Metaphysics Within Physics* (New York: Oxford University Press).

Penrose, R. (1989), *The Emperor's New Mind* (Oxford: Oxford University Press).

Pooley, O. (2003), 'Handedness, Parity Violation, and the Reality of Space', in K. Brading & E. Castellani (eds), *Symmetries in Physics: Philosophical Reflections* (Cambridge: Cambridge University Press), 250–280.

[24] I am grateful to Craig Callender, Jenann Ismael, and Tim Maudlin for comments on previous versions, and to the Australian Research Council and the University of Sydney, for research support.

Price, H. (1991), 'Agency and Probabilistic Causality', *British Journal for the Philosophy of Science,* 42: 15–76.

—— (1992a), 'Agency and Causal Asymmetry', *Mind,* 101: 501–520.

—— (1992b), 'The Direction of Causation: Ramsey's Ultimate Contingency', in D. Hull, M. Forbes and K. Okruhlik (eds.), *PSA 1992: Volume 2* (East Lansing, MI: Philosophy of Science Association), 253–267.

—— (1996), *Time's Arrow and Archimedes' Point* (New York: Oxford University Press).

—— (2001), 'Causation in the Special Sciences: the Case for Pragmatism', in D. Costantini, M. C. Galavotti and P. Suppes, eds., *Stochastic Causality* (Stanford: CSLI Publications), 103–120.

—— (2002), 'Boltzmann's Time Bomb', *British Journal for the Philosophy of Science,* 53: 83–119.

—— (2004), 'On the Origins of the Arrow of Time: Why There is Still a Puzzle About the Low Entropy Past', in Hitchcock, C., ed., *Contemporary Debates in the Philosophy of Science* (Malden, Mass.: Blackwell), 219–239.

—— (2006), 'The Thermodynamic Arrow: Puzzles and Pseudo-puzzles', in Bigi, I. and Faessler, M., eds., *Time and Matter,* Singapore: World Scientific, 209–224.

—— and Weslake, B. (2010), 'The Time-Asymmetry of Causation', in H. Beebee, C. Hitchcock & P. Menzies (eds), *The Oxford Handbook of Causation* (Oxford: Oxford University Press), 414–443.

Reichenbach, H. (1956), *The Direction Time* (Los Angeles: University of California Press).

Smith, N. J. J. (2010), 'Inconsistency in the A-Theory', *Philosophical Studies.* http://dx.doi.org/10.1007/s11098-010-9591-3.

Zimmerman, D. (2008), 'The Privileged Present: Defending an "A-Theory" of Time', in T. Sider, J. Hawthorne and D. Zimmerman (eds.), *Contemporary Debates in Metaphysics* (Malden, Mass.: Blackwell), 211–225.

CHAPTER 10

........................................................................................

# TIME IN THERMODYNAMICS

........................................................................................

JILL NORTH

# 1. INTRODUCTION

........................................................................................

Or better: time *asymmetry* in thermodynamics. Better still: time asymmetry in thermodynamic phenomena. "Time in thermodynamics" misleadingly suggests that thermodynamics will tell us about the fundamental nature of time. But we don't think that thermodynamics is a fundamental theory. It is a theory of macroscopic behavior, often called a "phenomenological science." And to the extent that physics can tell us about the fundamental features of the world, including such things as the nature of time, we generally think that only fundamental physics can. On its own, a science like thermodynamics won't be able to tell us about time per se. But the theory will have much to say about everyday processes that occur in time, and in particular, the apparent asymmetry of those processes. The pressing question of time in the context of thermodynamics is about the asymmetry of things *in* time, not the asymmetry *of* time, to paraphrase Price (1996, 16).

I use the title anyway, to underscore what is, to my mind, the centrality of thermodynamics to any discussion of the nature of time and our experience in it. The two issues—the temporal features of processes in time, and the intrinsic structure of time itself—are related. Indeed, it is in part this relation that makes the question of time asymmetry in thermodynamics so interesting. This, plus the fact that thermodynamics describes a surprisingly wide range of our ordinary experience. We'll return to this. First, we need to get the question of time asymmetry in thermodynamics out on the table.

# 2. THE PROBLEM

The puzzle that I want to focus on here, the puzzle that gets to the heart of the role of thermodynamics in understanding time asymmetry, arises in the foundations of physics, and has implications for many issues in philosophy as well. The puzzle comes up in connection with the project of explaining our macroscopic experience with micro-physics: of trying to fit the macroscopic world of our everyday lives onto the picture of the world given us by fundamental physics. The puzzle has been debated by physicists and philosophers since the nineteenth century. There is still no consensus on a solution.

Here is the problem. Our everyday experience is largely of physical processes that occur in only one direction in time. A warm cup of coffee, left on its own in a cooler room, will cool down during the day, not grow gradually warmer. A box of gas, opened up in one corner of a room, will expand to fill the volume of the room; an initially spread-out gas won't contract to one tiny corner. A popsicle stick left out on the table melts into a hopeless mess; the hopeless mess sadly won't congeal back into the original popsicle.[1]

While we would be shocked to see the temporally reversed processes, the familiar ones are *so* familiar that they hardly seem worth mentioning. But there is a problem lurking. The problem is that the physical laws governing the particles of these systems are symmetric in time. These laws allow for the time reversed processes we never see, and don't seem capable of explaining the asymmetry we experience.

Suppose I open a vial of gas in a corner of the room. The gas will spread out to fill the room. Take a film of this process, and run that film backward. The reverse-running film shows an initially spread-out gas contract to one corner of the room. This is something that we never see happen in everyday life. Yet this process, as much as the original one, evolves with the fundamental dynamical laws, the laws that govern the motions of the particles in a system like this. Consider it in terms of Newtonian mechanics. (For ease of exposition, I stick to classical mechanics. Assume this unless explicitly stated otherwise.) The dynamical law of this theory is $F = ma$. Now, the gas just *is* lots of particles moving around in accord with this law. And the law is time reversible: it applies to the reverse-running film as much as to the forward-playing one. Intuitively, the law does not contain any direction of time in it. We can see this by noting that each quantity in the law has the same value in both films: in the backward film, the forces between the particles are the same as in the forward film (these forces are functions of the particles' intrinsic features and their relative spatial separations); their masses are the same; and their accelerations (the second time derivatives of position) are the same. For any process that evolves with this law, the time reversed process—what we would see in a reverse-running film—will also satisfy the law.

---

[1]  A video, with background music: http://www.youtube.com/watch?v=tPFqBrC9ynE.

This is, roughly, what we mean when we say that the laws governing a system's particles are time reversal symmetric, or time reversal invariant. The movie-playing image brings out an intuitive sense of time reversal symmetry. On a time reversal invariant theory, the film of any process allowed by the laws of the theory, run backward, also depicts a process that obeys the laws. (A bit more on this in the next section.)

In this sense, the classical laws are time reversal symmetric. Run the film of an ordinary Newtonian process backward, and we still see a process that is perfectly in accord with the Newtonian dynamical laws.[2] This time reversal symmetry isn't limited to classical mechanics, either. For (exceptions and caveats aside for now[3]) it seems that all the plausible candidates for the fundamental dynamical laws of our world are temporally symmetric in just this way.

But then there is nothing in the relevant physics of particles to prohibit the spontaneous contracting of a gas to one corner of the room, or the warming up of a cup of coffee, or the reconstitution of a melted popsicle—what we would see in reverse-running films of the usual processes. And yet we never see these kinds of things happen. That's puzzling. Why don't we ever see boxes of gas and cups of coffee behave like this, if the laws governing their particles say that it is possible for them to do so?

It's not that we never see the temporally reversed phenomena, of course. We do see water turning to ice in the freezer, coffee heating up on the stove, melted popsicles recongealing in the fridge. But these processes are importantly different from the time reverse of the usual phenomena. These processes differ from what we would see in the reverse-running films of ice melting, coffee cooling, and popsicles dripping. For they require an input of energy. (We can formulate the asymmetry as the fact that energetically isolated systems behave asymmetrically in time.[4])

More, it seems to be a lawlike fact that popsicles melt and gases expand to fill their containers. These generalizations support counterfactuals, they are used in successful explanations and predictions, and so on. They seem to satisfy any criteria you like for lawfulness; they surely don't seem accidental. Think of how widespread and reliable they are.

And, in fact, there is a physical law that describes these processes: the second law of thermodynamics. This law says that a physical quantity we can define for all these systems, the entropy, never decreases. We'll return to entropy in section 4. For now, note that this is a time asymmetric law: it says that different things are possible in either direction of time. Only non-entropy-decreasing processes can happen in the direction of time we call the future.

It turns out that ordinary processes like the expansion of gases and the melting of popsicles are all entropy-increasing processes. All the processes mentioned so far are

[2]  See Feynman (1965, ch. 5) and Greene (2004, ch. 6) for accessible discussion.
[3]  Caveats in the next section; an exception in 6.2.
[4]  One approach to the puzzle, which I do not discuss here, exploits the fact that typical systems are not in fact energetically isolated. This approach faces similar questions about the asymmetry of the influences themselves: Sklar (1993, 250–254); Albert (2000, 152–153). But see Earman (2006, 422) for a recent suggestion along these lines.

characterized by the science of thermodynamics; their time asymmetry, in particular, is characterized by the second law of thermodynamics. This law seems to capture the very asymmetry we set out to explain. Does this then solve our initial puzzle?

No. If anything, it makes the problem starker. The question now is where the asymmetry of the second law of thermodynamics comes from, if not from the underlying physical laws. The macroscopic systems of our experience consist of groups of particles moving around in accord with the fundamental physical laws. Our experience is of physical processes that are due to the motions of the particles in these systems. Where does the asymmetry of the macroscopic behavior come from, if not from the motions of systems' component particles? Consider that if we watch a film of an ice cube melting, we can tell whether the film is being played forward or backward. But we won't be able to tell which way the film is playing if we zoom in to look at the motions of the individual molecules in the ice: these motions are compatible with the film's running either forward or backward.[5]

Where does the observed temporal asymmetry come from? What grounds the lawfulness of entropy increase, if not the underlying dynamical laws, the laws governing the world's fundamental physical ontology? Can we can explain the asymmetry of thermodynamics, and of our experience, by means of the underlying physics?

## 3.  INTERLUDE: TIME REVERSAL INVARIANCE

The puzzle of time asymmetry in thermodynamics stems from the temporal symmetry of the fundamental physical laws. Before moving on, a caveat and a related issue.

Caveat. We now have experimental evidence that there is a fundamental, lawful time asymmetry in our world. Given the CPT theorem, the observed parity violations in the decay of neutral (chargeless) $k_0$-mesons implies a violation of time reversal symmetry. Yet it is widely thought that these violations are too small and infrequent to account for the widespread macroscopic asymmetries of our experience and of thermodynamics.[6] I set this aside here.

A related issue, also interesting and important, but also to be set aside here. Whether the laws are time reversal invariant is the subject of recent debate, for two reasons.

First, what quantities characterize a system's fundamental state at a time is subject to debate. Take Newtonian mechanics. Think of a film of a Newtonian process, such as a baseball flying through the air along a parabolic trajectory. Run this film backward, and we seem to have a process that also evolves with Newton's laws. We see baseballs fly through the air in opposite directions, obeying the laws of

---

[5]  The example is from Feynman (1965, 111–112).

[6]  See Sachs (1987, ch. 9) for a summary of the empirical evidence. For dissent on the conclusiveness of this evidence, see Horwich (1987, 3.6). See Arntzenius (2010) for more on time and the CPT theorem.

physics, all the time. However, the reverse-flying ball only obeys the physics supposing that we invert the directions of the velocities at each instant in the time reversed process. Otherwise, we get a process in which, at any given instant, the ball at a later instant will have moved in the opposite direction to that in which its velocity had been pointing previously. If we simply take the time reversed sequence of instantaneous states—if we run the sequence of film frames in reverse temporal order—and those states (movie frames) include particle velocities, then not even Newtonian mechanics would be time reversal symmetric: the time reversed process would not be possible.

In other words, if our time reversal operator—the mathematical object we use to figure out whether a theory is time reversal symmetric—only inverts sequences of instantaneous states, and those states include velocities, then Newtonian mechanics will not be symmetric under the time reversal operation. In general, we apply an operator to a theory to learn about the symmetry of the theory and the world it describes. We compare the theory with what happens to it after undergoing the operation. If the theory is the same afterward, then it is symmetric under the operation: we say that the theory is invariant under the operation. And surely Newtonian mechanics is time reversal invariant, if any theory is. Newtonian mechanics, that is, should be symmetric under the time reversal operator.[7] This theory doesn't seem to indicate any asymmetric temporal structure in the worlds it describes.

Physics texts respond by allowing the time reversal operator to act on the instantaneous states that make up a given time reversed process. In Newtonian mechanics, for example, the standard time reversal operator does not just invert the time order of instantaneous states; it also flips the directions of the velocities within each instantaneous state. Then a time reversal invariant theory is one on which, for any process allowed by the theory, the reverse sequence of *time reversed* states is also allowed. This is a slightly different understanding of time reversal symmetry from the intuitive, movie-playing idea. Still, all the candidate fundamental theories (with an exception to come later, and aside from the caveat mentioned above) are time reversal symmetric, on this understanding.

This raises a question. Should we allow time reversal operations on instantaneous states, and if so, which ones? We cannot allow any old time reversal operators, else risk our theories' trivially coming out time reversal symmetric.[8]

An alternative view, advocated by Horwich (1987) and Albert (2000, ch. 1), holds that particle velocities aren't intrinsic features of instantaneous states to begin with. If velocities aren't included in the fundamental state of a system at a time, then when we line up the instantaneous states in reverse temporal order, we won't get the problem that we did above for Newtonian mechanics, the problem that motivated us to move to a slightly different time reversal operation. (Instead, velocities

---

[7] For disagreement on this view of classical mechanics, see Hutchison (1993, 1995); also Uffink (2002). Savitt (1994); Callender (1995) are replies.

[8] See Arntzenius (2004, 32–33).

will invert when we take the reverse sequence of position states.) On this view, for any sequence of states allowed by a time symmetric theory, the inverse sequence of the *very same states* is also allowed. Newtonian mechanics, for example, is time reversal invariant. But there are some radical consequences of this view; for example, it loses the time reversal invariance of other theories that are standardly taken to be invariant.[9]

(It is interesting to think of how the debate will go for other formulations of the dynamics. There has not been as much discussion of this.[10] Presumably, equivalent formulations of a theory should all be invariant, or not, with respect to any given operation. For a theory's (non-)invariance indicates (a)symmetries in the world it describes. And different, equivalent formulations of a theory should describe the same set of possible worlds. Thus, take the Hamiltonian formulation of classical mechanics. The equations of motion are $\frac{\partial H}{\partial p} = \frac{dq}{dt}$ and $\frac{\partial H}{\partial q} = \frac{-dp}{dt}$. On the standard view, this theory is time reversal invariant just in case $H$ is invariant when $p \rightarrow^T -p(-t)$ and $q \rightarrow^T q(-t)$, where $T$ is the time reversal operator. Why these time reversal properties for the momentum and position coordinates? A common suggestion is by analogy to the Newtonian case: there, we inverted velocities under time reversal; here, we invert the momenta. This suggestion, however, faces Arntzenius' (2004) challenge to justify the time reversal operators used in Newtonian mechanics. Note that it seems that someone with Albert's view can also say that Hamiltonian mechanics is time reversal invariant, but for interestingly different reasons. Albert can get this result by taking the second equation of motion above to be a definition of momentum rather than an additional fundamental law. Albert only allows time reversal operations to act on non-fundamental quantities that are the time derivatives of fundamental quantities; for example, in Newtonian mechanics, particle velocities invert because sequences of positions do. By taking momentum to be defined by this law, it becomes a non-intrinsic, non-fundamental quantity, which then inverts under time reversal because sequences of positions do.)

A second reason that the time reversal symmetry of the laws is subject to recent debate is that the proper action of the time reversal operator is under debate. (A third reason, whether a particular theory of quantum mechanics is correct, will be discussed in section 6.2.) Setting aside their differences on the intrinsic properties of instantaneous states, the above views all agree that the basic action of the time reversal operator is to invert the time order of a sequence of states. But whether this is the proper action for the time reversal operator is debatable. A different notion of time reversal, as an inverting of the temporal orientation, was recently proposed by Malament (2004), and is defended by North (2008).

---

[9] On this debate, see Earman (1974, 2002); Horwich (1987, ch. 3); Albert (2000, ch. 1); Callender (2000); Arntzenius (2000, 2003); Smith (2003); Malament (2004). See Arntzenius (1997a) for time reversal invariance of indeterministic theories. See Arntzenius and Greaves (2009) for time reversal and quantum field theory.

[10] Some discussion of this can be found in Arntzenius (2000); Uffink (2002).

# 4. TROUBLE IN THERMODYNAMICS

Thermodynamics was originally developed in the nineteenth century, by figures such as Carnot, Clausius, and Thomson (Lord Kelvin), as an autonomous science, without taking into account the constituents of thermodynamic systems and the dynamics governing those constituents.[11] The original developers of the theory didn't try to explain thermodynamics on the basis of anything more fundamental. In particular, the puzzle of thermodynamic asymmetry was not yet recognized.[12]

That changed with the advent of the atomic hypothesis, and the identification of various thermodynamic quantities with properties of systems' particles.[13] Given the atomic make-up of matter, and given the successful identification of macroscopic properties such as average temperature, pressure, and volume, with properties of groups of particles, the question arises as to where the asymmetry of macroscopic behavior comes from.

Statistical mechanics is the physical theory applying the fundamental dynamical laws to systems with large numbers of particles, such as the systems studied in thermodynamics. (Statistical mechanics adds some probability assumptions to the fundamental dynamics, more on which below.) So the question is whether we can explain thermodynamics on the basis of statistical mechanics, and in particular, whether we can locate a statistical mechanical grounding of the second law.[14]

The work of Maxwell, Boltzmann, and Gibbs, among others, led to key components of an answer. Each of them had their own approach to statistical mechanics and the explanation of entropy increase. Since there remains disagreement on the proper understanding of statistical mechanics, there remains disagreement on the proper grounding of thermodynamics. In what follows, I use a Boltzmannian

[11] Fermi (1956) is a readable book on thermodynamics. Here is Fermi on the autonomy: in thermodynamics, the "laws are assumed as postulates based on experimental evidence, and conclusions are drawn from them without entering into the kinetic mechanism of the phenomena" (1956, x).

[12] There are many issues that I set aside here. Throughout, I stick with the modern formulation of the second law in terms of entropy. For discussion of the historical development of thermodynamics and of other aspects of the theory, see Sklar (1993); Uffink (2001); Callender (2001, 2008b); and references therein. Later axiomatizations of thermodynamics, beginning with Carathéodory in 1909, rigorized the theory; Lieb and Yngvason (2000) explains a recent version. Sklar (1993); Hagar (2005); Uffink (2007); Torretti (2007) (see also references therein) contain surveys of the development of, and different approaches to, both statistical mechanics and thermodynamics. See Earman (1981, 2006); Liu (1994) on the question of a relativistic thermodynamics. On entropy in quantum statistical mechanics, see Hemmo and Shenker (2006); Campisi (2008).

[13] On whether this constitutes a reduction of thermodynamics, see Sklar (1993, ch. 9); Callender (1999); Hellman (1999); Yi (2003); Batterman (2005; 2010); Lavis (2005).

[14] Ehrenfest and Ehrenfest (2002) is a classic text on the development and foundations of statistical mechanics. See also Sklar (1973, 1993, 2000, 2001, 2007); O. Penrose (1979); Uffink (1996, 2007); Frigg (2008a,b). A different approach is discussed in Liu (2001). Wallace (2002); Emch (2007) discuss issues in the foundations of quantum statistical mechanics. Some textbooks on statistical mechanics, all with varying approaches: Khinchin (1949); Prigogine (1961); Tolman (1979); Landau and Lifshitz (1980); Pathria (1996); O. Penrose (2005a).

approach. Gibbs' statistical mechanics can be adapted to the discussion (and according to some, it must be); the main difference lies in the conception of entropy. I will have to stick to one version here, and choose Boltzmann's out of my own views on the matter.[15]

Boltzmann's key insights were developed in response to the so-called reversibility objections (of Loschmidt and Zermelo).[16] The objection, in a nutshell, is the one that we have already seen. If the laws governing the particles of thermodynamic systems are symmetric in time, then entropy increase can't be explained by these laws. Think of any system that has been increasing in entropy, such as a partly melted ice cube, and imagine reversing the velocities of all its particles. The determinism[17] and time reversal invariance of the dynamics entail that the system will follow the opposite time development: the system will decrease in entropy, becoming a more frozen ice cube. (Another reversibility objection, with a similar conclusion for the inability of the dynamics to ground entropy increase, employs Poincaré recurrence.) The time reversed, anti-thermodynamic behavior is just as allowed by the physics of the particles.

The first part of a reply is this. Boltzmann and others realized that, given the time reversal invariance and determinism of the underlying dynamical laws, the second law of thermodynamics can't be a strict law. It must instead be a *probabilistic* law. Entropy decrease is not impossible, but extremely unlikely.

To understand the move to a probabilistic version of the second law, Maxwell's thought experiment is illuminating.[18] Imagine that a demon, or a computer, controls a shutter covering an opening in a wall that divides a box of gas. The gas on one side of the divider is warmer than the gas on the other side. Now, the average temperature in the gas is a function of the mean kinetic energy of its molecules. Within the warmer portion of the gas, then, there will be molecules that are moving, on average, slower than the rest of the molecules. Within the cooler portion of the gas, there will be molecules that are moving, on average, faster than the rest of the molecules. Suppose that the shutter is opened whenever a slower-on-average molecule within the warmer gas moves near the opening, sending that molecule into the cooler gas, and whenever a faster-on-average molecule within the cooler gas moves near the opening, sending that

---

[15]  For more on Gibbs' approach, see Gibbs (1902); Ehrenfest and Ehrenfest (2002); Sklar (1993); Lavis (2005, 2008); Earman (2006); Pitowsky (2006); Uffink (2007). For arguments in favor of Boltzmann's, see Lebowitz (1993a,b,c, 1999a,b); Bricmont (1995); Maudlin (1995); Callender (1999); Albert (2000); Goldstein (2001); Goldstein and Lebowitz (2004); against the approach, see Earman (2006). On information-theoretic notions of entropy (especially in relation to Maxwell's demon), see Earman and Norton (1998, 1999); Bub (2001); Weinstein (2003); Balian (2005); Maroney (2005); Norton (2005); Ladyman et al. (2007, 2008); and references therein.

[16]  On the history of the debate over the reversibility objections, see Brush (1975).

[17]  I set aside the cases of indeterminism in classical mechanics: see Earman (1986); Norton (2008); Malament (2008).

[18]  More, the thought experiment suggests that Maxwell recognized the reversibility problem, and the probabilistic version of the second law, sooner than did Boltzmann: Earman (2006).

molecule into the warmer gas. The net result is that the two gases will become *more* uneven in temperature, against the second law of thermodynamics.[19]

Intuitively, this separation in temperature could happen. It could even happen on its own, just by accident. Imagine that there is no shutter, just an opening in the middle of the divider. The particles could just happen to wander through the hole at the right times to cause the gas to grow more uneven in temperature. This seems extremely unlikely; but it also seems possible. The fundamental laws governing the molecules of the gas do not prohibit this from happening.

Still, it would take a massive coincidence, an extremely unlikely coordination among the motions of all the molecules in the gas. In other words: the second law of thermodynamics—the tendency of systems to increase in entropy, such as the tendency of a gas to even out in temperature—holds probabilistically. Entropy decrease is possible, but unlikely. Indeed, given the huge numbers of particles in typical thermodynamic systems, and given the extent of the coordination among their motions that would be required, entropy decrease is *extremely* unlikely.

The probabilistic understanding of the second law is a first step toward a solution to the puzzle of time asymmetry in thermodynamics. But it is only a first step (albeit a very large one). For the reversibility objections apply just as much to the probabilistic version of the second law as to the non-probabilistic one. The probabilistic version of the law says that entropy decrease, while not impossible, is extremely unlikely. Given the time reversal symmetry of the underlying laws, though, where does *this* asymmetry come from? Why isn't entropy extremely unlikely to decrease in either direction of time? If it's so unlikely that my half-melted popsicle will be more frozen to the future, then why did it "unmelt" to the past? Remember, the temporally symmetric laws don't make any distinction between the past and future time directions. If I feed the current state of my half-melted popsicle into those laws, then how do the laws tell the particles to increase in entropy to the future and not to the past? The laws don't even pick out or *mention* the future as opposed to the past!

Where in the world *is* thermodynamic asymmetry?

# 5. STATISTICAL MECHANICS

In order to make progress at this point, we need a more detailed understanding of the statistical mechanical basis of the second law.

This takes some setting up. A typical thermodynamic system will have a large number of particles. We can specify the state of such a system by means of its macroscopic features, such as its average temperature, pressure, and volume. These features pick out the *macrostate* of the system. Another way to specify the system's state is by means

---

[19] This can arguably be done without doing any work. Whether or not a genuine Maxwell's demon is possible is a matter of continuing debate. See, for example, Earman and Norton (1998, 1999); Albert (2000, ch. 5); Callender (2002); Norton (2005); and references therein.

of the fundamental states of its constituent particles. This gives the system's *microstate*, its most precisely specified state, in terms of the positions and velocities and types of each of its particles. In general, corresponding to any macroscopically specified state, there will be many different compatible microstates, many different arrangements of a system's particles that give rise to the same set of macroscopic features.

Think of this in terms of phase space. The phase space of a system is a mathematical space in which we represent all its possible fundamental states. For a classical system with $n$ particles, the phase space has $6n$ dimensions, one dimension for the position and velocity of each particle, in each of the three spatial directions. (The phase space has dimension $2nr$, where $n$ is the number of particles and $r$ is the number of degrees of freedom, here assumed to be the three dimensions of ordinary physical space.) Each point in phase space picks out a possible microstate for the system; a curve through the space represents a possible micro-history. A macrostate corresponds to a region in phase space, each point of which picks out a microstate that realizes the macrostate.

In the phase space of a gas, for example, each point represents a different possible way for the particles in the gas to be arranged, with different positions and velocities. Think of a macrostate of the gas, say the one where it fills half the room. Think of all the different ways the particles could be arranged to yield a gas that fills this volume of the room. Swap some of their positions, or change a few of their velocities, and we still get a gas that fills this volume of the room. These changes amount to picking out a different point in the phase space of the gas, consistent with its filling this volume of the room. The region comprising all those points represents the macrostate in which the gas fills half the room.[20]

Now we can get more precise about Boltzmann's insight. Boltzmann showed that the thermodynamic entropy, $S$, of a given system, the same entropy appearing in the second law of thermodynamics, is a function of how many arrangements of the system's particles are compatible with its macrostate. He found that $S = k \log n$ (up to additive constant), where $n$ is the "number" of microstates consistent with the macrostate, and $k$ is a constant. Here $n$ is the "number" of particles in the sense of the size of the region in phase space that the corresponding macrostate takes up— the region's volume, on the standard measure.[21] This quantity has been empirically determined to track the thermodynamic entropy. Boltzmann thus arguably discovered a statistical mechanical correlate of thermodynamic entropy, just as had been done for other thermodynamic quantities, such as the identification of average temperature

---

[20]  I skirt over details about how to divide up, or coarse-grain, the phase space: see Sklar (1993); Albert (2000, ch. 3); Earman (2006); Uffink (2007).

[21]  The standard measure in classical statistical mechanics is the Liouville volume measure: the standard Lebesgue measure defined over the canonical coordinates. Boltzmann's equation is thus $S = k \log |\Gamma_M|$, where $|\Gamma_M|$ is the standard (normalized) volume of the phase space region corresponding to the macrostate $M$. Typically, the constant energy $E$ is one of the macro constraints; in which case we use the volume induced by the standard measure on the $6n - 1$-dimensional energy hypersurface. A bit more on this later.

with the mean kinetic energy of a system's particles, or of average pressure with the rate and force of particle collisions with a system's container.[22]

Boltzmann's equation tells us that macrostates compatible with many distinct microstates—macrostates that can be formed by many different arrangements of a system's particles, macrostates that correspond to large regions of a system's phase space—have higher entropy than macrostates compatible with fewer microstates. Basic combinatorics shows that these macrostates have *overwhelmingly* higher entropy. Thus, spread-out, uniform, even-temperature macrostates have overwhelmingly higher entropy than concentrated, unevenly distributed ones. Think of the gas in the room. Intuitively, there are many more ways for the gas' particles to be arranged so that the gas is spread out in the room than concentrated in one tiny part: there are many more microstates compatible with the spread-out macrostate. This is reflected in the difference in entropy. The state in which the gas has spread out to fill the room has a much higher entropy than the state in which it is concentrated in one corner. The equilibrium macrostate, the one in which the gas has stabilized to fill the volume of the room, has the overwhelmingly highest entropy.

So higher entropy macrostates have many, many more distinct possible microstates than do lower entropy macrostates. Boltzmann showed this to be the case for all thermodynamic systems.

This suggests that we can understand entropy increase as the progression toward more and more probable macrostates. Add to our theory a natural-seeming probability assumption, that a system is as likely to be in any one of its possible microstates as any other—that is, place a uniform probability distribution, on the standard measure, over the phase space region corresponding to the system's macrostate—and we get that high entropy, large-volume-occupying macrostates are overwhelmingly more *probable* than low entropy, small-volume ones.[23] At any time, a system is overwhelmingly likely to evolve to a microstate realizing a macrostate that takes up a larger phase space region (or to stay in its current macrostate if it is already at equilibrium), the very higher entropy macrostate that thermodynamics says it should evolve into. This is because, according to the particle dynamics and the uniform probability measure, there are overwhelmingly more such states for the system to be in.

Have we finally found the statistical mechanical grounding of thermodynamics? Have we managed to derive the probabilistic version of the second law from the

---

[22] Rather, for a system whose microstate realizes the equilibrium macrostate, Boltzmann saw that this quantity agrees with the thermodynamic entropy defined for systems at equilibrium. Boltzmann then extends this notion of entropy to systems not at equilibrium too; indeed, that is one of the reasons to prefer his notion to Gibbs': Callender (1999); Lebowitz (1999b); Goldstein and Lebowitz (2004). Though whether it can be so extended is a matter of continuing debate.

[23] Where this probability distribution comes from is another large subject of debate. See Sklar (1973, 1993, 2001, 2007); Jaynes (1983); Lebowitz (1993b); Bricmont (1995); Strevens (1998, 2003); Callender (1999); Albert (2000, ch. 4); Goldstein (2001); Loewer (2001, 2004); Wallace (2002); North (2004, ch. 3; 2010); (2010); Goldstein and Lebowitz (2004); Lavis (2005, 2008); Earman (2006); Maudlin (2007b); Frigg (2008a).

underlying dynamics and the statistical postulate (as it is often called) of uniform probabilities over the microstates compatible with a system's macrostate?

Well, no. But it turns out to be a big start.

We know that this can't be enough to ground the thermodynamic asymmetry, for the now-familiar reason that the dynamical laws are time reversible. Take any entropy-increasing microstate compatible with a system's macrostate—any microstate for which, if the system starts out in it, the dynamical laws predict that it will (deterministically, on the classical dynamics) increase in entropy—and there will be another, entropy-*decreasing* microstate compatible with that macrostate: just reverse all the particle velocities. There is a one-one mapping between microstates and their time reverses. And for any microstate that is compatible with a given macrostate, so is its time reverse. So there will be just as many entropy-increasing as entropy-decreasing ways for the system to evolve from its current state. But then entropy increase can't be any more likely than entropy decrease.

In other words, the uniform probability distribution, combined with a dynamical law like $\mathbf{F} = m\mathbf{a}$, will predict overwhelmingly likely entropy increase to the future of any thermodynamic system. That is all to the good. But the uniform probability distribution, combined with the dynamics, also says that any system is overwhelmingly likely to increase in entropy to the *past*. That is decidedly *not* to the good. It is contrary to the second law of thermodynamics—not to mention most of our ordinary experience. We remember the coffee having been warmer than the room, the gas having been more concentrated, the popsicle more frozen. These are the mundane observations that got us going along this puzzle-solving path in the first place! Although the elements of our theory thus far predict what we expect for these systems' future behaviors, they radically contradict what we take to be the case for their pasts. Hence the depth of our problem. Statistical mechanics seems to make predictions that are radically falsified by our ordinary experience and by the evidence we have for the second law of thermodynamics.

Our problem goes deeper still. If statistical mechanics says that the past was radically different from what our current evidence suggests, then this undermines the very evidence we have for the physics that got us into this mess! Take the current macrostate of the world, a uniform probability distribution over its compatible microstates,[24] and the dynamics governing the world's particles. These are the elements of our statistical mechanical theory as it now stands. In the overwhelming majority of these microstates, the world increases in entropy to the future; but likewise, in the overwhelming majority of these microstates, the world increases in entropy to the past. In other words, the overwhelming majority of possible micro-histories for our world are ones in which the records we currently have are not, in fact, preceded by the events that they seem to depict. It is overwhelmingly *more* likely that the current state of the world, apparent

---

[24] Alternatively, take the macrostate of any given sub-system and a uniform distribution over its possible microstates. Whether the theory can be applied to the world or universe as a whole remains contentious. For a particularly forceful contention, see Earman (2006).

records and all, spontaneously fluctuated out of a past equilibrium state, than that our current records are veridical accounts of the world's having evolved from an extremely unlikely low entropy state. Statistical mechanics deems it extremely unlikely that any of our evidence for the world's lower entropy past is reliable. And that is the very evidence we have for the dynamical laws and the uniform distribution in the first place. Not only does this undermine the asymmetry of thermodynamics, but it undermines all of the evidence we have for thermodynamics, not to mention the rest of physics. We have on our hands the threat of "a full-blown skeptical catastrophe" (Albert, 2000, 116).[25]

At this point, it may be surprising to hear that we have made any headway; but indeed we have. The work done by Boltzmann and others in the foundations of statistical mechanics suggests that we can conclude this much: for any given macrostate, the overwhelming majority of its compatible microstates are those for which, if the system were in it, the system would (deterministically) increase in entropy. So we can reasonably infer entropy increase to the future, just as our experience leads us to expect.

The problem is that we can just as reasonably infer entropy increase to the past. That is the problem we now have to solve.

## 6. DIFFERENT APPROACHES

In order to explain thermodynamics, we will need a temporal asymmetry somewhere in our fundamental theory. As Price (1996) emphasizes: no asymmetry in, no asymmetry out.[26] The question is where. Answers can generally be divided into one of two camps: an asymmetry in boundary conditions or an asymmetry in the dynamics. Within each of these camps, there are differing approaches. I survey here a few of the representative and, to my mind, most promising. For discussion of other approaches (interventionism, expansion of the universe, others), I refer the reader to the references cited here; for comprehensive overviews, see especially Sklar (1993); Price (1996); Uffink (2007); Frigg (2008b).

### 6.1 Boundary conditions

Recall the problem we have now gotten ourselves into. If Boltzmann's reasoning explains overwhelmingly likely entropy increase to the future, then why isn't entropy just as likely to increase to the past?

---

[25] See Albert (2000, ch. 6) for more on the problem of records and the solution to it discussed below. Earman (2006) argues against both the apparent problem and this solution.

[26] See Price (1996) for arguments that many accounts can be faulted for smuggling in unwarranted asymmetric assumptions.

Another way of seeing the problem is emphasized by Albert (2000, ch. 4) in order to motivate the move to asymmetric boundary conditions. Take a partly melted popsicle. A uniform probability distribution over the microstates compatible with its macrostate, when combined with $\mathbf{F} = m\mathbf{a}$, predicts that the popsicle is overwhelmingly likely to be more melted in five minutes. But a uniform distribution over the microstates compatible with the macrostate that obtains five minutes from now, combined with $\mathbf{F} = m\mathbf{a}$, predicts that the popsicle is overwhelmingly likely to have been more melted five minutes ago—contrary to our initial assumption, as well as to thermodynamics. Not only does our theory make false predictions about the past, but it cannot be consistently applied at more than one time in a system's history. Apply the theory at one time, and the theory itself predicts that it will fail at any other time.[27]

The solution in terms of boundary conditions goes like this. The basic idea is simple. *Assume* that entropy was lower to the past, as our records and memories suggest that it was, and take the uniform probability distribution over the compatible microstates then. The dynamical laws will predict overwhelmingly likely entropy increase to the future of that time. That is what we learned from the work of Boltzmann and Gibbs.

Think of our partly melted popsicle. Our theory as it currently stands predicts that the popsicle is extremely likely to be more melted five minutes from now, and also five minutes ago. But suppose we now *posit* that the popsicle was more frozen five minutes ago; suppose we keep the more-frozen macrostate fixed to the past. Relative to this posit, the uniform distribution (taken over the microstates compatible with the five-minutes-ago macrostate) and the dynamics predict overwhelmingly likely entropy increase for the popsicle's future. Of course, this won't help if we want to make inferences about the popsicle half an hour ago: the popsicle is overwhelmingly likely to have been more melted half an hour ago. So now move the low entropy posit to the thirty-minutes-ago macrostate. Relative to that posit, the popsicle is extremely likely to keep on melting to the future.

You see where this is going. In order to predict entropy increase for the entire history of the world, posit the low entropy macrostate at its very beginning. This past hypothesis, as Albert (2000) calls it, that soon after the big bang the entropy of our universe was extremely low, disallows the high entropy inferences that statistical mechanics makes about the past.[28] Add the past hypothesis to statistical mechanics, and plausibly, we can explain the fact that thermodynamic systems behave asymmetrically in time, even though the dynamical laws governing their particles are time reversible. It is because the world started out with extremely low entropy, and at any

[27] For disagreement on this point, see Earman (2006); another source of disagreement will be discussed in section 7.2. Note that this is not the case for any probabilistic theory. In Bohmian quantum mechanics, for example, the compatibility of the dynamics and the probabilities, at all times, can be demonstrated: Dürr et al. (1992a,b).

[28] The idea has been suggested in different ways by Boltzmann (1964) (see Uffink (2007); Goldstein (2001)); Feynman (1965, ch. 5); Penrose (1989, ch. 7), (2005b, ch. 27); Lebowitz (1993a; 1993b; 1993c; 1999b); Bricmont (1995); Price (1996; 2002a; 2002b; 2004); Albert (2000); Goldstein (2001); Goldstein and Lebowitz (2004); Callender (2004a,b); Wald (2006). Challenges to its account of thermodynamics are in Winsberg (2004a); Parker (2005); Earman (2006); a bit more of which soon.

given time entropy is overwhelmingly likely to go up. Don't posit a low entropy "future hypothesis" since the evidence suggests that entropy was lower to the past and not to the future. (That is, unless we found evidence to the contrary.) On this view, the time asymmetry of thermodynamics comes from an asymmetry in the boundary conditions of our universe.

In this way, amazingly, modern big bang cosmology seems to be getting at what we need in order to explain thermodynamics. The reasoning we get in statistical mechanics from the likes of Boltzmann and Gibbs, and the empirical evidence we get from cosmology, are converging on the same initial low entropy macrostate of our universe.[29] Although we must ultimately assume the past hypothesis, since without it our evidence of past low entropy is likely mistaken, this gives us a kind of justification for the assumption: the evidence we have from both foundations of statistical mechanics and cosmology, evidence empirical and theoretical, suggests that we can reasonably assume initial low entropy. That, plus the fact that all of this evidence would be self-defeating without the assumption of the past hypothesis. Without this assumption, remember, physics deems it overwhelmingly likely that the past was completely different from what we think, and that the laws are completely different as well. We would have no reliable evidence for what the physics of our world is really like, and no reason to infer anything in particular about the past or future. We could not even trust our belief in Newtonian mechanics. For what evidence we have deems it overwhelmingly likely that this evidence is radically misleading. If we assume the past hypothesis, however, we plausibly avoid getting into that muddle.

Some objections, replies, and clarifications, before moving on.

*What is the status of the past hypothesis?* Some (Albert, Feynman, Penrose, among others) regard the past hypothesis as a fundamental law. Whether you agree will depend on your view of laws. The past hypothesis does satisfy many of the generally accepted criteria of lawhood (counterfactual support, explanatory and predictive success), but for its being a non-dynamical generalization. Still, if successful, the past hypothesis yields a simple and unifying theory—no need to add anything to the laws we already have other than a simple statistical constraint on initial conditions—and this counts in favor of its law status.[30] Note that if we do treat the past hypothesis like this, then there is an asymmetry in the fundamental laws after all, albeit a non-dynamical one. If not, then the past hypothesis is a contingent generalization for

---

[29]    See Earman (2006) for sustained argument against this claim. Earman points out, for example, that not just any low entropy macrostate will be capable of grounding thermodynamics: it should be the kind of small, dense, hot, uniform state that big bang cosmology suggests it was. Even then, we need further details about the initial state to show that this should yield thermodynamics; and in Earman's view, no details likely to be forthcoming will do the job. More, the initial posit doesn't say anything about the rate at which entropy will increase. We need more details about the dynamics to show that the theory predicts a current macrostate of relatively low entropy. All of which leads Earman to conclude that this is no more than a "just-so" story, "a solution gained by too many posits and not enough honest toil" (2006, 412).

[30]    In particular on a Lewisian best-system account: Loewer (2001; 2004). But see Frigg (2008a, 2010); Winsberg (2008) for argument against this.

which we have empirical evidence, albeit evidence that is only reliable once we assume that it is.

*Why does the initial state have low entropy? Doesn't big bang cosmology say that the universe began in a uniform macrostate?* Although this has not been worked out rigorously, there is a rough answer that strikes many people as plausible.[31] Immediately after the big bang, the universe was in a uniformly hot "soup," with matter and energy uniformly distributed in thermal equilibrium. This state did have high thermodynamic entropy. The thought is that it had extremely low entropy due to gravity. Gravity is an attractive force: matter tends to clump up under this force, and then to stay clumped up. We know from thermodynamics that maximal entropy states are the equilibrium states toward which systems tend to evolve and then stay. For systems primarily under the influence of gravity, then, a clumped-up state has high entropy.[32] The early state of the universe, non-clumped-up and uniformly spread out, had extremely low entropy due to gravity.[33]

*Why did the universe start out in such an unlikely state?* This account tells us to assume something that is extremely unlikely by its own lights.[34] Price (1996, 2002*a,b*, 2004), who argues in favor of the boundary conditions strategy, says that more is needed to complete the story. In his view, the real puzzle about thermodynamic asymmetry is to explain the low entropy initial state itself.[35] We know that entropy increase is extremely likely from the work of Boltzmann and others, after all. The puzzle, according to Price, is why entropy was so low to begin with. On a view which thinks of this state as a fundamental law, though, this search for explanation will seem misguided. Even without that view of the past hypothesis, one might question the need to explain initial conditions.[36]

*How can the past hypothesis get us anywhere, when statistical mechanics says that all the evidence we have for it is extremely likely to be mistaken?* Without the past hypothesis, our theory says it's extremely likely that our current memories and records are mistaken, and radically so. If we don't start out assuming past low entropy, then the overwhelmingly most likely scenario is that our current records and memories—

[31] Such as Penrose (1989, 317–322), (2005*b*, ch. 27). For a recent account, see Wallace (2010). See Earman (2006) for disagreement. Earman argues that even the rough answer is implausible, for we do not have, and are unlikely to get, a theory of the entropy due to gravity, let alone such a theory that allows us to calculate the entropy of the entire universe. Wald (2006) is more optimistic. See Callender (2010) for recent discussion. See Ellis (2007) for recent discussion on this and other philosophical issues in cosmology.

[32] Another way of putting it is that the state will be spread out in momentum space, even though it will be relatively clumped up in position space.

[33] Though not completely uniform: enough non-uniformities are needed to start the clumping-up process that leads to the formation of stars and galaxies and so forth.

[34] How unlikely? See Penrose (1989, 343).

[35] A similar view is in Carroll (2008, 2010).

[36] As do, for example, Boltzmann (in Goldstein (2001)); Sklar (1993, 309–318); Callender (1998, 2004*a,b*); North (2002). Penrose (1989, ch. 7), (2005*b*, ch. 28) argues that there may be a dynamical explanation on which the initial state is not unlikely. Carroll and Chen (2004, 2005); Carroll (2008, 2010) attempt to explain the initial state by means of the large-scale structure of the multiverse. Wald (2006) argues against these ideas.

no matter how well-correlated they are with other records and alleged states of the world—spontaneously fluctuated out of past equilibrium. Yet once we assume the past hypothesis, the suggestion is, this will no longer be the case. For, given past low entropy, it is overwhelmingly more likely that a popsicle had been more frozen to the past than that it formed spontaneously out of a homogeneous soup. Relative to the assumption that the popsicle was frozen to the past, the overwhelming majority of micro-histories yielding its current state will have come by way of that lower entropy past state; for if that weren't the case, then the entropy of the popsicle would not have been increasing since then. See a footprint on a beach, and without the past hypothesis, the footprint is overwhelmingly likely to have spontaneously formed out of past equilibrium; assume past low entropy, and this is much less likely than a person's having walked on the beach.[37] Relative to the assumption of past low entropy, it is much more likely that the world is in a microstate, compatible with its current macrostate, which evolved through a lower entropy past state with a person on the beach.

In other words, plausibly, the low entropy initial posit makes it overwhelmingly likely that our usual causal accounts for how things got to be the way that they currently are, are correct: that there was a person to cause the footprint on the beach, a more frozen popsicle to cause my memory, and not just a homogeneous equilibrium soup. For the past hypothesis makes it overwhelmingly likely that the correlations among our current records and memories are due to past states of the world. This is not a rigorous argument. It is a plausibility claim that the theory should be able to ground our records in this way, given Boltzmann's reasoning in statistical mechanics, and given big bang cosmology's account of the formation of stars and galaxies, which in turn lead to the existence of beaches and people, who in turn lead to the existence of frozen popsicles, and so on.[38] Plausibly, given the past hypothesis, we can reconstruct a picture of the world on which our inferences, and the records that they rely on, come out successful in the way that we think they are.[39]

(Take the current macrostate of the world, the fundamental dynamics, and the probability postulate. Conditionalize on the past hypothesis, and we constrain the overwhelming majority of possible world-histories to those in which our records are by and large veridical. Hence the basis for Albert's (2000, ch. 6) claim that this explains why we know more about the past than about the future: initial low entropy restricts the possible pasts of our world more than its possible futures. Not that it restricts the set of possible past microstates more than possible future microstates. By Liouville's theorem, the volume in phase space taken up by the world's macrostate at any time will be the same. Rather, there is a "branching tree structure" to the world Loewer (2007), in

---

[37] See Earman (2006) for disagreement on both parts of this claim.

[38] See Penrose (1989, ch. 7).

[39] Note the different sense of "record" here from that of Lewis (1979). Rather than being a determinant, a record is something in the current state relative to which, conditional on the past hypothesis, it is overwhelmingly likely that the system passed through the state that the record appears to be a record of.

which the initial low entropy macrostate constrains the possible past macrostates of the world more than the possible future macrostates, relative to the current macrostate.[40])

*This only gives us records at the level of entire macrostates of the world. What about the localized records we're familiar with—footprints, photographs, and the like?* More needs to be done to suggest that the past hypothesis, even if necessary for thermodynamics, is sufficient as well, in particular for the localized records that make up our ordinary evidence for thermodynamics. More details about the initial state might help. Immediately after the big bang, everything in the universe was distributed relatively uniformly, in a dense, hot, equilibrium soup; the matter and fields were evenly distributed and particles were moving around randomly.[41] Some of these randomly moving particles will eventually collide, and under gravity, some clumps of matter will begin to form. These clumps contain accelerating particles and will be hotter on average than the surrounding space. As the universe expands, then, it evolves away from its initial homogeneous state into macrostates that consist of hotter clumps of matter in a cooler surrounding space. Relative to the assumption of initial homogeneity, these later states indicate that there had been particle collisions in the past. For if everything started out moving randomly in an even-temperature soup, the later states containing warmer masses within a cooler surrounding space indicate past particle collisions and not future ones. More, the clumps are relatively localized records of past collisions.[42]

*How can the past hypothesis explain entropy increase in the world's various subsystems, even if it can ground entropy increase for the universe as a whole?*[43] Given the deterministic dynamics, a probability distribution taken over the microstates compatible with the macrostate of the world at any time will induce a probability distribution over the world's possible microstates at any other time: conditionalize the initial distribution on the macroscopic constraints at the other time. By means of this conditionalizing procedure, the initial distribution will assign probabilities to the different possible fundamental states of the world at any time. But any microstate of the world includes a specification of the exact state of any sub-system. So the distribution

[40] Why this structure? Plausibly, because the microstates of the world compatible with its macrostate at any time are on trajectories that spread out, or "fibrillate," over more and more distinct macrostates to the future; see note 45. Objections to the explanation of the asymmetry of knowledge are in Parker (2005); Frisch (2005a, 2007, 2010); Earman (2006).

[41] We'll ultimately need quantum mechanics to describe an equilibrium state of matter and energy. Even then, you may be skeptical that such a description is possible: note 31. A very brief sketch of how that might go is at the end of North (2003).

[42] This is extremely rough, at best only a very beginning. See Elga (2007) for a more worked-out account; see also Albert (2000, ch. 6).

[43] Winsberg (2004a), for example, argues that we need a further posit to rule out local anti-thermodynamic behavior (since small, relatively isolated sub-systems will have randomized microstates as a result of past interactions with the rest of the universe), a posit which, moreover, we don't think is true; Earman (2006, 420) concurs. A similar criticism is in Reichenbach's "branching systems" objection Sklar (1993, 318–331). (See Winsberg (2004b) for an updated version of Reichenbach's idea.) Frigg (2008b) suggests that the standard measure cannot tell us the probabilistic behavior of an ordinary system, whose microstate is confined to an energy hypersurface, since any such lower-dimensional space gets zero measure on the standard volume measure taken over all of phase space.

taken over the phase space of the world, by assigning probabilities to its possible microstates at any time, will also assign probabilities to the possible microstates of any sub-system at any time. Restrict the initial distribution to the region representing the sub-system's macrostate; that is, conditionalize the initial universal distribution on the system's macroscopic features. More, this should yield relatively uniform probabilities over the sub-system's compatible microstates (see below).[44]

*Why should this yield the same probabilities as the empirically confirmed ones of ordinary statistical mechanics? Ordinary statistical mechanics takes the uniform distribution at any time we choose to call the initial one, without conditionalizing on the past.* Two reasons. The first is Boltzmann's combinatorics, which suggests that the overwhelming majority of microstates compatible with any given macrostate lie on trajectories that increase in entropy (both to the future and to the past). That is, think of the phase space region representing a system's macrostate. Boltzmann's reasoning suggests that the proportion of the volume of this region that is taken up by microstates leading to entropy increase is overwhelmingly large, and the proportion taken up by microstates leading to entropy decrease is overwhelmingly small. The second is a randomness assumption. Within any phase space region corresponding to a system's macrostate, the microstates leading to entropy decrease will be scattered, relatively randomly, throughout. Again, this is a reasonable, if unproven, assumption of ordinary statistical mechanics.[45] All of which suggests that the standard uniform distribution will yield the same probabilities of future thermodynamic behavior as the uniform distribution that is first conditionalized on the past hypothesis.[46] At the same time, the distribution conditionalized on the past hypothesis will improve upon the standard one with respect to inferences about past thermodynamic behavior.[47]

## 6.2 Dynamics

Another way of trying to solve our puzzle is with laws that aren't time reversal invariant. If the fundamental dynamical laws say that different things can happen to the past and to the future, then this might explain the asymmetry of thermodynamics.

---

[44] A standard assumption in statistical mechanics makes it plausible that, for any system, the initial distribution will be relatively uniform throughout any sub-space of the higher-dimensional phase space. So that when we conditionalize the initial distribution on the sub-space (and renormalize), we get another distribution that is relatively uniform. See Lebowitz (1993b,c, 1999b); and below.

[45] Thus Lebowitz: "for systems with realistic interactions the domain $\Gamma_{Mab}$ will be so convoluted that it will be 'essentially dense' in $\Gamma_{Mb}$" (1993a, 10); $M$ refers to the system's macrostate, $\Gamma$ the phase space, $\Gamma_M$ the phase space region corresponding to $M$, $M_a$ the system's initial macrostate, $M_b$ its later macrostate, and $\Gamma_{Mab}$ the region of $\Gamma_{Mb}$ that came via $\Gamma_{Ma}$ (the set of microstates within $\Gamma_{Mb}$ that are on trajectories that come from $\Gamma_{Ma}$). Again: "for systems with realistic interactions the domain $\Gamma_{Mab}$ will be so convoluted as to appear uniformly smeared out in $\Gamma_{Mb}$. It is therefore reasonable that the future behavior of the system, as far as macrostates go, will be unaffected by their past history" (1999b, S349).

[46] More, the conditionalized distribution arguably get, more inferences about the future correct than the ordinary, unconditionalized distribution does. See the discussion of Napoleon's boot in Albert (2000, ch.4).

[47] More on this is in North (2004, ch. 3).

You might wonder if even the classical laws, time reversal invariant though they be, could do the job. You might think that these laws have some property that will show entropy likely to increase over time; say, some chaotic property. You would not be alone. There is a history of trying to show just this, a history that continues to the present.[48] Yet no approach relying on these as the fundamental laws can suffice to explain the thermodynamic asymmetry, not without asymmetric boundary assumptions. This is for the usual reasons, namely, the time reversal symmetry and determinism that lead straight to the reversibility objections.[49]

That is why ergodic theory cannot do the job, as far as explaining thermodynamics goes. This approach to statistical mechanics uses mathematical theorems from ergodic theory to pinpoint features of the dynamics to explain entropy increase. In so doing, the approach also tries to explain the probability distribution that standard statistical mechanics, and its explanation of entropy increase, relies on.[50] This is a large approach to the foundations of statistical mechanics, which I cannot adequately address here; for survey and references, see Sklar (1993); Uffink (2004, 2007). Let me mention the reasons to be skeptical of its ability to ground the second law of thermodynamics. First, it has not been shown that ordinary systems are, in fact, ergodic. Although some notions of ergodicity have been demonstrated to hold of certain simple systems, results such as the KAM theorem suggest that most statistical mechanical systems will fail to satisfy any strict notion of ergodicity.[51] More generally, the results using ergodic theory do not seem necessary to the statistical mechanical grounding of thermodynamics. Goldstein (2001) argues that many of these results (such as the technique of Gibbs phase averaging, used to calculate the values of thermodynamic quantities at equilibrium) can be shown to hold regardless of whether a system is ergodic. Ergodic theory seems insufficient for this project as well, since it cannot avoid the need for

---

[48] Part of this history lies in Boltzmann's own H-theorem. For more on the H-theorem and on Boltzmann's later account—that the universe as a whole is almost always at maximum entropy, but we happen to be on the up-slope of one of its fluctuations out of equilibrium—see Ehrenfest and Ehrenfest (2002); Feynman (1965); Sklar (1993, ch. 2); Price (1996, ch. 2); Uffink (2004). For more on symmetric cosmological accounts, see Price (1996, ch. 4). For a recent version of a "symmetric on the whole" theory, see Carroll and Chen (2004, 2005); Carroll (2008, 2010); also discussed in Wald (2006).

[49] Nor does this depend on your view of time reversal (section 3). Even someone like Albert, who thinks that most theories other than Newtonian mechanics are non-time reversal invariant, won't for that reason explain the second law of thermodynamics. As Albert (2000, ch. 1) puts it, these theories are still symmetric with respect to the evolutions of particle positions, and that is enough to get the puzzle about thermodynamic systems going.

[50] The tradition of invoking ergodic theory in explanations of statistical mechanics goes back to Boltzmann (Ehrenfest and Ehrenfest, 2002). Boltzmann's original idea was that a system is ergodic if, for almost all (standard measure 1) initial conditions, its trajectory passes through every point in the available phase space. A version of Birkhoff's theorem then says that for such a system, the infinite time average of the phase function corresponding to a macroscopic property equals the function's average over the phase space. (Birkhoff's theorem says that infinite time averages exist for almost all initial conditions. A corollary says that if a system is ergodic, then those infinite time averages equal the standard (microcanonical) phase averages for almost all initial conditions. See Earman and Rédei (1996).)

[51] See Sklar (1993, ch. 5); Earman and Rédei (1996).

some initial probability assumption: the ergodic approach can't derive all probabilistic posits from the dynamics alone (as in the "measure zero problem" discussed in the literature), with laws that are deterministic and time reversal invariant, though that is one of its chief motivations. In particular, it can't avoid the need for an asymmetric boundary assumption like the one above.[52] (Though not enough to solve the puzzle here, note that ergodicity could help ground the randomness assumption (section 6.1; note 45): that the compatible microstates on entropy-decreasing trajectories will be scattered randomly throughout a system's phase space; that they will spread out over the phase space.[53])

A recent proposal based on non-time reversal invariant dynamics comes from Albert (1994, 2000). Here, too, the basic idea is simple, though we must now take into account quantum mechanics.[54] The suggestion is that a certain theory of quantum mechanics, the collapse theory of Ghirardi, Rimini, and Weber,[55] or GRW, is non-time reversal invariant in a way that can account for the thermodynamic asymmetry.[56]

In quantum mechanics, questions of time asymmetry are tricky, since there are different versions of the theory on the table. But there are some broad similarities and differences that are relevant to our question here.

All theories of quantum mechanics take the Schrödinger equation to be a fundamental law. This is a deterministic and time reversal invariant equation of motion.[57] It governs the evolution of a system's wavefunction, the mathematical object that describes a system's fundamental quantum state at a time. Different theories of quantum mechanics disagree on the scope of this law. Non-collapse theories of quantum mechanics, such as Bohm's theory or many worlds, posit the Schrödinger equation as the fundamental dynamical law governing the evolution of a system's wavefunction, at all times.[58] Collapse theories, on the other hand, say that Schrödinger evolution fails

---

[52]  See Sklar (1973), (1993, ch. 5); Friedman (1976); Leeds (1989); Earman and Rédei (1996); van Lith (2001) for presentations of the problem and various proposals for addressing it. For recent ergodic-based accounts that improve upon the traditional ones, see Malament and Zabell (1980); Vranas (1998); also Campisi (2005). Vranas, for example, suggests that something close enough to ergodicity might actually hold of ordinary systems. See Strevens (1998, 2003, 2005) for a different, non-ergodic-based approach to demonstrating that macroscopic generalizations, such as those of thermodynamics, come from chaotic properties in the micro-dynamics.

[53]  See Berkowitz et al. (2006) for recent work along these lines. See Earman (2006, 406) for such a suggestion based on a mixing property (stronger than ergodicity). Earman, however, argues that this suggestion undermines the Boltzmann apparatus.

[54]  A different approach, using quantum decoherence, is in Hemmo (2003); Hemmo and Shenker (2001, 2003, 2005). Bacciagaluppi (2007) is another. An overview of different theories of quantum mechanics is in Albert (1992), Uffink (2010) discusses others.

[55]  See Ghirardi et al (1985, 1986).

[56]  On the relevant sense of time reversal non-invariance, see Arntzenius (1997a).

[57]  It is time reversal invariant given the standard time reversal operator in quantum mechanics, which maps $t \mapsto -t$ and also takes the complex conjugate. Whether this is a legitimate time reversal operator is open to question: see section 2.

[58]  The guiding equation of Bohm's theory, which governs particle evolutions, comes from the Schrödinger equation plus some natural symmetry considerations: Dürr et al. (1992b, 852–854).

to hold whenever the wavefunction "collapses" onto one of its components, in accord with probabilities dictated by the theory. Collapses are non-unitary, indeterministic transitions, not governed by Schrödinger evolution.

GRW is a collapse theory. It posits a fundamental, probabilistic collapse law governing the evolution of the wavefunction, in addition to the Schrödinger equation. In GRW, the collapse law gives a probability per (small) unit time of a wavefunction collapse,[59] at which point the wavefunction is multiplied by a normalized Gaussian function. The result is that the wavefunction is localized to a small region within its phase space.[60] The probability that the multiplying Gaussian is centered on any given location in phase space depends on the wavefunction just before the collapse, in accord with the usual square amplitudes.

Not only is wavefunction collapse governed by a fundamental, indeterministic law on this theory, but by a fundamental, non-time reversal invariant law. GRW assigns probabilities to the different possible future wavefunctions that a system's current wavefunction could collapse into. (After which the wavefunction will evolve deterministically, in accord with the Schrödinger equation, until another collapse occurs.) The theory doesn't assign probabilities to different possible past wavefunctions, given a system's current wavefunction. The collapse law doesn't say anything about the chances of different past wavefunctions.[61] GRW then says that different things can happen in either direction of time: wavefunctions can collapse in accord with lawful probabilities to the future, not the past.

Time reversal invariant theories of quantum mechanics face the same problem of explaining the asymmetry of thermodynamics that classical theories do. As a result, in order to ground thermodynamics, these theories will need an asymmetric boundary assumption such as the one discussed above. This assumption is needed in order to make it unlikely that a system ever starts out in an entropy-decreasing quantum state; for if it did, then the deterministic and time reversible dynamics entails that the system will decrease in entropy to the future of that state.[62]

---

[59] Roughly per "particle." Recent empirical evidence suggests that a probability of collapse per atom isn't right, but there are other versions available. I say "particle" since there are no fundamental particles on this theory, and so arguably no particles at all: see Albert and Loewer (1995); Albert (1996). See Allori et al (2008) for different ways of understanding the theory's ontology.

[60] Except for the "tails": Albert and Loewer (1995).

[61] As Arntzenius (1997a) puts it, GRW is a theory of forward transition chances, with no backward transition chances. Against this, Price (1996, 2002a,b) argues that a theory like GRW might really be a symmetric theory, with backward transition chances in addition to the usual forward ones; the backward chances don't result in observed frequencies because they are subordinate chances that are overridden by the initial low entropy condition. One might wonder, though, why we should believe in the existence of lawful chances in that time direction, if they are never manifested in observable frequencies. More, it seems we can only add backward transition chances at the expense of empirical adequacy, since quantum phenomena don't display invariant backward transition frequencies, as argued by Arntzenius (1995, 1997a,b).

[62] There remains the question of how to define a uniform probability measure over the complex infinite-dimensional vector spaces of quantum mechanics. This is a large question for any version of quantum statistical mechanics.

Albert's suggestion is that this won't be the case for GRW. On this theory, there is a fundamental time asymmetry in the dynamics. So maybe this fundamental asymmetry can explain the macroscopic asymmetry of thermodynamics.

The reason to think that GRW might be able to do this stems from the structure of the entropy-decreasing microstates in phase space, combined with the nature of the theory's transition probabilities. Recall that, plausibly, the phase space regions consisting of the entropy-decreasing microstates (really, the microstates leading to any abnormal thermodynamic behavior) are scattered randomly, in extremely tiny clumps, throughout a system's phase space. Indeed, these entropy-deacreasing regions will be so scattered and tiny that a uniform probability distribution taken over just about any phase space region—any region that is not as tiny and scattered as the entropy-decreasing regions themselves—will deem it overwhelmingly unlikely that a system will decrease in entropy. That is, entropy decrease (to the future as well as to the past) is overwhelmingly unlikely, not only given a uniform distribution over the phase space region corresponding to a system's macrostate, but also given a uniform distribution over virtually any sub-region of that phase space region, however small, including the neighborhood of any single microstate, whether entropy-decreasing or not.[63] This is what Boltzmann made plausible.

If this is right, then the entropy-decreasing microstates are extremely unstable: any system that's in an entropy-decreasing state is extremely "close" to being in an entropy-increasing one.[64] Plausibly, therefore, on a theory of fundamental, indeterministic wavefunction collapses, any system will be overwhelmingly likely to increase in entropy. For it is overwhelmingly likely that, if the system is in an entropy-increasing quantum state at some time, then a wavefunction collapse to the future will keep its state within the entropy-increasing regions of its phase space—the regions containing the quantum states that will deterministically, in accord with the Schrödinger equation, increase in entropy to the future. And it is overwhelmingly likely that, if the system is in an entropy-*decreasing* state at some time, then a wavefunction collapse to the future will cause it to jump to a state within the entropy-increasing regions. Given the extent of the instability of the entropy-decreasing microstates, these wavefunction jumps should make it overwhelmingly likely that *any*[65] system will evolve in accord with the second law of thermodynamics, even if it starts out in an entropy-decreasing microstate (a microstate on a trajectory that takes the system to a lower entropy future macrostate).

Since the abnormal microstates take up non-zero phase space volumes, not just any kind of collapse will get this result. But there are reasons to think the collapses of a theory like GRW will.[66] The GRW collapse law assigns a probability per unit time to

---

[63]  See also note 43.

[64]  But see note 62.

[65]  That is, any large enough system, large enough to exhibit an entropy-increasing tendency: see Albert (1994), (2000, ch. 7); North (2002). And setting aside the possibility of Maxwell's demon type systems: Albert (2000, ch. 5).

[66]  Briefly here. See Albert (1994, 2000, ch. 7), also North (2004, ch. 1), for more.

a wavefunction collapse, in accord with the usual quantum mechanical probabilities, at which point the wavefunction is multiplied by a Gaussian. At any time, that is, the theory places a probability distribution over the different possible microstates that a system could evolve into after collapse, given its wavefunction at that time, in accord with the usual Born rule. This probability distribution is centered on the system's initial microstate. It is also uniform over a very small phase space region, smaller than any region representing a macrostate. Given the level of the instability of the abnormal microstates, and given the Gaussian width of the collapsed wavefunction (the width over which the probabilities are distributed), the region over which this probability distribution is taken should be at least as large as the smallest sub-region of phase space which yields the same probabilistic predictions as the standard distribution over all the microstates compatible with a system's macrostate. At the same time, the region should be small enough that any system, even after undergoing a wavefunction collapse, will remain in one of its possible microstates. Collapses won't cause the system to behave non-thermodynamically by carrying it to some far-off region in phase space, that is.

This is non-rigorous. Calculations are needed to show that the GRW distributions are of this order. But the conclusion seems plausible, given how tiny and scattered the entropy-decreasing portions of a system's phase space should be.

Of course, whether this works as a theory of thermodynamics depends on the truth of GRW as a theory of quantum mechanics. And the account relying on asymmetric boundary conditions arguably succeeds in grounding thermodynamics. So the question is whether GRW, should it turn out to be a true theory, could explain thermodynamics *better*.

There is reason to think that it can. In order to answer the reversibility objections, any time symmetric theory of quantum mechanics will require two fundamental probability distributions: the probabilities of quantum mechanics and the statistical mechanical probability distribution. We are thus left with "two utterly unrelated sorts of chance," as Albert puts it, "one (the quantum-mechanical one) in the fundamental microscopic equations of motion, and the other (the statistical-mechanical one) in the statistical postulate" (2000, 161).

A statistical mechanics based on GRW dynamics, on the other hand, does away with the latter distribution. On this theory, it is the probability per unit time of a wavefunction collapse that yields overwhelmingly likely entropy increase, not a probability distribution over possible initial wavefunctions. There is no need for an additional probability distribution, since no matter which microstate compatible with its macrostate a system starts out in—even an entropy-decreasing one—the stochastic dynamics predicts that it is overwhelmingly likely to evolve with the second law of thermodynamics. No need for an initial distribution to make entropy-decreasing quantum states unlikely: the dynamics takes care of this for us. (The past hypothesis is still needed for inferences about the past, but not as a correction to otherwise faulty retrodictions, as it was above.)

On this theory, there would be only one probability law underlying thermodynamics: the probabilistic law of wavefunction collapse. The probability distribution posited by GRW to yield a viable theory of quantum mechanics would also yield the probabilities of statistical mechanics and of thermodynamics. This is then a simpler, more unified theory. Other things being equal, it is preferable, assuming that an explanation is better to the extent that it is simpler and more unifying, relying on fewer independent assumptions. Although we must wait and see what the right theory of quantum mechanics is, and although asymmetric boundary conditions should do the job if GRW is not that theory, this is a better theory of thermodynamics, if GRW does turn out to be true.[67]

# 7. THE STATUS OF STATISTICAL MECHANICS

Another way to challenge the above attempts at a solution, whether via asymmetric boundary conditions or asymmetric dynamics, is to challenge the view of statistical mechanics as a scientific theory. So far, I've been assuming that statistical mechanics is a fundamental theory. Hence the problem with thermodynamics: if statistical mechanics is a fundamental theory and thermodynamics is not, then where does the asymmetry of the latter come from, if not an asymmetry in the former?

What, then, is the status of statistical mechanics? Could denying it fundamental status help solve the puzzle about thermodynamics?

## 7.1 Universal and fundamental

As Albert presents it, and as it tends to be treated in physics textbooks, statistical mechanics is a fundamental theory. Statistical mechanics consists of the fundamental dynamics, whether classical or quantum mechanical, for systems of large numbers of particles.[68]

Albert goes further than ordinary statistical mechanics books do. On his version of the theory, the probability distribution is pushed back to the initial state of the universe. This distribution is then updated, by conditionalizing, for use at all other times. Albert argues that this is the right thing to do in the face of the reversibility objections (that is, unless a GRW dynamics is true). But note the effect of this maneuver. Given the deterministic dynamics, the initial distribution yields a probability distribution over each possible microstate of the universe at any time. In so doing, it assigns a probability to anything that supervenes on the fundamental physical state of the

---

[67] See Albert (2000, ch. 7) for further considerations in its favor, Callender (1997) for argument against, and North (2004, ch. 1) for more discussion. See Price (2002a,b), Uffink (2002) for further argument against; North (2002) is a reply to Price.

[68] In Albert's presentation, statistical mechanics comprises the following three fundamental laws: the dynamics, the statistical postulate, and the past hypothesis (2000, 96). Remove the past hypothesis to get the version of the theory presented in textbooks.

universe at any time. This means that the theory makes probabilistic predictions for every physical event in the world's history—for every fundamental physical event, and for every event that supervenes on the fundamental physical state.[69] (Indeed, it assigns probabilities to the possible microstates of a system conditional on *any* less-precisely-specified state.[70]) Since this theory makes probabilistic predictions for everything that happens in the world, everything that happens must either conform to its predictions, or else disconfirm the theory.

If this theory is right, then statistical mechanics underlies not only the future behavior of gases, as it does in ordinary textbooks, but it underlies their past behaviors as well. It also underlies all sorts of macroscopic phenomena, even such things as the fact that people tend to keep spatulas in kitchen drawers rather than in their bathtubs, to use an example of Albert's. The idea that statistical mechanics can ground these phenomena might strike you as outlandish. But it follows naturally from the past hypothesis as a solution to the reversibility objections, combined with a realism about statistical mechanics (against a view such as Leeds', below) and a physicalism according to which everything supervenes on the world's fundamental physical state.

Of course, statistical mechanics is not ordinarily used to predict things like spatula locations. Nor should it be: the calculations involved would be much too complicated. Why then think it should do so in principle? Because the evidence we have so far supports this theory (at the least, it does not contradict it; more below), and this is the theory we end up with in reply to the reversibility objections. Or if GRW is correct, then because the statistical mechanical probabilities are the fundamental quantum mechanical probabilities.

## 7.2  A more limited theory

That's an awful lot to ask of statistical mechanics. Leeds (2003) argues that it's too much. Where Albert takes the reversibility objections to motivate a reformulation of the statistical postulate, Leeds suggests that we treat statistical mechanics instrumentally. Statistical mechanics is simply a successful instrument of prediction, and for just those phenomena we have evidence that it is successful for.

Ordinary statistical mechanics takes the uniform distribution over the macrostate of a system at any time we choose to call the initial one, regardless of its past behavior. And it is successful in doing so. According to Leeds, we should follow this ordinary practice, using the standard distribution to predict things such as the future behavior of gases and the values of thermodynamic quantities at equilibrium, and leave it at that. For we have no reason to think that statistical mechanics can (or should) yield successful inferences about the past, let alone where people tend to keep their

---

[69]  We can set aside here the question of how to spell out this supervenience relation. It suffices to assume that there is such a relation.

[70]  Hence Loewer's argument (2008; 2009) that this can account for the existence of the special sciences. See Albert(ms) for further discussion.

spatulas.[71,72] A more limited version of the theory is "all we need, and also the most we are likely to get" (Leeds, 2003, 126).

In Leeds' view, we needn't worry so much about the reversibility objections. For we can simply rely on our records, explaining the fact that an ice cube was more frozen ten minutes ago by describing its macrostate half an hour ago, for instance.[73] In particular, we needn't try to correct the predictions of ordinary statistical mechanics by pushing the statistical postulate back so far as to get a theory with claims to universality. Instead, we should use the standard postulate for making future thermodynamic predictions, since it has proven its mettle in that temporal direction. We should refrain from using it for the past, given its manifest failure in that direction. Empirical evidence shows that statistical mechanics only gives us rules for making inferences about the future, not about the past.

Likewise for other macroscopic phenomena, such as where people tend to keep their spatulas. Here, it's not that we have evidence that statistical mechanics fails, but that we lack any reason to think it can succeed. We certainly don't see any positive evidence in ordinary statistical mechanics textbooks: spatulas are a far cry from the systems that statistical mechanics ordinarily talks about, such as boxes of gas. So we should refrain from using statistical mechanics to predict these things, and not try to alter the theory so that we can.

The result is a statistical mechanics that is committed to less, and is correspondingly less prone to failure. This also means that we get less out of it, however: it can't be used to predict macroscopic phenomena other than the future behavior of ordinary thermodynamic parameters. And given that we don't yet have disconfirmation of the stronger theory, and given the success of statistical mechanics for other macroscopic systems made up of the same kinds of particles,[74] and given the reasoning of Boltzmann and Gibbs, it is not so crazy to hope that it could.

Leeds (and Callender, below) suggest that statistical mechanics, properly understood, doesn't make any inferences about the past; in particular, it doesn't make false inferences that need correcting with a revised statistical postulate. But is this the right view of the theory's range of predictions? Why did we entertain the idea that it makes these predictions? The reason is the time reversal invariance and determinism

---

[71]  A similar inductive skepticism could stem from a view like that of Cartwright (1999). Cartwright argues that we have no reason to infer that the physical laws will hold of ordinary systems outside the laboratory, even granting their truth in laboratory situations we set up.

[72]  Further, if we regard the distribution as no more than a successful instrument of prediction, we can consistently apply it at arbitrary times, avoiding the inconsistency with the dynamics mentioned at the beginning of 6.1. One of Leeds' motivations is a view of the statistical mechanical probabilities as subjective and epistemic; against, for example, the view of Albert (2000, ch. 4); Loewer (2001).

[73]  Cf. Earman (2006, 421–422).

[74]  Thus Pathria (1996, 1): "Statistical mechanics is a formalism which aims at explaining the physical properties of matter *in bulk* on the basis of the dynamical behavior of its *microscopic* constituents. The scope of the formalism is almost as unlimited as the very range of the natural phenomena, for in principle it is applicable to matter in any state whatsoever. It has, in fact, been applied, with considerable success, to the study of matter in the solid state, the liquid state or the gaseous state, matter composed of several phases and/or several components", and more (italics in the original).

of the dynamics. The classical dynamics (and the dynamics of no-collapse quantum mechanics), taken by itself, does yield inferences about the past. Plug in the state of a system at any one time, and the dynamics will predict its state at any other time—without taking into account the time of the initial state, and without making a distinction between past and future temporal directions. As far as the dynamics is concerned, the prediction could hold to the past or to the future of the state we plug in. Statistical mechanics, which takes this dynamics and applies it to large systems, should be likewise temporally symmetric, leading again to the reversibility objections and the past hypothesis as a means of responding to them. Without the past hypothesis, that is, statistical mechanics will yield these inferences of past high entropy, inferences that disconfirm the theory unless we do something to prevent them.

But Leeds has pointed to a trade-off. Either we accept a more limited version of the theory and evade the reversibility objections in that way, or end up with a theory that is committed to a lot, including much that it can get wrong. Since the stronger theory would be deeper and more unifying, explaining not only individual systems' behavior, but the success of thermodynamics as a whole, and other macroscopic phenomena besides, it seems worthwhile to aim for it, unless we get evidence otherwise.

## 7.3  Special science

Another take on statistical mechanics comes from Callender (1997, 2008a). Callender argues that the thermodynamic puzzle shows that statistical mechanics is a special science rather than a fundamental theory (let alone a theory with ambitions to universality). This is because it bears the hallmark of a special science, namely, the requiring of special initial conditions—in this case, initial low entropy—for its generalizations, such as the second law of thermodynamics, to hold.

Compare this with a special science generalization such as Fisher's fundamental theorem of natural selection, which says that the rate of evolution in a population is roughly equal to the variance in fitness. This law does not always hold of real organisms (as in artificial selection by breeders). But by regarding it as a special science law with an implicit ceteris paribus clause, we understand that it is only supposed to hold given initial conditions where natural selection is the lone force at work.

Similarly here. By regarding statistical mechanics as a special science, we understand that its predictions are only supposed to hold when the requisite initial conditions are in place. Statistical mechanics simply does not hold of models of the world with high entropy pasts.

This view of statistical mechanics has its own solution to the puzzle of thermodynamics: there was no problem to begin with. Statistical mechanics, properly understood, does not make any high entropy predictions about the past. Not once the requisite boundary conditions are in place.

There are reasons to disagree with this view of statistical mechanics, however. Take the past hypothesis as a fundamental law, and statistical mechanics is a fundamental theory which rules out past high entropy, without the need for special initial conditions; the initial state is itself a law that rules out models with high entropy pasts. Take the fundamental dynamics to be GRW, and there is no problem of past high entropy, because the theory does not say anything about the past. Both these accounts avoid the conflict with thermodynamics. Yet neither one requires statistical mechanics to be a special science.

Even without those views, one might deny that statistical mechanics is a special science, for the way in which statistical mechanics requires special boundary conditions is different from the way that ordinary special sciences do. First, the initial constraints of statistical mechanics, initial low entropy and a uniform probability distribution over the microstates compatible with that state, are very simple and natural. A special science generalization like Fisher's law posits constraints that are intuitively contrived: the initial state must be such that there are no artificial breeders, for instance. Second, statistical mechanics requires an initial state for which we arguably have independent evidence from cosmology. Fisher's law approximates what happens only by ignoring intervening factors that actually occur. Third, statistical mechanics makes extremely successful predictions about the future; it is in order to get it to succeed for the past that we need the initial constraint. None of the predictions of Fisher's theorem would come out absent its initial constraints. Fourth, without the past hypothesis, it is not just that statistical mechanics makes predictions that conflict with thermodynamics. Most of our inferences about the world would fail, and fail radically. If the initial conditions required of Fisher's law did not hold, though, we wouldn't lose the same handle on our evidence about the world.

Here's a different idea. Suppose that statistical mechanics comprises the fundamental dynamical laws and a statistical postulate. (Add the past hypothesis to get Albert's version.) The second law of thermodynamics is then a consequence of statistical mechanics, not a generalization of statistical mechanics itself. In that case, it is not the generalizations of statistical mechanics that fail absent special initial conditions; it is the generalizations of thermodynamics, and even then, only when applied to the past. Thermodynamics is the special science here, not statistical mechanics.

Of course, if statistical mechanics is fundamental and thermodynamics is not, then we will want an account of the latter on the basis of the former. Callender suggests that we elude this puzzle with a different conception of statistical mechanics. But the puzzle stems from the fact that statistical mechanics applies directly to the fundamental constituents of the world: it describes macroscopic systems in virtue of their comprising fundamental particles. That's why it is so puzzling that it should fail to ground the widespread and familiar macroscopic regularities. The generalizations of thermodynamics (or of a science like evolutionary theory), on the other hand, are stated independently of the fact that systems are composed of particles (see note 11). Ordinary special science generalizations hold without reference to the fundamental physical ontology; that is part of why they are special sciences. And if GRW is true, all the more reason to think that statistical mechanics is not a special science in the way that evolutionary theory is, for statistical mechanics would be a direct consequence of the fundamental dynamics, even without the initial constraint.

# 8. OTHER TIME ASYMMETRIES AND THE DIRECTION OF TIME

Thermodynamics covers a surprisingly wide range of the time asymmetric phenomena of our ordinary experience: the spreading of gases, the cooling of cups of coffee, the melting of popsicles, and more besides. This raises a tempting prospect. Perhaps whatever explains thermodynamics can explain all of the widespread macroscopic asymmetries we are familiar with—the asymmetry of knowledge, of counterfactuals, of causation, and more.

It might seem strange to hope that it could. Less strange, once we notice that these other asymmetries, like the thermodynamic one, all involve sequences of fundamental physical states that can occur in one temporal order and not the other. This raises the same question. Where do these macroscopic asymmetries come from, if not from asymmetries in the underlying laws?

As an example, take the wave asymmetry. Waves (water waves, electromagnetic waves) behave asymmetrically in time. Waves propagate away from their sources to the future and not to the past. We see waves spread out from their sources after those sources (a rock dropped in a pond, a light switch flipped on) begin to accelerate; we don't see waves converging on sources which then begin to accelerate. The puzzle is that the physical laws governing waves are symmetric in time. This is similar to the puzzle of thermodynamics, and solutions tend to fall into one of the two same camps: posit an asymmetry in boundary conditions or in the dynamics. Thus, Frisch (2000, 2005b, 2006) argues that the wave asymmetry is an additional fundamental dynamical law, whereas Price (1996, 2006, ch. 3) argues that it stems from special initial conditions. My own view (North, 2003) is that the wave asymmetry can be explained analogously to the thermodynamic one, by means of initial low entropy, so that it is not the additional law it is for Frisch, but neither is it the same explanation as the one for Price.[75] I argue that this is a reason to prefer the account: it can explain, in one simple and unified theory, both the asymmetry of thermodynamics and the asymmetry of wave phenomena.

More generally, if any account of the thermodynamic asymmetry is able to explain other macroscopic time asymmetries, then this would be a reason to prefer it. Indeed, if one such account could give a single, unified explanation for all the pervasive time asymmetries of our experience, then that would be a huge—perhaps decisive—mark in its favor. Some accounts of thermodynamics aim to do just this.[76]

[75]  See also Arntzenius (1993); North (2004, ch. 2); Atkinson (2006). Zeh (1999) is a different initial conditions approach. See Price (2006); Earman (2010) for more on the radiation asymmetry.

[76]  This is a current area of research in philosophy. On other time asymmetries and ways of accounting for them, see, among others: Reichenbach (1999); Horwich (1987); Savitt (1995, 1996); Price (1996); Callender (1998, 2008b); Zeh (1999); Albert (2000, ch. 6); Rohrlich (2000); Elga (2001); Huggett (2002); Kutach (2002, 2007, 2010); Rovelli (2004); Frisch (2005a); Eckhardt (2006); and references therein.

One reason that thermodynamics seems so central to questions about time and our experience is that thermodynamics covers such a wide range of the everyday processes we experience. The correct theory of thermodynamics might even account for the asymmetry of records, as well as the asymmetries of knowledge and memory (see section 6.1), all of which are particularly central to our ordinary experience in time.[77]

Another is that the best theory of the thermodynamic asymmetry may be able to tell us whether time itself has a direction: whether there is an objective distinction between past and future, a distinction that is intrinsic to the nature of time itself. (That is, whether there is a temporal orientation on the space-time manifold.) We can't directly observe whether time has this structure. Nor do the phenomena, however asymmetric they appear to be, suffice to tell us this. Things are asymmetrically distributed in space, but we don't conclude from this alone that space itself is asymmetric.

We can learn about the structure of time another way: from the fundamental dynamical laws. In general, we infer a certain structure to the world from features of the dynamics. If the fundamental dynamical laws can't be formulated without referring to some structure, then we infer that this structure must exist in order to support the laws—"support" in the sense that the laws could not be formulated without it. Thus, if these laws are non-time reversal invariant, then we couldn't state them without presupposing an objective distinction between the two temporal directions: a structural difference picking out which time direction things are allowed to evolve in and which they are not. This would then give us reason to infer that time has a direction. If the laws are time reversal invariant, on the other hand, then they do not presuppose a temporal direction. They "say the same thing" regardless of which direction things are evolving in. In that case, we would not infer a direction of time.

Non-time reversal invariant, fundamental laws would thus give us reason to believe that time has a direction. However, this inference won't be conclusive. There could instead be highly non-local laws or asymmetric boundary conditions, neither of which suggest a direction of time. How do we decide?

Any inference to fundamental structure in the world, whether a direction of time or some other, must take into account the best, most fundamental physical theory. And this is where the account of thermodynamics comes in. Thermodynamic phenomena are asymmetric in time; they encompass much of our everyday experience of asymmetric processes in time. Recall the two approaches to explaining thermodynamics: posit an asymmetry in boundary conditions or in the dynamics. If the former is the best account of the thermodynamic asymmetry, then we arguably would not have reason to infer that time has a direction.[78] If the latter is the best account of thermodynamics, then we arguably would have reason to infer that time is asymmetric, ultimately responsible for the asymmetries we observe.[79] On this view, GRW, for example, posits

---

[77] Against this, see Earman (2006).

[78] But see Maudlin (2007a) for argument that this asymmetry in boundary conditions is evidence for a direction of time.

[79] A different view says that time's passage, over and above any structural asymmetry between past and future, explains our experience in time: Maudlin (2007a).

non-time reversal invariant laws that would give us a reason to infer a direction of time.[80] A no-collapse theory such as Bohmian mechanics does not. (On that theory, the asymmetry in the phenomena is not a matter of fundamental dynamical law, but the result of asymmetric boundary conditions.[81])

The best account of our ordinary macroscopic experience in time, in other words, can give us insight into the nature of time itself. All of which is to say that thermodynamics, which covers such a wide range of the ordinary phenomena of our experience—including, perhaps, the fact that we have memories of the past and not the future—is central to our explanation of, and our experience in, time.

# REFERENCES

Albert, David and Loewer, Barry (1995), Tails of Schrödinger's Cat, in *Perspectives on Quantum Reality* (ed. R. Clifton), 181–192 (Kluwer).

Albert, David Z. (1992), *Quantum Mechanics and Experience* (Harvard University Press, Cambridge, Mass).

—— (1994), The Foundations of Quantum Mechanics and the Approach to Thermodynamic Equilibrium. *British Journal for the Philosophy of Science*, **45**, 669–677.

—— (1996), 'Elementary Quantum Metaphysics', in *Bohmian Mechanics and Quantum Theory: An Appraisal* (ed. J. T. Cushing. A. Fine, and S. Goldstein), 277–284 (Kluwer Academic Publishers, Dordrecht).

—— (2000). *Time and Chance* (Harvard University Press, Cambridge, MA).

—— 'Physics and Chance' (unpublished manuscript).

Allori, V., Goldstein, S., Tumulka, R., Zanghì, N. (2008), 'On the Common Structure of Bohmian Mechanics and the Ghirardi–Rimini–Weber Theory', *British Journal for the Philosophy of Science*, **59**(3), 353–389.

Arntzenius, Frank (1993), 'The Classical Failure to Account for Electromagnetic Arrows of Time', in *Scientific Failure* (ed. T. Horowitz and A. Janis), 29–48 (Rowman & Littlefield Savage, Maryland).

—— (1995), 'Indeterminism and the Direction of Time'. *Topoi*, **14**, 67–81.

—— (1997a), 'Mirrors and the Direction of Time', *Philosophy of Science (Proceedings)*, **64**, S213–S222.

—— (1997b), 'Transition Chances and Causation', *Pacific Philosophical Quarterly*, **78**, 144–168.

—— (2000), 'Are There Really Instantaneous Velocities?', *The Monist*, **83**, 187–208.

—— (2003), 'An Arbitrarily Short Reply to Sheldon Smith on Instantaneous Velocities', *Studies in History and Philosophy of Modern Physics*, **34B**, 281–282.

—— (2004), 'Time Reversal Operations, Representations of the Lorentz Group, and the Direction of Time', *Studies in History and Philosophy of Modem Physics*, **35**, 31–43.

—— (2010), 'The CPT Theorem', in *Oxford Handbook on Time*. (Oxford University Press, Oxford).

—— and Greaves, Hilary (2009), 'Time Reversal in Classical Electromagnetisn', *British Journal for the Philosophy of Science*, **60**(3), 557–584.

---

[80]  Price disagrees: see note 61.

[81]  See Arntzenius (1995). Which theories of quantum mechanics do, and which do not, indicate time's asymmetry is debatable: see Callender (2000).

Atkinson, David (2006), 'Does Quantum Electrodynamics Have an Arrow of Time?', *Studies in History and Philosophy of Modern Physics*, **37**, 528–541.

Bacciagaluppi, Guido (2007), 'Probability, Arrow of Time, and Decoherence', *Studies in History and Philosophy of Modern Physics*, **38**, 439–456.

Balian, Roger (2005), 'Information in Statistical Physics', *Studies in History and Philosophy of Modern Physics*, **36**, 323–353.

Batterman, Robert W. (2005), 'Critical Phenomena and Breaking Drops: Infinite Idealizations in Physics', *Studies in History and Philosophy of Modern Physics*, **36**, 225–244.

—— (2010), 'Reductions and Renormalization', in *Time, Chance and Reduction: Philosophical Aspects of Statistical Mechanics*, (ed. Gerhard Ernst and Andreas Hüttemann). (Cambridge University Press, Cambridge), 159–179.

Berkowitz, Joseph, Frigg, Roman, and Kronz, Fred (2006), 'The Ergodic Hierarchy, Randomness and Hamiltonian Chaos', *Studies in History and Philosophy of Modern Physics*, **37**, 661–691.

Boltzmann, Ludwig (1964), *Lectures on Gas Theory* (University of California Press, Berkeley), tr. S. G. Brush. Originally published in German in 1896.

Bricmont, Jean (1995), 'Science of Chaos or Chaos in Science?', *Physicalia*, **17**, 159–208.

Brush, Stephen G. (1975), *The Kind of Motion We Call Heat* (North-Holland, Amsterdam).

Bub, Jeffrey (2001), 'Maxwell's Demon and the Thermodynamics of Computation', *Studies in History and Philosophy of Modern Physics*, **32**, 569–579.

Callender, Craig (1995), 'The Metaphysics of Time Reversal: Hutchison on Classical Mechanics', *British Journal for the Philosophy of Science*, **46**, 331–340.

—— (1997), 'What is "the Problem of the Direction of Time"?', *Philosophy of Science (Proceedings)*, **2**, S223–S234.

—— (1998), 'Review: The view from no-when', *British Journal for the Philosophy of Science*, **49**, 135–159.

—— (1999), 'Reducing Thermodynamics to Statistical Mechanics: The Case of Entropy', *Journal of Philosophy*, **96**, 348–373.

—— (2000), 'Is Time "Handed" in a Quantum World?', *Proceedings of the Aristotelian Society*, 247–269.

—— (2001), 'Taking Thermodynamics (Too) Seriously', *Studies in History and Philosophy of Modern Physics*, **32**, 539–553.

—— (2002), 'Who's Afraid of Maxwell's Demon—and Which One?', in *Quantum Limits to the Second Law* (ed. D. Sheehan), 399–407, American Institute of Physics. Reprinted as "A Collision Between Dynamics and Thermodynamics" in *Entropy* 6 (2004), 11–20.

—— (2004a), 'Measures, Explanation and the Past: Should "Special" Initial Conditions he Explained?', *British Journal for the Philosophy of Science*, **55**, 195–217.

—— (2004b), 'There is No Puzzle about the Low-Entropy Past', in *Contemporary Debates in Philosophy of Science* (ed. C. Hitchcock), 240–255 (Blackwell, Oxford).

—— (2008a), 'The Past Histories of Molecules', in *Probabilities in Physics* (eds Claus Beisbart and Stephen Hatmann), Oxford University Press, Oxford, forthcoming.

—— (2008b), 'Thermodynamic Asymmetry in Time', Stanford Encyclopedia of Philosophy.

—— (2010), 'The Past Hypothesis Meets Gravity', in *Time, Chance, and Reduction* (ed. G. Ernst and A. Hüttemann), 34–58 (Cambridge University Press, Cambridge).

Campisi, Michele (2005), 'On the Mechanical Foundations of Thermodynamics: The Generalized Helmholtz Theorem', *Studies in History and Philosophy of Modern Physics*, **36**, 275–290.

—— (2008), 'Statistical Mechanical Proof of the Second Law of Thermodynamics Based on Volume Entropy', *Studies in History and Philosophy of Modern Physics*, **39**, 181–194.

Carroll, Sean M. (2008), 'Does Time Run Backward in Other Universes?', *Scientific American*, June, 48–57.

—— (2010), *From Eternity to here: The Quest for the Ultimate Theory of Time*. Penguin, New York.

—— and Chen, J. (2004), 'Spontaneous Inflation and the Origin of the Arrow of Time', <arXiv.org>hep-th>arXiv.hep-th/0410270>

———— (2005), 'Does Inflation Provide Natural Initial Conditions for the Universe?, <arXiv.org>gr.qc>arXiv.gr.qc/050503>

Cartwright, Nancy (1999), *The Dappled World: A Study of the Boundaries of Science* (Cambridge University Press, Cambridge).

Dürr, Detlef, Goldstein, Sheldon, and Zanghì, Nino (1992a), 'Quantum Chaos, Classical Randomness, and Bohmian mechanics', *Journal of Statistical Physics*, **68**, 259–270.

———— (1992b), 'Quantum Equilibrium and the Origin of Absolute Uncertainty', *Journal of Statistical Physics*, **67**, 843–908.

Earman, John (1974), 'An Attempt to Add a Little Direction to the Problem of "the Direction of Time"', *Philosophy of Science*, **41**, 15–47.

—— (1981), 'Combining Statistical-Thermodynamics and Relativity Theory: Methodological and Foundations Problems', in *Proceedings of the 1978 Biennial Meeting of the Philosophy of Science Association* (ed. P. Asquith and I. Hacking), Volume 2, 157–185.

—— (1986), *A Primer on Determinism*, Volume 32 (Reidel, University of Western Ontario Series in the Philosophy of Science).

—— (2002), 'What Time Reversal Invariance is and Why It Matters', *International Studies in the Philosophy of Science*, **16**, 245–264.

—— (2006), 'The "Past Hypothesis": Not Even False', *Studies in History and Philosophy of Modern, Physics*, **37**, 399–430.

—— (2010), 'The Electromagnetic Arrow of Time', in *Oxford Handbook on Time* (Oxford University Press, Oxford).

—— and Norton, John (1998), 'Exorcist xiv: The Wrath of Maxwell's Demon. Part i. From Maxwell to Szillard', *Studies in History and Philosophy of Modern Physics*, **29**, 435–471.

———— (1999), 'Exorcist xiv: The Wrath of Maxwell's Demon. Part ii. From Szilard to Landauer and Beyond', *Studies in History and Philosophy of Modern Physics*, **30**, 1–40.

—— and Rédei, Miklós (1996), 'Why Ergodic Theory Does Not Explain the Success of Equilibrium Statistical Mechanics', *British Journal for the Philosophy of Science*, **47**, 63–78.

Eckhardt, William (2006), 'Causal Time Asymmetry', *Studies in History and Philosophy of Modern Physics*, **37**, 439–466.

Ehrenfest, Paul and Ehrenfest, Tatiana (2002), *The Conceptual Foundations of the Statistical Approach in Mechanics* (Dover, New York). Translated by Michael J. Moravcsik. First edition in English published in 1959 by Cornell University Press. Originally published in 1912.

Elga, Adam (2001), 'Statistical Mechanics and the Asymmetry of Counterfactual Dependence', *Philosophy of Science (Proceedings)*, **68**, S313–S324.

—— (2007), 'Isolation and Folk Physics', in *Causation, Physics, and the, Constitution of Reality: Russell's Republic Revisited* (ed. H. Price and R. Corry), 106–119 (Oxford University Press, Oxford).

Ellis, George F. R. (2007), 'Issues in the Philosophy of Cosmology', in *Philosophy of Physics, Part B* (ed. J. Butterfield and J. Earman), 1183–1283 (North-Holland, Amsterdam).

Emch, Gérard (2007), 'Quantum Statistical Physics', in *Philosophy of Physics, Part B* (ed. J. Butterfield and J. Earman), 1075–1182 (North-Holland).

Fermi, Enrico (1956), *Thermodynamics* (Dover, New York). Originally published in 1937.

Feynman, Richard (1965), *The Character of Physical Law* (MIT Press, Cambridge, Mass).

Friedman, K. S. (1976), 'A Partial Vindication of Ergodic Theory', *Philosophy of Science*, **43**, 151–162.

Frigg, Roman (2008a), 'Chance in Boltzmannian Statistical Mechanics', *Philosophy of Science. (Proceedings)*, **75**(5), 670–681.

——(2008b), 'A Field Guide to Recent Work on the Foundations of Statistical Mechanics', in *The Ashgate Companion to Contemporary Philosophy of Physics* (ed. D. Rickles), 99–196 (Ashgate, London).

——(2010), 'Probability in Boltzmannian Statistical Mechanics', in *Time, Chance and Reduction: Philosophical Aspects of Statistical Mechanics* (ed. G. Ernst and A. Hüttemann), 92–118 (Cambridge University Press).

Frisch, Mathias (2000), '(Dis-)solving the Puzzle of the Arrow of Radiation', *British Journal for the Philosophy of Science*, **51**, 381–410.

——(2005a), Counterfactuals and the Past Hypothesis', *Philosophy of Science (Proceedings)*, **72**, 739–750.

——(2005b), *Inconsistency, Asymmetry, and Non-Locality: A Philosophical Investigation of Classical Electrodynamics* (Oxford University Press, Oxford).

——(2006), 'A Tale of Two Arrows', *Studies in History and Philosophy of Modern Physics*, **37**, 542–558.

——(2007), 'Causation, Counterfactuals, and Entropy', in *Causation, Physics, and the Constitution of Reality: Russell's Republic, Revisited* (ed. H. Price and R. Corry), 351–395 (Oxford University Press, Oxford).

——(2010), 'Does the Low-Entropy Constraint Prevent us from Influencing the Past? in *Time, Chance and Reduction: Philosophical Aspects of Statistical Mechanics*, ed Gerhard Ernst and Andreas Hüttemann (Cambridge University Press, Cambridge), 13–33.

Ghirardi, G. C., Rimini, A., and Weber, T. (1985), 'A Model for a Unified Quantum Description of Macroscopic and Microscopic Systems', in *Quantum Probability and Applications* (ed. L. Accardi) (Springer, Berlin).

——————(1986), 'Unified Dynamics for Microscopic and Macroscopic Systems', *Physical Preview*, D 34, 470.

Gibbs, W. (1902), *Elementary Principles of Statistical Mechanics* (Yale University Press, New Haven).

Goldstein, Sheldon (2001), 'Boltzmann's Approach to Statistical Mechanics', in *Chance in Physics: Foundations and Perspectives* (ed. J. Bricmont), 39–54 (Springer-Verlag). Available at arXiv:cond-mat/O105242vl.

——and Lebowitz, Joel L. (2004), 'On the (Boltzmann) Entropy of Non-Equilibrium Systems', *Physica D*, **193**, 53–66.

Greene, Brian (2004), *The Fabric of the Cosmos: Space, Time, and the Texture of Reality* (Knopf, New York).

Hagar, Amit (2005), 'The Foundations of Statistical Mechanics—Questions and Answers', *Philosophy of Science (Proceedings)*, **72**, 468–478.

Hellman, Geoffrey (1999), 'Reduction to *What?*', *Philosophical Studies*, 95, 203–214.

Hemmo, Meir (2003), 'Remarks on the Direction of Time in Quantum Mechanics', *Philosophy of Science (Proceedings)*, **70**, 1458–1471.

——and Shenker, Orly (2001), 'Can we Explain Thermodynamics by Quantum Decoherence?', *Studies in History and Philosophy of Modern Physics*, **32**, 555–568.

————(2003), 'Quantum Decohernce and the Approach to Equilibrium i', *Philosophy of Science*, **70**, 330–358.

————(2005), 'Quantum Decoherence and the Approach to Equilibrium ii', *Studies in History and Philosophy of Modern Physics*, **36**, 626–648.

————(2006), 'Von Neumann's Entropy Does Not Correspond to Thermodynamic Entropy', *Philosophy of Science*, **73**, 153–174.

Horwich, Paul (1987), *Asymmetries in Time: Problems in the Philosophy of Science* (MIT Press, Cambridge, Mass).

Huggett, Nick (2002), 'Review of David Albert', *Time and Chance Notre Dame Philosophical Reviews*.

Hutchison, Keith (1993), 'Is Classical Mechanics Really Time-Reversible and Deterministic?, *British Journal for the Philosophy of Science*, **44**, 307–323.

——(1995), 'Differing Criteria for Temporal Symmetry', *British Journal for the Philosophy of Science*, **46**, 341–347.

Jaynes, Edwin T. (1983), *Papers on Probability, Statistics, and Statistical Physics* (D. Reidel, Dordrecht, Holland).

Khinchin, A. I. (1949), *Mathematical Foundations of Statistical Mechanics* (Dover. New York). First English translation. Tr. by G. Gamow.

Kutach, Douglas N. (2002) 'The Entropy Theory of Counterfactuals', *Philosophy of Science*, **69**, 82–104.

——(2007), 'The Physical Foundations of Causation', in *Causation, Physics, and the Constitution of Reality: Russell's Republic Revisited* (ed. H. Price and R. Corry), 327–350 (Clarendon Press, Oxford).

——(2010), 'The Asymmetry of Influence', Ch.8 this volume.

Ladyman, James, Presnell Stuart, and Short, Anthony J. (2008), 'The Use of the Information-Theoretic Entropy in Thermodynamics', *Studies in History and Philosophy of Modern Physics*, **39**, 315–324.

——————and Groisman, Berry (2007), 'The Connection between Logical and Thermodynamic Irreversibility', *Studies in History and Philosophy of Modern Physics*, **38**, 58–79.

Landau, L. D. and Lifshitz, E. M. (1980), *Statistical Physics* (3rd edn, part 1) (Butterworth-Heineinann, Oxford).

Lavis, David A. (2005), 'Boltzman and Gibbs: An Attempted Reconciliation', *Studies in History and Philosophy of Modern Physics*, **36**, 245–273.

——(2008), 'Boltzmann, Gibbs, and the Concept of Equilibrium', *Philosophy of Science (Proceedings)*, **75**(5), 682–696.

Lebowitz, Joel L. (1993a, September), 'Boltzmann's Entropy and Time's Arrow', *Physics Today*, 32–38.

——(1993b), 'Macroscopic Laws, Microscopic Dynamics, Time's Arrow and Boltzmann's Entropy', *Physica A*, **194**, 1–27.

——(1993c), 'Microscopic Reversibility and Macroscopic Behavior: Physical Explanations and Mathematical Derivations', *Physics Today*, **46**, 1–20.

——(1999a), 'Microscopic Origins of Irreversible Macroscopic Behavior', *Physica A*, **263**, 516–527.

——(1999b), 'Statistical Mechanics: A Selective Review of Two Central Issues', *Reviews of Modern Physics*, **71**, 346–357.

Leeds, Stephen (1989), 'Malament and Zabell on Gibbs Phase Averaging', *Philosophy of Science*, **56**, 325–340.

Leeds, Stephen (2003), 'Foundations of Statistical Mechanics—Two Approaches', *Philosophy of Science*, **70**, 126–144.

Lewis, David (1979), 'Counterfactual Dependence and Time's Arrow', *Noûs*, **13**, 455–476.

Lieb, E. H. and Yngvason, J. (2000), 'A Fresh Look at Entropy and the Second Law of Thermodynamics', *Physics Today*, **53/4**, 32–37.

Liu, Chang (1994), 'Is There a Relativistic Thermodynamics? A Case Study of the Meaning of Special Relativity', *Studies in History and Philosophy of Modern Physics*, **25**, 983–1004.

—— (2001), 'Infinite Systems in SM Explanations: Thermodynamic Limit, Renormalization (Semi-) Groups, and Irreversibility', *Philosophy of Science (Proceedings)*, **68**, S325–S344.

Loewer, Barry (2001), 'Determinism and Chance', *Studies in History and Philosophy of Modern Physics*, **32**, 609–620.

—— (2004), 'David Lewis's Humean Theory of Objective Chance', *Philosophy of Science (Proceedings)*, **71**, 1115–1125.

—— (2007), 'Counterfactuals and the Second Law', in *Causation, Physics, and the Constitution of Reality: Russell's Republic Revisited* (ed. H. Price and R. Cony), 293–326 (Oxford University Press, Oxford).

—— (2008), 'Why There *Is* Anything Except Physics', in *Being Reduced: New Essays on Reduction, Explanation, and Causation* (ed. J. Hohwy and J. Kallestrup), 149–163 (Oxford University Press, Oxford).

—— (2009), 'Why Is There Anything Except Physics?', *Synthese*, **170**(2), 217–233.

Malament, David (2004), 'On the Time Reversal Invariance of Classical Electromagnetic Theory', *Studies in History and Philosophy of Modern Physics*, **35**, 295–315.

—— (2008), 'Norton's Slippery Slope', *Philosophy of Science (Proceedings)*, **75**(5), 799–816.

—— and Zabell, Sandy L. (1980), 'Why Gibbs Phase Averages Work—The Role of Ergodic Theory', *Philosophy of Science*, **47**, 339–349.

Maroney, O. J. E. (2005), 'The (Absence of a) Relationship Between Thermodynamic and Logical Reversibility', *Studies in History and Philosophy of Modern Physics*, **36**, 355–374.

Maudlin, Tim (1995), 'Review of Lawrence Sklar's *Physics and Chance* and *Philosophy of Physics*', *British Journal for the Philosophy of Science*, **46**, 145–149.

—— (2007a), On the Passing of Time', in *The Metaphysics Within Physics*, 104–142 (Oxford University Press, Oxford).

—— (2007b). 'What Could be Objective About Probabilities?', *Studies in History and Philosophy of Modern Physics*, **38**, 275–291.

North, Jill (2002), 'What is the Problem about the Time-Asymmetry of Thermodynamics?— A reply to Price', *British Journal for the Philosophy of Science*, **53**, 121–136.

—— (2003), 'Understanding the Time-Asymmetry of Radiation', *Philosophy of Science (Proceedings)*, **70**, 1086–1097.

—— (2004), *Time and Probability in an Asymmetric Universe*, Ph.D. thesis, Rutgers. New Jersey.

—— (2008), 'Two Views on Time Reversal', *Philosophy of Science*, **75**(2), 201–223.

—— (2010), 'An Empirical Approach to Symmetry and Probability', *Studies in History and Philosophy of Modern Physics*, **41**, 27–40.

Norton, John (2005), 'Eaters of the Lotus: Landauer's Principle and the Return of Maxwell's Demon', *Studies in History and Philosophy of Modern Physics*, **36**, 375–411.

—— (2008), 'The Dome: An Unexpectedly Simple Failure of Determinism', *Philosophy of Science (Proceedings)*, **75**(5), 786–798.

Parker, Daniel (2005), 'Thermodynamic Irreversibility: Does the Big Bang Explain What it Purports to Explain?', *Philosophy of Science (Proceedings)*, **72**, 751–763.

Pathria, R. K, (1996), *Statistical Mechanics* (2nd edn) (Butterworth Heinemann, Oxford). First edition published in 1972.

Penrose, Oliver (1979), 'Foundations of Statistical Mechanics', *Reports on Progress in Physics*, **42**(12), 1937–2006.

—— (2005a), *Foundations of Statistical Mechanics: A Deductive Treatment* (Dover, New York). First edition published in 1970.

Penrose, Roger (1989), *The Emperor's New Mind* (Oxford University Press, Oxford).

—— (2005b), *The Road to Reality* (Knopf, New York).

Pitowsky, Itamar (2006), 'On the Definition of Equilibrium', *Studies in History and Philosophy of Modern Physics*, **37**, 431–438.

Price, Huw (1996), *Time's Arrow and Archimedes' Point: New Directions for the Physics of Time* (Oxford University Press, Oxford).

—— (2002a), 'Boltzmann's Time Bomb', *British Journal for the Philosophy of Science*, **53**, 83–119.

—— (2002b), 'Burbury's Last Case: The Mystery of the Entropic Arrow', in *Time, Reality and Experience* (ed. C. Callender), 19–56 (Cambridge University Press).

—— (2004), 'On the Origins of the Arrow of Time: Why There is Still a Puzzle about the Low-Entropy Past', in *Contemporary Debates in Philosophy of Science* (ed. C. Hitchcock), 219–239 (Blackwell, Oxford).

—— (2006), 'Recent Work on the Arrow of Radiation', *Studies in History and Philosophy of Modern Physics*, **37**, 498–527.

Prigogine, Ilya (1961), *Introduction to the Thermodynamics of Irreversible, Processes* (2nd edn) (Interscience, New York).

Reichenbach, Hans (1999), *The Direction of Time* (Dover, New York). First edition published in 1956 by University of California Press, Berkeley, CA.

Rohrlich, Fritz (2000), 'Causality and the Arrow of Classical Time', *Studies in History and Philosophy of Modern Physics*, **31**, 1–13.

Rovelli, Carlo (2004), 'Comment on: "Causality and the Arrow of Classical Time", by Fritz Rohrlich', *Studies in History and Philosophy of Modern Physics*, **35**, 397–405.

Sachs, Robert G. (1987), *The Physics of Time Reversal* (University of Chicago Press, Chicago).

Savitt, Steven (1994), 'Is Classical Mechanics Time Reversal Invariant?', *British Journal for the Philosophy of Science*, **45**, 907–913.

—— (ed.) (1995), *Time's Arrows Today: Recent Physical and Philosophical, Work on the Direction of Time* Cambridge University Press, Cambridge.

—— (1996), 'The Direction of Time', *British Journal for the Philosophy of Science*, **47**, 347–370.

Sklar, Lawrence (1973), 'Statistical Explanation and Ergodic Theory', *Philosophy of Science*, **40**, 194–212.

—— (1993), *Physics and Chance: Philosophical Issues in the Foundations of Statistical Mechanics* (Cambridge University Press, Cambridge).

—— (2000), 'Interpreting Theories: The Case of Statistical Mechanics', *British Journal for the Philosophy of Science*, **51**, 729–742.

—— (2001), 'Philosophy of Statistical Mechanics', *Stanford Encyclopedia of Philosophy*.

—— (2007), 'Why Does the Standard Measure Work in Statistical Mechanics?', in *Interactions: Mathematics, Physics and Philosophy, 1860–1930 (Boston Studies in the Philosophy of Science)* (ed. V. F. Hendricks, K. F. Jørgensen, J. Lützen, and S. A. Pedersen), 307–320 (Springer, Dordrecht, Holland).

Smith, Sheldon (2003), 'Are Instantaneous Velocities Real and Really Instantaneous? An Argument for the Affirmative', *Studies in History and Philosophy of Modern Physics*, **34**, 261–280.

Strevens, Michael (1998), 'Inferring Probabilities from Symmetries', Noûs, **32**, 231–246.

—— (2003), *Bigger Than Chaos: Understanding Complexity through Probability* (Harvard University Press, Cambridge, Mass).

—— (2005), 'How Are the Sciences of Complex Systems Possible?', *Philosophy of Science*, **72**, 531–556.

Tolman, Richard C. (1979), *The Principles of Statistical Mechanics* (Dover, New York). Originally printed in 1938 by Oxford University Press.

Torretti, Roberto (2007), 'The Problem of Time's Arrow Historico-critically Reexamined', *Studies in History and Philosophy of Modern Physics*, **38**, 732–756.

Uffink, Jos (1996), 'Nought but Molecules in Motion', *Studies in History and Philosophy of Modern Physics*, **27**(3), 373–387.

—— (2001), 'Bluff your Way through the Second Law of Thermodynamics', *Studies in History and Philosophy of Modern Physics*, **32**, 305–394.

—— (2002), 'Essay Review: *Time, and Chance. Studies in History and Philosophy of Modern Physics*', **33**, 555–563.

—— (2004), 'Boltzmann's Work in Statistical Physics', *Stanford Encyclopedia of Philosophy*. Available at http://plato.stanford.edu/entries/statphys-Boltzmann/.

—— (2007), 'Compendium of the Foundations of Classical Statistical Physics', in *Philosophy of Physics, Part B* (ed. J. Butterfield and J. Earman), pp. 923–1074. North-Holland.

—— (2010), 'Irreversibility in Stochastic Dynamics', in *Time, Chance and Reduction: Philosophical Apects of Statistical Mechanics*, eds. Gerhard Ernst and Andreas Hüttemann, (Cambridge University Press, Cambridge), 180–207.

van Lith, Janneke (2001), 'Ergodic Theory, Interpretations of Probability and the Foundations of Statistical Mechanics', *Studies in History and Philosophy of Modern Physics*, **32**, 581–594.

Vranas, Peter B. M. (1998), 'Epsilon-Ergodicity and the Success of Equilibrium Statistical Mechanics', *Philosophy of Science*, **65**, 688–708.

Wald, Robert M. (2006), 'The Arrow of Time and the Initial Conditions of the Universe', *Studies in History and Philosophy of Modern Physics*, **37**, 394–398.

Wallace, David (2002), 'Implications of Quantum Theory in the Foundations of Statistical Mechanics', Available at http://philsci-archive.pitt.edu/.

—— (2010), *From Eternity to Here: The Quest for the Ultimate Theory of Time*. (Penguin, New York).

Weinstein, Steven (2003), 'Objectivity, Information, and Maxwell's Demon', *Philosophy of Science (Proceedings)*, **70**, 1245–1255.

Winsberg, Eric (2004a), 'Can Conditioning on the "Past Hypothesis" Militate Against the Reversibility Objections?', *Philosophy of Science (Proceedings)*, **71**, 489–504.

—— (2004b), 'Laws and Statistical Mechanics', *Philosophy of Science (Proceedings)*, **71** (707–718).

—— (2008), 'Laws and Chances in Statistical Mechanics', *Studies in History and Philosophy of Modern Physics*, **39**, 872–888.

Yi, Sang Wook (2003), 'Reduction of Thermodynamics: A Few Problems', *Philosophy of Science (Proceedings)*, **70**, 1028–1038.

Zeh, H. D. (1999), *The Physical Basis of the Direction of Time*, ($3^{rd}$ edn) (Springer, Berlin).

# PART III

# TIME, ETHICS, AND EXPERIENCE

CHAPTER 11

························································································

# PROSPECTS FOR TEMPORAL NEUTRALITY[1]

························································································

## DAVID O. BRINK

WE often assess actions and policies at least in part by how they distribute goods and harms across different people's lives. For example, utilitarians favor distributions that maximize welfare, egalitarians endorse equal distributions, and friends of maximin favor distributions that are to the greatest advantage of the worst off. In parallel fashion, we might assess actions and policies in part by how they distribute goods and harms across time. Intertemporal distribution has not been as extensively studied as interpersonal distribution. Whereas there are many competing conceptions of interpersonal distributive justice, there are not so many competing conceptions of intertemporal distribution. This may be in part because one view about intertemporal distribution has seemed uniquely plausible to many people. This traditional conception of intertemporal distribution is the demand of *temporal neutrality*, which requires that agents attach no normative significance per se to the temporal location of benefits and harms within someone's life and demands equal concern for all parts of that person's life. For example, this kind of temporal neutrality is reflected in the demands of prudence to undergo short-term sacrifice for the sake of a later, greater good, as when it requires us to undertake routine but inconvenient and unpleasant preventive dental care. Indeed, as we shall see, some have claimed that temporal neutrality is an essential part of rationality.

Despite its hegemony, temporal neutrality deserves philosophical scrutiny. We need to know what exactly temporal neutrality requires and why we should care about its dictates. Even if we can locate a rationale for temporal neutrality, it has several apparently controversial or counter-intuitive normative implications about our attitudes

[1] This chapter draws on but significantly extends the discussion in Brink (1997a) and (2003). Thanks to Craig Callender for encouraging me to write this chapter and to Theron Pummer for thoughtful comments on the penultimate version of this chapter.

toward the temporal location of goods and harms that must be addressed as part of any systematic assessment.

# 1. PRUDENCE AND TEMPORAL NEUTRALITY

Prudence demands that an agent act so as to promote her own overall good. It is usually understood to require an equal concern for all parts of her life. But one can also have an equal concern for all parts of the lives of others. So, while prudence requires temporal neutrality, temporal neutrality is not limited to prudence. Nonetheless in discussing temporal neutrality, I think it will often help to focus on the special case of temporal neutrality within the agent's own life, which prudence demands.

Consider Adam Smith's claims in *The Theory of Moral Sentiments* (1790), linking prudence and temporal neutrality with the approval of the impartial spectator.

> [I]n his steadily sacrificing the ease and enjoyment of the present moment for the probable expectation of the still greater ease and enjoyment of a more distant but more lasting period of time, the prudent man is always both supported and rewarded by the entire approbation of the impartial spectator, and of the representative of the impartial spectator, the man within the breast. The impartial spectator does not feel himself worn out by the present labour of those whose conduct he surveys; nor does he feel himself solicited by the importunate calls of their present appetites. To him their present, and what is likely to be their future situation, are very nearly the same: he sees them nearly at the same distance, and is affected by them very nearly in the same manner. He knows, however, that to the persons principally concerned, they are very different from being the same, and that they naturally affect *them* in a very different manner. He cannot therefore but approve, and even applaud, that proper exertion of self-command, which enables them to act as if their present and their future situation affected them nearly in the same manner in which they affect him [VI.i.11].

As Smith's appeal to an impartial spectator suggests, the demand for temporal neutrality need not be confined to a prudential concern with one's own well-being but can extend to concern for the well-being of *others*. This is why temporal neutrality is often an aspect, explicit or implicit, in conceptions of impartiality and benevolence, as well as prudence. Also, as Smith makes clear, he conceives of temporal neutrality as a *normative* requirement, not as a description of how people actually reason and behave. As Smith notes, it is an all too familiar fact that people are often temporally *biased*, investing short-term benefits and sacrifices with normative significance out of proportion to their actual magnitude and discounting distant benefits and harms out of proportion to their actual magnitude. This sort of temporal bias is sometimes thought to play a major role in various familiar human failings, such as weakness of will, self-deception, and moral weakness.[2] But it is almost always regarded as a mistake, typically a failure of rationality.

In *The Methods of Ethics* (1907) Henry Sidgwick recognizes the normative aspect of temporal neutrality in criticizing Jeremy Bentham for assigning normative significance to the temporal proximity of pleasures and pains.

> [P]roximity is a property [of pleasures and pains] which it is reasonable to disregard except in so far as it diminishes uncertainty. For my feelings a year hence should be just as important to me as my feelings next minute, if only I could make an equally sure forecast of them. Indeed this equal and impartial concern for all parts of one's conscious life is perhaps the most prominent element in the common notion of the *rational*—as opposed to the merely *impulsive*—pursuit of pleasure [124n; cf. 111].

Later, he elaborates on the demands of temporal neutrality and notes that it has broader application than its role in his own version of hedonistic egoism.

> Hereafter *as such* is to be regarded neither less nor more than Now. It is not, of course, meant, that the good of the present may not reasonably be preferred to that of the future on account of its greater certainty: or again, that a week ten years hence may not be more important to us than a week now, through an increase in our means or capacities of happiness. All that the principle affirms is that the mere difference of priority and posteriority in time is not a reasonable ground for having more regard to the consciousness of one moment than to that of another. The form in which it practically presents itself to most men is 'that a smaller present good is not to be preferred to a greater future good' (allowing for differences of certainty).... The commonest view of the principle would no doubt be that the present *pleasure* or *happiness* is reasonably to be foregone with the view of obtaining greater pleasure or happiness hereafter; but the principle need not be restricted to a hedonistic application, it is equally applicable to any other interpretation of 'one's own good', in which good is conceived as a mathematical whole, of which the integrant parts are realised in different parts or moments of a lifetime [381].

There are several aspects of Sidgwick's account of prudence and temporal neutrality that deserve discussion.

First, Sidgwick recognizes here that prudence's temporal neutrality is a structural constraint about the distribution of goods and harms over time within a single life. As such, it is neutral or agnostic about the content of the good. Though all conceptions of prudence are temporally neutral, different conceptions result from different conceptions of the good. Sidgwick's own conception of the good is hedonistic. Alternatively, one might understand the good in preference-satisfaction terms, as consisting in the satisfaction of actual or suitably informed or idealized desire. Hedonism and preference-satisfaction views construe the good as consisting in or

---

[2] Temporal bias plays an important role in Socratic and Aristotelian discussions of weakness of will. Compare Plato's *Protagoras* (356a-357e) and Aristotle's *Nicomachean Ethics* vii 2–10. The significance of temporal bias or discounting is explored in Ainslie (1992) and (2001).

depending upon an individual's contingent and variable psychological states. By contrast, one might understand the good in more objective terms, either as consisting in the perfection of one's essential capacities (e.g. one's rational or deliberative capacities) or as consisting in some list of disparate objective goods (e.g. knowledge, beauty, achievement, friendship).

Second, just as Sidgwick makes clear that temporal neutrality is not limited to hedonistic conceptions of prudence, so too we can notice that it is not limited to prudence. As Smith recognizes, temporal neutrality can be applied to concern for another, as well as oneself. So, for example, the two methods of ethics that form Sidgwick's dualism of practical reason—egoism and utilitarianism—are equally temporally neutral.

Third, Sidgwick is careful to claim that temporal neutrality insists only that the temporal location of goods and harms within a life has no *intrinsic* or *independent* significance. Prudence is intrinsically concerned with the magnitude of goods and harms, but not their temporal location. Temporal location can inherit significance when it is correlated with factors that do affect the magnitude of goods and harms. So if at some future point in time I will, for whatever reason, become a more efficient converter of resources into happiness or well-being, however that is conceived, then a neutral concern with all parts of my life will in one sense require giving greater weight to that part of my life. Perhaps, in the "prime of life" I have greater opportunities or capacities for happiness. If so, temporal neutrality will justify devoting greater resources to the prime of life. However, this is not a pure time preference for that future period over, say, the present, precisely because the same resources yield goods of different magnitudes in the present and the future. The rationality of this sort of discounting is an application of, not a departure from, temporal neutrality.

Furthermore, we may be differentially epistemically situated with respect to different points in time, and this will affect what temporal neutrality requires. Relative to events in the near future, events in the further future depend on more intervening events and are typically harder to predict and less certain. The most obvious case of this sort is the certainty or predictability of my continued existence. It is less certain or predictable that I will exist the further into the future I project. The probability that I will exist in 2030 is lower than the probability that I will exist in 2020. Presumably, rational planning can and should take this kind of uncertainty into account by discounting the significance of a future good or harm by its improbability. But, again, this seems to be an application of, rather than a departure from, temporal neutrality. Insofar as near and distant goods and harms are equally certain, I should have equal concern for them.

Another way to make this point is in terms of the important distinction, which Sidgwick draws, between *objective* and *subjective* reasons (1907: 207–08, 394–95). Claims of objective rationality are claims about what an agent has reason to do given the facts of the situation, whether he is aware of these facts or in a position to recognize the reasons that they support. Claims of subjective rationality are claims about what the agent has reason to do given his beliefs about his situation or what it would be reasonable for him to believe about his situation. Actions that are objectively rational can be subjectively irrational, and vice versa. Prudence can admit that the existence of my near future is more certain than the existence of my distant future and that this epistemic fact should affect what it is subjectively rational for me to do; it claims only that insofar as I have both present and future interests, they provide me with equally strong objective reasons for action.

This point reflects the fact that prudence is, at least in the first instance, a theory about an agent's objective reasons. This focus on objective reasons is worth elaborating. Subjective reasons are normatively important. In particular, it is common for those who make the distinction to think that we should tie praise and blame to subjective, rather than objective, reasons insofar as an agent's subjective reasons are accessible to her in a way that her objective reasons may not be. Insofar as praise and blame are constrained by what is within the agent's power to recognize and do, we have reason to tie praise and blame to an agent's conformity with her subjective reasons. But we can and should still recognize objective reasons. Objective reasons are independent of subjective reasons, as is reflected in the perspective of second-person and third-person evaluators, who distinguish between what was reasonable to do *tout court* and what was reasonable to do from the agent's perspective. But objective reasons are also essential to first-person evaluation in two ways. Objective reasons are central to the retrospective evaluation of one's own conduct and to learning from past successes and failures, even when these successes and failures are not appropriate objects of praise or blame. Moreover, objective reasons appear to be the object of prospective evaluation and deliberation. In practical deliberation, one aims at forming one's best judgment about what it is objectively rational to do, even if praise and blame are best apportioned in accordance with one's subjective reasons. Indeed, objective reasons have a kind of explanatory primacy insofar as we identify an agent's subjective reasons with the actions that would be objectively rational if only her beliefs about her situation, or the beliefs about her situation that it would be reasonable for her to hold, were true. These considerations give objective reasons an independence and theoretical primacy in discussions of practical reason. Prudence, is in the first instance, a theory about objective reasons, and that will be our primary, but not exclusive, focus in assessing its commitment to temporal neutrality.

We have now seen two ways in which Sidgwick thinks that temporally neutral concern can justify differential treatment of different periods in one's life. There is another way in which prudence might justify temporal discriminations that might initially seem incompatible with temporal neutrality, but which Sidgwick does not anticipate. On some views, a life is an organic whole whose value cannot be reduced to the sum of the values of its parts, or, at least, cannot be reduced to the sum of the values of its non-relational parts. It is possible to hold a version of this view that treats lives with certain narrative structure as being more valuable, all else being equal, than other lives (see, e.g., Velleman 1991). One could hold, for example, that it is intrinsically better for the value of one's life to display an upward trajectory, such that a life in which evils (e.g. misfortunes, pain, and failure) preceded goods (e.g. good luck, pleasure, and success) was, all else being equal, better than a life in which the goods came first. I do not want to defend this view, but it is, I think, coherent. Such a view says, in effect, that the distribution of goods and harms within a life is itself a good, improving the quality of the person's life. Such a view would require assigning normative significance to the temporal location of goods and harms within a life. But this unequal *treatment* of different periods in one's life would be justified by an equal *concern* for all parts of one's life.[3] Though such an agent is equally concerned about all parts of her life, she sees that by locating the goods later in life she actually makes a greater contribution to the value of her life overall. This sort of temporal bias does not assign normative significance to temporal location as such. It is compatible with and, indeed, required by temporal neutrality if and only if the temporal distribution of goods and harms within a life actually contributes to the value of that life.

This means that temporal neutrality should be understood to claim that the temporal location of goods and harms within a life has no normative significance except insofar as it contributes to the value of that life. We might say that on this view temporal location has no independent significance or no significance per se. The prudent person, concerned to advance his overall good, will be temporally neutral, assigning no independent significance to the temporal location of goods and harms within his life. There will often be diachronic intrapersonal conflicts of value in which what one does affects both the magnitude of goods and harms in one's life and also their temporal distribution. Temporal neutrality requires sacrificing a nearer good for a later, greater good. Call this *now-for-later sacrifice*. This aspect of temporal neutrality, Sidgwick thinks, is a central aspect of our concept of rationality. This claim is echoed by others—for instance by Frank Ramsey, who describes temporal bias as "ethically indefensible" (1928: 261), and by John Rawls who endorses Sidgwick's claim and describes the commitment to temporal neutrality as "a feature of being rational" (1971: 293–94).

However, this conception of temporal neutrality contrasts with a narrower conception that is suggested by some of Sidgwick's remarks. As he sometimes conceives the demand of temporal neutrality, all that the principle affirms is that the mere difference of priority and posteriority in time should not affect the normative significance of goods and harms (1907: 381). This may suggest that the principle is limited in its application to intrapersonal conflicts in which the only variable is temporal location. But that would be far too restrictive. In particular, that conception of temporal neutrality would limit its application to intrapersonal conflicts between goods of the same kind—for instance, smaller pleasure now versus greater pleasure later. The principle

---

[3]  In interpersonal contexts, we sometimes distinguish between equal *concern* and equal *treatment*. Cf. Dworkin (1977: 227). For instance, treating my two children, one of whom has a significant physical disability, with equal concern may require treating them unequally in terms of medical and other resources. We need to make the same distinction in the intrapersonal context. Prudence and temporal neutrality require equal concern, rather than equal treatment per se, for all parts of an agent's life.

would not apply to conflicts in which different kinds of goods are at stake. Sidgwick's focus on conflicts among homogeneous goods is, of course, reinforced by his sympathy for hedonism, which is a monistic theory of the good. Though he contemplates other conceptions of prudence, informed by non-hedonistic theories of the good, Sidgwick does not explore them in much detail, and he may assume that all significant rivals to hedonism would also be monistic. But there is no reason for us to make this assumption or to restrict the application of temporal neutrality to conflicts of homogeneous goods. We avoid this problem if we allow temporal neutrality to apply to conflicts with multiple variables insisting only that it prohibits assigning value to temporal location except insofar as this affects the value of the whole. If so, temporal neutrality can apply to conflicts of heterogeneous goods of the sort that would be recognized by suitably pluralistic theories of the good. Prudence will demand now-for-later sacrifice even when the goods at stake are of different kinds, provided that the plurality of goods is not an obstacle to commensurability.

Prudence requires temporal neutrality, which, in turn, requires now-for-later sacrifice. This sort of sacrifice provides us with some of our most compelling paradigms of rationality. It seems a mark of rationality to undertake actions, projects, and commitments to which we would otherwise be indifferent or averse for the sake of some later, greater good. This kind of rational planning is ubiquitous. We may not notice its more mundane applications, such as when we stand in line in order to get tickets to a movie, when we stop to refuel our cars, or when we go to the dentist for routine preventive dental care. We are more likely to recognize now-for-later sacrifice when the sacrifice is more significant. For instance, I engage in such sacrifice when I undergo a medical procedure that involves an extended and painful recovery in order to regain full range of motion and the ability to participate in a fuller range of physical activities than would otherwise be possible. The training required for success in many vocations and avocations often requires various non-negligible physical, financial, and personal sacrifices. Provided the later benefits genuinely do outweigh the near term costs, the sacrifices seem rational and failure to persevere, if understandable, nevertheless seems to be a form of weakness. Indeed, the evolution of the ability to recognize the rationality of now-for-later sacrifice and to regulate one's appetites, emotions, and actions in accordance with this recognition is arguably a significant part of the process of normative development that marks the progress from adolescence to responsibility and maturity.[4]

---

[4] Nagel (1970) defends temporal neutrality and interpersonal neutrality or altruism. Prudence insists that an agent's future interests provide her with reason for action now, and altruism insists that the interests of others provide her (now) with reason for action for their sake. Nagel argues that failure to recognize prudence involves temporal solipsism—failure to see the present as one time among others, equally real—and that failure to recognize altruism involves interpersonal solipsism—failure to see oneself as one person among others, equally real. I have often thought that the real value of Nagel's thesis lies in its adequacy as a description of developmental psychology. For it seems to me that the process of turning children into mature and responsible adults (a process that in some cases is never completed) is in significant part the process of overcoming temporal and personal solipsism.

## 2. COMPENSATION AND THE RATIONALE
## FOR TEMPORAL NEUTRALITY

As Sidgwick points out, the temporal neutrality of prudence seems to accord with assumptions most of us make about rationality. Failures to make now-for-later sacrifices for later, greater goods seems a paradigmatic form of irrationality, even if it is a common and familiar kind of weakness. So temporal neutrality enjoys intuitive support. But can we say more about why we should conform to the demands of temporal neutrality? Can we provide a rationale for temporal neutrality? This is important, because temporal neutrality requires sacrifice, and we should be able to justify sacrifices we demand. In this case, we should be able to justify sacrifices made at one point in an agent's life for the sake of some other period.

A traditional rationale appeals to *compensation*. Now-for-later sacrifice is rational, because the agent is compensated later for her earlier sacrifice. To see how this rationale works, it will help to consider a familiar interpersonal/intrapersonal analogy. Whereas prudence is temporally neutral, utilitarianism is person neutral. Prudence is temporally neutral and assigns no intrinsic significance to *when* a benefit or burden occurs in a person's life. It says that we should balance benefits and harms, where necessary, among different stages in a person's life and pursue the action or policy that promotes the agent's overall good best. Utilitarianism is interpersonally neutral; it assigns no intrinsic significance to *whom* a benefit or burden befalls. Just as temporal neutrality requires intrapersonal balancing, so too person neutrality requires interpersonal balancing. It requires that benefits to some be balanced against harms to others, if necessary, to produce the best interpersonal outcome overall. Utilitarianism's person neutrality thus effects a kind of interpersonal balancing akin to the intrapersonal balancing that prudence's temporal neutrality requires.

But many think that this sort of interpersonal balancing is unacceptable because it ignores the *separateness of persons*. For instance, Rawls famously makes this claim in *A Theory of Justice*.

> This view of social cooperation [utilitarianism's] is the consequence of extending to society the principle of choice for one man [i.e. prudence], and then, to make this extension work, conflating all persons into one.... Utilitarianism does not take seriously the distinction between persons [1971: 27].

Bernard Williams (1976: 3), Thomas Nagel (1970: 134, 138–42) and Robert Nozick (1974: 31–34) agree. They all accept prudence's intrapersonal balancing, at least for the sake of argument, but reject utilitarianism's interpersonal balancing. But perhaps the right reaction is not to deny the parity of intrapersonal and interpersonal cases but to extend intrinsic distributional considerations into intrapersonal contexts. Perhaps we should be concerned with the way in which we distribute goods and harms among the stages in a single life, as well as among lives, and not just with maximizing value over the course of one's life.

We can see how to deny the parity of intrapersonal and interpersonal cases and provide a rationale for the temporal neutrality of prudence by highlighting the role of compensation in the separateness of persons objection. Nozick's discussion is especially instructive here.

> Individually, we each sometimes choose to undergo some pain or sacrifice for a greater benefit or to avoid a greater harm.... Why not, *similarly*, hold that some persons have to bear some costs that benefit other persons more? But there is no *social entity* with a good that undergoes some sacrifice for its own good.... To use a person in this way does not sufficiently respect and take account of the fact that he is a separate person, that his is the only life he has. *He* does not get some overbalancing good from his sacrifice, and no one is entitled to force this upon him ... [1974: 32–33].

Like the others, Nozick is invoking claims about compensation to explain the asymmetric treatment of intrapersonal and interpersonal balancing. Whereas balancing benefits and harms is acceptable *within* a life, balancing benefits and harms *across* lives appears unacceptable. In the intrapersonal case, benefactor and beneficiary are the same person, so compensation is automatic. In the interpersonal case, benefactor and beneficiary are different people; unless the beneficiary reciprocates in some way, the benefactor's sacrifice will not be compensated. Whereas intrapersonal compensation is automatic, interpersonal compensation is not. This leads the critics of utilitarianism to defend the need for independent principles of interpersonal distribution that would be acceptable, in a way that needs to be specified, to *each* affected party.

# 3. RATIONALIZING THE HYBRID STRUCTURE OF PRUDENCE

This appeal to compensation also allows us to address a concern about the *hybrid structure* of prudence. Prudence or egoism is a hybrid theory, because it is temporally neutral, assigning equal importance to all parts of an agent's life, but agent-relative, because it assigns significance only to benefits and harms that accrue to the agent. As such, prudence can be contrasted with two pure-bred rivals. *Neutralism* is fully neutral; it holds that an agent has reason to do something just insofar as it is valuable, regardless of whom the value accrues to or when it occurs. *Presentism* is fully relative; it claims that an agent has reason to do something just insofar as that would promote his own present interest.[5]

---

[5] What I am calling presentism here is a normative theory about how an agent's reasons for action are grounded in her present interests. It is different from presentism as a metaphysical view about the

Time and person are parallel distributional dimensions; we need to decide where to locate goods and evils in time and among persons. Once we adopt this perspective, prudence may seem like an unstable hybrid. It says that it makes all the difference *on whom* a benefit or burden falls and none whatsoever *when* it falls. On reflection this may seem arbitrary. In *The Methods of Ethics* Sidgwick considers this issue in the context of his discussion of the proof of utilitarianism.

> I do not see why the axiom of Prudence [rational egoism] should not be questioned, when it conflicts with present inclination, on a ground similar to that on which Egoists refuse to admit the axiom of Rational Benevolence. If the Utilitarian [neutralist] has to answer the question, 'Why should I sacrifice my own happiness for the greater happiness of another?' it must surely be admissible to ask the Egoist, 'Why should I sacrifice a present pleasure for a greater one in the future? Why should I concern myself about my own future feelings any more than about the feelings of other persons?' [418]

The egoist asks the neutralist: Why should I sacrifice my own good for the good of another? The egoist doubts that concern for others is non-derivatively rational. But the presentist can ask the egoist: Why should I sacrifice a present good for myself for the sake of a future good for myself? The presentist doubts that concern for one's future is non-derivatively rational. These doubts may seem parallel. We must decide where among lives and when within lives to locate goods and harms. Because both are matters of position or location, we may think that they should be treated the same. Derek Parfit pushes this same worry about the hybrid structure of prudence, or the self-interest theory (S), as he calls it, in Part II of *Reasons and Persons* (1984).

> As a hybrid S can be attacked from both directions. And what S claims against one rival may be turned against it by the other. In rejecting Neutralism, a Self-interest Theorist must claim that a reason may have force only for the agent. But the grounds for this claim support a further claim. If a reason can have force only for the agent, it can have force for the agent only at the time of acting. The Self-interest theorist must reject this claim. He must attack the notion of a time-relative reason. But arguments to show that reasons must be temporally neutral, thus refuting the Present-aim Theory, may also show that reasons must be neutral between different people, thus refuting the Self-interest Theory [140].

If present sacrifice for future benefit is rational, why isn't sacrifice of one person's good for the sake of another's? In this way, the appeal to parity may support neutralism. This is roughly the view Thomas Nagel adopts in *The Possibility of Altruism* (1970). His primary aim is to argue against egoism's agent-bias and in favor of impartiality or altruism, and he relies on the parity of intertemporal and interpersonal distribution to do so. Just as the interests of an agent's future self provide him with reasons for action now, so too, Nagel argues, the interests of others can provide him with reason for action. Failure to recognize temporal neutrality involves temporal dissociation—

nature of time, according to which only the present, and neither the past nor the future, is real. For a discussion of this metaphysical version of presentism, see Mozersky's contribution to this volume.

failure to see the present as just one time among others—and failure to recognize impartiality or altruism involves personal dissociation—failure to recognize oneself as just one person among others (1970: 16, 19, 99–100).[6]

Alternatively, we might treat time and person as parallel and argue from the agent-bias that egoism concedes to temporal bias, in particular, present-bias. If my sacrifice for another is not rationally required, it may seem that we cannot demand a sacrifice of my current interests for the sake of distant future ones. If so, we will think that it is only the present interests of the agent that provide her with non-derivative reason for action. Though Parfit mentions Nagel's fully neutral response to parity, it is the fully biased response that he develops and thinks Sidgwick anticipated (1984: 137–44).

Whereas Parfit thinks that one cannot defend the hybrid character of prudence, Sidgwick thinks that this challenge to prudence is unanswerable only if we accept Humean skepticism about personal identity over time (1907: 418–19). Sidgwick thinks that prudence is defensible provided we recognize the separateness of persons.

> It would be contrary to Common Sense to deny that the distinction between any one individual and any other is real and fundamental, and that consequently "I" am concerned with the quality of my existence as an individual in a sense, fundamentally important, in which I am not concerned with the quality of the existence of other individuals: and this being so, I do not see how it can be proved that this distinction is not to be taken as fundamental in determining the ultimate end of rational action for an individual [498].

This appeal to the separateness of persons suggests a rationale for the hybrid structure of prudence. We saw that when the separateness of persons is invoked to discredit utilitarianism critics of utilitarianism appeal to the compensation principle. But the compensation principle and the metaphysical separateness of persons explain the asymmetry between intrapersonal and interpersonal distribution. We saw that there is automatic intrapersonal compensation but no automatic interpersonal compensation. Compensation requires that benefactors also be beneficiaries, and for compensation to be automatic benefactor and beneficiary must be one and the same. In the diachronic, intrapersonal case one's sacrifice of a present good for a (greater) future good is rational, because there is compensation later for the earlier sacrifice; benefactor and beneficiary are the same. This explains temporal neutrality. But in the interpersonal case, benefactor and beneficiary are different people; unless the beneficiary recipro-cates in some way, the agent's sacrifice will be uncompensated. This explains agent relativity or bias. So we have a rationale for the hybrid treatment prudence accords intertemporal and interpersonal distribution.

---

[6] Nagel's remarks about the "combinatorial problem" (1970: 134–42) show that he is skeptical of an impersonal interpretation of impartiality. Nonetheless his appeal to parity seems to require neutralism and not just impartiality. He appeals to parity to argue from egoism's temporal neutrality to non-derivative concern for others. But if intertemporal and interpersonal distribution must be isomorphic, and we accept a temporally neutral interpretation of intertemporal impartiality, then we seem forced to accept a person-neutral interpretation of interpersonal impartiality.

Or do we? Couldn't doubts about interpersonal balancing be extended to intrapersonal balancing? If the separateness of persons defeats interpersonal balancing, why
doesn't the separateness of different periods within a person's life defeat intrapersonal balancing? After all, me-now and me-later are distinct parts of me.[7] But then
it is hard to see how me-now is any more compensated for its sacrifices on behalf
of me-later than I am compensated by my sacrifices for you. Just as doubts about
interpersonal balancing lead to a distributed concern with each person, perhaps doubts
about intrapersonal balancing should support a distributed concern with each part
of a person's life. There are different interpretations of what this distributed concern
requires in the interpersonal context, such as equal distribution and maximin. Perhaps
we need to explore comparable interpretations of distributed concern in the intrapersonal context. (McKerlie 1989 explores some of these possibilities in interesting ways.)
However, this concern about temporal neutrality is not compelling, as it stands, for
several reasons.

First, we might distinguish between temporal impartiality and temporal neutrality.
Consider again the interpersonal case. Here, one norm might be called the norm of
impartiality; it insists that everyone be given equal concern. This norm of impartiality
admits of different interpretations, including a norm of substantive equality and maximin, among others. Indeed, utilitarianism's person neutrality is one interpretation
of interpersonal impartiality. Similarly, we might identify a more generic notion of
intertemporal impartiality that would admit of different interpretations, including that
of temporal neutrality. One way to read the separateness argument, then, is to see it
mandating a temporal impartiality. That would not vindicate temporal neutrality, as
such, but it would require a form of impartiality that was inconsistent with the sort of
temporal bias displayed in ordinary life by familiar forms of temporal discounting and
displayed theoretically in the pure-bred presentism.

Second, this challenge to temporal neutrality requires thinking that we can and
should adopt a sub-personal perspective when reckoning compensation. But there are
problems with this idea. Once we go sub-personal and appeal to full relativity, there
seems no reason to stop until we reach the sub-personal limit—a momentary time
slice of the person. But notions of compensation have no application to momentary
time slices, which do not persist long enough to act or receive the benefits of earlier
actions. Moreover, many of the goods in life, especially the pursuit and achievement
of worthwhile projects, seem to be realized only by temporally extended beings. But if
we stop short of momentary time slices and appeal to larger sub-personal entities, call
these person segments, other problems arise. One question is just where to stop. If we

[7] I intend talk about temporal parts of a person or person's life to be metaphysically ecumenical in
two ways. First, it is convenient to talk about persons and their temporal parts whether persons are
four-dimensional entities that literally have temporal parts (as three-dimensional entities have spatial
parts) or whether they are three-dimensional entities that have no temporal parts but do have lives,
histories, or careers that have temporal parts or stages. Talk about a person's temporal parts can refer
to temporal parts of persons or to parts of lives or careers of persons. Second, my talk of temporal parts
is neutral in the debate among those who treat persons as four-dimensional entities having temporal
parts about whether persons or their temporal parts are prior in order of explanation.

don't fully relativize, why relativize partially? Moreover, if we do relativize partially, we introduce indeterminacy. This is because the careers of person segments overlap, with the result that any one point in time is part of the career of indefinitely many different segments. To decide whether compensation has occurred, we need a determinate subject. But if we appeal to person segments, we seem to lack a determinate subject (for more details, see Brink 1997a).

Of course, persons are just maximal segments. They also seem to be the most salient segments. Many of the things we value and that structure our pursuits are certain sorts of lives. We aim to be certain sorts of people. Insofar as these ideals structure our beliefs, desires, and intentions, the correct perspective from which to assess success would seem to be the perspective of a whole life. Even when persons have more parochial aims and ambitions, the successful pursuit of these aims and ambitions requires interaction and cooperation among segments, much as persons must often cooperate with others to achieve individual, as well as collective, aims. They do interact and cooperate, much as distinct individuals interact and cooperate in groups, in order to plan and execute longterm projects and goals. They must interact and cooperate if only because they have to share a body and its capacities in order to execute their individual and collective goals, much in the way that individuals must sometimes interact and cooperate if they are to use scarce resources to mutual advantage (cf. Korsgaard 1989). Indeed, both the ease and necessity of interaction among person segments will be greater than that among persons, because the physical constraints and the reliability of fellow cooperators are greater in the intrapersonal case. But this means that person segments will overlap with each other; they will stand to each other and the person much as strands of a rope stand to each other and the rope.[8] Though we can recognize the overlapping strands as entities, the most salient entity is the rope itself. So too, the most salient entity is the person, even if we can recognize the overlapping person segments that make up the person.

In this way, person segments represent a rather arbitrary stopping place. If the appeal to full bias argues for agents with shorter lifespans than persons, then an appeal to *full* bias ought to argue for person slices as agents. But if, as I have argued, that conception cannot be maintained, then it seems arbitrary to settle on person segments. Once we extend the lifespan of the agent beyond that of a person slice, it seems we should keep going until we reach an entity with the most natural borders, viz. the person.

These appear to be reasons to preserve the normal assumption that it is persons that are agents. But is this assumption really coherent? I have identified the person with a temporally extended entity, some of whose parts lie in the future. But then the person is in one sense "not all there" at the time of deliberation and action. How then could the person be the agent who deliberates and acts and possesses reasons for action?

[8] The rope metaphor is familiar from Wittgenstein (1958: 87), though he does not apply the metaphor to personal identity or agency.

This raises difficult issues, but I doubt that they threaten the assumption that it is persons that are agents.[9] Notice, first, that person slices seem to be the only candidates for agency that avoid some form of this objection. For person segments extend from the instant of deliberation or action into either the future or the past (or both); so person segments are also entities with parts that are "not all there" at the time of deliberation or action. Only one person slice is "all there" at this time. But we've already seen that that conception of agency is indefensible. We might, therefore, wonder whether the agent or entity whose interests determine what rationally ought to be done need be "all there" at the time of action.

Consider an analogy with nations. We speak of nations as actors that enact legislation, start wars, and so on. We also think of nations as having interests and acting in their interests. But a nation is composed, at least in part, by its entire current population. And there is certainly some sense in which the entire population does not enact legislation or start wars. Instead, certain individuals or groups act as representatives of a larger spatially dispersed group of which they are members. We don't conclude that nations cannot be actors or the bearers of interests. Instead, we conclude that a nation can act when its deputies act on behalf of the national interest, that is, the interest of the spatially dispersed group. Similarly, the present self can act as representative of the temporally dispersed entity, the person, by acting in the interest of this being. If so, then the fact that the temporally extended person is "not all there" at the time of action is not a reason to deny that it is the actor or the entity whose interests determine what agents have reason to do. On this assumption, there is automatic diachronic, intrapersonal compensation and so compensation does justify temporal neutrality.

# 4. PERSONAL IDENTITY AND TEMPORAL NEUTRALITY

So far, we have explored a rationale for temporal neutrality that appeals to the separateness of persons and the unity within a life. But then the rationale for temporal neutrality may seem to rest on potentially controversial assumptions about personal identity. There is an important tradition of thinking about personal identity, dating back at least to John Locke (1690: II.xxvii), which analyzes personal identity into relations of psychological continuity and connectedness. Following Parfit, we might call this tradition *psychological reductionism*. Bishop Butler claimed that special concern for one's future and moral responsibility would be undermined by Lockean reductionism

---

[9] Perhaps the difficulty only arises if we are realists about temporal parts, and perhaps the proper moral of the difficulty is that we should reject realism about temporal parts. The defense of presentism that I am considering in this section presupposes a realism about temporal parts. If we reject realism about temporal parts, this hurts presentism, not prudence.

(1736: 267). In a similar way, Parfit argues that psychological reductionism, of the sort he defends, would undermine prudence's demand of temporal neutrality.

Parfit's version of psychological reductionism is similar to other views in the Lockean tradition of thinking about personal identity, including those of Shoemaker (1963, 1984), Wiggins (1967), and Nozick (1980: ch. 1). As a first approximation, psychological reductionism holds that two persons are psychologically connected insofar as the intentional states and actions of one influence the intentional states and actions of the other. Examples of intrapersonal psychological connections include A's earlier decision to vote Democratic and her subsequent casting of her ballot for the Democratic candidate, A's later memories of a disturbing childhood incident and her earlier childhood experiences, and A's later career change and her earlier re-evaluation of her priorities concerning work and family. Two persons are psychologically continuous insofar as they are links in a chain or series of people in which contiguous links in the chain are psychologically well connected. Both connectedness and continuity can be matters of degree. According to the psychological reductionist, it is the holding of many such relations of connectedness and continuity that unify the different stages in a single life. More specifically, personal identity consists in maximal (non-branching) psychological continuity.[10]

One of Parfit's arguments against temporal neutrality defends a discount rate as an apparent consequence of diminished connectedness.

> My concern for my future may correspond to the degree of connectedness between me now and myself in the future. Connectedness is one of the two relations that give me reasons to be specially concerned about my own future. It can be rational to care less, when one of the grounds for caring will hold to a lesser degree. Since connectedness is nearly always weaker over long periods, I can rationally care less about my further future [1984: 313].

As Parfit notes, this is a discount rate with respect to connectedness and not with respect to time itself. His discount rate should, therefore, be distinguished from the discount rate with respect to time that C.I. Lewis calls "fractional prudence" (1946: 493). Prudence is neutral with respect to time itself and so must deny fractional prudence. But prudence's temporal neutrality is also inconsistent with Parfit's discount rate, because temporal neutrality requires a kind of equal concern among parts of one's life. The magnitude of a good or harm should affect its rational significance.

---

[10]  Two qualifications are in order. (1) If we are to define identity in terms of relations of psychological continuity, these relations cannot themselves presuppose identity. Relations such as remembering one's earlier experiences and fulfilling one's prior intentions, which do presuppose identity, will have to be replaced by more general quasi-relations that are otherwise similar but presuppose causal dependence, rather than identity. See Shoemaker (1970) and Parfit (1984: 220–21). (2) If we are to define identity, which is a one-one relation, in terms of psychological continuity, which can take a one-many form, we must define it in terms of *nonbranching* psychological continuity. But the reasoning that leads us to this conclusion may also lead us to the conclusion that it is continuity (a potentially one-many relation), rather than identity per se, that is what has primary normative significance. See Parfit 1984: ch. 12 and Brink 1997b.

But temporal neutrality implies that the temporal location of a good or harm within a life should be of no rational significance per se. If so, then, all else being equal, an agent should be equally concerned about goods and harms at any point in his life. In particular, if near and more distant future selves are both stages in his life, then, other things being equal, an agent should have equal concern for each, even if the nearer future self is more closely connected with his present self.

Indeed, Parfit's claim about the discount rate seems too modest. He insists only that this discount rate of concern for one's future is not irrational; he does not claim that it is rationally required. Though the friend of temporal neutrality must deny the more modest claim as well, the reductionist argument, if successful, surely supports the stronger claim that a discount rate of concern is rationally appropriate where the relations that matter hold to a reduced degree. This is because concern should track and be proportional to the relations that matter.

However, reductionism justifies neither the permissibility nor the duty to discount. A symptom that something is amiss in this reductionist justification of a discount rate is that the same reasoning would imply that we lack prudential reason to improve ourselves in ways that involve significant psychological transformation (e.g. an addict giving up his addiction and the associated psychology and lifestyle or a neo-Nazi replacing hate with tolerance and sympathy). For if the improvement involves psychological change that diminishes connectedness, then we must have less prudential reason to undertake it. Improvements that diminish connectedness would be like benefiting another. But self-improvement is a paradigmatic demand of prudence. Something must be wrong with the reductionist case for discounting.

First, notice that diminished connectedness does not follow from psychological change or dissimilarity. Connectedness is defined in terms of psychological interaction and dependence. Sometimes psychological connectedness takes the form of maintaining beliefs and desires, which will ensure some degree of psychological similarity. But connectedness is also preserved in change, as when one changes one's career goals in light of a reassessment of one's opportunities, abilities, and responsibilities. This applies to character change as well. Provided one plays a suitable role in generating and shaping the change in his beliefs, desires, and ideals, his change in character is no obstacle to preserving connectedness over time.

Second, this reductionist argument for a discount rate appeals to diminished connectedness over time, but psychological reductionism needs to be formulated in terms of continuity, rather than connectedness. As Thomas Reid suggested in his criticism of Locke's account of personal identity in terms of memory connectedness, identity is, but psychological connectedness is not, a transitive relation (1785: III 357–58). Transitivity requires that if A = B and B = C, then A = C. But even if A is connected to B and B is connected to C, A need not be connected to C. Not so with continuity, which is defined as a chain the links of which are connected. Provided A is connected to B, and B is connected to C, A and C will be continuous, even if they are not well connected. But then diminished connectedness between A and C does not diminish the continuity between A and C. If reductionism is formulated in terms of continuity, rather than connectedness, then diminished connectedness over time does not justify a discount rate.

Third, even if connectedness did matter, the reductionist case for discounting confounds parts and wholes. The question is how a person should view different stages or periods in her life. This is a question about how a whole should view its parts. But the temporally dispersed parts of a person's life are equally parts of that person's life regardless of how the parts are related to each other. Consider, again, the person P and three different temporally successive periods in her life A, B, and C. The fact that A is more connected to B than A is to C does not show that C is any less part of P's life than B is. As long as it is the person who is the agent and whose interests are at stake, differences in connectedness among the parts of a person's life should not, as such, affect her reasons to have equal regard for all parts of her life.

At one point, Parfit considers a version of this appeal to the idea that parts of a person's life are equally parts of that life (1984: 315–16). He rejects this appeal with an analogy involving relatives. He claims that although all members of an extended family are equally relatives, this does not justify equal concern among them. For instance, it would not give my cousin as strong a claim to my estate as my children. But to focus on the division of my estate would be the intrafamily analog of asking about the interests of a person slice or segment in the intrapersonal case, which we have claimed is problematic. The intrafamily analog of the person would require focusing on the distribution of some asset that belonged to the entire extended family. But here equality or neutrality seems the right norm in light of the fact that all are equally parts of the family, even if some are more closely related to some relatives than they are to others.

These considerations undermine the reductionist case for a discount rate and show that the rationale for temporal neutrality is metaphysically robust.

## 5. INTRAPERSONAL CONFLICTS OF VALUE

Temporal neutrality can seem defensible when we restrict our attention to cases in which there is diachronic fixity of interests, because we can see how the agent is compensated later for the sacrifices she makes now. But what about cases in which there is significant change in an agent's character or ideals?

In *The Possibility of Altruism* Nagel claims that temporal neutrality is unproblematic when "preference changes" are regarded with indifference. However, he sees a potential problem when neutrality is applied to intrapersonal conflicts of ideals.

> It might happen that a person believes at one time that he will at some future time accept general evaluative principles—principles about what things *constitute* reasons for action—which he now finds pernicious. Moreover, he may believe that in the future he will find his present values pernicious. What does prudence require of him in that case? Prudence requires that he take measures which promote the

realization of that for which there *will* be reason. Do his beliefs at the earlier time give him any grounds for judging what he will have reason to do at the later [time]? It is not clear to me that they do, and if not, then the requirement of prudence or timeless reasons may not be applicable [74].

Parfit shares Nagel's worries about the application of temporal neutrality to intrapersonal conflicts of ideals (1984: 155). Later, he describes the case of the nineteenth-century Russian nobleman.

In several years, a young Russian will inherit vast estates. Because he has socialist ideals, he intends, now, to give the land to the peasants. But he knows that in time his ideals may fade. To guard against this possibility, he does two things. He first signs a legal document, which will automatically give away the land, and which can be revoked only with his wife's consent. He then says to his wife, 'Promise me that, if I ever change my mind, and ask you to revoke this document, you will not consent.' He adds, 'I regard my ideals as essential to me. If I lose these ideals, I want you to think that I cease to exist. I want you to regard your husband then, not as me, the man who asks for this promise, but only as his corrupted later self. Promise me that you would not do what he asks' [327].

Parfit uses the Russian nobleman example to argue that adoption of a reductionist view of personal identity should lead us to revise our views about promissory fidelity, especially in cases involving intertemporal conflicts of ideals. But we can also use it to raise questions about the plausibility of the demands of temporal neutrality in such cases.

The Russian nobleman example is supposed to derive some of its force against the norm of temporal neutrality from Parfit's reductionist conception of personal identity. He seems to think that reductionism justifies the Russian nobleman's claim that loss of his socialist ideals represents a substantial change, which he does not survive. This is what is supposed to justify the nobleman's wife in regarding his bourgeois successor as "another" who cannot revoke the nobleman's commitment.

But as our earlier discussion (§4) implies, there are several problems with this reductionist use of the Russian nobleman example. First, if this really were a substantial change, then prudence would not require neutrality between the socialist and bourgeois selves. Prudence requires intrapersonal neutrality but not interpersonal neutrality. If the example involves a substantial change, then it creates an interpersonal context. But then prudence does not demand concern and sacrifice for others. Absent some kind of reciprocation, these would be uncompensated sacrifices. So if the change of ideals produced a substantial change, prudence would not counsel the nobleman to moderate his socialist ideals out of concern for his bourgeois successor. Second, psychological reductionism does not justify regarding the change of ideals in this case as a substantial change. Even if such changes of ideals disrupted psychological connectedness, they would presumably not disrupt psychological continuity. But reductionism needs to be formulated in terms of continuity, rather than connectedness, to avoid Reid's transitivity concern. Moreover, psychological connections include ways an agent modifies his beliefs, desires, ideals, and intentions. So long as the nobleman plays a suitable role in generating and shaping his change of ideals (e.g. he is not the unwitting victim of psychological manipulation by another), character change of this sort is no obstacle to psychological connectedness. So in assessing the significance of the Russian nobleman example for the norm of temporal neutrality, we should resist any suggestion that socialist and bourgeois selves are literally different people. Both are equally parts of the nobleman's life, and, as such, prudence demands temporal neutrality.

But avoiding Parfit's reductionist gloss on intrapersonal conflicts of value does not itself remove the challenge that such cases pose to temporal neutrality. For we can still wonder if the demand of temporal neutrality makes sense in such cases. Should I be expected to moderate my pursuit of ideals I now hold dear for the sake of ideals I now reject but will or may later accept?

We should first notice something a little odd about the way intrapersonal conflicts of value are typically represented. Imagine that Before is at a crucial fork in the road of life and her prudential ideals speak in favor of route A, but she knows that she will later become After, whose prudential ideals will only be served if she now chooses route B. Should Before be true to her own ideals and choose route A, should she empathize with After and choose route B, or should she try to forge some third route C that compromises between A and B? This way of posing the problem assumes that there is a fact of the matter about the content of one's future character and ideals independently of the crucial choices one makes now. But often, perhaps typically, this is false. One's future character and ideals are very much influenced by crucial practical decisions one makes on the road of life. It is quite unlikely that a radical young socialist will turn into a complacent bourgeois regardless of the decisions he now makes. Who one becomes depends in part upon what one does now.[11] But then it may be possible to avoid many intertemporal conflicts of value by making choices now that preserve, rather than compromise, one's present ideals. Provided one's present ideals are worthwhile (about which more below), one can honor temporal neutrality by acting in accord with one's present ideals and thereby avoiding intertemporal conflict.

But perhaps some intertemporal conflicts are unavoidable. What then? Remember that prudence and its demand of temporal neutrality are, at least in the first instance, claims about what we have objective reason to do. The implications of temporal neutrality in situations involving unavoidable intrapersonal conflicts of ideals depend on the merits of the conflicting ideals. For present purposes, we can be quite ecumenical among different metaethical accounts of what makes some ideals meritorious and

---

[11] Lurking somewhere here is a relative of Parfit's Non-Identity Problem (1984: ch. 16). That problem makes it hard to assess the moral consequences of alternative actions in certain familiar ways (e.g. person-affecting ways) inasmuch as many alternatives affect not just how benefits and harms are distributed among a given set of people but also who exists to be benefitted or harmed. In the intrapersonal case, alternatives often determine which ideals exist to be promoted or hindered. Parfit takes the non-identity problem to support a form of interpersonal neutrality. I am unclear whether the corresponding intrapersonal problem about plasticity of ideals supports temporal neutrality. How far the parallels extend and what they show about the intrapersonal case deserve further consideration.

others meretricious. It will be helpful to divide unavoidable conflicts into *symmetrical* ones—those in which the merits of conflicting ideals are comparable—and *asymmetrical* ones—those in which the merits of conflicting ideals are very different.

The asymmetrical conflicts are perhaps more straightforward. There are two such cases. In the case of *Corruption*, Before's ideals are valuable, whereas After's are not. By contrast, in the case of *Improvement*, Before's ideals are worthless, whereas After's ideals are valuable. In cases of Corruption and Improvement, the demands of temporal neutrality are clear—act on the worthwhile ideal when you have it, not the worthless one. This is a claim about one's objective reasons, the reasons one has in virtue of the facts about the situation whether one is in a position to recognize them or not. In these cases, temporal neutrality does not require neutrality between current and future ideals.

It is fairly easy to see how the agent can and will act on these reasons in the case of Corruption, for this just requires acting on his current ideals. Here acting on one's current ideals is also what temporal neutrality requires.

However, matters are more complicated in the case of Improvement. Temporal neutrality's claim about one's objective reasons remains plausible. One has objective reason to act later on those valuable ideals that one will hold, rather than the worthless ideals that one now embraces. But can one act on this verdict if it is the worthless, rather than the valuable, ideal that one now embraces? Can temporal neutrality make plausible claims about subjective rationality? Could it be subjectively rational to act on valuable ideals that one does not now hold? The answer is Yes, provided that we understand subjective reasons as the reasons one has, not in virtue of what one now judges, but in virtue of what it would be reasonable for one to judge now if one gave the matter due attention. It is part of a theory of subjective rationality to specify more precisely what kind of idealization of the agent's epistemic situation is appropriate in determining her subjective reasons. As long as the worthlessness of Before's ideals and the merit of After's ideals do not transcend reasonable idealizations of the agent's epistemic situation, whatever those are, the comparative merits of earlier and later ideals will be ascertainable in the relevant way. If the comparative value of her current and future ideals is available to her in this way, we can ascribe to her a subjective reason to favor her future ideals. However, in cases of Improvement in which the comparative values of current and future ideals is a transcendent fact (transcending the relevant idealization), then the demands of objective and subjective rationality appear to diverge. The friend of prudence can and should defend temporal neutrality as a claim about the agent's objective reasons. Whether she is in a position to recognize it or not, she has no reason to act on her current ideals and will have reason to act on her future ideals. This can be a case where it may not be subjectively rational to do what is in fact objectively rational.

What about unavoidable conflicts whose merits are symmetrical? The *Minus-Minus* situation occurs when the conflicting ideals are similarly worthless. Here it seems right to agree with neutrality's claim that there is objective reason not to act on either ideal but to find, adopt, and act on some third ideal that has merit. Provided that the comparative merits of the meretricious and genuinely valuable ideals are reasonably ascertainable and are not (in the relevant sense) transcendent facts, this also yields a plausible claim about the agent's subjective reasons. The agent should act on neither meretricious ideal but adopt and act on the new valuable ideal.

Perhaps the most interesting case of unavoidable conflict is the symmetrical case in which the conflicting ideals are both valuable and comparably so. One example might be a conflict between excelling as a professional athlete early in life, which may require forgoing extended educational and professional training and may impose significant health costs later in life, and various forms of professional and personal success later in life. Another example might be familiar conflicts between success in professional and family life. We might call any such case a *Plus-Plus* case. By hypothesis, the conflict is unavoidable, so that After's ideals conflict with Before's no matter what the agent now does, and each ideal is valuable. Here, temporal neutrality recognizes a conflict of objective reasons and counsels a kind of neutrality among the competing ideals. On reflection, this seems right. If the agent can pursue Before's ideals unreservedly only by completely frustrating After's ideals (and vice versa), then there seems something objectively wrong with the unreserved pursuit of present ideals. Ideally, one would try to find a way to achieve substantial success in one's ideals both now and later, even if it required some moderation in or restrictions on the pursuit of one's ideals now or later. Neutrality's counsel of moderate or restricted pursuit of current ideals is an instance of the familiar adage "Not to burn one's bridges". Where such compromise and accommodation are possible, neutrality makes good normative sense. Call these cases of *Accommodation*. But accommodation may not always be possible. In cases of *Genuine Dilemma* there is no prospect of substantially accommodating both ideals. Here, neutrality seems compatible with two possibilities. On the one hand, one might achieve some less-than-substantial success along both ideals—neither a stellar success nor an abject failure at any time. Alternatively, one might engage in the unreserved and successful pursuit of ideals either now or later (but, by hypothesis, not both), provided that the process of selecting the favored ideal gave equal chances of success to both ideals (as in a coin flip). Neither alternative is attractive, but that seems to be a consequence of the situation being dilemmatic.[12] One consolation is that unavoidable conflicts are somewhat rare, and Genuine Dilemmas are even more exotic. Neutrality's claims about our objective reasons in such cases seem plausible enough. And, as before, provided the merits of the conflicting ideals are not transcendent facts, these claims about the agent's objective reasons apply to her subjective reasons as well.

We can reconcile the demands of prudence and fidelity to one's ideals if we remember that the agent is a person who is temporally extended. Her past, present, and future are equally parts of her and her life. To be true to herself, since she is a temporally extended person, she must be true to all of her reasonable ideals and cannot be selectively attentive to her current ideals. She must weigh her future reasonable ideals,

---

[12] These claims about intrapersonal dilemmas parallel claims we might make about moral dilemmas. See Brink (1994).

where these are fixed, against her current reasonable ideals, where this is necessary, in order to conform her behavior to all of her reasonable commitments. This sort of concern for one's whole life does not require forsaking one's current prudential ideals. But it does require conditioning their pursuit on recognition of the legitimate claims that one's reasonable future prudential ideals make on one.

## 6. THE SYMMETRY ARGUMENT

A very different concern about temporal neutrality can be seen in common responses to Epicurean arguments about why we should not fear death. The Epicureans saw the main aim of philosophy as confronting and, if possible, removing the fear of death, which, as hedonists, they regarded as bad insofar as it causes anxiety. They thought that fear of death was predicated largely on fear of retribution from anthropomorphic gods. They offered many different sorts of arguments for why we should not fear death. They argued that if the gods do exist we have reason to think that they do not interfere in human affairs and that even if they do exist and intervene in human affairs we are invulnerable to harm after death. Some of these arguments assume death brings nonexistence. Others do not. The argument that bears on temporal neutrality purports to show that we have no reason to fear death even if—indeed, because—it implies our nonexistence. In *De Rerum Natura* Lucretius gives expression to temporal neutrality in appealing to a parallel between our prenatal and postmortem nonexistence to counteract our fear of death.

> From all this it follows that death is nothing to us and no concern of ours, since our tenure of mind is mortal. In days of old, we felt no disquiet when the hosts of Carthage poured into battle on every side—when the whole earth, dizzied by the convulsive shock of war, reeled sickeningly under the high ethereal vault, and between realm and realm the empire of mankind by land and sea trembled in the balance. So, when we shall be no more—when the union of body and spirit that engenders us has been disrupted—to us, who shall then be nothing, nothing by any hazard will happen any more at all. Nothing will have the power to stir our sense, not though earth be fused with sea and sea with sky [III 830–51].

Later, he invokes the same symmetry between postmortem and prenatal nonexistence.

> Look back to see how the immense expanse of past time, before we were born, has been nothing to us. Nature shows us that it is the mirror-image of the time that is to come after we are dead. Is there anything there terrifying, does anything there seem gloomy? Is it not more peaceful than any sleep [III 972–77]?

The Symmetry Argument is wonderful. Here is its structure.

1. Death brings nonexistence.
2. Postmortem nonexistence is no different than prenatal nonexistence.

3. We do not regret our prenatal nonexistence.

4. Hence, we should not regret our death.

The Epicureans notice an asymmetry in our attitudes toward past and future nonexistence. They reject this asymmetry as irrational and propose to make our attitudes toward death consistent with our attitudes toward prenatal nonexistence. In so doing, they embrace temporal neutrality.

But symmetry is a two-edged sword. The parity of prenatal and postmortem nonexistence could be exploited to expand, as well as contract, regret. Consider this second appeal to symmetry.

1. Death brings nonexistence.

2. Postmortem nonexistence is no different than prenatal nonexistence.

3. We do regret our death.

4. Hence, we should regret our prenatal nonexistence.

This second appeal to symmetry may seem more compelling if we have no independent explanation of why death is not bad. Of course, the Epicureans also appeal to an *Existence Requirement*—one cannot be harmed if one does not exist—to explain why nonexistence is not to be feared (III 860–70). But the Existence Requirement does not explain why death is not bad. Even if one cannot be harmed *after* death, one can be harmed *by* death, because death deprives the person whom it befalls of the goods she would have enjoyed had she continued to exist and led a life worth living (Nagel 1979: 3; McMahan 1988; Feldman 1992). If this is what is bad about death, then symmetry suggests that we do have reason to regret our prenatal nonexistence. Had we existed earlier (and lived to the same date as we actually do), we would have enjoyed more goods than we will in fact. Either form of nonexistence deprives us of possible goods and so is a legitimate source of regret. Of course, to say that death or prenatal nonexistence is an appropriate object of regret is not to endorse preoccupation with it.

Some may regard either symmetry argument as a *reductio* of temporal neutrality. One common response to the second symmetry argument is to appeal to a metaphysical thesis about *the essentiality of origin* to defend asymmetry. The possible goods account of being harmed has an important counterfactual element—for something to harm me, it must make me worse off than I would otherwise have been. This allows us to explain how death can harm us, because it deprives us of goods we would have enjoyed if we had lived longer. But some think that the essentiality of origin implies that, though I could live longer than I actually will, I could not have been born earlier than I actually was. If this were true, then we couldn't make sense of the counterfactual that I would have enjoyed more goods if only I had been born earlier (Nagel 1979: 8; Parfit 1984: 175).

But the second symmetry argument is metaphysically robust. First of all, the essentiality of origin, as usually understood, does not establish the essentiality of time of birth. In his classic discussion of the essentiality of origin, Saul Kripke (1980: 110–15) suggests that individual humans essentially had their origins as the particular zygotes out of which they grew. Presumably, what's essential to a particular zygote, by the same criterion, is being the union of a particular sperm and a particular egg. But then the time of birth is not essential to a particular person or human. The gestation period for a fetus—the time from conception to birth—could be longer or shorter, and so one could well wish that it had been shorter and one had been born earlier. Neither is the time of conception essential to a particular individual. A particular zygote can be formed at different times, depending on when that sperm and that egg are joined. If so, then not only could one have been born earlier than one was but also one could have been conceived earlier.

Even if we assumed, contrary to fact, that one's time of birth was essential to one, it still wouldn't follow that one couldn't sensibly regret one's prenatal nonexistence. For while it would be true, on this assumption, that we couldn't have been born earlier than we in fact were, it doesn't follow that we know when this was. We need to distinguish between metaphysical and epistemic possibility here. Even if it was not metaphysically possible to have been born earlier than we were, it is still possible to discover that what we thought was our date of birth is incorrect and that we were in fact born earlier than previously thought.[13] For instance, I could know that I was adopted but imagine discovering that the adoption agency confused my records with those of Baby Doe who was born later than I was—a discovery that would imply that I was actually born earlier than I thought I was. Such a discovery could be a legitimate basis for being pleased that one's life contains more goods than one had previously realized. Correlatively, closing such epistemic possibilities could be a legitimate basis for regret that one was not in fact born earlier than one had thought. Of course, being a coherent and legitimate object of regret does not make it appropriate for me to be preoccupied with the possibility (metaphysical or epistemic) of my prenatal nonexistence any more than it follows from the fact that my death is a legitimate object of regret that I should be preoccupied with it.

These considerations show that the second symmetry argument is surprisingly robust. Of course, this conclusion (expanding regret from death to prenatal nonexistence) is not one the Epicureans would welcome. But it takes seriously and defends their appeal to temporal neutrality.

# 7. MINIMIZING FUTURE SUFFERING

A final challenge to temporal neutrality worth considering here is the claim that most of us would prefer learning that our suffering is past, even if this suffering is greater than would be an alternative future suffering. Parfit illustrates this claim with his

---

[13] In a wonderful paper, Philip Mitsis (1989) invokes this distinction between metaphysical and epistemic possibilities to defend the robustness of Epicurean assumptions about symmetry.

ingenious case of *My Past and Future Operations* (1984: §64). Imagine that there is a painful operation that requires the patient's cooperation and, hence, can only be performed without the use of anesthetic. But doctors can and do induce (selective) amnesia after the operation to block memories of these painful experiences, which are themselves painful. I knew I was scheduled for this procedure. I wake up in my hospital bed and ask my nurse whether I have had the operation yet. He knows that I am one of two patients, but doesn't know which. Either I am patient A, who had the longest operation on record yesterday (10 hours), or I am patient B, who is due for a short operation (one hour) later today. While I wait for him to check the records, I find that I have the strong preference and hope that I am patient A, even though A's suffering was greater than B's will be. Temporal neutrality would seem to imply that this preference is irrational. But that might not seem right. More generally, it might seem that we prefer to minimize future suffering, even if that is not a way to minimize total suffering.

I assume that when contemplating this example the preference for minimizing future suffering is common. But we can still ask if it is, on reflection, rational. We should put Parfit's example in proper context before deciding on the rationality of the preference in question.

First, notice that Parfit must appeal to a *double* sort of temporal relativity. The preference is not simply for earlier rather than later suffering. If we keep the time of the two possible procedures fixed, but ask whether we prefer the greater earlier suffering from a point in time that is either prospective or retrospective with respect to *both* possible procedures, then most people would prefer the later operation with shorter suffering. I prefer B to A if you ask me before I enter the hospital, as I do if you ask me as I leave. So it's not about preferring earlier pain to later pain; instead, it's about preferring past pain to future pain. This makes the bias in question more narrow or isolated.

But it also makes the preference unstable. When I view both procedures prospectively or retrospectively, I have the temporally neutral preference to minimize suffering. It is only when the greater suffering is past and the smaller suffering lies in the future that I display the temporally biased preference for greater past pain. To see why this kind of diachronic instability of preferences might be reason to think the bias is irrational, consider briefly a structurally similar sort of instability that Socrates addresses in his discussion of akrasia or weakness of will in the *Protagoras*.

In the *Protagoras* Socrates famously denies the possibility of akrasia, claiming that, appearances to the contrary notwithstanding, it is really not possible to act contrary to what one judges on balance best. He focuses on cases in which our judgment about what is best is overcome by pleasure, in particular, proximate pleasure (356a–357e). He suggests that our judgments about what is best are inappropriately influenced by the proximity of pleasures and pains. The proximity of pleasures and pains leads to inflated estimates of their magnitude. This kind of temporal bias, Socrates thinks, produces instability in the agent's beliefs about what is best. For instance, in a cool prospective moment an agent might judge that a short-term indulgence should be forsaken for the sake of a later greater good. But as the indulgence becomes imminent—in the heat of the moment—its proximity changes the agent's estimate of the magnitude of the pleasure associated with the indulgence, leading him to conclude that the indulgence is actually on balance better. But in a cool retrospective moment, when his passions no longer inflame his judgment, he sees that he purchased the indulgence at too high a cost and experiences regret. There is no genuine weakness of will, on this interpretation of events, because the agent's beliefs about what is best actually change and he acts in accord, rather than against, the beliefs about what is best that he holds at the time of action. Though Socrates denies that the agent acts akratically, he does think that he acts irrationally, allowing temporal proximity to affect his beliefs about the magnitude of the benefits and harms associated with his options. One sign of the irrationality of the bias is the instability temporal proximity induces in the agent's beliefs and preferences. The fact that the hot judgment is preceded and followed by contrary cool judgments is evidence that the hot judgment is not to be trusted.

One needn't accept Socratic skepticism about weakness of will in order to accept his attack on temporal bias. Socrates believes that we are optimizers and that our desires and passions reflect our beliefs about what is best. He interprets temporal proximity as inducing a change of belief about what is best, which means that the putative akrates does not act contrary to his practical beliefs at the time of action. But we might believe, instead, that agents are not always optimizers and can and do act on autonomous desires and passions that do not reliably track beliefs about what is best. On this alternative interpretation, we might think that temporal bias influences the agent's actions, not by changing her beliefs about what is best, but by triggering or inflaming good-independent desires or passions. Even so, we might still agree with Socrates that temporal proximity does not affect the magnitude of goods and harms and therefore should not affect their significance. Moreover, we might treat the diachronic instability of the bias as evidence of its irrationality: the brief hot judgment appears anomalous against the background of prospective and retrospective cool judgments. Similarly, we might think, the bias in favor of minimizing future suffering appears anomalous against the background of prospective and retrospective cool judgments that are temporally neutral. We might regard the diachronic instability of this bias as evidence of its irrationality.

Second, this preference does not generalize well. While I may have this preference for past over future pain, I don't have this preference, for example, about my own past and future *disgraces*. I might well prefer a smaller future disgrace to a larger past disgrace. Suppose that I drank too much at the firm's party last night and can't remember what I did. I overhear that someone made lewd remarks to the boss before vomiting on her dress in front of the whole gathering. I desperately hope that somebody wasn't me and would gladly commit a minor faux paux this evening in exchange for not being implicated in last night's huge disgrace. If the preference is limited to pains or perhaps a few bad things, then it may not challenge temporal neutrality per se. Again, the bias proves, on inspection, to be rather isolated in scope.

Third, notice that the preference only seems to hold for *one's own* pains. As Parfit concedes (§69), my preferences about the pain of others, including loved ones, seems to be temporally neutral. My daughter undertakes volunteer work in a remote and largely inaccessible part of the world. I receive a message from someone that traveled through her village that she has a terminal disease that has become quite painful and will soon kill her. I am depressed. When I am told later that this was substantially correct but mistaken about the timing so that my daughter has already died, I feel no relief that her pain is behind her. Yet again, this narrows the scope of the bias.

Fourth, all else being equal, prospective pain is worse than past pain that one cannot remember, because one can anticipate prospective pain and this anticipation is itself painful. But then there is a danger that our preference for past pain may not be a preference for the larger amount of suffering, as it needs to be to challenge temporal neutrality. For the comparison to be fair, we must do one of two things: (a) we must change the example so that the past suffering is something that one can recollect, just as prospective pain can be anticipated, or (b) we must change the example so that it involves administration of a drug that blocks anticipation of future pain, much as the doctors induce amnesia to block recollection of the pain of the operation. But if we modify the example in either of these ways, it is somewhat less clear that the preference for the past pain persists.

Finally, it seems quite possible that evolution might have favored a forward-looking bias that prioritizes the minimization of future pain, inasmuch as a concern with future pain could contribute to a creature's fitness in a way that a concern with past pain could not. By itself, an evolutionary bias for minimizing future pain would not undermine that bias. But if there are other reasons to question the robustness or the rationality of the bias, of the sort we have just canvassed, then the existence of an evolutionary explanation of that bias could help explain why we might be subject to this bias even if it is not rational. Together, these considerations would make it easier to reject this rather isolated form of a bias for the future as irrational.

I am not sure if these considerations completely undermine Parfit's defense of the bias for the future. The intuitions his example evokes are common and strong. But the example's handling of memory and anticipation is not fair, and it's not clear that our intuitions survive unchanged when the comparison is made fair. And, in various ways, any bias that can be uncovered turns out not to generalize well and to be unstable.

## 8. CONCLUDING REMARKS

This chapter has focused on issues about the intertemporal distribution of benefits and harms, especially within a single life. We have focused on prudence's demand of temporal neutrality, which is a traditional norm of intrapersonal distribution. It assigns no normative significance per se to the temporal location of benefits and harms within a person's life and demands equal concern for all parts of that life. After clarifying the commitments of temporal neutrality, we located its primary rationale in the principle of compensation. We saw that compensation provides a rationale for the hybrid structure of prudence—the fact that it is agent-biased but temporally neutral. This rationale appeals, in part, to assumptions about the separateness of persons. We saw that those assumptions are metaphysically robust and not upset by reductionist assumptions about personal identity. Even with this kind of support, temporal neutrality remains a controversial norm, in part because some of its implications seem counter-intuitive. Though it may seem to give controversial advice in cases involving intrapersonal conflicts of values or ideals, neutrality does seem defensible in light of the fact that ideals of equal value hold sway in periods of a person's life that are equally real and equally parts of her life. Though it might seem to be a philosophical liability to be committed to Epicurean ideas about the symmetry of death and prenatal nonexistence, that symmetry turns out to be surprisingly robust and defensible. Perhaps the most counter-intuitive implication of temporal neutrality is its rejection of our apparent preference for past over future pain, even when this means preferring more total pain. But this bias does not generalize well and remains limited in scope and unstable. Moreover, it may not survive once the example used to elicit the bias is corrected in certain ways.

Our discussion of legitimate worries about temporal neutrality has been selective, and my assessment of some of the worries has been sketchy and provisional. But we have seen a strong rationale for central features of temporal neutrality, and many of these worries have surprisingly good responses. A more systematic assessment of temporal neutrality would be comparative in nature—comparing it with alternatives, whose rationale and adequacy are explored in comparable detail. At this stage in the inquiry, the prospects for temporal neutrality still seem good.

## References

Ainslie, G. (1992), *Picoeconomics* (Cambridge).

—— (2001), *Breakdown of the Will* (Cambridge).

Aristotle. *Nicomachean Ethics*, trs. T. Irwin (Hackett, 1985).

Brink, D. (1994), "Moral Conflict and Its Structure", *Philosophical Review* 103: 215–47.

—— (1997a), "Rational Egoism and the Separateness of Persons" in *Reading Parfit*, ed. J. Dancy (Blackwell).

—— (1997b), "Self-love and Altruism", *Social Philosophy & Policy* 14: 122–57.

—— (2003), "Prudence and Authenticity: Intrapersonal Conflicts of Value", *Philosophical Review* 112: 215–45.

Butler, J. (1736), "Of Personal Identity" in *The Works of Joseph Butler* (William Tegg, 1847).

Dworkin, R. (1978), *Taking Rights Seriously* (Harvard).

Feldman, F. (1992), *Confrontations with the Reaper* (Oxford).

Korsgaard, C. (1989), "Personal Identity and the Unity of Agency: A Kantian Response to Parfit", *Philosophy & Public Affairs* 18: 101–32.

Kripke, S. (1980), *Naming and Necessity* (Harvard).

Lewis, C.I. (1946), *An Analysis of Knowledge and Valuation* (Open Court).

Locke, J. (1690), *An Essay Concerning Human Understanding*, ed. P Nidditch (Oxford, 1979).

Lucretius. *De Rerum Natura*, trs. R. Latham (Penguin, 1994).

McKerlie, D. (1989), "Equality and Time", *Ethics* 99: 475–91.

McMahan, J. (1988), "Death and the Value of Life", *Ethics* 99: 32–61.

Mitsis, P. (1989), "Epicurus on Death and Duration", *Proceedings of the Boston Area Colloquium in Ancient Philosophy* 4: 303–22.

Nagel, T. (1970), *The Possibility of Altruism* (Princeton).

—— (1979), *Mortal Questions* (Cambridge).

Nozick, R. (1974), *Anarchy, State, and Utopia* (Basic Books).

—— (1980), *Philosophical Explanations* (Harvard).

Parfit, D. (1984), *Reasons and Persons* (Oxford).

Plato. *Protagoras*, in *Plato's Complete Works*, ed. J. Cooper (Hackett, 1997).

Ramsey, F. (1928), "A Mathematical Theory of Saving" reprinted in *Foundations* (Routledge, 1978).

Rawls, J. (1971), *A Theory of Justice* (Harvard).

Reid, T. (1785), *Essays on the Intellectual Powers of Man*, ed. B. Brody (MIT, 1969).

Shoemaker, S. (1963), *Self-Identity and Self-Knowledge* (Cornell).

—— (1970), "Persons and Their Pasts", *American Philosophical Quarterly* 7: 269–85.

—— (1984), "Personal Identity: A Materialist Account" in S. Shoemaker and R. Swinburne, *Personal Identity* (Blackwell).

Sidgwick, H. (1907), *The Methods of Ethics*, 7th ed. (Macmillan).

Smith, A. (1790), *A Theory of the Moral Sentiments* (Oxford, 1976).

Velleman, D. (1991), "Well-Being and Time" *Pacific Philosophical Quarterly* 72: 48–77.

Wiggins, D. (1967), *Identity and Spatio-Temporal Continuity* (Blackwell).

Williams, B. (1976), "Persons, Character and Morality" reprinted in *Moral Luck* (Cambridge, 1981).

Wittgenstein, L. (1958), *The Blue and Brown Books* (Blackwell, 1958).

CHAPTER 12
......................................................................................................

# TIME, PASSAGE, AND IMMEDIATE EXPERIENCE

......................................................................................................

BARRY DAINTON

## 1. EXPERIENCE WITHOUT PASSAGE

......................................................................................................

FOR the past couple of minutes you have (let's suppose) been riding on an escalator; with little else to do, you've been gazing absent-mindedly at the advertisements on the wall as they slowly and smoothly slide by. How would your visual experience change if the escalator were to come to a stop? It's not difficult to predict: the advertisements would stop sliding by and you would find yourself staring at a single motionless patch of wall. And it would not be long before you would find yourself getting more than a little bored.

Let's consider a rather more dramatic variant of this scenario. One obvious difference between time and space is that the latter has three dimensions, whereas time has but one. A seemingly equally obvious, but arguably more profound, difference is that time *passes* but space does not. There is no spatial counterpart to the steady second-by-second advance of the present. Bearing this in mind, we can pose this question: what would it be like if time *ceased* to pass? How would things change if the future-directed movement of present were to come to a complete halt?

A natural first thought might be: 'It would be a bit like the elevator coming to a stop, but writ large: the whole *world* would come to a stop—it would be like someone hitting the pause button on the video player: everything would just *freeze*.' A little further reflection might bring this thought:

> Ah, it would be a lot worse than that. If the elevator were to stop I would continue to see my surroundings. I would no longer see the walls sliding by, but I would continue to see the wall as a thing there before me, albeit now motionless—this is because I would continue to have experiences, the flow of my stream of consciousness would be unaffected. But if time were no longer to pass, I would no longer have new thoughts or fresh experiences: if time were to stop flowing, so too would

my stream of consciousness. Consequently I wouldn't be *aware* of world's having stopped, I wouldn't *see* the people and things frozen in place. The lights would go out. Without the passage of time there wouldn't be experience, there wouldn't be anything that it is like to live *through* such a period.

While I would not want to suggest that this is the only 'natural' response to this 'What would it be like?' question, this response does I think have a fair measure of plausibility. For anyone unacquainted with (or, if you prefer, untainted by) philosophical theorizing about time, it *is* very natural to suppose that things would be very different if time did not pass. Narrowing the focus, it is very natural to think that our *experience* of the world would be very different if time did not pass.

That this is so is a significant obstacle to the acceptance of a conception of time to which many contemporary philosophers (and physicists) are drawn. According to the 'Block view' of time—also known as *eternalism*, or the *static* conception—the universe consists of a four-dimensional spread of objects and events. Time exists in this model (it is one of the four dimensions), but there is nothing corresponding to a single privileged and transient 'now': every event, irrespective of when it occurs, is fully and equally real, in the same way that events occurring at different *spatial* locations are fully and equally real. Just as people at different spatial locations pick out different places by their uses of 'here' without any one place being ontologically privileged, so too with 'now'. Clearly, if the Block view is correct, there is nothing in reality which corresponds to the steady future-directed advance of a privileged moment of presence. It follows that temporal passage, as commonly conceived, is unreal.

This gives rise to a question. If temporal passage *is* unreal, why do we have such a strong intimation that our experience would not be as it is if time were *not* to pass? This question has an obvious and quite plausible answer: we find passage in our experience, and we are naturally inclined to believe what our experience tells us. By way of an analogy, we naturally suppose that the objects we see are clothed in the colours we see. But for those who subscribe to the Lockean-style representational (or indirect) theory of perception, this is a mistake: colours as they feature in our immediate experience—*phenomenal colours*—are not to be found in external bodies, they exist only in our experience. But while there are many good reasons for supposing this view of perception is correct, it does not *seem* correct: it is not at all easy to believe that the objects we perceive do not really possess the colour-qualities we perceive them as possessing. If our experience of colour misleads us in this way, might not the same apply to our temporal experience? Might we not be *projecting* passage onto the world in the same way as we project colour?

For a Block theorist this line of thought holds a good deal of promise, but before we can begin to assess its merits, further clarification is needed.[1] Someone might object

---

[1] The relegation of the tensed or dynamic aspects of time to the mental realm has been a standard move for Block theorists for many years. Grünbaum has been a prominent defender of the 'mind-dependence of becoming', and proposed the analogy with colour experience (Gale 1968: 323);

to the Block view thus: 'I'm as certain as I can be of anything that this conception of time is wrong. Why? Because I am *directly aware* of the passage of time. What stronger evidence could there be?' If we were aware of time's passage, it might well be that some or all of the dynamic character of our experience derives from this awareness. But is the passage of time itself something we can perceive? It is certainly not *obvious* that we can. When travelling by car or train (or riding on escalators) we unquestionably see our surroundings passing by, but there is no obvious sign of *time's* passing us by. However, as William James observes, there are occasions when time's passing can seem discernible:

> Let one sit with closed eyes and, abstracting entirely from the outer world, attend exclusively to the passage of time, like one who wakes, as the poet says, to "hear time flowing in the middle of the night". There seems under such circumstances as these no variety in the material content of our thoughts, and what we notice appears, if anything, to be the pure series of durations budding, as it were, and growing beneath our indrawn gaze. Is this really so or not?
>
> (1890, *Principles*, 'We Have No Sense for Empty Time')

Are we to suppose we have (what James calls) a *special sense for pure time*? Or can we explain the appearances in more mundane terms? Perhaps the 'hearing' of time flowing is to be explained in terms of the hearing of something else. This was James' diagnosis:

> It takes but a small exertion of introspection to show that the latter alternative is the true one, and that *we can no more intuit a duration than we can intuit an extension, devoid of all sensible content.* Just as with closed eyes we perceive a dark visual field in which a curdling play of obscurest luminosity is always going on; so, be we never so abstracted from distinct outward impressions, we are always inwardly immersed in what Wundt has somewhere called the twilight of our general consciousness. Our heart-beats, our breathing, the pulses of our attention, fragments of words or sentences that pass through our imagination, are what people this dim habitat. . . . In short, empty our minds as we may, some form of *changing process* remains for us to feel, and cannot be expelled. And along with the sense of the process and its rhythm goes the sense of the length of time it lasts. Awareness of *change* is thus the condition on which our perception of time's flow depends.                (*ibid.*)

I think James was right to take this line: it seems very plausible to suppose that on those occasions when our consciousness is comparatively empty, and we might seem to be aware of nothing save time passing, what we are really aware of are various low-level sensory flows located in Wundt's 'twilight' regions of our consciousness. After all, if on such occasions our consciousness were *entirely* devoid of content we would, in effect,

---

for Russell 'past, present and future arise from time-relations of subject and object, while earlier and later arise from time-relations of object and object' (1915: 212).

be experiencing nothing at all, and if we are not experiencing *anything* we can hardly be experiencing the passage of time.[2]

The fact that *we* don't perceive time itself doesn't mean that time is necessarily unperceivable: there may be logically possible worlds where space-time is substantival, and as easily perceived by its inhabitants as any other part of the material furniture of their world. But our universe is not of this kind. It may well be that space-time in our world *is* substantival—if so, then space-time can rightly be regarded as a material object—but it is not an object that is readily or directly detectable by our senses or any other instrument. (If it were, the debate between substantivalists and relationists would have been settled long before now.)

Pulling these points together, the problem for the Block theorist is not explaining how we perceive the passage of time when in fact there is no such thing—we don't perceive the passage of time per se. We are, however, certainly aware of what James called the 'changing process' of our consciousness. To put it another way, the contents of our immediate experience flow: we are aware of sounds and sensations *continuing on*, we are aware of one content (one moment of sensation, or feeling, one fragment of thought or mental image) giving way to the next. Duration and succession are inherent in our experience—so much so, indeed, that it is tempting to think that this sort of dynamic character is an essential attribute of conscious states. More relevantly for present purposes, it also seems plausible to suppose that our impression that time itself passes or flows is heavily bound up with the dynamic, flowing, changing character of our ordinary everyday experience—the sort of experience we have during all our waking hours.

Hence the issue we need to address, and the issue I will be concentrating on in what follows: how can our immediate experience be as it is if time does not pass? Or to put it slightly differently: how can our experience have the flowing, stream-like character that it does in a passage-free universe? How can the calm, eternal character of the Block universe be reconciled with the turbulent, dynamic character of our immediate experience?[3]

---

[2]  For more on experiencing nothing, and the impossibility of a truly empty consciousness, see Dainton (2002).

[3]  The focus of Grünbaum's defence of the mind-dependence of becoming is the *now*: an event only qualifies as now (or present) if some '*mind-possessing organism* M is conceptually aware of experiencing the event at that time . . . For example, if I just hear a noise at a time *t*, then the noise does not qualify at *t* as *now* unless at *t* I am judgmentally aware of the fact of my hearing it at all and of the temporal coincidence of the hearing with that awareness' (1968: 332–3). So far as saving the (temporal) appearances goes, Grünbaum supplies some of what the Block theorist needs, but by no means all: he says nothing about the flowing, stream-like character of our experience. For the reasons just outlined, this very general and basic aspect of our experience may well be responsible for much of the most deep-seated resistance to the Block conception, the conviction—shared by many—that it omits what is most distinctively *timelike* about time. One further point here: these more elemental aspects of consciousness do not (obviously) require or depend on conceptualized awareness—couldn't the experience of animals or infants be stream-like?

# 2. OPTIONS

In responding to this question a number of options are open to the Block theorist. I will consider four of the more important.

(1)   *Reduction*: our ordinary experience may seem to flow or stream, but this poses no real problem. Since there is every reason to believe physicalism is true, there is every reason to believe that every feature of our consciousness can be accounted for in entirely physical terms. We may be beings with the capacity for conscious experience, but we are also composed of the same elementary ingredients as tables and chairs—ingredients that can be completely characterized in terms of the properties and relationships recognized by fundamental physics. Since fundamental physics requires no time other than tenseless block-time, the same must apply to our experience.

Many Block theorists may find this a tempting option, but they would be unwise to avail themselves of it. The main elements of this response may be correct—we may very well be wholly material beings—but accepting this does not, in itself, help with the problem we are engaged with here: what we are looking for is an explanation or elucidation of *how* our experience can have the dynamic character it clearly does have in a passage-free universe. Simply stating that there *must be* such an explanation will not satisfy or assist those who find it difficult (or impossible) to accept the Block theory precisely because they cannot see how their experience could be as it is in a universe of the Block type.

There is a second point to note here. From a particularly tumultuous couple of decades in the philosophy of mind only a few clear messages have emerged. Of these, the most obvious and important is that reducing the phenomenal to the physical is proving more difficult by far than many once hoped or assumed. Even the most sophisticated attempts to reduce the phenomenal (or experiential) to the material, causal, functional, or computational have encountered serious problems. Not surprisingly this has given renewed impetus to a variety of non-reductionist approaches. Some have opted for the dualist route. According to the doctrine of 'non-reductive functionalism' elaborated by Chalmers, experiences are immaterial particulars that are nomically correlated with information-processing activity in physical things (1996: ch.8). In an only slightly less radical vein, others have suggested that the conception of the physical realm to be found in physics is inadequate or incomplete, a proposal which opens the way for taking phenomenal properties to be previously unrecognized *physical* properties of a basic and irreducible kind—that is, properties of matter which exist in addition to mass, charge, momentum, and the other properties recognized by physics in its current (incomplete) form.[4] Of course, there are many who remain convinced that some form of reductionist account will one day prove viable, and those in this camp see no need or reason to embrace the more radical alternatives. But for

---

[4] In this camp see, for example, Lockwood (1989), Strawson (1995), Stoljar (2006).

present purposes what matters is simply this: since it is too early to tell which of these approaches will turn out to be closest to the truth, it would be a mistake for Block theorists aiming to convince a wide audience that their conception of time can accommodate our consciousness simply to assume that the problematic phenomenal features will simply disappear when one or other strong reductionist programme wins the day. And if it should turn out that phenomenal properties—in their familiar, unreduced form—are as much a part of the material world as mass or charge, the problem of reconciling their dynamic character with the eternal character of the Block universe will be all the more pressing.

Let us move on to a second line of response. Appealing to psycho-physical reductionism is one way to dissolve the problem, but it is not the only way:

(2)  *Sanitization*: descriptions of ordinary experience as 'flowing' or 'streaming' misrepresent the true character of our experience; when accurately characterized, our experience is static in character, and lacks any features which are in any way problematic from a Block-theoretical perspective.

Thomas Reid is one notable philosopher who fully recognized that it is extremely natural for us to talk as though we perceive change, but who also held that this talk should not be taken literally, at least not when we are engaged in serious philosophizing. Reid's reason for taking this stance might well seem plausible: 'if we speak strictly and philosophically, no kind of *succession* can be an object of either the senses or of consciousness; because the operations of both are confined to the present point of time' (1855: 235). Holding that our ordinary ways of talking are misleading (at least for strict philosophical purposes) solves one problem, but another remains: what should we say about the experiences we find it so natural to describe in dynamic terms?

Reid turns to the one form of consciousness which we know can provide us with access to the past, both distant and recent: 'philosophically speaking, it is only by the *aid of memory* that we discern motion, or any succession whatsoever. We see the present place of the body; we remember the successive advance it made to that place: the first can, then, only give us a conception of motion, when joined to the last (1855: 236–7).

Reid gets some things right. It does seem common sense to hold that our immediate experience is confined to the present, and that our experiential access to the past is via memory.[5] It is also true that we frequently employ the sort of memory-based reasoning Reid sketches. If I look out of the window and see that my neighbour's car is no longer parked where I remember seeing it this morning, I will probably infer that it has moved. More generally, from one day (or hour) to the next, we would have little or no sense of future events forever getting closer if we were unable to compare our current location on the time-line with memories of occupying earlier

---

[5]  Of course in the case of distant events—such as thunderstorms (as heard) or supernovae (as seen)—the fact that sound and light travel at a finite speed means we perceive the events in question some time after they actually occurred, but the resulting perceptual experiences (when they eventually occur) certainly seem to occur in the present.

locations. However, Reid's claim that the *only* way in which we become aware of change or motion is via memory-based inferences of this sort is not very plausible at all. As many philosophers have noted, from a phenomenological standpoint we seem to experience change or movement with the same immediacy as we experience colour or pain.[6] The wings of a flying sparrow (or hummingbird) are moving so quickly we see them only as a blur—or not at all—but the more sedate movements of the wings of an eagle in flight are clearly and *cleanly* visible, with no hint of a blur. When a saxophonist plays a swooping melody line, we can *hear* the rising and falling pitch; when we hear a melody we hear each successive tone flowing into the next. It is not only change which is directly perceivable, persistence is too: think of what it is to hear a single tone *continuing on*: isn't there a constant (and constantly *experienced*) renewal of auditory content? Quite generally, both change and persistence feature prominently in our immediate experience.

Of course Reid would insist that while it is natural to describe these sorts of experience in such terms, these terms do not reflect the strict truth of the matter. When we see an eagle falling from the sky in pursuit of its prey, our visual consciousness actually comprises a succession of momentary visual experiences, each possessing an entirely static content (of the eagle frozen at a particular location). The impression of motion derives from the fact that each of these static visual images is accompanied by memory-images of the eagle as it appeared at previous locations. Or so Reid would claim.[7] But this proposal suffers from a severe credibility problem. In the case of the eagle's descent, the phenomenological datum to be explained is motion at the sensory or impressional level—motion *as seen*. How could this be created by the addition of memory-images to perceptual experiences that are themselves duration-less? The phenomenal character of a memory-image of a visual experience, and a visual experience itself, are very different; we are never in any danger of confusing the two—roughly speaking, memory-images are far less vivid, far less detailed, than

---

[6] Russell tells us that '*Succession* is a relation which may hold between two parts of the same sensation' (1913: 65); for Broad (1923: 287) experienced motion and rest are basic features of visual experience on a par with colour; more recently and in a similar vein, Foster observes that 'duration and change through time seem to be presented to us with the same phenomenal immediacy as homogeneity and variation of colour through space (1982: 255).

[7] Reid is by no means alone in opting for this line: also see Mellor (1998: 114–5); for a more sophisticated treatment involving some additional ingredients, see Le Poidevin (2007: ch. 5). Drawing on Dennett's treatment of the phi phenomenon, Paul (forthcoming) offers an alternative diagnosis: 'More generally, when we have an experience of passage, we can interpret this as an experience that is the result of the brain producing a neural state that represents inputs from earlier and later temporal states and simply 'fills in' [fn: Not literally. It just gives the impression of being filled in. There is no 'figment', as Dennett would say . . .] the representation of motion or of changes. So on this account , , , there is no real flow or animation from one time to the next. Rather, our brains create the *illusion* of such flow'. Holding that our experience does not really possess the dynamic features it seems to possess, that we merely *believe* it does, may make life easier, at least in some respects, but the cost is high: we are being asked to endorse a highly revisionary account of what our experience is like. For a critical assessment of Le Poidevin and Dennett, see Dainton (2010: §4.4, §4.5 & §7.3).

the corresponding perceptual experiences. Combining memory-images (in the right sort of way) might conceivably generate a *memory* of seeing something in motion, but it can never amount to *seeing* motion first-hand. Might salvation lie in the fact that the posited momentary perceptual experiences form a gap-free continuum? It might if such a collection of experiences could secure continuity at the sensory or impressional level, but it is difficult to see how this could come about, given the assumptions currently in play. As James remarked, a succession of experiences (in and of itself) does not amount to an experience *of* succession. Unless the momentary experiences are *apprehended together*, they are doomed to remain experientially isolated from one another. But clearly, for them to be *experienced* together, consciousness must in some manner extend beyond the 'present point of time' to which Reid confines it.[8]

If the problematic phenomena cannot be reduced or re-described, the Block theorist has no option but to accommodate it somehow. Here is one quite radical way in which this can be done:

(3) *Exclusion*: our experiences are dynamic in a way the Block universe isn't, but our experiences are not parts of the Block universe.

Some famous lines by Weyl suggests a position of this sort: 'The objective world simply *is*, it does not *happen*. Only to the gaze of my consciousness, crawling upward along the life line of my body, does a section of this world come to life, as a fleeting image in space which continuously changes in time' (1949: 116).[9] Phenomenologically speaking Weyl's image is appealing: it can seem that we are apprehending the world through a narrow window that is steadily advancing into the future. However, taking this picture literally is not without its costs and consequences.

Since (ordinary) time makes up one of the Block universe's four dimensions, to make sense of a consciousness (or point of apprehension) *moving along* the block— and so moving through time—we need to posit an additional temporal dimension, a *meta-time*. The required mechanism is depicted in Figure 12.1.

Such a scheme may be metaphysically coherent, but it is far from unproblematic. First of all, we are being required to embrace an unusually radical form of psycho-

---

[8] As is well known, a succession of static images flashed onto a screen at a rate of around 15 frames per second (or above) can be seen by us as continuous, an effect (a variant of the phi phenomenon) which underlies cinema and TV technology. But it is a mistake to conclude from this that a continuous stream of consciousness can be formed merely by placing momentary experiences with static contents side-by-side, as it were. When we watch a movie the stimuli hitting our retinas may be discontinuous, but this discontinuity does not survive processing carried out in our brains' visual centres. The resulting visual experience is fully continuous. Furthermore, and contrary to what some philosophers (e.g., Dennett and Paul, see note 7) have maintained, the content of this experience is not itself static: it typically involves *movement* that is directly apprehended as such (there is all the difference in the world between watching a movie, and looking at a collection of still images). There is thus a strong case for holding that the 'still' images that are shown on-screen have no phenomenological reality: they function as *stimuli*, but the experiences to which they give rise are (typically) wholly dynamic in character, rather than static or motion-free.

[9] Or at least it does if Weyl's words are taken literally, which may well not have been his intention.

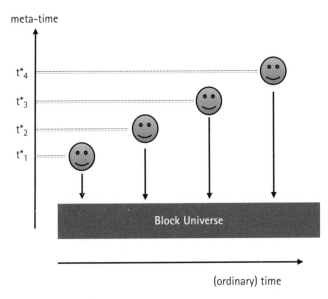

FIGURE 12.1  Gliding along the timeline requires a meta-time

physical dualism. If our states of consciousness are external to the four-dimensional Block universe, then on the natural assumption that the latter comprises the entirety of the physical universe, our states of consciousness must themselves be non-physical. Now, as just noted in connection with the Reductionist response, this sort of dualism has enjoyed something of a comeback in recent years, so this is not as damaging as would otherwise be the case. Even so, while Descartes and his contemporary followers are of the view that conscious states are immaterial, they do not think it necessary to posit an additional temporal dimension to house these immaterial entities. Weyl-type dualism is thus more radical than most. It goes without saying that introducing an additional temporal dimension is an ontologically costly move, and not one that should be made until lest costly alternatives have been explored and found wanting.

A further difficulty is perhaps less obvious. What *sort* of time is the proposed meta-time? If it too has the character of a Block universe then from the point of view of explaining the temporal appearances absolutely nothing has been gained. So meta-time must be a *dynamic* temporal dimension of some kind. This will not be acceptable to those Block theorists who doubt the intelligibility of dynamic conceptions of time. It also means that the Exclusionist has to supply a credible explanation of how the dynamic features of meta-time contribute to the dynamic character of our consciousness. And as we shall see in due course, this task is non-trivial.

None of the options surveyed thus far is very palatable. Is there nothing better on offer? As it happens, there *is* an alternative path open to Block theorists, and on the face of it a rather more promising one:

FIGURE 12.2 Integrating change: the seeing of a falling block in a Block universe. The resulting experience—unlike this static representation—would have a dynamic content: it would be of a block *moving*. According to the Integrationist, phenomenal contents with highly dynamic characteristics such as this can exist within Block universes

(4)   *Integration*: the claim that many of our experiences have a dynamic, flow-ing, character is correct, but these experiences are as fully integrated into the ordinary four-dimensional continuum as planets, bricks, earthquakes or hurricanes.

This is the preferred path of D. C. Williams who held that the 'whoosh of process' and 'the felt flow of one moment into the next' are part and parcel of the 'concrete stuff of the manifold' (1951: 466–7). Integrationists do not reject passage, but they do hold that it is confined to the contents of our consciousness. It follows that in one sense at least our talk of passage corresponds to a real feature of the world: even if times and events do not themselves undergo passage, such talk corresponds with how things *seem*, and these seemings are perfectly ordinary—perfectly real—ingredients of the four-dimensional manifold.

Those who believe experiences are themselves material in nature can of course follow this path, obviously, but so too can dualists. The defining claim of the Inte-grationist is not that streams of consciousness and streams of water are composed of the same sort of (physical) stuff, but that they belong to the same temporal framework.

In a slightly more precise and explicit vein, the Integrationist is forwarding the following combination of claims:

(1)   Our sense that time passes or flows derives (to a significant extent) from the dynamic features of our immediate experience.
(2)   Time in our universe is as the Block theorist maintains, and our experiences are themselves parts of a Block universe.
(3)   The dynamic character of our immediate experience does not require time itself to be dynamic.

If (2) is true, (3) obviously follows, but the latter is nonetheless worth stating explicitly. Many of those who reject the Block conception in favour of a dynamic conception of time do so because—in part at least—they believe that the dynamic character of our experience *derives from* the dynamic character of time. To put the point simplistically: it seems as though our experience is confined to a momentary present that is steadily advancing into the future because *there is* a momentary present that is doing precisely this. However, the situation is by no means so straightforward, for there are very different dynamic conceptions of time. To mention just a few of the leading contenders: some dynamists hold that there is a privileged point of presentness moving along a four-dimensional block, others hold that only the present moment exists, others hold that the present moment is but the most recent addition to a constantly expanding past. Given this divergence of opinion in the dynamist camp, there is no *one* way for the passage of time to generate (or contribute to) the appearance of passage in consciousness.

Clearly, assessing the merits of the Integrationist case will require us to examine the relationship between the dynamic character of experience and the various dynamic models of time. I will reserve this task for sections §6-8 below. But I will start with experience. It is one thing to recognize that our immediate experience has a dynamic character, it is another to explain how it is possible for it to have this character. Reconciling real time with time as it is manifest in our experience is one problem, but it is by no means the only one: the temporal features of immediate experience can themselves be baffling in the extreme.

## 3. THE EXPERIENCED PRESENT

For Integrationists our impression that time (or the world as a whole) is continually undergoing passage depends to a significant extent on our direct awareness of change and persistence. However, the claim that we can directly apprehend change and persistence can easily seem downright paradoxical. Change and persistence both take time. Accordingly, for us to apprehend change and persistence, our awareness must be able to embrace or encompass a temporal interval. But isn't our immediate awareness confined to the present? We can remember experiences we have enjoyed at earlier times, and anticipate what the future will bring, but we are only *directly* aware of what is occurring in the present. Since the present lacks duration, if our experience is confined to the present then—as Reid points out—it seems we cannot possibly directly experience change, or persistence.

Faced with this apparent paradox there is an obvious and appealing move available to philosophers who are determined to accommodate the phenomenological data. To circumvent the problem it suffices to draw a distinction between two forms of the present: the ordinary or *strict* (or mathematical) present, and the experiential or *phenomenal* present. Whereas the strict present is durationless, the present-as-experienced has a small but finite duration: it possesses just enough duration to allow us to apprehend change and persistence in the way that we commonly do. Hodgson referred to this interval as the 'experienced present' (1898: II.2). Husserl preferred to call it 'the living present' (or *lebendige Gegenwart*). That the phenomenal present became known as the 'specious' present in Anglo-American circles is largely due to William James' use of the term in his influential discussion in *The Principles of Psychology*. James regarded the strict (durationless or mathematical) present as 'an altogether ideal abstraction, not only never realized in sense, but probably never even conceived of by those unaccustomed to philosophic meditation' (1890: 608). Hence for James the only present we ever encounter in experience is the specious present: 'the original paragon and prototype of all conceived times is the specious present, the short duration of which we are immediately and incessantly sensible' (1890: 631). Despite some misgivings, to avoid complication I will stick with James' term.[10]

On the face of it, the doctrine of the specious present provides the Integrationist with much of what they need. Specious present possesses some (apparent) temporal depth, and their contents (by hypothesis) contain the change and persistence we find in our immediate experience. Further, if the span of our awareness is limited to a brief interval, it is hardly surprising that we have the impression that our consciousness is confined to the present: so far as the appearances go—so far as our experience goes—it *is*. But the fact that at any point during our waking lives our direct awareness extends only over a brief interval is perfectly compatible with our having experiences at other temporal locations which are just as real as our present experience (as the Block conception entails). If the span of our awareness were wider—if the specious present were broader—we would be aware of some (or even all) of these experiences, but since it isn't, we aren't.

However, for the specious present to perform these valuable services it must be possible to provide a coherent account of it, and there are those who are sceptical as to whether this is possible. Some of the difficulties can be traced back to James himself, and some of the claims made in the *Principles*. In a well-known passage he characterizes the specious present thus:

> the practically cognized present is no knife-edge, but a saddle-back, with a certain breadth of its own on which we sit perched, and from which we look in two directions into time. The unit of composition of our perception of time is a duration, with a bow and a stern, as it were—a rearward—and a forward-looking end. It is only as parts of this duration-block that the relation of succession of one end to the other is perceived. We do not first feel one end and then feel the other after it, and from the perception of the succession infer an interval of time between, but we seem to feel the interval of time as a whole, with its two ends embedded in it.
> (1890: 609–10)

There is nothing obviously absurd here. The claim that 'we look in two directions into time' *can* seem absurd if read literally: the ability to see into the past or future is an

---

[10]  From a purely phenomenological perspective there is nothing in the least illusory or deceptive (or specious) about the 'presence' of the contents appearing within a specious present: nothing could appear *more* present than the colours, sounds, thoughts, and feelings which we encounter in our immediate experience.

ability that is usually only ascribed to proud owners of crystal balls. But if the claim is simply taken to mean that as we go about our business we have some awareness (thanks to our memories) of what we have already seen and done, and an anticipatory awareness of what lies ahead, then it seems entirely innocuous.[11] Unfortunately, this innocuous reading sits uneasily with other things James says.

When James talks of the length (or duration) of the specious present he usually means the amount of change, as measured by normal clock-time, that we are able to apprehend as a whole. A few simple experiments suggest this duration is quite short. Rap a table several times in succession, trying for one tap per second. Ignoring any echoes, are the actual *sounds* made by earlier taps still present in your consciousness when the later sounds come along? Clearly not, which suggests that, in the auditory case, the specious present cannot embrace more than a second of change. Although the estimates James supplies vary a good deal, in general he is more accommodating: 'our maximum distinct *intuition* of duration hardly covers more than a dozen seconds (while our maximum vague intuition is probably not more than a minute or so)' (1890: 630). That James opts for a dozen seconds here—and at several other places—is baffling: what can he mean by 'intuition of duration'? This mystery aside, these longer estimates make for problems when combined with the saddle-back analogy. Are we to suppose that we are able to *see* six seconds into the past and future? Do we have a dim perceptual awareness of what will happen thirty seconds from now? If so, it must be so dim we aren't aware of it. And if we *were* aware of it some interesting questions would arise. If you dimly perceive yourself getting flattened by a truck while crossing a road half a minute from now, could you choose not to cross? It is considerations such as these which lead Plumer to conclude: 'consistently construing the sensory present as an interval would cause nothing less than a *riot* in our conceptualization' (1985: 4).

Rescuing the specious present from these riotous consequences is easy enough. The first step is to ignore James' puzzlingly high estimates for its temporal span: as just noted, considerably shorter estimates (a second or less) are more plausible by far. Next to go is any trace of mysterious clairvoyance: let us assume henceforth that the perceptual content of a specious present of a subject $S$ occurring (or starting) at a time $t$ is determined by the sensory data which falls on $S$'s sensory organs just prior to $t$—what occurs after $t$ contributes nothing. Taking these steps does not mean entirely abandoning James' saddle-back analogy, it simply means that we should construe it in the innocuous way: one doesn't need to be a clairvoyant to *anticipate* what the future will bring, and act accordingly.

Purging the specious present of these more dubious Jamesian ingredients helps with some problems, but not all. In the eyes of some, the doctrine of the specious present is irredeemably flawed. The contents of a single specious present are supposed to *seem present*; they are also supposed to have some apparent temporal depth: the contents are not compressed into a single instant, they are (seemingly) spread through a brief interval. Since one part of this interval will be experienced as occurring before the

---

[11]  See Gallagher [this volume] for a more detailed look at some aspects of this.

other, how can both parts also seem present? If both parts seem present, won't they be experienced as simultaneous? Here is Le Poidevin making this point in a recent discussion: 'If we have a single experience of two items as being present, then, surely, we experience them as *simultaneous*. Suppose we are aware of A as preceding B, and of B as present. Can we be aware of A as anything other than past?' (2007: 87). Le Poidevin fully appreciates the strength of the phenomenological considerations which motivate the postulation of the specious present, but since he can see no way around this difficulty, he falls back on Reid's position: 'What gives rise to the experience of pure succession . . . is the conjunction of the perception of E with the very recent memory of C' (2007: 92).

We can call this difficulty the 'simultaneity problem'. In fact, there is not one simultaneity problem, but two. This is because there are two fundamentally different conceptions of the specious present to be found in the literature. More specifically, there are two conceptions of the temporal character of *individual* specious presents. Proponents of both approaches agree on the character of the contents of an individual specious present—they house the change and persistence we encounter in our immediate experience—but they diverge markedly on the question of how these contents are related to ordinary objective time. On one view, call it the 'Extensionalist model', a specious present is a temporally extended episode of experiencing, just as its contents would suggest. On the alternative view—call this the 'Retentionalist' model— specious presents are *not* in fact extended through time, rather, they are momentary (or extremely brief) states of consciousness with a content which *appears* to be spread through a brief temporal interval.[12]

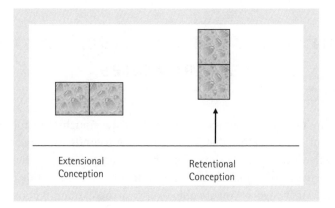

FIGURE 12.3 Two conceptions of the specious present

---

[12] Foster is a leading advocate of the Extensionalist approach (1982, 19 92), also Dainton (2000: ch. 7), (2003). Husserl (1991) is probably the most influential Retentional theorist—see Miller (1984) for a useful introduction. For further discussion see Gallagher and Varela (2003), Kelly (2005), Zahavi (2007).

We will be taking a closer and more detailed look at how these competing models can be developed shortly. But we have already seen enough for it to be clear that the relationship between presentness and simultaneity poses very different problems depending on which approach is adopted:

> **Extensional Simultaneity Problem:** how is it possible for contents which are all experienced together, and all experienced as present, to be experienced as successive rather than simultaneous?

> **Retentional Simultaneity Problem:** how is it possible for a collection of contents which occur simultaneously, at the same moment of time, to seem successive?

For either conception of the specious present to be viable, the associated Simultaneity Problem must be solved or circumvented. But before proceeding to take a closer look at these different approaches, it is worth noting that both face a further challenge.

The Simultaneity Problems arise in connection with the accounts being offered of the structure and composition of *individual* specious presents. We also need an account of how individual specious presents connect together to form continuous streams of consciousness. We can experience change and persistence over short intervals of time, of the order of a second or so, but we can also experience both change and persistence over much longer periods: recall your escalator trip, and the way each brief phase of your experience slides seamlessly into the next. More generally, many of us enjoy continuously streaming experience from when we awake in the morning until we fall asleep in the evening. How do individual specious presents—assuming such things exist—manage to combine in this way? Call this the *Connection Problem*. As we shall see, it too is non-trivial.

# 4. THE SPECIOUS PRESENT: EXTENSIONAL APPROACHES

We can start by taking a closer look at how the Extensionalist conceives of a single specious present. Let's suppose that for the past few seconds you have been watching a ball roll slowly down a gentle slope. Figure 12.4 depicts, in simplified form, a brief phase of your stream of consciousness during this period. Since this phase contains as much perceived change as you are able to apprehend at once, it constitutes a single specious present.

The vertical arrow on the left serves as a reminder that at any moment during this period, the diverse contents of your consciousness are unified: they are *synchronically co-conscious*, that is, they are experienced together, as parts of a single state of consciousness. So far as the diachronic aspect of your experience is concerned, there are several points to note. First, this specious present is extended a certain distance through ordinary time—as indicated by the lower time-axis. Second, it also *seems*

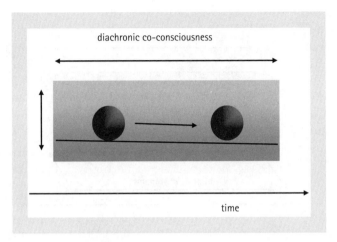

FIGURE 12.4  An Extensionalist specious present: a ball rolling along the floor

(subjectively) to have a certain temporal depth, and its contents are dynamic: in this case, they present a *ball in motion*. Third, although the successive phases of this specious present are experienced as occurring *in* succession—you see the ball move from one place to another—you also experience them together, as part of a temporally extended whole: the contents are thus *diachronically co-conscious*, as indicated by the upper arrow. By virtue of being thus unified, the specious present can be regarded as a *single* experience.

Is there anything incoherent here? Recalling the Extensional Simultaneity Problem it might seem so. If we regard the specious present as a single experience, its parts must all be *present* in the manner of any experience; if these parts are not only present but experienced together, how can they fail to be experienced as simultaneous? In fact, this reasoning is guilty of a conflation of two senses of 'present'. The Extensional theorist who takes 'A and B are experienced as present' to mean 'A and B are experienced as occurring at precisely the same instant' faces a problem: with 'present' construed in this way, A and B cannot fail to be simultaneous. But when Extensionalists talk of 'present' in this context they mean (or should mean) something very different. There is clearly a sense in which it is true to say that a remembered pain doesn't have the same *presence* as an actually experienced pain of the same type. Here 'presence' simply denotes the phenomenal immediacy and vivacity that all experiences possess at the time they occur—and which their remembered (or anticipated) counterparts possess to a lesser degree. The contents of a specious present (Extensionally construed) all possess presence in *this* sense. Is there anything puzzling or problematic about contents at different times possessing presence in this sense? Certainly not from the perspective of the Block conception of time, for according to the latter, experiences occurring at different times are equally real, and experiences of the same intrinsic type possess the same phenomenal properties, presence included, irrespective of where or when they occur.

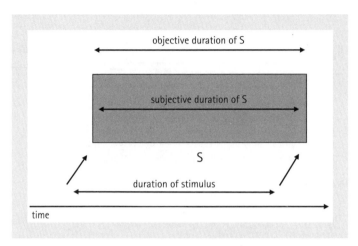

FIGURE 12.5 A single Extensional specious present and its relationship to objective time

The latter point has broader relevance. A potentially devastating objection to the notion that contents occurring at different times can be diachronically co-conscious runs thus: 'If A occurs before B, A can't be experienced *with* B, because by the time B occurs A has ceased to be present, and so has ceased to exist.' This objection may have considerable force for those who believe reality is confined to the present moment, but it has none whatsoever for the Block theorist.

Extensional specious presents may extend through time in the way their contents suggest, but there are nonetheless certain complications which are worth noting. Figure 12.5 depicts another simple specious present, S, this time of a single extended C-tone.

After the stimulus which triggers S commences, there is a short delay before S itself begins; this lag reflects the time required for processing the (in this case) auditory stimuli in the ear and brain. The *objective* duration of S is the amount of clock time the whole experience takes up. In this particular case the objective duration and the duration of the stimulus closely match, and no doubt this is usually the case—if it were not, our perceptual experience would not be as well synchronized with the world as it is—but the Extensional model can accommodate discrepancies between these magnitudes.[13] What of the rationale for distinguishing between the objective and *subjective* duration of S? The rationale for marking this distinction lies in the well-known fact that there are occasions when (as we say) time seems to pass more slowly (or quickly) than usual—for example, five minutes standing in a queue can seem an eternity. Of course if time doesn't pass—as we are currently assuming—such *façons de parler* must not be taken literally. Quite what underpins variations in apparent

[13] The divergence can be quite marked at smaller timescales: a visual stimulus of 1 msec can give rise to a visuals sensation lasting up to as much as 400 msec—a phenomenon known as 'visible persistence', without which fireworks would be a good deal less spectacular than they are. For more see Weichselgartner and Sperling (1985).

FIGURE 12.6 One conception of how Extensional specious presents combine to form streams of consciousness

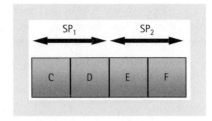

duration (or rate of flow) is not well understood, but for present purposes all that matters is that the Extensional theorist looks to have the resources to accommodate such fluctuations.[14]

So much for individual specious presents, but what of the Connection Problem: how do they combine to form streams of continuous consciousness? One option is to suppose they are lined up end-to-end, like a row of wooden blocks. This proposal is illustrated for a simple case in Figure 12.6, which represents the hearing of a sequence of tones C-D-E-F. The first two tones are experienced together in a first specious present $SP_1$, and the second two tones are experienced together in a second specious present $SP_2$. Taken together $SP_1$–$SP_2$ constitute a continuous, gap-free sequence of tones. Can we suppose our streams of consciousness have a structure of this kind?

Not given the assumptions currently in play. For the Extensional theorist, contents at different times are experientially unified only if they are connected by the relationship of diachronic co-consciousness—only then do they form parts of a single specious present. In the present case, C and D are related in this way, as are E and F, but D and E are not. Consequently, D and E do not belong to a single specious present, they are not experienced together, and a fortiori they are not experienced as successive. In effect, $SP_1$ and $SP_2$ constitute two entirely distinct (experientially speaking) streams of consciousness of short duration.

A suitable remedy is not hard to find. To secure the missing phenomenal continuity it suffices to introduce a third specious present, one which connects D and E, in the manner of $SP_3$ in Figure 12.7 below. With $SP_3$ in place, D is diachronically co-conscious with E and hence C-D-E-F form an experientially continuous stream of experiences, with each tone being experienced as flowing into its successor.

More generally, from an Extensional perspective, phenomenal continuity can be secured—and the Connection Problem solved—if neighbouring specious presents in a single stream of consciousness *overlap* in the manner of $SP_1$, $SP_2$ and $SP_3$. It is important to note that the mode of overlap consists of the *sharing of common parts*: the D-content which forms the second half of $SP_1$ and the D-content which forms the

---

[14]  There are at least two potentially relevant variables in this connection. A given objective interval might contain phenomenal contents with different subjective durations, but also, it may be possible for the breadth of the specious present to vary. Lockwood (2005: ch. 17) speculates that a diminution of the span of the specious present may be responsible for the apparent slowdown in the surrounding world which occurs during periods of extreme stress (e.g. in the midst of a car accident).

FIGURE 12.7 An alternative—and more plausible—conception of how Extensional specious presents combine to form streams of consciousness

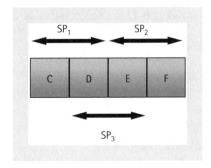

first half of SP$_3$ are numerically identical—and similarly for the E-tone which forms the second half of SP$_3$ and the first half of SP$_2$. This identity ensures that experienced continuity does not come at the cost of an unrealistic repetition (or duplication) of contents: D occurs in both SP$_1$ and SP$_3$, but—in virtue of the numerical identity—it is experienced but once.

More realistically, there is no reason to think specious presents in real-life streams of consciousness occur at regular, widely spaced intervals in the manner of SP$_1$ and SP$_2$. Indeed, if they were so spaced, content-patterns which are actually experienced—such as the content comprising the second-half of C, D and the first-half of E—would not be experienced as temporally extended wholes in the way they in fact are. To accommodate this fact we need simply hold that specious presents are more closely-packed, as shown in Figure 12.8.

This figure depicts a short stretch of a stream of consciousness, with each double-headed arrow representing a distinct specious present. In fact, the packing will be a good deal denser than shown on the page, with adjoining specious presents separated by barely discernible intervals. That neighbouring specious presents possess common parts means they can overlap almost completely without any risk of repeatedly experienced contents.

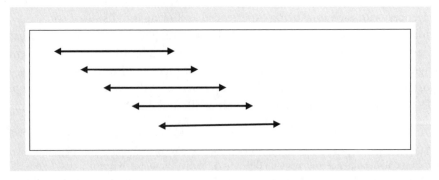

FIGURE 12.8 A more realistic depiction of the distribution of specious presents in an Extensional stream of consciousness

# 5. THE SPECIOUS PRESENT: RETENTIONAL APPROACHES

The Extensional approach is not obviously incoherent, and it promises to be able to do full justice to the dynamic and continuous character of our ordinary experience. What of the Retentional alternative? As with the Extensional approach, it will be useful to start by taking a closer look at an individual specious present. The basic elements of the Retentional picture are depicted in Figure 12.9.

Once again we are dealing with an artificially simple example: this time our subject is experiencing nothing but an extended C-tone. The diagram depicts our subject's consciousness at a single moment $t$. At this moment our subject is enjoying a full specious present, and the latter has an apparent duration of around one second. This momentary state of consciousness consists of a sequence of representations— Husserl called these *retentions*—of just-past occurrences (in this case, earlier phases of the C-tone), along with a momentary episode of fully present-seeming phenomenal content—a *primal impression* in Husserlian terminology.

Retentions bear some obvious similarities to ordinary memories, but as we saw earlier, in connection with Reid's proposal, the contents of ordinary memories are of the wrong sort to supply the sort of change we encounter in our immediate experience. Like memories, retentions are presently occurring representations of earlier experiences, but they are significantly more detailed and vivid: they are more like ordinary experience, they have greater *presence*. But although retentions possess a degree of presence, in standard Retentionalist schemes, they do not all possess it to the same

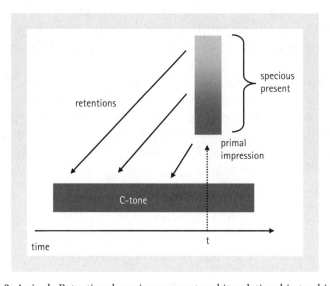

FIGURE 12.9  A single Retentional specious present and its relationship to objective time

degree. Within the confines of a single specious present, retentions possess varying degrees presence, dropping from a maximum to a minimum—as indicated in Figure 12.9 by the gradual diminution of colour-density, as we move from bottom-to-top. The phenomenal consequence of this variation is that contents within a single specious present appear as successive, but also as increasing *less present* (or as *more past*). Or to put it another way, the contents within a single specious present appear under different *temporal modes of presentation*: fully present, slightly in-the-past, slightly more in-the-past, and so on.

This appeal to temporal modes is by no means an ad hoc complication, it is crucial to the plausibility of the Retentional model, as we shall see shortly. But we have already seen enough of the Retentional specious present to assess the threat posed by the Simultaneity Problem. How can contents which occur at a single moment of time also possess temporal extension? It is surely impossible for anything to be both extended and non-extended at the same time. However, the Retentionalist can respond that there is no real difficulty here at all, for as a cursory glance at Figure 12.9 makes plain, the temporal extension of the Retentional specious present lies *orthogonal* to ordinary time: in effect, the Retentional theorist is offering a two-dimensional account of phenomenal temporality. Hence the fact that the specious present has zero extension in one temporal dimension—that of ordinary clock time—is perfectly compatible with its having non-zero extension in a second temporal dimension.[15]

Recalling the criticism laid at the door of the Weyl-inspired Exclusionist earlier, isn't this a wildly extravagant solution, one that is highly vulnerable to Occam's razor? Arguably not: there is no commitment to matter-consciousness dualism, and the additional temporal dimension is of comparatively small size, being no more than the breadth of a single specious present, and localized in individual brains. Also relevant here is how little we know about the precise relationship between matter and consciousness. Given this ignorance, we are hardly in a position—at least at present—to deem it impossible for the conscious states generated by momentary brain states to possess apparent (phenomenal) temporal depth. Or so the Retentionalist could reasonably maintain.

The next question to be addressed is the Connection Problem, and the manner in which specious presents, construed along Retentional lines, combine to form streams of consciousness. Are there gaps (objectively speaking) between neighbouring specious presents, or are they packed in a dense, gap-free manner? Retentionalists generally subscribe to the latter option, as approximated in Figure 12.10, where each vertical represents a distinct specious present.

Given the way our sensory experience can reflect sensory stimuli in a (more or less) continuous fashion, the close-packing of specious presents is certainly what the Retentional theorist requires. However, it potentially gives rise to a problem of *surplus content*. To illustrate: suppose each objective minute typically contains one thousand specious presents; if each of the latter contains around one second's worth of change,

---

[15]    Dobbs (1951) explicitly recognizes as much.

then a typical one minute of ordinary experiencing would contain 1,000 sec (around 17 minutes) of experienced change—surely an absurd result. This is where Retentionalists appeal to temporal modes of presentation. There would be a fatal problem here if all the content within a single specious present were as vivid and detailed as actual sensory experience; but as we saw earlier, this is not the case: the bulk of a specious present is composed of retentions, and these possess rapidly diminishing presence. This response may seem promising, but it does mean that proponents of the Retentional approach are treading a fine line. If retentions are too vivid, the surplus content problem returns; if they are too transparent, too diaphanous, the specious present will no longer be capable of housing our immediate impressional consciousness of change—the purpose for which it was introduced in the first place.

Figure 12.10 reflects one aspect of the Retentional model, but fails to capture another: it provides no indication of the interplay between contents occurring in successive specious presents. A better idea of this can be gleaned from Figure 12.11, which represents a short stream of auditory experience, deriving from hearing someone count from one to eight.

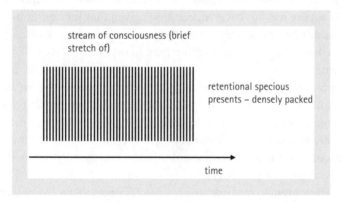

FIGURE 12.10  A more realistic depiction of a Retentional stream of consciousness

| | | | | 1 | 2 | 3 | 4 |
|---|---|---|---|---|---|---|---|
| | | | 1 | 2 | 3 | 4 | 5 |
| | | 1 | 2 | 3 | 4 | 5 | 6 |
| | 1 | 2 | 3 | 4 | 5 | 6 | 7 |
| 1 | 2 | 3 | 4 | 5 | 6 | 7 | 8 |
| $S_1$ | $S_2$ | $S_3$ | $S_4$ | $S_5$ | $S_6$ | $S_7$ | $S_8$ |

FIGURE 12.11  Retentions in action

Each of $S_1$-$S_8$ is a distinct specious present, each of which has a certain phenomenal depth. With regard to particular contents, larger font-size indicates greater presence. Note first the way '1' occurs as a primary impression in $S_1$, and is retained with gradually decreasing degrees of presence through $S_2$, $S_3$, $S_4$ and $S_5$, until it finally disappears from the sphere of immediate experience. What goes for '1' also applies to '2', '3' and so on. Note also the way the sequence '3–4' is progressively modified once it enters the retentional sphere: it features in each of $S_5$, $S_6$, $S_7$, and in each case '3' appears as more past (or less present) than '4'. It is not just individual tones (or sensation-slices) but tone-sequences and intervals which undergo progressive retentional modification. It is thanks to this systematic pattern of modification that we experience change and persistence in the way we do—or so the Retentional theorist maintains.

# 6. TIME AND CONSCIOUSNESS

There are further questions concerning the Extensional and Retentional approaches that this brief survey has not covered; certainly, given that they differ in fundamental ways, both cannot be correct. However, for present purposes—that of assessing the viability of the Integrationist stance with regard to experienced passage—two relevant points have emerged.

First, accepting the phenomenologically compelling doctrine that we have immediate experience of change and persistence may well mean accepting that our consciousness has (in one way or another) some temporal extension, but embracing this does not lead directly to a conceptual catastrophe in the way that some have alleged. The two main accounts of the specious present on offer may be very different, but neither is *obviously* incoherent or paradoxical.

Second, in attempting to make sense of our experience, neither approach to the specious present appeals, at least in any obvious way, to temporal passage. In both cases, the only players on stage are phenomenal contents and experiential interrelations. Consequently, each of these accounts looks to be capable of providing Integrationists with what they need—namely, an explanation of how the immediate experience of change can exist in a passage-free world. Extensional theorists view specious presents as unified experiential wholes which extend a short distance through time. Since in a Block universe all times and events are real, there is no difficulty in accommodating such experiences in such a universe. Retentional theorists hold that individual specious presents are momentary episodes of experiencing which possess *apparent* (or phenomenal) depth. Clearly, there is no problem whatsoever in accommodating experiences of this sort in a Block universe either. It seems as though any Block theorist who opts for the Integrationist stance will be standing on solid ground.

However, before reaching any firm conclusions as to just how solid this ground is, there are further avenues to be explored. When it comes to metaphysical conceptions of time itself—world-time, universe-time—the Block conception is by no means the

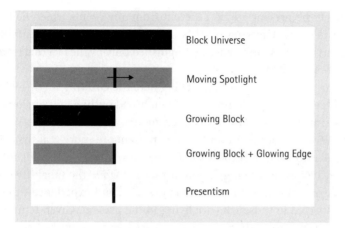

FIGURE 12.12 Five conceptions of the nature of time—and the large-scale structure of the universe

only game in town. There are also *dynamic* conceptions of time, conceptions which, in one way or another, find room for temporal passage, and these alternative conceptions also have their contemporary adherents. Might it be that the appearance of passage can be accounted for in an easier or more compelling way if passage is in fact real? This is a very natural thought, to say the least. Hence we need to consider the extent to which adding passage to the world might help to explain the dynamic character of our experience. This task is complicated by the fact that there are a number of very different dynamic conceptions of time; the main contenders are displayed in schematic form in Figure 12.12.

Advocates of the *Moving Spotlight* view agree with orthodox Block theorists that all times and events (whether past, present, or future) are real, but whereas the Block theorists hold that all parts of the four-dimensional universe are *equally* real, the Moving Spotlight theorists hold that events that are present are *more real* than those which are not. Most significantly, given our purposes, Moving Spotlighters also generally hold that *conscious experience* is confined to the present; the mode of existence enjoyed by subjects who inhabit the past and future does not permit them to be conscious. Needless to say, since the present is moving steadily forward—like a spotlight across a darkened stage—events are maximally real only for a moment. (In Figure 12.12 the lighter shade of grey used for the non-present regions of the universe is intended to indicate their lesser degree of reality.)

Advocates of the *Growing Block* model agree with their orthodox Block counterparts about the nature of the past—it is maximally or fully real—but they hold that the future is wholly unreal or non-existent. For Growing Block theorists, time is dynamic by virtue of new slices of reality continually coming into being, by virtue of a process of 'absolute becoming', as it is sometimes called. Despite these very significant differences, there are also some significant similarities between Growing and non-Growing Block universes. Intrinsically speaking, those events in a Growing Block universe which happen to be present are no different from any events which lie in the past—or any events in a *non*-Growing Block universe. Present events are simply the most recent additions to reality, and hence distinguished only by the fact that (momentarily) nothing lies ahead of them.

This takes us on to view what I am calling here the *Growing Block + Glowing Edge* model (hereafter abbreviated 'Grow-Glow'). In broad outline this version of temporal dynamism is similar to the Growing Block model, but there is one key difference: Grow-Glow theorists hold that events in the present are more real than those in the past. Most relevantly for our purposes, in Grow-Glow universes, *experience* is confined to the present. Just as in Moving Spotlight models, while the inhabitants of the past may enjoy a degree of reality, they do not enjoy conscious experience—this important privilege is restricted to those who are living in the present. Given temporal passage, this privilege is transitory and evanescent: as new slices of reality are added to the block it is *they* that acquire the mantle of maximal reality, as their predecessors lose it. (As with the Moving Spotlight view, this is indicated in Figure 12.12 by different shades of grey.)

As for Presentists, they deny reality to both the past and the future, and hold that only what is present is real. On this more radical view, past and future events do not enjoy *any* form of existence, no matter how attenuated. Presentism (in its most common guise) counts as a dynamic model of time by virtue of its recognizing a succession of presents: Napoleon's defeat at Waterloo isn't (now) present, but it *was*.

Some of these dynamic temporal models are more popular than others, and some (it is probably fair to say) are more metaphysically suspect than others.[16] What, precisely, does the lesser degree of reality ascribed to non-present things by the Moving Spotlight and Grow-Glow theorists amount to? It is incompatible with the existence of consciousness, but what implications does it have for the ordinary physical properties of an object? What is a less-than-fully-real *lump of rock* like, for instance? No obvious (or plausible) answer springs to mind. The difficulties do not end here. If a moving present bestows maximal reality in a transitory fashion, then all objects are both maximally real (when present) and less-than-maximally real (when the present has moved on). How can any object, at a given time, have different and inconsistent intrinsic properties in this manner? It would be absurd to suppose that a particular pane of glass, at a particular time *t*, is both wholly opaque and wholly transparent—so how can it be both maximally real and less-than-maximally real at *t*? But for present purposes challenges of this sort can be left to one side.[17] Let's just assume that each of these dynamic models is a way

---

[16] When McTaggart argued that time was unreal, his target was the Moving Spotlight conception (or something very much like it)—see Smith (2002) for a contemporary defence. Tooley (1997) has offered a recent defence of the Growing Block view, and Bourne (2006) of Presentism. Forrest defends a Grow-Glow model: 'Life and sentience are, I submit, activities not states. Activities only occur on the boundary of reality, while states can be in the past . . . It is then intuitively plausible that life and consciousness are causal activities. The past is then dead.' (2004: 359)

[17] And it is worth remembering in this connection that in paraconsistent logic there can be true contradictions.

time could actually be. What we need to consider is whether any form of temporal dynamism can help in explaining the dynamic character of our immediate experience.

The variety of dynamic conceptions of time on offer is not the only reason why this issue is less than straightforward: there is also more than one way in which temporal passage could contribute to the dynamic character of our experience. On both the Extensionalist and Retentionalist approaches the contents of individual specious presents have a dynamic character—they consist of movement, change, persistence, and so forth. How might the existence of such contents be related to passage? There seem to be two main options or possibilities. To put the distinction into a convenient shorthand form: passage could work *on* specious presents, or it could work *with* them. If the first obtains, passage contributes in an essential way to the creation or existence of the dynamic phenomenal contents which are housed in specious presents. Call this the *strong passage-dependence* thesis. If the second obtains, passage is not necessary for the existence of dynamic contents, but it nonetheless makes an essential contribution to the temporal character of our experience as it unfolds. The movement of the present might, for example, help bind neighbouring specious presents into continuous streams of consciousness, or it might contribute to the apparent *direction* of phenomenal flow. Call this second option the *weak passage-dependence* thesis. If the strong dependence thesis is true, without temporal passage there can be no direct experience of change, because the latter requires phenomenal contents possessing a dynamic character, and the latter in turn cannot exist in a passage-free universe. If only the weak dependence thesis is true, the immediate experience of change can exist in a passage-free universe, but only within the confines of single, isolated specious presents; in such a universe individual specious presents cannot combine to form continuous intervals of experiencing, of the sort we can enjoy.

These theses are by no means mutually exclusive. It could be that passage contributes essentially to both the existence of dynamic content *and* the way these contents combine to form continuous streams of consciousness. Of course, if it should turn out that both the strong and weak theses are false, we needn't consider this *combined* passage-dependence thesis separately. Since I will be arguing in what follows that the constituent theses *are* both false, I will not be devoting any further attention to the combined thesis.

## 7. WEAK PASSAGE-DEPENDENCE

Let's start by looking at the weak dependence doctrine. Is there any reason to suppose temporal passage could contribute to the *continuity* of our experience? Might it play a key role in the forging of continuous-seeming streams of consciousness from individual specious presents? Alternatively, is there any reason to suppose temporal (or metaphysical) passage contributes to the apparent *direction* of experienced passage?

The answers to these questions are connected, but we can make a useful start by starting with the latter.

As noted in §1, while it seems unlikely that we are aware of the passage of time itself, it seems a good deal more plausible to suppose that we are aware of flow, passage, and succession in our immediate experience. We see birds flying across the sky, we hear sounds continuing on, or tones succeeding one another, we are aware of one thought or mental image giving way to another—to mention just a few examples. Typically, all these diverse forms of experience exhibit an inherent directionality: one phase is experienced as *giving way* to the next, or as *being followed* by the next. More generally, if much of our experience has a flowing character, different in precise nature for different forms of consciousness, it is undeniable that all our experience seems to flow in a single direction: away from the present towards the past. Since dynamic models posit a privileged present which advances in a smooth, continuous manner towards the future, leaving the past behind, it may well seem natural to suppose these modes of passage are connected, in such a way that the direction of experienced passage can be explained by temporal passage.

However, for all that it may initially seem very plausible, this claim is difficult to defend in the context of the *weak* passage-dependence doctrine that we are currently exploring. According to the latter, although temporal passage contributes (in some as-yet unspecified way) to the temporal phenomenal features of our streams of consciousness, it is not itself responsible for the existence of the dynamic phenomenal contents within individual specious presents. Hence the difficulty. There is every reason to suppose that experiential flow, in all its myriad forms, is a feature of the phenomenal contents which exist within individual specious presents. To take a simple example, if a tone C is experienced—*directly* experienced—as giving-way to a tone D, then the succession (C–D) will of necessity exist within a single specious present. But if this experienced succession exists within a single specious present, then—assuming only weak-dependence obtains—temporal passage itself has nothing to contribute to the phenomenal characteristics of this episode of experiencing. It is only if the strong dependence thesis obtains that the contents within individual specious presents are essentially dependent upon temporal passage.

In fact, the situation is not quite so straightforward. In the context of the *Retentional* approach, the reasoning just outlined looks entirely valid. On this conception of the specious present, although the contents (C–D) seem to unfold over a brief interval of time, objectively speaking they exist at a single moment of time. Consequently, when they become present, they will become present simultaneously, irrespective of whether we construe the present in the manner of the Growing Block theorist, the Presentist, or any of the other dynamic temporal models. In contrast, if we assume the *Extensional* conception of the specious present is correct, there looks to be more room for a moving privileged present to exert an influence. To illustrate, if the envisaged experience occurs in a universe of the Growing Block variety, the contents (C–D) will not come into existence simultaneously, by virtue of the manner in which Extensional specious presents extend through ordinary time, content C will come into existence *before* content D. Might not this contribute to the phenomenal character of (C–D)? Might it explain why C *seems* to occur before, and flow into, D, rather than vice versa? It may be less obvious, but here again the limitations of the weak-dependence doctrine effectively rule this out. Since the proponent of the weak-dependence doctrine holds that the phenomenal character of an *individual* specious present is entirely independent of temporal passage, an Extensional specious present comprising or containing (C–D) will have precisely the same phenomenal features—apparent direction-of-flow-included—in a Block universe as it does in any dynamic universe.

There is more to be said on this, and further complications which need to be dealt with, but since these are of a general character, we may as well move on to consider the second way in which the weak-dependence doctrine might be true. Is there any reason to suppose temporal passage contributes to the moment-to-moment *continuity* that we find in a typical stream of consciousness?

Not if the Retentionalist approach is correct, and for a simple reason. Retentionalists view a stream of consciousness as composed of a series of momentary states; each of these states possesses apparent phenomenal depth, and—crucially for present purposes—each is *experientially encapsulated*. By which I mean: the contents within a single specious present are phenomenally unified (or diachronically co-conscious), but the phenomenal unity relationship does not extend to contents in neighbouring specious presents. (If it did, we would be switching from the Retentional to the Extensional conception: the defining feature of the latter is precisely the positing of direct experiential relationships over brief intervals.) Specious presents being thus encapsulated makes it difficult to see how phenomenal continuity at the stream-level could essentially depend on any form of passage. To illustrate: consider a short stretch of consciousness consisting of a series of two specious presents $SP_1$ and $SP_2$. For Block theorists both are equally real. For the Moving Spotlight theorist the situation is quite different: as the present hits $SP_1$ the latter briefly blossoms into full reality; when the present glides over to $SP2$ *it* gains full reality, while $SP_1$ loses this distinctive character. For this movement of the present to impact on phenomenal continuity, it would have to (in some manner) blend, connect, or unify successive specious presents. But no such connections *can* exist on the Retentional model: for as we have seen, phenomenally speaking, Retentionalist specious presents are entirely self-contained. The situation is much the same with the other models of passage. To illustrate with just one instance, suppose the universe is a Growing Block; when $SP_1$ comes into existence, $SP_2$ lies in the future, and so does not exist; a few moments later $SP_2$ comes into existence. Once again, given the encapsulated nature of Retentionalist specious presents, how can the coming-into-being of $SP_2$ have an impact on the phenomenal continuity of this stretch of experience?

Now, it is true that some critics of the Retentional approach have argued that if specious presents *are* experientially encapsulated, our streams of consciousness could not possess the moment-to-moment continuity they obviously do possess.[18] But for

---

[18] E.g. Dainton (2000: ch. 6, 2003); for counter-arguments see Gallagher (2003), Zahavi (2007).

present purposes this debate is irrelevant: all that matters is that *if* the Retentionalist approach is the right one, there is no obvious or compelling reason to suppose that temporal passage contributes to the continuity of our streams of consciousness.

From an Extensionalist standpoint the situation looks interestingly different, depending on which form of temporal dynamism one considers.

Let's start with the Growing Block. It might seem far from absurd to suppose that the continuous creation of new slices of reality (and hence fresh moments of experience) might contribute to the sort of moment-to-moment continuity that is characteristic of our consciousness. However, with a little further reflection it soon becomes apparent that this cannot easily be the case. Earlier, when considering whether passage could contribute to the *directional* features of our experience, I suggested that the limitations of the weak-dependence doctrine ruled this out: if specious presents in dynamic and non-dynamic universes have the same phenomenal features, they will have the same immanent or intrinsic directional features too. Although analogous reasoning applies with regard to the continuity issue, there is a more general point which can be made in this connection.

Consider a specious present $SP_1$ consisting of the experiencing of an extended C-tone of around one second's duration, which—we can suppose—occurs in a Growing Block universe we can call $U$. Given U's dynamic character, $SP_1$ comes into being in a succession of brief or momentary stages: (i) $[C_1]$, (ii) $[C_1+C_2]$, (iii) $[C_1+C_2+C_3]$, ... where each of the 'C's is a brief or momentary phase of a C-tone as-it-is experienced. Now consider the *non-dynamic counterpart* of this universe, $U^*$; the latter is indistinguishable from U in all respects save one: $U^*$ is a Block universe of the standard, non-growing sort. Let $SP_1^*$ be $SP_1$'s counterpart in $U^*$; given the sort of universe $SP_1^*$ inhabits, $SP_1^*$ does not come into being in a succession of stages in the manner of $SP_1$. Can $SP_1$ and $SP_1^*$ differ in their intrinsic phenomenal characteristics? No. As noted in the previous section, events which find themselves in the present of a Growing Block universe are not intrinsically different from any other event. More generally, for any event E in a Growing Block Universe and its counterpart $E^*$ in a standard Block universe, there are *no* intrinsic differences between E and $E^*$; the *only* difference between them is that E comes into existence in phases (assuming it is non-momentary), whereas $E^*$ does not. In which case, there can be no intrinsic or qualitative differences between $SP_1$ and $SP_1^*$; and this surely means there is no phenomenologically discernible difference between them either.

Evidently, if $SP_1$ and $SP_1^*$ are phenomenally indistinguishable, each must seem to possess the same degree of phenomenal *continuity*. And so we see that coming-into-existence-by-phases contributes nothing to the experienced continuity of individual specious presents. Since entire streams of consciousness are entirely composed of overlapping series of individual specious presents, it seems there is no reason for supposing that streams of consciousness in Growing Block universes could seem any *more* continuous than the streams in their non-dynamic Block-universe counterparts.

FIGURE 12.13 Experiencing a sequence of tones in a Moving
Spotlight universe

A-<u>**B**</u>-C-D-E-F-G
A-<u>**B-C**</u>-D-E-F-G
A-B-<u>**C-D**</u>-E-F-G
A-B-C-<u>**D-E**</u>-F-G
A-B-C-D-<u>**E-F**</u>-G
A-B-C-D-E-<u>**F-G**</u>

What of the Moving Spotlight model? Here we encounter a hurdle. Those who subscribe to this conception of time usually assume the present is momentary. This assumption does not sit easily with the Extensional specious present, which extends through an interval of time. It seems safe to assume that contents only possess maximal *phenomenal* presence when they fall under the spotlight of the *metaphysical* present. Since—as will be recalled from §4—all the parts or phases of an Extensional specious present *seem* fully present (in the phenomenal sense), we evidently need a metaphysical present which also extends through an interval. Perhaps this hurdle is not insurmountable: perhaps the Moving Spotlight needn't be momentary; perhaps it can match the temporal extension of the Extensional specious present. Without inquiring further, let us make this assumption. Can a temporally *extended* Moving Spotlight contribute to the continuity of our experience? This might not seem implausible, but once again, on closer scrutiny it is difficult to see how it could.

Figure 12.13 depicts the progression of an extended Moving Spotlight—indicated by **bold** + <u>underlining</u>—through a sequence of seven experienced tones, only two of which can fit into a single specious present. In reality, assuming the Spotlight moves in a continuous way, there would be many intermediate cases, but to simplify we will omit these.

The question to be addressed is whether positing this sort of structure better explains the continuity of our experience than the alternatives. In this case, the relevant alternative is the overlap mechanism exploited by the Extensionalist in the context of Block universes. Here is how a theorist of the latter persuasion would construe this episode of experiencing:

$$(A - B), (B - C), (C - D), (D - E), (E - F), (F - G)$$

We have here six specious presents, all of which are fully real, and which overlap by sharing common parts. (Here too the additional intermediate cases, of the sort illustrated in Figure 12.8, have been omitted so as to make it easier to focus on the essentials.) What is of interest is that if we compare the two cases, while restricting our attention to experiences that are fully real, there is no difference whatsoever between this pattern, and the pattern generated by the Moving Spotlight: either way we have precisely the same pattern of overlaps.[19] If we look to the broader surroundings of these specious presents then a difference does emerge, for on the Moving Spotlight view the fully-real specious presents are surrounded by less-than-fully-real experiences. But it is difficult to see how presence or absence of these pseudo-experiences contributes significantly to continuity as it exists at the level of genuine (fully real) experience. Surely, so far as the latter is concerned, relationships between genuine (fully real) experiences are all that count. In which case, the Moving Spotlighter is clearly failing to add anything of significance to the account of experiential continuity supplied by the orthodox Block theorist.

Similar remarks apply in the case of the Grow-Glow model. For this approach to be compatible with Extensionalism, the glowing edge must extend over an interval, rather than being momentary. The emboldened descending diagonal in Figure 12.13 can now be construed as the advancing glowing edge in a Grow-Glow universe. As is plain, the additional complexity of this conception brings no additional explanatory power.

We have one last version of temporal dynamism to consider: Presentism. Although it might not be immediately evident, the Presentist's predicament is similar to that of the Moving Spotlighter. In its standard form Presentism posits a *momentary* present, and hence—given the claim that only what is present is real—a momentary universe. Since the Extensionalist specious present possesses temporal extension, such a specious present can find no place in a Presentist universe. We could conclude that the Extensional approach and Presentism are simply incompatible, and that if the Extensionalist account of the temporal structure of our experience were to turn out to be correct, we could conclude that Presentism is false. Alternatively, and recalling the modification made to the Moving Spotlight, the Presentist could opt for a limited retreat, and hold that the present (or the sum total of what is real as of any given time) in fact has a brief duration, sufficient to house our specious presents. Let us suppose this move is made.

How should we suppose such presents are related? Should we hold that they occur in discrete extended blocks, e.g. [A-B], [C-D], etc.? Not if we want a version of Presentism which is compatible with phenomenal continuity: in the case just envisaged, there would be no experiential continuity linking B with C. No, we must opt once more for the overlap mechanism. To avoid the fragmentation of our streams of consciousness into isolated pulses the Presentist must hold that successive presents are related in this manner [A–B], [B–C], [C–D], [D–E], and so forth.[20] But since this is precisely the inter-experiential structure posited by Extensionally-inclined Block theorists, it is clear that the Presentist is not offering a better or richer account of phenomenal continuity.

---

[19] Of course, this is on the assumption that the specious presents in Figure 12.13 *do* overlap by sharing parts, but the consequences of abandoning this mode of overlap are catastrophic. On the one hand, the Extensionalist would no longer have a solution to the Connection Problem (of what links successive specious presents). On the other, there is a menace of repeatedly experienced contents: *unless* the **B**'s in the first two rows of Figure 12.13 *are* numerically identical, doesn't this tone get experienced twice rather than just once?

[20] For further discussion of this non-momentary version of Presentism see Dainton (2001: §6.11).

# 8. STRONG PASSAGE-DEPENDENCE

It is time to turn our attention to the *strong* passage-dependence thesis, the claim that dynamic phenomenal contents are impossible in a passage-free universe. How plausible is it to suppose that passage contributes to, or is responsible for, the dynamic characteristics we find in the contents of our experience? Proponents of passage might be inclined to argue thus: 'Since passage is found in reality, and passage is also found in our experience, isn't it just obvious that the two are linked? The latter is simply a reflection of the former!' But this is far too quick. First, there are several different conceptions of passage: are we to suppose they all leave the same mark on experience? Second, and more importantly, we saw in §1 that the claim that we directly experience temporal passage per se is open to serious question. Many of our experiences have a dynamic aspect to them, but the relevant flowing/streaming characteristics can plausibly be explained in terms of the flowing/streaming characteristics of many of the *contents* of our consciousness—there is no obvious place for passage *in addition* to these dynamic contents; phenomenologically speaking, passage seems surplus to requirements.

There remains the question of how experiences with such contents come to be, and in this regard the tales told by Block theorists who incline to materialism and dualism (or idealism or panpsychism, for that matter) will differ in familiar ways. It is also true, also for familiar reasons, that these various tales are problematic: the relationship between the phenomenal and physical realms remains mysterious and controversial. But does rejecting the Block conception in favour of a dynamic model of time do anything to reduce the matter-consciousness mystery? This is not obvious, to say the least. Furthermore, as we shall now see, the claim that passage can contribute *anything* to the explanation of dynamic phenomenal contents must cross or circumvent a number of significant hurdles.

Despite their divergence of opinion over the existence of the past, the Growing Block theorist and the Presentist work with a similar conception of what passage involves: it's all down to *existence*. Both agree that new times and events come into being by a process of absolute becoming. Whereas for the Growing Block theorist objects and events remain in existence after entering it, according to the Presentist they invariably depart from the scene (via *absolute annihilation*) after the briefest of visits. Of key importance with regard to the strong dependency thesis is this point: absolute becoming and annihilation contribute nothing that is in any way distinctive to the *intrinsic or qualitative* characteristics of the objects and events which undergo these processes (for want of a better expression). We noted this point earlier in connection with the Growing Block model (recall U and U*), but it also applies in the case of Presentism. An experience which is brought into being by the process of absolute becoming is qualitatively indistinguishable from its counterpart in a non-dynamic Block universe. It follows at once that *if* passage consists of becoming and/or annihilation, and nothing more, then passage cannot contribute to, or create, the dynamic features of our experience—at least on the (very) plausible assumption that these features are intrinsic or qualitative. Since this point applies equally directly on both the Retentionalist and Extensionalist conceptions of the specious present, it seems we can dismiss the strong passage-dependence thesis without further ado, at least with regard to these conceptions of passage.

Lest we be accused of moving too swiftly, a brief clarification is in order. A proponent of the strong dependence doctrine might argue along these lines:

> How can you say that passage isn't playing a creative role vis-à-vis dynamic phenomenal contents? For this is clearly wrong. In *all* the various temporal models on offer, passage is responsible for bringing such contents into being, albeit in different ways: in Presentist and Growing Block universes the process of absolute becoming does the work, in other models a Moving Present fulfils the same role. What more do you want?

While this objection is correct as far as it goes, it is also misguided, for more *is* required. To appreciate this, if it is not already obvious, it suffices to distinguish two ways in which phenomenal contents could be passage-dependent. Let us say that contents are *trivially* passage-dependent if temporal passage brings them into existence just as it brings every other ingredient of concrete reality into existence. Let us say that contents are *non-trivially* passage-dependent if temporal passage not only brings them into existence, but *also* contributes essentially to their dynamic character, in such a manner that these contents can *only* exist in worlds where passage is to be found. In Growing Block and Presentist universes, the objects and properties which exist at any moment, or over any interval, only exist because the process of temporal becoming has brought them into existence, and because of this *everything* in such universes is trivially passage-dependent, dynamic phenomenal contents included. But with regard to their intrinsic or qualitative properties, the contents of these universes are indistinguishable from their counterparts in passage-free Block universes. It is for this reason that it is difficult to see how a case for a *non-trivial* version of the passage-dependence doctrine can be made for universes of this kind.

So much for the Growing Block and Presentism, what of the other main ways of making sense of passage? Despite their other differences, proponents of the Moving Spotlight and Grow-Glow conceptions agree that passage essentially involves objects and events becoming maximally real when they become (transiently) present. Assessing the extent to which this process (for want of a better term) could contribute to the dynamic character of experiential contents is hindered by obscurities in the doctrine under consideration. Being fully real (or present) is comparatively unproblematic: this sort of reality—and this sort of experience—is the sort we're acquainted with from moment to moment, but what precisely is involved in being *less*-than-maximally real is a good deal less clear.[21] Even so, we can make some progress with the issue at hand.

---

[21] Quentin Smith, in a recent defence of this sort of doctrine, explains the difference thus: 'Past (or future) particulars do not have nonrelational, monadic properties, but only stand in relations or have

To begin, let's suppose individual specious presents are as the Retentional theorist claims: objectively speaking momentary, but with apparent phenomenal depth. Consider a brief segment of a stream of consciousness consisting of three neighbouring specious presents $SP_1$, $SP_2$ and $SP_3$. As the present advances—and here the differences between the Moving Spotlight and the Grow-Glow models are of no import—each of these in turn acquires full reality before passing this boon onto its successor(s). The question we need to consider, of course, is whether this sort of passage contributes to the dynamic character of phenomenal contents. It may not seem impossible: as the (metaphysical) present *sweeps through* these specious presents, mightn't it inject a degree of dynamism or animation as it goes? Think of the turbulence produced by dragging a plank of wood through a placid pool of water—might not the movement of the present create something analogous within the confines of a specious present?

This simplistic analogy is misleading. Objectively speaking, like any other Retentional specious present, $SP_1$ is *momentary*, and consequently the boon of maximum reality is granted to all its parts at once. These simultaneously existing parts comprise a primal impression and a spread of retentions, each of which presents the recent past under a different temporal mode of presentation. It is the combination of these ingredients which—according to the Retentionalist—creates both the temporal depth we find in experience, and the dynamic contents which exploit this depth. Of course, for any specious present to be genuine full-blooded *experience* it will have to fall under the spotlight of the present—or so the Moving Spotlighter will insist—hence the Present is certainly playing a role in this scheme. But its role is the limited one of granting experiences maximal reality: for once a specious present acquires this status, the work of generating an appearance of temporal depth (and hence dynamism) is done by the interplay between primal impressions and retentions which lie *inside* objectively momentary specious presents. It seems that a moving Present may bring an experience possessing dynamic temporal content into (full) being, but it makes no essential contribution to this dynamic content. Once again, we have a form of passage-dependency, but it seems clearly to be of the trivial rather than the non-trivial variety (in the sense introduced earlier).

While this may seem cut-and-dried, there are a couple of further points to note. From a metaphysical standpoint, what is currently being proposed differs markedly from what the Block theorist has to offer. According to the latter, each of $SP_1$, $SP_2$, and $SP_3$ is fully real, whereas according to subscribers to the Moving Spotlight and Grow-Glow models this status is fleetingly bestowed upon each in turn. Can *this* difference contribute in a non-trivial way to the dynamic character of the contents within individual specious presents? It's difficult to see how. For it to do so, the subjective

---

relational properties. They are thus "bare particulars" in the sense that they lack nonrelational, monadic properties. This "bareness" is due to the fact that these particulars are only partly real; they are partly unreal in the sense (among other senses) that they are bare in this respect.' (2002: 132) But can an object lacking *any* monadic (or intrinsic) properties really be considered even partly real?

character of $SP_2$ would have to be affected by whether or not $SP_1$ and/or $SP_3$ are fully real. But since, as we have already seen, Retentional specious presents are experientially encapsulated, there seems no room for such an influence. Someone might object: 'Yes, but the motion of the present still makes an all-important difference. If $SP_2$ were Present for more than an instant it would not have the short duration that it actually has; rather than seeming to last only for a second or so, it would seem to go on and on!' This line of thought may have some intuitive appeal, but Block theorists will plausibly insist that it is rooted in a confusion. If $SP_2$ has an apparent duration of one second, that is *all* the duration it has; it does not need to undergo absolute annihilation to have its experienced duration curtailed, its internal features and boundaries within the four-dimensional space-time manifold accomplish that.

Let us move on to consider things from the Extensional standpoint. If the specious present possesses objective temporal extension it may seem there is more scope for the motion of the present to stir things up, so to speak, and thus contribute to the dynamic character of phenomenal contents. In fact, there are serious difficulties here too.

For the Extensional theorist, the specious present is a temporal spread of content which all seems fully present (in the phenomenal sense of 'maximally clear and vivid'). For proponents of the Moving Spotlight and Grow-Glow models, it is only contents which fall under the present (in the metaphysical sense) which possess full reality, and so possess maximal presence (in the phenomenal sense). Putting these points together gives us the conclusion we arrived at earlier: for these temporal models to be compatible with the Extensional approach, the (metaphysically significant) present must possess some temporal depth, sufficient to accommodate the specious present. So let us suppose this is the case. The strong passage-dependence thesis is immediately imperilled, and for a simple reason. It looks very much as though we have arrived at a conception of the specious present which is, in effect, indistinguishable from that of the orthodox Block theorist: a temporally extended spread of content, all of whose parts possess maximal phenomenal presence. If so, it seems that switching to the Spotlight (or Grow-Glow) conception of time is adding nothing to our understanding of how dynamic phenomenal content is possible.

The story does not quite end here. 'But no', it might be objected, 'there is a crucial difference: by virtue of the present's *motion*, the successive phases of a single specious present gain maximal reality *in succession*; there is no counterpart to this in a Block universe'. This is true—assuming a succession of the kind being envisaged can exist at all—but again it is not obvious that this helps. The reason for this is by now familiar: the experiential structures the alleged succession would create would be indistinguishable from the experiential structures posited by the Extensionalist in a Block universe. If this is not already obvious, a glance at Figure 12.14 should make it so. We are concerned once more with the experience of the rolling ball; shown here are two ways of conceiving of this experience.

In the upper part of the figure is a succession of specious presents, illuminated in turn by a temporally extended moving present. Each of these specious presents consists

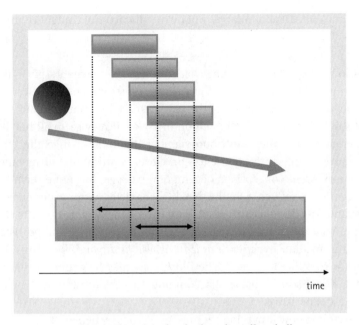

FIGURE 12.14  Another look at the rolling ball

of a spread of (dynamic) content; each part or phase of each specious present possesses a maximal degree of (phenomenal) presence. Neighbouring specious presents
overlap by sharing common parts.[22] The lower part of the figure shows how the Block
theorist (of the Extensional persuasion) construes the same episode of experiencing.
Here also there is a series of specious presents, their boundaries shown by the dotted lines, with the double-headed arrows indicating the reach of the diachronic co-
consciousness relationship. The contents of each of these specious presents are all
experienced together, they are dynamic, and their parts all possess presence to the
maximal degree. These specious presents also overlap by sharing common parts. As
can be seen, the specious presents in the upper part of the figure map onto those in
the lower part perfectly (to avoid clutter only two mappings are shown). Given these
profound similarities, isn't it to be expected that these experiential structures would be
subjectively indistinguishable?

As per usual, there is one significant difference between the two cases. In the Block
universe all the parts or phases of this stream-segment are equally real, and all possess
presence to the same degree; in the temporally dynamic universes (of the Moving
Spotlight or Grow-Glow varieties) this is not so: when the contents of any one specious
present are maximally present, the contents of its neighbours possess presence to a
lesser degree. But while this difference is real enough, there is no reason to suppose
this difference could be responsible for a subjectively discernible difference between
the two cases. The subjective character of a single specious present is surely determined

[22] For the reasons given in connection with the example considered earlier—see Figure 12.13.

by the contents *within* its boundaries, not by the character of contents lying *outside* its boundaries.

————————————

The discussion latterly has been somewhat convoluted—inevitably, given the number of options in play—but the result is clear. Both the strong and weak versions of the passage-dependence thesis fail—or at the very least are in serious difficulties. And this result is of some significance. Although on first view it may seem all but self-evident that the passage of time must contribute significantly to the change, flux, and flow we encounter in our immediate experience, on closer scrutiny this turns out not to be so. There are different conceptions of temporal passage, but none seems capable of exerting a significant influence on the dynamic character of our experience.

It goes without saying that this result supplies a considerable boost to the Integrationist's case. That the Block theorist can accommodate change and persistence as we find it in our immediate experience *at least as well as anyone else* means there is one less reason for rejecting this view. The Block view may be counterintuitive in some respects, but it does seem capable of accommodating the most basic dynamic aspects of our everyday experience.

Or at least, this is how things look at present. Perhaps when we have a better understanding of the relationship between material systems and consciousness—or of the temporal structures *within* consciousness—it will become clear how and why temporal passage *can* leave its mark on phenomenal contents. Perhaps an alternative account of temporal passage will emerge. We will have to wait and see. In the meantime, the conclusion stands: when it comes to immediate experience, the Block theorist has less to fear than is sometimes thought.

## REFERENCES

Bourne, C. (2006), *A Future for Presentism* (Oxford: Oxford University Press).
Broad, C. D. (1923), *Scientific Thought* (London: Routledge and Kegan Paul).
Chalmers, D. (1996), *The Conscious Mind* (Oxford: Oxford University Press).
Dainton, B. (2000), *Stream of Consciousness* (London: Routledge).
—— (2001) (2nd edn 2010), *Time and Space* (Chesham: Acumen).
—— (2002), 'The Gaze of Consciousness', *Journal of Consciousness Studies*, Vol. 9, issue 2, 31–48.
—— (2003), 'Time in Experience: Reply to Gallagher', *Psyche* 9(10).
—— (2010), 'Temporal Consciousness', *Stanford Encyclopedia of Philosophy*.
Dobbs, H. A. C. (1951), 'The Relation between the Time of Psychology and the Time of Physics Part I'. *The British Journal for the Philosophy of Science*, Vol 2: 6.
Forrest, P. (2004), 'The Real but Dead Past: Reply to Braddon-Mitchell', *Analysis*, 64.4: 358–62.
Foster, J. (1982), *The Case for Idealism* (London: Routledge and Kegan Paul).
—— (1991), *The Immaterial Self* (London: Routledge).
Gallagher S. (2003), 'Sync-ing in the Stream of Experience: Time-Consciousness in Broad, Husserl, and Dainton', *Psyche* (9)10.

Grünbaum, A. (1968), 'The Status of Temporal Becoming', in Gale (ed.) *The Philosophy of Time*, Sussex: Harvester.

Husserl, Edmund (1991), *On the Phenomenology of the Consciousness of Internal Time* (1893–1917), edited and translated by J. B.Brough (Dordrecht: Kluwer).

James, William (1890), *The Principles of Psychology* (New York: Dover).

Kelly, Sean D. (2005), 'Temporal Awareness', in *Phenomenology and Philosophy of Mind*, ed. Woodruff Smith and Thomasson (Oxford: Oxford University Press).

Le Poidevin, R. (2007), *The Images of Time* (Oxford: Oxford University Press).

Lockwood, M. (1989), *Mind, Brain and the Quantum* (Oxford: Blackwell).

—— (2005), *The Labyrinth of Time* (Oxford: Oxford University Press).

Mellor, H. (1998), *Real Time II* (London: Routledge).

Miller, I. (1984), *Husserl, Perception and Temporal Awareness* (MIT: Cambridge, Mass.).

Paul, L. A. (forthcoming), 'Temporal Experience', *Journal of Philosophy*.

Plumer, G. (1985), 'The Myth of the Specious Present', *Mind*, 94, 373: 19–35.

Reid, T. (1855), *Essays on the Intellectual Powers of Man*. ed. Walker (Derby: Boston).

Smith, Q. (2002), 'Time and the Degrees of Existence', in *Time, Reality and Experience*, ed. Callender (Cambridge University Press: Cambridge) 119–36.

Stoljar, D. (2006), *Ignorance and Imagination* (Oxford: Oxford University Press).

Strawson, G. (1995), *Mental Reality* (Cambridge: MIT Press).

Tooley, M. (1997), *Time, Tense and Causation* (Oxford: Oxford University Press).

Russell, B. (1913/1984), 'On the Experience of Time', in *The Collected Papers of Bertrand Russell*, Vol. 7 (London: Allen and Unwin).

Weichselgartner, E. and Sperling, G. (1985), 'Continuous Measurement of Visible Persistence', *Journal of Experimental Psychology, Human Perception and Performance*, 11(6): 711–25.

Weyl, H. (1949), *The Philosophy of Mathematics and Natural Science* (Princeton: Princeton University Press).

Williams, D. C. (1951), 'The Myth of Passage', *Journal of Philosophy*. 48 (15): 457–72.

Zahavi, D. (2007), 'Perception of Duration Presupposes Duration of Perception—or does it? Husserl and Dainton on Time', *International Journal of Philosophical Studies*, 15/3: 453–71.

CHAPTER 13

........................................................................................................

# TIME IN ACTION

........................................................................................................

SHAUN GALLAGHER

# 1. INTRODUCTION

........................................................................................................

WHEN we look at infants younger than three months of age our impression is that their movements lack proper coordination. When they move their arms and legs they seem to be flailing about, attempting, perhaps, to gain control over their movements as they adjust to their newly found gravity (Hopkins and Prechtl 1984; Prechtl and Hopkins 1986). For this reason, in part, developmental studies have traditionally argued that body schemas (understood as mechanisms of motor control) are absent at birth, and that their development depends on prolonged experience. Video studies, however, have shown that there is more organization in these movements than the casual glance reveals. Close to one-third of all arm movements resulting in contact with any part of the head lead to contact with the mouth, either directly (14%) or following contact with other parts of the face (18%) (Butterworth and Hopkins 1988; Lew and Butterworth 1995). Moreover, a significant percentage of the arm movements that result in contact with the mouth are associated with an open or opening mouth posture, compared with those landing on other parts of the face. In these movements the mouth *anticipates* arrival of the hand.[1] This kind of coordination can be traced to even earlier points in development. Ultrasonic scanning on fetuses shows that similar hand–mouth movements occur between 50–100 times per hour from 12 to 15 weeks gestational age (DeVries, Visser and Prechtl 1984).[2] This suggests that a

---

[1]  There is no evidence, however, that the movement is guided by sight: eyes are no more likely to be open than closed when the hand finds the mouth directly. Importantly, tests rule out the possibility that these movements are the result of reflex responses such as the Babkin reflex where the infant's mouth opens when the palm is pressed, and no instance of the rooting reflex is observed in relation to these movements.

[2]  The ultrasonic scans were sufficiently fine-grained to show jaw openings, yawns, and even movements of the tongue. De Vries et al. state: 'There was a striking similarity between prenatal and postnatal movements, although the latter sometimes appeared abrupt because of the effect of gravity'

basic hand–mouth coordination may be an aspect of early, centrally organized and organizing processes that come to involve proprioceptive input even prior to birth.

The anticipation involved in hand–mouth coordination suggests that at the very least, from early post-natal life onwards, human (and most likely animal) movement involves an apparent timing that reflects an intrinsic or inherent temporality. I note this distinction between timing and temporality. Timing is something that we can see and measure. Timing, however, can be accidental or merely coincidental. The fact of a more consistent timing, the fact that the mouth almost always anticipates the hand, for example, suggests deeper temporal processes involved in bodily systems capable of such timing. In other words, it is not just a matter of the system carrying or processing temporal information; rather, the important thing is that the system is capable of organizing its processing and its behavior in a temporal fashion. For the system to have this anticipatory aspect in its movement, it needs to have a practical orientation towards what is just about to happen. Throughout its movement the system also needs to keep track of how previous movement has brought it to its current state, and this is especially true if the movement is intentional, and if a conscious sense of movement is generated.

The kind of time that I want to discuss in this chapter is not the objective time that can be measured by a clock, although action certainly does take place in time, and it may be important in various contexts that its duration can be measured. Rather, I want to discuss the intrinsic temporality that is found in both bodily movement and action, and that manifests itself at both the subpersonal and the personal levels of analysis. Some philosophers distinguish objective time from lived time (e.g. Husserl 1991; Merleau-Ponth 1962; Strauss 1966). The latter includes psychological or phenomenological time, that is, time as we experience it passing, sometimes seeming to pass slowly and sometimes rapidly. The intrinsic temporality I will discuss includes more than phenomenological time: it includes temporal structures that shape action and experience, but might not be experienced as such.

## 2. THE DYNAMICS OF INTRINSIC TEMPORALITY IN THE BODY SCHEMA

This intrinsic temporality is expressed in Henry Head's definition of the body schema. Head noted that the body schema dynamically organizes sensory-motor feedback in such a way that the final sensation of position is 'charged with a relation to something that has happened before' (Head 1920: 606). He uses the metaphor of a taximeter, which computes and registers movement as it goes. Merleau-Ponty borrows this metaphor from Head and suggests that movement is organized according to the 'time

(p. 48). Lack of sufficient viewing angles made it impossible to tell whether the same anticipatory mouth opening occurs in the fetus, however.

of the body, taximeter time of the corporeal schema' (1968: 173). This includes an incorporation of past moments into the present: 'At each successive instant of a movement, the preceding instant is not lost sight of. It is, as it were, dovetailed into the present ... [Movement draws] together, on the basis of one's present position, the succession of previous positions, which envelop each other' (Merleau-Ponty 1962: 140). This kind of effect of the past on the present is a rule that applies more generally on the level of neural systems: a given neural event is normally encoded in the context of preceding events (Karmarkar and Buonomano 2007), and, as we'll see below, not necessarily on a linear model.

These retentional aspects of movement are integrated into a process that includes the ubiquitous anticipatory or prospective aspects already noted in the hand–mouth coordination in infants. Empirical research has shown that anticipatory or prospective processes are pervasive in low-level sensorimotor actions. Eye-tracking, for example, involves moment-to-moment anticipations concerning the trajectory of the target. Our gaze anticipates the rotation of our body when we turn a corner (Berthoz 2000: 126). Similar to the mouth's anticipation of the hand, when I reach down to the floor to grab something, my body angles backward in order to adjust its center of gravity so it doesn't go off balance and fall over when I bend forward (Babinski 1899). Reaching for an object involves feed-forward components that allow last-minute adjustments if the object is moved. On various models of motor control, for example, a copy of the efferent motor command (efference copy) is said to create 'anticipation for the consequences of the action' (Georgieff and Jeannerod 1998) prior to sensory feedback, allowing for fast corrections of movement. Forward control models involve an anticipatory character so that, for example, the grasp of my reaching hand tacitly anticipates the shape of the object to be grasped, and does so according to the specific intentional action involved (see Jeannerod 2001; MacKay 1966; Wolpert, Ghahramani and Jordan 1995). My grasp moves in a teleological fashion. In this sense, anticipation is 'an essential characteristic' of motor functioning, and it serves our capacity to reorganize our actions in line with events that are yet to happen (Berthoz 2000: 25). Similar anticipations characterize the sensory aspects of perception (see Wilson and Knoblich 2005 for review). Since these prospective processes are present even in infants, the 'conclusion that [anticipatory processes] are immanent in virtually everything we think or do seems inescapable' (Haith 1993: 237).

What is inescapable, ubiquitous, and pervasive for human experience and action is not just the anticipatory aspect, but the full intrinsic temporality of the processes involved. A good model for this, as Berthoz suggests, is the Husserlian analysis of the retentional-protentional structure of experience (Berthoz 2000: 16; Husserl 1991; also see Gallagher 2005, Ch. 8). Husserl finds phenomenological evidence for what he calls the 'retention' of the just past, and the 'protention' or anticipation of that which is just about to occur, and considers these to be structural features of consciousness. The general structure of this temporality, however, can also be applied to movement and to motor processes that are not conscious.

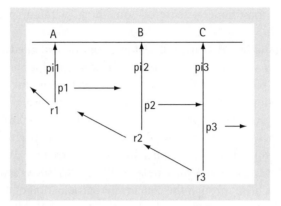

FIGURE 13.1  Husserl's model of time-consciousness (from Gallagher 1998)

Here is a brief summary and diagram (Figure 13.1) of Husserl's model (for more discussion, see Dainton, this volume). Husserl uses the example of the perception of a melody. The horizontal line ABC represents a temporal object such as a melody of several notes. The vertical lines represent abstract momentary phases of an enduring act of consciousness. Each phase is structured by three functions:

- *primal impression* (pi), which allows for the consciousness of an object (a musical note, for example) that is simultaneous with the current phase of consciousness;
- *retention* (r), which retains previous phases of consciousness and their intentional content;
- *protention* (p), which anticipates experience which is just about to happen.

In the current phase there is a retentioning ($r_3$) of the previous phase, and this just-past phase includes its own retentioning of the prior phase. This means that there is a retentional continuum—$r_3(r_2[r_1])$, and so forth—that stretches back over prior experience. The protentional aspect provides consciousness with an intentional sense that something more will happen. Although Husserl provides an exhaustive explication of retention, he says very little of protention, except that it is like retention in the direction of the future. Husserl does point out that the protentional aspect allows for the experience of surprise. If I am listening to a favorite melody, there is some sense of what is to come, a primal expectation of the notes to follow, and the best indication of this is that if someone hits the wrong note, I am surprised or disappointed. If a person fails to complete a sentence, I experience a sense of incompleteness. This kind of perceptual disappointment is based on a lack of fulfillment of protention: what happens fails to match my anticipation. A similar protentional feature is found in the phenomenon of representational momentum where movement or implied movement results in the extrapolation of a trajectory that goes beyond what was actually perceived (Wilson and Knoblich 2005). The content of protention is never completely determinate, however. Indeed, to the extent that the future itself is indeterminate, the content of protention may approach the most general sense of 'something (without specification) has to happen next'.

This model, or its broad structural features, can be extended to non-conscious motor processes as well, and is reflected in Head's description of the retentional aspect of the body schema and the anticipatory aspects of motor control. The same retentional-protentional structure can act as the organizing principle for proprioceptive processes that give rise to a phenomenal sense of movement and agency.

A number of theorists have proposed to capture the subpersonal processes that would instantiate this Husserlian model by using a dynamical systems approach (Thompson 2007; van Gelder 1996; Varela 1999). On this view, action and our consciousness of action arise through the concurrent participation of distributed regions of the brain and their sensorimotor embodiment (Varela et al. 2001). The integration of the different neuronal contributories involves a process that is understood as an integration of three different scales of duration (Pöppel 1988, 1994; Varela 1999), the first two of which are said to be directly relevant to protentional-retentional processes.

(1) the elementary scale (the 1/10 scale varying between 10-100 milliseconds)
(2) the integration scale (the 1 scale, varying from 0.5 to 3 seconds)
(3) the narrative scale involving memory (the 10 scale)

Neurophysiologically, the elementary timescale corresponds to the intrinsic cellular rhythms of neuronal discharges within the range of 10 milliseconds (the rhythms of bursting interneurons) to 100 milliseconds (the duration of an excitatory postsynaptic potential (EPSP) / inhibitory postsynaptic potential (IPSP) sequence in a cortical pyramidal neuron). Neuronal processes on this scale are integrated into the second scale, which, at the neurophysiological level, involves the integration of cell assemblies, distributed subsets of neurons with strong reciprocal connections (see Varela 1995; Varela et al. 2001). Phenomenologically, the second scale corresponds to the experienced living present, the level of a fully constituted, normal cognitive operation; motorically, it corresponds to a simple action, for example, reaching, grasping.

The important point in this analysis is the integration of the close-to-momentary processing events of the elementary scale into the extended present of the integration scale. The neuronal-level basic events that have a duration on the 1/10 scale synchronize (via phase-locking) and form aggregates that manifest themselves as incompressible but complete acts on the 1 scale.[3] The completion time is not dependent on a fixed

[3] This currently has the status of a working hypothesis in neuroscience. Thompson summarizes: 'integration happens through some form of temporal coding, in which the precise time at which individual neurons fire determines whether they participate in a given assembly. The most well-studied candidate for this kind of temporal coding is *phase synchrony*. Populations of neurons exhibit oscillatory discharges over a wide range of frequencies and can enter into precise synchrony or phase-locking over a limited period of time (a fraction of a second). A growing body of evidence suggests that phase synchrony is an indicator (perhaps a mechanism) of large-scale integration ... Animal and human studies demonstrate that specific changes in synchrony occur during arousal, sensorimotor integration, attentional selection, perception, and working memory' (Thompson 2007: 332).

integration period measurable by objective time, but rather is dynamically dependent on a number of dispersed assemblies. Moreover, the integration does not necessarily preserve an objective linear sequence in the events. For example, as Karmarkar and Buonomano (2007, p. 432) show, 'a 50 ms interval followed by a 100 ms interval is not encoded as the combination of the two. Instead, the earlier stimulus interacts with the processing of the 100 ms interval, resulting in the encoding of a distinct temporal object. Thus, temporal information is encoded in the context of the entire pattern, not as conjunctions of the component intervals'. The temporal order that manifests itself at the integration level is the product of a retentional function that orders information according to a pragmatic pattern (a pattern that is useful to the organism), rather than according to some internal or external clock. One result may be something like the temporal binding that occurs when subjects are asked to judge the timing of their voluntary movements and the effects of those movements (see Engbert 2007; Haggard et al. 2002). In addition, the temporal window of the integration scale is necessarily flexible (0.5 to 3 seconds), depending on a number of factors: context, fatigue, physical condition, age of subject, and so on. Furthermore, the integrating synchronization is dynamically unstable and will constantly and successively give rise to new assemblies, such transformations defining the trajectories of the system.

It is suggested that this integration process at the 1 scale level corresponds to the experienced present, describable in terms of the protentional-retentional structure discussed above (Varela 1999; Thompson 2007). Whatever falls within this window counts as happening 'now' for the system, and this 'now', much like James' (1890) notion of the specious present, integrates (retains) some indeterminate sequence of the basic 1/10 scale neuronal events that have just happened. The system dynamically parses its own activity according to this intrinsic temporality. Each emerging present bifurcates from the previous one determined by its initial and boundary conditions, and in such a way that the preceding emergence is still present in (still has an effect on) the succeeding one as the trace of the dynamical trajectory (corresponding to, or causally constituting *retention* on the phenomenological level). The initial conditions and boundary conditions are defined by embodied constraints and the experiential context of the action, behavior, or cognitive act. They shape the action at the global level and include the contextual setting of the task performed, as well as the independent modulations (i.e. new stimuli or endogenous changes in motivation) arising from the contextual setting where the action occurs (Gallagher and Varela 2003: 123; see also Varela 1999: 283). The outcome of this neuronal integration manifests itself at a global level as an experience, action or behavior (Thompson 2007; Thompson and Varela 2001; Varela and Thompson 2003).

That the dynamical analyses offered by Varela and others are not a perfect fit for Husserl's analysis of temporality has been critically pointed out by Grush (2006). It is not clear, however, that such analyses confuse vehicle and content, as Grush further maintains. Grush thinks of dynamic processes as the neural vehicles, and he equates retention with 'aspects of the current contents of awareness'. Retention, however, should be thought of, not as the content of awareness, or an aspect of that content, but as a structural feature of experience. The important correlation here is a structural one. One could say that the dynamical processes described by Varela and others constitute the vehicles that carry information about what has just been happening in the system, but that is just to say that the dynamical processes are retentional. There is general agreement, however, that whatever the dynamical explanation is, it has to be more than simply a claim that the neuronal processes are isomorphic to experience (see Varela 1996; Thompson 2007).

Whatever the limitations of this kind of dynamical analysis, it is generally consistent with Aleksandr Luria's notion of 'kinetic melody', which suggests that movement is a flowing, dynamic process involving a temporal dimension. Developed motor skills are '*integral kinaesthetic structures* or *kinetic melodies* . . . a single impulse is sufficient to activate a complete dynamic stereotype of automatically interchanging elements' (Luria 1973). As Sheets-Johnstone (2003: 85) puts it,

> A kinetic melody is not a *thing* in the brain (or in the central nervous system) but a particular neurological and experiential dynamic. Each melody is in fact a *neuromuscular dynamic* whose innervations and denervations, together with the constantly changing muscle tone they generate, constitute a particular temporal organization.

Sheets-Johnstone also agrees that the dynamic nature of this movement can be modeled by Husserl's analysis of time-consciousness: 'A coordinated series of movements whose dynamics are engrained in kinesthetic memory is run off and recognized kinesthetically. As it runs off, it is unified by retentions and protentions (Husserl 1991) until the series and its familiar and unique dynamics come to an end' (Sheets-Johnstne 2003: 75). The kinds of dynamics involved in body schemas, however, involve not just the flowing sequence of change of movement or experience, but also the forces and the temporal structures implicit in that change. The diachronic aspects of dynamics are accompanied by synchronies that constrain and structure movement. Body schematic processes are not simply about change of position over time, or the pure kinetic flow, but about the complex interactions of different parts of the body that are changing relationships throughout the movement,[4] but that also impose limits on that movement.

On such views, the notion of body or motor schema is understood as a dynamical entity. This is not always clearly the case, however. Marc Jeannerod, for example, offers a critique of the concept of schema put forward by Michael Arbib (1985; Arbib and Hesse 1986) on just this point (see Jeannerod and Gallagher 2002). Central to Jeannerod's view is the cognitive science notion of representation, taken to mean the complex neuronal firing patterns responsible for planning and carrying out bodily movement and action. The concept of motor schema is a way of describing the lower

---

[4]  That body schemas are not fixed but constantly adapting to movement conditions is shown clearly in recent work in neuroscience which demonstrates that the connections between cortical areas and muscles are dynamically self-organizing, 'not fixed but fluid, changing constantly on the basis of feedback from the periphery' and providing the flexibility needed to encode behaviorally relevant actions (Graziano 2006: 127).

levels of neuronal activity of a motor representation. But action, Jeannerod insists, involves something more than a snapshot activation of some neuronal population. For a correct characterization of actions, the correct level of discourse involves the complex and dynamic processes of representations and networks. Networks, in contrast to schemas, are things you can see using brain-imaging techniques. Networks are physical processes and one can actually see the activation of specific neural ensembles, and the possible overlaps and distinctions among such networks. For Jeannerod, then, the notion of schema is an abstract and static theoretical construct that doesn't capture the dynamic structure of representations and networks:

> here is the problem of the temporal structure of nervous activity. There are no real attempts to conceive the temporal structure in the schema. The schema is a static thing, ready to be used. You take one schema, and then another, and another, and they add up to an assembly or a larger schema (Jeannerod and Gallagher 2002: 12).

Arbib responds to Jeannerod's critique by emphasizing both the complex synchronic features of schemas, and their dynamic assemblage. Schemas are constantly being combined in order to deal with novel situations. Arbib calls this 'schema assemblage'.

> At any particular time there is a network of interacting schemas pulled together to represent the situation. It's possible to provide a microanalysis of how schemas are integrated into abilities for recognizing objects and acting on them. This kind of integration gives you a wide ability to cope with novel situations in their complexity.                    (Arbib and Gallagher 2004: 55)

Thus, in contrast to Jeannerod's characterization of schemas as lower-level static structures, Arbib emphasizes the dynamical nature of schema systems as well as their hierarchical complexity which cuts across motor, perceptual, and cognitive aspects of experience.

> Elementary motor schemas are stored for 'automated actions', but they should be distinguished from *dynamic coordinated control programs* which can recursively define new schemas as a network of previously defined schemas which includes the ability to activate and deactivate these subschemas as the situation demands. The point is that something like higher-order or intentional 'deliberation' may require explicit construction of a symbolic model (but that's still schemas!) to guide construction of the *executed* coordinated control program—which may then need to be restructured in the face of unexpected contingencies—and so we put stress on *dynamic planning.*                    (Arbib and Gallagher 2004: 55–56)

Complex motor schemas, then, are not snapshots, but dynamical processes that involve assimilation and accommodation (Piaget 1970). Schematic combinations that define a certain action may become stabilized, but then may be restructured in a way that overrides their original structure. Jeannerod and Arbib are in general agreement on the importance of the dynamic and temporal nature of motor control processes; they disagree only on the proper terminology for describing the brain processes important for action—schemas, representations, networks.

## 3. Intentional Action and the Narrative Scale

Intentional action often involves prior intentions, and these may involve conscious and thoughtful deliberations about what to do and what goals to aim at. In these respects intentional action involves a larger, but still intrinsic, time frame, that is, the narrative scale, in contrast to the elemental or integrative scales. I may, for example, form my intention to purchase a car a week or a month, or even a year before I actually begin any kind of action that could be considered shopping for or purchasing the car. In some sense, however, that intention is carried through to the action itself, and one's present action is guided by that past (or still present) intention.

Elizabeth Pacherie (2006; 2007) offers a model of how the various intentional aspects of action interconnect. She identifies three cascading 'stages' of action specification. The first corresponds to the formation of future-directed intentions (F-intentions); the second corresponds to present-directed intentions (P-intentions) (also see Bratman 1987). This distinction follows Searle's (1983) distinction between 'prior intentions' and 'intentions-in-action'. Pacherie also introduces a third concept: motor intentions (M-intentions).

> F-intentions are formed before the action and represent the whole action as a unit. They are usually detached from the situation of action and specify types of actions rather than tokens. Their content is therefore conceptual and descriptive. F-intentions are also ... subject to distinctive normative pressures for consistency and coherence: in particular, they should be means-end coherent, consistent with the agent's beliefs and consistent with other intentions he or she may have. P-intentions serve to implement action plans inherited from F-intentions. They anchor the action plan both in time and in the situation of action and thus effect a transformation of the descriptive contents of the action plan into perceptual-actional contents constrained by the present spatial as well as non-spatial characteristics of the agent, the target of the action, and the surrounding context. The final stage in action specification involves the transformation of the perceptual-actional contents of P-intentions into sensorimotor representations (M-intentions) through a precise specification of the spatial and temporal characteristics of the constituent elements of the selected motor program.                                   (Pacherie 2007: 3)

Much of the discussion in the previous section focused on the temporality involved in M-intentions and the generation of movement at the micro level. F-intentions and P-intentions, and the larger-scale temporality involved in them, are not reducible to M-intentions or the intrinsic temporality that we described above. This is an important point when we consider the question of free will (see below).

Both F-intentions and P-intentions, *pace* the 'P' for 'present', are future-directed. They are about the future, since they are about goals, and goals are things to be accomplished. I can, of course, talk about goals that I have attained, but to say that

I have an intention to do something necessarily means that it is something that is still to be done. It would be odd and nonsensical for me to say, while writing this essay, that I intend to meet Napoleon some day (unless I believed in some kind of heavenly, or hellish, afterlife). To intend to do something is to be oriented to the future. Even as I start to perform my intentional act and carry it out, my P-intention is oriented to moving my present action in the direction of the future, not yet achieved goal. Once the goal is achieved (or once I have given up on the goal or forgotten about it), the intention dissipates. I can, of course, remember and talk or write about my past intentions and actions. I may also have a long-standing intention. Someday I will visit Dharamsala. I'm not sure when this will happen, but I certainly intend to do this, and I formed this intention several years ago. My intention is still a present intention, directed at the future, but it has a past, a history, in the sense that I can tell you when I formed it, and perhaps even when I enhanced it with various qualifications. Someday I will visit Dharamsala to meet the Dalai Lama.

This tells us something about the temporality of intentions. On the one hand, and unremarkably, an intention, if it exists, exists in the present. At least on one conception of time this is true of everything and anything. Furthermore, once the intended goal is attained (or we give up on the goal), the intention ceases to exist, except in memory or narrative. Again, this is unremarkable. On the other hand intentions belong to a small group of phenomena (like desires, hopes, expectations, and predictions) that are essentially directed at the future.

Intentional action takes time: it begins and ends, and takes up some duration in between. The time frame of intentional action may vary from very short to very long, depending on the degree of complexity involved in the action. But again, this is an unremarkable observation in terms of objective measurable time. What is remarkable is the temporal structure of action which derives from intention. An occurrent action is, per se, ongoing towards the future, specifically towards its future end, and this feature is not reducible to the fact that this action requires more time to be complete. As Heidegger (1962: 236) would put it, action is always 'ahead-of-itself'. Moreover, as a way of being-in-the-world, my action is always and already situated in a particular set of circumstances, and these circumstances are shaped by what has gone before, which includes my own action up to this point. What I *can* do (what my possibilities are) is shaped by those circumstances. Yet my actions always transcend these circumstances in so far as I act for the sake of something other than the action itself. At the same time, at the same stroke, my action incorporates the situation that has been shaped by *past* actions, and the projected *future* toward which it is moving, in the *present* circumstances that can both limit and enable it. This is a temporal structure that is not captured by objective time. It's not enough to say that action takes time; there is a time *in* action, an intrinsic temporality or a temporal structure in action.[5]

---

[5] To be clear, in the previous section I indicated that there is an intrinsic temporality at the subpersonal level of motor system processes; here I am indicating that there is an intrinsic temporality at the personal or existential level of intention and action. By calling each of these temporalities intrinsic I do not mean to indicate that they are identical. The term 'intrinsic' signifies a characteristic

This intrinsic temporal structure is tied to the meaning of the action, which is tied to its goal, and reflects its significance to the agent. It also contributes to my ability to perceive meaning in the actions of others. The meaning of an action is not determined simply by how long it takes, as measured by the clock. For example, person A is five feet away from a door; person B is 15 feet away. There comes a knock on the door. Person A strolls over to the door to answer it; person B dashes to the door arriving at the same time as A. In one respect A and B's actions are equivalent actions; both of them answered the knock. Both actions took exactly the same amount of time. I can see, however, that B's action was hurried in a way that A's was not. The kinematics of this movement conveys meaning (see Berthoz 2000: 137). What the meaning is, of course, depends on many other circumstances to which I may or may not have access. But the hurriedness, and its meaning, are not simply a matter of covering more ground in less than normal time. The meaning of B's action in contrast to A's, is shaped by how motivated B is by what he already knows, or by what has already happened, by his specific expectation of who is knocking, by the present circumstances that either enable him or hinder him from reaching the door, and by his intention to reach the door as fast as possible. These factors are ordered in the intrinsic temporality of the action, and they manifest themselves in that action as meaningful.

If we treat action as a physical event stretched out and confined to objective time, we are naturally led to questions about causality, a concept that is defined in the framework of objective time. A cause always precedes its effect; an effect always follows its cause. What causes me to perform this particular action? To answer this question we naturally look to something that precedes the action. Perhaps I have a certain belief and a desire that 'cause' me (or motivate me) to act the way I do, or that at least give me reasons to act the way I do (and if I do act on this basis, this is often referred to as mental causation). Perhaps, however, my physical (brain) state or the social circumstances cause me to act the way I do (accordingly we can easily arrive at some version of physical or social determinism). These are the terms of the traditional discussion. If, however, we reframe the question in terms of the intrinsic temporality of action, it is not something in the past that causes or determines my action; it is some possibility of the future, some goal that draws me out of my past and present circumstances and allows me to transcend, and perhaps to change, all such determinations. Kant's antinomy concerning freedom and determinism may be deflated by this distinction between two different time frames. The scientific analysis of physical causes in objective time draws the conclusion that our actions are always determined by prior events. The practical analysis of possibilities as they are outlined in the intrinsic temporality of action opens the door to the possibility of freedom.

---

rather than a name for a certain kind of time. How the intrinsic temporal processes at the subpersonal motor control level relate to the intrinsic temporal structure of action is an open question that I don't try to answer here.

# 4. TEMPORALITY, AGENCY, AND FREE WILL

In contrast to the micro-analysis of action in terms of motor control, body schemas, and brain processes, which are framed on the elementary (1/10) and integration (1) scale of retentions and protentions, the analysis of action in terms of F-intentions and P-intentions is framed on the narrative (10) scale of time. An intention to perform a specific act, or generally to act in a certain way may be long standing or it may have been formed yesterday or an hour ago. In such cases I may have well-defined reasons, or I may know that my reasons are not quite clear to me. I may also simply be going along with normal expectations, letting my intention to do some action be determined by what others expect me to do, for example. I can also act spontaneously and decide, without deliberation, to answer the door I'm standing next to when I hear the knock. In this case the formation of a P-intention borders on the short time frame of the integration scale and there is not much more to it than the workings of body-schematic processes.

It seems unreasonable to claim that all intentional actions are in some way free actions, or that they have the same degree of freedom. A spontaneous action that is unaccompanied by an F-intention may have a degree of freedom that is less than an action that is guided by a deliberated and well-planned F-intention. It is also the case that a complex action (or set of actions) may have a degree of freedom that is lacking in any particular part of it (or any action that is part of it). In any of these cases I want to argue that the question of freedom, or the free use of the will, is a matter that is most appropriately decided on the narrative timescale (the scale of intentions and actions), and that it is a mistake to think that it can be decided at the elementary scale (the scale of motor control). Moreover, in a discussion of free will that is framed in terms of the integration scale, the issue will remain ambiguous. I think this is very clear from the

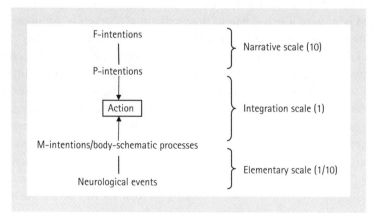

FIGURE 13.2 Intentions and timescales

much-discussed experiments by Benjamin Libet. Before I address this issue, however, it will be important to get clear on the concept of agency.

There is a renewed interest in the concept of agency in the cognitive neurosciences and the philosophical discussion of recent developments in that area. One claim is that even though we may have a sense of agency for a particular action, this sense of agency is an illusion (e.g. Wegner 2002). The experience of agency, however, is a complex phenomenon (Gallagher 2008; 2010). On one definition, the sense of agency is the first-order (pre-reflective) phenomenal awareness of causing an action, an experience that is generated in neural efferent processes that control bodily movement—motor processes that are measured on the elemental timescale of milliseconds (Gallagher 2000; Haggard and Magno 1999; Marcel 2003; Tsakiris and Haggard 2005). Other definitions focus on intentional aspects of the action—that is on the effects of one's action in the world. If I accomplish some kind of task, then I have a first-order (pre-reflective) phenomenal sense of agency for doing so (e.g. Chaminade and Decety 2002; Farrer and Frith 2002). On this definition, the sense of agency is generated at a level where P-intentions get fulfilled in action, and are best conceived in terms of the integration or narrative timescales. Finally, the sense of agency can be defined in terms of a second-order reflective or introspective attribution of agency (Stephens and Graham 2000), and this is best conceived on the narrative scale.

It is easy to confuse these different senses of the sense of agency, and in fact I've argued that these neat conceptual distinctions may or may not show up in the phenomenology of action. If they do, they may be ambiguous and this ambiguity is part of the phenomenology of action that needs to be acknowledged (Gallagher, 2010). It is also possible to find confusions among these various conceptions in the experimental literature. For example in Farrer and Frith (2002) the experimenters clearly think of the sense of agency as something tied to the intentional aspect of action (the second definition listed above) and not to mere bodily movement. Yet, when it comes to *explaining why* the anterior insula should be involved in generating the sense of agency, they frame the explanation in terms of motor control (i.e. the first definition listed above) (see Gallagher 2007 for discussion).

A related confusion about different timescales in action can be found in the interpretation of various experiments, including Libet's experiments on free will (Libet et al. 1983; Libet 1985, 1992; also see discussion in Wegner 2002: 52ff, 70ff). As Wegner notes, many researchers have tried to find the will 'somewhere in the link between brain and muscles' (2002: 36). On some interpretations this is what Libet's experiments set out to do. Libet shows that motor action and the sense of agency (or of willing or of the urge to move) depend on neurological events that happen before our conscious awareness of deciding or moving. In one of Libet's experiments subjects with their hands on a tabletop are asked to move their index finger whenever they want to. Their brain activity is monitored with special attention given to the time course of brain activity leading up to the movement, between 500–1000 ms. Just before the movement, there is 50 ms of activity in the motor nerves descending from motor cortex to the wrist. But this is preceded by several hundred (up to 800) ms of brain activity known as the readiness potential (RP). Subjects report *when* they were first aware of their decision (or urge or intention) to move their finger by referencing a large clock that allows them to report fractions of a second. It turns out that on average, 350 ms before they are conscious of deciding (or of having an urge) to move, the RP is initiated and their brains are already working on the motor processes that will result in the movement.

The conclusion is that voluntary acts are 'initiated by unconscious cerebral processes before conscious intention appears' (Libet 1985). The brain seemingly decides and then enacts its decisions in a nonconscious fashion, on a subpersonal level. It then creates the illusion that we consciously decide matters and that our actions are under our personal control. 'The initiation of the freely voluntary act appears to begin in the brain unconsciously, well before the person consciously knows he wants to act. Is there, then, any role for conscious will in the performance of a voluntary act?' (Libet 1999: 51). The epiphenomenalist interpretation of these results is that what we call 'free will' is nothing more than a false sense or impression, an illusion. Libet himself answers in the positive: consciousness can have an effect on our action, and free will is possible, because there is still approximately 150 ms remaining after we become conscious of our intent to move, and before we move. So, he suggests, we have time to consciously veto the movement (1985, 2003).

This challenge to the concept of free will, however, is cast at the wrong timescale. I suggested above that F-intentions and P-intentions, and the larger-scale, narrative temporality involved in them, are not reducible to M-intentions, which are about initiation and control of bodily movement, and involve the elemental timescale measured in hundreds of milleseconds. Free will is a concept that belongs to the realm of F-intentions and P-intentions; it is not a concept that relates to motor system mechanisms that carry out those intentions. The best answers we have to the question of motor control indicate that most control processes happen at a sub-personal, unconscious level in the elemental timescale, which may generate a sense of agency at the integrative timescale. This correlates well with the phenomenology of action. As we move through the world we do not normally monitor the specifics of our motor action in any explicitly conscious way. Rather, we are aware of the world and of our projects, we are caught up in thoughtful conversations, we are enjoying a meal with friends, and so on. Body schematic processes that involve proprioception, efference copy, forward comparators, ecological information, etc., keep me moving in the right direction, but the direction comes from the intentional dimensions that develop on the larger timescales.

The kinds of processes associated with free actions do not occur in the spur of the moment—they are not momentary and cannot fit within the thin phenomenology of the milliseconds between RP and movement. Consider the following example (from Gallagher 2006).

> At time $T$ something moves in the grass next to my feet. At $T+150$ ms the amygdala in my brain is activated, and before I know why, at $T+200$ ms I jump and move

several yards away. Here, the entire set of movements can be explained purely in terms of non-conscious perceptual processes, neuronal firing and muscles contracting, together with an evolutionary account of why our system is designed in this way, etc. My behavior, of course, motivates my awareness of what is happening and at T+1000 ms I see that what moved in the grass was a small harmless lizard. My next move is not of the same sort. At T+5000 ms, after observing the kind of lizard it is, I decide to catch it for my lizard collection. At T+5150 ms I take a step back and *voluntarily* make a quick reach for the lizard. My choice to catch the lizard is quite different from the reflex behavior. What goes into this decision involves awareness of what has just happened (I would not have decided to catch the lizard if I had not become conscious that there was a lizard there) plus recognition of the lizard as something I could appreciate.

At T+5150 ms I take a step back and reach for the lizard. One might focus on this movement and claim that at T+4650 ms processes in my brain were already under way to prepare for my reaching action, before I had even decided to catch the snake. Thus, what seemed to be my free decision was actually predetermined by my brain. This, however, ignores the context defined by the larger timeframe involving previous movement, a conscious recognition of the lizard, and a narrative-level reference to my lizard collection. There are likely newly initiated P-intentions at work as this action unfolds. I may not be able to go as fast as I've portrayed in picking up the lizard. Waiting for the strategic moment, I may not actually reach for the lizard until ten seconds after I made the decision to catch it. Libet might suggest that an extra decision would have to be made to initiate bodily movement precisely at that time. But any such decision about moving is already under the influence of the initial conscious decision to catch the lizard. My action, in this case, is not well described in terms of making bodily movements, but of attempting to catch the lizard for my collection, and this is spread out over the larger timescales than the elemental timescale of milliseconds.

Libet's results, then, are pitched at the wrong level of temporality to tell us anything about free will. Free will cannot be squeezed into timeframes of 150–350 milliseconds; it is a phenomenon that emerges only at larger timescales. Issues pertaining to free will do not apply primarily to motor control processes or even to bodily movements that make up intentional actions. Rather, they apply to intentional actions themselves, with their own intrinsic temporality.

# 5. CONCLUSION

Action is characterized by intrinsic lived temporalities at both the micro-level of motor control and the macro-level of F-intentions and P-intentions. Neither of these lived temporalities is reducible to objective time as measured by the clock, although they do fall into specific (elemental, integrative, narrative) timescales. The intrinsic temporality at work at the micro-level (elemental timescale) is shaped in part by the fact that action is dynamically embodied and situated, and that initial and boundary conditions are not determined simply by neural parameters but include a variety of conditions of the acting body (fatigue, physical condition, age of subject, starting posture, etc.) and the circumstances of action. This intrinsic temporality, which can be captured in the dynamic model of the retentional-protentional structure outlined by Husserl and other phenomenologists, characterizes the integration of body schematic activity and is expressed (on the integrative scale) in our first-order (pre-reflective) conscious experience of action and the sense of agency. That there is an intrinsic temporality at work at the macro-level (narrative scale) means that actions are always existentially situated in a world of meaning, that they are future oriented (toward possibilities), and that they allow us to transcend present circumstances. It is at this level and in this time frame that the concept of free will becomes relevant.

# REFERENCES

Arbib, M. A. (1985), 'Schemas for the Temporal Organization of Behaviour', *Human Neurobiology* 4: 63–72.

—— and Gallagher, S. (2004), 'Minds, Machines, and Brains: An Interview with Michael Arbib', *Journal of Consciousness Studies* 11/12: 50–67.

Arbib, M. A. and Hesse, M. B.(1986), *The Construction of Reality* (Cambridge: Cambridge University Press).

Babinski, J. (1899), 'De l'asynergie cérébelleuse', *Revue de Neurologie* 7: 806–816.

Berthoz, A. (2000), *The Brain's Sense of Movement* (Cambridge, MA: Harvard University Press).

Bratman, M. E. (1987), *Intention, Plans, and Practical Reason* (Cambridge, MA: Cambridge University Press).

Butterworth, G. and Hopkins, B. (1988), 'Hand-Mouth Coordination in the Newborn Baby', *British Journal of Developmental Psychology* 6: 303–314.

Chaminade, T. and Decety, J. (2002), 'Leader or Follower? Involvement of the Inferior Parietal Lobule in Agency', *Neuroreport* 13/1528: 1975–78.

De Vries, J. I. P., Visser, G. H. A. and Prechtl, H. F. R. (1984), 'Fetal Motility in the First Half of Pregnancy', in H. F. R. Prechtl (ed.), *Continuity of Neural Functions from Prenatal to Postnatal Life* (Spastics International Medical Publications), 46–64.

Engbert, K., Wohlschläger, A., Thomas, R., Haggard, P. (2007), Agency, Subjective Time, and Other Minds', *Journal of Experimental Psychology, Hum Percept Perform* 33/6:1261–8.

Farrer, C. and Frith, C. D. (2002), 'Experiencing Oneself vs. Another Person as Being the Cause of an Action: The Neural Correlates of the Experience of Agency', *Neuroimage* 15: 596–603.

Gallagher, S. (1998), *The Inordinance of Time* (Evanston: Northwestern University Press).

—— (2000), 'Philosophical Conceptions of the Self: Implications for Cognitive Science', *Trends in Cognitive Science* 4/1: 14–21.

—— (2005), *How the Body Shapes the Mind* (Oxford: Oxford University Press).

—— (2006), 'Where's the Action? Epiphenomenalism and the Problem of Free Will', In W. Banks, S. Pockett, and S. Gallagher (eds.), *Does Consciousness Cause Behavior? An Investigation of the Nature of Volition* (Cambridge, MA: MIT Press), 109–124.

Gallagher, S. (2008), 'Agency, Free Will, and Psychopathology', in J. Parnas and Kenneth S. Kendler (eds.), *Philosophical Issues in Psychiatry: Natural Kinds, Mental Taxonomy and the Nature of Reality* (286–312) (Baltimore: Johns Hopkins University Press).

—— (2007), 'The Natural Philosophy of Agency', *Philosophy Compass* 2 (http://www.blackwell-synergy.com/doi/full/10.1111/j.1747-9991.2007.00067.x).

—— (2010), Multiple Aspects of Agency, *New Ideas in Psychology*. In press, Available online 31 March 2010, DOI: 10.1016/j.newidea.psych.2010.03.003.

—— and Varela, F. (2003), 'Redrawing the Map and Resetting the Time: Phenomenology and the Cognitive Sciences', *Canadian Journal of Philosophy*, Supplementary Volume 29: 93–132.

Georgieff, N. and Jeannerod, M. (1998), 'Beyond Consciousness of External Events: A Who System for Consciousness of Action and Self-Consciousness', *Consciousness and Cognition* 7: 465–477.

Graziano, M. (2006), 'The Organization of Behavioral Repertoire in Motor Cortex', *Annual Review of Neuroscience* 29: 105–134.

Grush, R. (2006), 'How to, and How *not* to, Bridge Computational Cognitive Neuroscience and Husserlian Phenomenology of Time Consciousness', *Synthese* DOI 10.1007/s11229-006-9100-6.

Haggard, P., Aschersleben, G., Gehrke, J., and Prinz, W. (2002), 'Action, Binding and Awareness', in W. Prinz and B. Hommel (eds.), *Common Mechanisms in Perception and Action: Attention and Performance*, Vol. XIX (Oxford: Oxford University Press), 266–285.

Haggard, P. and Eimer, M. (1999), 'On the Relation between Brain Potentials and the Awareness of Voluntary Movements', *Experimental Brain Research* 126: 128–133.

Haggard, P. and Magno, E. (1999), 'Localising Awareness of Action with Transcranial Magnetic Stimulation', *Experimental Brain Research* 127: 102–107.

Haith, M. M. (1993), 'Future-Oriented Processes in Infancy: The Case of Visual Expectations', in C. Granrud (ed.), *Carnegie-Mellon Symposium on Visual Perception and Cognition in Infancy* (Hillsdale, NJ: Lawrence Erlbaum Associates), 235–264.

Head, H. (1920), *Studies in Neurology*, Volume 2 (Oxford: The Clarendon Press).

Heidegger, M. (1962), *Being and Time*, trans. J. Macquarrie and E. Robinson (Oxford: Blackwell).

Hopkins, B. and Prechtl, H. F. R. (1984), 'A Qualitative Approach to the Development of Movements during Early Infancy', in H. F. R. Prechtl (ed.), *Continuity of Neural Functions from Prenatal to Postnatal Life* (Oxford: Blackwell).

Husserl, E. (1991), *On the Phenomenology of the Consciousness of Internal Time (1893–1917)*, Collected Works IV, trans. J. Brough (Dordrecht: Kluwer Academic).

James, W. (1890), *The Principles of Psychology* (New York: Dover, 1950).

Jeannerod, M. (2001), 'Neural Simulation of Action: A Unifying Mechanism for Motor Cognition', *Neuroimage* 14: S103–S109.

—— and Gallagher, S. (2002), 'From Action to Interaction: An Interview with Marc Jeannerod', *Journal of Consciousness Studies* 9 (1): 3–26.

Karmarkar, U. R. and Buonomano, D. V.(2007), 'Timing in the Absence of Clocks: Encoding Time in Neural Network States', *Neuron* 53: 427–438.

Libet, B. (1985), 'Unconscious Cerebral Initiative and the Role of Conscious Will in Voluntary Action', *Behavioral and Brain Sciences* 8: 529–566.

—— (1992), 'The Neural Time-Factor in Perception, Volition, and Free Will', *Revue de Métaphysique et de Morale* 2: 255–272.

——(1999), 'Do We Have Free Will?' *Journal of Consciousness Studies* 6/8–9: 47–57.

Libet, B., Gleason, C. A., Wright, E. W. and Pearl, D. K. (1983), 'Time of Conscious Intention to Act in Relation to Onset of Cerebral Activity (Readiness Potential): The Unconscious Initiation of a Freely Voluntary Act', *Brain* 106: 623–642.

Lew, A. and Butterworth, G. E. (1995), 'Hand-Mouth Contact in Newborn Babies before and after Feeding', *Developmental Psychology* 31: 456–463.

Luria, A. R. (1973), *The Working Brain*, tr. Basil Haigh (Hammondsworth, Middlesex: Penguin Books).

MacKay, D. (1966), 'Cerebral Organizatioin and the Conscious Control of Action', in J. C. Eccles (ed.), *Brain and Conscious Experience* (New York: Springer), 422–445.

Marcel, A. J. (2003), 'The Sense of Agency: Awareness and Ownership of Actions and Intentions', in J. Roessler and N. Eilan (eds.), *Agency and Self-Awareness* (Oxford: Oxford University Press), 48–93.

Merleau-Ponty, M. (1962), *Phenomenology of Perception*, trans. C. Smith (London: Routledge and Kegan Paul).

Pacherie, E. (2007), 'The Sense of Control and the Sense of Agency', *Psyche* 13/1 (http://psyche.cs.monash.edu.au/).

——(2006), 'Towards a Dynamic Theory of Intentions', in S. Pockett, W. P. Banks and S. Gallagher (eds.), *Does Consciousness Cause Behavior? An Investigation of the Nature of Volition* (Cambridge, MA: MIT Press), 145–167.

Piaget, J. (1970), *The Science of Education and the Psychology of the Child* (New York: Grossman).

Pöppel, E. (1988), *Mindworks: Time and Conscious Experience* (Boston: Harcourt Brace Jovanovich).

——(1994), 'Temporal Mechanisms in Perception', *International Review of Neurobiology* 37: 185–202.

Prechtl, H. F. R. and Hopkins, B. (1986), 'Developmental Transformations of Spontaneous Movements in Early Infancy', *Early Human Development* 14: 233–283.

Searle, J. (1983), *Intentionality: An Essay in the Philosophy of Mind* (Cambridge: Cambridge University Press).

Sheets-Johnstone (2003), 'Kinesthetic Memory: Theoria et Historia Scientiarum', *International Journal for Interdisciplinary Studies* (Poland), 7, 1, 69–92.

Stephens G. L. and Graham G. (2000), *When Self-Consciousness Breaks: Alien Voices and Inserted Thoughts* (Cambridge MA: MIT Press).

Straus, E. (1966), *Philosophical Psychology* (New York: Basic Books).

Thompson, E. (2007), *Mind in Life: Biology, Phenomenology, and the Sciences of Mind* (Cambridge, MA: Harvard University Press).

——and Varela, F.J. (2001), 'Radical Embodiment: Neural Dynamics and Consciousness', *Trends in Cognitive Sciences* 5: 418–425.

Tsakiris, M. and Haggard, P. (2005), 'Experimenting with the Acting Self', *Cognitive Neuropsychology* 22/3–4: 387–407.

Van Gelder, T. (1996), 'Wooden Iron? Husserlian Phenomenology Meets Cognitive Science', *Electronic Journal of Analytic Philosophy* 4; reprinted in J. Petitot, F. J. Varela, B. Pachoud, and J-M. Roy (eds.),*Naturalizing Phenomenology: Issues in Contemporary Phenomenology and Cognitive Science* (Stanford: Stanford University Press, 1999), 245–265.

Varela, F. J. (1999), 'The Specious Present: A Neurophenomenology of Time Consciousness', in J. Petitot, F. J. Varela, B. Pachoud, and J.-M. Roy (ed.), *Naturalizing Phenomenology:*

*Issues in Contemporary Phenomenology and Cognitive Science* (Stanford: Stanford University Press, 1999), 266–314.

Varela, F. J. (1995), 'Resonant Cell Assemblies: A New Approach to Cognitive Functioning and neuronal synchrony', *Biological Research* 28: 81–95.

—— (1996), 'Neurophenomenology: A Methodological Remedy for the Hard Problem', *Journal of Consciousness Studies* 3/4: 330–349.

—— Lachaux, J. P., Rodriguez, E. and Martinerie, J. (2001), 'The Brainweb: Phase-Synchronization and Long-Range Integration', *Nature Rev. Neuroscience* 2: 229–239.

Varela, F. J. and Thompson, E. (2003), 'Neural Synchrony and the Unity of Mind: A Neurophenomenological Perspective', in A. Cleeremans, ed., *The Unity of Consciousness: Binding, Integration and Dissociation* (New York: Oxford University Press), 266–287.

Wegner, D. (2002), *The Illusion of Conscious Will.* Cambridge, MA: MIT Press.

Wilson, M. and Knoblich, G. (2005), 'The Case for Motor Involvement in Perceiving Conspecifics', *Psychological Bulletin* 131/3: 460–473.

Wolpert, D. M., Ghahramani, Z. and Jordan, M.I. (1995), 'An Internal Model for Sensorimotor Integration', *Science* 269/5232: 1880–1882.

# CHAPTER 14

# TIME IN COGNITIVE DEVELOPMENT

## CHRISTOPH HOERL AND TERESA MCCORMACK

## 1. INTRODUCTION

THE topic of this chapter is the development of temporal understanding, and in particular the question as to when children can be said to be able to grasp temporal concepts such as 'before' and 'after'. One specific idea we wish to look at is that the development of temporal understanding, and the emergence of a grasp of temporal concepts, is closely linked to developments in children's understanding of causal relationships. There are, of course, substantive theories dealing with the general question of what concepts are, or what it is to possess a concept (see, e.g. Peacocke 1992, Fodor 1998). However, a detailed discussion of these issues is beyond the scope of this chapter. Instead, we will start with a fairly rough, intuitive understanding of the explanatory project at hand that locates it somewhere in between two other projects familiar from the philosophical literature on time.

One such project is exemplified by attempts to provide what is typically referred to as a causal theory of time. As usually understood, a causal theory of time has it that there is a sense in which temporal notions can be defined in terms of causal notions. Key to causal theories of time such as Reichenbach's (1956, 1957) is the thought that we can give an account of what makes one event a cause and another its effect, rather than vice versa, without using temporal notions. This, in turn, then allows us to explain, for example, what it is for one event to happen before, rather than after another, by reference to the relation in which a cause stands to its effect.

Even though causal theories of time are often characterized as claiming that temporal concepts can be analysed in terms of causal ones, they are not intended to provide a descriptive psychological account of the origins of temporal concepts.[1] Indeed, it has been argued that, if they were to be read as attempting the latter, causal theories of time would clearly be false. Mackie takes this view for reasons he describes as follows:

> Our concept of time is based on a pretty simple, immediate, experience of one event's following straight after another…Our experience of earlier and later, on which our concept of time direction is based, itself remains primitive, even if it has some unknown causal source.
>
> (Mackie 1977, quoted in Sklar 1995: 218; see also Sklar 1981)

Mackie's words here point to a second type of philosophical project—that of giving an account of the phenomenology of experiences involving, for example, movement and change. Part of his claim seems to be that there are direct experiences of events unfolding in time that a subject can have without any grasp of the particular types of causal facts at issue in causal theories of time. This is perhaps difficult to deny. However, Mackie also seems to think that our possession of temporal concepts can somehow be straightforwardly explained by appealing to such experiences, which is much less obvious (see also Hoerl 1998).

In what follows, we will describe a number of ways in which young children can be sensitive to, learn about, or keep track of the temporal order of events in a sequence that, arguably, can be explained without invoking a grasp, on the part of the child, of temporal concepts such as 'before' and 'after'. In each case, we will argue that the sensitivity to temporal relations shown by the children is tied very closely to their directly experiencing events in sequence, and that it is the sequence in which events are experienced, rather than any ability to reason about sequences, which might actually explain children's competence in the early stages of development. Now, put very schematically, at least some of these temporal abilities either do already involve 'experiences of earlier and later', to use Mackie's words, or they do not (where one stands on this issue will typically depend on whether one adopts a form of non-conceptualism or a form of conceptualism about temporal experience). If they do, then arguably there is more involved in the possession of temporal concepts than can be explained by a mere appeal to such experiences. If they do not, then it appears that we might need to explain children's ability to have experiences of earlier and later, at least in part, in terms of their grasp of temporal concepts, rather than vice versa (or the two explanatory tasks may actually be of a piece with one another).

Thus, it may be true that possession of temporal concepts does not require anything like a grasp of a causal theory of time, but neither, it seems, can we account for the acquisition of temporal concepts simply by gesturing at experiences of temporal relationships. From a developmental perspective, the crucial question we need to ask is: in what kinds of contexts do children first need to engage in reasoning about temporal relationships using concepts such as 'before' and 'after', rather than just relying on

---

[1]  For a similar distinction between normative and descriptive approaches, in the case of causal concepts, see Woodward (2007).

some more primitive sensitivity to temporal relations? This is the question we will focus on in this chapter.

## 2. THE ROLE OF TEMPORAL PRIORITY IN CHILDREN'S CAUSAL JUDGEMENTS

One of the most immediately obvious areas of research to turn to in considering whether young children can reason about before and after relationships is research that has been done on whether or not children's causal judgements respect the 'temporal priority principle', that is, the principle that causes must precede their effects (e.g. Bullock et al. 1982, Kun 1978). When it comes to selecting one amongst several potential causes for an effect, do children restrict themselves to candidate causes that temporally precede the effect, or do they sometimes judge that an event that occurred after the effect was its cause? Bullock and Gelman's thorough study (1979; see Bullock et al. 1982), which we will discuss below, is often cited as providing definitive evidence that children as young as three do make use of a temporal priority principle in their causal judgements. Indeed, following their study, very little subsequent research was conducted on this issue, perhaps because many researchers believed that the issue had been empirically resolved. Yet, others have viewed this principle as one that is adopted relatively late in development (Piaget 1930, White 1988). For example, Shultz, Altmann, and Asselin (1986) argued that while use of generative transmission rules is pervasive in 3-year-olds, such children do not reliably exploit temporal priority (but see also Shultz and Kestenbaum 1985). Other experimental studies that suggest that children's causal judgements do not always accord with the temporal priority principle include Shultz and Mendelson's (1975) study and that of Sophian and Huber (1984), which we will also discuss briefly below (see also Das Gupta and Bryant 1989).

Some of the existing theoretical debate in this area gives the impression that the issue at stake in the experiments is children's ability to reflect on the relationships between two orders: the temporal order and the causal order. In this section, though, we will argue that there is another way of interpreting the experimental results. In particular, we will argue that some of the results of studies on the use of temporal priority in causal selection can be understood in terms of the idea that, whilst young children are sensitive to temporal structure in making causal judgements, they may not initially have a reflective grasp of the relationship between causal and temporal order.

Experimental studies in this area typically involve children seeing an event A, then an effect E, and then a further event B, after which they are asked whether A or B caused E. Thus, in Bullock and Gelman's (1979) study children saw the experimenter drop first one ball down a chute (event A), after which a Jack-in-the-box jumped up (event B), and then another ball was dropped down a different chute. They found that even 3-year-olds tended to judge that A was the causally efficacious event. In Sophian and Huber's (1984) study, by contrast, 3-year olds were not above chance, on certain trial types, at choosing amongst two candidate events the one that was temporally prior to the effect. In their study, the effect was a toy animal doing a trick, and events A and B were two different kinds of sounds (clicking vs. buzzing). Importantly, their study involved an initial training phase, in which the effect was shown to covary only with B, and not with A. It was when children were subsequently shown a sequence in which A occurred before the effect and B after it, that some of them were prepared to judge that B had caused the effect.

How should we interpret these apparently conflicting findings? Consider, first, Bullock and Gelman's (1979) experiment. In a scenario in which children see A, then E, and then B, there are two ways in which children may end up judging that A is the cause of E. The first is that they actually recall the order in which the three events occurred (AEB) and then base their judgements on their memory for the event sequence, bringing to bear a grasp of the general principle that causes must precede effects. We will label this the *reasoning about order* account. The second is that children make such a judgement without actually remembering or reflecting on the order in which the events have occurred. Rather, causal learning processes may usually operate according to a default along the lines of: 'ignore any events that occur after the relevant effect'. Thus, once E has occurred, children may simply stop encoding any further events as causally relevant to E. We will call this the *encoding default* explanation of children's success in selecting a temporally prior event as the cause of an effect. Use of such a default is non-insightful, since it does not involve thinking about event order and considering its causal significance. Yet it would normally be sufficient for accurate judgements because, under most circumstances, learning order is identical to causal order. That is, the order in which children find out about events is normally the order in which the events themselves occur (see also Lagnado, Waldmann, Hagmayer, and Sloman 2007 for this point).

The important contrast between the two types of accounts we are outlining is that on the encoding default account, children's causal judgements in the early stages of development obey the temporal priority principle as a result of the way in which the temporal relationships between observed events themselves affect how they are processed and encoded, whereas the reasoning about order account assumes that the principle itself is embodied in children's reasoning. In fact, on the former type of account, children need not even be able to remember and report or comment on event order. On the encoding default account, the order in which the events occur does indeed determine how the events are represented (i.e. as causal or non-causal), but order information itself need not even be part of that representation.

Note, however, that even the encoding default account assumes that children's encoding processes go beyond what is often assumed by some standard theories of animal causal learning. For example, this account assumes that children do more than simply encode associations between events or information about the co-occurrence of events. Rather, it assumes that there must be some selective encoding of events in terms of their causal relevance. Nevertheless, this temporally-sensitive selectivity need not involve an explicit grasp of *why* it is that only some events should be considered causally relevant. Indeed, when Bullock and Gelman (1979) directly asked children why they had chosen A rather than B as the cause, the youngest children (3-year-olds) very rarely gave an explanation that mentioned the order in which events had occurred, even using fairly lenient criteria for categorizing their explanations as temporal. This suggests that although their judgements may have respected the principle of temporal priority, they appear not to have an explicit grasp of the grounds of their judgements. In contrast, the majority of the 5-year-olds referred to the order in which the events had occurred in explaining their decisions, which suggests that, by this stage in development, a reasoning about order account may provide a better description of the basis of children's causal judgements.

How do the two accounts we have outlined fare in light of the findings of Sophian and Huber's (1984) study? At least on the face of it, those findings seem difficult to explain on a reasoning about order account. Arguably, in adult common-sense causal reasoning, the idea that causes precede their effects has the status of an inviolable principle (although there is of course considerable debate in philosophy on what, if anything, rules out the possibility of backward causation). If young children's causal judgements were also governed by the grasp of such a principle, it would therefore be surprising to find that they sometimes seem to tolerate violations of it. And even if we thought that somehow children don't treat the principle as inviolable, there would still be a prima facie contradiction between the belief that, generally, causes precede their effects, and their contrary judgement in a particular situation. On the encoding default account, by contrast, we need not credit the child herself with a grasp of the temporal priority principle to explain why her causal judgements are typically in accord with it. Thus, if the child were to judge, in a particular situation, that an effect was caused by an event that in fact succeeded it, this judgement would not be in contradiction with any general principle that the child also endorses. To show how such a judgement might arise, all we need to do is explain how there can be situations in which the normal encoding default is not operative.

Arguably, just such an explanation is available in the case of Sophian and Huber's (1984) study. Recall that this study included a training phase, in which event B covaried with the effect, and event A failed to do so. After this training phase, Sophian and Huber then examined whether children would judge that event A was the cause if shown an évent sequence in which B occurred after its typical effect, but A occurred before it, and they found that 3-year-olds were not above chance at choosing A over B.

At least one way of explaining this result is that Sophian and Huber's (1984) training phase, which is likely to have led children to judge that only B had the causal powers to make the effect occur and that A was causally inefficacious, ensured that the normal encoding default was not operative during the test phase. In other words, at test, A had already been encoded as causally inefficacious, so that when A occurred, followed by the effect, children were still looking for a cause. Under these circumstances, it is perhaps not surprising that, in at least some cases, their prior knowledge that B can bring about the effect won out. Thus, what we have called an encoding default account seems able to resolve the apparent conflict between Bullock and Gelman's (1979) results and those of Sophian and Huber (1984). A reasoning about order account, on the other hand, would have to provide some explanation as to why, if children grasp the general principle that causes precede their effects, they do not seem to apply such a principle consistently.

Further empirical work is needed to establish whether the encoding default account does indeed give the best description of young children's initial sensitivity to temporal order in causal selection. Our main aim in this section was to outline just how children's sensitivity to temporal order, in the context of selecting between a number of potential causes for an effect, might be explained without invoking an ability, on the part of the child, to grasp temporal concepts. We have also pointed out that, on the encoding default account we have sketched, such judgements may not even require memory for the order in which events happened. This is not to say, though, that such memory would necessarily involve a grasp of temporal concepts. In the next section, we will suggest that there might also be a basic form of memory for sequences that does not require an ability to reason about temporal relationships. Again, our suggestion will be that what makes this basic form of memory possible is the fact that children directly experience events in the order in which they happen.

## 3. TWO ACCOUNTS OF SCRIPT KNOWLEDGE

In *The Principles of Psychology* Williams James writes: 'A succession of feelings, in and of itself, is not a feeling of succession' (1890: 629). Almost identical words can be found in Edmund Husserl's *On the Phenomenology of Consciousness of Internal Time*. As Husserl puts it, 'The succession of sensations and the sensation of succession are not the same' (Husserl 1991: 12). James and Husserl are here concerned primarily with the question as to how time figures in perceptual experience, whereas our focus in this section will be on a particular kind of memory for familiar sequences found in young children, which is sometimes characterized by saying that they have acquired a script for the relevant sequence. However, the upshot of our discussion might also be put in terms of a variation on James' slogan, namely, that a succession of representations, in and of itself, does not amount to a representation of succession (at least if the latter is supposed to involve a grasp of concepts such as 'before' and 'after'). As we will argue, a central question raised by research on children's acquisition of scripts is whether it can show that children possess the latter, rather than merely being able to entertain the former.

In the mid-1970s, Katherine Nelson and her colleagues began a groundbreaking programme of research demonstrating that 3- to 4-year-olds seem to possess surprisingly robust and detailed knowledge about routine event sequences, such as what happens when you go to a fast-food restaurant, or at a visit to the doctor's. As Nelson explains, at the time, 'the term *event* was not widely used in psychology, and when it was it was usually interpreted as a conditioned response to a stimulus. Objects, object concepts, object perception and object categories were the focus of main-stream psychology, as well as developmental psychology, as had always been the case' (Nelson 1997: 1). Developmentalists usually attributed only 'limited and fragmented action-object connections' (ibid.: 2) to children. By contrast, the children in Nelson's studies seemed to show relative complex knowledge about whole sequences of events and their temporal structure (see esp. Nelson and Gruendel 1981, 1986).

Nelson and her colleagues adopted the term 'script', originally introduced by Schank and Abelson (1977) as a theoretical construct in cognitive psychology, to describe both children's verbal reports of such event sequences and the representations underlying them. Their suggestion was that scripts in fact constitute a developmentally basic form of representation, geared specifically to the acquisition and retention of knowledge of recurring event sequences.

The following snippet from a conversation might perhaps provide a flavour of the basic idea behind the notion of a script. In it, an adult experimenter (E) wants to find out what a child (C) did during a recent camping holiday.

E:   You slept outside in a tent? Wow! That sounds like a lot of fun.
C:   And then we waked up and had dinner. First we eat dinner, then go to bed, and then wake up and eat breakfast. (Fivush and Hamond 1990: 231)

If an adult was asked about a recent camping holiday, we would normally expect the answers to concentrate on events specific to that holiday. In retrieving memories of such specific events, though, she might be helped by general knowledge of the sequence of events that happens during a normal day. For instance, her attempts to recall might be guided by questions such as what she had for dinner and where she slept during that holiday. This is one way in which the notion of a script, and the role scripts have in memory, is sometimes understood in the adult literature. Now, it is of course possible to interpret the responses of the child in the above example along similar lines, that is, in terms of the idea that the child is using his knowledge of the sequence of events that happen during a normal day to think back to events during the camping holiday, but failing to remember anything distinctive about those events. Yet, insofar as the child does indeed fail to remember anything distinctive about those events, an alternative interpretation is that he does not actually manage to think back to those events at all, but falls back on a memory capacity of a quite different, more primitive type that involves only the ability to rehearse the order in which a sequence of familiar events usually happens. Understood along these lines, scripts constitute a distinctive form of memory for sequences in their own right, a form of memory which encodes the order in which a certain type of event sequence typically unfolds, but without locating occurrences of that sequence at a specific location in time (see also Hoerl 2007).

We should stress that we have quoted the example above only as a possible illus-tration, and our aim is not to try to argue for one kind of interpretation rather than another of this specific example. Moreover, we are not trying to argue that young children in general only ever produce such script-like responses to questions about past events. It is widely accepted that 2-to-3-year-olds can verbally retrieve, as Cordón, Pipe, Sayfan, Melinder, and Goodman (2004: 108) put it, 'bits and pieces' of specific past experiences, particularly when prompted to do so by adults (for discussion see, for example, Fivush 1993; Fivush & Schwarzmueller 1998). Yet, it is unclear to what extent doing so involves having an idea of particular events as located in the past and organized along a linear time series (see McCormack & Hoerl 1999, 2008). By contrast, script knowledge clearly does seem to involve a capacity of some kind to encode temporal relationships. Thus, the issue for us is how we should characterize the temporal abilities of a child at a stage of development at which she may be adept at acquiring script knowledge, but perhaps still lack a number of other temporal abilities. We can contrast two approaches to this issue.

John Campbell has recently suggested that considerations about the semantics of tensed expression might help us understand the difference between script-based thought about events and what he calls 'ordinary thinking about time', that is, the mature ability to locate particular events in a linear timeline. Following Reichenbach (1947), Campbell conceives of tensed expressions such as 'now' as being governed by token-reflexive rules—for example, in the case of 'now', the rule that any token of 'now' refers to the time at which it was produced. However, as Campbell also claims, two terms can share the same token-reflexive rule, in this sense, and still differ in their semantics, because the underlying domain of times in each case is different. This, he believes, is how we should think of the difference between script-based thinking and ordinary, mature thinking about time. In the case of scripts,

> [t]he domain of times over which [the token-reflexive] rule is defined will not, of course, be times drawn from our ordinary range of linearly organised times; they will themselves be times defined in terms of the temporal framework provided by the script. Within each script times are temporally related; but we cannot express temporal relations between times identified in different scripts.
>
> (Campbell 2006: 6)

Another way of characterizing Campbell's view is that, in ascribing to a child knowledge of, say, a fast-food restaurant script, we credit the child both with an overall representation of the order in which a sequence of events are arranged, and with the ability to orient herself within this order using tensed notions.[2] This might involve, for instance, the child's using her knowledge of the script to frame the thought that it is *now* time to tell the person behind the counter what she would like to order, and that the items she orders will be put on her tray *in a little while*. The critical sense in which, on Campbell's account, the child's cognitive abilities nevertheless fall short of those of mature thinkers is that her ability to give significance to these temporally token-reflexive terms is exhausted by her ability to orient herself within the relevant script.

Campbell's view may be contrasted with one that is perhaps hinted at in the following passage from Nelson:

---

[2] For the notion of temporal orientation, see Friedman et al. (1990), and Campbell (1994).

> The infant or young child does not consciously try to 'master' the script of a birthday party, for example (although the mother may have mastered it through deliberate planning), nor does the child try to remember how the bath sequence goes. Rather, through repeated occurrences, the pattern finds its place in the child's repertoire of event knowledge, a repertoire that provides the basis for action in repeated and in new events. (Nelson 1999: 242)

One way of understanding Nelson's words, here, is in terms of the idea that learning a script, at least at an early age, is not so much a matter of coming to think of a sequence of events *as* unfolding in a certain order, but rather a matter of coming to think of those events *in the right order* (see also Hoerl 2008). On this view, there is no overall representation of the sequence in which the child then orientates herself with the help of token-reflexive notions. Rather, recounting a script, at least initially, actually involves representing those different events *in* sequence.

If this latter suggestion is along the right lines, it might also help shed some light on a controversy in developmental linguistics, where it has been claimed that, when children first come to use tense morphology, they don't actually use it to mark tense, but instead use it to mark aspectual distinctions. The specific empirical form this claim takes in what has become known as the 'aspect before tense' or 'aspect first' hypothesis is that children's use of the past tense is initially restricted to verbs describing events that result in a change of state or have a natural completion point (e.g. 'broke' or 'built'), whereas they use present tense or imperfective morphology with verbs that describe events that can go on for an indefinite amount of time (e.g., 'dancing'). Our intention here is not to assess whether or not the linguistic evidence supports this specific empirical prediction. Rather, we introduce the claim in order to point out that it can also be interpreted as a cognitive one about limitations in children's thought about events, according to which children initially think of events only in terms of aspectual notions such as 'ongoing' and 'completed', and perhaps also 'yet to start' (i.e. an aspectual notion focusing on inception rather than completion), before they are capable of employing tensed notions such as 'past', 'present', and 'future'.[3]

What exactly does the cognitive difference at issue here come to? One basic feature of tense, which is implicit in Campbell's approach as described above, is that tensed notions serve to locate an event within a wider domain of times by marking its relationship with the present time. By contrast, if a young child is thinking of events only in terms of aspectual notions, there is a sense in which her focus is restricted entirely to the present itself. That is, she does not think of past and future events in their own right, as located at other points in time, but only about their current status as completed, ongoing, or yet-to-start. To transcend the perspective of the present, she would also have to be capable of thinking, say, of events she now thinks of as

---

[3] The notion of aspect of particular relevance here is that of grammatical aspect. See, e.g. Wagner (2001) for an overview of the existing debate in developmental linguistics and further discussion. See also McCormack & Hoerl (1999).

completed as having once been ongoing, or of events she thinks of as ongoing as completed in the future.

We want to suggest that the idea that children first come to think of events in terms of aspectual notions, before becoming capable of genuinely employing tenses, actually fits in well with the type of alternative to Campbell's account of script knowledge that we have sketched above. According to that alternative, it is wrong to think of script knowledge in the very early years as involving one unified representation of a sequence of events and the temporal relationships in which they stand to each other, within which the child then orientates herself with the help of tensed notions. Instead, going through the script involves entertaining a sequence of different representations.[4] Rather than encompassing the idea of different times across which the events making up the sequence are spread out, there is a sense in which each of these representations will simply be concerned with what is the case at the time it is entertained. Yet, that does not mean that such representations might not include aspectual notions marking the fact that, at that time, some events are ongoing, others completed, and others yet to come.

# 4. A KANTIAN INTERLUDE

If what we have said above is at least roughly along the right lines, there are potentially interesting parallels between developmental questions regarding children's script knowledge and some of the concerns that motivate Kant's Second Analogy of Experience in the *Critique of Pure Reason* (which is arguably a historical source for the claims made by James and Husserl that we quoted at the beginning of the preceding section).[5]

There is considerable debate as to what Kant's argument in the Second Analogy actually is, or indeed whether it is one or several arguments, and there is no scope here to engage with many of the exegetical and substantive questions that have been raised in this debate. Also, we need to stress that Kant's project is a specifically philosophical one, which is ultimately concerned with the very possibility of empirical thought and knowledge. There is no suggestion that it should have straightforward developmental implications. What might nevertheless make it fruitful to draw a comparison between at least one strand of thought in the Second Analogy and the kind of account we have

---

[4] Interestingly, the dispute between tensed and tenseless theories in the metaphysics of time is also sometimes put in terms of the question as to whether there is a more global sense in which, even on a mature understanding of time, any representation of things in time is ultimately only a representation from one amongst a number of different perspectives, or whether a unified representation of time is possible (Dummett 1960, Fine 2005, Moore 2001).

[5] In what follows, quotations from the *Critique of Pure Reason* are taken from the Norman Kemp Smith translation (London: Macmillan and Co. Ltd., 1929). We have followed the usual convention of referring to the standard paginations of the first ('A') and second ('B') editions. Our account draws heavily on expositions of Kant's argument in the Second Analogy given in Bennett (1966); Guyer (1987); and Strawson, (1966).

given of young children's script knowledge is that both seem to deal with basic types of mental phenomena which themselves display a temporal order or organization that is determined by the order in which events or states of affairs are or were perceived.

Kant's main focus is on the *sequence of perceptions*, or perceptual experiences, that occurs when we observe a change in the states of an object, say a ship moving from one location on a river to another. He writes:

> That something happens, i.e that something, or some state which did not previously exist, comes to be, cannot be perceived unless it is preceded by an appearance which does not contain in itself this state. [...] Every apprehension of an event is therefore a perception that follows upon another perception.          (A191/B236)

As he notes, however, we can have a succession of different perceptions also when there is no objective succession to be perceived. An example he uses is that of looking up and down the façade of a house. In Kant's words,

> [t]he apprehension of the manifold of appearance is always successive. The representations of the parts follow upon one another. Whether they also follow one another in the object is a point which calls for further reflection.     (A189/B234)

The issue, then, for Kant is what makes it possible for us to apply time-determinations such as 'before' and 'after' in empirical judgement. To do so, it seems, we need to have a sequence of different perceptions; yet, this alone is clearly not enough, since such a sequence can also occur when the perceived states of affairs in fact obtain concurrently. Put very crudely, Kant's solution is that we can distinguish the 'subjective sequence of apprehension from the objective sequence of appearances' (A193/B238) only if we think of the latter as governed by causal laws, which, in turn, determine the order of the former in the case of the perception of changes. In other words, the possibility of using time determinations such as 'before' and 'after' turns on thinking, when there is a succession of two different perceptions, that the relation between the different states of affairs perceived is causally determined, such that one of those states must follow, rather than precede, the other.

Kant's starting point, as we have seen, is the idea that perceiving a sequence of events involves having a sequence of perceptions. Similarly, the account of young children's script knowledge we have given above turns on the idea that a basic form of retaining knowledge of the order in which certain sequences of events happen simply consists in retaining an ability to entertain a sequence of representations. Kant also says that merely having a sequence of perceptions, by itself, cannot explain the ability to apply time-determinations in empirical judgement. On our account of script-knowledge, a similar point might be seen to apply, though in a slightly different guise, as the following might help to bring out.

Kant's point is sometimes put by saying that the possibility of time-determination, the ability to apply concepts such as 'before' and 'after' in empirical judgement, requires an implicit recognition that some of our perceptions are order-indifferent, but that there are also cases in which our perceptions necessarily happen in a certain order, because what is being perceived is an objective succession of events or states of affairs (Strawson 1966: 83). Kant's further claim, then, is that drawing this distinction requires a certain form of causal understanding. Now, on our account, script knowledge involves the ability to entertain a succession of representations, just as, according to Kant, the perception of succession involves a succession of perceptions. However, the issue of order-indifference vs necessary order does not seem to arise in the same way in the case of scripts. Rather, the right thing to say seems to be that the child does implicitly recognize that the representations making up the script have to occur in a particular sequence, but that this recognition is a purely practical matter of knowing how to go on, for instance, when recounting or re-enacting the script, rather than a matter of reflectively forming judgements involving time-determinations such as 'before' and 'after'.

Setting this difference to one side, for the moment, it may be thought that something like Kant's appeal to causal understanding as a necessary ingredient in the possibility of time-determination might nevertheless be right, and might also explain what is involved in children moving on from a mere capacity to acquire script-type knowledge to a grasp of temporal concepts such as 'before' and 'after'. How exactly might a grasp of certain causal relations be involved in a grasp of such concepts? If we take our inspiration from Kant, or at least from some of his commentators, we are likely to focus on two types of causal relationship.

Kant himself seems to have taken the conclusion of his argument to be that the possibility of time-determination requires thinking of perceived events as falling under causal laws that determine the order in which those events happen. Such laws might determine a temporal order among events, either by determining that a certain type of event necessarily follows, rather than precedes another, because the former is the cause, the latter the effect, or by determining that, given a certain initial condition, events necessarily follow one another in one temporal order rather than another (see Guyer 1987: 239f.).

According to one influential objection to Kant, however, his argument involves an illicit move from the idea of a necessary order of perceptions, in the case of an experience of an objective succession of events, to the idea that the objective succession of events is itself made necessary by a causal law. In Strawson's words, Kant confuses 'causal transactions or dependencies relating objects of subjective perception to one another . . . with the causal dependencies of subjective perceptions themselves upon their objects' (Strawson 1966: 84). Thus, for Strawson, the focus of the argument should be on the role played in our grasp of temporal concepts by the idea that the order in which two events happen can have a causal impact on our psychology, by determining the order in which the two events are perceived.

For our purposes, we can set aside whether Strawson's objection is to the point or misconstrues the Kantian project, as some have argued (see, e.g. Guyer 1987: 255ff.). For, both of the two lines of thought just sketched seem only to be of limited help from a developmental perspective. Seen from such a perspective, a natural way of interpreting the first one is in terms of the idea that the emergence of a grasp of temporal concepts is connected to a type of causal understanding that involves a grasp of general scientific laws. And a natural way of interpreting the second one is in terms of the idea that a grasp of temporal concepts is connected to certain aspects of what is sometimes called a 'theory of mind' that have to do with the causal dependence of the temporal order of perceptions on the order of events perceived. In each case, the relevant type of causal understanding is invoked to explain a sense in which the subject can appreciate that there is a necessary order to (some of) her perceptions. As we have seen, though, script knowledge can involve an implicit recognition of a necessary order in a sequence of representations that does not seem to rest on any causal understanding. Thus, what both of the above suggestions must come down to, in effect, is that making judgements involving temporal concepts requires a more explicit way of making sense of the type of causal necessity in question, although they differ on how the latter should be conceived. Yet, an obvious question is how we can make more concrete what the relevant implicit/explicit distinction comes to, that is, what can count as an explicit grasp in each case that can provide the required basis for the ability to apply temporal concepts. Even more importantly, though, we also have to ask what could count as a demonstration that children possess the relevant explicit grasp.

From a developmental point of view, a key question we need to ask is: what are circumstances that would provide clear evidence as to whether or not children have a form of understanding of sequences that cannot be explained by the possession of a script? And it is in this context that we should address the question as to how a grasp of causal relations might be involved in such understanding.

From this perspective, though, one particular set of causal connections seems absent from the discussion so far. The Kantian considerations mentioned above are concerned, either, with the idea that causal laws can determine the sequence in which two events happen, or the idea that the order in which two events happen, in turn, can make a difference to our psychology, in the form of determining a certain sequence of perceptions. Arguably, however, this leaves out a third idea, namely that the order in which two objective events happen cannot just have psychological consequences, but can also itself have causal consequences in the mind-independent world.

In what follows, we suggest that one way in which children might manifest a sensitivity to temporal relationships that cannot be explained in terms of the capacities underlying, for example, script learning, is in contexts in which the fact that two events happen in one order rather then another also has causal consequences outside the psychological realm and in which children can't rely on direct experiences of those events to work out those consequences.

## 5. THINKING OUTSIDE THE SCRIPT: TEMPORAL-CAUSAL REASONING

The kind of account of early script knowledge that we have given above turns crucially on the idea that scripts are acquired through direct (and usually, though perhaps not necessarily, repeated) experience of the relevant sequence of events. In particular, the thought is that the sequence in which the child witnesses the events itself determines the sequence in which they will later be recalled, and this can explain a basic form of knowledge for sequences which does not require a reflective grasp of temporal relations between the events in those sequences. The idea that children's early script recall may not be underpinned by such a reflective grasp is not a new one. For example, Catellani (1991: 100) cautions that 'the fact that young children's reports have an accurate temporal-causal sequence does not imply that they are able to use such temporal-causal connections explicitly'. Researchers have tried to address this issue by examining, for example, how young children deal with misordered sequences or with requests to recall in backwards order (e.g. Catellani 1991; Hudson and Nelson 1983; Fivush and Mandler 1985). Typically, the findings suggest that there are notable developmental changes in performance on such tasks. The intuition behind such studies is broadly similar to the one being articulated here: that young children's ability to recall sequentially, while impressive, is tied very closely to their having experienced events in a certain sequence, and that they will have difficulty on tasks that require any sort of manipulation of the temporal relationships between events.

We note also that psycholinguistic research that has examined children's comprehension of the terms 'before' and 'after' has come to a similar conclusion. Although it is generally accepted that there are large developmental changes in the extent to which children spontaneously employ such terms (e.g. Orbach and Lamb 2007), studies examining children's production of such terms in the context of script recall have found that even 3-year-olds will sometimes use them appropriately, and appear to interpret them accurately when used in connection with familiar event sequences (Carni and French 1984, French and Nelson 1985, see also Nelson 1996). Nevertheless, it is quite clear that the type of understanding thus manifested is highly limited. In particular, children of this age do not seem to reliably understand sentences in which events are mentioned in an order that does not correspond to the order of event occurrence (for discussion, see Harner 1982, Weist 1989, Winskel 2003). As with the work on children's ability to manipulate sequences, this suggests that young children have difficulties thinking about the temporal relationships between events when the order in which they have to consider the events differs from their actual order.

We have recently carried out a series of studies in which passing the task at hand required children to reason about a series of events that they had not directly witnessed (McCormack and Hoerl 2005, 2007; see also McColgan and McCormack 2008, Povinelli et al. 1999). In particular, in each of these studies, what children had to realize was that a particular outcome was only possible if two events happened in a certain order rather than another. These studies yield further evidence in favour of the view that young children's sensitivity to the order in which events happen is tied closely to the child's having observed the relevant events in succession. However, they may also, in turn, be seen to give some indication as to the types of circumstances in which children might first start employing temporal concepts.

One of the studies (McCormack and Hoerl 2007) involved two doll characters, John and Peter, and a doll's house. The doll's house had a bathroom with a door that could be closed, so that children could not look inside, although the experimenter could still reach into the bathroom through the back of the doll's house. The children were told that the dolls' hair had got messy when they were playing outside, and that they were going to go into the bathroom to brush their hair. Three items in the bathroom were pointed out to the children: the hairbrush, which was sitting by the sink, and two differently coloured cupboards. The two dolls then went into the bathroom, and the door was closed behind them. After this, the experimenter said, 'You can't see John right now, but he goes first and gets the hairbrush and now he is brushing his hair. Now he puts the hairbrush in one of the cupboards. Peter goes last. You can't see him now, but he gets the hairbrush out and now he is brushing his hair. Now he puts the hairbrush into the other cupboard.' Then bathroom door was opened, and each of the dolls could be seen standing beside one of the cupboards. It was explained to children that each doll was standing beside the cupboard that he had placed the hairbrush in, and they were asked two control questions to confirm that they could recall the order in which the two dolls had brushed their hair. Finally, they were asked the test question 'So, where do you think the brush is right now?' Five-year-olds could answer this question correctly, but 4-year-olds performed at chance.

Four-year-olds' poor performance in this task contrasts sharply with the ease with which even 3-year-olds passed a modified version of the task. In this modified version, an identical procedure was used, except that the bathroom door was left open, so that children were actually able to see the relevant actions of the dolls.

What might explain the striking difference in children's performance in these two tasks? Once again, it seems that the 3-year-olds' ability to pass the modified version of the task turns crucially on the fact that they witnessed the relevant events in sequence. Success, in this case, can be explained in terms of what McCormack and Hoerl (2005) call temporal updating. In temporal updating, the child simply has a model of the world (e.g. of where certain objects are located) that changes as and when she receives information about changes in the world. In other words, there will be a sequence of changes to the child's model over time as a result of the child receiving information about each event in turn, without the child herself having to reason about the order in which those events happened. Thus, in the modified version of the task, children could provide the right answer without having to consider the order in which the two dolls had acted. All they needed to do was retrieve a representation of the hairbrush's location that had been appropriately updated in the course of the child's observing it being moved first to one cupboard and then to the other.

By contrast, in the version of the task in which the bathroom door is closed, successful performance does seem to require reasoning about the order in which events have happened. In particular, children need to appreciate that the actions of the doll that went second undid the consequences of the actions of the doll that went first in order to make use of the retrospectively provided information as to which doll had put the hairbrush into which cupboard.

This study (along with the others reported in McCormack and Hoerl 2005, McColgan and McCormack 2008, and Povinelli et al. 1999) thus serves to reinforce the idea, already discussed above in connection with the role of temporal priority in children's causal judgements and children's script-knowledge, that young children's sensitivity to temporal relations is tied to them finding out about events in the order in which they happen. Precisely because this type of sensitivity is not sufficient to pass the task, though, this study may also be seen to point to the type of context in which older children first start to reason about temporal relations using concepts such as 'before' and 'after'.

The general suggestion we want to make is that children first make proper use of concepts such as 'before' and 'after' in the context of reasoning about situations in which the order in which two or more events happen makes a difference to the overall outcome, that is, situations that involve what we might call temporal-causal relationships. Such relationships might be seen to play a crucial role in children's grasp of temporal concepts because they allow children to give empirical significance to the idea that events happened in a certain order, in a way that goes beyond just thinking of them in that order. The task described above exemplifies one such temporal-causal relationship, because the actions of the doll that goes second undo the consequences of the actions of the doll that goes first. There are also other types of temporal-causal relationships, however. For instance, a certain outcome may depend on A happening before B, rather than B before A, with neither A nor B being able to produce that outcome on its own. Or B may produce a certain outcome either on its own or if followed by A, but not if preceded by A. Children's ability to reason about the whole range of such relationships has yet to be explored systematically. If what we have argued is at least broadly along the right lines, and a grasp of temporal concepts first emerges in such contexts, in which taking account of the order in which events happened is required to make causal judgements, further empirical research on this issue seems merited.

## 6. THE ROLE OF OBJECTS IN A GRASP OF TIME AS LINEAR

In the preceding sections, we have argued that there is a range of basic ways in which young children can be sensitive to, learn about, or keep track of the order in which a sequence of events happens if they directly experience the events in that order. In each case, we have suggested that it is the sequence in which events are experienced, rather than an ability to reason about such sequences, that might actually explain children's competence in the early stages of development. Yet, we have pointed out that the order in which two events happen can also have causal consequences outside of experience, and we have suggested that situations in which this is the case play a crucial role in the emergence of a grasp of temporal concepts. That is to say, children first give significance to the thought of one event occurring before, rather than after another event, in the context of situations in which the order in which two events happen can not just make a difference to the order in which they are experienced and perhaps later remembered, it also makes a difference to the state the world is left in.

In this final section, we want to discuss the relationship between the account we have offered and what might be seen as a rival account given by John Campbell.

Campbell's main focus is on autobiographical memory and on the question as to what it takes for a subject to give empirical significance to the linear structure of time in memory. Some theorists hold that there is a developmental dissociation between the ability to remember past events and the ability to think of them as being arranged in a linear order, so that individual memories start off as unconnected 'islands in time' (Friedman 2005: 151). Yet, even if we don't take that view and instead think that the idea of events as arranged in a linear order is integral to autobiographical memory, we need to give some account of how we give substance to that idea. Crudely speaking, Campbell's thought is that we give substance to the idea of remembered events as being arranged in a linear temporal order by thinking of those events as involving a common set of objects, and grasping the role objects play in transmitting causal influence across space and time. This is how Campbell summarizes this idea:

> [T]he various narratives constituting the autobiographical memory of a single individual will be thought of as organised around a single linear time so long as there is some overlap in the persisting things which figure in the various narratives. These objects will ensure the temporal connectedness of all the times remembered by ensuring the potential causal connectedness of all the events remembered.                                          (Campbell 1997: 116f.)

Straight off, though, it is at least not obvious that there couldn't also be a way of picking up on the role that concrete objects have in transmitting causal influence over time that does not entail grasp of time as linear. Consider, for instance, the case of a child with a teddy that has been left bruised and battered by a variety of events, in the way teddies are prone to. Suppose we credit the child herself with some insight into how this has happened. For instance, the child might think of Teddy's leg as having been chewed by the dog, the ear as having been torn on a fence, and the nose as having been squashed by sitting on it. At least on the face of it, there might be a way for the child to grasp all this without thereby being able to attach any significance, say, to the idea that the dog chewed the leg before, rather than after, the ear got torn. At the same time, though, we might nevertheless want to say that the child has some sort of grip on the fact that the reason Teddy has all of those marks is that it's still the same teddy that got into all of those scrapes. Certain kinds of planning abilities—for example, putting Teddy in the suitcase before a holiday—might also show that the child has a grip on objects as being capable of transmitting causal influence over space and time.

To make good Campbell's claims about the role that thought about objects plays in a grasp of time as linear, it seems the idea that objects transmit causal influence over time is perhaps not enough. Instead, we may have to look specifically at cases in which the fact that things happen to an object in a certain order rather than another makes a difference to what the object ends up being like as a result. But this, of course, is just the idea that there can be temporal-causal relationships between different things that happen to the object, in the sense described in the previous section, and that it is a grasp of such relationships that allows us to give substance to the idea of a linear order of past events. In other words, as we saw in the case of the teddy, it seems we can make sense of a basic grasp of the idea of an object having been affected by certain events that doesn't entail grasp of the idea of those events having occurred in a particular order. By contrast, what does require the idea of a temporal order, in the case of the task described in McCormack and Hoerl (2007) is the thought of a particular kind of causal relationship between the two dolls' actions, in that each of them has the power to undo the effects of the other.

Where exactly does that leave Campbell's proposal? We can distinguish between a weaker and a stronger line one might take here. According to the weaker one, we need to draw some sort of distinction between a more implicit and a more explicit grasp of the causal role of objects. Campbell himself is possibly quite sympathetic to this kind of line (see, e.g. Campbell 1993: 92). The upshot of what we have been arguing would then be that part of what having a more explicit grasp comes to, in this context, needs to be spelled out in terms of the idea that the subject can make sense of the idea that how an object is at one time depends not just on what has happened to it before, but also on the order in which things have happened to it. Once we take that line, however, it seems there's also a stronger one in the offing, because the real explanatory weight in accounting for a subject's grasp of the linear structure of time is carried by her understanding of such temporal-causal relationships. And this might make us wonder whether thought of the causal role of objects is indeed essential to a grasp of the linear structure of time, as Campbell claims. It seems that it can be so only if it is essential to a grasp of temporal-causal relationships.

# ACKNOWLEDGEMENTS

Work on this chapter was carried out within the context of the AHRC Project on Causal Understanding, and while the first author held a Research Leave Grant from the AHRC. We are very grateful to the AHRC for its support.

# REFERENCES

Bennett, J. (1966), *Kant's Analytic* (Cambridge: Cambridge University Press).
Bullock, M. and Gelman, R. (1979), 'Preschool Children's Assumptions about Cause and Effect: Temporal Ordering', *Child Development* 50: 89–96.

Bullock, M. and Baillargeon, R. (1982), 'The Development of Causal Reasoning', in W. Friedman (ed.), *The Developmental Psychology of Time* (New York: Academic Press), 209–254.

Campbell, J. (1993), 'The Role of Physical Objects in Spatial Thinking', in N. Eilan, R. McCarthy and B. Brewer (eds.), *Spatial Representation: Problems in Philosophy and Psychology* (Oxford: Basil Blackwell), 65–95.

—— (1994), *Past, Space, and Self* (Cambridge, MA: MIT Press).

—— (1997), 'The Structure of Time in Autobiographical Memory', *European Journal of Philosophy* 5: 105–118.

—— (2006), 'Ordinary Thinking About Time', In F. Stadler, and M. Stoeltzner (eds.), *Time and History: Proceedings of the 28th international Wittgenstein symposium 2005* (Frankfurt: Ontos Verlag), 1–12.

Carni, E. and French, L.A. (1984), 'The Acquisition of Before and After Reconsidered: What Develops?', *Journal of Experimental Child Psychology* 37: 394–403.

Catellani, P. (1991), 'Children's Recall of Script-Based Event Sequences: The Effect of Sequencing', *Journal of Experimental Child Psychology* 52: 99–116.

Cordón, I. M., Pipe, M.-E., Sayfan, L., Melinder, A., and Goodman, G. S. (2004), 'Memory for Traumatic Experiences in Early Childhood', *Developmental Review* 24: 101–132.

Das Gupta, P. and Bryant, P. (1989), 'Young Children's Causal Inferences', *Child Development* 60: 1138–1146.

Dummett, M. (1960), 'A Defense of McTaggart's Proof of the Unreality of Time', *The Philosophical Review* 69: 497–504.

Fine, K (2005). 'Tense and Reality', in: *Modality and Tense: Philosophical Papers* (Oxford: Oxford University Press).

Fivush, R. (1993), 'Developmental Perspectives on Autobiographical Recall', in G. Goodman and B. Bottoms (eds.), *Child Victims, Child Witnesses: Understanding and Improving Testimony* (New York: Guilford Press) 1–24.

—— and Hamond, N. R. (1990), 'Autobiographical Memory Across the Preschool Years: Toward Reconceptualizing Childhood Amnesia', in R. Fivush and J. A. Hudson (eds.), *Knowing and Remembering in Young Children* (Cambridge: Cambridge University Press), 223–248.

Fivush, R., & Mandler, J. M. (1985), 'Developmental Changes in the Understanding of Temporal Sequence', *Child Development* 56: 1437–1446.

Fivush, R. and Schwarzmueller, A. (1998), 'Children Remember Childhood: Implications for Childhood Amnesia', *Applied Cognitive Psychology* 12: 455–473.

Fodor, J. A. (1998), *Concepts: Where Cognitive Science Went Wrong* (Oxford: Oxford University Press).

French, L. A. and Nelson, K. (1985), *Young Children's Understanding of Relational Terms: Some Ifs, Ors and Buts* (New York: Springer-Verlag).

Friedman, W. J. (1990), *About Time: Inventing the Fourth Dimension* (Cambridge, MA: MIT Press).

—— (2005), 'Developmental and Cognitive Perspectives on Humans' Sense of the Times of Past and Future Events', *Learning and Motivation* 36: 145–158.

Gopnik, A., Glymour, C., Sobel, D. M., Schulz, L. E., Kushnir, T., and Danks, D. (2004), 'A Theory of Causal Learning in Children: Causal Maps and Bayes Nets', *Psychological Review* 111: 3–32.

Guyer, P. (1987), *Kant and the Claims of Knowledge* (Cambridge: Cambridge University Press).

Harner, L. (1982), 'Talking About the Past and the Future', in W. J. Friedman (ed.), *The Developmental Psychology of Time* (New York: Academic Press) 141–170.

Hoerl, C. (1998), 'The Perception of Time and the Notion of a Point of View', *European Journal of Philosophy* 6: 156–171.

—— (2007), 'Episodic Memory, Autobiographical Memory, Narrative: On Three Key Notions in Current Approaches to Memory Development', *Philosophical Psychology* 20: 621–640.

—— (2008), 'On Being Stuck in Time', *Phenomenology and the Cognitive Sciences* 7: 485–500.

Hudson, J. A. and Nelson, K. (1983), 'Effects of Script Structure on Children's Story Recall', *Developmental Psychology* 19: 625–635.

Husserl, E. (1991), *On the Phenomenology of Consciousness of Internal Time (1893–1917)*, trans. J. B. Brough (Dordrecht: Kluwer Academic Publishers).

James, W. (1890), *The Principles of Psychology*, Vol. I (London: Macmillan).

Kant, I. (1929), *The Critique of Pure Reason*, trans. N. Kemp Smith (London: Macmillan and Co. Ltd).

Kun, A. (1978), 'Evidence for Preschoolers' Understanding of Causal Direction in Extended Causal Sequences', *Child Development* 49: 218–222.

Lagnado, D. A., Waldmann, M. R., Hagmayer, Y., and Sloman, S. (2007), 'Beyond Covariation: Cues to Causal Structure', in A. Gopnik and L. Schulz (eds.), *Causal Learning: Psychology, Philosophy, and Computation* (Oxford: Oxford University Press), 154–172.

Mackie, J. L. (1977), 'Causal Asymmetry in Concept and Reality', unpublished paper presented at the 1977 Oberlin Colloquium in Philosophy, quoted in Sklar (1995), p. 218.

McColgan, K. and McCormack, T. (2008), 'Searching and Planning: Young Children's Reasoning about Past and Future Event Sequences', *Child Development* 79: 1477–1497.

McCormack, T. and Hoerl, C. (1999), 'Memory and Temporal Perspective: The Role of Temporal Frameworks in Memory Development', *Developmental Review* 19: 154–182.

—— (2005), 'Children's Reasoning About the Causal Significance of the Temporal Order of Events', *Developmental Psychology* 41: 54–63.

—— (2007), 'Young Children's Reasoning About the Order of Past Events, *Journal of Experimental Child Psychology* 98: 168–183.

—— (2008), 'Temporal Decentering and the Development of Temporal Concepts', in: P. Indefrey and Marianne Gullberg (eds.): *Time to Speak. Cognitive and Neural Prerequisites of Time in Language* (Oxford: Blackwell Publishers), 89–113.

Moore, A. W. (2001), 'Apperception and the Unreality of Tense', in C. Hoerl and T. McCormack (eds.): *Time and Memory: Issues in Philosophy and Psychology* (Oxford: Clarendon Press), 375–391.

Nelson, K. (1996), *Language in Cognitive Development: The Emergence of the Mediated Mind* (Cambridge: Cambridge University Press).

—— (1997), 'Event Representations: Then, Now, and Next', in P. W. van den Broek, P. J. Bauer, and T. Bourg (eds.), *Developmental Spans in Event Comprehension and Representation: Bridging Fictional and Actual Events* (Mahwah, NJ: Lawrence Erlbaum), 1–26.

—— (1999), 'Event Representations, Narrative Development and Internal Working Models', *Attachment & Human Development* 1: 239–252.

—— and Gruendel, J. (1981), 'Generalized Event Representations: Basic Building Blocks of Cognitive Development', in M. Lamb & A. Brown (eds.), *Advances in Developmental Psychology*, Vol. 1 (Hillsdale, NJ: Lawrence Erlbaum), 131–158.

—— —— (1986), 'Children's Scripts', in K. Nelson (ed.), *Event Knowledge: Structure and Function in Development* (Hillsdale, NJ: Lawrence Erlbaum), 21–46.

Orbach, Y. and Lamb, M. E. (2007), 'Young Children's References in Temporal Attributes of Allegedly Experienced Events in the Course of Forensic Interviews', *Child Development* 78: 110–1120.

Peacocke, C. (1992), *A Study of Concepts*, (Cambridge, Massachusetts: MIT Press).

Piaget, J. (1930), *The Child's Conception of Physical Causality* (London: Kegan Paul).

Povinelli, D. J., Landry, A. M., Theall, L. A., Clark, B. R., and Castille, C. M. (1999), 'Development of Young Children's Understanding that the Recent Past is Causally Bound to the Present', *Developmental Psychology* 35: 1426–1439.

Reichenbach, H. (1947), *Elements of Symbolic Logic* (London: Macmillan).

—— (1956), *The Direction of Time*, ed. by Maria Reichenbach (Berkeley: University of California Press).

—— (1957), *The Philosophy of Space and Time*, trans. by Maria Reichenbach and John Freund (New York: Dover Publications).

Schank, R. C. and Abelson, R. P. (1977), *Scripts, Plans, Goals, and Understanding* (Hillsdale, NJ: Erlbaum).

Shultz, T. R., Altmann, E., and Asselin, J. (1986), 'Judging Causal Priority', *British Journal of Developmental Psychology* 4: 67–74.

Shultz, T. R, and Kestenbaum, N. R. (1985), 'Causal Reasoning in Children', *Annals of Child Development* 2: 195–249.

Shultz, T. R., and Mendelson, R. (1975), 'The Use of Covariation as a Principle of Causal Analysis', *Child Development* 46: 394–399.

Sklar, L. (1981), 'Up and Down, Left and Right, Past and Future,' *Nous* 15: 111–129.

—— (1995), 'Time in Experience and in Theoretical Description of the World', in S. Savitt (ed.), *Time's Arrow Today* (Cambridge: Cambridge University Press), 217–229.

Sobel, D. M. and Kirkham, N. Z. (2007), 'Interactions Between Causal and Statistical Learning', in A. Gopnik and L. Schulz (eds.), *Causal Learning: Psychology, Philosophy, and Computation* (Oxford: Oxford University Press), 139–153.

Sophian, C., and Huber, A. (1984), 'Early Developments in Children's Causal Judgments', *Child Development* 55: 512–526.

Strawson, P. F. (1966), *The Bounds of Sense: An Essay on Kant's Critique of Pure Reason* (London; Methuen & Co., Ltd.).

Wagner, L. (2001), 'Aspectual Influences on Early Tense Comprehension', *Journal of Child Language* 28: 661–681.

Weist, R. M. (1989), 'Time Concepts in Language and Thought: Filling the Piagetian Void from Two to Five Years', in I. Levin and D. Zakay (eds.), *Time and Human Cognition: A Life-Span Perspective* (Amsterdam: Elsevier) 63–118.

White, P. A. (1988), 'Causal Processing: Origins and Development', *Psychological Bulletin* 104: 36–52.

Winskel, H. (2003), 'The Acquisition of Temporal Event Sequencing: A Cross-Linguistic Study Using an Elicited Imitation Task', *First Language* 23: 65–95.

Woodward, J. (2007), 'Interventionist Theories of Causation in Psychological Perspective', in A. Gopnik and L. Schulz (eds.), *Causal Learning: Psychology, Philosophy, and Computation* (Oxford: Oxford University Press), 19–36.

CHAPTER 15

...........................................................................................

# TEMPORAL EXPERIENCE

...........................................................................................

JENANN ISMAEL

# 1. INTRODUCTION

...........................................................................................

WE are temporal beings. We have histories, we keep a running record of our histories
as they unfold, and we act with an eye to the future. Time as we encounter it in
experience is very different from time as conceived by physics. Time as conceived
by physics is very simple. There is no intrinsic difference between past and future.[1]
Change and movement are represented as static relations between different parts of
time. All of the parts of time exist in a fixed set of relations to one another. As we
encounter it in experience, by contrast, time is intrinsically directed and in continuous
flux. There are differences between past and future in how much we know about them,
in whether we can affect them, and other ways that have come under examination
in this volume. The past seems fixed, but there is a sense of openness about the
future. Change and movement are the rule rather than the exception. We are almost
irresistibly inclined to describe time in dynamical terms. We say that one event gives
rise to the next, that time passes or flows, that we cannot stop the fleeting moment from
being incorporated irretrievably into the past. Some of this dynamical terminology is
the product of misleading mental pictures, but it arises so naturally and spontaneously
that one suspects it captures something about the way we experience time. What
are the psychological sources of the temptation to speak of time as flowing? Why
does it seem to have a direction? What leads us to regard the past as fixed and the
future as open? Any attempt to reconcile the physical conception of time with the
way that time is encountered in experience has to begin with an analysis of temporal
experience.

The experience of time has been a mainstay of discussion in the phenomenological tradition, but has received relatively little attention in the analytic tradition. But
every aspect of our psychological lives is pervaded by the fact that we experience our

---

[1] And this is linked to the fact that the dynamical laws are symmetric under reflection in time.

histories in stages, remembering the past and anticipating the future. A good part of the complexity of temporal experience has to do with the interaction between temporal perspectives. I'll begin with a schematic description of the history of an historically extended consciousness when its parts are plotted in a temporal sequence and then switch to a temporally embedded point of view, asking what things look like from the perspective of particular moments in that history and how they differ from the perspective of one moment to the next.

## 2. PHENOMENOLOGY AND PERCEPTION

Phenomenological analysis takes it for granted that there is some neurocognitive story to be told, but studies only the structures that arise at the personal level, that is, structures that are present to consciousness, introspectively available to the subject.[2] Consciousness has many elements, from sensory experiences and bodily sensation, to non-sensory aspects such as volition, emotion, memory, and thought. At any waking moment we are aware of patterns of sound, light, color, sound, kinesthetic sensations, and internal moods and emotions. We are also aware of a way the world is presenting itself to us perceptually: we see and hear events occurring in the space around us, we see objects arranged in and moving through the space around us, we feel the motion of our own bodies, and experience some of that motion as governed by volition. We also have memories in the form of recollected images of past events, as well as knowledge of our own histories, and a body of semantically structured belief that can be accessed more or less on demand.

If we focus for the moment just on perceptual consciousness, a simple and natural view would have it that the sensory surfaces register information about the environment and relay it to the mind where it produces experience, in the way that a video camera registers and relays information to a screen, so that we have real-time covariation of states of the world and states of the screen. One representational state replaces another, each reflecting the more or less occurrent state of the environment. Although a person watching the screen will remember the passing images and piece them together to arrive at an idea of how the screen changed over time, there is no representation of time on the screen itself, and no accumulation of information on the screen over time. If perception were like that, the representational content of perceptual experience at any given moment would be an instantaneous state of the world at, or immediately before, the moment that the experience occurs.[3] The content

---

[2]   The notion of personal level representation, first proposed by Dennett in 1969 and quickly became an entrenched distinction in the cognitive science literature. It is used here to refer to the level of representation that is available to consciousness.

[3]   Representational content is what would be reported in the that-clause in sentences of the form "I see that there are three cars in the road/that there is an apple on the table . . .".

would ordinarily have three spatial dimensions—it would depict material objects in a spatial configuration—but it would have no temporal dimension.

The first conspicuous challenges to this idea came from James and Husserl. Both of them defended versions of the Doctrine of the Specious Present.

> The Doctrine of the Specious Present (SP): says that if we consider a particular temporal cross section of experience at a point t in time (call it a t-section), the content carried by the t-section has temporal breadth. It spans a finite interval of time centered on t.[4]

The two primary texts are James' *Principles of Psychology* and Husserl's *Lectures on Internal Time Consciousness*.[5] James attributes the term 'the specious present' to the psychologist E. R. Clay. But he introduced it to the philosophical literature, and his own discussion is so vividly written that it is still the classic text on the specious present. Husserl's is perhaps deeper, but it is exceedingly hard to read. Exegetical difficulties stemming from the complexity of his view are compounded by the fact that the text was not published by Husserl himself, but culled by his secretary (Edith Stein) and student (Heidegger) from notes on time consciousness penned between 1901 and 1917, a period throughout which his own views were in flux.[6]

The most common misunderstanding of SP is to fail to realize that it is a claim about content, and does *not* entail that sensation comes in discrete pulses. James was quite explicit that aside from periods of unconsciousness and sleep, there is no discontinuity in experience at the level of phenomenology. We experience trajectories as smooth and change as continuous.[7] A ball moving across a table from point A to point B appears to pass through all points in between. A bowl of soup doesn't go from hot to cold without passing through all temperatures in between.[8] As James says:

> [experience] does not appear to itself chopped up in bits. Such words as 'chain' or 'train' do not describe it fitly as it presents itself in the first instance. It is nothing jointed; it flows. A 'river' or a 'stream' are the metaphors by which it is most naturally described.

It was well known even in James' time that the phenomenological continuity is partly the product of extrapolation by the brain. Much in the same way that the mind artificially glosses over the blind spot we have in the vision-field of each eye (created by the break in the sheet of photoreceptors where the optic nerve enters the eye) the brain extrapolates a temporally continuous stream of events out of a well-timed

---

[4]  Whether the doctrine was clearly and distinctly conceived in exactly this manner by either is a question I'll bracket. I've sharpened up the basic insight and given it the most defensible expression. See also LePoidevin references in Poidevin for more traditional ways of understanding the specious present.

[5]  From volume, *On the phenomenology of the consciousness of internal time*, translated by John Brough of *Husserliana Brand X* (Rudolph Boehm, ed.).

[6]  I'm indebted to Rick Grush here for an especially lucid exposition of Husserl's views (Grush 2006).

[7]  For models that take seriously the hypothesis that experience is not continuous, but discrete, see Dainton (2000) and Grush (2006).

[8]  See Grush (to appear) for a survey and comparison of neurocognitive models of temporal perception empirical support.

set of discrete events. There is a certain frequency of experience at which distinct events blur into a single continuous duration. This frequency of events is known as the Continuous Flicker Frequency (CFF), at which the experience of a flickering light becomes an experience of a continuously burning light. Depending upon variables such as size of the light source and the characteristics of the observer, the CFF can vary between 2 and 80 cycles per second, but the standard recognized CFF, often used in cinema to turn many still images into an illusion of a motion picture, is 60 cycles per second. Brain activities are, at their basis, coordinations of action potentials, and action potential firings have beginnings and ends. Each flicker of a movie projector sets off a complex perception event, in which many neurons have discrete moments of action and then inaction. The experience, however, is one of continuous motion on the movie screen with much longer duration than any action potential's firing.

SP is not always clearly distinguished from a claim about the minimal duration *occupied by* an episode of perceptual awareness.[9] To think otherwise is to confuse semantic levels—to mix up what's true of the representational content with what's true of the representational vehicle. This is a confusion we're especially prone to in the case of time. We are not apt to suppose that the brain represents spatially separated objects by means of spatially separated perceptions, or red surfaces by red perceptions. But there is a long tradition of thinking that time is special precisely in that the temporal relations between events are represented in experience by the temporal relations between the events that represent them. Helmholtz expresses such a view when he writes,

> Events, like our perceptions of them, take place in time, so that the time-relations of the latter can furnish a true copy of those of the former. The sensation of the thunder follows the sensation of the lightning just as the sonorous convulsing of the air by the electric discharge reaches the observer's place later than that of the luminiferous ether.[10]

There is some degree, clearly, to which temporal subdivisions of the perceptual stream correspond to temporal subdivisions of its content. If you eat breakfast before you get dressed, you experience breakfast eating before you experience getting dressed. SP does not deny this. It simply places a limit on the correspondence, holding that we can't go on subdividing the stream of perceptual consciousness into components that correspond to parts of time up to the level of points. SP claims that every perceptual content, even at the finest level of resolution, is awareness of a finite temporal interval. Even if we consider the content of instantaneous temporal cross section of experience, the representational content of that cross-section will span a finite interval of time. It is

---

[9]  When Wittgenstein asks "If I see the picture of a galloping horse, … Is it superstition to think I see the horse galloping in the picture?—And does my visual impression gallop too?" (Wittgenstein (1999: 202). The answer is that the galloping is part of the representational content of the picture along with the three-dimensionality of the horse, and the space that contains it. And the same should be said about the experience of a galloping horse.

[10]  Helmholtz (1910: 40).

analogous in this respect, as James remarks, to spatial perception. We are never aware of an instant of time but only of some finite interval, just as we are never aware of a point in space but only of some finite spatial volume. The minimal unit of perceptual awareness has both spatial and temporal breadth. This doesn't mean that we can't arrive at the concept of an instantaneous state. It means simply that we get that idea by carrying the process of subdivision to its limit, and what we have left when we do so is empty of any experiential content.[11]

This is connected to another way of misunderstanding SP. To say that the specious present represents an interval with past, present, and future parts does not mean that the specious present has temporal parts lying in the past, present, and future, but that the content *represents* an interval of time as a temporally ordered whole centered on the present. To see the difference here, consider the spatial analogue. When you perceive a cathedral. Although the cathedral itself is composed of stones laid out in different parts of space, your *percept* of the cathedral is not composed of percepts of stones that are located in different parts of space. That would leave the spatial relations outside the scope of any percept. To see them arranged in cathedral configuration, in that case, there would have to be a further seeing that spans those parts and relates them to one another. SP asserts that the most elementary contents incorporate lower-level elements that might be separated by analysis, but are themselves highly structured. Husserl refers to the past, present, and future components of the specious present, respectively, as retention, primal impression, and protention. James describes the structure with a memorable image,

> The unit of composition of our perception of time is a duration, with a bow and a stern, as it were—a rearward—and a forward-looking end.

He continues, emphasizing the synthetic character of the content;

> It is only as parts of this duration-block that the relation of succession of one end to the other is perceived. We do not first feel one end and then feel the other after it, and from the perception of the succession infer an interval of time between, but we seem to feel the interval of time as a whole, with its two ends embedded in it. The experience is from the outset a synthetic datum, not a simple one.[12]

The best introspective evidence for SP is the perception of motion or change. When you see a ball thrown across a room, you don't see instantaneous representations of the ball's position, you see movement. The motion, which is not present in any instant of the series, falls in the scope of your percept. You may be able to break it into smaller components, but even the smallest includes some motion, and so even

---

[11] 'And here again we have an analogy with space. The earliest form of distinct space-perception is undoubtedly that of a movement over some one of our sensitive surfaces, and this movement is originally given as a simple whole of feeling, and is only decomposed into its elements—successive positions successively occupied by the moving body—when our education in discrimination is much advanced.' James (1890: 622).

[12] James (1890: 610).

the smallest has temporal breadth. When you hear a descending pitch, you don't just have a descending series of impressions of notes, you *hear the descent*. Which is to say that the descent, which is not present in any instantaneous part of the series, falls in the scope of your percept. And notice that when you perceive motion, you don't just perceive motion in a certain direction (a ball was first here, then there), you see *how fast* it occurs, that is, how long it took to get from here to there. This perception of the speed involves perception of *quantity* of time. It imposes not just an order, but a metric on the perceived process, and the order of the parts and metric are part of the content of the impression. The same goes for other modalities. A potter at the wheel feels the motion of the clay in his fingers. The passenger on a train feels the vibration of the rails as they pass under his car. This perception of movement includes both order and quantity.

In all of these cases, the movement and the speed are both part of the content of the experience. This bears emphasis. In order to have experiences of succession, movement, or duration, the contents of those experiences must have temporal breadth. It is not enough for experience itself to be extended in time, there has to be a temporal dimension in the representational content. This is a generalization of Kant's oft-cited observation that successive experiences are not an experience of succession. To have an apprehension of temporal order, it is not enough to apprehend instants of time individually in succession. That leaves the relation of succession outside the scope of apprehension of any experience. We have to apprehend them together, rather, as an ordered collection. And to get a measure of amount of time, it is not enough to apprehend instants of time individually over some period. We have to apprehend the period itself in a single act. This primitive perception of a minimal unit of time can then serve as a yardstick in terms of which we conceive of longer units of time.

It's not hard to incorporate SP into a modern conception of the mind. We are blind to the subpersonal processes that generate perceptual awareness, but this doesn't mean that the brain is passively conveying information from the sensory surfaces to the conscious part of the mind.[13] When simple stimulus response mechanisms in the human brain incorporate more complex forms of mediation between input and output, we start to talk about sensorimotor loops. Collections of these get cobbled together, sometimes in a manner that is regulated by a superloop, and these in their turn are collected under the partial supervision of further loops. At every stage, there is filtering, transformation, integration, and what is *given* at one level, is *constructed* or *restructured* by the levels below.[14] The emergence of personal level representational states is a late development on the phylogenetic scale that involves the emergence of a new kind of superloop that selectively integrates information from lower-level sensorimotor loops to generate an overarching conception of a spatiotemporally ordered world. What one is consciously aware of at the personal level—that is, what is given immediately and without inference in the contents of personal level perceptual states—is the product

---

[13] Personal level representation is representational content that is introspectively available to consciousness.

[14] By 'constructed' we mean arranging, assembling, imposing a new order on.

of that integration. The lower-level processing supplies the embedding structure that organizes the complex, cross-modal patterns of sensory qualities into a conception of an orderly, three-dimensional reality viewed from the spatial vantage point of an embodied subject. A unified frame of spatial and temporal reference is supplied; temporally separated and qualitatively distinct images are strung together into the world-lines of places and objects that are reidentified across experiences, viewed from different angles, and apprehended through different modalities.

The movie projector model leads us to believe that structures that are present in our perceptual state at a time are a simple mirror-like reflection of structure that is present in the stimulus. This picture suggests something quite different. When you stand on a street corner looking out at the world, you may be aware of a multi-dimensional pattern of light and color, sound and smell, but *what you see* is cars whizzing past, people walking by, speaking to one another, a streetlight changing color. What you see—that is, the representational content of your experience—is an evolving three-dimensional space in which lights, sounds, and smells are related to one another and to the vantage point of your own eyes. The spatiotemporal structuring of experience, which is given in the representational content, presupposes an embedding structure that imposes strong constraints over vast tracts of experience. The unification of the sensory manifolds, the separation of space and time, and the conception of oneself as a material presence in the landscape are all parts of the embedding structure, but they impose more structure than is present in the occurrent stimulus. It's a very Kantian idea that concepts of self and world and space and time all get sorted out together as part of a categorial framework that brings order to experience.[15] But it's one that to this extent is borne out by what we know about the way the brain processes sensory information.[16]

One might accept that the representational content of any given temporal cross section of experience has both spatial and temporal dimensions, and still wonder whether there is reason for thinking that perceptual contents have a *protentive* component. This is one of the most interesting implications of SP. Both James and Husserl asserted it, but neither of them spends much time on it. Rick Grush has argued that there is empirical evidence for the existence of a protentive component coming from experimental work on temporal illusions, and has developed a neurocognitive model for perceptual processing that incorporates SP. He writes

> The basic idea that perception involves constructing representations that are based, in part, on sensory information, is fairly standard, and has been for some time. But part of this standard view has been that the job of the perceptual system is to produce representations of states of the environment. I want to suggest, though, that we should reconceive the job of the perceptual system as producing representations

---

[15] These concepts aren't definable in sensory terms: they function rather as primitives whose application imposes order on experience.

[16] Questions about how the cognitive mind is implemented in the brain are unsettled. Analysis of the structures that arise at the personal level can mostly avoid them.

that attempt to capture temporally extended *processes* (or, synonymously, *trajectories*) in the environment.[17]

One of the most innovative aspects of Grush's account is that trajectory estimates don't just represent parts of processes that are already completed, but anticipate the direction at the next moment. His reasoning is that we need to initiate behavior designed to react to future states of the world before we actually receive information that we are reacting to. Perceptually guided behavior exhibits sensitivity to the temporal features of trajectories: it can adjust to the speed and anticipated duration of processes as they unfold. When Santonio Holmes caught the winning touchdown in the 2009 Superbowl, there was no conscious calculation of where the ball would be. His brain was moving his body to where the football would descend *before* his senses registered its presence there. If Grush is correct, Holmes didn't see where the ball was and *infer* where it would be, he literally saw both where it had been and *where it was going*. There was no time for inference, and no conscious awareness of having made any inference. The forward and backward looking part of the trajectory was all part of the instantaneous content of his visual state—it was given to him immediately in the content of the experience. George Bush was able to duck out of the path of a flying shoe *before* he registered its presence where he had been standing because he *saw* where it was going. Your hand is able to catch the bag of peas as it falls out of the freezer not only because you see where it is, but where it will be. We are always in this sense *reacting* to what we *foresee*, acting to fend off, forestall, divert trajectories in process, but uncompleted.[18] If SP is correct, those actions are guided by the protentive component of experience.

# 3. MEMORY

So far, I've spoken only about perceptual consciousness. Perceptual consciousness, however, isn't the whole story. To fill out our portrait of a psychological history, we need to embed perceptual consciousness in a psychological stream whose full description includes the contents of memory. What we do at the personal level in forming a conception of history to some extent mirrors what perceptual processing does on a very small scale in forming a conception of change over the interval of a specious present. The term 'memory' covers quite a large variety of phenomena. There is what is sometimes called 'habit memory' or 'procedural memory', a label for embodied skills such as typing, playing golf, using a knife and fork, or solving jigsaw puzzles. These are not directly representational forms of memory. They do not represent the world as being a certain way. Among representational forms of memory, we distinguish between short-term and long-term memory. Short-term memory acts

---

[17] Grush (to appear), p. 15.

[18] That is, the anticipatory component is learned rather than hard-wired and gives us the practical flexibility we require in new environments. Repeated exposure to different patterns of events leads to new behavioral expectations.

as a scratch-pad for temporary recall of the information under process. For instance, in order to understand a sentence you need to hold in your mind the beginning of the sentence as you read the rest. Short-term memory has a limited capacity and decays rapidly (200 ms.). Long-term memory is intended for storage of information over a long period. There is little decay for information in long-term memory. Information from the working memory is transferred to it after a few seconds. Among the forms of long-term memory is semantic memory which is memory of fact, the accumulated fund of particular and general belief acquired through book-learning, hearsay, and all of the other ways that we pick up information about the world. It is usually impersonally expressed and stored in a propositionally structured form, a vast internal encyclopedia of knowledge that includes the fact that water contains hydrogen, that Wittgenstein was Viennese, and that elephant tusks are made of ivory.

Personal memories of past experience hold special interest. These come in two forms. When I think about the first time I visited Cairo, I can remember what it felt like stepping into the desert air. I remember the smells and sounds, the date palms right next to the taxi rank. I recall the sensory field almost as I experienced it. These kinds of recollected images of past experiences are *episodic* memories. They are representational, but not propositional in form. They are singular and image-like. They have qualitative properties that resemble the experiences they represent, and they don't involve any explicit representation of time or self. Like a photograph taken at a particular time, they represent the view of a space at the time at which they were taken, but neither the time nor photographer (ordinarily) appear in the image. I also remember that the trip was in 1987, that we stayed the first few nights at the Hilton, that there was a restaurant at the Hilton that became our haven, that I visited Minia, then the Sinai, then Dhahab ... These memories, by contrast, are propositional in form: they explicitly portray me as subject and ascribe certain experiences to me in a particular order. They are the products of autobiographical memory, whose function it is to weave the collection of episodic memories into a portrait of personal history.[19]

Episodic and autobiographical memory work together. Episodic memory allows information from past experience to collect in the mind by making records of past experiences, and autobiographical memory gives that information form summarizing, constructing, interpreting, and condensing life experiences, to produce a coherent narrative sense of a personal past. Autobiographical memory is the psychological source of the conception of self as temporal continuant. The psychological sense of continuity depends on the fact that I remember my past and expect in the future to remember my present. One's sense of self extends as far into the past as one's memories, and as far into the future as one expects to remember the present. Autobiographical memory opens up the psychological space for a conception of self that spans a whole life. Without autobiographical memory, psychological life would consist of a series of psychological episodes—one thought or experience and then another—with

---

[19]  To say that a representation is explicit is to say that it falls within the scope of the representational content of a state. No further explication is possible without a full-blown theory of content.

a temporal horizon no longer than a specious present. A being with autobiographical memory, by contrast, has the capacity to survey its past from the earliest recollected moment in childhood at any point in the course of its life.

There is no one better than Proust at evoking the thin-ness and ephemerality of the sense of self supported by the specious present, and the role that memory plays—as he says 'like a rope let down from heaven'—integrating specious presents into a personal history. Early in *Swann's Way*, for example, he writes:

> When a man is asleep, he has in a circle round him the chain of the hours, the sequence of the years, the order of the heavenly host. Instinctively, when he awakes, he looks to these, and in an instant reads off his own position on the earth's surface and the time that has elapsed during his slumber; but this ordered procession is apt to grow confused, ... [there were times when] I lost all sense of the place in which I had gone to sleep, and when I awoke in the middle of the night, not knowing where I was, I could not even be sure at first who I was; I had only the most rudimentary sense of existence, such as may lurk and flicker in the depths of an animal's consciousness; ... but then the memory not yet of the place in which I was, but of various other places where I had lived and might now very possibly be— would come like a rope let down from heaven to draw me up out of the abyss of not-being, from which I could never have escaped by myself; in a flash I would traverse centuries of civilization, and out of a blurred glimpse of oil-lamps, then of shirts and turned down collars, would gradually piece together the original components of my ego.[20]

There's the lone thought, which first situates itself as part of the community of connected memories that form a single life. This jumble of memories then shakes itself into an order that is embedded in the larger narrative of history. And all of this structure— the occurrent thoughts and experiences, the episodic memories, the personal history, and the impersonal history in which it is embedded—are all present—in a more or less definite, more or less explicit form—in every momentary part of the psychological life of a consciousness with autobiographical memory. This structure is not always part of the foreground of thought, but it is present in a form that allows it to be accessed more or less on demand. The contents of memory are like psychological time capsules, providing each momentary cross section of an evolving consciousness with a compact, backward-looking representation of its own past.

## 4. THE STREAM OF CONSCIOUSNESS

Intuitive understanding of perspective is strongly shaped by the spatial case, which suggests the need for an *owner* of perspective, a spatially extended occupant of space that retains its identity across changes in spatial location and whose movement corresponds to changes in spatial perspective. Carrying the analogy over to the case of time

[20] Proust (1988: 56).

would require a *temporally extended occupant of time that moves through time, retaining its identity across changes in temporal location.* This picture is rife with confusion,[21] but most of us nevertheless retain some version of it when we think about time. Either we think of time as forming a fixed background and ourselves as moving through it, or we think of time as, in some objective way flowing past us, bringing our ends ever nearer. The schematic structure above gives us a way of thinking of transitions between temporally embedded perspectives inside a life, without slipping into the idea that we move through our lives occupying now one, now another temporal perspective in it, and allows us to begin to explore the psychological structures that underpin temporal experience. In what follows I'll review abstract description of what the history of a normal human consciousness looks like from the outside, and then we'll turn to temporally embedded perspectives within that history.

I've said that the stream of perceptual contents is embedded in a psychological context lined with memory. The contents of memory grow by 'accretion of fact' moving up the temporal dimension of a psychological history with the addition of new memories. The contents associated with each temporal cross section of that history include a backward looking portrait of its past. The result is an asymmetric arrangement, with information accumulating in memory along the temporal dimension in an almost profligate reification of structure, representation, and re-representation of the same events in every momentary cross section of experience. Rehashing, reevaluation, reorganization occurs at each stage. That rehashing and reorganization is the conscious counterpart to the subpersonal processing that generates perceptual contents. It is an ineliminable part of practical reasoning. Whether I decide to take another drink depends not only on how many I've already had, but on whether I believe I'm slipping into an unhealthy pattern, and that is a judgment that takes some consideration. Whether I decide to abandon a partner or friend depends on my understanding of my history with him or her, the loyalties, resentments, and affections that have been formed, memories of expectations realized or relinquished, fears and hopes and aspirations recalled as they were experienced and viewed through the lenses of later events. All of this is woven into the history of the relationship itself, and plays into decisions about how to act in the here and now. And it is something that requires constant rethinking. Each momentary content of consciousness contains, alongside information coming in from observation, a remembered image of the preceding state. And that image of the preceding state contains an image of its predecessor nested in it. And that one, likewise, and so on like a string of Chinese boxes, each containing a reproduction of its predecessor.

Autobiographical memory doesn't extend indefinitely into past or future: the sequence is bounded by birth on one end and death on the other. And everything we know about the transformative effects of both memory and self-observation should caution us against a naive presumption of either accuracy or completeness. Memory is notoriously reconstructive and this business of representing one's past is not

---

[21] See Price, this volume.

necessarily veridical.[22] Each moment is only very partially and selectively reified in the next, and reification doesn't merely copy, but transforms its objects: filtering, shifting, and sometimes distorting in an attempt to bring them into sharper focus.[23] But we do represent our pasts, and re-represent them with very passing moment, reexamining, reevaluating, and reorganizing them in an ongoing process of self-definition. Some of us do this more than others, but all of us do it to some degree. No other animal so far as we know has the cognitive infrastructure to support a reflexive conception of its autobiographical history with anything that approaches human complexity.[24] And no animal so far as we know engages in this complex process of reflexive self-definition.

Compare this structure with the representational history of a system without a memory—a robot, for instance, navigating by an internal map of space, but not representing its past. The epistemic states of such a system follow one another in an ordered sequence, but there is no retention of information. Each replaces the next: there is no representation of time at any point in its history, no representational state that spans the contents of these specious presents and integrates them into a history in which they are simultaneously represented in a temporally ordered form, and no internal *point of view* whose temporal horizon includes past, present, and future. And compare both of these to the psychological life of an even simpler, sentient system that reacts to stimulus in ways designed to produce adaptive behavior, but represents neither time nor space. All that exists for such a system at any point in time is the occurrent contents of consciousness. The concept of a world distinct from experience extended beyond the boundaries of that state is not provided for. There is change—each is different from the preceding state—but no preservation of information across change. If there are causal connections between one state and the next, there will be continuity that is visible from the outside—that is, to a perspective from which multiple states are simultaneously visible—but there is no retention of information in the explicit content of the states. The continuity won't be visible from the embedded point of view within such a life. The system itself does not have a point of view that spans its temporal parts. At no point in its own psychic history are its temporal parts present simultaneously to consciousness.

Because we represent both space and time, our psychological states have both spatial and temporal content. In the first case described above, we have a system whose states have spatial content but no temporal content. And in the second case, we have a system whose states have neither spatial nor temporal content. I'll leave it open whether those states are properly ascribed objective content at all. We know for a fact that we

---

[22]  The relevant notion of veridicality is something more than forensic accuracy. An honest or fair representation employing thick ethical concepts is a more subtle matter than a bare transcription of fact (if such there be).

[23]  'Observation' and 'memory' are not used here as success terms. There is no assumption of veridicality. We can replace them, respectively, with 'process which generates representations of the environment', and 'process which generates representations of the past'.

[24]  See the old, but still excellent collection by David Rubin (1986).

can construct systems whose cognitive lives conform to these descriptions which do remarkably well at navigating complex environments and completing practical tasks necessary to survival. So the practical advantages of explicit mental map-keeping and calendar-making (i.e. of explicit representation of space and time) are subtler than we might think.[25] But it is an innovation that makes all the difference in the world to the internal character of a psychological life.

# 5. LOOKING FORWARD

I spoke earlier of the anticipatory component of the specious present. There's a much more far-reaching anticipatory representation of the future, a forward-looking analogue to autobiographical memory that represents events both closely connected to the here and now, and events that are far away in space and time. I'll bundle our representations of the future together under the heading 'expectations', though this term conceals a great deal of variety. Some representations of the future are predictions, but there are also hopes, aspirations, fears, and intentions, . . . and each one of these has its own epistemic caste. There are great differences between our representations of the past and our representations of the future that bear more careful analysis. They're not very well understood, though in psychological terms, the asymmetry between past and future is embodied in the differences between memory and expectation.

When we pull all of the pieces together, the schematic picture we get of the contents of an evolving consciousness is this. Experience is continuous, which means that we can consider the contents of any temporal cross section of experience at any moment in the conscious waking life of a cognitively normal human subject. If the subject is attending perceptually to the world, those contents will include a sensory field carrying a perceptual content spanning a finite interval centered on the present. And it will be embedded in a psychological context containing a backwards-looking self-image on one side and anticipatory representations of the future on the other. The backwards-looking self-image will be a collection of episodic memories woven into an autobiographical history, and the forward-looking representation of the future will be the mixture of predictions, hopes, fears, aspirations, and intentions I have called 'expectation'.

There is a great deal of variability among persons, and over time, in the history of a person in how the schema gets realized. How much of the past and future is represented, in how much detail, in what terms, and how faithfully? How much thought is given to the past and future, and how does it figure in practical reasoning? But any complete account of the contents of a normal human consciousness at any point in time would have a temporal dimension organized in roughly this way. Memory would

---

[25]  Ismael (ms) is an examination of these advantages. See also Ismael (2006: last chapter).

have the recursive structure described above (each containing a partial, not necessarily accurate reproduction of the contents of memory at the preceding time, with all of the nested images of preceding contents that it contains). If we look up the temporal dimension, comparing the contents of consciousness at different moments in the history of a consciousness organized in this way, we see the whole structure centered on different moments of the internal timeline we provide for our own histories.[26]

Because we update both memory and expectation as we learn and evolve, what we remember and what we expect varies from one moment to the next, as does the line between events represented in memory and events represented in expectation. So if we move up the sequence, keeping our eye on an event that starts out as expectation (say the wedding day of a daughter), it will get progressively closer to the line, eventually crossing and passing into memory. The slow shift of balance within a life that starts out light in memory and heavy in expectation and ends relatively heavy in memory and light in anticipation is a poignant and inevitable feature of growing old.

Now that we have some sense of what the history of a self looks like from the outside, we can switch points of view and ask what things seem like from the embedded perspective of a moment within a life. And we can also ask how things change with changes in perspective, in the same way we can ask what things look like from a particular point in space, and how that changes with changes in spatial perspective.[27] The temporal perspective here is given by the moment on the internal timeline on which the temporal content is centered, that is, given by the moment on our internal timeline that we call 'now'.

Let's consider a very simple example and try to describe the phenomenology. It's common to use auditory experience, and the perception of music in particular, in discussion of temporal phenomenology, because it provides a highly simplified setting So imagine yourself immersed in a warm bath, stop up all your senses except for your ears, make your mind a blank slate, attach an iPod to the stereo and hit 'play'. Suppose that as it happens, what comes on is a recording of Bach's Cello Suite #4 in E flat. Before the first sound emerges from the earphones, you don't expect silence over sound, Janis Joplin over John Cage. Once the first note is sounded, registered, and recorded, even if you have no conscious memory of having heard the piece before, you have at least some memory and some new expectation. You have probably increased your expectation of hearing more cello and lowered your expectation of hearing Janis Joplin in the next moment. A second note is registered and added to memory. Your earlier expectation is confirmed. A new note is registered, compared against expectation from previous cycle, added to memory, new expectation is generated, and new, more definite expectations begin to take shape.

---

[26] Since we conceive of our histories in objective terms, this will also be the objective timeline of history.

[27] The analogy holds perfectly so long as we are careful not to slip into thinking of transitions between epistemic perspectives as suggesting that there is any *thing* that undergoes those transitions. See Ismael (2006).

The cycle repeats, with memories accumulating, and expectations becoming more definite at every stage. The mind begins to discern patterns, recognize motifs. It jumps ahead and completes a theme before the notes register. It is either satisfied, or surprised by what it hears, delighted, or disappointed. At the first stage, the mind registers a note and forms a very indefinite sort of expectation. There's nothing at this stage yet in memory. At the next stage, the note and expectation registered at the first stage are incorporated into memory and form the psychological backdrop against which the second note is heard. A newer, more definite expectation is formed that draws both on the note that is being currently registered and the contents of memory. And so it goes, at each stage, the contents of the previous stage being incorporated into memory, a new note being registered, and a new expectation formed that draws on the whole accumulating stock of information being registered perceptually and incorporated into memory.

The sort of system that keeps an evolving record of its past and forms expectations for the future encounters every note as a partial revelation of an extended structure that will be eventually apprehended in its entirety. It encounters each note essentially as part of a melody in progress, a partially recollected and partly anticipated whole. The notes themselves occur one at a time. They are not co-present in physical space, or co-represented in auditory experience, except on the very small dimensions afforded by the specious present. It is in the memory and expectation of the subject that they are brought together on the larger scale of the piece as a whole, setting up the cross-temporal pattern of resonance and reverberation that makes them musical.[28] It's not simply that the parts of the song need to be simultaneously represented in memory to permit apprehension of patterns and recurrences. That is one part of it, but that is available also to the person looking at a musical score. And notice that it doesn't matter, for purposes of perception of these regularities whether he reads the score front to back or back to front. But it is essential to the musical experience that listening *itself* is a process, that is, that the song is revealed in stages, and in stages that follow a particular order. And that is because it is essential that each note is encountered from a different temporal perspective, in a psychological context lined with different memories and different expectations. Changes are wrought in the listener at each stage in the listening process. In physics, we would say that the listener doesn't 'return to his ground state' after each observation, but that memory serves like a cognitive ratchet, saving changes wrought by experience and propelling the listener into an ever new frontier. And these changes don't affect representational content. They make a difference to the *quality* of the experience. The mind that confronts a theme for the third or fourth time hears it differently than a mind that confronts it for the first. Surprise,

---

[28]   The meeting of an expectation is a kind of consonance between a remembered expectation and an observation; surprise, or disappointment, is a kind of dissonance. To have suspense resolved or to recognize a repeated theme, to see a theme developed, these are all cross-temporal relations. Surprise, disappointment (what you expect doesn't come to pass); pleasing or unpleasing cross-temporal dissonance. Resolution, satisfaction, repetition (what you expect comes to pass); pleasing or unpleasing cross-temporal consonance.

recognition, disappointment…these epistemic attitudes have a phenomenology of their own. That phenomenology is not available to a system without a memory, and it is as much a part of the musical experience as the sounds emerging from the bow. You have to think of the quality of the experience as determined not just by input from perception, but jointly by the input from perception and memory. Now, consider the mind that confronts that final note. It has a high degree of internal complexity. It is very much less innocent than the mind that crawled into the bath. It has memories of the view from all perspectives that preceded as constituents. That structure has to be built up by passing through those stages: it can't be bypassed by simultaneous apprehension of the notes in a higher dimensional medium, as, for example, by looking at the score. When you look at a score, you see in two dimensions all at once what you hear as a temporally ordered sequence.

Such—at a much higher degree of complexity—are the epistemic states of a normal human adult. We add the full complement of experiences, with all of their internal complexity, and we extend the sequence to cover a lifetime.[29] The complexity of these states is appreciated by Husserl,[30] and is vividly in Velleman's discussion of reflexive memory in *Self to Self*. He writes:

> I don't just anticipate experiencing the future; I anticipate experiencing it as the payoff of this anticipation, as the cadence resolving the present, anticipatory phase of thought…Within the frame of my anticipatory image, I glimpse a state of mind that will include a memory of its having been glimpsed through this frame—as if the image were a window through which to climb into the prefigured experience."[31]

Not just our epistemic states, but also (and perhaps especially), our emotional responses are closely tied to these cross-temporal patterns of resonance and reverberation, consonance and dissonance, not only among remembered experiences, but among our memories of expectations and expectations of memories reproduced—partially, at least implicitly—inside each momentary part of our lives. Think, for example, of the complexity of sadness at the memory of years of regret attached to expectations for a relationship in light of what actually came to pass. You have only a little difficulty attaching both a phenomenology and a content to that state, but it has an exceedingly complex temporal structure of iterated nesting. States with the complexity to support these epistemic and emotional attitudes have the nested structure that arises from autobiographical memory. And again, because they have as constituents, representations of the view from different epistemic perspectives, they have to be built up in stages by passing through those perspectives, one at a time, in a particular order.[32] In a world

---

[29] The example is too simple to support the embedding structure of a spatiotemporally organized world, so the kind of memory here is not yet autobiographical, but merely reflexive.

[30] What I have said is in agreement with the central elements of Husserl's view, but I've refrained from explicit discussion because there's much that I'm not confident of having understood.

[31] Velleman (2005: 198).

[32] The view from a particular perspective just is the view with a particular set of memories at one's back, and so the events stored in memory have an intrinsic order.

like ours, building those states is an attenuated process. It requires nothing less than the laborious process of living.[33]

And that brings us to a very salient aspect of the phenomenology of a life lived in time: suspense, not knowing *what comes next*. There is a tension set up in the mind of the subject that represents her own life, in the gaps between anticipation and resolution, when the mind prompted by history has formed an expectation and awaits its resolution. From the perspective of any moment in a life, there's always a space between what is known and what awaits revelation, between what's been stored in memory and what lingers in expectation. We live our lives in that space, perpetually poised between expectation and resolution in the limbo between what is and what might yet be. And the transitions between temporal perspectives are accompanied by impression of possibilities melting away. As we look back over our pasts and forward to the impending end, we have the impression moving away from our pasts and towards the future. In the beginning of life, we are separated from the end by a yawning gap full of possibilities awaiting resolution. That gap is narrowed as we move up the temporal dimension of our lives and possibilities give way to actualities. The space between what is known and what is still to be revealed is closed at the end of life,[34] but we spend our lives in a state of suspended cognitive animation, representing ourselves as captured in the middle of a cycle that has been repeated as far back—quite literally—as time immemorial. It is worth remarking on how to understand this without incoherence. We shouldn't think of ourselves as moving through our lives occupying different temporal perspectives, or thinking of time as flowing past us, as we stand fixed. We should think of the mind as looking back over the changing temporal perspective it has represented in memory and seeing the shift over time.[35] There is nothing illusory about that shift, and nothing illusory about the differences between past and future from an embedded perspective.

In the discussion of music, I have emphasized memory because the temporal breadth of the specious present is small relative to the length of a musical piece, and I wanted to bring out the cross-temporal patterns that stretch over the piece as a whole. They arise within the wider temporal context provided by memory, because it's there that the parts of the song are pieced together and represented simultaneously. But that's not to say that the specious present isn't important to the phenomenology of music. We don't just experience instantaneous parts of notes and piece them together in memory. When I hear one note, the preceding note still lingers in experience and the experience carries an expectation about what will follow. The content of the experience stretches over a temporal interval, and it is this that allows us to hear melody as descending and to experience the quickness of a tempo. If it were not so, that is, if SP were not true, the descent or the quickness could not be part of the content.

---

[33] In a world like the one that Russell once envisioned, in which the world is created *ex nihilo* with all memories and records of the past intact, things would be different, but in a world of the sort we take ours to be, the process takes time.

[34] Which is not to say that everything gets revealed, but that what ignorance is left, forever remains.

[35] See Ismael (2007, ch. 11).

Someone 'looking from the outside', without the particular set of epistemic limitations that characterize the view from within time, doesn't undergo the cycle of expectation and resolution, doesn't experience the accompanying emotional tension and release. For God, as surely as for the cow that has no sense of its past or future, there is no uncertainty, no nostalgia, no anticipation, discovery, anxiety, or expectation. There is no cycle of suspense and resolution; there is only the set of events laid out in time.[36] What this brings out is that it is the combination of autobiographical memory (in its truly reflexive form) and the fact that each of our momentary selves has an epistemic horizon that is essential to the epistemic phenomenology. Things seem different from different perspectives, and there seems to be a definite direction of movement, only because each perspective has a view that is both partial and asymmetric.[37] There is a special phenomenology that arises for a system that represents time, a whole cluster of cognitive, emotional and epistemic attitudes that are essential to the felt character of human life. Those attitudes are not available either to a system without memory (in a form that involves the representation of time) or to an all-knowing god. They are not available to a system without a memory because such a system doesn't have the states with the complexity to support those attitudes. (Recall what those states look like: just one representation of the occurrent state of the environment after another, like pearls on a string.) And they are not available to an all-knowing god, because the psychological history of an all-knowing god does not evolve. There is no development, no change, no difference in how things seem at different times.

## 6. TIME AND THE SELF

Memory lengthens the range of temporal vision from the very small interval afforded by the specious present to the much wider expanse stretching from early childhood to the present. Even though much of the literature on time perception in cognitive psychology has focused on the specious present, phenomenologists like Husserl and Ricouer, and moral philosophers, particularly in the Lockean tradition, recognize that the most interesting and most distinctively human features of temporal experience have to do with the larger dimensions provided by memory, and specifically memory in its autobiographical form. This provides us with a conception of self that stretches across the years, and is a necessary condition for personhood in the legal, political, and moral sense. It is what provides for relationships that grow and evolve, personal

---

[36] Nor does your dog, if his representational states don't have a temporal dimension, experience the passage of time. Which means that there are certain experiences—agency might be one as well—that depend on a certain kind of representational setting.

[37] This contrasts with the spatial case, in which there is limitation, without asymmetry: what one sees from a given spatial perspective is a proscribed region of space centered on one's body, but one doesn't see more from one perspective than from another (ignoring contingencies like obstacles and so on that limit one's field of vision).

commitments that stretch into the future, plans and projects that can structure a life. There's been a lot of discussion in recent years calling attention to the role of autobiographical memory and the rendering of one's history in narrative terms in constituting one's awareness of oneself as a temporally extended being. In more radical incarnations of the view, it is said that what it is to *be* (or have) a self is to possess a narrative self-identity. The suggestion is that the very activity of piecing together an inner biography constitutes one as the temporally extended bearer of the biography.

This is an extremely interesting suggestion. It is offered as a potential alternative to the ontological obscurities of traditional Cartesianism and the skepticism of Hume, Nietzsche, or contemporary anti-realists like Dennett. Dan Zahavi, in a recent review of narrativist accounts of self, writes:[38]

> Ricoeur, who has frequently been regarded as one of the main proponents of a narrative approach to the self, has occasionally presented his own notion of narrative identity as a solution to the traditional dilemma of having to choose between the Cartesian notion of the self as a principle of identity that remains the same throughout the diversity of its different states and the positions of Hume and Nietzsche, who held an identical subject to be nothing but a substantialist illusion.[39]

The view, however, suffers from a lack of clarity or consensus about what a narrative self-identity is.[40] Proponents of the view seem to have something more in mind than the minimal form of psychological temporality involved in representing one's past. Narrative structure is usually linked with having a 'story-like character'. The impulse to organize the pieces of one's life into a story requires, at the very least, their co-presentation to consciousness and rendering in a temporally ordered form. So the minimal form of psychological temporality that I have described is a necessary condition for narrative structure. It opens up the space within which the narrative impulse finds expression.[41]

I myself am skeptical of monolithic accounts of selfhood, that is, accounts that suppose that there is a single notion of self that will cover all of its uses. But we don't have to adopt narrativism as a monolithic account of selfhood to acknowledge its insights. We clearly do engage in the construction and elaboration of an inner biography. And piecing together an inner biography leaves us with an internal point

---

[38] For discussion of the narrativist thesis. See Bruner (2002), Carr (1991), and against them Strawson (2005), and Zahavi (2008).

[39] Zahavi, p. 2. Although he's careful to note Ricouer's ambivalence.

[40] Aristotle and Plato both had accounts of narrative structure and it saw renewed popularity as a critical concept in the mid- to late-twentieth century in literary theory when structuralists argued that all human narratives have certain universal, deep structural elements in common, but that view fell out of fashion, and there's still no generally agreed definition of narrative.

[41] The idea that narrative unity is a requirement of selfhood has been challenged most loudly and persistently by Galen Strawson (2005). He is attacking something stronger than the claim that to be a temporally extended self requires the possession of a point of view that ranges over one's past and future. He is challenging the descriptive claim that we live our lives with our pasts and futures always in full view, acting for the sake of narrative unity. And he is challenging the normative claim that we should strive for narrative unity because a good life is one that is narratively unified.

of view that ranges over the temporal parts of a life. The construction of this point of view is a cognitive achievement, and the idea that it literally brings into existence the self as occupant of that point of view is one that is especially congenial to naturalists because it makes understandable how selves could arise in a natural world.[42]

# 7. METAPHYSICS

Let me close with some remarks about metaphysics. It is sometimes said that physics represents space and time as seen by an all-knowing god 'looking' at time from the 'outside'. The metaphor is misleading in innumerable ways, but we can dispense with it. When we talk about the contrast view of time presented in physics with the view from within time presented in experience, what we really mean to contrast is the embedded view of time from the perspective of a particular perceptual encounter with the world, and a representation that is invariant under transformations between such perspectives. The shift between these two ways of looking at a time—that is, from a point of view that spans perspectives and from the point of view of the various perspectives embedded in it—corresponds to the shift from the view of space presented in a map to the view presented in the visual field or in a coordinate-dependent description of space. They represent the same facts, but the latter in a manner that is relativized to a position in it. Each position (characterized fully enough to provide a frame of reference)[43] corresponds to a (distinct) perspective, and *shifts* in perspective induce shifts in appearance even though nothing in the *field* of representation is actually changing. The visual field changes as you walk around an object—say a table, sitting motionless in the center of a room—even though the object itself remains the same. Here the part of space in which the object is located constitutes the field of representation, and the perspective is given by the position and orientation of the viewer. Part of knowing how to *interpret* the visual field—that is, how to distill out its objective content, how to separate what it is telling you about the world from what it is telling you about your position in it—is knowing to anticipate and account for changes in appearance due to perspective. The representational content of a non-perspectival representation, by contrast, is invariant under changes in perspective.[44] It tells you nothing about your position vis à vis the object of representation, and is unaffected by changes in your position. The formal relations among representations from different perspectives and between perspectival and non-perspectival (equivalently, frame-dependent and non-frame-dependent) representations are of special importance in trying to understand temporal experience. Formally, a space S is a set of elements with a relation defined over it. A frame of reference for S is a set of points or elements of S

---

[42] And indeed, *why* they might arise, for the practical importance of the perspective-spanning viewpoint is easy to see. It allows for planning and practical reasoning.

[43] In spaces of n dimensions, n points are needed to specify a frame of reference. In the spatial case, for example, it's 3. Since time is one dimensional, only a single point is needed.

[44] See the discussion below in 'Metaphysics'.

relative to which descriptions of other elements are relativized (either explicitly or implicitly). A perspectival representation of S is one that is relativized to a frame of reference. In an n-dimensional space an n-dimensional frame of reference is needed to fully characterize a perspective.[45] An invariant representation of S is that one that is unchanged by transformations between perspectives.

In physics,[46] time is conceived as one dimension of a four-dimensional manifold whose other dimensions are spatial, and whose structure we aim to describe in invariant terms. We can *represent* space-time in terms that are invariant under transformations between spatiotemporal perspectives, but we can't *experience* it as such. The world as encountered in experience is the world as encountered from a spatially and temporally embedded perspective, that is, from the *here-now* of a particular perceptual encounter with it. The reconciliation of time as conceived in physics with time as encountered in experience is the central problem in the metaphysics of time. A big part of that problem is the reconciliation of time as represented in invariant terms with time as presented to consciousness from different perceptual perspectives.[47] I opened with a catalogue of some of the differences between time as represented by physics and time as encountered in experience. Time as conceived by physics is one dimension of a four-dimensional manifold of events. There is no intrinsic difference between past and future;[48] there is no change or movement; the parts of time exist together, eternally, in a fixed configuration. Time as encountered in experience, by contrast, exhibits a cluster of well-known past/future asymmetries; change and movement are the rule rather than the exception; the world is in the process of becoming, new facts are constantly coming into existence; the past is fixed, but the future remains to be decided.

I have not been concerned with the metaphysical problem directly. I have been concerned, rather, to elucidate the psychological structures that arise in the mind of a being that encounters time from different perspectives and remembers those encounters. The discussion bears on the metaphysical problem, however, in the following way. It makes available to the metaphysician the resources of psychological explanation of elements of phenomenology that don't correspond in any obvious way to features of the spatiotemporal manifold. One of the things that is especially intriguing about temporal experience is how much psychological complexity it presupposes and how much of it is not generated by interaction with the environment, but generated by internal interaction among representational contents. Start with a system that is receiving perceptual input from the environment, add a temporal dimension to its representational states, and allow memory to work recursively on those states, and the result is an internal environment that is a virtual hothouse for the cultivation of

---

[45] These notions can be generalized to allow for incomplete perspectives and implicit perspectivity. There may be ambiguity among perspectives if the space possesses symmetries.

[46] The physics of time is unsettled. The problems that beset the understanding of time in quantum gravity are wholly new. I am speaking here entirely within the setting of a relativistic understanding of time.

[47] And it is very much complicated by the fact that the representation of time from multiple perspectives are always co-present in consciousness in the form of memories.

[48] And this is linked to the fact that the dynamical laws are symmetric under reflection in time.

increasingly complex psychological structures. We have the specious present with its internal structure, with the long line of nested memories and expectations (and memories of memories, and memories of expectations and expectations of memories...) superimposed over the specious present in a resonating interaction. All of this structure is co-present in every temporal cross section of consciousness, and can provide the basis for yet further higher level states. It's not that this phenomenology is not perceptual: it clearly is. It arises from continued interaction with an external environment. It's rather that it doesn't fit the over-simple film projector, or property-tracking model of perception. Over time, as one viewpoint is exchanged for another, we get an emergent phenomenology involving the experience of movement and change, and eventually suspense, and the more complex narratively structured emotions: suspense, nostalgia, excitement, regret.

Understanding the different elements of this phenomenology is not, or not purely, a matter of finding objective correlates in the invariant picture of time. Some elements of the phenomenology *do* have objective correlates. A lot of progress has been made in the philosophical literature, for example, on explaining the cluster of past/future asymmetries in terms of the thermodynamic gradient. But others do not. The fixity of the past and openness of the future are real but perspectival effects. From an embedded perspective, it is right to think of ourselves as perpetually in the process of becoming, transitioning from one perspective to the next. It's right to think of the past as fixed and the future as open: transitions between perspectives always close off possibilities that were open in the past. There is nothing illusory about these asymmetries. To say that they are perspectival is to say that they are represented in the temporally unembedded view of reality (i.e. the representation of time that is invariant under transformations between temporal perspectives) by relativization to a point (or frame).[49]

What does all of this teach us about the fraught issue of the experience of flow? We can reject outright the incoherent pictures criticized in other papers in this volume, but link the feeling of flow to the experience of movement and change. It was a penetrating phenomenological insight of both James and Husserl to recognize that movement and change are part of the immediate content of the most basic kind of experience. We arrive at the idea of stasis, of an instant of time, or of a point of space only by analysis, by carrying the process of subdivision to its limit, but the idea of a static state is phenomenologically empty. Of all of the ink that's been spilled over the question of whether, and in what sense, time flows, perhaps what we're really trying to get at when we speak of the flow of time, is the phenomenologically basic experience of a world in perpetual flux. It's appropriate that James should have the last word here:

> Empty our minds as we may, some form of changing process remains for us to feel, and cannot be expelled...Awareness of change is thus the condition on which our perception of time's flow depends.[50]

---

[49]  And, perhaps, to add that the relevant point (or frame) is undistinguished from others of the same kind in the invariant representation.

[50]  James (1890).

## REFERENCES

Bruner, J. (2002), *Making Stories* (Farrar, Strauss, and Giroux).
Carr, D. (1991), *Time, Narrative, and History* (Indiana University Press).
Dainton, B. (2000), *Stream of Consciousness: Unity and Continuity in Conscious Experience*, Routledge (International Library of Philosophy).
Dennett, D. (1969), *Content and Consciousness* (London: Routledge & Kegan Paul).
Grush, R., 'Time and Experience' (forthcoming), in *Philosophie der Zeit*, Thomas Müller, ed.
—— 'How to, and How not to, Bridge Computational Cognitive Neuroscience and Husserlian Phenomenology of Time Consciousness', *Synthese* 2006 153(3): 417–450.
—— 'Brain Time and Phenomenological Time', draft, at http://mind.ucsd.edu/papers/!papers.html
Helmholtz, H. (1910), *Physiol Optik*, 1910 (English translation by Optical Society of America (1924–5)).
Husserl, E., *On the Phenomenology of Internal Time Consciousness*, (English Translation by John Brough Springer 2008).
James, W. (1890), *The Principles of Psychology* (Classics in the History of Psychology, an internet resource developed by Christopher D. Green of York University, Toronto, Ontario).
Ismael, J. (2007), *The Situated Self* (Oxford University Press).
—— 'Self-Organization and Self-Governance', MS.
LePoidevin, R., 'The Experience and Perception of Time', The Stanford Encyclopedia of Philosophy (Winter 2009 edition), Edwin N. Zalta (ed.), URL=<http://plato.stanford.edu/archives/win2009/entries/time>.
McTaggart, J., 'The Unreality of Time', in Le Poidevin, Robin, and McBeath, Murray (eds.), *The Philosophy of Time* (Oxford University Press, 1993), 23–34.
Price, H., 'The Flow of Time' (this volume).
Proust, M. (1988), *Swann's Way* (Cambridge University Press).
Rubin, D. (1986), *Autobiographical Memory* (Cambridge University Press).
Strawson, G. (2004), 'Against Narrativity', in *Ratio*, 17: 428–452.
Velleman, D. (2005), *Self to Self* (Cambridge University Press).
Wittgenstein, L. (1999), *Philosophical Investigations* (SUNY Press).
Zahavi, D (2008), *Subjectivity and Selfhood* (MIT Press).

## ADDITIONAL READING

McInerney, P. (1991), *Time and Experience* (Temple Univ. Press: Philadelphia).
Merleau-Ponty, M. (1945), *Phénoménologie de la Perception* (Gallimard, Paris).
Sutton, J., 'Memory', *The Stanford Encyclopedia of Philosophy* (Spring, 2004 edition). Edward N. Zalta (ed.), URL = <http://plato.stanford.edu/entries/memory/>
Varela, F. E. Thompson, and E. Rosch (1991), *The Embodied Mind: Cognitive Science and Human Experience*, MIT Press, Cambridge.

# PART IV

# TIME IN CLASSICAL AND RELATIVISTIC PHYSICS

# CHAPTER 16

## SHARPENING THE ELECTROMAGNETIC ARROW(S) OF TIME

### JOHN EARMAN

## 1. INTRODUCTION

THE various "arrows of time" and their interrelations are the subject of a seemingly never ending discussion in the physics and the philosophy of science literature. While the discussion in recent decades has undoubtedly produced numerous advances in the details of our understanding, it is hard not to be discouraged by the overall lack of progress in reaching a consensus on key issues.[1] In addition to ruing the lack of progress, one could also make two general complaints about the arrows of time literature. First, one could complain (as several authors have) that the talk of "arrows of time" suggests that what is at issue is the directionality of time, whereas what is often at issue is not directionality but temporal asymmetries. This is a defect that could be corrected by a change of terminology, but since talk of "arrows" has become well entrenched, I will continue to use it. What is not a matter of terminology, however, is whether the various asymmetries are merely asymmetries *in* time, or whether they constitute asymmetries *of* time itself. That there are differences of opinion on this matter is not surprising. But what is disconcerting is the lack of agreement on how this matter is to be decided.[2] Second, it is especially vexing that the typical way of stating the puzzle about various temporal arrows or asymmetries rests on a false presupposition. The general form of the puzzle is supposed to be: "Since the fundamental laws of physics exhibit a symmetry $X$, why is the world we see around us so $X$-asymmetric?" For

---

[1] The reader may find it instructive to consider five benchmarks for recent decades: Davies 1976 Zeh (1989 2001), Savitt 1995, and Savitt 2006.

[2] For remarks on the considerations that go into such a decision, see Sklar (1993: 378–384).

a continuous symmetry, such as spatial translation or spatial rotation, it is typically true that for laws as expressed in terms of differential equations, generic solutions of $X$-symmetric equations are $X$-asymmetric.[3] There are (as far as I know) no general results to this effect for discrete symmetries, such as time reversal invariance. Nevertheless, such results often do hold. For example, Einstein's gravitational field equations are time reversal invariant, but within the class of Friedmann-Walker-Robertson (FRW) models used in contemporary cosmology to describe the large-scale features of our universe, the subclass of models that are time symmetric about a time slice $t = const$ has "measure zero" (see Castagnino et al. 2003).[4] In short, not only is it not surprising that we find ourselves in an $X$-asymmetric world even though the laws that govern this world are $X$-symmetric, it would be surprising if we *didn't* find ourselves in an $X$-asymmetric world! Still, some actually observed asymmetries seem so striking and/or pervasive that they call for explanation. But I am unaware of a persuasive analysis of how these privileged asymmetries are to be identified and, thus, am left with the nagging feeling that the choice of the asymmetries that get promoted to arrows of time is a matter of fashion rather than principle.

Turning from the general picture to the electromagnetic (EM) arrows in particular, several complaints could be raised about the state of the debate. First, and foremost, there is the lack of a clear and unproblematic statement of what the EM arrows are. Second, there are conflicting claims about a number of issues: the status of retarded/advanced fields; the time reversal invariance, or lack thereof, of the equations of motion of charged particles that incorporate radiation reaction; and the linkage between the EM arrow and the other arrows of time, especially the cosmological arrow. Third, there is a feeling of *déjà vu* all over again about the debate. The modern phase of the debate can be dated to the Einstein-Ritz controversy of 1908–1909. The predominant opinion had been that Einstein prevailed. But recently neo-Ritzian points of view have been expressed, not only in the philosophy literature but the physics literature, as well (see, for example, Frisch 2000, 2005, and Rohrlich 1998, 1999, 2000).

There is no hope that one review can clear up all of the unresolved issues about the EM arrow(s). But it should be possible to separate the genuine from the pseudo-problems, and to put to rest the latter while sharpening the former. That is the goal of this chapter.

In section 2, the search for EM arrows is focused on the space of solutions of Maxwell's equations for a specified charge distribution and the initial/boundary conditions used to pick out a particular solution or a subclass of solutions. In section 3 the focus shifts to the equations of motion of accelerating charges that radiate and experience a damping force. One has to be prepared to find that one or both of these searches may turn up different results depending upon whether the setting is Minkowski spacetime or a cosmological model whose local or global structure differs

---

[3]  The technical result here assumes that the symmetry is codified as a Lie group of transformations.
[4]  Here $t$ is the time coordinate for which the FRW line element takes the form $ds^2 = a(t)d\sigma^2 - dt^2$. Here $a(t)$ is called the scale factor, and the spatial line element $d\sigma^2$ can take one of three forms corresponding to $t = const$ slices which have zero curvature, constant positive curvature, or constant negative curvature.

from Minkowski spacetime. Most of the discussion is devoted to classical electrodynamics, but the implications of quantum electrodynamics (QED) are examined at the end of section 3. Conclusions are presented in section 4.

# 2. SEARCHING FOR THE EM ARROW IN THE SOLUTIONS TO MAXWELL'S EQUATIONS

## 2.1 Time reversal invariance

The search for an EM arrow would be brought to a swift and successful conclusion if Albert (2000) were right, and Maxwell's equations failed to be time reversal invariant.[5] However, Albert's claims rest on a non-standard and, arguably, unilluminating analysis of time reversal invariance (see Earman 2002 and Malament 2004).[6] This matter will not be rehearsed here, and I will operate with the standard version of time reversal invariance for electromagnetism.

In covariant notation, Maxwell's equations read

$$\nabla_\mu F^{\mu\nu} = J^\nu \tag{1}$$

$$\nabla_{[\sigma} F_{\mu\nu]} = 0 \tag{2}$$

where $F^{\mu\nu}$ is the Maxwell tensor, $J^\nu$ is the charge-current field, and $\nabla_\mu$ is the covariant derivative operator.[7] These equations apply to curved as well as flat spacetime; but until further notice the application will be to Minkowski spacetime, in which case the covariant derivative can be replaced by ordinary differentiation with respect to inertial coordinates. In inertial coordinates the charge-current field takes the form $J^\nu = (\mathbf{j}, \rho)$, where $\mathbf{j}$ is the three-vector current and $\rho$ is the charge density. Further, the contravariant components of the Maxwell tensor are related to the electric and magnetic fields, $\mathbf{E}$ and $\mathbf{B}$, as follows[8]

$$F^{\mu\nu} = \begin{pmatrix} 0 & \mathbf{B}_z & -\mathbf{B}_y & \mathbf{E}_x \\ -\mathbf{B}_z & 0 & \mathbf{B}_x & -\mathbf{E}_y \\ \mathbf{B}_y & -\mathbf{B}_x & 0 & -\mathbf{E}_z \\ \mathbf{E}_x & \mathbf{E}_y & \mathbf{E}_z & 0 \end{pmatrix} \tag{3}$$

---

[5] "[C]lassical electrodynamics is *not* invariant under time-reversal" (Albert 2000: 20).

[6] In the cosmological context where the spacetime structure is not invariant under time reflection, Malament's (2004) analysis of time reversal invariants needs to be employed. However, for the sake of simplicity and familiarity, I will operate here with the standard textbook analysis, and will simply issue the promisory note that the main points I make below survive the translation to his analysis.

[7] Greek indices and Latin indices run respectively from 1 to 4 and 1 to 3. The signature convention for the spacetime metric is $(+ + +-)$. Units are chosen so that $c \equiv 1$.

[8] More generally, the electric and magnetic fields as measured by an observer whose (normed) four-velocity is $V^\mu$ are defined respectively by $E^\mu := F^{\mu\nu} V_\nu$ and $B^\mu := \frac{1}{2} \epsilon^{\mu\nu\alpha\beta} V_\nu F_{\alpha\beta}$, where $\epsilon^{\mu\nu\alpha\beta} = \epsilon^{[\mu\nu\alpha\beta]}$ is the volume element of the spacetime. Note that the electric and magnetic fields measured by the observer $V^\mu$ are spatial vectors in that they lie in the spacelike plane orthogonal to $V^\mu$.

And in three-vector notation the Maxwell equations (1)–(2) take their familiar forms

$$\nabla \cdot \mathbf{E} = \rho \quad \nabla \mathbf{x} \mathbf{B} - \frac{\partial \mathbf{E}}{\partial t} = \mathbf{j} \tag{1'}$$

$$\nabla \cdot \mathbf{B} = 0 \quad \nabla \mathbf{x} \mathbf{E} + \frac{\partial \mathbf{B}}{\partial t} = 0 \tag{2'}$$

One approach to time reversal invariance involves the literal reversal of time orienta-tion. A time orientation is given by a continuous non-vanishing timelike vector field $t^\nu$ (or more properly by an equivalence class of such fields, where two such fields $t^\nu$ and $t'^\nu$ are counted as equivalent just in case at every spacetime location $x$, $t^\nu(x)$ and $t'^\nu(x)$ point into the same lobe of the null cone at $(x)$.[9] Under literal time reversal (denoted by '$T$'), $^T t^\nu = -t^\nu$. To check whether the equations of a given theory are time reversal invariant it is necessary to define the resultant action of this operation on the basic variable of the theory, and verify that the solution set of the equations remains invariant under the defined action.[10] The obvious drawback of this approach to time reversal invariance is the lack of a direct connection to experiment—at least for those experimenters who do not have control of a switch by means of which they can reverse the orientation of time.

A second approach that lends itself better to experimental test works with a fixed time orientation. A model for a given theory is an assignment $x \mapsto S(x)$ to spacetime points $x$ of a state description $S(x)$ at $x$ appropriate for said theory. A theory is deemed to be time reversal invariant just in case whenever $x \mapsto S(x)$ satisfies the laws of the theory, so does $x \mapsto (^T S)(x)$ where $(^T S)(\mathbf{x}, t) = \,^R S(\mathbf{x}, -t)$ for an inertial coordinate system $(\mathbf{x}, t)$. (Nothing depends on the choice of the inertial system since it is assumed that the laws are Poincaré invariant.) Here, '$^R S$' denotes the "reversed" state, which is supposed to describe, relative to the fixed time orientation, the analogue of the corresponding state in the world with the literally reversed time orientation. In particle mechanics the "reversed" state is standardly defined by reversing the three-velocities of the particles. For Maxwell, electromagnetism, the standard definition of the rever-sal operation is $^R(\mathbf{E}(\mathbf{x}, t), \mathbf{B}(\mathbf{x}, t), \mathbf{j}(\mathbf{x}, t), \rho(\mathbf{x}, t)) = (\mathbf{E}(\mathbf{x}, t), -\mathbf{B}(\mathbf{x}, t), -\mathbf{j}(\mathbf{x}, t), \rho(\mathbf{x}, t))$. It is easy to see that, whenever the history $(\mathbf{x}, t) \mapsto (\mathbf{E}(\mathbf{x}, t), \mathbf{B}(\mathbf{x}, t), \mathbf{j}(\mathbf{x}, t), \rho(\mathbf{x}, t))$ satisfies (1')-(2'), so does $(\mathbf{x}, t) \mapsto \,^T(\mathbf{E}(\mathbf{x}, t), \mathbf{B}(\mathbf{x}, t), \mathbf{j}(\mathbf{x}, t), \rho(\mathbf{x}, t)) = (\mathbf{E}(\mathbf{x}, -t), -\mathbf{B}(\mathbf{x}, -t), -\mathbf{j}(\mathbf{x}, -t), \rho(\mathbf{x}, -t))$. In terms of the Maxwell tensor, $^T F^{4\nu}(x') = F^{4\nu}(x)$ and $^T F^{mn}(x') = -F^{mn}(x)$, where $x = (\mathbf{x}, t)$, $x' = (\mathbf{x}, -t)$, and $m, n = 1, 2, 3$. The four-velocity $V^\mu$ transforms as $^T V^4 = V^4$ and $^T V^m = -V^m$. In the alternative approach, where time reversal involves the literal reversal of the time orientation, $^T F^{\mu\nu}(x) = -F^{\mu\nu}(x)$ and $^T V^\mu = -V^\mu$.

---

[9]  Any simply connected spacetime admits a time orientation. Thus, if a given spacetime is not time orientable, one can obtain a time orientable spacetime that is locally the same as the given spacetime by passing to a covering spacetime. It is assumed that all of the spacetimes at issue here are time orientable.

[10]  See Malament (2004) for an elegant application of this approach to classical electromagnetism.

While both of the above approaches take for granted the existence of a time orientation, it is fair to ask whether it would be justified to posit such an object if all of the fundamental laws of physics were time reversal invariant. The suggestion behind the question is that, when the supposition obtains, the time reversal symmetry should be treated as a gauge symmetry in the sense that it connects equivalent descriptions of the same physical state of affairs. And the further suggestion would be that our perception of the time order of events is not due to our communion with an orientation defining vector field $t^\nu$, but rather to our reaction to, say, local entropy gradients. I will not attempt to tackle these issues here, and will simply assume the existence of a time orientation.[11]

## 2.2 Explaining electromagnetic asymmetries: the Einstein-Ritz controversy

For phenomena governed by time reversal invariant laws $L$ (such as Maxwell's laws of electromagnetism), at least two strategies are available for explaining observed asymmetries. First, additional laws $L'$ can be postulated such that the combined laws $L \& L'$ are not time reversal invariant. Second, the asymmetry can be attributed to the fact that, although certain time developments are allowed by the laws $L$, they are vastly more "improbable" than their time reversed counterparts. A familiar but still controversial example of the latter is the modern form of Boltzmannian statistical mechanics. The time reversal invariant laws of Newtonian mechanics are cast in Hamiltonian form; a measure on the phase space is adopted and is used to gauge the probability of macroscopic outcomes as identified with regions of phase space; and the tendency of (coarse-grained) entropy to increase with time is explained by the tendency of the microstate of system to evolve to regions of phase space that correspond to ever more probable macrostates. However, as is well known, the time reversal invariant character of the laws of mechanics entails that this explanation of the asymmetry of entropic behavior requires the help of a posit of special, low entropy initial conditions (see Albert 2000). Different positions can be taken with regard to the issue of whether these special initial conditions have a de facto or a lawlike character. In the latter case, the distinction between the two strategies is somewhat blurred, but a crucial difference remains in that in the first strategy the additional laws $L'$ are supposed to be non-probabilistic.

Commentators have interpreted the Einstein-Ritz controversy (see Ritz 1908 and 1909, Einstein 1909, and Ritz and Einstein 1909) as exemplifying the competition between these two strategies, with Ritz opting for the first strategy and Einstein for the second. Commentators typically quote the joint Einstein-Ritz declaration in which they agreed to disagree:

---

[11] For a discussion of these issues, see Earman (2002).

> [E]xperience compels one to consider the representation by means of retarded potentials as the only one possible, if one is inclined to the view that the fact of the irreversibility of radiation must already find its expression in the fundamental equations. Ritz considers the restriction to the form of retarded potentials as one of the roots of the second law [of thermodynamics], while Einstein believes that irreversibility is exclusively due to reasons of probability.
>
> (Ritz and Einstein 1909: 324)

Superficially, this quotation does seem to count as an exemplification of the competition between the two strategies. But first appearances are deceiving. Before trying to get behind the appearances, a digression is in order.

The second sentence of the joint declaration reveals a wholly implausible feature of Ritz's position, namely, the idea that the electromagnetic asymmetries explain the thermodynamic asymmetries. This explanatory linkage is hard to forge for at least two reasons. First, thermodynamics works for electrically neutral matter for which, trivially, there is no electromagnetic EM arrow. Second, the talk of retarded potentials invokes relativistic considerations, whereas thermodynamics does not cease to be valid in the Newtonian limit where velocities are small in comparison with the velocity of light. Curiously, Einstein did not make these obvious points. They are made, over and over again, by commentators on the Einstein-Ritz controversy.

Of course, Ritz could be mistaken about the relation between the electromagnetic and the thermodynamic asymmetries, but right about the basis of the former. But the first sentence of the joint declaration seems to reveal a confusion that Einstein noted in an earlier and more sharply critical assessment of Ritz's position. Ritz proposed to find the expression of the asymmetry of electromagnetic radiation in the fundamental equations of electromagnetism by positing "the representation by means of retarded potentials as the only one possible." But Einstein (1909) claimed that the representation by means of retarded potentials is not more special than the representation by, say, a linear combination of retarded and advanced potentials, both being representations of the same solution. In fact, Ritz and Einstein were at cross purposes: although Einstein's claim is correct if it refers to orthodox classical electromagnetism, there is a sense in which Ritz held a scientifically respectable, if not ultimately defensible, position. This position exemplifies a third strategy, which involves changing the theory. For Ritz, the fundamental equations of electrodynamics were to be formulated in the context of a theory that postulates particles acting at-a-distance without the mediation of fields. For such a theory, the restriction to retarded action has a natural expression, one which leads to time reversal non-invariant laws.

By contrast, the recent proposals that I label as neo-Ritzian—because they also invoke a restriction to retarded fields—are supposed to be implemented in the field theoretic setting of orthodox classical electrodynamics. As a result, they do not have, I contend, a scientifically respectable expression in terms of physical laws, but require chanting incantations about "causation." In order to lay the ground work for this negative judgment, I must first explain the nomenclature of advanced and retarded representations/potentials/fields. After explaining the difference between the Ritzian and neo-Ritzian proposals, I concentrate mainly on the orthodox field-theoretic formulation of electromagnetism, but the inability of one widely discussed action-at-a-distance theory of classical electrodynamics—the Wheeler-Feynman theory—to explain the asymmetry of radiation is discussed.

## 2.3 The Kirchhoff representation theorem for Minkowski spacetime

In a simply connected spacetime, the Maxwell equation (2) guarantees the existence of a global potential $A_\mu$ for $F_{\mu\nu}$:

$$F_{\mu\nu} = \nabla_{[\nu} A_{\mu]} \tag{4}$$

Written in terms of the potentials, Maxwell's equations do not have a well-posed initial value problem since values of the potentials and their time derivatives at a given time do not fix a unique solution, a not unexpected result, since the introduction of the potentials injects gauge freedom. Fixing the gauge by imposing the Lorentz gauge condition

$$\nabla_\mu A^\mu = 0 \tag{5}$$

turns the Maxwell equation (1) into the equation

$$\Box A^\nu + R^\nu{}_\mu A^\mu \equiv g^{\alpha\beta}\nabla_\alpha\nabla_\beta(A^\nu) + R^\nu{}_\mu A^\mu = J^\nu \tag{6}$$

where $R_{\mu\nu}$ is the Ricci tensor of the spacetime metric $g_{\mu\nu}$. In Minkowski spacetime, which is the focus of this section, (6) reduces to the inhomogeneous wave equation

$$\Box A^\nu = J^\nu \tag{7}$$

In inertial coordinates (7) takes the familiar form

$$\frac{\partial^2 A^\nu}{\partial x^2} + \frac{\partial^2 A^\nu}{\partial y^2} + \frac{\partial^2 A^\nu}{\partial z^2} - \frac{\partial^2 A^\nu}{\partial t^2} = J^\nu \tag{7'}$$

which does have a well-posed initial value problem.

Consider a given electric current distribution $J^\nu$ satisfying the law of conservation of charge

$$\nabla_\nu J^\nu = 0 \tag{8}$$

which is implied by the Maxwell equation (1). Then the Kirchhoff representation theorem shows that any solution of (7) can be written in either retarded or advanced form as a sum of a volume integral and a surface integral:

$$A^\nu(x) = \int_{\Omega^-} ret + \int_{\partial\Omega^-} ret \equiv {}_{ret}A^\nu(x) + {}_{in}A^\nu(x) \tag{9a}$$

$$= \int_{\Omega^+} adv + \int_{\partial\Omega^+} adv \equiv {}_{adv}A^\nu(x) + {}_{out}A^\nu(x) \tag{9b}$$

Here $x$ is a spacetime location in a compact spacetime volume $\Omega$ with orientable boundary $\partial\Omega$, $\Omega^-$ and $\Omega^+$ are respectively the past and future light cones of $x$ in $\Omega$, and $\partial\Omega^- \equiv \Omega^- \cap \partial\Omega$ and $\partial\Omega^+ \equiv \Omega^+ \cap \partial\Omega$. In *four*-dimensional Minkowski spacetime the propagation of the field is clean-cut so that the supports of the volume integrals in (9a)–(9b) are restricted to the surface of the light cones; this result does not hold if space has dimension one or two. In the retarded representation, the volume integral gives the contribution of the sources in $\Omega^-$, while the surface integral gives the contribution of the incoming radiation that is either associated with sources lying outside of the chosen volume or else is truly source-free radiation that is not tied to any sources. The interpretation of the advanced representation is analogous.

To make gauge-free statements, the Maxwell tensor can be computed from the potentials in (9a)–(9b) to give the retarded and advanced representations of the field

$$F^{\mu\nu}(x) = {}_{ret}F^{\mu\nu}(x) + {}_{in}F^{\mu\nu}(x) \tag{10a}$$

$$= {}_{adv}F^{\mu\nu}(x) + {}_{out}F^{\mu\nu}(x) \tag{10b}$$

The fields ${}_{ret}F^{\mu\nu}(x)$ and ${}_{adv}F^{\mu\nu}(x)$ are solutions of the inhomogeneous Maxwell equations with the specified sources, whereas ${}_{in}F^{\mu\nu}(x)$ and ${}_{out}F^{\mu\nu}(x)$ are solutions of the homogeneous Maxwell equations ($J^\nu \equiv 0$). Taking advantage of the linearity of the Maxwell equations, a general solution can be written in the mixed form

$$F^{\mu\nu}(x) = \lambda[{}_{ret}F^{\mu\nu}(x) + {}_{in}F^{\mu\nu}(x)] + \tag{11}$$

$$(1-\lambda)[{}_{adv}F^{\mu\nu}(x) + {}_{out}F^{\mu\nu}(x)]$$

where $0 \le \lambda \le 1$.

Commentators' talk of "retarded and advanced fields" and "retarded and advanced solutions" invites confusion if the following distinction is not kept in mind. The fields ${}_{ret}F^{\mu\nu}(x)$ and ${}_{adv}F^{\mu\nu}(x)$ derived from the potentials defined by the volume integrals in (9a) and (9b) respectively are generally *different solutions* of the inhomogeneous Maxwell equations—rather special time symmetric cases  in which these fields are equal, are the exceptions that prove the rule. By contrast, as is evident from (10a)–(10b), the total retarded and total advanced fields, ${}_{ret}F^{\mu\nu}(x) + {}_{in}F^{\mu\nu}(x)$ and ${}_{adv}F^{\mu\nu}(x) + {}_{out}F^{\mu\nu}(x)$ respectively, are *not* different solutions but merely different representations of the *same solution*, as noted by Einstein (1909).

Further, it is worth noting that the distinctions between ${}_{ret}F^{\mu\nu}(x)$ and ${}_{ret}F^{\mu\nu}(x) + {}_{in}F^{\mu\nu}(x)$ on one hand, and between ${}_{adv}F^{\mu\nu}(x)$ and ${}_{adv}F^{\mu\nu}(x) + {}_{out}F^{\mu\nu}(x)$ on the other, are somewhat artificial since they depend on the chosen volume $\Omega$. It is only in cases where the volume and surface integrals in (9) have well-defined limits as $\partial\Omega^-$ and $\partial\Omega^+$ "go to infinity," that is as $\Omega$ tends to the entire volume of spacetime, that an absolute meaning can be assigned to statements about *the* advanced and retarded fields $_{ret}F^{\mu\nu}(x)$ and $_{adv}F^{\mu\nu}(x)$ and *the* incoming and outgoing fields $_{in}F^{\mu\nu}(x)$ and $_{out}F^{\mu\nu}(x)$.

The retarded and advanced fields for the special case of a single point charge are referred to as the Liénard-Wiechert fields, the general expressions for which can be found in standard texts (see, for example, Jackson 1998: section 14.1). The actual values of these fields have to be worked out on a case-by-case basis, once the details of the motion of the charge are specified.

## 2.4 Retarded fields: "causality" and the "retardation condition"

Some textbooks on electromagnetism work out the expressions for the advanced Liénard-Wiechert potentials and fields, only to discard them for reasons of "causality." Thus, Heald and Marion (1995) opine that

> This so-called advanced potential [the time component of the advanced four-potential] appears to have no physical significance because it corresponds to an anticipation of the charge distribution (and current distribution for the case of the vector potential [the space components of the four-potential]) at a future time. Such a potential does not satisfy the requirement that causality must be obeyed by a physical system.                                                              (260)

This requirement of "causality" must involve something in addition to satisfying Maxwell's equations, but the something more is not elucidated, much less justified, in the text.

Frisch's (2000) proposal seemed to promise an implementation of the first strategy mentioned in section 2.2—that is, add additional laws to the original set of time reversal invariant laws such that the augmented set of laws is time asymmetric—that bypasses issues about "causality" in favor of two neo-Ritzian posits. The first is that, in addition to satisfying Maxwell's equations, "electromagnetic fields associated with charges satisfy the retardation condition" (405); and the second is that the retardation condition is to be regarded as a law of classical electromagnetism that is just as fundamental as Maxwell's equations, in that it makes no more sense to ask for an explanation of why the retardation condition obtains than it does to ask why Maxwell's equations obtain (405–406).

Unfortunately, Frisch's proposed new law reintroduces "causality" under another name: his retardation condition is the condition that each charge "physically contributes" a fully retarded component to the total field. But this proposed law does not pass muster that a potential law of physics must satisfy even prior to empirical testing. In the soft sciences, where it is difficult, if not impossible, to find any precise and exceptionless lawlike generalizations with broad scope, it is apparently acceptable to use escape clauses—for example, *ceteris paribus*—or to gesture to wantabe laws by using suggestive but imprecise terminology—for example, X produces (causes, contributes, ...) Y. In physics this is not an acceptable practice. A putative fundamental law of physics must be stated as a mathematical relation without the use of escape clauses or words that require a PhD in philosophy to apply (and two other PhDs to referee the application, and a third referee to break the tie of the inevitable disagreement of the first two).

In his later book, *Inconsistency, Asymmetry, and Non-Locality* (2005), Frisch refers to the retardation condition not as a law but as a "causal constraint," which suggests that the original quest of providing additional laws to ground electromagnetic asymmetries has been abandoned in favor of a preferred interpretation of the existing laws of electromagnetism. The retardation condition as a causal constraint comes to this: the notion that a charge "physically contributes" a fully retarded component to the total field is parsed by saying that that component would be absent if the charge were absent (Frisch 2005: 153ff). The exercise of trying to divine the truth value of such counterfactual assertions, even when it is agreed at the outset what the basic laws are, is an invitation to a contest of conflicting intuitions about cotenability of conditions and the closeness of possible worlds. This is a contest that may generate many learned philosophical articles, but I am skeptical that such a contest or, more generally, the philosophical exercise of interpreting the equations of physics by performing incantations using the phrases "physically contribute," "cause," and the like will reveal an electromagnetic asymmetry that was not perceptible when the equations were allowed to speak for themselves. By "speak for themselves" I do not mean that the equations are taken as uninterpreted mathematical squiggles, but rather that only minimalist interpretations are allowed, for example, '$\rho$' denotes the electric charge density, '$\mathbf{E}$' denotes the electric field strength, etc. *Until proven otherwise, my assumption is that an EM arrow worth having is one that only requires the equations of physics to speak for themselves under such a minimalist interpretation.*

It is worth understanding why Ritz himself had access to a scientifically respectable version of the retardation condition, albeit one that ultimately is not tenable by the standard criteria of theory evaluation. This is the task of the next subsection; the example given there involves action-at-a-distance electrodynamics. An example of how to achieve a scientifically respectable, time asymmetric variant of orthodox electrodynamics that does not involve the resort to incantations of "causality" will be given in section 2.8.

## 2.5 The retardation condition for particle theories

Because he advocated a particle theory of electrodynamics rather than a field theory, Ritz was able to express his conviction that "experience compels one to consider the representation by means of retarded potentials as the only one possible" in the form of a law of physics requiring only minimalist interpretation and no philosophy-speak (produce, cause, contribute, ...). But because Ritz's own theory was a bit of a mess, and because the claims I want to make are general conceptual claims that are independent of details of Ritz's theory, I will illustrate them using a different and much cleaner toy theory that allows the relevant points to shine forth.

Consider a pure particle theory of classical electrodynamics, by which I mean that all of the basic variables of the theory are particle variables. Electromagnetic fields may be introduced, but only in an auxiliary role for purposes of calculation. The goal is to illustrate how, in this setting, Ritz's retardation condition can be implemented in terms of a Poincaré covariant form of Newton's $\mathbf{F} = m\mathbf{a}$ governing the motion of a system consisting of a finite number $N$ of charged particles. Here is one way to proceed. For each particle $j$ calculate its (auxiliary) retarded Liénard-Wiechert field $_{ret}F^{\mu\nu}_{(j)}$. Then postulate that each particle $k$ with worldline $z^{\mu}_{(k)}(\tau_{(k)})$, parameterized by proper time $\tau_{(k)}$, obeys the equation of motion

$$m_{(k)}a^{\mu}_{(k)} = q_{(k)} \sum_{\substack{j=1 \\ j \neq k}}^{N} {}_{ret}F^{\mu\nu}_{(j)} u_{(k)\nu} \tag{12}$$

where $m_{(k)}$ is the mass of particle $k$, $u^{\mu}_{(k)} = \dot{z}^{\mu}_{(k)} := dz^{\mu}_{(k)}/d\tau_{(k)}$ and $a^{\mu}_{(k)} = \ddot{z}^{\mu}_{(k)} := d^2z^{\mu}_{(k)}/d\tau^2_{(k)}$ and are respectively the four-velocity and four-acceleration of particle $k$, and $q_{(k)}$ is the electric charge of particle $k$. Once the calculation of the auxiliary fields that appear on the right-hand side of (12) is completed, (12) can be restated using only particle variables.

When the equations of this theory (call it $T_1$) are allowed to speak for themselves, they entail an EM arrow, for these equations are not time reversal invariant. To see why, compute the (auxiliary) electromagnetic fields associated with a finite system of charges obeying the equations of motion (12). The computed fields will satisfy the inhomogeneous Maxwell equations for the prescribed sources. Thus, the Kirchhoff representation theorem applies. When the volume $\Omega$ is chosen large enough to include all of the charges, $_{ret}F^{\mu\nu}(x)$ in (10a) will be $\sum_{all\ j}^{N} {}_{ret}F^{\mu\nu}_{(j)}(x)$ and $_{in}F^{\mu\nu}(x)$ will be zero at all $x$. Except in very special cases where the motions of the charges are time symmetric, $\sum_{all\ j}^{N} {}_{ret}F^{\mu\nu}_{(j)}(x)$ and $\sum_{all\ j}^{N} {}_{adv}F^{\mu\nu}_{(j)}(x)$ will not be equal and, thus, $_{ret}F^{\mu\nu}(x)$ in (10a) will not equal $_{adv}F^{\mu\nu}(x)$ in (10b), which means that, except for said special cases, $_{out}F^{\mu\nu}$ in 10b will not be zero everywhere. But under time reversal, $_{ret/adv}F^{4\nu}_{(j)}(x') \to _{adv/ret}F^{4\nu}_{(j)}(x)$ and $_{ret/adx}F^{mn}(x') \to - _{adv/ret}F^{mn}(x)$ with $x = (\mathbf{x}, t)$, $x' = (\mathbf{x}, -t)$, and similarly for $_{in/out}F^{\mu\nu}$. This implies that if $_{in}F^{\mu\nu} \equiv 0$ in a dynamically possible history, then $_{out}F^{\mu\nu} \equiv 0$ in the time reversed history. Thus, a contradiction ($_{out}F^{\mu\nu} \equiv 0$ and $\neg(_{out}F^{\mu\nu} \equiv 0)$) would result if time reversal invariance held.

An equally time reversal non-invariant theory ($T_2$) would be produced by substituting advanced for retarded interactions of the charges, yielding the rival equations of motion

$$m_{(k)}a^{\mu}_{(k)} = q_{(k)} \sum_{\substack{j=1 \\ j \neq k}}^{N} {}_{adv}F^{\mu\nu}_{(j)} u_{(k)\nu} \tag{13}$$

where $_{adv}F^{\mu\nu}_{(j)}$ is the advanced Liénard-Wiechert field of particle $j$. This theory would be the embodiment of an anti-Ritz principle of advanced action.

Or, following Fokker (1929) and Wheeler and Feynman (1945, 1949), a time reversal invariant theory ($T_3$) can be produced by using a symmetric combination of retarded and advanced interactions of the charges, resulting in the equations of motion

$$m_{(k)}a_{(k)}^{\mu} = q_{(k)} \sum_{\substack{j \neq k}}^{N} \frac{1}{2}[\,_{ret}F_{(j)}^{\mu\nu} + \,_{adv}F_{(j)}^{\mu\nu}]u_{(k)\nu} \tag{14}$$

$T_1 - T_3$ are distinct theories that make distinct predictions. Since they embody respectively the principles of retarded action, advanced action, and symmetric retarded-plus-advanced action, these principles are seen to have real theoretical and empirical bite in the setting of pure particle theories of electrodynamics. Choosing among these theories is not a matter to be settled by appeal to considerations of "causality" or to intuitions about counterfactual conditionals, but rather by appeal to the standard criteria of theory evaluation—empirical adequacy and theoretical fruitfulness.

Unfortunately, all three of these theories are seen to be wanting by the lights of the standard criteria. One of Einstein's (1909) sharpest criticisms of Ritz's theory was that it does not uphold the validity of the "energy principle." The criticism applies to any pure particle theory in the setting of Minkowski spacetime: without a field to mediate the interactions of the particles, the conservation of energy-momentum cannot hold in the usual form as a statement about the constancy of the instantaneous values of energy-momentum. A sharp form of this negative result can be found in van Dam and Wigner (1966): compute the kinetic energy and the linear momentum of each of the particles in some inertial coordinate system $(\mathbf{x}, t)$; require that the sum of the computed energies and the sum of the computed momenta are each the same for all $t$; then Poincaré covariance entails that there can be no interaction among the particles, that is, the particle world-lines are geodesics of Minkowski spacetime. This is not a fatal objection to relativistic pure particle theories. In some such theories it may be possible to maintain a weakened, asymptotic version of the instantaneous form of conservation of energy-momentum as a statement about the equalities of the sums of the energies and momenta of the particles as $t \to \pm\infty$. Alternatively, conservation of energy-momentum may be expressible in an integral form rather than an asymptotic form (see van Dam and Wigner 1966).

A more troubling problem with pure particle theories is specific to electrodynamics; namely, such theories do not offer any natural explanation of the phenomenon of the radiation reaction experienced by accelerating charges. Wheeler and Feynman's (1945, 1949) heroic attempt to provide an explanation in their action-at-a-distance electrodynamics uses an argument that is of dubious validity and that relies on an "absorber condition" which may well fail for the actual universe; and even if the argument goes through, it cannot explain the observed asymmetry of radiation reaction in flat spacetime. And, finally, in curved spacetime, Wheeler-Feynman action-at-a-distance electrodynamics does not produce the same radiation reaction force as standard classical electrodynamics, even if the absorber condition holds. These points will be taken up in sections 3.4 and 3.6 respectively.

A third strike against the pure particle theories comes from the criterion of theoretical fruitfulness. The best current theory we have of electrodynamics is a quantum field theory—QED—which arises as the quantization of a classical field theory—Maxwell's theory—rather than the quantization of a particle theory. Philosophers have an obsessive fascination with the Wheeler-Feynman theory. They do not balance this obsession with the remark that one of the authors of this theory—Feynman—was also a principal architect of QED. A large part of the motivation for the Wheeler-Feynman theory was the desire to avoid the infinities that arise in classical theories with a mixed ontology involving particles that create fields that act back on the particles. Wheeler and Feynman explored the escape route of eschewing fields in favor of a pure particle ontology. But the other route is to promote the field concept and demote the particle concept. Arguably, this is exactly what relativistic quantum field theory does by treating local fields as the basic entities, and explaining particle-like behavior in terms of the behavior of the fields (see Wald 1994). The infinities that arise in the classical theory of electrodynamics formulated in terms of a mixed particle-field ontology are cured by the quantum field theory treatment, at least in the sense that QED is a renormalizable theory.

Despite the fact that theories $T_1 - T_3$ are found wanting, I repeat that they serve to make the conceptual point that a Ritzian retardation condition has a clear meaning and function in a pure particle theory. That this is not so for standard classical electrodynamics is emphasized in the next section.

## 2.6 The retardation condition for field theories

After the excursion into particle theories, I return to orthodox classical relativistic electromagnetism. How might someone who insists on trying to find some role in this context for the neo-Ritzian retardation condition—the condition that each charge "physically contributes" a fully retarded component to the total field—proceed? For a system consisting of a finite number $N$ of charged particles, the closest analogue $\bar{T}_1$ for the retarded action-at-a-distance theory $T_1$ of the preceding section would consist of the conjunction of Maxwell's laws plus the posit

$$F_1^{\mu\nu}(x) = \sum_{j=1}^{N} {}_{ret}F_{(j)}^{\mu\nu}(x) + {}_{hom}F_1^{\mu\nu}(x) \tag{15}$$

where $F_1^{\mu\nu}$ stands for the (total) electromagnetic field in a physically possible history, the ${}_{ret}F_{(j)}^{\mu\nu}$ are the retarded Liénard-Wiechert fields of the particles, and ${}_{hom}F_1^{\mu\nu}$ is a homogeneous solution of Maxwell's equations. Allowance for the homogeneous solution must be made on pain of restricting the range of validity of the theory.

Those who insist, *contra* the neo-Ritzians, that each charge physically contributes a fully advanced component to the total field will endorse $\bar{T}_2$, consisting of Maxwell's equations plus the posit

$$F_2^{\mu\nu}(x) = \sum_{j=1}^{N} {}_{adv}F_{(j)}^{\mu\nu}(x) + {}_{hom}F_2^{\mu\nu}(x) \tag{16}$$

where the ${}_{adv}F_{(j)}^{\mu\nu}$ are the advanced Liénard-Wiechert fields of the particles. And those who insist that each charge physically contributes a half-retarded-half advanced field to the total field will endorse $\bar{T}_3$, consisting of Maxwell's equations plus the posit

$$F_3^{\mu\nu}(x) = \frac{1}{2}\sum_{j=1}^{N}[\,{}_{ret}F_{(j)}^{\mu\nu}(x) + {}_{adv}F_{(j)}^{\mu\nu}(x)] + {}_{hom}F_3^{\mu\nu}(x) \tag{17}$$

But how exactly are the proponents of these three theories $\bar{T}_1 - \bar{T}_3$ disagreeing? Measurements can in principle fix the actual value ${}_{act}F^{\mu\nu}(x)$ of the electromagnetic field at every spacetime location $x$, and by the Kirchoff theorem the homogeneous solutions in (16)–(18) can be chosen so that $F_1^{\mu\nu}(x) = F_2^{\mu\nu}(x) = F_3^{\mu\nu}(x) = {}_{act}F^{\mu\nu}(x)$.[12] With this choice the three theories $\bar{T}_1 - \bar{T}_3$ will agree on experimental outcomes even if we allow ourselves access to the results of thought experiments about what would happen if a hypothetical test charge were used to probe the value of the field. Indeed, in contrast to $T_1 - T_3$, which undoubtedly are distinct theories, $\bar{T}_1 - \bar{T}_3$ seem more like different modes of presentation of the same theory—Maxwell's theory written in different ways and anointed with different philosophy-speak about "physically contributes."

Nevertheless, depending on the circumstances, one mode of presentation may seem more pleasing than the others. Consider, for example, the case of a single charged particle that has been in inertial motion from time immemorial to the present and then is set into hyperbolic motion by a non-electromagnetic force (see Fig. 16.1). [Hyperbolic motion (aka. uniform acceleration) means that $\dot{a}^{\mu} := u^{\nu}\nabla_{\nu}a^{\mu} = a^{\beta}a_{\beta}u^{\mu}$ where $u^{\mu}$ is the (normed) four-velocity of the particle. Differentiating $u^{\mu}u_{\mu} = -1$ gives $a^{\mu}u_{\mu} = 0$. Using this fact, hyperbolic motion is seen to imply that $\dot{a}^{\mu}a_{\mu} = 0$, and that the magnitude of acceleration $a := (a^{\mu}a_{\mu})^{1/2}$ is constant.] Suppose that at all spacetime locations $x$, ${}_{act}F^{\mu\nu}(x)$ is given by the retarded Liénard-Wiechert field of the charge. Then theory $\bar{T}_1$ provides a simple description that comes from setting ${}_{hom}F_1^{\mu\nu}$ in (15) to zero, that is, $\bar{T}_1$ simply has to be supplemented by the condition that there is no incoming source-free radiation. Consider, by contrast, the description that $\bar{T}_2$ provides. For any spacetime location $x$ in the sectors $I$ and $IV$ of Figure 16.1 the advanced Liénard-Wiechert field of the particle is identically zero. Thus, $\bar{T}_2$ must invoke a homogeneous solution ${}_{hom}F_2^{\mu\nu}(x)$ that, for $x$ in sectors $I$ and $IV$, exactly mimics the retarded field Liénard-Wiechert of the particle. In addition, the advanced Liénard-Wiechert field of the particle involves a delta-function field on the null surface separating sectors $I$ and $IV$ from $II$ and $III$ (see Boulware 1980), so to reproduce

---

[12] This also follows from the facts that $\sum_{j=1}^{N} {}_{ret}F_{(j)}^{\mu\nu}(x)$ and $\sum_{j=1}^{N} {}_{adv}F_{(j)}^{\mu\nu}(x)$ are solutions to the inhomogeneous Maxwell equations, and that any two such solutions differ only by a homogeneous solution.

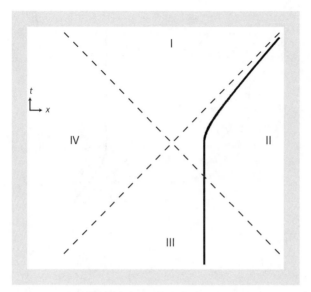

Figure 16.1 A charge that is hyperbolically accelerated in the future

the hypothesized total field, $_{\text{hom}}F_2^{\mu\nu}$ must be arranged to cancel out this delta-function field. These features of $\bar{T}_2$'s description seem contrived. And for similar reasons $\bar{T}_3$'s description will also seem contrived.[13]

But note that, in the hypothesized case, no guardian angel of Maxwell's theory is needed to step in to generate the $_{\text{hom}}F_2^{\mu\nu}$ required by $\bar{T}_2$; for as long as we are not contemplating contra-nomological scenarios, Kirchhoff's representation theorem guarantees the existence of the required $_{\text{hom}}F_2^{\mu\nu}$. And again, barring magic meters, only the total field—in this case, $_{act}F^{\mu\nu} = {}_{ret}F^{\mu\nu} = {}_{adv}F^{\mu\nu} + {}_{\text{hom}}F_2^{\mu\nu}$—is measurable. Furthermore, the circumstances are the tail that wags the dog, for with a reversal of circumstances comes a reversal of fortunes of the theories.

Consider the time reversed motion of the particle in the above scenario (see Fig. 16.2)[14] and suppose that at all spacetime locations $x$, $_{act}F^{\mu\nu}(x)$ is given by the advanced Liénard-Wiechert field of the particle. Now $\bar{T}_2$'s description will seem natural while $\bar{T}_1$'s will seem contrived. And between the extremes of these two hypothetical cases—the one favoring $\bar{T}_1$, the other favoring $\bar{T}_2$—are all the messy cases where no one of the theories $\bar{T}_1$, $\bar{T}_2$, or $\bar{T}_3$ offers a markedly more simple description.

Moving from the hypothetical to the actual, one can ask: Is it in fact the case in the actual world that the retarded representation is always (mostly, typically, ...) simpler and more natural than the advanced or mixed representations? And if so, what is the explanation of the asymmetry? These and related issues will be taken up

---

[13] I take it that North (2003) is proposing that the judgment of what electromagnetic field is produced by a charged source is to be formed relative to the representation that has the most natural source-free field. I do not object to this as long as 'produce' means just this. I *do* object if 'produce' has a metaphysically charged meaning.

[14] Again, it is supposed that the acceleration of the particle is due to non-electromagnetic forces.

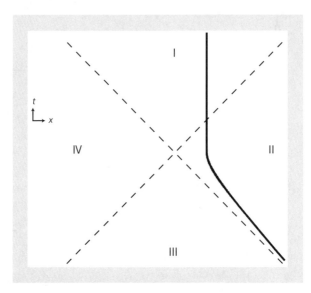

FIGURE 16.2  A charge that is hyperbolically accelerated in the past

in the following subsection. But before turning to these issues there is some unfinished business generated by the present discussion.

The alert reader will have noticed a gap in the above analysis which concentrates exclusively on fields. When it comes to the motion of charged particles, $\bar{T}_1 - \bar{T}_3$ do seem to yield different results, at least if the Lorentz force law is operative, and if the Lorentz force on a charge is computed from the fields due to the *other* charges plus the source-free field. For then the electromagnetic force acting on charge $k$ at point $x$ on $k$'s world line, according respectively to $\bar{T}_1 - \bar{T}_3$, is

$$q_k[\sum_{j \neq k} {}_{ret}F^{\mu\nu}_{(j)}(x) + {}_{hom}F^{\mu\nu}_1(x)]u_{(k)\nu} \qquad (18)$$

$$q_k[\sum_{j \neq k} {}_{adv}F^{\mu\nu}_{(j)}(x) + {}_{hom}F^{\mu\nu}_2(x)]u_{(k)\nu} \qquad (19)$$

$$q_k[\sum_{j \neq k} \frac{1}{2}({}_{ret}F^{\mu\nu}_{(j)}(x) + {}_{adv}F^{\mu\nu}_{(j)}(x)) + {}_{hom}F^{\mu\nu}_3(x)] \qquad (20)$$

In general these forces are not equal and, thus, there would seem to be a decisive test to decide which of $\bar{T}_1 - \bar{T}_3$ is the true theory. However, any attempt to carry out such a crucial experiment would end in failure, for none of the equations of motion (18)–(20) is empirically adequate since they all neglect the radiation damping force experienced by accelerating charges.[15]

---

[15]  Note that the difference between any two of the forces (18)–(20) is proportional to $q_k({}_{ret}F^{\mu\nu}_{(k)} - {}_{adv}F^{\mu\nu}_{(k)})u_{(k)\nu}$. According to Dirac's (1938) analysis, the value of the radiation reaction force experienced by charge $k$ is $\frac{q_k}{2}({}_{ret}F^{\mu\nu}_{(k)} - {}_{adv}F^{\mu\nu}_{(k)})u_{(k)\nu}$ (see section 3.3).

But if they are unable to win by dint of a crucial experiment, the ever resourceful neo-Ritzians claim to find a new purchase in radiation damping since, they claim, this phenomenon requires a posit of retarded action. This matter will be taken up in section 3.5.

## 2.7  Some electromagnetic arrows, real and alleged

One of the most oft-cited EM arrows involves spherical electromagnetic waves: we commonly experience such waves diverging from a center, but rarely if ever do we experience such waves converging on a center. This innocent seeming dictum disguises a number of potential misunderstandings and pitfalls. In the first place, it is essential to distinguish collective from individual phenomena. As an instance of the former, consider an antenna in which the electrons are induced to oscillate in unison, producing an outgoing radio signal, all parts of which are eventually absorbed. The time reverse of this process, in which the materials that played the role of absorbers in the original scenario now emit in a coordinated fashion so that the antenna receives an anti-broadcast signal, is never experienced, save for contrived situations which are hard to engineer except on small scales.

This is undoubtedly a real asymmetry, but it does not reveal anything novel about electromagnetism per se since analogous asymmetries are common to water waves and sound waves. The predominant view about such asymmetries is in line with that of Einstein in the Einstein-Ritz debate, namely, the asymmetries are to be traced, not to a failure of time reversal invariance of a fundamental law, but to statistical considerations that are of a piece with those that lie at the origin of the thermodynamic arrow and other arrows involving collective phenomena of non-charged matter. As mentioned above, the currently favored way of implementing this approach in order to explain the thermodynamic arrow is to supplement classical statistical mechanics with the posit of a low Boltzmann entropy state for the very early universe ("past hypothesis").[16] Exactly how the thermodynamic arrow links to the EM arrow under discussion is an important issue that will not be tackled here.[17]

---

[16]  For some skepticism about the ability of the past hypothesis to explain the thermodynamic arrow, see Weisberg (2004) and Earman (2006); for skepticism about the ability of the past hypothesis to explain EM arrows, see Frisch (2006). For sake of completeness it should also be acknowledged that there are alternative approaches, such as that championed by Penrose and Percival (1962) who posit a time asymmetric statistical law. Their proposed law, called the "law of conditional independence," is supposed to explain several of the arrows of time, including the thermodynamic arrow and the EM arrow. This law is inconsistent with distant correlations between relatively spacelike regions in a cosmology that has particle horizons; see section 2.9 below.

[17]  Price (2006) sees the linkage forged in the following way: the observed asymmetry at issue depends on the contrast between a few large outgoing waves vs many small incoming waves; that contrast is explained by the thermodynamics of the environment which derives large additions of energy but few large subtractions; and the second contrast is explained by the low entropy past. I find the first two links plausible but am suspicious of the third.

Notice that for the asymmetry just discussed, nothing about retarded vs advanced fields or representations need enter the discussion. The key question is simply whether the actual total field $_{act}F^{\mu\nu}$ is in the form of an outgoing or an incoming wave. Whether a retarded or an advanced decomposition of $_{act}F^{\mu\nu}$ is preferable is a side issue that might, but need not, be raised.[18]

This situation appears to change when the focus shifts from the collective phenomena associated with groups of charges to the phenomena associated with individual charges, for the typical way of describing the latter invokes retarded and advanced fields. The retarded field of a charge corresponds (it is said) to emission or a wave spreading from the charge, whereas the advanced field of the charge corresponds to absorption or a wave collapsing on the charge. Since (it is said) we experience waves spreading from a charge but not waves collapsing on a charge, the impression is left that an explanation of the asymmetry between incoming and outgoing radiation calls for a quashing of advanced fields. A number of clarifications and cautions need to be attached to these dicta if a muddle is to be avoided.

The retarded Liénard-Wiechert field for a point charge in Minkowski spacetime can be covariantly separated into a velocity (or generalized Coulomb) field $_{ret}F^{\mu\nu}_{Coul}$ and an acceleration (or radiation) field $_{ret}F^{\mu\nu}_{rad}$. The velocity field is so-called because it is independent of the acceleration of the charge. In the instantaneous rest frame of the charge, the retarded velocity field takes the familiar form of the Coulomb field found in elementary textbooks, and in a moving frame it is the Lorentz transform of the rest of the Coulomb field. The strength of this field falls off inversely as $d^2_-$, where $d_-$ is the spatial distance between the field point and the retarded position of the charge as measured in the Lorentz frame in which the charge is instantaneously at rest at the retarded time. An analogous story holds for the advanced Liénard-Wiechert velocity or Coulomb field with the distance $d_+$ between the field point and advanced position of the charge in place of the retarded distance $d_-$. So far, nothing about outgoing or incoming waves.

That interpretation comes from the acceleration or radiation part of the Liénard-Wiechert field, which depends linearly and homogeneously on the acceleration of the charge. The spacetime support of the retarded radiation field of a point charge (and, thus, of the stress-energy tensor[19] calculated from this field) consists of a sequence of future null cones whose vertices lie on the world line of the charge and coincide with those points at which the charge is accelerating. The value of the retarded radiation field at a point on one of these future null cones is inversely proportional to the spatial distance $d_-$ to the retarded position of the charge. Similarly, the advanced radiation field (and, thus, the stress-energy tensor calculated form this field) of a point charge has support on a series of past null cones whose vertices lie on the world line of the

---

[18]  Thus, I agree with Price (2006) that the asymmetry at issue is not captured by the condition that $_{in}F^{\mu\nu} = 0$ leading to a purely retarded description $_{act}F^{\mu\nu} = {}_{ret}F^{\mu\nu}$. However, unlike Price, I do not identify the Sommerfield radiation condition with $_{in}F^{\mu\nu} = 0$ for local systems (see below).

[19]  The stress-energy tensor $T^{\mu\nu}_{em}$ associated with the electromagnetic field $F^{\mu\nu}$ is defined by $T^{\mu\nu}_{em} := \frac{1}{4\pi}(F^{\mu\beta}F^{\nu}_{\beta} - \frac{1}{4}g^{\mu\nu}F^{\alpha\beta}F_{\alpha\beta})$.

charge and coincide with those points at which the charge is accelerating, and the value of the field at point one of these past null cones is inversely proportional to $d_+$.

So far so good. But what exactly is the observed asymmetry here? Let us agree to use "Observe" to indicate a liberal sense of observation that includes not only the data gathered from the immediate deliverances of our senses or measuring instruments but also inferences drawn from this data by means of Maxwell's equations and the background theories of the measuring instruments. Is it the case that we commonly Observe radiation diverging from an electron but never (or hardly ever) Observe radiation collapsing on an electron? There are two senses of the latter question that need to be distinguished. The first is: Do we never (or hardly ever) Observe cases of an actual total field $_{act}F^{\mu\nu}$ that is collapsing on an electron? I don't know that the answer is Yes, but nothing absurd results from granting a Yes. The second sense of the question is: Do we never (or hardly ever) Observe the advanced Liénard-Wiechert radiation field $_{adv}F^{\mu\nu}_{rad}$ of an electron? Here a Yes answer cannot be granted under conditions where the Kirchhoff theorem is valid since otherwise the advanced representation would be invalidated. The only non-muddled message about advanced and retarded fields that could come out of a Yes answer to the first question is not that advanced Liénard-Wiechert radiation fields have to be quashed but only that it is simpler to use the retarded representation of the actual total field $_{act}F^{\mu\nu}$ to describe the hypothesized Observations.

These cautionary remarks designed to ward off muddles about advanced/retarded fields do nothing positive to sharpen and explain the asymmetry of radiation reaction of accelerating charges. That job will be tackled below in section 3. In anticipation, it is worth remarking here that there is a reason not to try to quash or ignore the advanced Liénard-Wiechert field of a charge. As we will see in section 3.4 below, the Dirac expression for the radiation damping force experienced by accelerating charges involves evaluating at the world line of the charge the difference between the retarded and advanced Liénard-Wiechert fields of the charge, offering an explanation of the origin of the radiation reaction force. A positive value for the difference indicates that more energy-momentum is radiated away by the charge than is absorbed, and energy-momentum balance requires a compensating damping force. An immediate implication—accepted as correct on all analyses of radiation reaction—is that a charged particle that is (always) uniformly accelerated does *not* experience a damping force since the difference between the advanced and retarded Liénard-Wiechert fields is zero on the world line of the charge.

With these lessons in mind, let us turn to other attempts in the literature to specify some electromagnetic asymmetries.

One is posed in the form of a question by Zeh (2001: 21):

Why does the condition $_{in}F^{\mu\nu} = 0$ (in contrast to $_{out}F^{\mu\nu} = 0$) approximately apply in most situations?[20]

---

[20] I have replaced the potentials in Zeh's formulation with the Maxwell field tensor in order to assure gauge independence.

Here "situation" refers to some local system, and $_{in}F^{\mu\nu}$ and $_{out}F^{\mu\nu}$ are fields corresponding respectively to the potentials $_{in}A^\nu(x)$ and $_{out}A^\nu(x)$ in (9a)–(9b) evaluated for domains of integration appropriate to the local system.

The most obvious difficulty with this formulation concerns the "most" qualification. It would seem that in a natural sense of 'most', $_{in}F^{\mu\nu}$ is not approximately zero in the visible part of the electromagnetic spectrum for most of the systems of which we are aware, since otherwise we would not be aware of them. And the ubiquity of the cosmic background radiation makes one think that in a natural sense of 'most', $_{in}F^{\mu\nu}$ is not approximately zero in the microwave spectrum for most systems, whether we are aware of them or not (see North 2003).

Seeking to preserve the idea behind Zeh's radiation asymmetry, Frisch (2005: 108) reformulates it as

> (RADASYM) There are many situations in which the total field can be represented as being approximately equal to the sum of the retarded fields associated with a small number of charges (but not as the sum of the advanced fields associated with the charges), and there are almost no situations in which the total field can be represented as being approximately equal to the sum of the advanced fields associated with a small number of charges.[21]

The counterexamples to Zeh's formulation are avoided by shifting from 'most' to the vaguer 'many', and by restricting attention to systems consisting of a small number of charges. The condition that $_{in}F^{\mu\nu}\approx 0$ can be assured by using an absorber to keep the incoming radiation from impinging on the system of interest, and by noting that, per Frisch's stipulation, the absorber (which necessarily uses many charges) cannot be regarded as part of the system. From $_{in}F^{\mu\nu}\approx 0$ and Kirchhoff's representation theorem it follows that $_{out}F^{\mu\nu}\approx {}_{adv}F^{\mu\nu}-{}_{ret}F^{\mu\nu}$, where $_{ret}F^{\mu\nu}$ and $_{adv}F^{\mu\nu}$ are respectively the sums of the retarded and advanced Liénard-Wiechert fields associated with the charges of the system. The explanation of why $_{out}F^{\mu\nu}$ is typically not approximately zero in this situation (and, thus, why the total field cannot be represented purely as the sum of the advanced Liénard-Wiechert fields associated with the charges) is then quite straightforward: the result follows from the fact that typically $_{ret}F^{\mu\nu}-{}_{adv}F^{\mu\nu}$ is not approximately zero. This fact follows in turn from a remark in the Introduction; namely, typical solutions of time reversal invariant laws are not time symmetric. And so it is in electromagnetism: typically the motions of the charges in the system are not time symmetric, which implies that typically the retarded and advanced Liénard-Wiechert fields of the charges will not be equal. For a charge that is always and forever in hyperbolic motion (see Fig. 16.3), the retarded and advanced Liénard-Wiechert fields of the charge are the same near the charge and, indeed, throughout sector II to which the charge is confined (see Boulware 1980). But this is the exception that proves the rule.

[21] A somewhat different version (labeled (R)) is given in Frisch (2006: 546).

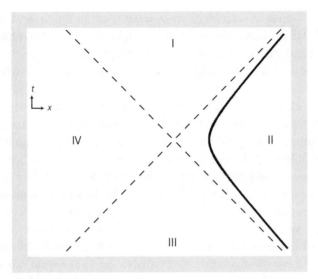

FIGURE 16.3  A charge that is hyperbolically accelerated for all times

Buried in this explanation sketch is a suppressed premise about collective behavior. The assurance that the system at issue is shielded from incoming radiation is merely probabilistic, for there is no inconsistency with the laws of electromagnetism that the material intended to function as an absorber in fact radiates into the system it was intended to shield.[22] The improbability of such behavior is presumably of a piece with the improbability of the anti-thermodynamic behavior of non-charged matter. The need to refer to such behavior is even more evident in the more general case where an absorber is not used to keep incoming radiation from impinging on the system. Here $_{out}F^{\mu\nu} = [_{adv}F^{\mu\nu} - _{ret}F^{\mu\nu}] + _{in}F^{\mu\nu}$ with the last term on the right-hand side typically being non-zero. That $_{out}F^{\mu\nu}$ is typically not approximately zero in the general case is partly explained by Frisch's definitional move: an absorber that captures $_{out}F^{\mu\nu}$ cannot be considered to be part of the system which, by definition, contains only a small number of charges. But the explanation in the general case must also invoke the improbability of the incoming radiation being configured so as to cancel out $_{adv}F^{\mu\nu} - _{ret}F^{\mu\nu}$ (which, by the same argument as in the special case, is typically non-zero).

Frisch (2005) does not reject outright these explanation sketches of RADASYM but he favors an explanation in which "the brunt of the explanatory work is done by the retardation condition—the assumption that the field physically contributed by a charge is fully retarded" (152). My skepticism about explanations which do not allow the theory to speak for itself and which require philosophy-speak (causes,

---

[22]  As Price (2006) puts it, unless we posit an EM arrow, we have no right to assume that ordinary matter will act as an absorber.

physically contributes, ...) can be given concrete form in the present instance: if the explanation sketches offered above are on the right track, then the philosophy-speak of the retardation condition not only is not needed, but it covers up the need to fill in the details of the sketches by specifying the nature and source of the improbabilities involved.

The sorts of asymmetries to which Zeh and Frisch point have two disturbing features. First, they are vague and hedged, requiring qualifiers like 'most', 'many', 'typically', or 'approximately' to ward off counterexamples. Second, these asymmetries are formulated in terms of the quantities $_{ret}F^{\mu\nu}$, $_{in}F^{\mu\nu}$, $_{adv}F^{\mu\nu}_{out}$, and $_{out}F^{\mu\nu}$ as evaluated for local systems; and as what counts as "the system" is expanded or contracted and, thus, as the domains of integration $\Omega^{\pm}$ and $\partial\Omega^{\pm}$ implicit in these quantities change, the asymmetries can come and go. This is not to say that the vagueness and the relativity of the asymmetries means that they are not genuine asymmetries. But it does underscore the need to come to grips with an issue noted in the Introduction. Even though Maxwell's equations are time reversal invariant, one would expect that among the solutions of these equations, the subset of time symmetric solutions is "measure zero" (in a sense that needs to be made precise). Which of the unlimited number of temporal asymmetries that are present in a generic solution should be viewed as interesting and fundamental enough to merit being promoted to "arrows of time" and to merit the search for an explanatory account? The comparison with thermal physics is revealing. The asymmetries encapsulated in the Second Law of thermodynamics certainly do merit promotion to the status of thermodynamic "arrows": they are systematic and pervasive, and although there are exceptions—the use of a microscope of resolving power great enough to observe Brownian motion will reveal some of them—no Maxwell's demon that produces at will macro-scale exceptions has ever been constructed and, arguably, could ever be constructed. Of the EM asymmetries discussed in this section, the only one that comes close to matching these features of the thermodynamic arrow is the non-muddled version of expanding vs contracting waves. But this version involves collective behavior of emitters and absorbers, and as such it does not reveal a distinctively electromagnetic arrow; indeed, the most plausible line of research is to try to explain this arrow in terms of the same considerations that explain temporal asymmetries in the collective behavior of non-charged matter.

One obvious idea for finding cleaner and more robust electromagnetic asymmetries would be to look for asymmetries that are independent of the choice of the volume $\Omega$. One safe and sure way to guarantee such independence is to formulate the asymmetries in terms of the limiting behavior of $_{in}F^{\mu\nu}$ and $_{out}F^{\mu\nu}$ as $\Omega$ is enlarged to encompass all of spacetime. Perhaps the asymmetries that emerge in this limit will reveal an arrow of time that is peculiar to electromagnetism. To return to the opening example of this section, suppose that the radio antenna broadcasts into empty space so that the outgoing radio waves are not absorbed but travel to spatial infinity. It would seem nearly miraculous if the time reverse of this scenario were realized in the form of anti-broadcast waves coming in from spatial infinity and collapsing on the antenna. The absence of such near miracles might be explained by an improbability in the coordinated behavior of incoming source-free radiation from different directions in space. Or it might be explained non-probabilistically by a prohibition against any truly source-free incoming radiation. The latter is one motivation for the Sommerfeld radiation conditions.

## 2.8  Sommerfeld radiation conditions for Minkowski spacetime

There is no a priori guarantee that the volume and surface integrals in the Kirchhoff representation (9a)–(9b) have well defined limits when $\Omega$ tends to the entire spacetime. But suppose for sake of discussion that the limits (denoted by a superscript $\infty$) exist. Then $\overset{\infty}{\underset{in}{}} F^{\mu\nu}$ gives the truly source-free incoming radiation. The retarded Sommerfeld radiation condition is the statement that

$$\overset{\infty}{\underset{in}{}} F^{\mu\nu} = 0 \tag{21}$$

If this condition is treated as an additional law of electromagnetism, then the reduced set model of classical electromagnetism is not closed under time reversal. The proof is just a variant of the argument given in section 2.5 to show that the Ritzian theory $T_1$ is not time reversal invariant. Here then is the promised example of how to achieve a time asymmetric variant of orthodox classical electromagnetism without using incantations about causation.

Similarly, if Maxwell's theory is augmented, not by postulating (21) as an extra law, but rather by postulating the advanced Sommerfeld radiation condition

$$\overset{\infty}{\underset{out}{}} F^{\mu\nu} = 0 \tag{22}$$

then again the resulting theory is not time reversal invariant. If both (21) and (22) are promoted to the status of extra laws, then the resultant theory is time reversal invariant.

A motivation for promoting (21) (respectively, (22)) to law status might stem from the conviction that electromagnetic fields must have sources (respectively, must have sinks). It is hard to credit such convictions if they are supposed to stand alone. Buttressing for these convictions might come from the further conviction that electromagnetic fields are merely mathematical devices for describing what are actually direct interactions among charges. But this is a view that was rejected in section 2.5. A further problem for motivating the promotion of (21) (or (22)) to law status is that, in conjunction with Maxwell's laws, the posit of (21) (or (22)) as a de facto condition that holds in the actual universe would seem to do just as much work in explaining features of the actual universe as positing (21) (or (22)) as an additional law.

Continuing this line of thought, one can wonder whether (21) or (22)—qua laws or qua de facto truths—can do any useful explanatory work at all with regard to the local electromagnetic asymmetries alluded to by Zeh and Frisch (section 2.7). These asymmetries concern the behavior of systems that typically comprise small finite chunks of the universe; and for such systems asymmetries between $_{in}F^{\mu\nu}$ and $_{out}F^{\mu\nu}$ need not have anything to do with asymmetries between $_{in}^{\infty}F^{\mu\nu}$ and $_{out}^{\infty}F^{\mu\nu}$ since, for example, $_{in}F^{\mu\nu}$ may not represent truly source-free incoming radiation but rather radiation associated with sources outside the volume $\Omega$ that is implicit in the expression for $_{in}F^{\mu\nu}$.

I will return below to the last point. But even if the point is well taken, there are three reasons why it is worth inquiring about $_{in}^{\infty}F^{\mu\nu}$ and $_{out}^{\infty}F^{\mu\nu}$. First, an asymmetry between $_{in}^{\infty}F^{\mu\nu}$ and $_{out}^{\infty}F^{\mu\nu}$ would give a clean and distinctively electromagnetic asymmetry; second, this asymmetry may admit linkages with cosmological arrows; and third, even if an asymmetry between $_{in}^{\infty}F^{\mu\nu}$ and $_{out}^{\infty}F^{\mu\nu}$ is irrelevant to the Zeh-Frisch asymmetries, it may be relevant to explaining the asymmetry of radiation reaction for accelerated charges. The first two reasons will be taken up in the immediately succeeding subsections, and the third will be discussed in section 3.5.

## 2.9 Sommerfeld radiation conditions in cosmology

The threshold issue to be faced in discussing the Sommerfeld radiation conditions in cosmology is that the Kirchhoff representation theorem in the form (9a)–(9b) for Minkowski spacetime does not generalize in any straightforward way to arbitrary cosmological models.[23] If the spacetime is not globally hyperbolic[24] the global existence and uniqueness of advanced and retarded Green's function is not assured, and without this assurance the theorem does not get off the ground. (Restricting to a globally hyperbolic neighborhood[25] does not suffice for present purposes since the Sommerfeld radiation conditions are global in nature.) Next, a condition is needed to assure that the electromagnetic field propagates cleanly without "tails" that trail along behind the wave front (as would be the case if the Green's functions do not vanish in the interior of the light cones).[26] For this assurance it is sufficient that the (four-dimensional)

---

[23] For an overview of the problems encountered, see Ellis and Sciama (1972).

[24] See Wald (1984: 210–209) for various equivalent definitions, one of which is the existence of a Cauchy surface, a spacelike hypersurface that is intersected exactly once by every endless timelike curve.

[25] Let $\mathcal{M}$, $g_{\mu\nu}$ be an arbitrary relativistic spacetime. Then for any $p \in \mathcal{M}$ there is a neighborhood $\mathcal{N}(p)$ of $p$ such that the spacetime $\mathcal{N}$, $g_{\mu\nu}|_{\mathcal{N}}$ is globally hyperbolic.

[26] In the non-conformally flat case, the retarded (respectively, advanced) representation contains an additional "tail" term consisting of an integral over the interior as well as the surface of the past (respectively, future) light cone of the field point.

spacetime be conformally flat.[27] The FRW cosmological models currently used to describe the large-scale structure of the actual universe provide examples of spacetimes that are both globally hyperbolic and conformally flat.

Suppose then that we are working in a cosmological model where the Kirchhoff representation of the electromagnetic field holds in its usual form, and suppose that the volume and surface integrals in the retarded and advanced Kirchhoff representations have finite limits as the volume $\Omega$ tends towards the entire spacetime of the model. Nevertheless, it may not be physically consistent to impose the retarded Sommerfeld condition (21) (respectively, the advanced Sommerfeld condition (22)) if the model has particle horizons (respectively, event horizons).[28] The standard hot big bang models (which belong to the family of FRW models) have particle horizons, regardless of whether the model contains an early inflationary epoch and regardless of whether $k = 0$ (flat space sections), $k = +1$ (space sections of constant positive curvature), or $k = -1$ (space sections of constant negative curvature). Furthermore, there is strong evidence that "dark energy" is driving the currently observed accelerated expansion of the universe (see Carroll 2004), and if this "dark energy" is due to a positive cosmological constant, the future asymptotic structure of spacetime becomes that of de Sitter spacetime, which has event horizons.

To see the difficulty with the retarded Sommerfeld radiation condition in the presence particle horizons, consider the toy example of past truncated Minkowski spacetime (see Fig. 16.4a) where all the spacetime points on or below some $t = const.$ hypersurface (with $t$ an inertial time coordinate) have been deleted. Particle #2 is outside of #1's particle horizon at $x$, and so at $x$ the retarded volume integral in (10a) will be zero, no matter how large the volume $\Omega$ is taken to be. Nevertheless, at $x$ particle #1 will feel particle #2's Coulomb field[29] and, thus, (10a) implies that $_{in}F^{\mu\nu}(x) \neq 0$.[30] Analogously, in future truncated Minkowski spacetime illustrated in Fig. 16.4b the advanced Sommerfeld radiation condition cannot be satisfied. Particle #4 is outside of particle #3's event horizon at $z$, and so at $z$ the advanced volume integral in (10b) will be zero, no matter how large the volume $\Omega$ is taken to be. Nevertheless, at $z$ particle #3 will feel particle #4's Coulomb field and, thus, (10b) implies that $_{out}F^{\mu\nu}(z) \neq 0$.

Aichelberg and Beig (1977) have studied the compatibility between Sommerfeld type radiation conditions and cosmological structure for an interesting toy model involving a one-dimensional, non-relativistic harmonic oscillator coupled to a scalar field $\Phi$. Because the model is completely soluble, Aichelberg and Beig are able to

[27] A metric $g_{\mu\nu}$ is conformally flat just in case there is a scalar field $\phi$ such that at all points $x$ of the spacetime manifold, $g_{\mu\nu}(x) = \phi^2(x)\eta_{\mu\nu}$ where $\eta_{\mu\nu}$ is the Minkowski metric.

[28] For an introduction to horizons in cosmology, see Ellis and Rothman (1993).

[29] This is a result of the fact that two of the Maxwell equations are constraint equations—in the case of Minkowski spacetime, these equations are the first of the equations in (1')–(2'). These elliptic equations constrain the joint values of the fields and the charge-current on a spacelike hypersurface. See Penrose (1964) for a more precise presentation of this point.

[30] An exception occurs when the charges are symmetrically arranged around the point $x$ so that their Coulomb fields cancel out.

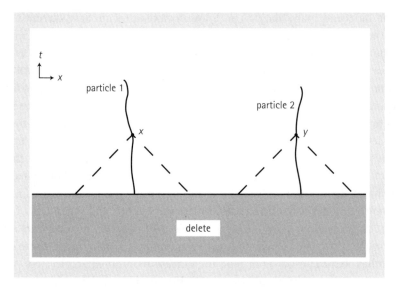

FIGURE 16.4a  Past truncated Minkowski spacetime

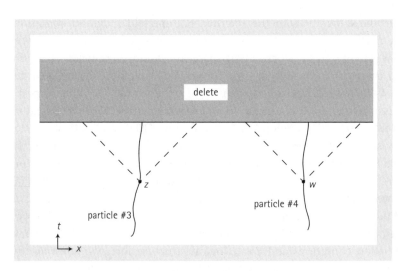

FIGURE 16.4b  Future truncated Minkowski spacetime

determine whether there are solutions of the coupled oscillator-field equations of motion[31] with $\overset{\infty}{\underset{in}{}}\Phi = 0$, $\overset{\infty}{\underset{out}{}}\Phi = 0$, or $\overset{\infty}{\underset{in}{}}\Phi + \overset{\infty}{\underset{out}{}}\Phi = 0$ for a range of different $k = 0$ (flat

---

[31]  The Aichelberg-Beig model assumes that the scalar field $\Phi$ obeys a conformally invariant wave equation. Thus, they use a conformal coupling to the spacetime, and $\Phi$ satisfies

$$(\Box_g + \tfrac{1}{6}R)\Phi(\mathbf{x}, t) = \lambda \frac{\delta^3(\mathbf{x})}{a^3(t)} Q(t), \text{ where } \Box_g := \frac{1}{\sqrt{-g}} \frac{\partial}{\partial x^\mu}(\sqrt{-g}g^{\mu\nu}\frac{\partial}{\partial x^\nu}), R \text{ is the Ricci curvature}$$

scalar, and $Q$ is the oscillator amplitude obeying the equation of motion $\ddot{Q}(t)+\omega_o^2 Q(t) = \lambda\Phi(0, 0, 0, t)$, where $\omega_o$ is the spring constant.

space sections) FRW cosmologies. The line element in these cosmologies takes the form $ds^2 = g_{\mu\nu}dx^\mu dx^\nu = a(t)(dx^2 + dy^2 + dz^2) - dt^2$, where $a(t)$ is the scale factor. In expanding Friedmann models for which $a(t) \sim t^n$ and $n < 1$, and also in expanding models for which $a(t) \sim t^n$ and $n \geq 1$, Aichelberg and Beig found that it is consistent to impose the condition $\overset{\infty}{\underset{in}{}}\Phi = 0$ but not $\overset{\infty}{\underset{out}{}}\Phi = 0$ or $\overset{\infty}{\underset{in}{}}\Phi + \overset{\infty}{\underset{out}{}}\Phi = 0$. In the time reversed contracting versions of these cosmological models the asymmetry is reversed. But in the case of an expanding de Sitter model with $a(t) \sim \exp(t/a)$, $a$ a positive constant, and also in Minkowski spacetime with $a(t) \sim 1$, it is not consistent to impose any of these conditions.

To return to electromagnetism, the fact that, say, $\overset{\infty}{\underset{in}{}}F^{\mu\nu} = 0$ is incompatible with a cosmological model does not entail that this cosmology does not enforce an interesting asymmetry between $\overset{\infty}{\underset{in}{}}F^{\mu\nu}$ and $\overset{\infty}{\underset{out}{}}F^{\mu\nu}$ but only that the asymmetry cannot be stated in the most obvious way, e.g. $\overset{\infty}{\underset{in}{}}F^{\mu\nu} = 0$ while $\overset{\infty}{\underset{out}{}}F^{\mu\nu} \neq 0$. The most dramatic such cosmologically enforced asymmetry would be a case where, say, the volume and surface integrals of the retarded representation have finite limits as the volume $\Omega$ tends to the entire spacetime of the considered cosmological model, whereas in the advanced representation the volume and/or surface integrals diverge. Sciama (1963) has worked out the details of just such a case for an expanding Einstein-de Sitter model, which is a $k = 0$ FRW model with $a(t) \sim t^{2/3}$. For a plausible distribution of sources, Sciama found that both the volume and surface integrals in the advanced Kirchhoff representation diverge as $\Omega$ tends to the entire spacetime. By contrast, the retarded volume integral converges and, thus, the demand for a finite solution means that the amount of incoming source-free radiation is such that the retarded surface integral converges. Discounting the possibility that the divergences in the advanced volume and surface integrals cancel to give a finite sum, the demand for a finite solution uniquely singles out the retarded representation from the general class of representations (11). This asymmetry is reversed in the time reversed contracting Einstein-de Sitter model, emphasizing that the EM asymmetry at issue is enslaved to the cosmological arrow.

That the cosmological based electromagnetic asymmetries may well be independent from the local electromagnetic asymmetries discussed in section 2.7 does not show that the former are not interesting nor that they are not fundamental. By way of analogy, the failure of time reversal invariance that has been demonstrated for neutral kaon decay[32] does not seem to be connected with any of the more familiar arrows of time. But arguably this failure is of fundamental importance, not only in providing a fundamental lawlike asymmetry *in* time, but also in providing a basis for an asymmetry *of* time by supporting the conclusion that the actual universe is time orientable and possesses a time orientation (see Earman 2002).

[32]  The experimental results reported in Angelopoulos (1998) give a direct demonstration of time reversal invariance in the sense that they do not (as earlier results did) appeal to observed $CP$ violation and the $CPT$ theorem.

# 3. RADIATION DAMPING

## 3.1 From the well-defined to the not-so-well-defined

Section 2 was concerned with a well-defined situation; namely, a charge-current distribution $J^\mu$, satisfying $\nabla_\mu J^\mu = 0$, is specified, and one seeks to find the corresponding solution to the Maxwell equations. One could add a second well-defined problem: an external electromagnetic field $_{ext}F^{\mu\nu}$ and a non-electromagnetic force $_{nem}F^\mu$ are specified, and one seeks to find the world line $z^\mu(\tau)$ of a point mass carrying charge $q$ by using the Lorentz force law and solving the resultant equation of motion

$$ma^\mu = q\left(_{ext}F^{\mu\nu}u_\nu\right) + {}_{nem}F^\mu \tag{23}$$

And combining the two also leads to a harmonious situation in that (i) the coupled Maxwell-Lorentz equations have a well-posed initial value problem and (ii) the total stress-energy tensor, consisting of the sum of the stress-energy tensor of the electromagnetic field and the mechanical stress-energy tensor of the particle,[33] is conserved.

Unfortunately, equation (23) is empirically inadequate, for accelerated charges radiate and, as a result, they experience a damping force that is not reflected in (23).[34] The generally accepted equation of motion describing this phenomenon for a point charge in Minkowski spacetime is called the Lorentz-Dirac equation:

$$ma^\mu = q\left(_{ext}F^{\mu\nu}u_\nu\right) + {}_{nem}F^\mu + \frac{2}{3}q^2(\dot{a}^\mu - a^2u^\mu) \tag{24}$$

$$= q\left(_{ext}F^{\mu\nu}u_\nu\right) + {}_{nem}F^\mu + \frac{2}{3}q^2(\eta^{\mu\nu} + u^\mu u^\nu)\dot{a}_\nu$$

where $a := a^\mu a_\mu$ is the magnitude of the four-vector acceleration and $\eta^{\mu\nu}$ is the Minkowski metric. The second equality in (24) follows from the first by differentiating the identity $u_\mu a^\mu = 0$, and using the result to eliminate $a^2$. Note that the crucial radiation reaction term (the third term on the right-hand side of (24)) vanishes when the acceleration of the charge is hyperbolic, suggesting that such a charge must be absorbing as much electromagnetic energy as it radiates.

Radiation reaction is observed, for example, in synchrotron radiation, but high precision tests of the predictions of (24) for the orbits of charged particles are lacking. A summary of the current experimental situation and proposed high precision tests can be found in Spohn (2004: section 9.3).

Note that under time reversal, the spatial component of $(\dot{a}^\mu - a^2u^\mu)$ changes sign. Thus, if (24) is regarded as a fundamental law, then time reversal invariance is broken. The issue of whether radiation reaction does indeed indicate such a breaking will be approached in stages.

---

[33] $T^{\mu\nu}_{mech} := m \int d\tau u^\mu u^\nu \delta(x - z)$.
[34] An exception to the rule will be mentioned shortly.

## 3.2  The reduced order equation

The Lorentz-Dirac equation is beset by several well-known difficulties. The most often mentioned is the existence of "run away" solutions in which the acceleration of the charge increases without bound (and consequently $|\mathbf{u}/c| \to 1$) even in the absence of an external force. These pathological solutions can be quashed by narrowing the class of solutions with the demand that the acceleration of the charge vanishes as the proper time along its world line approaches $+\infty$, but the price to be paid is a "pre-acceleration" effect in which the charge begins to accelerate before the external force is applied.

In response one can adopt the pragmatic attitude that these difficulties result from pushing the Lorentz-Dirac equation beyond its domain of validity. The Lorentz-Dirac equation was intended to describe the reaction of a point charge to an external force that causes the charge to accelerate and, thus, to radiate. But only a glance at (24) is needed to realize that this intent is not fulfilled since the self-interaction ($q^2$) term can be non-zero even when $_{ext}F^{\mu\nu}$ and $_{nem}F^{\mu}$ in (24) vanish (as emphasized by Rohrlich 2001). This suggests that the domain of validity of the Lorentz-Dirac equation is circumscribed by a value for the self-interaction term that is small in comparison with the external force (see Teitelboim et al. 1980). That the equation (24) yields pathological results when the external force is zero is then no surprise. Adding a note of optimism to the pragmatism, the Lorentz-Dirac equation can be expected to yield empirically adequate results when its domain of validity is not exceeded.

The pragmatic attitude towards the Lorentz-Dirac equation (24) sketched above may strike some as too casual. It is not unreasonable to demand that the pragmatism be justified by finding an equation of motion that is just as accurate as the Lorentz-Dirac equation within its domain of validity, but which is not subject to the pathologies of run-away solutions and pre-acceleration effects. A clue to finding the desired equation lies in the fact that these pathologies can be traced to the presence of the third-order time derivative in (24). This suggests that a more satisfactory equation can be found by finding a technique to reduce the highest order of the time derivatives. Such a reduction of order procedure was first proposed by Landau and Lifshitz (1975) and subsequently has been extensively discussed and refined.

Suppose that the external force is due to the external electromagnetic field $_{ext}F^{\mu\nu}$. Since the self-interaction term in (24) is supposed to be small compared to the external force, the acceleration $a^{\nu}$ appearing in the $\dot{a}^{\nu}$ term on the right-hand side of the second equality of (24) can be approximated by $_{ext}F^{\nu\beta}u_{\beta}/m$. When the proper time derivative of this expression is taken and substituted for $\dot{a}^{\nu}$ the resulting equation of motion takes the form

$$ma^{\mu} = q\left(_{ext}F^{\mu\nu}u_{\nu}\right) + \tag{25}$$

$$\frac{2}{3}\frac{q^3}{m}\left[u^{\beta}\nabla_{\beta}(_{ext}F^{\mu\nu})u_{\nu} + \frac{q}{m}F^{\mu\nu}\,_{ext}F_{\nu\beta}u^{\beta} + \frac{q}{m}u^{\mu}\,_{ext}F^{\alpha\beta}\,_{ext}F_{\beta\gamma}u_{\alpha}u^{\gamma}\right]$$

This heuristic derivation of the reduced order equation can be replaced by a more rigorous argument, as given in detail by Spohn (2000, 2004).[35]

Like the Lorentz-Dirac equation in the form (24), the reduced order equation (25) is not time reversal invariant (*pace* Rohrlich 2001). To see this it suffices to focus on a special case that reduces the formidable clutter of super- and sub-scripts in (25). Work in an inertial system and take the case where (in the chosen system) the external magnetic field vanishes and the external electrostatic potential $\phi(\mathbf{r})$ varies only along, say, the $y$-axis. Setting $\mathbf{r} = (0, y, 0)$, $\mathbf{u} = (0, \dot{y}, 0)$, and $\phi(\mathbf{r}) = V(y)$, (25) reduces to

$$m\frac{d}{dt}(\gamma\dot{y}) = -V'(y) - \frac{2}{3}\frac{q^3}{m}V''(y)\gamma\dot{y} \tag{26}$$

where $\gamma := \sqrt{1 - \mathbf{u} \cdot \mathbf{u}}$ and the prime and the over-dot denote respectively differentiation with respect to $y$ and with respect to inertial time (see Spohn 2000). When $V''(y) > 0$ the charge experiences a frictional damping force. But under time reversal, there is a flip in the sign of the term describing this force and, thus, in the time reversed history the charge experiences an anti-damping force.

Because equation (25) avoids the pathologies of the Lorentz-Dirac equation and is just as accurate as the Lorentz-Dirac equation within the domain of validity of the latter, Rohrlich (2001) has dubbed equation (25) as "the exact" classical equation of motion covering radiation reaction of a point charge. But these virtues of (25) do not constitute a proof that (25) is anything more than a phenomenological equation that encapsulates the observed asymmetry of radiation—that accelerating charges radiate and experience a damping force rather than absorbing radiation and experiencing an anti-damping force—but does not provide an explanation of the origins of this asymmetry. Before turning to possible explanations of the observed asymmetry of radiation, it will be helpful to look at another form of the Lorentz-Dirac equation that, unlike (24) and (25), is time reversal invariant.

## 3.3 The Dirac expression for radiation reaction

The retarded and advanced Liénard-Wiechert fields of a point charge are singular on the world line of the charge, but since both solutions have the same singularity structure near the world line of the charge, the *difference* between the two is non-singular on the world line. Following Dirac (1938), postulate that the radiation reaction force $\Gamma^\mu$ experienced by a point particle with charge $q$ is given by

$$\Gamma^\mu := \frac{q}{2}[_{ret}F^{\mu\nu} - {}_{adv}F^{\mu\nu}]u_\nu \tag{27}$$

---

[35] When the external force is due to a non-electromagnetic force $_{nem}F^\mu$, a reduced order equation can be produced by approximating the $\dot{a}^\nu$ term on the right-hand side of the second equality of (24) by $\frac{1}{m}d(_{nem}F^\nu)/d\tau$.

where $_{ret}F^{\mu\nu}$ and $_{adv}F^{\mu\nu}$ are respectively the retarded and advanced Liénard-Wiechert fields of charge in question. By performing a power series expansion in the proper time of the worldline of the charge and dropping terms of third and higher order, Dirac (1938) showed that $\Gamma^{\mu}$ is equal to the radiation reaction term in (24).[36] As far as phenomenology goes, (27) has just as good a claim to providing a correct description of radiation reaction as the expression in (24).

Now consider a closed system consisting of a finite number $N$ of point charges subject to no external forces. Using the retarded representation, the external electromagnetic force acting on one of the charges will be the sum of the retarded Liénard-Wiechert fields of the other charges plus the incident source-free radiation. Using the Dirac expression (27) for the radiation reaction and setting $_{nem}F^{\mu}$ in (24) to zero, the equation of motion of the $k$th charge is

$$ma_{(k)}^{\mu} = q_k \sum_{j \neq k}^{N} {}_{ret}F_{(j)}^{\mu\nu}u_{(k)\nu} + q_k({}_{in}F^{\mu\nu}u_{(k)\nu}) + \frac{q_k}{2}[{}_{ret}F_{(k)}^{\mu\nu} - {}_{adv}F_{(k)}^{\mu\nu}]u_{(k)\nu} \qquad (28)$$

To get the equation of motion describing the time reverse of the history described by (28), replace each of the quantities in (28) by its time reverse counterpart (per section 2.1). The result is

$$ma_{(k)}^{\mu} = q_k \sum_{j \neq k}^{N} {}_{adv}F_{(j)}^{\mu\nu}u_{(k)\nu} + q_k({}_{out}F^{\mu\nu}u_{(k)\nu}) + \frac{q_k}{2}[{}_{adv}F_{(k)}^{\mu\nu} - {}_{ret}F_{(k)}^{\mu\nu}]u_{(k)\nu} \qquad (29)$$

where the quantities in (28) are evaluated at the proper times $\tau_{(k)}$ and the corresponding quantities (29) are evaluated at the proper times $\tau'_{(k)} = -\tau_{(k)}$. At this juncture appeal can be made to the Kirchhoff theorem in the form

$$\sum_{j \neq k}^{N} {}_{adv}F_{(j)}^{\mu\nu} + {}_{out}F^{\mu\nu} = \sum_{j \neq k}^{N} {}_{ret}F_{(j)}^{\mu\nu} + {}_{in}F^{\mu\nu} + [{}_{ret}F_{(k)}^{\mu\nu} - {}_{adv}F_{(k)}^{\mu\nu}] \qquad (30)$$

Substituting (30) into (29) shows that the equation describing the time reversed history can be brought into the form (28) for the original history but with the terms evaluated at the proper times $\tau'_k = -\tau_k$ appropriate to the time reversed history.

The conclusion that the Lorentz-Dirac equation in the form (28) is time reversal invariant might seem to break down in the special case where $_{in}F^{\mu\nu} = 0$ and $N = 1$. But taking into account the inadequacies of the Lorentz-Dirac equation (recall sections 3.2 and 3.3), we can reason informally as follows. Since the impressed force acting on the lone charge is zero by hypothesis, the acceleration of the charge is zero; and since the radiation reaction force arises only when a charge is accelerating, it too is zero in the present case (see Rohrlich 1990: 251).

However, it does seem that the failure of time reversal invariance announces itself in the $N = 1$ case when there is a *non*-electromagnetic force that drags along the lone charge. But postulating by brute force a $_{nem}F^{\mu}$ term to be added to the right-hand side

---

[36] For a user-friendly presentation of this result, see Poisson (1999).

of (28) amounts to treating the system as an open system and to allowing ourselves to play God by "poking" the system with an arbitrary force. The point remains that for a closed electromagnetic system that is fully described by the Lorentz-Dirac equation in the form (28), no violation of time reversal invariance can be detected. It remains to investigate the time reversal invariance properties of a closed system that couples electromagnetism to another force, once the source equations of that additional force field are specified.

That the Lorentz-Dirac equation in the form (28) is time reversal invariant seems to generate a puzzle since, taken at face value, (28) seems to entail the temporally asymmetric consequence that an accelerating charge experiences a damping rather than an anti-damping force. The latter impression is incorrect.

## 3.4 Attempts to explain the asymmetry of radiation reaction

The Lorentz-Dirac equation in the form (28) cannot explain the observed asymmetry of radiation reaction. Rather than applying the time reversal transformation to (28) to get the equation describing the time reversed history, simply rewrite (28) using the advanced representation. The result is an equation (29′) with the same form as (29) but with the terms evaluated at the same proper times $\tau_k$ as in (28). Thus, if (28) licenses the inference that an accelerating charge experiences a damping force (because the last term on the right-hand side of this equation represents a damping force), then (29′) licenses the inference that an accelerating charge experiences an anti-damping force (because the last term on the right-hand side of this equation represents an anti-damping force). But since (28) and (29′) are equivalent, neither inference can be correct (unless (28)–(29′) is self-contradictory) since otherwise an accelerating charge would experience both a damping and an anti-damping force. Thus, why it is that we observe radiation damping rather than radiation anti-damping remains to be explained.

Here is where cosmological considerations may be relevant. With the volume $\Omega$ implicit in the expressions $_{in}F^{\mu\nu}$ and $_{out}F^{\mu\nu}$ pushed to the limit, these expressions become $_{in}^{\infty}F^{\mu\nu}$ and $_{out}^{\infty}F^{\mu\nu}$ respectively. Suppose then that the Sommerfeld radiation condition $_{in}^{\infty}F^{\mu\nu} = 0$ obtains. Then as a consequence of the Kirchoff theorem, $_{out}^{\infty}F^{\mu\nu} = \sum_{all\ j}^{N}[_{ret}F_{(j)}^{\mu\nu} - _{adv}F_{(j)}^{\mu\nu}]$. Substituting this latter relation into (29′) shows that (29′) reduces to (28) (with the domains of integration chosen so that $_{in}F^{\mu\nu} = _{in}^{\infty}F^{\mu\nu}$ and $_{out}F^{\mu\nu} = _{out}^{\infty}F^{\mu\nu}$), which might be seen as an explanation of why we observe radiation damping rather than anti-damping.

However, we know from section 2.9 that the Sommerfeld radiation condition is inconsistent with various cosmologies, and even when the Sommerfeld condition obtains, it may also be the case that $_{out}^{\infty}F^{\mu\nu} = 0$, in which case symmetry is restored and the proffered explanation fails. In addition, even in cases where $_{in}^{\infty}F^{\mu\nu} = 0$ but $_{out}^{\infty}F^{\mu\nu} \neq 0$ it is not clear that the proffered explanation is cogent. How does showing that (29′) reduces to (28) license the inference that radiation damping rather than radiation anti-damping is to be expected? After all, one could equally well argue that the Sommerfeld condition $\substack{\infty \\ in} F^{\mu\nu} = 0$ shows that (28) reduces to (29′). The best good hope for using cosmology to explain the asymmetry of radiation reaction in standard classical electrodynamics appears to lie in cases where the cosmology is such that only the retarded representation leads to a finite solution (see section 2.9).

The attempt to explain the asymmetry of radiation reaction does not fare any better in the Wheeler-Feynman action-at-a-distance version of classical electrodynamics. Their basic equation of motion for a point charge is (14). It is postulated that the universe consists of an island of matter containing $N$ charges surrounded by empty space. Because of destructive interference effects, Wheeler and Feynman conclude that in empty space

$$\sum_{all\ j}^{N} [_{ret}F^{\mu\nu}_{(j)} + {}_{adv}F^{\mu\nu}_{(j)}] = 0 \tag{31}$$

Then by means of a not entirely convincing argument, they conclude that the absorber condition

$$F^{\mu\nu}_{rad,tot} =: \sum_{all\ j}^{N} [_{ret}F^{\mu\nu}_{(j)} - {}_{adv}F^{\mu\nu}_{(j)}] = 0 \tag{32}$$

holds everywhere, in which case their equation of motion (14) reduces to the Lorentz-Dirac equation in the form (28). [The Wheeler-Feynman definition of the total radiation field $F^{\mu\nu}_{rad,tot}$ differs from the sum of the retarded radiation fields $\sum_{all\ j}^{N} {}_{ret}F^{\mu\nu}_{rad(j)}$, with $_{ret}F^{\mu\nu}_{rad(j)}$ as defined above in section 2.7. At a field point $x$ sufficiently far from the charges and at a time sufficiently long after the charges have ceased to accelerate, the two expressions for the total radiation field will agree because the Coulomb part of $_{ret}F^{\mu\nu}_{(j)}(x)$ will be approximately zero as will $_{adv}F^{\mu\nu}_{(j)}(x)$. But one could complain that the explanation of radiation reaction is not concerned solely with such locations.]

Of course, the actual universe is not an island universe. This inconvenient fact can be overcome if the universe is opaque to electromagnetic radiation, but there is evidence that it is not (see Partridge 1973).[37] But the main point for the present discussion is that, even if all of the qualms about the Wheeler-Feynman argument are waived, their theory, supplemented by the absorber condition (32), does not explain the observed asymmetry of radiation reaction since (32) equally well shows that their equation of motion (14) reduces to (29′) (as emphasized by Zeh 2001: 35).

It seems that conjuring with the time reversal invariant form of the Lorentz-Dirac equation (28) or with the time reversal invariant Wheeler-Feynman direct particle interaction equations is not going to lead to a satisfactory explanation of the observed asymmetry of radiation. Non-time reversal invariant forms of the equation, such as the reduced order equation (25), would provide the basis for an explanation if they

---

[37] Detailed critical discussions on the Wheeler-Feynman absorber theory of radiation can be found in Davies (1972); Davies and Twamley (1993); Price (1966); and Zeh (2001).

stood for fundamental laws rather than phenomenological descriptions of what is be explained, but affirming the 'if' seems highly dubious.

## 3.5  Neo-Ritzian explanations of the asymmetry of radiation

The rather dreary accounting in the preceding subsection of the failures in explaining the observed asymmetry of radiation and radiation reaction might tempt one to listen again to the Siren song of neo-Ritzian posits. For those who are seduced by this song, one way to proceed is to reject the Dirac analysis of radiation reaction for point charges and to provide an alternative analysis that appeals only to retarded fields. A conceptually clear and elegant derivation along these lines of the Lorentz-Dirac equation in the form (24) has been given by Teitelboim (1970, 1971). As mentioned above, the retarded Liénard-Wiechert field for a point charge in Minkowski spacetime can be covariantly separated into a velocity (or generalized Coulomb) field $_{ret}F^{\mu\nu}_{Coul}$ and an acceleration (or radiation) field $_{ret}F^{\mu\nu}_{rad}$. This splitting of the field induces a splitting of the stress-energy tensor $T^{\mu\nu}_{em}$ of the field into a "bound part" $T^{\mu\nu}_{Coul}$ and an "emitted part" $T^{\mu\nu}_{rad}$, each of which is separately conserved off the world line of the charge. The Lorentz-Dirac equation (24) is then obtained as a consequence of the assumption of energy-momentum balance in the form

$$\nabla_\nu T^{\mu\nu}_{bare} + \nabla_\nu T^{\mu\nu}_{Coul} = -\nabla_\nu T^{\mu\nu}_{rad} - {}_{ext}f^\mu \tag{33}$$

where $_{ext}f^\mu$ is the force density associated with the external force, and $T^{\mu\nu}_{bare}$ is the mechanical stress-energy tensor for a point particle of "bare" mass $m_b$. The observed mass $m$ of the particle is assumed to result from the "renormalization" of the total mass $m_b + m_{em}$ where the electromagnetic mass $m_{em}$ is the self-energy $\dfrac{q^2}{\varepsilon}$ of the charge, with $\varepsilon$ being a parameter that is zero for a point particle. Thus, a subtraction of an infinity is needed to produce a finite value for $m$. This is is not a blemish on the Teitelboim derivation since mass renormalization is a feature of any derivation of (24) from energy-momentum balance considerations.[38]

Such a derivation using only retarded fields is not sufficient to explain the asymmetry of radiation reaction. The explanation could be completed by using a neo-Ritzian invocation of "causality" to block any competing analysis of the radiation reaction that appeals to advanced fields. Rohrlich, who has heeded the Siren song (see Rohrlich 1998, 1999, 2000), is not opposed to this form of explanation; but because he finds a point charge to be an unacceptable idealization in classical electrodynamics, his analysis starts with the equations of motion for charges of finite size. His own neo-Ritzian position is based on two claims. The first is that the correct classical equation of motion for a finite sized charge is the Caldirola-Yaghjian equation (see

---

[38]  It might be noted, however, that the Teitelboim derivation also relies on the assumption that the world line of the charge tends to a straight line as the proper time along the world line approaches $-\infty$, an assumption that limits the domain of validity of (24) and introduces a temporal asymmetry.

Yaghjian 1992), and that this equation is not time reversal invariant. The latter part of this first claim has been disputed by Zeh (1999) and Rovelli (2004). I will not enter this dispute because I want to concentrate on Rohrlich's second claim which provides a diagnosis of the (alleged) failure of time reversal invariance for the equations of motion for radiating charges of finite size. The diagnosis goes as follows: the self-interaction of a finite sized charge involves the interaction of one element of the charge on another; this interaction "takes place by the first element emitting an electromagnetic field, propagating along the *future* light cone and then interacting with the other element of charge" (2000: 9); the choice of the future light cone over the past light cone is dictated by "causality" which "requires *retarded* rather than advanced self-interaction" (2000: 1); in sum, it is causality which is "ultimately responsible for the [EM] arrow of time" (ibid.).

Several comments are in order. 1. Rohrlich thinks that it follows from his analysis that in the limit of a point particle, the equation of motion ought to be time reversal invariant:

> In the point limit, the retarded and advanced actions can no longer be distinguished because the interaction distance between the charge-mass elements shrinks to zero. Therefore, in that limit the equations of motion are time reversal invariant
> (Rohrlich 1999: 5)

Thus, he is committed to the (I think) incorrect position that the reduced order form of the Lorentz-Dirac equation for a point charge is time reversal invariant (see Rohrlich 2005). 2. It also seems to follow from Rohrlich's analysis that the behavior of the equations of motion under time reversal is irrelevant to the asymmetry of radiation reaction; for even if the equations of motion are time reversal invariant (as Rohrlich wrongly says is the case with reduced order form (25) of the Lorentz-Dirac equation (24) and rightly says is the case for Lorentz-Dirac equation in form (28)), the appeal to "causality" still provides a basis for the asymmetry of radiation—which is a consequence that Rohrlich (2005) embraces. 3. I have already railed enough against the invocation of "causality" in lieu of genuine scientific theorizing, and at this juncture I only want to note that if Rohrlich's causality based account of the origins of the arrow of electromagnetic radiation is correct, then this arrow is non-contingent, in that it does not depend on initial/boundary conditions.[39] One persuasive reason for resisting this consequence—and thus for rejecting causality based accounts of the asymmetry of radiation—derives from the facts that QED is widely accepted as the correct theory of electrodynamics and that the basic laws of this theory are time reversal invari-ant. So either taking the classical limit of QED introduces an temporal asymmetry (implausible) or else asymmetry must drive from contingency of initial/boundary conditions.

This last thought suggests that a fruitful strategy for finding an explanation of the asymmetry of radiation is to study how the Lorentz-Dirac expression for radiation

---

[39] Unless, of course "causality" itself depends on initial/boundary conditions, a notion which the neo-Ritzians seem to reject.

reaction emerges from QED in an appropriate classical limit. But before pursuing this line of inquiry, it is worth making a final comment about the relevance of cosmology for the asymmetry of radiation reaction in classical electrodynamics.

## 3.6  The equation of motion of a point charge in a cosmological setting

The above discussion of radiation reaction in classical electrodynamics assumed a flat spacetime background. The generalization from flat to curved spacetime of the Lorentz-Dirac equation in the form (24) replaces $a^\mu$, $\dot{a}^\mu$, and $a^2$ by their counterparts that use the covariant path derivative $D/D\tau$ along the world line of the charge in place of $d/d\tau$. However, a more important modification is necessitated by the presence of spacetime curvature since the interaction between the electromagnetic field of a charge and the curvature produces additional terms in the expression for the radiation reaction. Specifically, in a conformally flat but non-Ricci flat spacetime (such as a FRW spacetime), the addition is

$$-\frac{q^2}{3}(R^\mu_\beta u^\beta + u^\mu R_{\beta\gamma} u^\beta u^\gamma) \tag{34}$$

(see Hobbs 1968 and Quinn and Wald 1997). Since odd powers of the four-velocity are involved in (34), the presence of the Ricci curvature seems to give rise to additional time asymmetry that is distinct from the asymmetry of radiation reaction in flat spacetime.

This curvature based asymmetry has received virtually no attention in the philosophical literature on the arrows of time. But it deserves attention if for no other reason than that it provides a decisive (in-principle) test of standard classical electrodynamics vs. Wheeler-Feynman action-at-a-distance electrodynamics, since in the natural generalization of the latter to curved spacetimes, the Ricci curvature term (34) is absent even when the Wheeler-Feynman absorber condition is imposed (see Unruh 1976).

## 3.7  Radiation reaction in QED

It is high time to conduct the search for the origins of asymmetry of radiation reaction by appealing to the theory that is currently thought to be the correct theory of electrodynamics—QED—and to the assumption that the valid core of classical electrodynamics is to be identified by what emerges from QED in an appropriate classical limit.

The threshold question is whether the classical limit of QED for a point charge reproduces the Lorentz-Dirac equation. Unfortunately, there is no clean-cut answer because of technical issues concerning the implementation of the classical limit (see Higuchi 2002). I will suppress these technicalities as far as possible, and will

outline the approach of Higuchi and Martin (2004, 2005) which studies the radiation reaction of a wave packet of a charged scalar field moving in an external
potential.[40]

In the classical model, a point particle of charge $q$ is linearly accelerated by an
external potential $V = V(z)$ such that $V(z) = V_o = const > 0$ for $z < Z_1$ and $V(z) = 0$
for $z > -Z_2$, where $Z_1 < Z_2$ are positive constants. Assume that the particle initially
moves in the positive $z$-direction and that at $t = 0$ it has passed through the region
$(-Z_1, -Z_2)$ where it is accelerated. Let $z_o$ denote the position the particle would have
at $t = 0$ if there there were no radiation reaction force. And let $z$ be the actual position
at $t = 0$ when the radiation reaction force—per the Lorentz-Dirac equation—is acting.
The classical position shift due to radiation reaction is then $\delta z_C := z - z_0$.

The goal is to compare the classical position shift $\delta z_C$ to the position shift $\delta z_{QED}$
calculated for a charged scalar field $\hat{\varphi}$ coupled to the electromagnetic field and subject
to the same external potential $V$ as in the classical model.[41] Towards this end, define
the charge density operator $\hat{\varrho}$ per usual as

$$\hat{\varrho}(x) := \frac{i}{\hbar} : \hat{\varphi}^\dagger \partial_t \hat{\varphi} - \partial_t \hat{\varphi}^\dagger \cdot \hat{\varphi} : \tag{35}$$

where : : indicates normal ordering. And define the expectation value of the position
of the particle by

$$< \hat{z} >:= \int z < \hat{\varrho}(\mathbf{x}, t) > d^3\mathbf{x} \tag{36}$$

Using these definitions, calculate the expectation value $< \hat{z} >^{off}$ of position at $t =$
0 with the electromagnetic field turned off using an initial state $|i\rangle$ in which the
momentum of the particle is strongly peaked about a value pointing in the positive
$z$-direction:

$$< \hat{z} >^{off} = \int z \langle i|\hat{\varrho}(\mathbf{x}, t)|i\rangle d^3\mathbf{x} \tag{37}$$

Next, use the WKB approximation to calculate to lowest non-trivial order in $q$ the
expectation value of position $< \hat{z} >^{on}$ at $t = 0$ with the electromagnetic field turned
on. To this order, time dependent perturbation theory gives a final state $|f\rangle$ of the
form

$$|f\rangle = |1\varphi, 0\gamma\rangle + |1\varphi, 1\gamma\rangle \tag{38}$$

where the elements of the superposition are interpreted as follows. The first element
$|1\varphi, 0\gamma\rangle$ is a state with one scalar particle and no photon, and is equal to $|i\rangle + |s\rangle$,
where $|s\rangle$ arises from the the forward scattering of the wave packet. The second
element $|1\varphi, 1\gamma\rangle$ is a state with one scalar particle and one photon. The expectation

---

[40]  For other approaches, see Munitz and Sharp (1977) and Johnson and Hu (2002).

[41]  The Lagrangian density for the QED model is taken to be
$$\mathcal{L} = [(D_\mu + iq A_\mu)\varphi]^\dagger [(D_\mu + iq A_\mu)\varphi] - (m/\hbar)^2 \varphi^\dagger \varphi - \tfrac{1}{4} F_{\mu\nu} F^{\mu\nu} - \tfrac{1}{2}(\partial_\mu A^\mu)^2,$$
where $D_\mu := \partial_\mu + i V_\mu/\hbar$ and $V_\mu := V(z)\delta_{\mu 0}$.

value $< \hat{z} >^{on}$ calculated from $|f\rangle$ is the sum of three terms $< \hat{z} >^{off} + < \hat{z} >^{(s)} + < \hat{z} >^{(1)}$, where the last two terms are respectively the contributions of the forward scattering and the one photon state. Since the former contribution arises without photon emission, Higuchi and Martin deem it to be irrelevant to radiation reaction, and they define the QED position shift by $\delta z_{QED} := < \hat{z} >^{on} - < \hat{z} >^{(s)} - < \hat{z} >^{off} = < \hat{z} >^{(1)}$. (If subtracting off the $< \hat{z} >^{(s)}$ term seems like hocus-pocus, two responses can be given. First, one could impose additional conditions as part of the classical limit to assure that $< \hat{z} >^{(s)}$ is small in comparison with the other terms. Second, one could take the attitude that classical electrodynamics is wrong because it does not include the effect codified in $< \hat{z} >^{(s)}$, which is of essentially quantum origins.) Higuchi and Martin (2004, 2005) show that in the $\hbar \to 0$ limit, $\delta z_{QED} = \delta z_C$. The same result is demonstrated for a time-dependent but position-independent potential.

What light does this derivation cast on the time asymmetry of radiation reaction? QED is a time reversal invariant theory (see Atkinson 2006). So if an initial state $|I\rangle$ evolves to the (exact) final state $|F\rangle$ over $\Delta t$, then $^R|F\rangle$ evolves to $^R|I\rangle$ over the same $\Delta t$. In the above model calculation $|I\rangle = |i\rangle$, but the state $|f\rangle$ calculated from time dependent perturbation theory differs from the exact final state $|F\rangle$, and so $^R|f\rangle$ won't evolve to $^R|i\rangle$.[42] But to the extent that the approximation procedure outlined above is to be trusted, calculating the QED position shift from the exact final state $|F\rangle$ which $|f\rangle$ approximates should not change the conclusion that in the $\hbar \to 0$ limit $\delta z_{QED} = \delta z_C$. Thus, from the perspective of QED the observed asymmetry of radiation and radiation reaction is to be traced to the fact that, in the circumstances we find ourselves, it is overwhelmingly more "probable" that $|I\rangle$ $(= |i\rangle)$ type states will be realized than it is that $^R|F\rangle$ type states will be realized. But if circumstances were different, and $|I\rangle$ type and $^R|F\rangle$ type states are equally "probable," then the observed asymmetry would disappear. Scare quotes were used because the relevant sense of probability is not supplied by QED in particular or quantum field theory in general. The mystery of the asymmetry of radiation reaction is thus kicked upstairs to quantum statistical mechanics. Reducing one mystery to another does not count as a solution of the first, but progress has been made in the sense that, despite first appearances gained from classical electrodynamics, the time asymmetry of radiation and radiation reaction is shown not to be different in kind from other time asymmetries having a statistical origin.

As for the Ricci curvature asymmetry, I can only offer opinion and conjecture. As with radiation reaction, I would maintain that the curvature effect given by equation (34) is valid only to the extent that it emerges in the $\hbar \to 0$ limit of QED done on the background of a globally hyperbolic, conformally flat, but not Ricci flat spacetime. Linear quantum field theory on a curved, globally hyperbolic spacetime is well under-

---

[42] Suppose that $|f\rangle$ and $|F\rangle$ are close in the Hilbert space norm. Since the norm is preserved by the reversal operation $R$ and by a unitary transformation, it follows that the unitary time evolutes of $^R|f\rangle$ and $^R|R\rangle$ remain close in the Hilbert space norm.

stood, at least for stationary spacetimes (see Wald 1994).[43] Presumably QED can be generalized to this setting, and presumably this generalization is time reversal invariant. If the presumptions hold, then as with radiation reaction, the time asymmetry of the Ricci curvature effect (if valid) has to be due to the asymmetry of probabilities of realization of initial and reversed final states.[44]

There are uncomfortably many promissory notes left to be redeemed. But if the suggested line of analysis is on the right track, then the neo-Ritzianism can find no purchase in the observed asymmetry of radiation and radiation reaction. Einstein was correct: the asymmetry is "due to reasons of probability."

# 4. CONCLUSION

One overarching conclusion that emerges from the above discussion is that the siren song of neo-Ritzian posits to supplement classical relativistic electrodynamics should not be heeded. These posits are not needed to explain the classical EM asymmetries. Furthermore, in the setting of a pure particle theory of electrodynamics—the type of theory Ritz hankered after—a Ritzian "retardation condition" makes sense as a scientific hypothesis that speaks for itself, but in the setting of orthodox classical relativistic electrodynamics, such a condition requires philosophy-speak (cause, produce, contribute) to gain any traction. And such traction as is gained not only does not produce any genuine scientific explanation, but by offering soothing words in place of scientific theorizing, it retards the search for scientific understanding.

The task of tracing the origins of EM asymmetries would ideally start with a clear formulation of the asymmetries. But some of the formulations that are found in the literature are vitiated by muddles about retarded and advanced solutions/representations. Others have a distressingly vague and hedged character, requiring the use of qualifiers like 'most', 'many', 'typically', as well as 'approximately'. Additionally, some of the asymmetries that are formulated in terms of the quantities $_{ret}F^{\mu\nu}$, $_{in}F^{\mu\nu}$, $_{adv}F^{\mu\nu}$, and $_{out}F^{\mu\nu}$ can come or go as the domains of integration implicit in these quantities change. It is far from clear which of the asymmetries exhibiting these characteristics of vagueness and relativity to domains deserve to be promoted to the status of arrows of time.

A clean EM asymmetry worthy of promotion to an arrow of time may emerge when the domains of integration implicit in the quantities $_{ret}F^{\mu\nu}$, $_{in}F^{\mu\nu}$, $_{adv}F^{\mu\nu}_{out}$, and $_{out}F^{\mu\nu}$ are pushed to their limits and cosmological considerations are brought into

---

[43]  Intuitively, a stationary spacetime is one whose metric $g_{\mu\nu}$ is time independent. The precise mathematical statement is that there exists a timelike killing vector field $V^{\mu}$, which implies that $\nabla_{(\alpha}V_{\beta)} = 0$.

[44]  When the background spacetime is not stationary, even linear quantum field theory becomes problematic due to lack of a natural way to separate positive and negative frequencies and to define a vacuum state.

play. The large-scale structure of the spacetime, together with the distribution of the sources, may allow one but not another of the conditions $\overset{\infty}{\underset{in}{}} F^{\mu\nu} = 0$, $\overset{\infty}{\underset{out}{}} F^{\mu\nu} = 0$, or $\overset{\infty}{\underset{in}{}} F^{\mu\nu} + \overset{\infty}{\underset{out}{}} F^{\mu\nu} = 0$ on incoming and outgoing radiation. Additionally, the demand for a finite solution may, for example, uniquely single out the retarded as opposed to the advanced Kirchhoff representation. But although clean, the resultant EM arrow is clearly enslaved to the cosmological arrow.

The case of radiation reaction of a point charge appears to offer an EM asymmetry that is pervasive enough and unequivocal enough to be promoted to an arrow of time and that is distinctively electromagnetic in origin, being neither enslaved to the cosmological arrow nor due to the probability considerations that underlie the temporal asymmetries of collective phenomena of non-charged matter. However, when the investigation is carried into QED, the initial impression about the status of this arrow changes: the arrow of radiation reaction is of a piece with other arrows that derive from asymmetries of the probabilities of initial and reversed final states. It was conjectured that a similar conclusion will hold for the Ricci curvature asymmetry that arises for point charges moving in a conformally flat but not Ricci flat spacetime. But confirming this conjecture will involve difficult calculations in quantum field theory on curved spacetime.

I will mention a more general conjecture: any EM asymmetry that is clean and pervasive enough to merit promotion to an arrow of time is enslaved to either the cosmological arrow or the same source that grounds thermodynamic arrow (or a combination of both). But much more work would be needed before I would be willing to make this conjecture with any confidence.

## REFERENCES

Aichelberg, P. C. and Beig, R. (1977), 'Radiation Damping and the Expansion of the Universe', *Physical Review D* 15: 389–401.

Albert (2000), *Time and Chance* (Cambridge, MA: Harvard University Press).

Angelopoulos, A. et al. (1998), 'First Direct Observation of Time-Reversal Non-Invariance in the Neutral Kaon Decay System', *Physics Letters B* 444: 43–51.

Atkinson, D. (2006), 'Does Quantum Electrodynamics Have an Arrow of Time?', *Studies in the History and Philosophy of Modern Physics* 37: 528–541.

Boulware, D. G. (1980), 'Radiation from a Uniformly Accelerated Charge', *Annals of Physics* 124: 169–188.

Carroll, S. (2004), 'Why is the Universe Accelerating?' In W. L. Friedman (ed.), *Measuring and Modeling the Universe* (Cambridge: Cambridge University Press). astro-ph/0310342

Castagnino, M., Laura, L., and Lombardi, O. (2003), 'The Cosmological Origin of Time Asymmetry', *Classical and Quantum Gravity* 20: 369–391.

Davies, P. C. W. (1972), 'Is the Universe Transparent or Opaque?', *Journal of Physics A* 5: 1722–1737.

—— (1976), *The Physics of Time Asymmetry* (Berkeley, CA: University of California Press).

—— and Twamley, J. (1993), 'Time-Symmetric Cosmology and the Opacity of the Future Light Cone', *Classical and Quantum Gravity* 10: 931–945.

Dirac, P. A. M. (1938), 'Classical Theory of Radiating Electrons', *Proceedings of the Royal Society (London) A* 167: 148–169.

Earman, J. (2002), 'What Time Reversal Invariance Is and Why it Matters, *International Studies in Philosophy of Science* 16: 245–264.

—— (2006), 'The "Past Hypothesis" Not Even False', *Studies in the History and Philosophy of Modern Physics* 37: 339–430.

Einstein, A. (1909), 'Zum genenwärtigen Stand des Stralungsproblems', *Physikalische Zeitschrift* 10: 185–193. English translation in A. Beck and P. Havas, *The Collected Papers of Albert Einstein*, Vol. 2 (Princeton, NJ: Princeton University Press, 1989), 357–359.

Ellis, G. F. R. and Sciama, D. W. (1972), 'Global and Non-Global Problems in Cosmology', in L. O'Raifeartaigh (ed.), *General Relativity: Papers in Honor of J. L. Synge* (Oxford: Clarenden Press), 35–59.

—— and Rothman, T. (1993), 'Lost Horizons', *American Journal of Physics* 61: 883–893.

Fokker, A. D. (1929), 'Ein invarianter Variationssatz für die Bewegung mehrerer elektrisher Massenteichen', *Zeitschrift für Physik* 58: 386–396.

Frisch, M. (2000), '(Dis)Solving the Puzzle of the Arrow of Radiation', *British Journal for the Philosophy of Science* 51: 381–410.

—— (2005), *Inconsistency, Asymmetry, and Non-Locality: A Philosophical Investigation of Classical Electrodynamics* (Oxford: Oxford University Press).

—— (2006), 'A Tale of Two Arrows', *Studies in the History and Philosophy of Modern Physics* 37: 542–558.

Hawking, S. W. (1965), 'On the Hoyle-Narlikar Theory of Gravitation', *Proceedings of the Royal Society (London) A* 286: 313–319.

Heald, M. A. and Marion, E. J. B. (1995), *Classical Electromagnetic Radiation*, 3rd edn. (Fort Worth, TX: Sonders Publishing Co.).

Higuchi, A. (2002), 'Radiation Reaction in Quantum Field Theory', *Physical Review D* 66: 105004-1-12.

—— and Martin, G. D. R. (2004), 'Lorentz-Dirac Force from QED for Linear Acceleration', *Physical Review D* 70: 081701-1-5.

—— (2005), 'Classical and Quantum Radiation Reaction for Linear Acceleration', *Foundations of Physics* 35: 1149–1179.

Hobbs, J. M. (1968), 'Radiation Damping in Conformally Flat Universes', *Annals of Physics* 47: 166–172.

Jackson, J. D. (1998), *Classical Electrodynamics*, 3rd edn. (New York: John Wiley and Sons).

Johnson, P. R. and Hu, B. L. (2002), 'Stochastic Theory of Relativistic Particles Moving in a Quantum Field: Scalar Abraham-Lorentz-Dirac-Langevin Equation, Radiation Reaction, and Vacuum Fluctuations', *Physical Review D* 65: 065015-1-24.

Landau, L. and Lifshitz, L. (1975), *Classical Theory of Fields* (New York: Pergammon Press).

Malament, D. B. (2004), 'On the Time Reversal Invariance of Classical Electromagnetic Theory', *Foundations of Physics* 35: 295–315.

Monitz, E. J. and Sharp, D. H. (1977), 'Radiation Reaction in Nonrelativistic Quantum Electrodynamics', *Physical Review D* 10: 2850–2865.

North, J. (2003), 'Understanding the Time-Asymmetry of Radiation', *Philosophy of Science* 70: 1086–1097.

Partridge, R. B. (1973), 'Absorber Theory of Radiation and the Future of the Universe', *Nature* 244: 263–265.

Penrose, O. and Percival, (1962), 'The Direction of Time', *Proceedings of the Physical Society* 79: 605–616.

Penrose, R. (1964), 'Conformal Treatment of Infinity', in C. DeWitt and B. De Witt (eds.), *Relativity, Groups and Topology* (New York: Gordon and Breach), 563–584.

Poisson, E. (1999), 'An Introduction to the Lorentz-Dirac Equation', gr-qc/9912045.

Price, H. (1996), *Time's Arrow and Archimedes' Point: New Directions for the Physics of Time* (Oxford: Oxford University Press).

—— (2006), 'Recent Work on the Arrow of Radiation', *Studies in the History and Philosophy of Modern Physics* 37: 498–527.

Quinn, T. C. and Wald, R. M. (1997), 'Axiomatic Approach to Electromagnetic and Gravitational Radiation Reaction of Particles in Curved Spacetime', *Physical Review D* 56: 3381–3394.

Ritz, W. (1908), 'Über die Grundlagen der Elektrodynamik und Theorie der schwarzen Strahlung', *Physikalische Zeitschrift* 9: 903–907.

—— (1909), 'Zum genenwärtigen Stand des Stralungsproblems', *Physikalische Zeitschrift* 10: 224–225.

—— and Einstein, A. (1909), 'Zum genenwärtigen Stand des Stralungsproblems', *Physikalische Zeitschrift* **10**: 323–324. English translation in A. Beck and P. Havas, *The Collected Papers of Albert Einstein*, Vol. 2 (Princeton, NJ: Princeton University Press, 1989), 376.

Rohrlich, F. (1990), *Classical Charged Particles*, 2nd edn. (Redwood City, CA: Addison-Wesley).

—— (1998), 'The Arrow of Time in the Equations of Motion', *Foundations of Physics* 28: 1045–1056.

—— (1999), 'Classical Self-Force', *Physical Review D* 60: 084017-1-5.

—— (2000), 'Causality and the Arrow of Classical Time', *Studies in the History and Philosophy of Modern Physics* 31: 1–13.

—— (2001), 'The Correct Equation of Motion of a Classical Point Charge', *Physics Letters A* 283: 276–278.

—— (2005), 'Time Reversal Invariance and the Arrow of Time in Classical Electrodynamics', *Physical Review E* 72: 057601-1-3.

Rovelli, C. (2004), 'Comment on: "Causality and the Arrow of Classical Time"', *Studies in the History and Philosophy of Modern Physics* 35: 397–405.

Savitt, S. (ed.) (1995), *Time's Arrow Today: Recent Physical and Philosophical Work on the Direction of Time* (Cambridge: Cambridge University Press).

—— (ed.) (2006), Special issue: 'The Arrows of Time', *Studies in History and Philosophy of Modern Physics* 37 (no. 3).

Sciama, D. W. (1963), 'Retarded Potentials and the Expansion of the Universe', *Proceedings of the Royal Society A* 273: 484–495.

Sklar, L. (1993), *Physics and Chance: Philosophical Issues in the Foundations of Statistical Mechanics* (Cambridge: Cambridge University Press).

Spohn, H. (2000), 'The Critical Manifold of the Lorentz-Dirac Equation', *Europhysics Letters* 50: 289–292.

—— (2004), *Dynamics of Charged Particles and their Radiation Field* (Cambridge: Cambridge University Press).

Teitelboim, C. (1970), 'Splitting of the Maxwell Tensor: Radiation Reaction Without Advanced Fields', *Physical Review D* 1: 1572–1582.

—— (1971), 'Splitting of the Maxwell Tensor. II. Sources', *Physical Review D* 3: 297–298.

—— Villarroel, D. and van Weert, Ch. G. (1980), 'Classical Electrodynamics of Retarded Fields and Point Particles', *Revisita del Nuovo Cimento* 3 (no. 9): 1–64.

Unruh, W. G. (1976), 'Self-Force on Charged Particles', *Proceedings of the Royal Society (London) A* 348: 447–465.

van Dam, H. and Wigner, E. P. (1966), 'Instantaneous and Asymptotic Conservation Laws for Classical Relativistic Mechanics of Interacting Point Particles', *Physical Review* 142: 838–843.

Wald, R. M. (1984), *General Relativity* (Chicago, IL: University of Chicago Press).

—— (1994), *Quantum Field Theory on Curved Spacetime and Black Hole Thermodynamics* (Chicago, IL: University of Chicago Press).

Weisberg, E. (2004), 'Can Conditioning on the "Past Hypothesis" Militate Against the Reversibility Objection?, *Philosophy of Science* 71: 489–504.

Wheeler, J. A. and Feynman, R. P. (1945), 'Interaction with the Absorber as the Mechanism of Radiation', *Reviews of Modern Physics* 17: 157–181.

—— —— (1949), 'Classical Electrodynamics in Terms of Direct Particle Interaction', *Reviews of Modern Physics* 21: 425–433.

Yaghjian, A. D. (1992), *Relativistic Dynamics of a Charged Sphere* (Berlin: Springer Verlag).

Zeh, D. (1989), *The Physical Basis of the Direction of Time*, 1st edn. (New York: Springer-Verlag).

—— (1999), 'Note on the Time Reversal Symmetry of Equations of Motion', *Foundations of Physics Letters* 12: 193–196.

—— (2001), *The Physical Basis of the Direction of Time*, 4th edn. (New York: Springer-Verlag).

CHAPTER 17

························································································

# TIME, TOPOLOGY,
# AND THE TWIN PARADOX

························································································

JEAN-PIERRE LUMINET

## SUMMARY

························································································

THE twin paradox is the best known thought experiment associated with Einstein's theory of relativity. An astronaut who makes a journey into space in a high-speed rocket will return home to find he has aged less than his twin who stayed on Earth. This result appears puzzling, since the situation seems symmetrical, as the homebody twin can be considered to have done the travelling with respect to the traveller. Hence it is called a 'paradox'. In fact, there is no contradiction, and the apparent paradox has a simple resolution in Special Relativity with infinite flat space. In General Relativity (dealing with gravitational fields and curved space-time), or in a compact space such as the hypersphere or a multiply connected finite space, the paradox is more complicated, but its resolution provides new insights about the structure of space-time and the limitations of the equivalence between inertial reference frames.

## 1. PLAY TIME

························································································

The principle of relativity ensures equivalence between inertial reference frames in which the equations of mechanics hold. Inertial frames are spatial coordinate systems together with some means of measuring time so that observers attached to them can distinguish uniform motions from accelerated motions.

   In classical mechanics as well as in Special Relativity, such privileged frames are those moving at a constant velocity, that is, in uniform rectilinear motion. Their rest states are equivalent, as every passenger in a train that is slowly starting relative to a neighbouring one at a train station can check. Without feeling any acceleration, the passenger cannot decide which train is moving with respect to the other one.

Classical mechanics makes the assumption that time flows at the same rate in all inertial reference frames. As a consequence, the mathematical transformations between inertial systems are just the usual Galilean formulae, which preserve space intervals $\Delta d$ and time intervals $\Delta t$. As invariant quantities, lengths and durations are independent of the positions and speeds of the reference frames in which they are measured. This corresponds to Newton's concepts of absolute space and absolute time.

Special Relativity makes a different assumption, namely that the speed of light in vacuum, $c$, remains the same for every observer, whatever his state of motion. This assumption was confirmed by the famous Michelson and Morley experiments (1887). The mathematical transformations between inertial systems are given by the Lorentz formulae, which allow us to reformulate the laws of mechanics and electromagnetism in a coherent way. Their most immediate consequence is that space and time are not absolute but 'elastic', in the sense that space intervals $\Delta d$ and time intervals $\Delta t$ now depend on the relative velocity between the observer and the system he measures.

However, the Lorentz transformations preserve the space-time interval, an algebraic combination of space and time intervals given by $\Delta s = \sqrt{c^2 \Delta t^2 - \Delta d^2}$ . According to the Lorentz formulae, the clock of a system in motion appears to tick more slowly than that of a system at rest, while distances in the moving system appear to be shortened. In effect, for an observer at rest with his clock $\Delta d = 0$, $\Delta s$ measures his so-called *proper time*. But if an observer moves relative to a clock, he will measure a time interval $\Delta t$ longer and a space interval $\Delta d$ shorter than the observer at rest. These rather counter-intuitive effects are called *apparent time dilation* (moving clocks tick more slowly) and *length contraction* (moving objects appear shortened in the direction of motion).

The more the relative velocity $v$ increases, the more the clock appears to slow down. Due to the expression of the coefficient of time dilation, $\sqrt{(1 - v^2/c^2)}$, the phenomenon is noticeable only at velocities approaching that of light. At the extreme limit, for a clock carried by a photon, which is a particle of light, time does not flow at all. The photon never ages, because its proper time $\Delta s$ is always zero.

Special Relativity is one of the best verified theories in physics. The reality of apparent time dilation has been tested experimentally using elementary particles that can be accelerated to velocities very close to the speed of light. For instance, muons, unstable particles which disintegrate after 1.5 microseconds of proper time, were accelerated until they reached $0.9994 c$. Their apparent lifetime (as measured in the rest frame of the laboratory) extended to 44 microseconds, which is thirty times longer than their real lifetime, in complete agreement with special relativistic calculations.

In order to avoid a misunderstanding, it is very important to make a distinction between the apparent time and the proper time. To illustrate the difference, let us compare two identical clocks consisting of light impulses travelling between two parallel mirrors. One of the clocks is in uniform rectilinear motion at velocity $v$ relative to the other one (in a direction parallel to the line joining the mirrors). At moment $t = 0$, both clocks are at the same location, and the light impulses are sent to each of them. At time $t$, the observer of the clock at rest checks that the beam of light reaches the second mirror—this moment corresponds to the first tick of the clock. The second clock moved during this time, and the beam of light has yet to reach its second mirror. Thus it seems to run more slowly, because its ticks are not synchronized with those of the clock at rest. But, as the notion of uniform motion is a relative one, *these effects are totally symmetric.* The observer bound to the clock in motion considers himself at rest, and he sees the other system moving. He thus sees the clock of the other system slowing down.

In other words, the observers make observations which apparently contradict one another, each seeing the other clock beating more slowly than his own. However, their points of view are not incompatible, because this *apparent* dilation of time is an effect bound to observation, and the Lorentz transformation formulae ensure the coherence of both measurements. Indeed, in the case of uniform rectilinear motion, the *proper* times of both clocks remain perfectly identical; they 'age' at the same rate.

## 2. THE TWIN PARADOX IN SPECIAL RELATIVITY

Now consider two clocks brought together in the same inertial reference frame and synchronized. What happens if one clock moves away in a spaceship and then returns? In his seminal paper on Special Relativity, Albert Einstein (see Einstein 1905) predicted that the clock that undergoes the journey would be found to lag behind the clock that stays put. Here the time delay involves the proper time, not the apparent one. To emphasize this, in 1911 Einstein restated and elaborated on this result with the following statement: 'If we placed a living organism in a box... one could arrange that the organism, after any arbitrary lengthy flight, could be returned to its original spot in a scarcely altered condition, while corresponding organisms which had remained in their original positions had already long since given way to new generations. For the moving organism the lengthy time of the journey was a mere instant, provided the motion took place with approximately the speed of light' (in Resnick and Halliday 1992).

The same year, the French physicist Paul Langevin (see Langevin 1911) picturesquely formulated the problem using the example of twins aging differently according to their respective space-time trajectories (called *worldlines*). One twin remains on Earth while the other undertakes a long space journey to a distant planet, in a spaceship moving at almost the speed of light, then turns around and returns home to Earth. There the astronaut discovers that he is younger than his sibling. That is to say, if the twins had been carrying the clocks mentioned above, the traveller's clock would be found to lag behind the clock that stayed with the homebody brother, meaning that less time has elapsed for the traveller than for the homebody. This result indeed involves *proper times* as experienced by each twin, since biological clocks are affected in the same way as atomic clocks. The twins' ages can also be measured in terms of the number of their heartbeats. The traveller is really younger than his homebody twin when he returns home.

However, due to apparent time dilation, each twin believes the other's clock runs more slowly, and so the paradox arises that each believes the other should be younger at their reunion. In other words, the symmetry of their points of view is broken. Is this paradoxical?

In scientific usage, a paradox refers to results which are contradictory, that is, logically impossible. But the twin paradox is not a logical contradiction, and neither Einstein nor Langevin considered such a result to be literally paradoxical. Einstein only called it 'peculiar', while Langevin explained the different aging rates as follows: 'Only the traveller has undergone an acceleration that changed the direction of his velocity.' He showed that, of all the worldlines joining two events (in this example the spaceship's departure and return to Earth), the one that is not accelerated takes the longest proper time.

The twin paradox, also called the Langevin effect, underlines a limitation of the principle of relativity: points of view are symmetrical only for inertial reference systems, that is, those that aren't undergoing any acceleration. In the twin experiment, the Earth and the spaceship are not in a symmetrical relationship: the ship has a 'turnaround' in which it undergoes non-inertial motion, while the Earth has no such turnaround. Since there is no symmetry, Special Relativity is not contradicted by the realization that the twin who left Earth is younger than his sibling at the time of their reunion. The subject has been widely discussed for pedagogical purposes (see e.g. Taylor and Wheeler 1992).

# 3. AN EXAMPLE WITH NUMBERS

Let us call the twins Homebody and Traveller. At time $t = 0$ they synchronize their clocks in the Earth's inertial reference frame. Then Homebody stays on Earth whereas Traveller leaves towards a star E situated 10 light years away, travelling at $v = 0.9\,c$, that is 270 000 km/s. Next, he returns to Earth with speed $-v$. For convenience, the ship is assumed to have instaneous accelerations, so it immediately attains its full speed upon departure, turnaround, and arrival.

What would each twin observe about the other during the trip? The (x–t) space-time diagrams below (Figs. 17.1–17.5) allow us to solve the problem without any numerical calculation. We can choose the light-year as the unit of distance and the year as the unit of time. Then the paths of light rays are lines tilted at 45° (the dotted lines). They carry the images of each twin and his age-clock to the other twin. The vertical thick line is Homebody's path through space-time, and Traveller's trajectory (thin line) is necessarily tilted by less than 45° with respect to the vertical. Each twin transmits light signals at equal intervals according to his own clock, but according to the clock of the twin receiving the signals they are not being received at equal intervals.

In this example the coefficient of time dilation is $\sqrt{(1 - v^2/c^2)} = 0.436$, that is, when Homebody reads '1 second' on his clock, he reads '0.436 second' on Traveller's clock which is moving away from him at $0.9\,c$, and vice versa. (See Figs 17.1, 17.2, 17.3, 17.4 and 17.5.)

Both aspects of the paradox are solved in an obvious way by these space-time diagrams.

1) Why is the global situation not symmetric?

During the outward journey, the situations are perfectly symmetric because the inertial frames of both Traveller and Homebody are in uniform motion with relative speed $v$ (Figs 17.1 and 17.2). Also, during the return journey, the situations are perfectly symmetric because the inertial frames of both Traveller and Homebody are in uniform motion with relative speed $-v$ (Figs 17.3 and 17.4). But if one considers the complete journey (Fig. 17.5), the trajectories are physically asymmetric because at E, Traveller— having modified his speed, that is, having undergone an acceleration—changes his inertial frame.

2) Why is Traveller's proper time shorter than that of Homebody?

One can consider that it is because of the accelerations and the decelerations that Traveller has to undergo to leave Homebody at O, turn back at E and rejoin Homebody

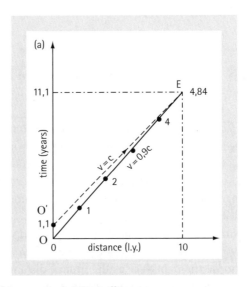

FIGURE 17.1 **Outward journey: what Traveller measures**
In principle, 11.1 years are required to cover 10 light years at the speed of 0.9 $c$. However, according to his clock, Traveller reaches E after only 4.84 years (11.1 x 0.436). Besides, once he arrives at E Traveller sees the Earth such as it was at O', which is 1.1 years after the departure according to Homebody's clock.
**Conclusion: Traveller sees Homebody's clock beating 4.36 times more slowly.**

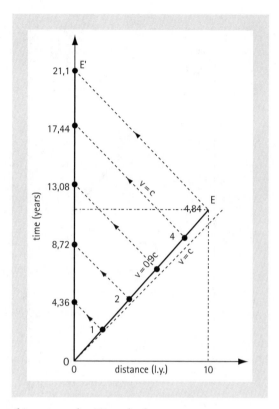

FIGURE 17.2  **Outward journey: what Homebody measures**
Homebody knows that, after 11.1 years, Traveller should arrive at E. However, the light rays sent from E take 10 years to reach him at E'. Homebody thus sees Traveller arriving at E only after 21.1 years.
**Conclusion: Homebody sees Traveller's clock beating 4.36 times more slowly.**

at R. Let us note, however, that the phases of acceleration at O and deceleration at R can be suppressed if one assumes that the trajectories of Traveller and Homebody cross without either observer stopping, their clocks being compared during the crossing. Nevertheless the necessary change of direction at E remains, translated as the acceleration of Traveller.

According to a more geometrical point of view, it is the particular structure of the relativistic space-time that is responsible for the difference of proper times. Let us see why. In classical mechanics and ordinary space, the Pythagorean theorem indicates that $DZ^2 = DX^2 + DY^2$, as in any right-angled triangle, which implies that $DZ < DX + DY$. But Special Relativity requires the introduction of a four-dimensional geometrical structure, the *Poincaré-Minkowski space-time*, which couples space and time through the speed of light. The Pythagorean theorem becomes $DS^2 = DX^2 - c^2DT^2$, and a straightforward algebraic manipulation allows us to deduce that DS is always longer than DX + cDT. As said previously, DS measures the proper time. *In Poincaré-Minkowski geometry, the worldlines of inertially moving bodies maximize the proper*

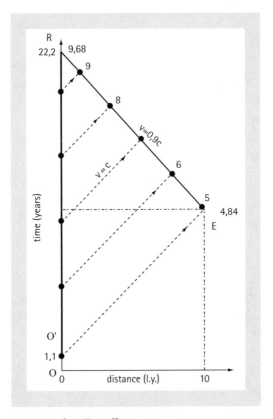

FIGURE 17.3  **Return journey: what Traveller measures**
Traveller returns to Earth at R 4.84 years after arriving at E. But during this time, he observes
21.1 years elapse on Earth.
**Conclusion: Traveller sees Homebody's clock beating 4.36 times faster.**

*time elapsed between two events.* One can also see that the proper time vanishes
for DX = cDT, in other words for $v=$ DX/DT$=c$. As already pointed out, a photon
never ages.

## 4. THE TWIN PARADOX IN GENERAL RELATIVITY

General relativity deals with more realistic situations, including progressive accelerations, gravitational fields, and curved space-time. The inertial frames are now systems
in free-fall, and the equations which allow us to pass between inertial systems are no
longer the Lorentz transformations, but the Poincaré transformations. The complete
treatment of the problem of the twins within this new framework was first described
by Einstein (1918); see also Perrin (1970). As in Special Relativity, the situation is

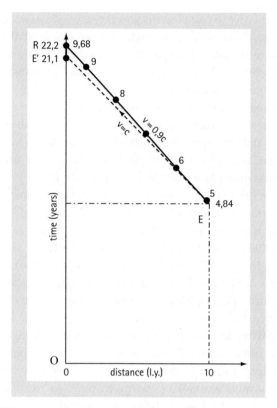

**FIGURE 17.4 Return journey: what Homebody measures**
Homebody sees the entire return journey of Traveller take place in 1.1 years, and meets him at R 22.2 years after the initial departure.
Conclusion: Homebody sees Traveller's clock beating 4.36 times faster.

**FIGURE 17.5 Complete journey**
When Homebody and Traveller meet each other at R, Homebody's clock has measured 22.2 years and Traveller's clock has measured 9.68 years.

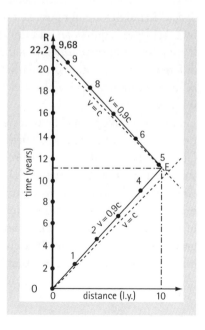

never symmetric. In order to complete his journey, Traveller necessarily experiences a finite and varying acceleration; thus he switches from one inertial reference frame to another, and his state of motion is not equivalent to that of Homebody. The rule stays the same: the twin who travelled through several inertial frames will always have aged less than the twin who stayed in the same inertial frame.

Assume for instance that the spaceship has a constant acceleration with respect to its instantaneous inertial reference frame, equal to the acceleration due to gravity at the Earth's surface and thus quite comfortable for Traveller. The spaceship velocity will rapidly increase and approach the speed of light without ever reaching it. On board, time will pass much more slowly than on Earth. In 2.5 years as measured by his own clock, Traveller will reach the closest star (Alpha Centauri) which is 4 light-years from Earth, and after about 4.5 years he will have travelled 40 light-years, while his homebody twin will have died of old age. The centre of the Galaxy will be reached in 10 years, but 15,000 years would have passed on Earth. In about thirty years of his proper lifetime, the Traveller would be able to cross once around the observable Universe, a distance of one hundred thousand million light years! It would be better therefore not to return to Earth, since the Sun would have extinguished long ago, after having burnt the planet to a cinder...

This shows in passing that, contrary to popular belief, although the theory of relativity prevents us from travelling faster than the velocity of light, it does facilitate the exploration of deep space. This fantastic journey is, however, impossible to realize because of the enormous amount of energy required to maintain the spaceship's acceleration. The best method would be to transform the material of the ship itself into propulsive energy. With perfectly efficient conversion, upon arrival at the centre of the Galaxy, only one billionth of the initial mass would remain. A mountain would have shrunk to the size of a mouse!

The full general relativistic calculations, although less straightforward than those in Special Relativity, do not pose any particular difficulty, but must take into account the fact that time acquires an additional elasticity: gravity also slows down clocks. Thus there is an additional *gravitational time dilation*, given by $(1 + \Phi/c^2)$, where $\Phi$ is the difference in gravitational potentials. For instance, a clock at rest on the first floor beats more slowly than a clock at rest on the second floor (although the difference is tiny).

Physicists have been able to design clocks precise enough to experimentally test the twin paradox in a gravitational field as weak as that of the Earth. In 1971, the US Naval Observatory placed extremely precise cesium clocks aboard two planes, one flying westward and the other eastward. Upon their return, the flying clocks were compared with a third (i.e. initially synchronized) clock kept at rest in a lab on Earth. In this experiment, two effects entered the game: a Special Relativistic effect due to the speed of the planes (about 1000 kph), and a General Relativistic effect due to the weaker gravity on board the planes. The clock which had travelled westward advanced by 273 billionths of a second, the one that had travelled eastward delayed by 59 billionths of a second—in perfect agreement with the fully relativistic calculations (see Hafele and Keating 1972).

Nevertheless, it is pointless to dream of extending one's lifetime by travelling. If a human being spent 60 years of his life on board a plane flying at a velocity of 1,000 kph, he would gain only 0.001 second over those who remained on the ground (and would probably lose several years of his life due to stress and sedentarity!).

# 5. PLAY SPACE

Is acceleration, which introduces an asymmetry between the reference frames, the only explanation of the twin paradox? The answer is no, as many authors have pointed out. See, for example, Peters (1983), who considers the example of non–accelerated twins in a space with a compact dimension (the closure being due to non-zero curvature or to topology). In such a case, the twins can meet again without either of them being accelerated and yet they age differently!

Before revisiting the question in such a framework, let us get some insight into the global properties of space. Topology is an extension of geometry that deals with the nature of space by investigating its overall features, such as its number of dimensions, finiteness or infiniteness, connectivity properties, or orientability, without introducing any measurement.

Of particular importance in topology are functions that are continuous. They can be thought of as those that stretch space without tearing it apart or gluing distinct parts together. The topological properties are just those that remain insensitive to such deformations. With the condition of not cutting, piercing, or gluing space, one can stretch it, crush it, or knead it in any way, and one will not change its topology, for example, whether it is finite or infinite, whether it has holes or not, the number of holes if it has any, and so on. For instance, it is easy to see that, although continuous deformations may move the holes in a surface, they can neither create nor destroy them.

All spaces that can be continuously deformed from one into another have the same topology. For a topologist, a ring and a coffee cup are one and the same object, characterized by a hole through which one can pass one's finger (although it is better not to pour coffee into a ring). On the other hand, a mug and a bowl, which may both serve for drinking, are radically different on the level of topology, since a bowl does not have a handle.

To speak only about three-dimensional spaces of Euclidean type (with zero curvature), there are eighteen different topologies. Apart from the usual, infinite Euclidean space, the seventeen others can be obtained by identifying various parts of ordinary space in different ways.

| Name | Fundamental domain and identifications | Shape | Closed | Orientable |
|---|---|---|---|---|
| Cylinder |  | 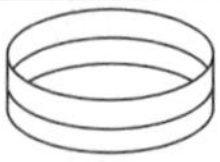 | NO | YES |
| Möbius band |  | 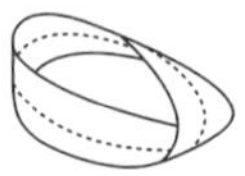 | NO | NO |
| torus | 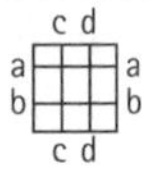 | 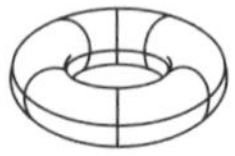 | YES | YES |
| Klein bottle | 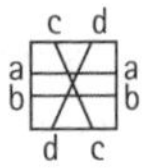 | 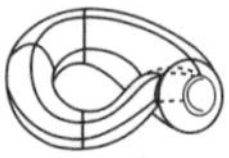 | YES | NO |

FIGURE 17.6 **The four multiply connected topologies of the two-dimensional Euclidean plane**
They are constructed from a rectangle or an infinite band (the fundamental domain) by identification of opposite edges according to allowable transformations. We indicate their overall shape, compactness, and orientability properties.

To make the description easier, it is useful to consider the more visualizable context of two-dimensional spaces (i.e. surfaces). Besides the usual infinite Euclidean plane, there are four other Euclidean surfaces (Figure 17.6). The cylinder is obtained by gluing together the opposite sides of an infinite strip with parallel edges, and Möbius band by twisting an edge through 180° before gluing the edges in the same way. The torus is obtained by gluing the opposite edges of a rectangle, and Klein's bottle by twisting one pair of edges before gluing.

All these surfaces have no intrinsic curvature—the sum of the angles of a triangle is always equal to 180°. They are only bent in a third dimension, which cannot be perceived by a two-dimensional being living on the plane. Such surfaces are said to be locally Euclidean.

The rectangle we start with is called the 'fundamental domain'. The geometrical transformations which identify the edge-to-edge points define the way objects move continuously within this space, leaving the rectangle by an edge immediately to reappear at the other edge (Figure 17.7).

One can visualize the metric properties of the space by duplicating the fundamental domain a large number of times. This generates the *universal covering space*, in which every point is repeated as often as the domain itself. One can draw in the universal covering space the various paths connecting a point to itself, called loops, either by going out of the fundamental domain to join a duplicate, in which case it is a loop which 'goes around' space, or by returning towards the original point in the fundamental domain, in which case it is a loop which can be continuously shrunk to a point (Figure 17.8).

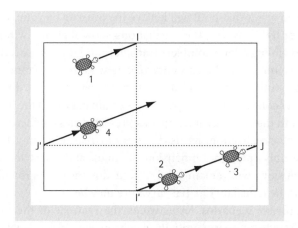

FIGURE 17.7 **Walking on a torus** As in those video games where characters who leave at an edge return at the opposite edge, the tortoise crosses the top of the square at $I$, reappears at the bottom at the equivalent point $I'$, continues to travel in a straight line, reaches the right-hand edge at $J$, reappears at $J'$, and so on. The torus is thus equivalent to a rectangle with the opposite edges 'glued together'.

FIGURE 17.8 **From multiply connected space to the universal covering space**
The fundamental domain of the torus is a rectangle. By repeatedly duplicating the rectangle, one generates the universal covering space—here the Euclidean plane $R^2$. The paths 2, 3, and 4 all connect the point 1 to itself. Loop 2 can be shrunk to a point; loops 3 and 4 cannot because they go around the space.

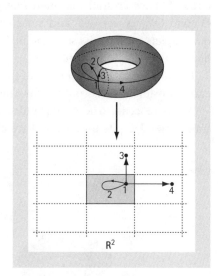

# 6. THE TWIN PARADOX IN FINITE SPACE

Now we can revisit the twin paradox whatever the global shape of space may be (Barrow and Levin 2001; Uzan et al. 2002). The travelling twin can remain in an inertial frame for all time as he travels around a compact dimension of space, never stopping or turning. Since both twins are inertial, both should see the other suffer a time dilation.

The paradox again arises that both will believe the other to be younger when the twin in the rocket flies by. However, the calculations show that the twin in the rocket is younger than his sibling after a complete transit around the compact dimension.

The resolution hinges on the existence of a new kind of asymmetry between the space-time paths joining two events, an asymmetry which is not due to acceleration but to the multiply connected topology. As we explain below, all the inertial frames are not equivalent, and the topology introduces a preferred class of inertial frames.

For the sake of visualization, let us develop our reasoning in a two-dimensional Euclidean space only (plus the dimension of time), and select the case of the flat torus. Our conclusions will remain valid in four-dimensional space-times, whatever the topology and the (constant) spatial curvature may be.

To span all possible scenarios, let us widen the example of the twins to a family of quadruplets (strictly speaking, initially synchronized clocks) labelled 1, 2, 3 and 4 (Figure 17.9). 1 stays at home, at point O, and his worldline can be identified with the time axis, so that he 'arrives' at O' on the space-time diagram; 2 leaves home at time t = 0, travels in a rocket, turns back and joins his sibling 1 at O'; 3 and 4 also leave at time t = 0, but travel in different directions along non-accelerated, straight worldlines issuing from O: 3 travels around the small axis, while 4 travels along a circumference around the main line of the torus. After a while they reach points O'' and O''' respectively, and since space is closed and multiply connected, all the quadruplets meet at the same point O'. Now, one wants to compare the ages of the quadruplets when they meet.

The motion of 2 corresponds to the standard paradox. Since he followed an accelerated worldline, he is younger than his sibling 1.

But there seems to be a real paradox with 1, 3, and 4, who all followed strictly inertial trajectories. Despite this, 3 and 4 are also younger than 1. In fact, the homebody 1 is

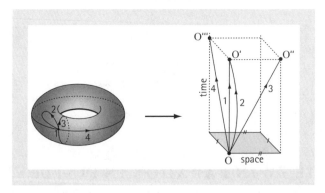

FIGURE 17.9 **From space to space-time**
On the right plot, quadruplets in a space-time with toroidal spatial sections leave O at the same time. While 1 remains at home, 2 goes away and then comes back to meet 1 at O' (corresponding to the standard case), 3 goes around the universe in a given direction from O to O'', and 4 goes around the universe along another direction from O to O'''. On the left plot, we depict the spatial projections of their trajectories on the torus. The space-time events O, O', O'', O''' are projected onto the same base point 1.

always older than any traveller, because his state of motion is not symmetrical with respect to those of non–accelerated travellers.

What kind of asymmetry is to be considered here? The only explanation lies in a global breakdown of symmetry due to the multiply connected topology. Let us investigate the case more closely. If one draws closed curves (i.e. loops) on a given surface, there are two possibilities. First, the loop can be tightened and continuously reduced to a point without encountering any obstacle. This is the case for all loops in the Euclidean plane or on the sphere, for instance, and such surfaces are called simply connected. Second, the loop cannot be tightened because it goes around a 'hole', as in the case of the cylinder or the torus. Such surfaces are said to have a *multiply connected* topology (multiple connectivity appears as soon as one performs gluings, or identifications of points, in a simply connected space).

Two loops are *homotopic* if they can be continuously deformed into one another. Homotopy allows us to define classes of topologically equivalent loops. In our example, the trajectories of brothers 1 and 2 are homotopic to {0}, because they can both be deformed to a point. However, they are not symmetrical because only 1 stays in an inertial frame. Here the asymmetry is due to acceleration. One can show that among all the homotopic curves from O to O′, only one corresponds to an inertial observer, and it is he who will age most, as expected in the standard twin paradox.

Now, 3 and 4 travel once around the handle and once around the hole of the torus respectively. From a topological point of view, their paths are not homotopic; they can be characterized by a so–called winding index. In a cylinder, the winding index is just an integer that counts the number of times a loop goes around the surface. In the case of a torus, the winding index is a couple $(m, n)$ of integers where $m$ and $n$ respectively count the numbers of times the loop goes around the hole and the handle. In our example, 1 and 2 have the same winding index $(0, 0)$, whereas 3 and 4 have winding indices of $(0, 1)$ and $(1, 0)$ respectively. The winding index is a topological invariant for each traveller: neither change of coordinates nor of reference frame can change its value.

To summarize, we have the following two situations:

1. Two brothers have the same winding index (1 and 2 in our example), because their loops belong to the same homotopy class. Nevertheless only one (twin 1) can travel from the first meeting point to the second without changing inertial frame. The situations relative to 1 and 2 are not symmetrical due to local acceleration, and 1 is older than 2. Quite generally, between two twins of the same homotopy class, the oldest one will always be the one who does not undergo any acceleration.

2. Several brothers (1, 3, and 4 in our example) can travel from the first meeting point to the second one at constant speed, but travel along paths with different winding indices. Their situations are not symmetrical because their loops belong to different homotopy classes: 1 is older than both 3 and 4 because his path has a zero winding index.

For observers to have the same proper times it is not sufficient that their movements are equivalent in terms of acceleration, their worldlines should also be equivalent in terms of homotopy class. Among all the inertial travellers, the oldest sibling will always be the one whose trajectory is of homotopy class {0}. The spatial topology thus imposes privileged frames among the class of all inertial frames, and even if the principle of relativity remains valid locally, it is no longer valid at the global scale. This is a sign that the theory of relativity is not a global theory of space-time.

This generalizes previous works, for example Brans and Stewart (1973), Low (1990), Dray (1990), by adding topological considerations that hold no matter what the shape of space is.

# 7. THE COMPLETE SOLUTION

In order to exhaustively solve the twin paradox in a multiply connected space, one would like not only to separately compare the ages of the travellers to the age of the homebody, but also to compare the ages of the various travellers when they meet each other. It is clear that only having knowledge of the winding index or the homotopy class of their loops does not allow us, in general, to compare their various proper time lapses. The only exception is that of the cylinder, where a larger winding number always corresponds to a shorter proper time lapse. But for a torus of unequal lengths, for instance when the diameter of the hole is much larger than the diameter of the handle, a traveller may go around the handle many times with a winding index (0, n), and yet be older than the traveller who goes around the hole only once with a winding index (1, 0). The situation is still more striking with a double torus, a hyperbolic surface rather than a Euclidean one (see e.g. Lachièze-Rey and Luminet 1995). The winding indices become quadruples of integers and, as in the case of the simple torus, they cannot be compared to answer the question about the ages of the travellers. As we shall now see, this problem can only be solved by acquiring additional metric information.

The torus is built from a rectangle by gluing together its opposite sides. If one repeatedly duplicates this rectangle so as to cover the plane, one generates the universal covering space, which is infinite in all directions. It is a fictitious space that represents space as it appears to an observer located at O. All points O are, however, identical. In this representation, the trajectory of 2 appears as a loop which returns to the inital point O, without passing through one of its duplicates, whereas the trajectories of 3 and 4 are straight lines which connect the point O to a duplicate with winding numbers (1, 0) and (0, 1) respectively. There are many ways to describe a loop in a closed space, and one could consider trajectories 5 and 6 with winding numbers (1, 1) and (2, 1), for example.

As mentioned above, the homotopy classes only tell us which twin is aging the fastest: the one who follows a straight loop homotopic to {0}. They do not provide a ranking of the ages (i.e. proper time lengths) along all straight loops. To do this,

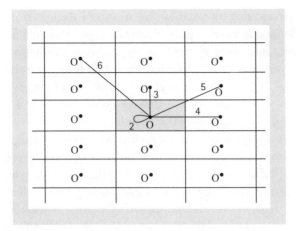

FIGURE 17.10  **Straight paths in the universal covering space of a (2 + 1)–space-time with flat, torus-like spatial sections**
Path 2 is an accelerated, curved loop with winding index (0, 0). Paths 3, 4, 5, and 6 are straight loops with respective winding indices (0, 1), (1, 0), (1, 1) and (1, 2), allowing the travellers to leave and return to the homebody at O without accelerating. The inertial worldlines are clearly not equivalent: the longer the spatial length in the universal covering space, the shorter the proper time traversed in space-time.

some additional information is necessary, such as the various identification lengths. Indeed there exists a simple criterion that works in all cases: a shorter spatial length in the universal covering space will always correspond to a longer proper time. To fully solve the question, it is therefore sufficient to draw the universal covering space as tessellated by duplicates of the fundamental domain, and to measure the lengths of the various straight paths joining the position of sibling 1 in the fundamental domain to his ghost positions in the adjacent domains (Figure 17.10). As usual in topology, all reasoning involving metrical measurements can be solved in the simply–connected universal covering space.

## 8. The Twin Paradox and Broken Symmetry Groups

With the homotopy class, we have found a topological invariant attached to each twin's worldline that accounts for the asymmetry between their various inertial reference frames. Why is this? In Special Relativity theory, two reference frames are equivalent if there is a Lorentz transformation from one to the other. The set of all Lorentz transformations is called the Poincaré group—a ten-dimensional group that combines translations and homogeneous Lorentz transformations called 'boosts'. The loss of equivalence between inertial frames is due to the fact that a multiply connected spatial topology globally breaks the Poincaré group.

The preceding reasoning involved Euclidean spatial sections of space-time. In the framework of General Relativity, general solutions of Einstein's field equations are curved space-times admitting no particular symmetry. However, all known exact solutions admit symmetry groups (although less rich than the Poincaré group). For instance, the usual 'big bang' cosmological models—described by the Friedmann–Lemaître solutions—are assumed to be globally homogeneous and isotropic. From a geometrical point of view, this means that spacelike slices have constant curvature and that space is spherically symmetric about each point. In the language of group theory, the space-time is invariant under a six-dimensional isometry group. The universal covering spaces of constant curvature are either the usual Euclidean space $R^3$, the hypersphere $S^3$, or the hyperbolic space $H^3$, depending on whether the curvature is zero, positive, or negative. Any identification of points in these simply-connected spaces via a group of continuous transformations lowers the dimension of their isometry group; it preserves the three–dimensional homogeneity group (spacelike slices still have constant curvature), but it globally breaks the isotropy group (at a given point there are a discrete set of preferred directions along which the universe does not look the same).

Thus in Friedmann–Lemaître universes, (i) the expansion of the universe and (ii) the existence of a multiply connected topology for the constant time hypersurfaces both break the Poincaré invariance and single out the same 'privileged' inertial observer who will age more quickly than any other twin—the one comoving with the cosmic fluid—although aging more quickly than all his travelling brothers may not be a real privilege!

## References

Barrow, J. D., and Levin, J. (2001), *The Twin Paradox in Compact Spaces*, Physical Review A63: 044104.

Brans, C. H., and Stewart, D. R. (1973), *Unaccelerated Returning Twin Paradox in Flat Spacetime* , Physical Review D8: 1662–1666.

Dray, T. (1990), 'The Twin Paradox Revisited', *American Journal of Physics* 58 (9): 822–825.

Einstein, A. (1905), *Zur Elektrodynamik bewegter Körper*, Annalen der Physik (Leipzig) 17: 891–921 (English translation: 'On the Electrodynamics of Moving Bodies').

——(1918), *Dialog über Einwände gegen die Relativitätstheorie*, Die Naturwissenschaften 48: 697–702 (English translation: 'Dialog about Objections against the Theory of Relativity').

Hafele, J. C., and Keating, R. E. (1972), 'Around-the-World Atomic Clocks: Observed Relativistic Time Gains', *Science*, 177: 168–170.

Lachièze-Rey, M. and Luminet, J.-P. (1995), *Cosmic Topology*, Physics Reports 254: 135–214.

Langevin, P. (1911), *L'évolution de l'espace et du temps*, Scientia X: 31–54.

Low, R. J. (1990), *An Acceleration Free Version of the Clock Paradox* , European Journal of Physics 11: 25–27.

Michelson, A. A., and Morley, E. W. (1887), 'On the Relative Motion of the Earth and the Luminiferous Aether', *Philosophical Magazine* 24: 449–463.

Perrin, R. (1970), ' Twin Paradox: 'A Complete Treatment from the Point of View of Each Twin', *American Journal of Physics* 44: 317–319.

Peters, A. C. (1983), 'Periodic Boundary Conditions in Special Relativity', *American Journal of Physics* 51: 791–795.

Resnick, R. and Halliday, D. (1992), *Basic Concepts in Relativity* (New York: Macmillan).

Taylor, E. F. and Wheeler, J. A. (1992), *Spacetime Physics*, Second Edition (Freeman).

Uzan, J.-P., Luminet, J.-P., Lehoucq, R., and Peter, P. (2002), 'Twin Paradox and Space Topology', *European Journal of Physics*, 23: 277–284.

CHAPTER 18

...................................................................................

# TIME IN THE SPECIAL THEORY OF RELATIVITY

...................................................................................

STEVEN SAVITT

## 1. INTRODUCTION

...................................................................................

PRIOR to the revolutions in physics in the early decades of the twentieth century, reflective individuals would likely have agreed with the following remarks about time by Roberto Torretti (1983: 220):

> In the Aristotelian philosophy that still shapes much of our common sense, the present time, called "the now" (*to nun*), separates and connects what has been (*to parelthon*) and what is yet to be (*to mellon*). Though the now, as the link of time (*sunekhei khronou*), the bridge and boundary between past and future, is always the same formally, materially it is ever different (*heteron kai heteron*), for the states and occurrences on either side of it are continually changing. Indeed, the so-called flow or flight of time is nothing but the ceaseless transit of events across the now, from the future to the past. If an event takes some time, then, while it happens, the now so to speak cuts through it, dividing that part of it which is already gone from that which is still to come. Two events which are thus cleaved by (materially) the same now are said to be *simultaneous*. Simultaneity, defined in this way, is evidently reflexive and symmetric, but it is not transitive. For a somewhat lengthy event— e.g. the French Revolution—can be simultaneous with two shorter ones—such as Faraday's birth in 1791 and Lavoisier's death in 1794—although the latter are not simultaneous with each other. However, if we conceive simultaneity as a relation between (idealized) *durationless* events, we automatically ensure that it is transitive and hence an equivalence. Thus conceived, simultaneity partitions the universe of such events into equivalence classes, and time's flow can be readily thought of as the march of those classes in well-aligned squadrons past the now. The succession of events on a given object—a clock—linearly orders the classes to which those events belong. The linearly ordered quotient of the universe of events by simultaneity is what we mean by physical time.

A philosopher might have added that the colorful language of classes "marching" past the now could have been replaced without loss by bland talk of (equivalence classes of) events happening successively.[1] The logician Kurt Gödel captured this idea in the pithy but not wholly unambiguous sentence: "The existence of an objective lapse of time, however, means (or, at least, is equivalent to that fact) that reality consists of an infinity of layers of 'now' which come into existence successively" (1949: 558).

It is not clear whether Gödel believed that these layers of 'now' come into existence and then remain in existence, being the foundation for a common-sense distinction between the past and future, or whether he thought that they come into existence then immediately cease to exist, as one might expect of events that are very brief or even durationless. This chapter will try to sidestep questions like this concerning 'reality' and 'existence' whenever possible, focusing instead on issues concerning the nature of simultaneity and the passage of time raised directly by the special theory of relativity.[2] We will begin with a brief presentation of the theory itself, before examining the philosophical issues it raises.

## 2. THE SPECIAL THEORY OF RELATIVITY

In his landmark 1905 paper Albert Einstein presented the special theory of relativity as the consequence of two postulates.[3] The first we shall call *the relativity principle*: "The laws by which the states of physical systems undergo change are not affected, whether these changes of state be referred to the one or the other of two systems of co-ordinates in uniform translatory motion" (1905: 41).

Each system of co-ordinates is four-dimensional, consisting of three spatial dimensions and one temporal dimension. We may label the co-ordinates in one system $(x, y, z, t)$ and those in another system $(x', y', z', t')$. Einstein specifies that these systems are to be such that in them "the equations of Newtonian mechanics hold good" (1905: 38) at least to first approximation. We shall call such co-ordinate systems *frames of reference* or *inertial frames*.[4]

---

[1] A prescient philosopher might also have added a *metaphysically neutral* indexical account of 'now' (Markosian 2001: 622). According to such an account, 'now' picks out one's temporal location without committing one to a particular account of the nature of that location, just as 'here' picks out one's spatial location without committing one to a particular geometry or to a position in (say) the relationalist/substantivalist debate. I am not, by the way, suggesting that Torretti is unaware or opposed to such refinements. In the quoted paragraph he is simply summarizing the classical view without endorsing it.

[2] A reader who wishes to engage these questions is directed to Putnam (1967), Stein (1968, 1991), Clifton and Hogarth (1995), Callender and Weingard (2000), Saunders (2000), Dorato (2006), Savitt (2006), and references therein.

[3] Most popular introductions to the special theory develop the theory in this manner. Two of the most readable books of this kind are Mermin (1968, 2005).

[4] See Norton (1993, §6.3) for a discussion of the difference between coordinate systems and frames of reference and DiSalle (2002, 2008) for further information regarding inertial frames.

When there are two such reference frames under consideration, it is often helpful (though it is not necessary) to arrange them so that the origins of the two systems, O and O′ (that is, the points that have co-ordinate values (0,0,0,) in each system), coincide, that the corresponding pairs of axes $y$ and $y'$, $z$ and $z'$ are parallel, that the $x$ and $x'$ axes coincide, and that the second system is moving in the positive $x$-direction in respect to the first with some constant velocity $v$.[5] We can therefore consider the first system to be "stationary", though we could with equal justice (given the relativity principle) regard the second as "stationary" and the first as moving in the negative $x$'-direction with velocity $-v$. Designating one or other of the two systems as "stationary" is a matter of expository convenience. We can now state Einstein's second postulate, *the light principle*: "Any ray of light moves in the 'stationary' system of co-ordinates with the determined velocity c, whether the ray be emitted by a stationary or by a moving body" (1905: 41).

From these principles Einstein derived *the Lorentz transformations*,[6] mathematical rules expressing the relations of the co-ordinates in two such systems. If the "stationary" system has un-primed co-ordinates, the "moving" system primed co-ordinates, and the relative velocity is $v$, then

$$x' = \frac{x - vt}{\sqrt{1 - v^2 / c^2}},$$

$$y' = y,$$

$$z' = z,$$

$$t' = \frac{t - vx / c^2}{\sqrt{1 - v^2 / c^2}}.$$

For convenience, the factor $1 \Big/ \sqrt{1 - {v^2}/{c^2}}$ is frequently abbreviated by '$\gamma$'. Note that when $0 < \|v\| < c, \gamma > 1$.

Now consider two inertial frames in standard configuration, the system with unprimed co-ordinates being chosen to be "at rest", and the system with primed co-ordinates moving with some non-zero speed $v$ relative to it along the $x$-axis. We can choose a pair of distinct points or events in spacetime, say the origin (the co-ordinates of which in the "stationary" co-ordinate system are of course (0,0,0,0)) and an arbitrary point on the $x$-axis at time $t=0$ (the co-ordinates of which would then be $(x, 0, 0, 0)$ in that same system). Since these two events have the same fourth, or time, co-ordinate,

---

[5] Two co-ordinate systems so aligned are said to be *in standard configuration*.

[6] Purists will note that the transformations indicated below are only a subset of the full group of Lorentz transformations, even allowing for the convenience of aligning the two co-ordinate systems in standard configuration. The "Lorentz transformations" exhibited in the text reverse neither spatial nor temporal axes: they are *proper* and *orthochronous*. For a discussion from a modern perspective of what is necessary for the derivation of the Lorentz transformations see Friedman (1983; chapter 4, §2).

we can say that they are *simultaneous*—at least relative to the unprimed co-ordinate system.[7]

The co-ordinates of these two points or events in the "moving" system may be found by applying the Lorentz transformations to their unprimed co-ordinates. We see that the co-ordinates of the origin remain $(0,0,0,0)$ in the moving system, which should not be surprising given that the two co-ordinate systems were stipulated to be in standard configuration. The co-ordinates of $(x, 0, 0, 0)$ in the moving system, however, must be $\left(\gamma x, 0, 0, \gamma\left(-vx/c^2\right)\right)$, given the Lorentz transformations above. Since neither $v$ nor $x$ nor $\gamma$ is equal to 0, the fourth primed co-ordinate $t'$ is not equal to 0. That is, relative to the moving co-ordinate system the two points or events that were chosen to be simultaneous in the "stationary" co-ordinate system are *not* simultaneous, if sameness of fourth co-ordinate in a given co-ordinate system remains our indicator of simultaneity. This result is quite general and is a startling but uncontroversial feature of the special theory of relativity known as *the relativity of simultaneity*.

Three years after Einstein introduced the special theory, Hermann Minkowski (1908) re-presented the theory in a different manner, developing it as a kind of four-dimensional geometry.[8] Choose again the coincident origins O and O' for two inertial frames in standard configuration. Imagine a burst of light at O (or O'). In time $t$, a particular light-ray will travel spatial distance $ct$ to reach a point P that we will call $(x, y, z, t)$.[9] By the Pythagorean theorem the spatial distance of that point from the origin can be written as $\sqrt{x^2 + y^2 + z^2}$. We then have

$$x^2 + y^2 + z^2 = (ct)^2, \quad \text{or}$$

$$x^2 + y^2 + z^2 - (ct)^2 = 0.$$

But if we consider the point P in terms of the primed co-ordinates, $(x', y', z', t')$, the same reasoning (including the light principle) will convince us that

$$(x')^2 + (y')^2 + (z')^2 - (ct')^2 = 0$$

and therefore that

$$(x')^2 + (y')^2 + (z')^2 - (ct')^2 = x^2 + y^2 + z^2 - (ct)^2. \tag{1}$$

This last equation shows us that, while many quantities, like simultaneity, turn out rather surprisingly to be frame-dependent or relative to a chosen inertial frame, there is at least one quantity that is independent of, or invariant between, frames. This quantity is usually called the *spacetime interval*. In the example given, the spacetime interval

---

[7] In the debate concerning the conventionality of simultaneity, which we will discuss in section 3, this notion of simultaneity—identity of time coordinate in an inertial frame—is in older literature sometimes called *metrical simultaneity*. By way of contrast, two events are said to be *topologically simultaneous* iff they are spacelike separated. This latter concept will be explained below. By 'simultaneity' *tout court* I will mean metrical simultaneity, unless otherwise specified.

[8] Most books on the general theory of relativity introduce the special theory geometrically. The classic introductory special relativity text in this vein is Taylor and Wheeler (1963).

[9] Here and throughout we ignore any complications that might arise from quantum theory.

is an invariant quantity determined by two points, the origin and the point P. But any point in the spacetime may be chosen as origin of a co-ordinate system, so the spacetime interval is an invariant quantity characterizing a relation between any two points in spacetime. But what relation is it?

The set of points or events that a light-ray from O can reach form an expanding spherical shell about O. If we suppress one spatial dimension, say $z$, then the points that a light-ray from O can reach form a cone, which Minkowski called *the back cone* but which is now usually called *the future light cone*. All these points by the reasoning above have spacetime interval 0 from the origin and have $t$ co-ordinate greater than 0. Events with $t$ co-ordinate less than 0 but spacetime interval 0 from the origin are events such that a light-ray from that event can just reach the origin. These points form what Minkowski called *the front light cone* but which is now usually called *the past light cone*. Since the spacetime interval is invariant, exactly the same set of events constitutes the light cone structure in the primed co-ordinate system.[10] The light cones are invariant structures in the special theory, and events on the light cones of O are said to be *lightlike* or *null separated* from O.

One can easily see that, given the finite specified speed of light c, there must be events that are so far from O yet occur so soon after O (in the "stationary" co-ordinate system, say) that no light-rays from O can reach them. For such events $\sqrt{x^2 + y^2 + z^2}$ must be greater than $ct$, and this relation must hold for the spatial and temporal co-ordinates in any inertial frame. In this case the invariant spacetime interval is greater than 0, and such events are said to be *spacelike separated* from O.

Conversely, there are evidently events such that $\sqrt{x^2 + y^2 + z^2}$ is less than $ct$. These are events are sufficiently near to O that some object traveling at a speed less than light speed c can reach them from O. For such events the spacetime interval from O is less than 0, and they are said to be *timelike separated* from O.

Consider now an event (or, as Minkowski calls it, a *world-point*) that is spacelike separated from O. Minkowski notes that (1908: 84)

> Any world-point between the front and back cones of O can be arranged by means of a system of reference [a co-ordinate system] so as to be simultaneous with O, but also just as well so as to be earlier than O or later than O.

We have, then, another way of expressing the relativity of simultaneity (for spacelike separated world-points or events). By choosing an appropriate value for $v$, the constant relative velocity of a "moving" frame with respect to the "stationary" frame, a frame can be specified in which any given point P′ that is spacelike separated from O′ (that is, O, when the two frames are in standard configuration) will be simultaneous with the origin O′ in that frame.

Pre-relativistically, whether in (classical, Newtonian) physics or in our ordinary, common-sense way of thinking, simultaneity is by no means relative in this way. To

---

[10] If this all sounds a bit pedestrian, bear in mind that the light and relativity principles imply that the single expanding spherical shell from the burst of light at O must stay centered on both O and O′ as they move apart.

our ordinary way of thinking, throughout the universe there are events taking place or occurring right now, as opposed to those that have taken place or are yet to take place at these various distant locations. There is one and only one now or present extended across the breadth of the universe. Aristotle captured part of this idea when he said (*Physics* 220$^b$5): "there is the same time everywhere at once". Newton may well be expressing the same thought when he says (1962: 137):

> [W]e do not ascribe various durations to the different parts of space, but say that all endure together. The moment of duration is the same at Rome and at London, on the Earth and on the stars, and throughout all the heavens.

In classical or Newtonian physics, the transformation laws that connected co-ordinates in one inertial frame to another (in standard configuration) are the *Galileo transformations*:

$$x' = x - vt,$$

$$y' = y,$$

$$z' = z,$$

$$t' = t.$$

Evidently, when co-ordinates are transformed according to the Galileo transformations, events that are simultaneous in one inertial system are simultaneous in all, exactly as common sense indicates that they should be.

Pre-relativistically, the successive occurrence of global nows or presents constitutes the passage of time or temporal becoming, the dynamic quality of time that distinguishes it from space and that seems to be essential to its nature. The relativity of simultaneity challenges not only the uniqueness of the now but also our understanding of the passage of time as well. If each inertial frame has its own sets of simultaneous events, and if the principle of relativity states that no physical experiment or system (and we human beings are physical systems too) can distinguish one such frame or another as (say) genuinely at rest, then we are able to discern no particular set of simultaneous events as constituting *the* now or *the* present. If the passage of time is the succession of global nows or presents, then the notion of passage threatens to become unintelligible. Yet what phenomenon seems more important or more intuitively evident to us than the passage of time?

Late in his life, looking back on his scientific achievements and their philosophical importance, Albert Einstein wrote the following (1949: 61):

> We shall now inquire into the insights of definite nature which physics owes to the special theory of relativity.
> (1)    There is no such thing as simultaneity of distant events.

This chapter is an elaboration of Einstein's remark in three distinct but related areas: the conventionality of simultaneity, the relativity of simultaneity, and the passage of time (or temporal becoming). Einstein's presentation of the special theory of relativity in 1905 initiated two decades in which new theories in physics (the general theory of relativity, quantum mechanics) arose to challenge many philosophical preconceptions. Questions raised by these theories are still the subject of intense debate, but questions raised by the special theory of relativity regarding the nature of time are also still deeply puzzling. Minkowski elegantly modeled a world with no privileged distant simultaneity, but integrating this model with our intuitive understanding of time is still—more than a century after the advent of the special theory of relativity—no mean feat.

## 3. THE CONVENTIONALITY OF SIMULTANEITY

In order to introduce the special theory of relativity—in particular the relativity of simultaneity—to the reader, the presentation in section 2 skipped over the most famous and arguably the most important (both physically and philosophically) insight in Einstein's paper. The discussion in section 2 concerned ideas developed in §§2–3 of Einstein (1905), but we need now to consider §1, "Definition of Simultaneity".[11]

The light postulate introduced above states that light (*in vacuo*) has a certain determinate constant velocity $v$. How could one determine what exactly this velocity is? It seems that what one must do is start a light ray at some point (Call it "$A$") at some time $t_A$ and then see at what time $t_B$ the ray reaches a distinct point $B$. If one knows the distance from $A$ to $B$, which can be in principle determined by a measuring rod, one can determine the velocity of the ray, which is the distance from $A$ to $B$ divided by the elapsed time $t_B - t_A$.

One can determine the relevant times $t_A$ and $t_B$ if there are *synchronized* clocks at the points $A$ and $B$. The clocks must be synchronized if their readings are to indicate a definite time difference. So we next ask, how does one synchronize clocks at distinct points $A$ and $B$? One obvious way to do this would be to determine the distance from $A$ to $B$, note the time at $A$, and send a signal to $B$. One could then set the clock at $B$ to the reading of the clock at $A$ plus $v$ times the distance from $A$ to $B$—*if* one knew the velocity $v$ of the signal. But, alas, determination of this velocity seems to require that we have synchronized clocks (as we saw above). We have landed in a circle—synchronizing distant clocks requires knowing signal velocities, but determining signal velocities requires synchronized distant clocks.

The existence of this circle explains the following otherwise surprising remark of Einstein's regarding the comparison of times at distinct locations:

[11] In this section of my paper, all quotes from Einstein (1905) will come from the translation of that paper in Stachel (1998). The reason for the switch is that in the classic Perrett and Jeffery translation in *The Principle of Relativity* there is a notorious mistranslation at a key point. The (mis)translation is discussed in detail in Jammer (2006: 111–15).

If there is a clock at point $A$ in space, then an observer located at $A$ can evaluate the time of events in the immediate vicinity of $A$ by finding the positions of the hands of the clock that are simultaneous with these events. If there is another clock at point $B$ that in all respects resembles the one at $A$, then the time of events in the immediate vicinity of $B$ can be evaluated by an observer at $B$. But it is not possible to compare the time of an event at $A$ with one at $B$ without further stipulation.

(126)

What sort of "stipulation" does Einstein think is required? He continues the above train of thought:

So far we have defined only an "$A$-time" and a "$B$-time," but not a common "time" for $A$ and $B$. The latter can now be determined by establishing *by definition* that the "time" required for light to travel from $A$ to $B$ is equal to the "time" it requires to travel from $B$ to $A$. For, suppose a ray of light leaves from $A$ for $B$ at "$A$-time" $t_{A_{,,}}$ is reflected from $B$ towards $A$ at "$B$-time" $t_{B_{,,}}$ and arrives back at $A$ at "$A$-time" $t'_A$. The two clocks are synchronous by definition if

$$t_B - t_A = t'_A - t_B.$$

That is, common time is arrived at by setting the clocks at $A$ and $B$ so that the time taken for light to travel from $A$ to $B$ is the same as the time it takes for the same signal to travel back from $B$ to $A$. This way of synchronizing distant clocks is called *standard* or *Einstein* or*Poincaré-Einstein synchronization*.

It is useful to write the displayed equation above in a slightly different form:

$$t_B = t_A + \frac{1}{2}\left(t'_A - t_A\right). \tag{2}$$

Equation (2) tells us that the event that occurs at $A$ at the same time as the event $t_B$,—that is, the event at $A$ that is *simultaneous* with the event $t_B$, the event at $B$ at which the light signal from $A$ is reflected—is, according to Einstein synchronization, the event that occurs at $A$ exactly halfway between the time of the emission and the time of the reception of the reflected light signal. This way of synchronizing clocks is so natural that one is apt to overlook the fact that there is a choice to be made.

To see how choice enters the picture, recall that the light principle says that the speed of light in any inertial frame is $c$. As the special theory of relativity is usually understood, this speed is taken to be an upper limit for the speed of propagation of *any* causal process.[12] In classical physics there is no upper limit to the speed of causal propagation. In Roberto Torretti's words:

Before Einstein...nobody appears to have seriously disputed that any two events might be causally related to each other, regardless of their spatial and temporal distance. The denial of this seemingly modest statement is perhaps the deepest innovation in natural philosophy brought about by Relativity.[13]    (1983: 247)

[12]  A useful discussion of this idea may be found in Grünbaum (1973: chapter 12 (C)).

[13]  The Lorentz-Fitzgerald contraction hypothesis had been invoked to explain the null result of the Michaelson-Morley experiment since the early 1890s. (See Lorentz (1895) and references therein. See

Hans Reichenbach argued that temporal order is fixed by causal order. (1958: §21) That is, if event $e'$ is the (or an) effect of event $e$, then it is later than $e$. Applied to Einstein's example above, Reichenbach's principle (or "axiom," as he calls it) implies that $t'_A$ is later than $t_B$, which in turn is later than $t_A$, but it leaves indeterminate with respect to $t_B$ the time ordering of all events at $A$ later than $t_A$ (since no causal process or signal leaving $A$ later than $t_A$ can reach $t_B$,) and earlier than $t'_A$, (since no causal process leaving $t_B$ can reach any event at $A$ earlier than $t'_A$). Any event in that interval, according to Reichenbach, may be chosen to be simultaneous with $t_B$. According to Reichenbach, then, one may replace (2) by

$$t_B = t_A + \varepsilon \left( t'_A - t_A \right), \tag{3}$$

where $0 < \varepsilon < 1$. Reichenbach's thesis is called *the conventionality of simultaneity*.

The conventionality of simultaneity is quite different from the relativity of simultaneity. Given an "observer" or (in the example most used nowadays) a spaceship far from any planets or stars that is not accelerating, its path in spacetime is a straight line. Given that particular straight line, the conventionality thesis claims that clocks distant from it can be synchronized with the spaceship clocks in many ways—in fact, an infinite number of ways—although one way does seem simpler than the others and so is usually preferred. That preference is just that, conventionalists say, a preference. It reflects no matter of fact as to what distant events are simultaneous with the events in the spaceship There is in their view no such matter of fact.

The relativity of simultaneity, on the other hand supposes that in a given inertial frame clocks are synchronized according to the Einstein convention.[14] It then asserts that in any inertial frame moving with respect to the first frame, there are pairs of events that are simultaneous according to its clocks, but not simultaneous according to the clocks of the first frame (and vice versa, of course). As Adolf Grünbaum expressed it:

> [I]f *each* Galilean observer adopts the particular metrical synchronization rule adopted by Einstein in Section 1 of his fundamental [1905] paper and if the spatial separation of P$_1$ and P$_2$ has a component along the line of the relative motion of the Galilean frames, then that relative motion issues in their choosing as metrically simultaneous *different pairs of events* from within the class of topologically simultaneous events at P$_1$ and P.    (1973: 353)

Grünbaum is careful to emphasize in his discussion the difference between the conventionality and the relativity of simultaneity.

If one takes the special theory of relativity seriously, then one must take the relativity of simultaneity seriously since, as we saw above, it follows from the fundamental

---

also §1.2 of Brown (2005) on Fitzgerald's neglected 1889 letter and article.) Robert Rynasiewicz pointed out to me that if in light of this hypothesis one thinks about what happens to any massive object as its speed approaches that of light, one might well surmise that $c$ is a limiting speed for it and even for any form of causal influence.

[14]  Actually, any value of $\varepsilon$ may be chosen, as long as the same value is chosen in all inertial frames.

postulates of the theory. The status of the conventionality thesis is more controversial, and it has evoked a wide range of reactions. Indeed, the bulk of Jammer's (2006) monograph, *Concepts of Simultaneity,* is given over to discussion of the conventionality of simultaneity. In this chapter, I will be able only to sketch a few of the most important battle lines.

Hans Reichenbach (1958: §19), along with his notable students Adolf Grünbaum (1973: Chapter 12) and Wesley Salmon (1975: Chapter 4), vigorously defended the conventionality thesis. One job of the philosopher of science, as they saw it, is to separate factual from conventional elements in scientific theories, and they thought it was a triumph of modern physics cum philosophical analysis to have discovered that simultaneity—long thought to have some deep physical and even metaphysical reality—is (merely) conventional.

Since the conventionality thesis (at least in Reichenbach's classic presentation in §19 of his 1958 book) rests on the circularity argument given above (synchronizing clocks at distant points requires knowing the speed of the signal sent from one clock to the other, but finding the speed of anything, including a signal, requires synchronized clocks at distant points), it is tempting to think that the conventionality thesis can be undermined by showing that the claim of circularity is unsound. Why must one rely, for example, on a *signal* traveling (to revert to Einstein's notation) from A to B? Perhaps one could synchronize two clocks at A and then transport one of the clocks to B. Would we not then have synchronized clocks at both A and B? We would, Reichenbach wrote (1958, §20) *if* we could assume that the rate of the clock that traveled from A to B was not affected by its speed or path during its journey, but the special theory implies that neither assumption is true.

To take a slightly different tack, then, suppose that we synchronized many clocks at A and then transported them to B ever more slowly. Could we not find a limit to the series of times indicated by the slow-transported clocks and use the limit to synchronize distant clocks (Bridgman 1962: 64–67, Ellis and Bowman 1967)? It turns out that the limit exists and the synchronization agrees with standard synchrony. This proposal is considerably more complex and controversial than the first. The interested reader should look at the discussion of the Ellis and Bowman paper by Grünbaum, Salmon, van Fraassen, and Janis in the March, 1969 issue of *Philosophy of Science,* Friedman (1977, 1983), and Chapter 13 of Jammer (2006).[15] The members of the 1969 "Pittsburgh Panel" all argue, one way or another, that

> Ellis and Bowman have not proved that the standard simultaneity relation is non-conventional, which it is not, but have succeeded in exhibiting some alternative conventions which also yield that simultaneity relation.     (van Fraassen 1969: 73)

---

[15] Although Jammer's book is remarkable in scope, it fails to discuss the argument in Friedman (1983: 309–317) that the standard simultaneity relation can be fixed by slow clock transport in a way that avoids any circularity or question-begging assumptions.

There have also been numerous ingenious attempts to break out of the second half of the circle and argue that one can determine "one-way" light speeds (that is, the speed of a light signal from A to B, as opposed to a "round-trip" light speed in which the travel time of a light signal emitted and received at one location but reflected from a distant place is measured by one stationary clock) without the use of synchronized clocks. Indeed, it is not at all obvious that the first determination of the speed of light by Ole Rømer in 1676 is not a one-way determination.[16] Salmon (1977) is a comprehensive discussion of all such proposals then known. He concludes:

> I have presented and discussed a number of methods which have been proposed for ascertaining the one-way speed of light, and I have given references to others. Some of these approaches represent methods which have actually been used to measure the speed of light. Others are obviously "thought experiments." Some are quite new; others have been around for quite a while. In all of these cases, I believe, the arguments show that the methods under discussion do not provide convention-free means of ascertaining the one-way speed of light (although some of them are excellent ways of measuring the round-trip speed). I am inclined to conclude that the evidence, thus far, favors those who have claimed that the one-way speed of light unavoidably involves a non-trivial conventional element.                      (288)

By an odd coincidence, Salmon's extended defense of the conventionality thesis was followed in the same journal issue by a short article by David Malament (1977) that is widely thought to be the definitive refutation of the view. Malament proves the following result:

*Proposition 2* Suppose S is a two-place relation on $R^4$ where

   i.   S is (even just) implicitly definable from $\kappa$ and $O$;
  ii.   S is an equivalence relation;
 iii.   S is non-trivial in the sense that there exist points p ∈ O and q ∉ O such that S(p,q);
 iv.   S is not the universal relation (which holds of all points),

   Then S is $Sim_O$.

In Proposition 2 '$\kappa$' designates the relation of causal connectibility. Two points are *causally connectible* iff they are either lightlike or timelike separated. $O$ is a timelike line representing an inertial observer. $Sim_O$ is standard Einstein simultaneity relative to $O$.

   What is the significance of Proposition 2?[17] Malament begins his paper by observing that one of the major defenders of conventionalism, Grünbaum, is committed to the following two assertions:

---

[16] See Holton and Brush 2001: 343–344 for a brief account of the measurement. Salmon (1977) of course offers an account of the measurement and an argument that it is not one-way. Bridgman (1962: 68–9) offers a completely different explanation for the measurement's not being one-way.

[17] The reader is urged to look at the details in Malament's paper or, failing that, the semi-popular account in Norton (1992).

(1)   The relation [of simultaneity relative to an inertial observer] is not uniquely definable from the relation of causal connectibility [that is, from invariant causal relations].

(2)   Temporal relations are non-conventional if and only if they are so definable. (1977: 293).

Malament's Proposition 2 shows that (1) is false and therefore by (2) that temporal relations, in particular standard simultaneity, are *non*-conventional.

Malament's result has clarified but not ended the discussion. One can always raise doubts about the reasonableness of the required conditions. Indeed, Norton 1992: 225) worries about the delicate dependence of the result of Proposition 2 on its conditions.[18] He also reports (226) that Grünbaum has pointed out to him that condition (ii) might not be as innocent as it initially appears, and the extended discussion of the symmetry and transitivity of the simultaneity relation in Jammer (2006: Chapter 11) gives this concern some weight. Grünbaum (forthcoming) reiterates this point but also questions Malament's first condition, a topic to which we will now turn.

The very brief explication of the conditions of Proposition 2 omitted any discussion of 'definition'. While this might seem over-fussy, Sarkar and Stachel (1999) raise serious concerns over the precise form of definition employed by Malament. Roughly, a structure is defined if it is invariant under a class of transformations. $Sim_O$ is the unique, non-trivial simultaneity relation left invariant under a class of transformations that Malament calls the $O$ *causal automorphisms*. Amongst the $O$ causal automorphisms are temporal reflections, mappings of a spacetime to itself by reflection about a hyperplane that take the given inertial line $O$ to itself and preserve the relation $\kappa$. It should be intuitively clear that such mapping will leave the hyperplane in place (invariant) but flip a past light cone into a future light cone and vice versa.

Sarkar and Stachel (1999: 213–215) argue that it is physically unreasonable to include such mappings in the set used to define simultaneity. "Since reflections are not physically implementable as active transformations,[19] it is not physically reasonable to demand that all relations between events be universally preserved under them" (215). If one removes all reflections from the set of $O$ causal automorphisms (resulting in a set Sarkar and Stachel call the $O$ causal automorphisms), then, they claim, $Sim_O$ is no longer uniquely definable. They claim that, given an event $e$, its past light cone and future light cone are each definable by $\lambda$, the relation of lightlike or null separation, alone (and so a fortiori are definable from $\kappa$), and claim that the definability of these structures is a clear counterexample to Malament's uniqueness result.[20]

In a response to Sarkar and Stachel's argument Rynasiewicz (2000) offers, amongst other things, a brief primer on the varieties of definition. At the end of his discussion of definition he writes:

---

[18]  Regarding this point, one should see the discussion in Budden (1998).

[19]  That is, while we can translate and rotate objects, we cannot reverse time.

[20]  A different criticism of Malament's construction may be found in §2.2 of Anderson et al. (1998).

> The punch line is now this. No matter what notion of definability is in question, the preservation of a relation under automorphisms of the structure or structures in question is a necessary condition for the definability of a relation. In the case of Minkowski spacetime, the automorphisms include temporal reflections and thus render futile any attempt to define an individual half cone in the absence of temporal orientation.    (S355)

That is, if temporal reflections about $e$ are permitted, then (say) the future light cone of $e$ is not mapped to itself by a temporal reflection but to the past light cone of $e$, and so is not preserved under the class of maps, and so cannot be defined (in *any* of the relevant senses), contrary to Sarkar and Stachel's claim.

Why not, then, consider the question of conventionality in the presence, rather than the absence, of a temporal orientation?[21] Rynasiewicz raises this question and answers it himself. "This is what Sarkar and Stachel do in effect by proposing an alternative 'definition' of definability. It is unclear, though, what they think is the upshot of this move" (S355). Given the foregoing, one might think that Sarkar and Stachel would respond that adding the extra structure of a temporal orientation allows them to produce a reasonable special relativistic spacetime in which simultaneity is no longer uniquely definable from $\kappa$ and $O$. We seem to be back to squabbling over the conditions of Malament's theorem, but Rynasiewicz adds an unexpected and disquieting final paragraph to his paper.

> The most serious question ... is this. Described as neutrally as possible, what Malament establishes is that the only (interesting) equivalence relation definable from $\kappa$ and $O$ is that of lying on the same hypersurface spacetime-orthogonal to $O$. Now, as silly or contentious as it may sound, we should ask, what does spacetime orthogonality have to do physically with simultaneity? The force of the question is more easily recognized if reframed as follows. Suppose an inertial observer emits a light pulse in all directions. Consider the intersection of the resulting light cone with some subsequent hypersurface orthogonal to the observer. Does causal connectibility (plus $O$ if you like) completely determine the spatial geometry of the light pulse on the hypersurface in the absence of some stipulation as to the one-way velocity of light? If not (and I urge you to think not), then relative simultaneity does involve a conventional component corresponding to a degree of freedom in choosing a (3+1)-dimensional representation of an intrinsically four-dimensional geometry.    (S357)

In this light, Malament's result, far from being a decisive refutation of conventionalism, looks to be nearly irrelevant to the thesis! Even granting the unique definability of the hypersurface orthogonal to $O$ from $\kappa$ and $O$ itself, why would one suppose that the light sphere generated by a pulse of light from $O$ at an earlier time intersects that hyperplane at a set of points equidistant from $O$? In the absence of some way to break

---

[21]  That is, in the absence, rather than presence, of the temporal reflections in the supposedly appropriate class of automorphisms.

Reichenbach's circle, only a stipulation that one-way light speeds are the same in all directions will do. What is going on?

A partial answer may be found in a distinction drawn in Friedman (1977: 426).

> It seems to me that there are at bottom only two arguments for the conventionality of simultaneity in the literature: Reichenbach's and Grünbaum's. Reichenbach argues from an epistemological point of view; he argues that certain statements are conventional as opposed to "factual" because they are unverifiable in principle. Grünbaum argues from an ontological point of view; he argues that certain statements are conventional because there is a sense in which the properties and relations with which they purportedly deal do not really exist, they are not really part of the objective physical world.

Insofar as Grünbaum admits the existence of causal structure and insofar as Malament proves the unique specification of simultaneity in terms of causal structure, Grünbaum's version of conventionalism seems to be untenable.[22] Reichenbach's version seems, in contrast, to remain untouched by Malament's argument, though Friedman argues against Reichenbach's verificationism on other grounds, claiming that it rests on a dubious semantics[23] (1977: 426–428).

Another possible source of some of the perplexity here is the notion of general covariance. The conventionality of simultaneity is supposed to be an exciting thesis. One (naively, perhaps) supposes that light travels with some definite speed to a distant location and with a definite speed back from there to its origin. Conventionalists say that there is no fact of the matter with respect to these speeds, either (as we have just noted) on verificationist or straightforwardly ontological grounds. The physicist Peter Havas (1987) thought that conventionalism is true, but for essentially formal and unsurprising reasons. The general theory of relativity (and Minkowski spacetime is one particular general relativistic spacetime) is generally covariant, permitting "the formulation of the theory using arbitrary space-time coordinates" (444). According to Havas, "What Malament has shown…is that in Minkowski space-time…one can always introduce time-orthogonal coordinates…, an obvious and well-known result which implies $\varepsilon = \frac{1}{2}$." (444). A straightforward reading of Malament's and Rynasiewicz's papers indicates, however, that they are about the definability of the standard simultaneity relation in terms of causal connection, rather than its mere

---

[22] A useful way of looking at this matter was pointed out to me by Rynasiewicz. Grünbaum's propositions (1) and (2) suffice for conventionalism, and Malament's construction shows that Grünbaum's argument is unsound. But conventionalism itself does not entail (1) and (2), so their falsity does not entail the falsity of conventionalism. Yet another way to look at the matter is to be found in Stein (2009), which appeared after this chapter was written. Stein carefully argues that the major differences between Malament and Grünbaum result from different ways that they understand the issues. "I think we should acknowledge," writes Stein, "side by side, the points made by Grünbaum and the points made by Malement et al"(437). There is much else in Stein's article to enlighten the reader.

[23] Others, like Dieks (forthcoming) would argue that Reichenbach's argument fails because it relies on an overly restrictive epistemology. Following this up would take us too far afield, but it does indicate the entanglement of the conventionality issue with other major philosophical issues.

introduction in a given spacetime. If there is some reason that this straightforward reading cannot be sustained, it is not to be found in Havas's paper, though there is much of interest there on general covariance and the range of permissible coordinate systems for Minkowski spacetime.

One final consideration should be brought to bear on the conventionalism debate. Of the seven basic units in the SI (Système Internationale) system of measurement, time can be measured the most precisely. In 1983 the Conférence Générale des Poids et Mesures (CGPM), the highest authority in definitions of units, defined the meter as "the length of the path traveled by light in vacuum during the time interval of 1/299,792,458 of a second" (Jones 2000: 156–160; Audoin and Guinot 2001: 287–289). Then *by definition* the speed of light in *any* direction is 299,792,458 meters per second. As Roberto Torretti (1999: 275) remarks in the only philosophical discussion of this episode known to me, "The definition of the meter ratifies Reichenbach's view of the one-way speed of light as conventional but also undercuts his claim that its two-way speed is factual."

What is surprising is not that there is choice to be made when it comes to fundamental units, but the breadth and variety of theoretical and practical considerations that constrain the "convention". But if there is a line to be drawn between statements that are factual and those that are conventional, then my conclusion in this section must be rather an odd one. Whether or not the conventionality of simultaneity was refuted by Ellis and Bowman in 1967 or by Malament in 1977, it has been true since the CGPM defined the meter in 1983.

# 4. THE RELATIVITY OF SIMULTANEITY AND THE PASSAGE OF TIME

Let us turn now to some problems raised by the relativity of simultaneity. If one synchronizes distant clocks in various inertial frames using standard Einstein synchrony, whether conventional or not, then we find (as noted in section 2) that different frames disagree as to which events happen simultaneously. This basic result raises questions about nowness and about passage, questions that seemed reasonably straightforward in the classical view sketched at the beginning of this chapter. Here is one well-known version of the problem:

> Change becomes possible only through the lapse of time. The existence of an objective lapse of time, however, means (or, at least, is equivalent to the fact) that reality consists of an infinity of layers of "now" which come into existence successively. But, if simultaneity is something relative in the sense just explained, reality cannot be split up into such layers in an objectively determined way. Each observer has his own set of "nows," and none of these various layers can claim the prerogative of representing the objective lapse of time.                    (Gödel 1949: 558)

If the passage of time has something to do with the advance of *the* now or with (as Torretti preferred to put it) the advance of events (from future to past) through *the* now, then passage is undermined by the fact that there is no longer a unique now to serve as *the* now. There are (at least) as many nows as inertial frames, and there are a non-denumerable infinity of such frames. On this view, if there is no unique now, then there can be no objective lapsing of time or passage.

A second problem for passage raised by the relativity of simultaneity is that (recalling the quote from Minkowski in section 2) events that are spacelike separated from a given event O have no definite time order in respect to it. If e is spacelike separated from O, then in some frames it precedes O temporally, while in others it follows O. In precisely one frame e is simultaneous with O. It is claimed, though, that if there is no objective ordering of events as past, present, and future, then there is no passage (that which turns future events into past events) either. As the physicist Olivier Costa de Beauregard wrote:

> In Newtonian kinematics the separation between past and future was objective, in the sense that it was determined by a single instant of universal time, the present. This is no longer true in relativistic kinematics: the separation of space-time at each point of space and instant of time is not a dichotomy but a trichotomy (past, future, elsewhere). Therefore there can no longer be any objective and essential (that is, not arbitrary) division of space-time between "events which have already occurred" and "events which have not yet occurred." There is inherent in this fact a small philosophical revolution.                                    (1981: 429)

The upshot of this argument, the result of the revolution, is that the special theory of relativity is supposed to show that we live in a static or "block"[24] universe. Here is Costa de Beauregard's depiction of it:

> This is why first Minkowski, then Einstein, Weyl, Fantappiè, Feynman, and many others have imagined space-time and its material contents as spread out in four dimensions. For those authors, of whom I am one . . . relativity is a theory in which everything is "written" and where change is only relative to the perceptual mode of living beings.                                    (1981: 430)

These two arguments carry the weight and prestige of an important scientific theory and are endorsed, we learn, by physicists of the highest order. Yet it may be possible to find, within the confines of the special theory of relativity (or in the geometrical model of it called *Minkowski spacetime*) and within Einstein's stricture, that there is no distant simultaneity, enough remnants of the pre-relativistic notion of becoming that one might hesitate to call the resulting picture of time in the special theory "static".

---

[24]  This unfortunate expression originated in William James's brilliant characterization of determinism in James (1897: 569–570), but it has become a standard designation for a "static" or non-dynamic universe, one that supposedly lacks passage. The two ideas are distinct but they are continually conflated. The confusions leading to their conflation are effectively exposed in Grünbaum (1971: section VI; 1973: Chapter 10), but no amount of pesticide seems able to kill this weed.

The first step in this direction is to note that in Minkowski spacetime the concept of time bifurcates. We have so far been discussing *coordinate time*, time spread from the origin of an inertial system throughout the rest of space. One can also define what is called *proper time* along the world line of a material particle. The path or world line of a material particle consists of points that are mutually or pair-wise timelike separated. Such a path is called a *timelike world line* or a *timelike curve*. We can write the invariant spacetime interval that we found in equation (1) in its infinitesimal form as

$$ds^2 = -dt^2 + dx^2 + dy^2 + dz^2.$$ (4)

If we multiply (4) by −1, which of course still leaves it invariant, we can write

$$d\tau^2 = dt^2 - dx^2 - dy^2 - dz^2.$$ (5)

The quantity '$\tau$' is the *proper time*.

Timelike curves or world lines can be parameterized by proper time, $\tau$. We can define proper time lengths between two points A and B on a timelike curve, $\tau_{AB}$ as:

$$\tau_{AB} = \int_A^B d\tau = \int_A^B [dt^2 - (dx^2 + dy^2 + dz^2)]^{1/2}.$$ (6)

If we choose some point on the timelike line and assign it proper time 0, we can then define the proper time function along the timelike line by:

$$\tau_A = \tau_{0A}$$ (7)

We can begin to see the importance of proper time in that, according to the clock hypothesis (Naber 1992: 52; Brown 2005: 94–95), ideal clocks measure proper time.[25] Suppose we then conjecture the following (Arthur 1982: 107): "It is this proper time which is understood to measure *the* rate of becoming for the possible process following this timelike line (or *worldline*)." It is this idea of becoming along a timelike line, local becoming, that underlies my negative evaluation of the two anti-passage arguments presented above.

It is important to note one more fact about proper time. Since proper time is a function of all four variables $x$, $y$, $z$, $t$, if two ideal clocks are transported from event A to event B along different paths, they will in general indicate different proper times. For example, a clock that moves inertially from A to B will yield proper time change equal to co-ordinate time change, since $dx^2 + dy^2 + dz^2 = 0$. If the clock is not inertial, then $dx^2 + dy^2 + dz^2 > 0$ (at least for some portions of the journey from A to B), and the change in its proper time will be less than the co-ordinate time, the change in proper time for the inertial clock.[26]

---

[25] The standard clocks of inertial observers indicate proper time, but in general clocks during their history may be accelerated. It is a non-trivial demand on a real clock that even when accelerated it read proper time (Sobel 1995).

[26] Hence, in the so-called *twin paradox*, the traveling twin must return younger than the stay-at-home (inertial) twin. For further discussion of this feature of the special theory see the chapter by Luminet in this volume and Marder (1971).

Let us now return—slowly—to the two anti-passage arguments based on the special theory of relativity presented above by considering a classic article (Grünbaum 1971) that takes them—or at least the conclusion that "change is only relative to the perceptual mode of living beings" very seriously. The thesis of Grünbaum's article is this: "Becoming is mind-dependent because it is not an attribute of physical events per se but requires the occurrence of certain conceptualized conscious experiences of the occurrence of certain events" (1971: 197). But what, according to Grünbaum, is this becoming that is mind-dependent?

> In the common-sense view of the world, it is of the very essence of time that events occur now, or are past, or future. Furthermore, events are held to change with respect to belonging to the future or the present. Our commonplace use of tenses codifies our experience that any particular present is superseded by another whose event-content thereby 'comes into being'. It is this occurring now or coming into being of previously future events and their subsequent belonging to the past which is called 'becoming' or 'passage'. Thus, by involving reference to present occurrence, becoming involves more than mere occurrence at various serially ordered clock times. The past and future can be characterized as respectively before and after the present.                                                        (1971: 195)

There are two elements in this passage that I wish to separate. The first is the common-sense idea indicated at the outset that events naturally sort themselves out into those that are present, past, and future. This aspect of time is typically called *tense* by philosophers.[27] Second is the distinction Grünbaum makes between becoming and the "mere occurrence [of events] at various serially ordered clock times". Grünbaum has no objection to (and, in fact, insists upon) the mind-independence of the latter.

> [T]o assert in this context that becoming is mind-dependent is not to assert that the obtaining of the relation of temporal precedence is mind-dependent. Nor is it to assert that the mere occurrence of events at various serially ordered clock times is mind-dependent.                                                        (1971: 197)

I suggested briefly in the introduction to this chapter that a philosopher might wish to consider the "mere occurrence [of events] at various serially ordered clock times" to be becoming, and I have defended the idea at greater length elsewhere (Savitt 2002). I argue there that this usage captures the mainstream, metaphysically unobjectionable content of the concept of passage. Of course, this is essentially a terminological matter, but given the way I believe the term *becoming* has been used traditionally, if we do agree to use the term in that way, then Grünbaum and I *agree* that becoming is mind-*in*dependent.

We also agree that something else is mind-*de*pendent, and that something else is tense. Grünbaum claims this directly. I claim it indirectly by claiming directly that the terms for tense, like 'now', 'past', and 'future', are indexical terms. Without minds there could not be language-users.[28] Without language-users there could not be languages. Without languages there could not be indexical terms.

Whatever the differences between these two views, both agree to the extent that they entail the following key claim (Grünbaum 1971: 206):

> [W]hat qualifies a physical event at a time $t$ as belonging to the present or as now is not some physical attribute of the event or some relation it sustains to other purely[29] physical events.

Once one disjoins passage from tense, one can see that the first argument against passage from the relativity of simultaneity, the one above presented by Gödel, is invalid. From the fact that in different inertial frames different events are simultaneous (or, to put it more tendentiously, from the fact that in different inertial frames different events merely share the same $t$-coordinate), there is no conclusion to be drawn regarding passage, the successive occurrence of events. The failure of the first argument can be seen even more clearly if one recalls Einstein's stricture, the guiding insight of this chapter, that there is no distant simultaneity. The successive occurrence of events need not rely on distant simultaneity. It can be a *local* process, confined to a world line, possessing a proper time that is measured by a clock (Arthur 2008: section 2).

Is this intelligible? Do we *really* have time if we have only this local process? That is, do we *really* have time if there is no way to say of events not on a world line that they are past, present, or future (relative to events on that world line)? That is, can there be passage without tense? This question, however, has an incorrect presupposition, as we shall see in our examination of the second anti-passage argument described above.

If the passage of time involves the future, "events which have not yet occurred" becoming the past, "events which have already occurred", how can there be passage if there is no objective way of separating events into these two classes (as well as "events that are occurring now")? The second argument, Costa de Beauregard's argument, notes that there is no objective (that is, frame independent) way to make this division for spacelike separated events. In the absence of tense, it concludes there can be no passage.

Does it follow, if the time ordering of spacelike separated events is frame dependent, that there is no tense? There are events, those that are timelike or lightlike separated, for which the time ordering is not frame dependent (Čapek, 1976). These are precisely the events that, given an event or spatio-temporal location O, are within or on O's past light cone and so can have some effect on it, or are within or on O's future light cone and so can be effected by it. Given the limiting speed $c$ of any causal process, the events spacelike separated from O and so not temporally ordered with respect to it in some

---

[28] Perhaps we will come to believe that machines can use language. Perhaps then we'll come to believe that those machines have minds.

[29] 'Purely' was doubtless added to avoid begging any questions with respect to the mind-body problem. To put this point in terms of a distinction introduced by Meehl and Sellars (1956, 252), one can deny that some item or process is "purely" or narrowly physical (that is, physical$_2$)—roughly, describable in physics–without being committed to its being broadly non-physical (or not physical$_1$)—that is, outside the spacetime network.

frame-invariant way are also causally irrelevant to anything that happens at O. Why should one, then, despair of "tense" in Minkowski spacetime if there is an absolute (that is, not frame dependent) past and an absolute future at each event, even if there are some events that are neither?

To put the same thought more formally, in the classical view of time as described by Torretti at the beginning of this chapter, there is a relation of *earlier than* ($<$) that completely orders events. The relation '$<$' is irreflexive, anti-symmetric, and transitive, and for any two events a and b, either a $<$ b or b $<$ a or a=b. In Minkowski spacetime there is still a relation of earlier than that is irreflexive, anti-symmetric and transitive, but the clause "either a $<$ b or b $<$ a or a=b" is no longer true. The ordering imposed by '$<$' is only partial rather than complete. Is complete ordering an essential feature of tense? Or should one rather say that in the shift to the special theory of relativity we learned that tense in fact was a partial rather than a complete ordering of events in spacetime? If one takes this latter course, then the second argument against passage in Minkowski spacetime, that the absence of tense implies the absence of passage, fails. The premise that there is no tense is incorrect.

There is a residual oddity in this view (Putnam 1967: 246). If being present is tied to being simultaneous with (no matter how distant) and if there is no distant simultaneity, then in the history of some material object or person an event not on its world line can at some earlier times be in its future and at some later times be in its past without ever being present.

It is tempting to take a high-handed approach to this complaint. The empirical evidence for the special theory (as opposed to classical mechanics) is overwhelming. If the evidence supports a theory that forces us to an odd conclusion, common sense must bow to the evidence. It might be worth noting, however, that there are two ways in which one might attempt to mitigate the oddity of the conclusion.

First of all, one might simply identify the entire region of Minkowski spacetime that is spacelike separated from some event O, its "elsewhere," as its present.[30] One motivation for such a thought was provided by Minkowski himself, since he noted in his original paper (1908: 77–79) in effect that if the speed c (of electromagnetic radiation *in vacuo*) is allowed to increase without bound, then the region of spacelike separated points approaches as a limit the flat plane of events orthogonal to O's world line. If one then thinks of the elsewhere as a relativistic counterpart of the classical present in virtue of this reduction relation, it is then true that any event that is once future must be present before it is past.

Nevertheless, it is worth noting that two events e and e' that are both spacelike separated from some event O and so both "present" in this sense to or for O may themselves be timelike separated and so invariantly time ordered. In fact, there is no upper bound on the proper time between e and e'. Identifying the elsewhere with the present does not seem to decrease oddity.

---

[30] As noted above, the "elsewhere" of O has also been called the *topological present*.

What, then, is the second way that might mitigate the oddity? To approach this idea, let us recognize that when we use indexical terms like 'now' and 'here' to indicate temporal or spatial location, the exact temporal or spatial extent or boundaries are context-dependent. When I say 'here', I might mean *in this room*, or *in British Columbia*, or even *on Earth*, since there is water here but not on Neptune. Similarly, with temporal terms I might wish to indicate a very short period of time ("Go to your room *now*.") or a much longer one ("Since we now have cell phones, public pay phones are disappearing."). All these *heres* are more-or-less spatially extended. All the *nows* are more-or-less temporally extended.

The *now* or *present* of experience is also extended. The extent may vary—estimates for normal human experience put the range from about .3 seconds to 3 seconds. That's the (varying) duration that we typically perceive as present or happening now— the period of time, say, that it takes to hear a sentence or a musical phrase. This period is called the *specious* or *psychological present*. To make the following discussion simpler (but without compromising any matters of principle, I hope), I will take the psychological present to be of one fixed convenient length, 1 second.[31]

The second proposal specifies a relativistic counterpart of the present by employing a period of (proper) time represented by a specious present. Suppose one chooses two events on a given timelike curve, say $e_0$ and $e_1$, that are one second apart. Then consider the region of spacetime that is the intersection of the future light cone of $e_0$ and the past light cone of $e_1$. This is the set of events that, at least in principle, can be reached by a causal process from $e_0$ *and* are also able then to reach $e_1$. In Savitt (2009) I call this structure the *Alexandroff present* for the interval $e_0$ to $e_1$ along the given timelike curve (assuming that it is parameterized by proper time) in honor of the Russian mathematician who first investigated these sets.[32]

To return, finally, to the claim that it is odd that a future event can become past without ever being present, one can note that an Alexandroff present is about 300,000 km wide at its waist, given the convention adopted above that the interval between beginning and end is one second. This means that at least events in one's vicinity cannot become past without being present, if Alexandroff presents are deemed to be reasonable relativistic stand-ins or counterparts of the classical present. The oddity can happen with events on Saturn but not events in Sydney. Perhaps that helps.[33]

---

[31] This period of time is a completely objective matter, though the extent of its duration is set by subjective and pragmatic considerations.

[32] Others prefer to call it the *Stein present*, in honour of Howard Stein, since those of us who have been toying with this structure lately were all more-or-less independently and more-or-less at the same time inspired to it by a remarkable set of reflections at the end of Stein (1991). Still others call it a *diamond present* given its shape when one or two spatial dimensions are suppressed and when units are adjusted such that the numerical value of $c$ is 1. It might in fact best be called the *causal present*, since it is the set of all events that can causally interact with any pair of events on the given world line between $e_0$ and $e_1$.

[33] The arguments in this section were influenced by the writings of and by conversations with Richard Arthur, Dennis Dieks, and Abner Shimony. The relevant papers will be found listed amongst

# 5. CONCLUDING REMARKS

In section 4 of this chapter I tried to show that a certain common view concerning time and the special theory of relativity is not true. The view is that the special theory mandates a "block" or static universe. I interpret this view to mean that there is no becoming, no passing or lapsing time, in Minkowski spacetime.

Since 'time' is an ambiguous term,[34] I tried to disambiguate it in this context into two strands, passage and tense. When these two strands are clearly separated, the two main arguments for the "Special Relativity implies block universe" view fail. Moreover, we saw that there are two concepts of time in the special theory itself, co-ordinate time and proper time, the latter being a kind of time perfectly apt for becoming. What is surprising about the special theory (at least in this regard; there are other surprises of course) is that time *qua* passage is a *local* phenomenon, tied to a world line. For eons we have tied passage to an advancing *global* now, and this idea is buried deep in our worldview. It is an idea that we must transcend.

## REFERENCES

Anderson, R., Vetharaniam, B, and Stedman, G. E. (1998), 'Conventionality of Synchronisation, Gauge Dependence and Test Theories of Relativity', in *Physics Reports* 295: 93–180.

Arthur, R. (1982), 'Exacting a Philosophy of Becoming from Modern Physics', in *Pacific Philosophical Quarterly* 63: 101–110.

—— (2008), 'Time Lapse and the Degeneracy of Time: Gödel, Proper Time and Becoming in Relativity Theory', in D. Dieks (ed.) *The Ontology of Spacetime* II (Elsevier): 207–228.

Audoin, C. and Guinot, B. (2001), *The Measurement of Time: Time, Frequency and the Atomic Clock* (Cambridge University Press). This is an English translation of *Les Fondements de la Mesure du Temps* (1998).

Bridgman, P. (1962), *A Sophisticate's Primer of Relativity* (Wesleyan University Press). Second edition (1983). References in this chapter are to the second edition.

Brown, H. (2005), *Physical Relativity: Space-time Structure from a Dynamical Perspective* (Oxford University Press).

Budden, T. (1998), 'Geometric Simultaneity and the Continuity of Special Relativity', in *Foundations of Physics Letters* 11: 343–357.

Callender, C. (2000), 'Shedding Light on Time', in *Philosophy of Science* 67: S587–S599.

the references. One can find a related view in Maudlin (2002), and Dorato (2006: 107) writes that he defends in his way the same view as Maudlin. I have been helped throughout by the advice of Robert Rynasiewicz, who has saved me from numerous errors and who no doubt wishes that he had saved me from even more. I wish also to thank the editor and Christian Wüthrich for helpful suggestions.

[34] Rovelli (1995: 81) exhibits "ten distinct versions of the concept of time", and he is restricting his attention only to time in physical theories.

Čapek, M. (1976), 'The Inclusion of Becoming in the Physical World' in M. Čapek (ed.), *The Concepts of Space and Time: Their Structure and Their Development* (D. Reidel Publishing Company): 501–524.

Clifton, R. and Hogarth, M. (1995), 'The Definability of Objective Becoming in Minkowski Spacetime' in *Synthese* 103: 355–387.

Costa de Beauregard, O. (1981), 'Time in Relativity Theory: Arguments for a Philosophy of Being' in J. T. Fraser (ed.) *The Voices of Time,* second edition (The University of Massachusetts Press).

Dieks, D. (1988), 'Discussion: Special Relativity and the Flow of Time' in Philosophy of Science 55: 456–460.

——(1991), 'Time in Special Relativity and its Philosophical Significance', in *European Journal of Physics* 12: 253–259.

——(2006), 'Becoming, Relativity and Locality' in D. Dieks (ed.), *The Ontology of Spacetime* I (Elsevier): 157–175.

——(forthcoming), 'The Adolescence of Relativity: Einstein, Minkowski, and the Philosophy of Space and Time'.

DiSalle, R. (2002, 2008), 'Space and Time: Inertial Frames', The Stanford Encyclopedia of Philosophy (Fall 2008 Edition), Edward N. Zalta (ed.), URL = <http://plato.stanford.edu/archives/fall2008/entries/spacetime-iframes/>

Dorato, M. (2006), 'The Irrelevance of the Presentist/Eternalist Debate for the Ontology of Minkowski Spacetime' in D. Dieks (ed.), *The Ontology of Spacetime* I (Elsevier): 93–109.

Earman, J. (2006), 'The "Past Hypothesis": Not Even False' in *Studies in History and Philosophy of Modern Physics* 37: 399–430.

Einstein, A. (1905), 'On the Electrodynamics of Moving Bodies', in *The Principle of Relativity* (Dover Publications, Inc.). This is an English translation by W. Perrett and G. B. Jeffery of 'Zur Elektrodynamik bewegter Körper' *Annalen der Physik* 17: 891–921.

——(1949), 'Autobiographical Notes' in P. A. Schilpp (ed.) *Albert Einstein: Philosopher-Scientist* (Open Court), 2–94.

Ellis, B, and Bowman, P. (1967), 'Conventionality in Distant Simultaneity', in *Philosophy of Science* 34: 116–136.

Friedman, M. (1977), 'Simultaneity in Newtonian Mechanics and Special Relativity', in J. Earman, C. Glymour, and J. Stachel (eds.), *Foundations of Space-Time Theories* (University of Minnesota Press), 403–432.

——(1983), *Foundations of Space-Time Theories: Relativistic Physics and Philosophy of Science* (Princeton University Press).

Gödel, K. (1949), 'A Remark about the Relationship between Relativity Theory and Idealistic Philosophy' in P. A. Schilpp (ed.), *Albert Einstein: Philosopher-Scientist* (Open Court), 557–562.

Grünbaum, A. (1969), 'Simultaneity by Slow Clock Transport in the Special Theory of Relativity', in *Philosophy of Science* 36: 5–43.

——(1971), 'The Meaning of Time' in E. F. Freeman and W. Sellars *Basic Issues in the Philosophy of Time* (Open Court): 195–228.

——(1973), *Philosophical Problems of Space and Time*. Second, enlarged edition (D. Reidel Publishing Company).

——(forthcoming), 'David Malament and the Conventionality of Simultaneity: a Reply', forthcoming in *Philosophy of Science in Action* (Oxford University Press). The article is currently available as http://philsci-archive.pitt.edu/archive/00000184/00/malament.pdf.

Havas, P. (1987), 'Simultaneity, Conventionalism, General Covariance, and the Special Theory of Relativity', in *General Relativity and Gravitation* 19: 435–453.

Holton, G. and Brush, S. (2001), *Physics: the Human Adventure: from Copernicus to Einstein
and Beyond* (Rutgers University Press).

James, W. (1897), 'The Dilemma of Determinism', in *The Will to Believe and Other Essays in
Popular Philosophy* (Longmans, Green & Co.). Rpt. in *William James: Writings 1878–1899*
(The Library of America): 566–594.

Jammer, M. (2006), *Concepts of Simultaneity* (The Johns Hopkins University Press).

Janis, I. (1969), 'Synchronism by Slow Transport of Clocks in Noninertial Frames of Reference', in *Philosophy of Science* 36: 74–81.

Jones, T. (2000), *Splitting the Second: the Story of Atomic Time* (Institute of Physics
Publishing).

Lorentz, H. (1895), 'Michelson's Interference Experiment' in *The Principle of Relativity*
(Dover Publications, Inc.). This is an English translation by W. Perrett and G. B. Jeffery
of 'Versuch einer Theorie der elektrischen und optischen Erscheinungen in bewegten
Körpern" (Leiden, 1895): §§ 89–92.

Malament, D. (1977), 'Causal Theories of Time and the Conventionality of Simultaneity', in
*Noûs* 11: 293–300.

Marder, L. (1971), *Time and the Space Traveller* (University of Pennsylvania Press).

Markosian, N. (2001), 'Review of Questions of Time and Tense', in *Noûs* 35: 616–629.

Maudlin, T. (2002), 'Remarks on the Passing of Time', in *Proceedings of the Aristotelian
Society* 52: 237–252.

McKeon, R. ed. (1941). *The Basic Works of Aristotle.* (Random House).

Meehl, P. and W. Sellars (1956), 'The Concept of Emergence', in H. Feigl and M. Scriven
(eds.), *The Foundations of Science and the Concepts of Psychology and Psychoanalysis*
(University of Minnesota Press), 239–252.

Mermin, N. D. (1968), *Space and Time in Special Relativity* (McGraw-Hill).

——(2005), *It's About Time: Understanding Einstein's Relativity* (Princeton University
Press).

Minkowski, H. (1908), 'Space and Time', in *The Principle of Relativity* (Dover Publications,
Inc.). This is an English translation by W. Perrett and G. B. Jeffery of an address delivered
at the 80th Assembly of German Natural Scientists and Physicians at Cologne on 21
September 1908.

Naber, G. (1992), *The Geometry of Minkowski Spacetime* (Springer-Verlag).

Newton, I. (1962), 'De Gravitatione et aequipondio fluidorum', in A. R. Hall and M. B. Hall
(eds.) *Unpublished Scientific Papers of Isaac Newton* (Cambridge University Press), 89–156.

Norton, J. (1992), 'Philosophy of Space and Time', in M. H. Salmon *et al. Introduction to the
Philosophy of Science* (Prentice-Hall): 179–231. Rpt. by Hackett Publishing Company.

Norton, G. (1993), 'General Covariance and the Foundations of General Relativity: Eight
Decades of Dispute', in *Reports on Progress in Physics* 56: 791–858.

Putnam, H. (1967), 'Time and Physical Geometry', in *The Journal of Philosophy* 64: 240–247.

Reichenbach, H. (1958), *The Philosophy of Space and Time* (Dover Publications, Inc.).

Rovelli, C. (1995), 'Analysis of the Distinct Meanings of the Notion of "Time" in Different
Physical Theories', in *Il Nuovo Cimento* 110 B: 81–93.

Rynasiewicz, R. (2000), 'Definition, Convention, and Simultaneity: Malament's Result and its
Alleged Refutation by Sarkar and Stachel', in *Philosophy of Science* 68: S345–S357.

Salmon, W. (1969), 'The Conventionality of Simultaneity', in *Philosophy of Science* 36: 44–63.

Salmon, W. (1975), *Space, Time, and Motion: A Philosophical Introduction* (Dickenson Publishing Co., Inc.).

——(1977), 'The Philosophical Significance of the One-Way Speed of Light', in *Noûs* 11: 253–292.

Sarkar, S. and Stachel, S. (1999), 'Did Malament Prove the Non-Conventionality of Simultaneity in the Special Theory of Relativity?', in *Philosophy of Science* 66: 208–220.

Saunders, S. (2000), 'Tense and Indeterminateness', in *Philosophy of Science* 67: S600–S611.

Savitt, S. (2002), 'On Absolute Becoming and Myth of Passage', in C. Callender (ed.) *Time, Reality and Experience* (Cambridge University Press): 153–167.

——(2006), 'Presentism and Eternalism in Perspective', in D. Dieks (ed.), *The Ontology of Spacetime* I (Elsevier): 111–127.

——(2009), 'The Transient *nows*', in W. Myrvold and J. Christian (eds.), *Quantum Reality, Relativistic Causality, and Closing the Epistemic Circle: Essays in Honour of Abner Shimony* (Springer-Verlag): 339–352.

Shimony, A. (1993), 'The Transient *now*', in *Search for a Naturalistic World View, volume II: Natural Science and Metaphysics* (Cambridge University Press): 271–287.

Sobel, D. (1995), *Longitude: The True Story of a Lone Genius who Solved the Greatest Scientific Problem of His Time* (Penguin Books).

Stachel, J. (1998), *Einstein's Miraculous Year: Five Papers that Changed the Face of Physics* (Princeton University Press).

Stein, H. (1968), 'On Einstein-Minkowski Space-Time', in *The Journal of Philosophy* 65: 5–23.

Stein, H. (1991), 'On Relativity Theory and the Openness of the Future', in *Philosophy of Science* 58: 147–167.

Stein, H. (2009), ' "Definability," "Conventionality", and Simultaneity in Einstein-Minkowski Space-Time' in W. Myrvold and J. Christian (eds.), *Quantum Reality, Relativistic Causality, and Closing the Epistemic Circle: Essays in Honour of Abner Shimony* (Springer-Verlag): 403–442.

Taylor, E. and Wheeler, J. (1963), *Spacetime Physics* (W. H. Freeman and Co.).

Torretti, R. (1983), *Relativity and Geometry* (Pergamon Press). (Reprinted, with minor corrections and a new preface by Dover in 1996.)

Torretti, R. (1999), *The Philosophy of Physics* (Cambridge University Press).

Van Fraassen, B. (1969), 'Conventionality in the Axiomatic Foundations of the Special Theory of Relativity', in *Philosophy of Science* 36: 64–73.

# CHAPTER 19

# TIME IN CLASSICAL DYNAMICS

## LAWRENCE SKLAR

## 1. INTRODUCTION

IT starts with Newton in his famous "Scholium to the Definitions" of the *Principia*:

> Absolute, true and mathematical time, of itself and from its own nature, flows equably without relation to anything external, and by another name is called duration: relative, apparent, and common time, is some sensible and external (whether accurate or unequable) measure of duration by means of motion, which is commonly used instead of true time; such as an hour, a day, a month, a year.

Now this is in the same "Scholium" that takes up absolute space and our empirical ability to determine at least some kinds of motion, accelerations, relative to "space itself" by means of the consequent inertial forces such absolute accelerations engender. And with that comes the still ongoing debate about the appropriate metaphysics of space or spacetime—in particular substantivalisms of various sorts versus relationisms of equally varied sorts. Parallel to that debate there has to be a metaphysical dispute about time. Here the notion of time as "substance" seems even more peculiar that it does in the case of space as substance—and even Newton thought that peculiar enough. In the light of relativistic spacetimes it seems as though the two metaphysical issues become inseparably intertwined.

But one can put these metaphysical quandries to the side and still explore a very important way in which the place of time in classical mechanics serves as a paradigmatic example of a kind of "conceptual refinement" that appears again and again in the historical evolution of physics. To see this let us first ask the question as to why Newton needs his "absolute" time. Well a very crucial aspect of classical dynamics is the free motion of particles. And, developing out of impetus theories, with improvements by Galileo and with further refinement at the hands of Descartes, we have what becomes

Newton's "First Law of Motion" in the *Principia*. Particles free of forces follow straight-line paths at constant speed. But the very notion of what is a straight-line path requires a preferred spatial reference frame (for Newton absolute space, for later refinements the inertial frame structure of Galilean (neo-Newtonian) spacetime). And constant speed also requires an absolute standard for the equality of temporal intervals. And that is absolute time.

## 2. REFINING THE CONCEPT OF TIME

There is a pattern of conceptual refinement in physics that goes like this: First we have some rather vague concept founded in our subjective experience as generated by aspects of our bodily interaction with the physical world and, perhaps, by some physical goings-on internal to us bodily. A way of making this concept more amenable to inter-subjective comparison, and to the kind of precise repeated applicability necessary for framing physical generalizations using the concept, is the resort to standard "measuring instruments" that allow us to construct an "objective" surrogate of greater precision and inter-subjective applicability than the concept we started off with. At the final stage, the concept becomes integrated into a fundamental physical theory by means of its place in the fundamental laws of that theory. At this point the very meaning of the concept becomes fixed by its place in those laws. As a corollary of this framing of the concept by foundational laws, we develop an account of why the measuring instruments that we used at the intermediate stage really were "good" measuring instruments for the concept in question. And we begin to gain at least some insights into how the property characterized by the concept played a role in that initial subjective experience that first brought it to our attention.

Consider, for example, temperature. Start with the subjective awareness of hot and cold. Eventually find an "objective correlative" for a measure of hot and cold by using thermometers, here relying implicitly on the fact that many substances (gases are best) have linear coefficients of volume change with temperature. Eventually discover the laws of thermodynamics, giving empirical temperature as an index of co-equilibrium for systems in thermal contact, and absolute temperature as a measure of ability to extract mechanical work from heat in an ideal engine. Finally with the kinetic theory of matter's picture of the micro-components of a macro-object in motion, and with the rich resources of full-blooded statistical mechanics, we have the ability to explain why it was that good thermometers were reliable indicators of temperature. At this stage the concept of temperature is principally tied no longer to subjective experience or to the results of thermometric measurement, but to its role in fundamental thermodynamics and statistical mechanics.

Now look at time. Its appearance in subjective experience is so pervasive and so primitive that we find it hard to accurately characterize it. We have an intuitive experience of events ordered in time. Such basic features as the transitivity of "later than" seem primitive and immediate. Aspects of our experience give us the inbred notion of an asymmetry of time, of the basic distinction between past and future. And we have some primitive, subjective experience of magnitudes of intervals, and of the notion of some one time interval in a metric comparison of length with another. Now all of these notions need the kind of regimentation into objective science we have described above. But let us focus on the idea of the measure of the length of a time interval.

Here the stage of finding measuring instruments to objectify the concept, in this case the concept of length of time interval, is thrust upon us by our experience of the astronomical phenomena in which we are embedded. The rotation of the earth gives us the solar day (but with a need to accommodate seasonal variation) and, even better, the sidereal day (but which itself is the subtle correction of the equation of time as Newton points out). On top of this we have the regular heavenly clockwork of the repetition of the seasons, and the less obvious regular temporal pattern of the motion of the planets. So profound and so immediately impressive are these already given temporal metric measuring devices that independent discovery of the temporal regularities of the heavens and use of them for framing clocks occurs across numerous cultures.

The process of finding handier clocks than the sundial and the mural quadrants of the astronomers is a slow one. From water clocks to escapements to the installation of the pendulum into clocks is a process of centuries. In each case, though, there is the constant attempt to find "good" clocks, where steadiness of rate always looks back to the astronomical clocks as the standards.

## 3. TEMPORAL METRIC AND DYNAMICAL LAWS

But now there must be the second stage in objectification of the concept. The concept of a metric length of time, fundamentally the concept of equality of time intervals, must be embedded into a foundational lawlike structure of physics. It is this that Newton suggests is accomplished by the inertial law of classical dynamics.

A free particle travels a straight line path at constant speed. Constant speed means equal spatial intervals covered in equal times. A good clock is one that rates equality of time intervals in such a way that it credits free particles with constant speed when the lapse of time is measured by the clock's standard. Of course there are many very deep conceptual issues hidden in all of this. How do we know when a particle is moving freely? For early Newtonians this was taken to be given by its sufficient distance from other particles. But we now know, in the light of general relativity with its notion of gravity as determining timelike geodesics and with the possibilities of cosmic curvatures, that the notion of "free" or "inertial" is not so easy to get rigorous. And to make the time standard work we first need a distance standard. For constant speed means covering equal *spatial* intervals in equal times. And where does the standard for equality of spatial intervals come from? We can resort to such "operational" notions as the use of "rigid rods" here. But to go further and figure out how equality of spatial intervals is to be fixed by proper embedding of the concept of spatial equality into the foundational physical laws is another story—and not either a simple one nor one yet fully understood.

But this is the beginning. The temporal metric has received its deep objective clarification by the embedding of the notion of equal temporal interval into the fundamental inertial law. Once having done this, and we can then take on the task of evaluating ordinary clocks for their goodness or lack of goodness: good clocks properly measure dynamic time. And we can seek for explanations as to why clocks are good or bad within the classical dynamical theory itself. The Earth's rotation is a good clock because it isn't too badly interfered with by other bodies, and the conservation of angular momentum gives it a steady rotation rate. But the rotating Earth isn't a perfect clock, and dynamics (say the dynamics of tidal friction by the Earth's interaction with the moon) explains that as well.

This new characterization of the absolute measure of time receives its practical application in celestial mechanics. The time measures provided by the best available terrestrial clocks, or even by the rotating Earth, were inadequate in the nineteenth century for the precision needs of the astronomy of the solar system. Instead, one constructed one's model of the behavior of the planets taking into account all the perturbations the components of the system induced on one another, assuming, of course, Newton's inverse square law of gravitation as the correct law of the planets' interacting forces. Then one sought the best time parameterization that gave the correct motion of the components of the system through time assuming the truth of the classical dynamical laws. This is so called "ephemeris time."

In pure theory this notion of time appears, for example, in Barbour-Bertotti Machianism. Here one takes as the system in question the entire cosmos, assumed as being both finite in extent and, crucially, with null absolute rotation as a system taken as a whole. A new dynamical principle that generalizes from the standard principles of least action will then pick out the inertial frames and will characterize the motion of the components of the system as those predicted by classical dynamics, with the forces given by the standard laws and the classical equations holding in the now relationistically characterized inertial frames. And how is time to be measured in such a theoretical framework? It is determined as the ephemeris time of the cosmic system as a whole.

Incidentally, Poincaré's famous claim that the scale of time is conventional, since we could pick a different way of scaling things if we were willing to pay the price of a more complicated formulation of the dynamical laws, is irrelevant to these considerations. It is an objective fact that the scaling we do pick is the one that puts the laws into their familiar, simple form. And it is a fact of nature that it is this scaling that is, to greater or lesser degrees of accuracy, picked out by the dynamical clocks, celestial or terrestrial, that had been used to give our objective measures of time prior to Newton. And it is these facts that make the standard scaling of time the "absolute" measure of equality of time intervals.

## 4. Time: From Subjective Perception to Place in Foundational Physics

But, of course, even astronomers don't now use ephemeris time in their practical activities. Contemporary atomic-based clocks are easier to use and stunningly accurate. At the theoretical level what has happened is that the metric of time has entered into many laws of foundational physics besides the laws of dynamics. It enters into the laws of electromagnetism and electromagnetic radiation. And we have, correspondingly, the possibility of optical clocks, say that of a light ray bouncing back and forth between two mirrors at rest with respect to one another and both in some inertial frame of motion. And time enters into the foundational laws of quantum mechanics, the laws that govern such things as the frequencies correlated to the state transitions of the atoms in the resonant chamber that form the guts of the atomic clock.

With each extension of physics into which the notion of time is embedded, the concept of "equal time interval" becomes implicitly framed by the new foundational phenomena. For example, the Michelson-Morley experiment pushes physics from Newton's absolute time to the frame-dependent time intervals of special relativity. One of the crucial facts of nature, though, is that the equality of time intervals as picked out by a properly reconstructed (relativistic) dynamics, the descendant of Newton's absolute time as picked out by inertial motion, *agrees* with the standards for equality of time intervals picked out by an idealized optical clock or by an idealized quantum oscillator. Indeed, it is this universality of the notion of equality of time interval across these physical phenomena that legitimizes our very talk of one and the same "time" in all these aspects of our fundamental theories.

Some fundamental physical properties of the world are unavailable to us either through any kind of direct perception or surface features of our physical environment. The concepts characterizing these properties are introduced for the first time by the role they play in foundational physical laws. Physicists deal with them symbolically (or, nowadays, give them silly names such as "strangeness"). But other features (temperature, temporal interval) are present in our everyday, pre-scientific experience. Here the process of conceptual refinement described above plays its role. We start with a loosely characterized "subjective" notion (hotter/colder, shorter or longer duration). We move on to an objectified (often metric) notion by means of measuring instruments (thermometers, clocks of the astronomical or terrestrial sort). Finally we begin to understand the role played by some feature of the world at the most basic level, by embedding the concepts describing that feature into our foundational physics (thermodynamics and statistical mechanics, classical dynamics).

It is the last stage of this process that is exemplified by Newton's characterization of equality of time intervals in the "Scholium" to the *Principia*. And this is the first profound step in the beginning of the scientific process of trying to decipher just what time *is*. It is the role played by time in classical dynamics that illustrates for us the role played by classical dynamics in deepening our understanding of time.

CHAPTER 20

# TIME TRAVEL AND TIME MACHINES

CHRIS SMEENK AND
CHRISTIAN WÜTHRICH

## ABSTRACT

THIS chapter is an enquiry into the logical, metaphysical, and physical possibility of time travel understood in the sense of the existence of closed worldlines that can be traced out by physical objects. We argue that none of the purported paradoxes rule out time travel either on grounds of logic or metaphysics. More relevantly, modern spacetime theories such as general relativity seem to permit models that feature closed worldlines. We discuss, in the context of Gödel's infamous argument for the ideality of time based on his eponymous spacetime, what this apparent physical possibility of time travel means. Furthermore, we review the recent literature on so-called *time machines*, that is, of devices that produce closed worldlines where none would have existed otherwise. Finally, we investigate what the implications of the quantum behaviour of matter for the possibility of time travel might be, and explicate in what sense time travel might be possible according to leading contenders for full quantum theories of gravity such as string theory and loop quantum gravity.

## 1. INTRODUCTION

The general theory of relativity allows an abundant variety of possible universes, including ones in which time has truly bizarre properties. For example, in some universes the possible trajectory of an observer can loop back upon itself in time, to form what is called a closed timelike curve (CTC). In these universes time travel is possible, in the sense that an observer traversing such a curve would return to exactly the same point in spacetime at the "end" of all her exploring. Such curves and other similarly exotic possible structures illustrate the remarkable flexibility of general relativity (GR) with regard to the global properties of spacetime. Earlier theories such as Newtonian mechanics and special relativity postulate a fixed geometrical and topological spacetime structure. In contrast, GR tolerates a wide variety of geometries and topologies, and these are dynamical rather than fixed *ab initio*. This toleration does have bounds: the theory imposes weak global constraints on possible universes, such as requiring four-dimensionality and continuity, along with local constraints imposed by the basic dynamical laws (Einstein's field equations) and the requirement that locally the spacetime geometry approaches that of the special theory of relativity. But within these bounds flourish an embarrassingly rich collection of possible topologies and geometries that depart quite dramatically from the tame structures apparently compatible with our experience.

What does the existence of solutions with such exotic structures imply regarding the nature of space and time? Below we will assess different answers to this question before offering our own. But the question itself has to be disambiguated before we can sketch answers to it. First, what do we mean by the "existence of solutions" with exotic structures? At a minimum we require that these are solutions of the field equations of GR, and in that sense "physically possible" models according to the theory. But one recurring theme of the discussion below is that this is a very weak requirement, due to the possibility of constructing "designer spacetimes" that satisfy the field equations only, in effect, by stipulating the right kind of matter and energy distribution to produce the desired spacetime geometry. Second, much of the discussion below will focus on GR, our best current spacetime theory. But part of the reason for interest in these solutions is the light their study may shed on the as yet unformulated theory of quantum gravity, so we will also discuss hybrid theories that include quantum effects within GR and briefly touch upon candidates for a full quantum theory of gravity. Third, our focus below will be on one kind of exotic structure—namely, CTCs. We expect that arguments roughly parallel to those below could be run for other kinds of exotic structure. Finally, we hope to isolate the *novel* consequences of the existence of exotic spacetimes, distinct from other lessons of special and general relativity.[1]

One answer to our question simply denies the relevance of exotic spacetimes entirely. Some physicists and many philosophers have argued that solutions with CTCs are logically or metaphysically impossible, based on paradoxes of time travel such as the grandfather paradox. Suppose Kurt travels along a CTC and has Grandpa in his rifle sight, finger at the trigger. Either outcome of this situation seems wrong: either Kurt succeeds in killing Grandpa, preventing the birth of his father Rudolf and his

[1] Two other chapters in this volume address these other implications of relativity theory for understanding the nature of time: Savitt focuses on the implications of special relativity, and Kiefer discusses the problem of time in quantum gravity.

own conception, or something such as a well-placed banana peel mysteriously prevents Kurt from fulfilling his murderous intentions. In §2 below we will argue that this paradox (and others) does not show that spacetimes with CTCs are logically incoherent or improbable: what they show instead is that in spacetimes with CTCs, questions of physical possibility—for example, whether it is possible for Kurt to kill Grandpa—depend upon global features of spacetime. We further argue against the idea that one should rule out exotic structures a priori by imposing stronger restrictions on causal structure than those imposed by GR. The ideas of causal structure in GR that give content to this debate are reviewed in §3.

A second answer delimits the opposite extreme. On this view, the existence of models with CTCs is taken to imply directly that there is no "objective lapse of time" according to GR. Gödel famously offered an argument for this conclusion based on his discovery of a spacetime with the property that a CTC passes through every point. We will discuss Gödel's solution and his arguments based on it in detail in §4. The problem we will focus on in assessing Gödel can be stated more generally: if we do not take the exotic spacetimes to be physically viable models for describing the entire universe or particular systems within it, what does their existence reveal regarding the nature of space and time?

Our own answer falls between the two extremes. The question proves to be a fruitful and difficult one due to our lack of understanding of the large-scale dynamics of GR and the space of solutions to the field equations. Suppose we allow that the existence of alternative cosmological models such as Gödel's that are not viable descriptions of the observed universe does not have direct implications for the nature of time in our universe. We can still ask the following question: what is the nature of time in a class of solutions that are directly accessible from our universe, in the sense that an arbitrarily advanced civilization could reach them by locally manipulating matter and energy? If it were possible to *create* CTCs via local manipulations—in effect, to operate a *time machine*—then we could argue much more directly than Gödel against the existence of an objective lapse of time. In §5 below we offer a definition of a time machine along these lines. However, even if a time machine so defined proves to be impossible, it is tremendously important to understand *why* it is impossible. Our general approach to this issue is familiar from John Earman's (1986) treatment of determinism: pushing a theory to its limits often reveals a great deal about its content, and may lead to refinement of core principles or even provide stepping stones to further theory. In this case, a proof that TMs are impossible could take the form of delimiting some range of solutions of the field equations as "physically reasonable," and showing that these solutions do not give rise to CTCs or other exotic structures. This way of formulating the problem brings out the parallels with Roger Penrose's "cosmic censorship conjecture," as we will discuss in §6. Finally, the vast physics literature on time travel and time machines has been inspired by intriguing connections with quantum field theory and quantum gravity (the topics of §7 and §8).

## 2. THE PARADOXES OF TIME TRAVEL

So what is time travel? The standard answer among philosophers, given by David Lewis (1976, 68), is that time travel occurs in case the temporal separation between departure and arrival does not equal the duration of the journey. However, this is not a necessary condition for time travel. Presumably, Lewis and everyone else should want to include a case when the time lapse between departure and arrival equals the duration of the journey, but the arrival occurs *before* the departure.

More significantly, we also claim that Lewis's definition does not state a sufficient condition for an interesting sense of time travel within the context of modern physics. Readers familiar with special relativity may have already asked themselves what Lewis might mean by temporal separation between arrival and departure. Due to the relativity of simultaneity, observers in relative motion will generally disagree about the temporal separation between events. We could try to skirt this difficulty by defining the temporal separation as the maximal value measured by any observer (corresponding to the proper time elapsed along a geodesic connecting the two events). This proposal would allow us to assign an objective meaning to Lewis's temporal separation between arrival and departure. But the resulting definition of time travel is far too promiscuous. Everyone who departs from geodesic motion—due to the slightest nudge from a non-gravitational force—counts as a time traveler. Admittedly, Lewis's definition does seem to capture an intuitive sense of "time travel" that is useful for some purposes. But it is too broad to capture a useful distinction within relativity, given that nearly every observer would qualify as a time-traveler.

Thankfully, an alternative conception of time travel that avoids these problems is close at hand in GR. There is a sense in which GR permits time travel into the past: it allows spacetimes containing closed timelike curves (CTCs), that is spacetimes with unusual causal structures.[2] Loosely speaking, a CTC is a path in space and time that can be carved out by a material object and is *closed*, that is returns to its starting point not just in space, but also in time. A curve is *everywhere timelike*, or simply *timelike*, if the tangent vectors to the curve are timelike at each point of the curve. A timelike curve represents a possible spatio-temporal path carved out by material objects, a so-called *worldline*. Of course, we also presuppose that the curves representing observers are continuous. GR declares as compatible with its basic dynamical equations spacetimes that contain closed worldlines that could be instantiated by material objects. It is evident that the presence of worldlines that intersect themselves is a sufficient condition

---

[2] Strictly speaking, as we will see in §3, spacetimes with CTCs do not allow a global time ordering and thus there is no global division into past and future. But it is always possible to define a local time ordering within a small neighborhood of a given point, and a CTC passing through the point would connect the point with its own past according to this locally defined time ordering.

for time travel to take place. For the rest of this essay, we shall also assume that it is a necessary condition.[3]

Both the popular and the philosophical time travel literature contain vivid debates regarding whether time travel in this sense is logically impossible, conceptually or metaphysically incoherent, or at least improbable. Let us address these three issues in turn.

## 2.1 Logical impossibility: the grandfather paradox

Although less prevalent than a decade or two ago, the belief that various paradoxes establish the logical or metaphysical impossibility of time travel is still widespread in philosophy. The grandfather paradox introduced above is no doubt the most prominent of these paradoxes. It allegedly illustrates either how time travel implies an inconsistent past and is thus ruled out by logic,[4] or that time travel is extremely improbable. Other time travel paradoxes include the so-called *predestination* and *ontological* paradoxes. A paradox of predestination arises when the protagonist brings about an event exactly by trying to prevent it. These paradoxes are not confined to scenarios involving time travel, although they add to the entertainment value of the latter. Just imagine a time traveler traveling into her own past in an attempt to prevent the conception of her father, whose actions instead kindle the romance between her grandparents. The related *ontological paradox* can be exemplified by the story of the unpainted painting. One day, an older version of myself knocks on my door, presenting a wonderful painting to me. I keep the tableau until I have saved enough money to be able to afford a time machine. I then use the time machine to travel back in time to revisit my younger self, taking the painting along. I ring the doorbell of my earlier apartment, and deliver the painting to my younger self. Who has painted the picture? It seems as if nobody did since there is no cause of the painting. All the events on the CTC have

---

[3] This might seem to be overly restrictive, as it would appear to rule out a scenario in which the time traveler follows a nearly closed trajectory rather than a CTC. First, in such scenarios there may also be CTCs, even if these are not instantiated by material objects, and then our definition holds. Second, we could liberalize the definition slightly to take the existence of near-CTCs as a necessary condition for time travel, which would amount to moving to a strictly weaker condition on the causal structure of spacetime, called "stable causality" (see §3 below). Monton (2009) argues that CTCs should not be taken as a necessary condition for time travel, but we believe that Monton's argument fails. If one rules out discontinuous worldlines and similarly unphysical constructs, then CTCs are arguably the only Lorentz-invariant way of implementing time travel. Cf. Arntzenius (2006, Sec. 3) for an alternative transposition of a Lewis-like understanding of time travel into the context of GR. We don't see, however, how this understanding can be extended to cover non-time orientable spacetimes, as Arntzenius seems to think (2006, 604f).

[4] In a *dialethic logic*, i.e. a logic in which contradictions can be true, and perhaps in other paraconsistent logics, such contradiction need not imply the impossibility of time travel. A possible reply to the grandfather paradox is thus the rejection of classical logic. This price is considered too high in this article, particularly also because the contradiction can be resolved by other means, as will be argued shortly.

just the sort of garden-variety causes as events not transpiring on CTCs do. The causal loop *as a whole*, however, does not seem to have an originating cause. For all these reasons, the popular argument goes, causal loops cannot exist.

Lewis (1976) has argued that although such scenarios contravene our causal intuitions, it is not in principle impossible that uncaused, and thus unexplainable, events in fact occur. According to Lewis, there are such unexplainable events or facts such as the existence of God, the big bang, or the decay of a tritium atom. True. Who would have expected that time travel scenarios will be easily reconcilable with our causal intuitions anyway? The fact that phenomena transpiring in a time-travel universe violate our causal intuitions, however, is no proof of the impossibility of such a world. Analogously, predestination paradoxes can be rejected as grounds for believing that time travel is impossible: although they undoubtedly exude irony, the very fact that it was the time traveler who enabled her grandparents' union is not in any way logically problematic. What is important as far as logic is concerned is that the time traveler has timelessly been conceived at some point during the year before her birth and has not been "added" or "removed" later. If it occurred, it occurred; if it didn't, it didn't. So despite their persuasiveness, the ontological and the predestination paradoxes don't go far in ruling out time travel.

The grandfather paradox cannot be dismissed so easily. Grandpa cannot simultaneously sire and not sire the parent of the time traveler. The central point is that the grandfather paradox does not rule out time travel *simpliciter*, but only inconsistent scenarios. In fact, all self-contradictory scenarios are forbidden, regardless of whether they involve time travel or not. Various options can be pursued in attempts to resolve the grandfather paradox. Apart from the costly rejection of bivalent logic, one can, following Jack Meiland (1974), postulate a two-dimensional model of time such that every moment entertains its own past which is distinct from the times that preceded that moment. According to this proposal, at a given moment there are two branches, one containing the actual events that preceded it, and the other representing an alternative past into which time travel can lead. If one travels back in time, then, one doesn't arrive at a time that preceded the departure, but rather at a time in the past of the moment when one departed. Time, on this understanding, is represented by a two-dimensional plane rather than a one-dimensional line. Following Lewis (1976, 68), we do not find this resolution particularly attractive, primarily because the time traveler would, on this conception, never be able to revisit the very past moment when Grandpa first met Grandma. She would only be able to reach a "copy" of this moment on the past line of the moment of when the time machine is switched on. The event reached would thus be different from the one steeped in history that the intrepid traveler intended as the goal of her journey. Whatever travel this is, it is not the time travel characterized above.

An obvious, but rarely seriously entertained option tries to make sense of time travel by allowing the universe to bifurcate each time consistency would otherwise be violated. The instant the time traveler arrives in her past, the spacetime splits into two "sheets." (Unlike Meiland's proposal, the branches are "created" by time travel, they

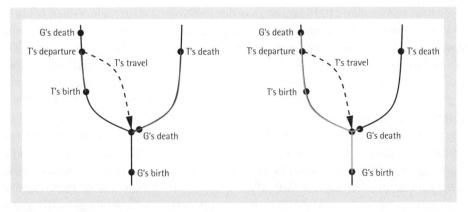

FIGURE 20.1    The worldlines of the time traveler $T$ (dark grey) and of Grandpa $G$ (light grey) according to the multiverse proposal. Note that both figures are of the same multiverse; they are just highlighting different worldlines

are not already in existence.)[5] This branching does not happen in time or space alone, but in the overall causal structure of the spacetime in which the journey takes place. In particular, the causal future of the event where the traveller arrives must permit "two-valuedness." In the case of such a "multiverse," the adventurous traveler not only journeys in time, but also to a branch distinct from the one in which she departed. A multiverse with more than one actual past history does timelessly contain the killing of Grandfather, but only in one of the branches (cf. Lewis 1976, 80). Interaction between the co-existing branches is solely possible by time travel, which does arguably not deserve to qualify for time travel as it is not a journey back in the traveler's "own" time. But the threat of inconsistency is surely banned if history along any given branch is consistent. This would, for example, mean that everybody's worldlines have an unambiguous beginning and end points in all branches (see Figure 20.1).

Does such travel in a multiverse change the past? Only in the sense that through the traveling activity, more and more branches of past histories seem to pop into existence. If this is the picture, then time traveling necessitates an inflation of branches as it becomes more popular. But since if it is possible to change the past, we run into the same difficulties as with the grandfather paradox again: these branches must in fact eternally co-exist with the sheet we are actually living in. Thus, if time travel is physically possible, then there will be an infinitude of branches corresponding to all the possible ways in which time travel could occur. Thus, there will be an infinity of actual past histories of the multiverse timelessly containing all time traveling activity. Even though such a construction does not live up to an ideal of metaphysical austerity,

[5] A further contrast between the proposals is that on Meiland's view the time traveler will not have complete freedom as to how to affect the past since, presumably, both pasts must lead to the same present moment located at the bifurcation point. This constraint seems to be absent in scenarios with branching structures into the future, at least if one grants the causal fork asymmetry (cf. Horwich 1987, 97–99).

logic does not preclude it. However, in order to accommodate multi-valued fields in physics—which would be necessary in such a multiverse—a radical rewriting of the laws of physics would be required. Although topology offers manifolds which could potentially deal with multi-valuedness,[6] these new types of laws would also have to tolerate it. But we do not know of a dynamical theory which could deliver this.

We concur with Earman (1995) (and, unsurprisingly, with Earman et al. (2009)) that the grandfather paradox only illustrates the fact that time-travel stories, just like any other story, must satisfy certain consistency constraints (CCs) that ensure the absence of contradictions. In other words, only one history of the universe is to be told, and this history had better be consistent. GR mandates that spacetimes satisfy what Earman dubbed a *global-to-local property*, that is, if a set of tensor fields satisfy the laws of GR globally on the entire spacetime, then they do so locally in every region of spacetime.[7] This property is shared by spacetimes with CTCs. The reverse *local-to-global property* would imply that any local solution could be extended to a global solution of the field equations. But this property need not hold in spacetimes with CTCs: situations that are admissible according to the local dynamical laws may lead to inconsistencies when evolved through a region containing CTCs. CCs are imposed to prevent such inconsistencies. John Friedman et al. (1990) encode the demand that CCs are operative in their *principle of self-consistency*, which "states that *the only solutions to the laws of physics that can occur locally in the real Universe are those which are globally self-consistent.*"[8] This principle guarantees the validity of the local-to-global property, at the cost of introducing non-trivial CCs.

How should we think of CCs? We can think of them as consisting of restrictions imposed on the initial data of, say, a matter field for point mass particles at a given point. Assume a single particle that moves along an inertial worldline in accordance with the dynamical laws that apply for the particle, and assume further that the worldline is a CTC. The CCs would then have to restrict the choice of the initial velocity of the particle such that its trajectory smoothly joins itself after one loop. More generally, however, the CCs for any macroscopic object involving more complex physical processes would become very complicated indeed if spelt out explicitly. Consider a more concrete example involving macroscopic objects, such as a spacecraft venturing out to explore deep space only to discover that it in fact traces out a CTC. Here, the spacecraft would have to go over into its earlier state smoothly, including restoring the "original" engine temperature and settings of all onboard computers, refueling to exactly the same amount of propellant, and so forth. If the scenario included humans, it would become trickier still. The time traveler would have to rejoin exactly his

---

[6] Cf. Visser (1996, 250–255). The concerned manifolds have to be non-Hausdorff in order to permit branching, as discussed in Douglas (1997). For a thorough critique of branching spacetimes, cf. Earman (2008).

[7] Cf. Earman (1995, 173) for a more mathematically rigorous account.

[8] Friedman et al. (1990, 1916f.) emphasis in original. For more advocacy of CCs, see Malament (1985b, 98f.) and Earman (1995, passim). They both see the emergence of CCs as the one and only lesson to be learnt from the grandfather paradox.

worldline, wearing the same clothes, with the same shave, with each hair precisely in the same position, with his heart beat cycle exactly coinciding, his memory reset to the state when he entered the CTC etc. The world is rich in variety and complexity, and such strong constraints *appear* to conflict with our experience. However, it is not clear how exactly such a conflict could arise: if the relevant dynamical laws have the local-to-global property in a given spacetime with CTCs, then the CCs would be enforced regardless of their apparent improbability. In any case, regions of causality violations are found beyond horizons of epistemic accessibility of an earth-bound observer in realistic spacetimes. Hence, if taken as an objection against the possibility of CTCs, the difficulty of accommodating complex scenarios has little theoretical force. But it surely shatters the prospect of sending humans on a journey into their own past in a way that has them instantiate the totality of a CTC.

Since CCs seem to mandate what time travelers can and cannot do once they have arrived in their own past, the CCs' insistence that there is only one past and that this past cannot be changed appears to give rise to a kind of *modal paradox*. Either John Connor's mother is killed in 1984 or she isn't. In case she survives, the deadliest Terminator with the highest firepower cannot successfully assassinate her. This inability stands in a stark contrast to the homicidal capacities that we would normally ascribe to an armed and highly trained cyborg. The modal paradox arises because the terminator can strike down Connor's mother—he has the requisite weapons, training of many years, and a meticulous plan, etc.—but simultaneously he cannot do it as Sarah Connor actually survived 1984, and the Terminator would thus violate CCs were he to kill her successfully. Lewis (1976) has resolved the looming modal inconsistency by arguing that "can" is ambivalently used here and that the contradiction only arises as a result of an impermissible equivocation. "Can" is always relative to a set of facts. If the set contains the fact that Sarah has survived 1984, then the terminator will not be able to kill her (in that year). If this fact is not included, however, then of course he can. The contradiction is only apparent and Lewis concludes that time travel into one's own past is not logically impossible.

Thus, the paradoxes invoked do not establish that logic precludes time travel, although they exhibit how they constrain the sort of scenarios that can occur. Although logic does not prohibit it, time travel still faces stiff resistance from many philosophers. The resistance typically comes in one of two flavors: either it turns on the alleged improbability of time travel or on an argument barring the possibility of backward-in-time causation. Let us address both complaints in turn.

## 2.2 Metaphysical impossibility: improbability and backward causation

The first philosophical objection originated in Paul Horwich (1987) and was arguably given the most succinct expression in Frank Arntzenius (2006). This popular objection admits that the logical arguments fail to establish the *impossibility* of time travel, but insists that what they show is its *improbability*. The main reason for this improbability is the iron grip that consistency constraints exert on possible scenarios involving time travel. Imagine the well-equipped, highly trained, lethally determined Terminator who consistently fails to kill Connor's mother again and again, in ever more contrived ways. The first time the trigger is jammed, another time a rare software bug occurs, the third time the Terminator slides on banana peel, the fourth time Reese shows up in time to spectacularly save her life, etc. The more often the Terminator attempts the murder and fails, the more improbable the ever more finely tuned explanations of his failure appear to us.

Apart from philosophical objections that have been raised against this argument,[9] which do not concern us here, this line of reasoning also faces difficulties, both concerning the exact formulation and interpretation of the probabilistic thesis to be defended as well as the interpretation of the consistency constraints. What exactly could be meant by the claim that scenarios involving CTCs are "improbable"? If the relevant probabilities should be interpreted as objective, then we should be in the position to define an adequate event space with a principled, well-defined measure. If we were handed such a space, we could then proceed to isolate the subspace of those events which include time travel. But this is hardly possible in a principled and fully general manner. If the generality is restricted to GR, the event space would arguably consist of general-relativistic spacetime models. In this context, the claim that time travel is highly "improbable" could then be given concrete meaning by identifying the models of GR that contain CTCs, and then arguing that these models-cum-CTCs represent only a very small fraction, perhaps of measure zero, of all the physically possible or physically realistic models of GR in the sense of some natural measure defined on the space of all general-relativistic models. Such a construction would allow us to conclude that CTCs are indeed very rare in physically realistic set-ups. It would, however, presuppose significantly more knowledge about the space of all solutions to the Einstein field equations than we currently have. Although there is some exploratory material in this direction in the literature, it is certainly neither in Horwich (1987) nor in Arntzenius (2006). The latter invites us to consider the issue in the context of spacetimes where "later" CTCs may impose bizarre constraints on data in "earlier" regions of spacetime.

Here is how the Horwich-Arntzenius argument that time travel is improbable may get some traction. They might insist that the choice of measure ultimately depends on the *actual* frequencies.[10] The reason why the Liouville measure comes out on top in statistical mechanics is because it is the one measure that returns the observed transitions from non-equilibrium to equilibrium states as typical. In analogy, given that we don't observe time travel, we ought to choose a measure such that time travel

[9] Cf. e.g. Smith (1997).
[10] Of course, for this imagined response to succeed, it must be clarified what these are frequencies *of*. Presumably, the frequencies are of observing time travel, or of observing phenomena that are reasonably interpreted as signatures of the existence of CTCs. This is vague, and it is the onus of the responder here to give a more precise formulation of what it is exactly that these frequencies are of.

comes out as highly atypical.[11] But while we also harbor Humean sympathies, this move still does not deliver the space on which to slap the measure and it ignores the possibility that there may be systematic reasons why we don't observe time travel. Thus, at the very least, this response must explicate what an observation of CTCs would involve. We will return to the problem of finding objective probabilities in the setting of a spacetime theory in §5.

Alternatively, the probabilities involved can be interpreted as subjective. Of course, it is plausible that systematically tested betting behavior of maximally rational human agents would ascribe a low probability to time-travel scenarios. But since we are unfamiliar with such phenomena, our untutored intuitions will not serve as reliable indicators in situations with CTCs. Our resulting betting behavior may thus yield nothing more than additional confirmation that time-travel scenarios appear bizarre to us. This is a potentially interesting psychological point, but hardly qualifies as a serious statement about the possibility or probability of time travel. So either way, regardless of whether we interpret the probabilities involved objectively or subjectively, we are at a loss. Either we propose a novel interpretation of probability that does not run into these difficulties, or we settle for a non-probabilistic likelihood in the claim that time travel is "improbable."

A second difficulty for Horwich's argument arises when we regard the consistency constraints as laws of nature, as we can with some justification. Whether or not consistency constraints really are laws will undoubtedly depend upon our analysis of laws of nature, as Earman (1995) has argued. In case the consistency constraints turn out to qualify as laws, it would be amiss to infer the improbability of time travel from their existence. After all, we do not conclude from the fact that Newton's law of universal gravitation is (at least approximately) a law of nature that a long sequence of bodies that have been released near the surface of the Earth and moved *toward* the Earth, rather than, say, away from it, is highly improbable. Of course, we can extend the set of relevant possible worlds, or of possible worlds simpliciter, in such a way that, despite the nomological status of Newton's law of gravitation in the actual world, there exists a subset of worlds in which objects move away from the center of the Earth and that this subset is much larger than the subset of worlds in which objects fly toward Earth's center. Such an extension, however, would involve such modal luxations that the argument would lose much of its force. Of course, all of this would in no way imply that time travel must occur with a fairly high probability, but only that from the necessity of consistency constraints, one cannot infer its improbability.

One man's modus ponens is another woman's modus tollens, a defender of the improbability might object. Thus, she might simply point out that these considerations drawing on the nomological status of consistency constraints only show, if anything, that we ought not to think of them as fully deserving of promotion to lawhood. The move, thus, is to deny the analogy to the case of Newton's law of universal gravitation. True, our point only commands any force *if* consistency constraints are awarded nomological status. But if we have independent reasons to so award them,

---

[11]  We are indebted to Craig Callender for suggesting this retort on behalf of Horwich and Arntzenius.

then nothing about the probability or improbability of time travel can be inferred from the fact that consistency constraints obtain. If we are right, then Horwich's claim must at least be strongly qualified or amended.

The second philosophical motivation for resisting time travel maintains that the presumed impossibility of backward causation disallows time travel. While most philosophers today would accept that time travel is not ruled out on grounds of paradoxes of consistency, they argue that it necessarily involves backward causation, and since backward causation is conceptually impossible, they believe, so is time travel. Both premises of this argument, however, can defensibly be rejected. The reason why time travel as we conceive it does not involve backward causation, at least not locally, will become clear in the next section when we sharpen the concepts by reconfiguring the issues in the context of spacetime physics. Most philosophers, however, beg to differ as they insist that although a time traveler cannot *change* her past, she must still *affect* it. A causal link between antecedent conditions prior to departure and consequent conditions upon the earlier arrival is required to ascertain the personal identity and thus the persistence of the time traveler. But if such a causal relation is necessary for genuine time travel, they argue, there must be backward causation, that is, causal relations where the effect precedes the cause. This argument, however, does not succeed if causation is conceptualized as a purely local phenomenon, connecting only events of adjacent spacetime regions or, more precisely, if causal propagation occurs only along smooth curves in spacetime. We will return to this in §3. The second premise—that backward causation is incoherent or impossible—is genuinely metaphysical and shall be dealt with here. We do not wish to commit ourselves to the metaphysical possibility of backward causation here, but we want to elucidate the dialectical grounds on which backward causation is ruled out of bounds.

The basic idea behind the ruling is captured by the so-called *bilking argument*.[12] Consider the following "experimental set-up." We plan an experiment in which we attempt to produce (prevent) the subsequent cause $C$ whenever we have previously observed the absence (presence) of the potential effect $E$. The experiment is then performed repeatedly, with sufficient runs to gain statistically meaningful results. It will be found that either one of two possibilities obtains: Either (i) $E$ often transpires despite the absence of $C$, and the occurrence of $C$ is often unaccompanied by $E$; or (ii) whenever $E$ does not occur, our attempts to bring about $C$ consistently fail, and whenever $E$ occurs, we cannot prevent the subsequent occurrence of $C$. In case (ii), our ability to produce $C$ depends upon the previous presence of $E$. Advocates of the bilking argument insist that in this situation, we ought to interpret $E$ as a necessary causal antecedent condition for the occurrence of $C$, rather than a consequent condition. The causal relation between $E$ and $C$ obtains, but should be taken to hold in the opposite direction: $E$ is the cause, not the effect, of $C$. In case (i), on the other hand,

---

[12]   For a classic formulation, cf. Mellor (1981). The bilking argument was first formulated by Black (1956). Dummett (1964) reacts to Black, defending backward causation. The argument as presented here is grossly simplified and neglects additional factors such as the informational state of the epistemic agents involved, the possibly statistical nature of certain causal relations, etc.

the hypothesis of backward causation is simply false, as the factors do not stand in any causal relation. In either of the two cases, we don't have backward causation.

Does the bilking argument rule out the possibility of backward causation? It is not so clear that it does. First, the experimental design may not be implementable in all situations of potential backward causation. Huw Price (1984) has argued that, while it seems reasonable to assume that backward causation is impossible in instances that could in fact be bilked, there is no guarantee that this is always possible. The conditions for bilking fail to obtain, for instance, if we cannot discover whether a supposed earlier effect has in fact occurred. Price (1984) has forcefully argued that this may indeed be the case for quantum-mechanical systems when the observation of the candidate effect can only be achieved at the price of disturbing the system in such a way that is itself causally relevant for the occurrence of $E$. Alternatively, it is at least conceivable that it may be nomologically impossible to determine whether $E$ has indeed occurred prior to $C$. Second, even if the experimental design can be implemented in all relevant situations, the argument may not succeed because bilking may be frustrated not by a reversed causal relation between $E$ and $C$, but simply by cosmic coincidences. We can run very long experimental series in an attempt to minimize the probability of such coincidences, but we never seem able to fully rule out their possibility.

Without delving into these matters here, we conclude that the case against backward causation is not closed. We would like to repeat, however, that the next section will show how the attack against the possibility of time travel based on the bilking argument against backward causation misses the point, because there is a perfectly respectable sense in which time travel as it arises in the context of spacetime theories does not involve backward causation.

We conclude the section by admitting that the discussed paradoxes reveal that unusual consistency constraints are indeed required in time travel spacetimes. We do not, however, grant that the presence of such consistency constraints indicates the "improbability" of time travel scenarios or even offers the basis for rejecting time travel altogether. Most importantly, we conclude that time travel is neither logically nor metaphysically impossible, as all arguments attempting to establish inconsistency have failed and the philosophical considerations adduced against the possibility or probability of time travel are inconclusive at best. But even if logic and metaphysics do not rule out time travel, physics might. Let us thus now turn to the question of whether physics permits time travel.

# 3. CAUSAL STRUCTURE OF RELATIVISTIC SPACETIMES

Formulating our question—whether physics permits time travel—precisely requires an understanding of the causal structure of relativistic spacetimes. The study of causal structure developed as physicists freed themselves from the study of particular solutions and began to prove theorems that would hold generically, most importantly the singularity theorems. The local constraints imposed by the dynamical laws of GR are compatible with solutions with a wide variety of different global properties, which are a rich source of counterexamples to proposed theorems. The ideas surveyed briefly in this section make it possible to characterize solutions based on their "large-scale" or global features, and to bar counterexamples by imposing conditions on the causal structure. We will see that the causality conditions can be thought of roughly as specifying the extent to which a spacetime deviates from the causal structure of Minkowski spacetime.[13]

We begin by reviewing definitions.[14] A *general-relativistic spacetime* is an ordered pair $\langle \mathcal{M}, g_{ab} \rangle$, where $\mathcal{M}$ is a connected four-dimensional differentiable manifold without boundary, and $g_{ab}$ is a Lorentz-signature metric defined everywhere on $\mathcal{M}$, such that $g_{ab}$ satisfies Einstein's field equations $R_{ab} - \frac{1}{2} R g_{ab} + \Lambda g_{ab} = 8\pi T_{ab}$ for some energy-momentum tensor $T_{ab}$.[15] The metric $g_{ab}$ fixes a light cone structure in the tangent space $\mathcal{M}_p$ at each point $p \in \mathcal{M}$. A tangent vector $\xi^a \in \mathcal{M}_p$ is classified as timelike, null, or spacelike, according to whether $g_{ab} \xi^a \xi^b > 0, = 0$, or $< 0$ (respectively).[16] Geometrically, the null vectors "form the light cone" in the tangent space with the timelike vectors lying inside and the spacelike vectors lying outside the cone. The classification of tangent vectors extends naturally to curves: a *timelike curve* is a continuous map of an interval of $\mathbb{R}$ into $\mathcal{M}$, such that its tangent vector is everywhere timelike. A spacetime is *time orientable* if and only if there exists a continuous, nowhere-vanishing vector field, which makes a globally consistent designation of one lobe of the null cone at every point as "future" (in the which-is-which sense) possible. In a time-orientable spacetime, one can then define *future (past)-directed* timelike curves as those whose tangent vectors fall in the future (past) lobe of the light cone at each point. Future-directed *causal* curves have tangent vectors that fall *on or inside* the future lobe of the light cone at each point.

---

[13]   Our approach in this section may seem to imply a substantivalist view of spacetime, given that we treat the manifold $\mathcal{M}$ as the basic object of predication and apply various mathematical structures to it. We do not argue for this implication here or implicitly endorse it, and we take the clarification of causal structure to be prior to the relational–substantival debate.

[14]   We can recommend nothing better than Geroch and Horowitz (1979) for a clear and self-contained introduction to the global structure of relativistic spacetimes; see also Hawking and Ellis (1973) or Wald (1984) for textbook treatments, or Earman (1995) for a more philosophically oriented discussion.

[15]   The *energy-momentum tensor* $T_{ab}$ is a functional of the matter fields, their covariant derivatives, and the metric $g_{ab}$ that satisfies the following conditions: (i) $T_{ab}$ is a symmetric tensor, (ii) $T_{ab} = 0$ in some open region $\mathcal{U} \subset \mathcal{M}$ just in case the matter fields vanish in $\mathcal{U}$, and (iii) $T^{ab}_{;b} = 0$. This third condition is a generalization of the special-relativistic conservation of energy and of linear momentum in the non-gravitational degrees of freedom, and is implied by the diffeomorphism invariance of the theory. Our discussion here is "kinematical," in the sense that we do not yet impose any further constraints on the energy-momentum tensor, such as the energy conditions discussed in §8 below. The symmetric tensor $R_{ab}$, called the *Riemann tensor*, is a functional of the metric, $R$ is the *Riemann scalar*, and $\Lambda$ is the cosmological constant. Throughout this article, we use natural units, i.e. $c = G = 1$.

[16]   In this article we use a $(1, 3)$ signature, which means that the metric assigns a length $+1$ to one of the four orthonormal basis vectors that can be defined in each tangent space, and $-1$ to the other three.

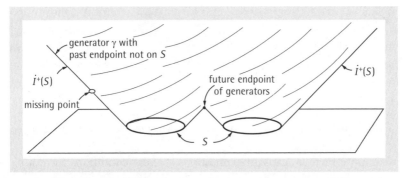

FIGURE 20.2   The generators of the boundary may have endpoints not in $S$

We can now precisely characterize the set of points in $\mathcal{M}$ that are causally connected to a given point $p \in \mathcal{M}$. The *chronological future* $I^+(p)$ is defined as the set of all points in $\mathcal{M}$ that can be reached from $p$ by (non-trivial) future-directed timelike curves; the *causal future* $J^+(p)$ includes all points that can be reached by a future-directed causal curve. (The past sets $I^-(p)$ and $J^-(p)$ can be defined analogously, and the definitions extend straightforwardly to sets $S \in \mathcal{M}$ rather than points.) More intuitively, $I^+(p)$ includes the points in $\mathcal{M}$ that can be reached by a "signal" emitted at the point $p$ traveling below the speed of light. (The causal future $J^+(p)$ includes, in addition to those in $I^+(p)$, points connected to $p$ by null geodesics; although light signals propagate along such curves, using $J^\pm(p)$ rather than $I^\pm(p)$ in what follows leads to unnecessary complications.) The boundary of the chronological future, $\dot{I}^+(S)$, is the set of points lying in the closure of $I^+(S)$ but not its interior.[17] It is easy to see that $\dot{I}^+(S)$ must consist of null surfaces (apart from the set $S$ itself).[18] In fact, $\dot{I}^+(S)$ is an achronal boundary, i.e. no point can be connected to another point in the surface via a timelike curve. Any given point $q \in \dot{I}^+(S)$ lies either in the closure of $S$ or in a null geodesic lying in the boundary, called a *generator* of the boundary. In Minkowski spacetime, the generators of the boundary can have *future* endpoints on the boundary, beyond which they pass into $I^+(S)$, but their *past* endpoints all lie on the set $S$. However, this does not hold in general, given the possibility that null geodesics making up part of the boundary may encounter "missing points" rather than reaching $S$ (as illustrated in Figure 20.2).

More generally, the properties of the sets $I^\pm(p)$ in Minkowski spacetime hold locally in any general-relativistic spacetime, but to insure that they also hold *globally*

---

[17]   The closure of a set $S$, denoted $\bar{S}$ is the smallest closed set containing $S$, whereas the interior is the largest open set contained in $S$.

[18]   Suppose the boundary is timelike. For an arbitrary point $p$ just outside the boundary $\dot{I}^+(S)$, there will be points in the interior of $I^+(S)$ that can be connected to it by a timelike curve—and hence $p$ lies in $I^+(S)$, contrary to our assumption. Similarly, suppose $\dot{I}^+(S)$ is spacelike and consider a point $p$ lying to the future of the boundary. But there are timelike curves connecting $p$ to points in $I^+(S)$, again contradicting the assumption. For further information regarding the properties of the boundary, see, e.g. Hawking and Ellis (1973, Prop. 6.3.1).

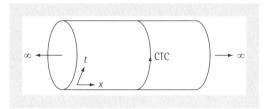

FIGURE 20.3   Rolling up a slice of (1 + 1)-dimensional Minkowski spacetime

one must impose further "causality conditions." These conditions form a hierarchy in terms of strength. The lowest condition in the hierarchy, *chronology*, rules out the existence of CTCs. The existence of a CTC passing through a point $p$ implies that $p \in I^+(p)$; chronology requires that there are no such points. One can construct "artificial" examples of relativistic spacetimes that fail this condition, such as the spacetime defined by "rolling up" (1 + 1)-dimensional Minkowski spacetime (see Figure 20.3). Although this example is obviously quite artificial, there are a number of more interesting chronology-violating spacetimes, such as Gödel spacetime (discussed in the next section).[19] Even this simple example serves to illustrate the point that *locally* chronology violating spacetimes do not involve backward causation or anything of the sort. The description of physical processes in a region of rolled-up Minkowski spacetime will not differ from that in ordinary Minkowski spacetime, except that the presence of CTCs may lead to consistency constraints (we return to this topic below). Chronology is a weak condition and does not imply other properties that one might hope that the chronological future and past satisfy. For example, Minkowski spacetime is *past* and *future distinguishing* in the sense that different points have different chronological pasts and futures (that is, for every $p, q \in \mathcal{M}$, $I^-(p) = I^-(q) \to p = q$ and $I^+(p) = I^+(q) \to p = q$). It is easy to construct spacetimes in which chronology holds that fail to be past and future distinguishing, and it is also the case that the disjunction (past *or* future distinguishing) does not imply the conjunction.

Before moving up the hierarchy, consider the following question: what conditions do we need to impose in order to reconstruct a spacetime based on causal relations encoded in the sets $I^\pm(p)$ for all $p \in \mathcal{M}$? In other words, given two spacetimes $\langle \mathcal{M}, g_{ab} \rangle$ and $\langle \mathcal{M}', g'_{ab} \rangle$ and a mapping between them $\phi : \mathcal{M} \to \mathcal{M}'$ that preserves the causal structure, in that $q \in I^+(p) \leftrightarrow \phi(q) \in I^+(\phi(p))$), what can we further claim regarding their full spacetime structures—including topology and geometry? Intuitively, if the causality conditions fail, then requiring that such a map exists is less informative; for example, one can define such a map between two chronology-violating spacetimes with quite different topological and geometric structures.[20] But at what

---

[19]   Discussions of causal structure often use simple "artificial" constructions, based on the conviction that the features they illustrate show up in more complicated guise in spacetimes that are more physically reasonable.

[20]   Given that the $I^\pm(p)$ are open sets, for a suitably well-behaved spacetime these can be used to define a topology (called the *Alexandrov topology*) equivalent to that of the manifold topology. In the presence of closed timelike curves or other acausalities, a topology defined using $I^\pm(p)$ is too coarse,

point on the hierarchy do the sets $I^{\pm}(p)$ encode sufficiently rich information about the spacetime structure to make it possible to reconstruct $\langle \mathcal{M}, g_{ab} \rangle$ from $\langle \mathcal{M}', g'_{ab} \rangle$? David Malament (1977) proved that if the spacetimes are past *and* future distinguishing, then $\langle \mathcal{M}, g_{ab} \rangle$ and $\langle \mathcal{M}', g'_{ab} \rangle$ have the same geometrical structure up to a conformal factor.[21] The import of this result is that any philosophical program to give a reductive analysis of spacetime structure in terms of causal relations has a fighting chance only in past and future distinguishing spacetimes.

At the top of the hierarchy of causality conditions we find *global hyperbolicity*. Define the *future domain of dependence* $D^+(\Sigma)$ for a global time slice $\Sigma$ to be the set of points in $\mathcal{M}$ such that every past inextendible causal curve through $p$ intersects $\Sigma$ (with the obvious analogous definition for $D^-(\Sigma)$). The "slice" $\Sigma$ is a spacelike hypersurfaces with no edges such that no causal or timelike curve intersects $\Sigma$ more than once. (The *edge* of an achronal surface $S$ is the set of points $p$ such that every open neighborhood $O$ of $p$ includes points in $I^+(p)$ and $I^-(p)$ that can be connected by a timelike curve that does not cross $S$.) Thus, a global time slice $\Sigma$ is an inextendible, smooth spacelike hypersurface which trisects $\mathcal{M}$: $\Sigma$ itself, as well as the "past" and the "future" of $\Sigma$. The future boundary of $D^+(\Sigma)$ is called the *future Cauchy horizon* $H^+(\Sigma)$.[22] The *domain of dependence* $D(\Sigma)$ is the union of the past and future domains of dependence. Global hyperbolicity requires that the spacetime possesses a *Cauchy surface*, that is a slice $\Sigma$ such that $D(\Sigma)$ is the entire spacetime.[23] Global hyperbolicity implies that the manifold $\mathcal{M}$ is topologically $\Sigma \times \mathbb{R}$. It also ensures that the generators of the boundary $\dot{I}^+(S)$ share the properties of boundaries in Minkowski spacetime, in particular that $\dot{I}^+(S)$ consists of the future-directed null geodesics with past endpoints on $S$—it rules out the possibility that there are incomplete null geodesics lying in the boundary of the chronological future (and past). A spacetime which possesses a Cauchy surface is safe for determinism, in that initial data specified on the Cauchy surface determine a unique solution (up to diffeomorphisms) to Einstein's field equations throughout the spacetime. Furthermore, there are typically existence and uniqueness theorems showing that data specified on a partial Cauchy surface $\Sigma$ uniquely determines a solution throughout the domain of dependence $D(\Sigma)$ for matter fields coupled to Einstein's field equations. (See §5 for further discussion of the initial value problem in GR.)

---

lacking open sets corresponding to every open set of the manifold topology. For example, there is a trivial map preserving causal structure between Gödel spacetime (discussed below) and a four-dimensional "rolled up" Minkowski spacetime (since in both cases $\forall p(I^+(p) = \mathcal{M})$), despite their quite different topological and metrical structures.

[21] Two spacetimes $\langle \mathcal{M}, g_{ab} \rangle$ and $\langle \mathcal{M}', g'_{ab} \rangle$ are said to have the *same geometrical structure up to a conformal factor* just in case $g_{ab} = \Omega^2 g'_{ab}$ where $\Omega$ is a smooth, strictly positive (and thus non-zero) function. A conformal transformation induced by $\Omega$ preserves local angles and ratios of magnitudes. Thus, the local light cone structure is preserved under conformal transformations.

[22] The *future boundary* of $D^+(\Sigma)$ is defined as $\overline{D^+(\Sigma)} - I^-(D^+(\Sigma))$, where the overbar denotes topological closure.

[23] There are several different ways of defining global hyperbolicity that are provably equivalent; see, e.g. Hawking and Ellis (1973, 206–212) for further discussion.

We will be interested below in *chronology-violating* spacetimes, that is, spacetimes which contain CTCs. In some cases, such as the "rolled up" Minkowski cylinder mentioned above, a CTC passes through *every* point in the spacetime. But this is not generally the case, and we can define the *chronology-violating region* $V \subset M$ as including all points $p \in M$ such that a CTC passes through $p$; in other words, $V$ is the region containing CTCs. If $V \neq \emptyset$ then it is an open region. The local-to-global property introduced in §2 can now be expressed more rigorously: A general-relativistic spacetime $\langle M, g_{ab} \rangle$ has the *local-to-global property* just in case for any open neighborhood $\mathcal{O} \subset M$, if $\langle \mathcal{O}, g_{ab}|_{\mathcal{O}} \rangle$ solves Einstein's field equations for some (admissible) energy-momentum tensor $T_{ab}|_{\mathcal{O}}$, then $\langle M, g_{ab} \rangle$ does so as well for a $T_{ab}$ that is identical to $T_{ab}|_{\mathcal{O}}$ in $\mathcal{O}$. Chronology-violating spacetimes typically do not exemplify the local-to-global property. In other words, locally well-behaved solutions to Einstein's field equation (perhaps coupled to some other dynamical equations) are not in general extendible to global solutions on all of $M$. This is not surprising, of course, since in chronology-violating spacetimes there will in general be local solutions that do not satisfy the consistency constraints. However, it is quite surprising that there are several chronology-violating spacetimes that *do* exhibit the local-to-global property for some dynamical equations. The discovery of several systems with this property spurred interest in this spacetimes with CTCs, although there is still not a general characterization of which chronology-violating spacetimes admit well-posed initial value problems for an interesting set of dynamical equations.[24]

The causality conditions make it possible to classify spacetimes according to how much their causal structure diverges from various natural features of Minkowski spacetime, and hence to prove theorems for "well-behaved" spacetimes. The conditions themselves are not consequences of the dynamics—we have already mentioned simple cases of chronology-violating spacetimes and we will see more interesting cases below. We have argued against the idea that something akin to causality conditions should be imposed as a further law of nature in order to avoid alleged time-travel paradoxes. But granting this point obviously does not settle the question of what the existence of chronology-violating spacetimes implies regarding the nature of time in GR. One response to this question, widespread in the physics literature until fairly recently, has been to simply dismiss chronology-violating spacetimes as mathematical curiosities without physical relevance. Advocates of this line of thought are faced with the task of articulating a clear set of constraints on what qualifies as a "physically reasonable" spacetime such that all the chronology-violating spacetimes are ruled out, a task that turns out to be surprisingly difficult (as we will see in §6 below). Admittedly, we are hard pressed to produce an example of a chronology-violating spacetime whose success in modeling a particular physical phenomena depends upon the presence of CTCs. On the other hand, since Kerr-Newman spacetimes are physically very important and afford a natural analytical extension containing CTCs, we feel that it would

---

[24]  See Friedman (2004) for a survey of the initial value problem in spacetimes with CTCs and references to earlier results.

be equally hasty to rule out spacetimes-cum-CTCs a priori. Furthermore, chronology-violating spacetimes need not be viable models of observed phenomena in order to be worthy of study, or to shed light on the conceptual structure of GR. In the next sections we turn to assessing the implications of the existence of chronology-violating spacetimes.

# 4. IMPLICATIONS OF TIME TRAVEL

Given that time travel cannot be straightforwardly ruled out as incoherent or logically impossible, we now face the following difficult questions: In what sense is time travel physically possible, and what does this imply regarding the nature of time? More precisely, what are the *novel* consequences of time travel, that is, ones that do not follow already from more familiar aspects of special or general relativity? As a first step towards answering these questions, we will consider Kurt Gödel's (in)famous argument for the ideality of time.[25]

Gödel (1949a) was the first to clearly describe a relativistic spacetime with CTCs.[26] Gödel's stated aim in discovering this spacetime was to rehabilitate an argument for the ideality of time from special relativity within the context of GR. In special relativity, Gödel asserts that the ideality of time follows directly from the relativity of simultaneity. He takes as a necessary condition for the existence of an objective lapse of time the possibility of decomposing spacetime into a sequence of "nows"—namely, that it has the structure $\mathbb{R} \times \Sigma$, where $\mathbb{R}$ corresponds to "time" and $\Sigma$ are "instants," three-dimensional collections of simultaneous events. But in special relativity the decomposition of the spacetime into "instants" is relative to an inertial observer rather than absolute; as Gödel puts it, "Each observer has his own set of 'nows,' and none of these various systems of layers can claim the prerogative of representing the objective lapse of time" (Gödel 1949b, 558).

This conclusion does not straightforwardly carry over to GR, because there is a natural way to privilege one set of "nows" in a cosmological setting. The privilege can be conferred on a sequence of "nows" defined with respect to the worldlines of galaxies or other large-scale structures. It is natural to require the surfaces of simultaneity to be orthogonal to the worldlines of the objects taken to define the "cosmologically preferred frame." The question is then whether one can extend local surfaces of simultaneity satisfying this requirement to a global foliation for a given

---

[25] The following papers, which we draw on below, discuss aspects of Gödel's argument: Stein (1970), Malament (1985b), Savitt (1994), Earman (1995), Dorato (2002), Belot (2005). Ellis (1996) discusses the impact of Gödel's paper.

[26] Although von Stockum (1937) discovered a solution describing an infinite rotating cylinder that also contains CTCs through every point, this feature of the solution was not discussed in print, to the best of our knowledge, prior to Tipler (1974). Gödel does not cite von Stockum's work. Others had noted the possibility of the existence of CTCs without finding an exact solution exemplifying the property (see, e.g. Weyl 1921, 249).

set of curves. For the cosmological models usually taken to be the best approximation to the large-scale structure of spacetime, the *Friedmann-Lemaître-Robertson-Walker (FLRW) spacetimes*, the answer is yes. These models have a natural foliation, a unique way of globally decomposing spacetime into a one-dimensional "cosmic time" and three-dimensional surfaces $\Sigma$ representing "instants," orthogonal to the worldlines of freely falling bodies. (Cosmic time in this case would correspond to the proper time measured by an observer at rest with respect to this privileged frame.) Thus Gödel's necessary condition for an objective lapse of time is satisfied in the FLRW cosmological models, and in this sense the pre-relativistic concept of absolute time can be recovered.

But in Gödel's spacetime one cannot introduce such a foliation. The spacetime represents a "rotating universe," in which matter is in a state of uniform rigid rotation.[27] Due to this rotation it is not possible to define a privileged frame with global "instants" similar to the frame in the FLRW models.[28] An analogy due to Malament (1995) illustrates the reason for this. One can slice through a collection of parallel fibers with a single plane that is orthogonal to them all, but if the fibers are twisted into a rope there is no way to cut through the rope while remaining orthogonal to each fiber. (The "twist" of the fibers is analogous to the rotation of worldlines in Gödel's model.) The construction of global "instants" described above can be carried out if and only if there is no "twist" (or rotation) of the worldlines used to define the cosmologically privileged frame. Demonstrating that such rotating models exist by finding an explicit spacetime model solving Einstein's field equations was clearly Gödel's main aim. But the welcome discovery that in his rotating universe there is a CTC passing through every point further bolstered his argument for the ideality of time.[29] It is noteworthy that many chronology-violating spacetimes resemble Gödel's solution in the following sense: they contain rotating masses and CTCs wind around the masses against the orientation of the rotation.[30]

What, then, is Gödel's argument? The crucial problem is how to get from discoveries regarding the nature of time in this *specific* spacetime to a conclusion about the nature of time *in general*. Gödel could avoid this problem if his spacetime, or a spacetime with similar features, were a viable candidate for representing the structure of the observed universe. Then his results would obviously have a bearing on the nature of time in

---

[27]   More precisely, in Gödel's universe a congruence of timelike geodesics has non-zero twist and vanishing shear. Defining rotation for extended bodies in general relativity turns out to be a surprisingly delicate matter (see, especially, Malament 2002).

[28]   As John Earman pointed out to us, Gödel does not seem to have noted the stronger result that Gödel spacetime does not admit of *any* foliation into global time slices.

[29]   Malament observed that the existence of CTCs is not mentioned in three of the five preparatory manuscripts for Gödel (1949a) and it appears that Gödel discovered this feature in the course of studying the solution. In addition, in lecture notes on rotating universes (from 1949), Gödel emphasizes that he initially focused on rotation and its connection to the existence of global time slices in discovering the solution. See Malament (1995) and Stein (1995, 227–229).

[30]   Cf. Andréka et al. (2008). That rotation may be responsible for the formation of CTCs is also suggested by Bonnor's (2001) result that stationary axially symmetric solutions of Einstein's field equations describing two spinning massive bodies under certain circumstances include a non-vanishing region containing CTCs.

*our* universe. Gödel apparently took this possibility quite seriously, and subsequently discovered a class of rotating models that incorporate the observed expansion of the universe (Gödel 1952). In these models, one can construct suitable "instants" as long as the rate of rotation is sufficiently low, and recent empirical work places quite low upper limits on the rate of cosmic rotation.[31] Gödel goes on to argue that even if his model (or models with similar features) fails to represent the actual universe, its mere existence has general implications (p. 562):[32]

> The mere compatibility with the laws of nature of worlds in which there is no dis-
> tinguished absolute time, and, therefore, no objective lapse of time can exist, throws
> some light on the meaning of time also in those worlds in which an absolute time
> *can* be defined. For, if someone asserts that this absolute time is lapsing, he accepts
> as a consequence that, whether or not an objective lapse of time exists...depends
> on the particular way in which matter and its motion are arranged in the world.
> This is not a straightforward contradiction; nevertheless, a philosophical view lead-
> ing to such consequences can hardly be considered as satisfactory.

Despite disagreement among recent commentators regarding exactly how to read Gödel's argument, there is consensus that even this modest conclusion is not war-ranted. The dynamical connection between spacetime geometry and the distribution of matter encoded in Einstein's field equations ensures that, in some sense, many claims regarding spacetime geometry depend on "how matter and its motion are arranged." Nearly any discussion of the FLRW models highlights several questions regarding the overall shape of spacetime—for example, whether time is bounded or unbounded and what is the appropriate spatial geometry for "instants"—that depend on apparently contingent properties, such as the value of the average matter density. What exactly is unsatisfactory about this? What does the mere possibility of space-times with different geometries imply regarding geometrical structure in general? Earman (1995, Appendix to Chapter 6) challenges the implicit modal step in Gödel's argument. How can we justify this step on Gödel's behalf, and elucidate what is unsatisfactory about objective time lapse in general, without lapsing back into pre-GR intuitions?

---

[31] These instants are not surfaces orthogonal to timelike geodesics, as there is still rotation present, but Gödel (1952) establishes that surfaces of constant matter density can be used to define a foliation that satisfies his requirements for an objective lapse of time. For recent empirical limits on global rotation based on the cosmic microwave background radiation, see, for example, Kogut et al. (1997).

[32] As Sheldon Smith pointed out to us, if this is taken to be Gödel's main argument then it is not clear why the mere existence of Minkowski spacetime, regarded as a vacuum solution of the field equations, does not suffice. Why did Gödel need to go to the effort of discovering the rotating model granted that there is no distinguished absolute time in Minkowski spacetime? Although we do not find a clear answer to this in Gödel (1949b), we offer two tentative remarks. First, Gödel may have objected to classifying Minkowski spacetime as physically reasonable because it is a vacuum spacetime. Second, and more importantly, Gödel took the prospect of discovering a rotating and expanding model consistent with observations more seriously than most commentators allow. This suggests that the argument in the quoted passage is a fall-back position, and that Gödel put more weight on the claim that he had discovered a viable model for the observed universe that lacks an objective lapse of time.

Perhaps the argument relies on an implicit modal assumption that lapsing, in the sense described above, must be an essential property of time. Then (given that $\Diamond(\neg P) \leftrightarrow \neg(\Box P)$), the demonstration that $\Diamond(\neg P)$ (where $P$ is the existence of an objective lapse of time) via finding the Gödel spacetime would be decisive. But what is the basis for this claim about the essential nature of time, and how can it be defended without relying on pre-relativistic intuitions? Earman (1995) considers this and several replies that might be offered on Gödel's behalf, only to reject each one. Steve Savitt (1994) defends a line of thought (cf. Yourgrau 1991) that is more of a variation on Gödelian themes than a textual exegesis. On Savitt's line, Gödel's argument rests not on essentialist claims regarding the nature of time, but instead on a claim of local indistinguishability. Suppose that it is physically possible for beings like us to exist in a Gödel spacetime, and (1) that it is possible for these denizens to have the "same experience of time" as we do. Assume further that (2) the only basis for our claim that objective time exists in our universe is the direct experience of time. Then the existence of the Gödel universe is a defeater for our claim to have established objective time lapse on the basis of our experience, because (for all we know) we could be in the indistinguishable situation—inhabiting a Gödel universe in which there is no such lapse. While this variation does not require a modal step as suspect as the original version, neither (1) nor (2) are obviously true—and it is unclear how they can be established without begging the question.[33]

One response to the challenge is simply to abandon Gödel's modal argument and formulate a different argument to the same effect. Consider an alternative argument that adopts a divide and conquer strategy rather than relying on a shaky modal step (suggested to us by John Earman). Divide the solutions of Einstein's field equations into (1) those that, like Gödel spacetime, lack a well-defined cosmic time, and (2) solutions that do admit a cosmic time.[34] The considerations above show that the spacetimes of type (1) lack an objective lapse of time in Gödel's sense. The spacetimes of type (2) have, by contrast, an embarrassment of riches: there are *many* well-defined time functions, and in general no way to single out *one* as representing the objective lapse of time. The definition of the cosmologically preferred reference frame in the FLRW models takes advantage of their maximal symmetry. Thus we seem to have an argument, without a mysterious modal step, that generic solutions of the field equations lack an objective lapse of time.

A different approach spelled out by Gordon Belot (2005) offers a *methodological* rather than *metaphysical* response to Earman's challenge. Belot concedes to Earman's challenge given a "natural-historical" construal of Gödel's argument, according to which the nature of time can be established based on empirical study of "how matter and its motion are arranged." On this reading, time in *our* universe is characterized by

---

[33]  See Belot (2005) and Dorato (2002) for further discussion.

[34]  In terms of the causality conditions in §3, a global time function exists for "stably causal" spacetimes—a condition slightly weaker than global hyperbolicity.

the appropriate spacetime of GR that is the best model for observations—and the mere existence of alternative spacetimes is irrelevant. But on a "law-structural" construal, questions regarding the nature of time focus on the laws of nature rather than on contingent features of a particular solution. Belot makes a case that a law-structural construal of the question is more progressive methodologically, in that it fosters deeper insights into our theories and aids in the development of new theories.[35] If we grant that understanding the laws may require study of bizarre cases such as Gödel's spacetime alongside more realistic solutions, then we have the start of a response to Earman's challenge.

It is only a start, because this suggested reading remains somewhat sketchy without an account of "laws of nature," which is needed to delineate the two construals more sharply. Even if we had a generally accepted account of the laws of nature, the application of "laws" to cosmology is controversial: how can we distinguish nomic necessities from contingencies in this context, granting the uniqueness of the universe? Setting this issue aside, Earman's challenge can be reiterated by asking which spacetimes should be taken as revealing important properties of the laws. Why should Gödel spacetime, in particular, be taken to reveal something about the nature of time encoded in the laws of GR? Suppose we expect that only a subset of the spacetimes deemed physically possible within classical GR will also be physically possible according to the as-yet-undiscovered theory of quantum gravity. How would we argue that Gödel spacetime should fall within that subset, and that it should be taken to reveal a fundamental feature of the laws of GR that will carry over to quantum gravity? The features Gödel used to establish the lack of absolute time in his model are often taken to support a negative answer to this question that does not appear to be ad hoc. Many approaches to quantum gravity simply rule out spacetimes with CTCs *ab initio* based on the technical framework adopted.[36] As we will discuss below, much of the physics literature on spacetimes with CTCs seeks clear physical grounds to rule them beyond the pale; insight into the laws of a future theory of quantum gravity would come from showing why the laws do *not* allow CTCs. But we agree with Belot that what is more unsatisfying regarding Gödel's argument, even on the "law-structural" construal, is that an argument by counter-example does little to illuminate deeper connections between the nature of time and the laws of the theory.

Assessing the implications of Gödel's spacetime clearly turns on rather delicate issues regarding modality and the laws of nature. Perhaps our failure to articulate a clear Gödelian argument indicates that the properties of such bizarre spacetimes can

[35]  Belot finds inspiration for this position in several brief remarks regarding the nature of scientific progress in manuscript precursors to Gödel (1949a); however, he does not take these considerations to be decisive (see p. 275, fn. 52).

[36]  Gödel's solution might be ruled out due to the symmetries of the solution, as Belot notes: symmetric solutions pose technical obstacles to some approaches to quantization, and it seems precarious to base assertions regarding features of quantum gravity on properties of special, symmetric solutions. But this argument seems too strong, in that it also would rule out the FLRW models, which are currently accepted as the best classical descriptions of the large-scale structure of the universe.

be safely ignored when we investigate the nature of time in GR. Tim Maudlin (2007) advocates a dismissive response to CTCs, which would otherwise pose a threat to his metaphysical account of the passage of time: "It is notable in this case that the equations [Einstein's field equations] do not force the existence of CTCs in this sense: for any initial conditions one can specify, there is a global solution for that initial condition that does not have CTCs." He anticipates a critic's response that his metaphysical account of passage boldly stipulates that the nature of time is not compatible with the existence of CTCs, and replies: "But is it not equally bold to claim insight into the nature of time that shows time travel to be possible *if we grant that it is not actual and also that the laws of physics, operating from conditions that we take to be possible, do not require it*" (Maudlin 2007, 190). These assertions would follow from the proof of the following form: CTCs do not arise from "physically possible" initial states under dynamical evolution according to Einstein's equations. Below we will consider a more precise formulation of this "chronology protection conjecture" (in §6). But at this point we wish to emphasize that this is still a *conjecture*, and that there are a number of subtleties that come into play in even formulating a clear statement amenable to proof or disproof.[37] Perhaps a claim like Maudlin's, suitably disambiguated, will prove to be correct, but part of the interest of the question is precisely due to the intriguing technical questions that remain open.[38]

In any case, Maudlin's remarks usefully indicate a fruitful way of addressing the importance of solutions with exotic causal structure. Arguments by counterexample—displaying a solution to Einstein's field equations with exotic causal structure—are unsatisfying because it is usually not clear how the solution in question relates to solutions used to model physical systems or how it is related to other "nearby" solutions. For example, given a solution with CTCs, is it an element of an open set of solutions that also have CTCs? Or does the presence of CTCs depend upon a symmetry or some other parameter fixed to a specific value? Rather than considering a solution in isolation, we are pushed towards questions about the space of solutions to the field equations. We can ask, for example, what Einstein's field equations imply for the dynamical evolution of some class of initial data we decide to treat as "physically possible." One advantage of framing the question this way is that we can exploit the initial value formulation of GR to address it, as we will see below. But there is also an important disadvantage: we can only address the existence of chronology-violating spacetimes indirectly, given that they lack surfaces upon which initial data can be specified. By framing the question this way we would avoid controversial questions regarding modalities in cosmology, and instead focus on whether it is possible according to GR to manipulate matter and energy in a local region such that, *contra* Maudlin, CTCs are the inevitable result. In more vivid language, is it possible in principle to build a time machine? Formulating this idea precisely is the task of the next section.

---

[37] And we should note that absent some further qualifications, Maudlin's first claim is false—as established by Manchak's theorem discussed in the next section.

[38] Our treatment here is influenced by the clear discussion of these issues in Stein 1970.

# 5. Time machines

In the usual parlance of science-fiction authors, time machines are simply devices that enable time travel in roughly Lewis's sense (discussed in §2)—the time elapsed during the journey does not match some appropriately external measure of the time interval between departure and arrival. We have already argued in favor of one departure from this usage, in that we define time travel in terms of the existence of CTCs. Time travel is possible in our sense only in chronology-violating spacetimes, and we will leave aside the issue of whether any observer or particle in fact instantiates time travel by following a CTC. But in a departure that will be more disappointing for science-fiction fans, we will define a "time machine" as a device which produces CTCs where none would have existed otherwise. This is more demanding than merely requiring the existence of CTCs. In Gödel spacetime, for example, one cannot claim that the CTCs are produced by local manipulation of matter and energy within some finite region, as we require for the existence of a time machine. Our immediate goal is to show how, following Earman et al. (2009) (cf. Earman and Wüthrich 2004), to flesh out this idea in terms of spacetime geometry. Science-fiction fans will be disappointed by this analysis because, as we will see, our definition rules out the possibility of using a time machine to travel into one's own past, that is, to times prior to the operation of the time machine. It will merely allow one to ride along CTCs in a spatiotemporal region *after* the time machine is switched on.

Our starting point is the idea that a "time machine" is a region of spacetime in which local manipulation of the fields can produce CTCs. As a first step, we turn to the initial value formulation of GR as a way of clarifying the sense in which manipulating fields in a finite region can be said to determine their values elsewhere. The initial value problem for a theory consists of proving theorems establishing, given a set of initial conditions and the dynamical laws of the theory, the existence and uniqueness for the dynamical evolution of a physical system that falls under the purview of the theory. A physical theory has an *initial value formulation* provided that one can prove that there exists a unique solution for a given set of appropriate initial data—intuitively, given a specification of the state of the system at one time, the dynamical laws determine a unique state at other times. A theory's initial value formulation must satisfy further conditions to be viable physically (and qualify as "well-posed"): "small changes" in the initial data lead to "small changes" in the evolved states, and changes to the initial data in a region $\Sigma$ should not effect causally disconnected regions. For GR, the dynamical equations are Einstein's field equations, coupled with additional equations for the matter fields. GR possesses a well-posed initial value formulation (up to gauge freedom).[39] However, this formulation applies only to globally hyperbolic spacetimes,

---

[39] The qualification is essential, for if one neglects gauge freedom, then the initial data appear to *underdetermine* the dynamical evolution—the solution is only fixed up to diffeomorphism. Furthermore, the initial data for GR must satisfy constraints. Once a gauge condition is imposed,

which form a proper subset of all spacetimes that solve Einstein's field equations. This restriction seems natural, however, since only globally hyperbolic spacetimes admit a foliation by Cauchy surfaces. And a Cauchy surface, as stated above, is a spacelike slice $\Sigma$ such that $D(\Sigma)$ is the entire spacetime. Thus, even without studying the initial value problem in its full mathematical detail, it ought to be clear that spacetimes admitting Cauchy surfaces are good candidates for a general-relativistic initial value formulation. What this also means, however, is that there is a large class of spacetimes, namely those which fail to admit a Cauchy surface, that lack a natural dynamical interpretation. For instance, Gödel spacetime has no three-dimensional, spacelike surfaces without boundary, and thus, a fortiori, no Cauchy surfaces.

Let us restrict our attention to globally hyperbolic spacetimes, for the moment. These spacetimes always admit a foliation by Cauchy surfaces $\Sigma_t$ and thus a global time function $t$, that is a function such that each surface of constant $t$ is a Cauchy surface of the spacetime.[40] The spacetime then has topology $\mathbb{R} \times \Sigma$, where the topology of the three-spaces $\Sigma_t$ is arbitrary but must be the same for all $\Sigma_t$.[41] Conversely, however, not every spacetime with topology $\mathbb{R} \times \Sigma$ automatically affords a global time function or a Cauchy surface. As a matter of fact, the spacetime may be foliated using a "flow of space" rather than a "flow in time." The topology of the spacetime is in both cases $\mathbb{R} \times \Sigma$, yet only the latter case would be amenable to the introduction of a global time function.

The analysis of time machines bears a close relationship to the initial value problem. Let us explicate this relationship by conceptualizing time machines starting out from some globally hyperbolic spacetime $\langle \mathcal{M}, g_{ab} \rangle$. Its Cauchy surfaces $\Sigma_t$ trisect $\mathcal{M}$: $\Sigma_t$ itself, and the "past" and the "future" of $\Sigma_t$.[42] Consider the chronological past $I^-(\Sigma)$ of $\Sigma$ as fixed. It is of considerable importance for dynamical formulations of GR to determine the "causal stability" of the dynamical evolution of globally hyperbolic spacetimes. The natural approach to an analysis of this issue is to consider globally hyperbolic spacetimes and study the causal structure of their maximal extensions. The original globally hyperbolic spacetime $\langle \mathcal{M}, g_{ab} \rangle$ can thus be thought of as the past and

Einstein's field equations take the form of quasi-linear, second order, hyperbolic partial differential equations, which admit a well-posed local initial value formulation. See Hawking and Ellis (1973, Ch. 7), Wald (1984, Ch. 10, particularly §10.2), and Rendall (2008) for discussions of the initial value formulation of GR.

[40] We thus assume that the spacetimes at stake afford a time orientation, and that this time orientation is encoded in the global time function.

[41] At least for spatially compact Hausdorff spacetimes $\langle \mathcal{M}, g_{ab} \rangle$; this is essentially Geroch's Theorem (Geroch 1967) which states that for a compact spacetime $\langle \mathcal{M}, g_{ab} \rangle$ whose boundary is the disjoint union of two closed spacelike three-manifolds, $S$ and $S'$, $S$ and $S'$ are diffeomorphic if $\langle \mathcal{M}, g_{ab} \rangle$ admits a time orientation and does not contain closed timelike curves. Intuitively, one can think of mapping points of $S$ into $S'$ using a congruence of timelike curves; for the topology to differ, at least some of these curves must fail to connect the boundaries, instead forming CTCs confined to the compact region bounded by $S$ and $S'$. The theorem does not apply to spatially open spacetimes. For a penetrating discussion of topology change in general and of Geroch's Theorem in particular, see Callender and Weingard (2000).

[42] For ease of notation, we drop the index "$t$" in what follows.

the present of some time slice $\Sigma$.[43] As a part of a globally hyperbolic spacetime, $I^-(\Sigma)$ does not contain CTCs. The important foundational issue is how much assurance does the theory give us that any dynamical evolution of $\langle \mathcal{M}, g_{ab} \rangle$, that is, any maximal extension of $\langle \mathcal{M}, g_{ab} \rangle$ in accordance with the dynamical equations, is causally well-behaved. An analysis of whether CTCs can arise is part of a systematic study of this foundational issue. We will discuss how it fits into the wider problem in the next section, §6.

The question of whether time machines can be had in a spacetime theory thus amounts to asking whether suitable maximal extensions of a globally hyperbolic space-time $\langle \mathcal{M}, g_{ab} \rangle$ contain CTCs. A spacetime $\langle \mathcal{M}, g_{ab} \rangle$ can be *extended* if it is isometric to a proper subset of another spacetime.[44] An extension of a spacetime is *maximal* if it is *inextendible*, that is, it is not isometric to a proper subset of another spacetime. The initial state of a physical system prior to the operation of the time machine is defined on a subset of $\Sigma$. In this setting, whether a time machine can be operated in a given spacetime depends on whether it admits a time slice $\Sigma$ such that $I^-(\Sigma)$ does not contain CTCs and $\langle \mathcal{M}, g_{ab} \rangle$ can be analytically extended to a spacetime $\langle \mathcal{M}', g'_{ab} \rangle$ with a non-vanishing chronology-violating region $\mathcal{V} \subset \mathcal{M}'$. One can distinguish three cases: either none, some, or all of the suitable maximal extensions of $\langle \mathcal{M}, g_{ab} \rangle$ contain CTCs. In the first case, $\langle \mathcal{M}, g_{ab} \rangle$ is maximally causally robust and no time machine can be operated. Let's discuss the two other cases in turn, starting with the latter. Thus, is it possible that all suitable maximal extensions of a globally hyperbolic spacetime contain CTCs?

Before we answer this question, two difficulties become immediately apparent. First, as should be obvious, time machines as conceptualized in this way will not be amenable to an avid science-fiction lover's desire to return from $\Sigma$ into the causal past $J^-(\Sigma)$ of $\Sigma$. In our present set-up, time travel is confined to the future of $\Sigma$. This is the price we pay in order to make the analysis more relevant to foundational concerns. And we are happy to pay it.

Second, and more importantly, in what sense could the emergence of a non-vanishing region $\mathcal{V}$ be appropriately said to be the "result of" the operation of the time machine? Within the domain of dependence of a surface $\Sigma$, the initial value formulation gives clear content to claims such as "wiggling the value of a field on $\Sigma$ is causally responsible for a corresponding wiggle elsewhere in $D(\Sigma)$." But by construction, $\mathcal{V}$ must lie outside of $D^+(\Sigma)$, since the past-inextendible, past-oriented

---

[43]   The immediate question that arises is what could guarantee that $\Sigma$ is the "latest" of all time slices, i.e. how can we assume that some part of the original globally hyperbolic spacetime isn't to the future of $\Sigma$? In general, we would surely expect that $I^+(\Sigma) \neq \emptyset$. However, even if $I^+(\Sigma) \neq \emptyset$ for a given $\Sigma$, it may be that there is no "later" Cauchy surface $\Sigma'$, where "later" is defined by the total ordering relation induced by the global time function $t$. Thus, we are interested in the one Cauchy surface $\Sigma_\tau \subset \mathcal{M}$ such that for all values $t$ of the global time function, $t \leq \tau$. Or at least in one reasonably close to it.

[44]   That is, it can be extended if there exists a spacetime $\langle \mathcal{M}', g'_{ab} \rangle$, $\mathcal{M} \subsetneq \mathcal{M}'$, and an isometric embedding $\phi : \mathcal{M} \to \mathcal{M}'$ such that $\forall p \in \mathcal{M}', \phi^*(g_{ab}(\phi^{-1}(p))) = g'_{ab}(p)$.

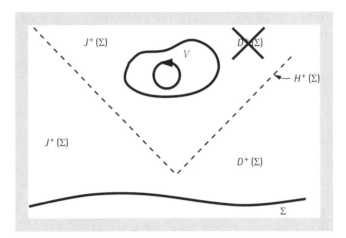

FIGURE 20.4   The causal future $J^+(\Sigma)$ contains a "chronology-violating region" $V \neq \emptyset$, i.e. the set of all points through which there exists a CTC

curves through points in $V$ don't intersect $\Sigma$ (cf. Figure 20.4).[45] This means that not all causal influences on the region $V$ stem from the conditions set on $\Sigma$. Some, perhaps decisive, influences on $V$ may come from outside of $\Sigma$. It also means that the original spacetime manifold $\mathcal{M}$ that we were seeking to extend may be thought of, without loss of generality, as the region $I^-(\Sigma) \cup D(\Sigma)$.[46]

The predicament is the following: in order to elucidate what it means to be "causally responsible" for CTCs, we have to abandon our best understanding of causal responsibility in GR—that provided by the initial value formulation.[47] That said, we claim that one can still give content to the idea of a time machine as follows. Although the conditions set on $\Sigma$ have no prayer of being fully causally responsible for the emergence of a region $V$, the third case where all maximal extensions of the original spacetime $I^-(\Sigma) \cup D(\Sigma)$ contain a non-vanishing $V$ is the next best thing to full causal responsibility. We say that in this case, the spacetime satisfies the

**Condition 1 (Potency Condition)** Every suitable smooth, maximal extension of $I^-(\Sigma) \cup D(\Sigma)$ contains CTCs.

The question, of course, is which smooth, maximal extensions ought to be deemed "suitable." The stronger the restrictions on which extensions qualify as suitable, and thus the fewer extensions required to contain CTCs in order for the Potency Condition

---

[45]   This means, of course, that $\Sigma$ is no longer a Cauchy surface of $\langle \mathcal{M}', g'_{ab} \rangle$, i.e. $\Sigma$ is no longer a spacelike hypersurface which every inextendible non-spacelike curve intersects exactly once. The Cauchy surfaces of $\langle \mathcal{M}, g_{ab} \rangle$ become *partial Cauchy surfaces* of $\langle \mathcal{M}', g'_{ab} \rangle$, i.e. spacelike hypersurfaces which no inextendible non-spacelike curve intersects *more than once*. Non-spacelike curves in $V$, for instance, will not intersect $\Sigma$.

[46]   This alleviates the worry expressed in footnote 43. For a more thorough discussion of these points, and for the details of the construction, cf. Earman et al. (2009).

[47]   This predicament was described clearly in Robert Geroch's lecture at the Second International Conference on Spacetime Ontology (Montreal, June 2006).

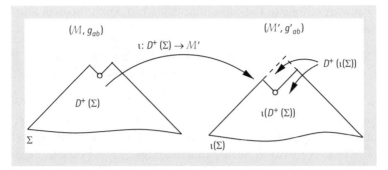

FIGURE 20.5  An illustration of the definition of hole freeness

to be satisfied, the weaker the potency. It would be natural to require a suitable extension to be a solution of the dynamical equations of the relevant theory. But at this stage
we are offering, in effect, a kinematical definition of a time machine, characterized
in terms of causal structure, that is not tied to a specific choice of dynamics such as
Einstein's field equations. In any case, for the remainder of this section we assume that
the satisfaction of Einstein's field equations is one necessary condition for a suitable
extension.

More fundamentally, however, and as also argued in Earman et al. (2009), we
impose a slightly modified version of Robert Geroch's (1977) definition of *hole freeness*.
This condition disbars artificial cut-and-paste manoeuvres in a principled manner.[48]

**Definition 1 (Hole freeness)**  A spacetime $\langle \mathcal{M}, g_{ab} \rangle$ is said to be *hole free* just in case
for any spacelike $\Sigma \subset \mathcal{M}$, there is no isometric imbedding $\iota : D^+(\Sigma) \to \mathcal{M}'$ into a
spacetime $\langle \mathcal{M}', g'_{ab} \rangle$ such that $\iota(D^+(\Sigma))$ is a proper subset of $D^+(\iota(\Sigma))$ (cf. Figure 20.5).

The demand for hole freeness has some bite, as can be seen from considering a theorem
by Serguei Krasnikov (2002):

**Theorem 5.1 (Krasnikov)** *Any spacetime* $\langle \mathcal{U}, g_{ab} \rangle$ *has a maximal extension*
$\langle \mathcal{M}_{max}, g'_{ab} \rangle$ *such that all closed causal (and, a fortiori, timelike) curves in* $\mathcal{M}_{max}$ *(if*
*they exist there) are confined to the chronological past of* $\mathcal{U}$.

The construction that Krasnikov gives in his proof of Theorem 5.1 allows any local
conditions on the metric, such as the satisfaction of Einstein's field equations or of
energy conditions, to be carried over to $\mathcal{M}_{max}$. If it weren't for the (global) condition
of hole freeness, Theorem 5.1 would imply that there are no spacetimes that satisfy the
Potency Condition. But the proof no longer goes through once the demand for hole
freeness is imposed, so the question remains open.

Does any general-relativistic spacetime satisfy the Potency Condition amended with
a demand of hole freeness (and of complying to Einstein's field equations)? A recent
result by John Manchak (2009b) answers this question in the affirmative, thereby

---

[48]  For a recent discussion on whether hole freeness is a physically plausible condition, cf.
Manchak (2009a).

offering a counterexample to the conjecture that Krasnikov's theorem also holds for hole-free spacetimes.

**Theorem 5.2 (Manchak)** *There exists an ESW time machine, where an* ESW *time machine is a hole-free spacetime satisfying the Potency Condition.*

As Manchak points out, this existence theorem establishes that we face a trilemma, in that some initial conditions force us to give up at least one of the following: either (i) the spacetime is inextendible, (ii) the spacetime is hole free, or (iii) the spacetime does not contain CTCs. We wish to qualify this trilemma.[49] It is possible, arguably even likely, that the general-relativistic spacetime that most accurately describes the large-scale structure of our actual universe is inextendible, hole free, and entirely free of CTCs. The initial conditions which are *known* to force us into this trilemma, encoded in Misner spacetime, are not physically realistic. Characterizing more precisely what qualifies as physically realistic initial data, and then demonstrating that the trilemma is avoided, is related to the censorship theorems discussed in the next section. Having said that, however, there seems to be a vast class of spacetimes $I^-(\Sigma) \cup D(\Sigma)$ such that *some*, but not all, suitable smooth, maximal extensions contain CTCs. We expect that it will typically be possible, by simple cut-and-paste strategies, to construct suitable smooth, maximal extensions of spacetimes with Cauchy horizons which are infested with CTCs.[50] It is to this case—the second as listed above—that we now turn.

GR is not a deterministic theory *simpliciter*. As with other dynamical theories, GR is deterministic insofar as existence and uniqueness theorems can be proven for its dynamical equations, and these theorems typically presuppose that imposing "energy conditions," which are assumptions concerning the energy-momentum tensor (discussed in more detail below). But the standard initial value formulation of general relativity only fixes dynamical evolution up to the Cauchy horizon of the spacetime (up to diffeomorphisms), and not beyond. As far as the evolution beyond the Cauchy horizon $H^+(\Sigma)$ is concerned, the data on $\Sigma$, together with the dynamical equations, does not uniquely determine the evolution into $J^+(H^+(\Sigma))$. This failure of determinism has motivated work on the censorship theorems discussed in the next section, but it is what makes it possible to find multiple, inequivalent extensions beyond $H^+(\Sigma)$— including extensions with CTCs.

While it is certainly the case that the causal stability of spacetimes that can be extended beyond their Cauchy horizon to include CTCs is limited, this does not imply

---

[49]  The first—very minor—qualification is that there is really a fourth option: the spacetime is not smooth.

[50]  Essentially, a spacetime is expected to be extendible in a way that the extension contains a Deutsch-Politzer gate in the sense of Deutsch (1991) and Politzer (1992). Consider an extendible spacetime, i.e. one with a Cauchy horizon. It is always possible to extend such a spacetime in a way that the resulting maximal extension contains a neighborhood $\mathcal{N}$ which is Minkowskian. Within $\mathcal{N}$, cut two achronal, timelike related strips and identify the lower edge of one strip with the upper edge of the other, and the lower edge of the other strip with the lower edge of the one. This creates a "handle region" in which CTCs are present (cf. Figures 5 and 6 in Wüthrich 2007). We wish to thank David Malament and especially John Manchak for discussions concerning this point.

that a time machine can be easily operated. First, these time machines would not operate perfectly in that they would sometimes fail to produce a non-vanishing region of CTCs. But why would one want to speak of a time machine if a certain fraction of suitable extensions contains CTCs while the rest doesn't and it is a matter of pure chance as to whether the would-be time machine operators will end up living in a spacetime that evolves to contain CTCs or not? We could only appropriately speak of a time machine if the local manipulation of energy and matter distribution would somehow *increase the chance* that there will be a CTC-containing region to the future of the Cauchy horizon. We will call such a device an *incremental time machine*.

The obvious manner in which talk of chance can be made respectable is in terms of probabilities. An incremental time machine would then be a device that would, by means of local operations, increase the probability that there emerge CTCs to the future of $H^+(\Sigma)$. This would be given a clear meaning if we had a probability measure defined over the set of suitable extensions. Thus, an incremental time machine would be operative just in case the probability of there being CTCs to the future of $H^+(\Sigma)$ conditional on the local manipulation performed by the incremental time machine is higher than the probability of there being CTCs to the future of $H^+(\Sigma)$ not so conditionalized, where the probabilities are given by the measure defined over the set of suitable extensions. If this can be done, we say that the spacetimes satisfies the

**Condition 2 (Mitigated Potency Condition)** The operation of an incremental time machine increases the measure of those extensions of $I^-(\Sigma) \cup D(\Sigma)$ containing CTCs among the set of all suitable smooth, maximal, and hole-free extensions of $I^-(\Sigma) \cup D(\Sigma)$.

It is not trivial to fill out this proposed condition. Earlier we confined our analysis to an investigation of whether all suitable extensions of a given, particular globally hyperbolic spacetime contained CTCs. Here, an analogous strategy faces the additional difficulty of getting a grip on what it could mean to "increase the measure of extensions" with certain properties. The notion of "increasing" straddles us with counterfactual discourse that is difficult to parse out. The best way to escape the counterfactual morass is to focus on two problems in the neighborhood of the original one: Can we get a principled handle on defining probability measures over the set of suitable extensions of $I^-(\Sigma) \cup D(\Sigma)$; and can we gain some understanding concerning the physical mechanisms that might be responsible for the emergence of causally unusual structures such as CTCs? We will not broach the second of these issues here.[51] And we have only very little to say concerning the first.

The most principled way of addressing the first issue of introducing probability measures over the set of suitable extensions is to start out from the set of all admissible spacetimes of the theory, define a probability measure over this set, and conditionalize on the subset of spacetimes in which the spacetime-to-be-extended at stake can be isomorphically embedded. This would result in a probability distribution over the set

---

[51] For a step into this direction, see Andréka et al. (2008). Cf. also footnote 30.

of suitable extensions. With such a probability distribution at hand, we could then determine the relative frequency of extensions-cum-CTCs in terms of their measure in the space of suitable extensions. This sounds all very principled and rigorous, but we face unresolved, and perhaps unresolvable, difficulties at every turn of this path.

First, at least for GR, the set of admissible spacetimes, that is, the space of solutions to Einstein's field equations, is not known. Second, and a fortiori, we cannot define a probability measure over this set. But even if we pretend that we know the set, or at least some significant subset of it, it is not trivial to endow it with a canonical measure and not much of help can be found in the literature. Most of the attempts to define such measures over sets of solutions focus on causally well-behaved spacetimes such as the FLRW cosmological models.[52] These measures have primarily been designed to deal with the "flatness problem" in standard cosmology in an attempt to avoid inflationary scenarios. Extant results in this field, however, are hardly of much use to our present purposes since they only extend to a particular parameter family of comparatively well-understood spacetimes. Without a more general measure defined over larger classes at hand, the prospect of this research program seems daunting, if not downright hopeless.

There is an alternative way to obtain a sense of how generic maximal extensions containing CTCs are for given initial spacetimes. Rather than measuring extensions with CTCs, we might be able to count them. Presumably, a theorem analogous to Theorem 5.1 may be found which shows that it is always possible to find an extension which respects certain local conditions while displaying CTCs. Perhaps we could then establish a theorem of "parallel existence" according to which there exists, for every causally virtuous extension, an extension with CTCs. Take, for instance, any "clean" extension received with the help of Theorem 5.1.[53] It seems that any of these extensions could be infected for example, by a Deutsch-Politzer gate, thus producing CTCs. In all fairness, such limited theorems could at best serve to strengthen our intuitions. In order to obtain more conclusive statements concerning the genericity of causal and acausal extensions, one would have to establish theorems asserting the open density of one of the two families of extensions, causal or acausal, thus showing that it is of measure one, while its complement is "nowhere dense" and therefore of measure zero. Such a proof, however, would again require a well-defined measure on the space of extensions.

There is thus little hope that much more can be said about the second case where some, but not all, extensions contain CTCs. Before we close this section, a brief word concerning the first case, the one where none of the suitable extensions harbors CTCs

---

[52] Cf. Hawking and Page (1988); Cho and Kantowski (1994); and Coule (1995). As it turns out, there really is a third problem: the "natural" measure may not be unique. In the FLRW case, some measures imply that flatness is generic, while others hold that it is special. We wish to thank Craig Callender for drawing our attention to this.

[53] Exploiting thus Krasnikov's construction entails that we drop the demand for hole freeness. If we drop this constraint, however, we cannot prove any more that there are any (strict) time machines. Clearly, it is a desperate move to sacrifice strict time machines in order to clarify the meaning of incremental time machines.

and the initial spacetime is maximally causally robust. It seems that if hole freeness is not required of the admissible extensions, then there always exists a smooth, maximal extension with CTCs that can be gained by the scissors-and-glue method.[54] Trivially, those spacetimes which already are maximal and thus admit of no non-trivial extension beyond the Cauchy horizon will not be extendible in a way that includes a non-vanishing region $V$. Thus, the conjecture has to be reformulated as claiming that for all spacetimes *that permit a non-trivial extension beyond their Cauchy horizon*, there exists a suitable smooth, maximal extension containing CTCs. Furthermore, once one adds the constraint of hole freeness, as we did above, it seems as if the scissors-and-glue method is no longer possible in general. It is thus an open question whether there always exists a *hole-free*, smooth, maximal extension of $I^-(\Sigma) \cup D(\Sigma)$ containing CTCs.

# 6. CENSORSHIP THEOREMS

Given this understanding of what a time machine requires in terms of spacetime structure, we can now return to the question of whether time machines are physically possible according to GR. The discussion above shows that time machines are physically possible in the weak sense that there are spacetime geometries that instantiate a time machine. But we have set aside until now the question of whether such spacetimes, or more generally spacetimes with CTCs or other causal pathologies, are physically possible in the stronger sense that there are solutions to Einstein's field equations exhibiting these features that are "physically reasonable." There are several conditions one might impose to delineate the subset of spacetimes that qualify as reasonable:

1. *Causality Conditions*: treat one of the causality conditions (e.g. global hyperbolicity) as a law of nature not derived from the field equations.
2. *Conditions on Source Terms*: impose energy conditions on the matter fields, or limit consideration to $T_{ab}$ derived from "fundamental fields."
3. *Generic*: rule out "special" spacetimes (e.g. those of measure zero) as possible models for real systems.
4. *Quantum Considerations*: impose conditions that the spacetime must satisfy to admit a QFT, or to be the classical limit of a quantum gravity solution.

The first option is sometimes motivated by the paradox-mongering we have criticized above. There is a further objection to simply imposing causality constraints, namely that one sets aside the possibility that they might be enforced by the dynamics combined with some combination of the other conditions. As Stephen Hawking puts it (in Hawking and Penrose 1996, 10),

---

[54]  Presumably including the satisfaction of local conditions such as the dynamical equations and energy conditions; cf. also footnote 50.

> [M]y viewpoint is that one shouldn't assume [global hyperbolicity] because that
> may be ruling out something that gravity is trying to tell us. Rather, one should
> deduce that certain regions of spacetime are globally hyperbolic from other physi-
> cally reasonable assumptions.

Attempts to decipher what gravity is trying to tell us have led to what we will generally "censorship hypotheses," and in some cases censorship theorems. Such a theorem
shows that specific features such as causal pathologies either do not develop under
dynamical evolution from suitable initial conditions, or, failing that, that these features
are "censored"—that is, hidden safely behind event horizons. The name derives from
Penrose's "cosmic censorship hypothesis" (Penrose 1969, 1979), which in slogan form
holds that nature abhors a naked singularity. Turning this slogan into a precise claim
amenable to proof is no easy task, and the proper formulation and status of cosmic
censorship remains one of the central open problems in classical GR. Here we will
briefly survey how conditions (2) and (3) come into play in attempts to state clearly
and then prove censorship theorems (leaving 4 until the next sections), and focus on
"chronology protection" results which aim to demonstrate the impossibility of time
machines.

Penrose formulated the cosmic censorship hypothesis on the heels of his groundbreaking work (along with Hawking and Geroch) on the singularity theorems. These
theorems show that a large class of physically reasonable spacetimes, relevant for
cosmology and for the gravitational collapse of stars, must be singular, in the sense
of geodesic incompleteness.[55] However, although all globally hyperbolic spacetimes
resemble one another, each singular spacetime is singular in its own way, and the singularity theorems reveal little about the nature of the singularity. A successful proof of
cosmic censorship would rule out all but the relatively benign singularities. Specifically,
one would hope to rule out *nakedly singular* spacetimes. These are defined, intuitively,
as spacetimes containing point(s) $p$ such that $I^-(p)$ includes an entire inextendible
timelike curve with finite proper length, representing the trajectory $\gamma$ of a point particle
that "falls into the singularity"—the singularity that $\gamma$ encounters is "visible" from
such points.[56] Strong cosmic censorship asserts that there are no such points, and
Penrose shows this holds iff a spacetime is globally hyperbolic. Strong censorship can
be formulated in slightly different terms as follows: strong cosmic censorship holds
if initial data specified on $\Sigma$ (for a suitable hypersurface)[57] have a maximal Cauchy
development $D(\Sigma)$ that is inextendible. There are counterexamples to the conjecture

---

[55] A spacetime is *geodesically incomplete* just in case there exist geodesics in it which are
inextendible in at least one direction (timelike, null, or spacelike) yet run only over a finite range of
their affine parameter.

[56] Here we follow Penrose's characterization of cosmic censorship in terms of detectability
(Penrose 1979); his formulation is based on the ideas of "indecomposable past sets" and "terminal
indecomposable past sets," a way of defining ideal or boundary points. See Earman (1995) and Geroch
and Horowitz (1979) for discussions of alternative approaches to formulating cosmic censorship.

[57] The qualifier is required to rule out surfaces such as an achronal surface $\Sigma$ for which
$\exists p : \Sigma \subset I^-(p)$ in Minkowski spacetime; for such a surface, $D(\Sigma)$ is not the entire manifold, but this
just reflects the poor choice of $\Sigma$.

if there are not further qualifications, however. One such counterexample (based on numerical simulations by Choptuik and analytical work by Christodoulou) prompted Hawking to pay up on a famous bet in favor of cosmic censorship against Kip Thorne and John Preskill. (Hawking wagered in favor of the claim that "When any form of classical matter or field that is incapable of becoming singular in flat spacetime is coupled to general relativity via the classical Einstein equations, the result can never be a naked singularity," at 2 to 1 odds.)[58]

One response to such counterexamples is to give a more refined (and weaker) formulation of cosmic censorship, as Hawking did in immediately placing a new wager with Thorne and Preskill regarding the claim that: "When any form of classical matter or field that is incapable of becoming singular in flat spacetime is coupled to general relativity via the classical Einstein equations, then dynamical evolution from generic initial conditions (i.e. from an open set of initial data) can never produce a naked singularity (a past-incomplete null geodesic from [future null infinity] $\mathscr{I}^+$)". The clarification of what is meant by a "naked singularity" is based on a distinction between "local observers" and "observers at infinity," and requires that only the latter are safely shielded from the singularity by the event horizon of a black hole. This is usually called weak cosmic censorship.[59] The formulation of weak cosmic censorship relies on a precise notion of infinity developed in the study of gravitationally isolated systems. Roughly speaking, a spacetime is *asymptotically flat* at late times if it can be conformally embedded into a spacetime with a boundary $\mathscr{I}^+$ composed of the endpoints of null geodesics that propagate to arbitrarily large distances.[60] The black hole region $\mathcal{B}$ is then defined as the region of the manifold from which light cannot escape to $\mathscr{I}^+$—that is, the complement of $I^-(\mathscr{I}^+)$—and the event horizon is the boundary $\dot{\mathcal{B}}$ (see Figure 20.6). One can demand that an asymptotic region $\mathscr{I}^+$ be "complete," that is, roughly, that it is "as large as" the asymptotic region of Minkowski spacetime.[61] We then have the following formulation of weak cosmic censorship: the maximal Cauchy development of *generic*, asymptotically flat initial data for Einstein's field equations with *suitable* matter fields has a complete $\mathscr{I}^+$. If weak cosmic censorship holds, it would imply that even if it is in principle possible to produce a singularity (even one that is not "benign" locally) by rearranging matter and energy within a finite region of spacetime, the effects of this singularity will not reach distant observers. The region

---

[58]   The bets are posted outside Thorne's office at Caltech, and are discussed in Thorne (2002).

[59]   There are other ways of formulating a weaker condition (Earman 1995, 74–75), such as adding further requirements related to the curve $\gamma$ in the definition of a naked singularity—e.g. that it is a geodesic curve, or that curvature invariants blow up along $\gamma$.

[60]   In slightly more detail, we require that there is a spacetime $\langle \tilde{\mathcal{M}}, \tilde{g}_{ab} \rangle$, the conformal completion of $\langle \mathcal{M}, g_{ab} \rangle$, consisting of $\mathcal{M} \cup \mathscr{I}$ (where $\mathscr{I}$ is the asymptotic region consisting of a past and a future part, $\mathscr{I}^+$ and $\mathscr{I}^-$ respectively, along with spatial infinity $i^0$), and a conformal isometry such that $\tilde{g}_{ab} = \Omega^2 g_{ab}$ on $\mathcal{M}$; see, e.g. Wald (1984, Ch. 11).

[61]   More precisely, an asymptotic region $\mathscr{I}^+$ is said to be *complete* just in case $\tilde{\nabla}^a \Omega$ is complete on $\mathscr{I}^+$, where $\tilde{\nabla}^a$ is the covariant derivative with respect to $\tilde{g}_{ab}$ and $\Omega$ is the conformal factor with which infinity is "brought into the finite" (cf. footnote 60).

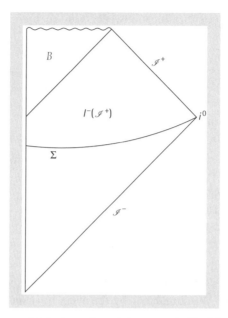

FIGURE 20.6    A conformal diagram of a black hole

outside the black hole is "safe for determinism" in the sense that the data on an appropriate slice $\Sigma$ uniquely determines the evolution throughout the asymptotic region.

Proving weak cosmic censorship requires first giving some precise content to the qualifications inserted in the above formulation (regarding *generic* initial data, and *suitable* matter fields). The initial wager above explicitly limited suitable matter fields to those that do not themselves develop singularities in flat spacetime. This restriction is motivated by the desire to separate singularities due to the matter fields themselves from those due to gravity, and it rules out fluids because shocks and other singularities do occur in flat spacetime.

The restriction that is more significant for our purposes is an implicit imposition of energy conditions (ECs). These are restrictions on $T_{ab}$, the energy-momentum tensor appearing on the right-hand side of Einstein's field equations, sometimes taken to characterize "reasonable" classical fields. Einstein's field equations can be solved for specific sources—that is, a specific choice regarding the matter distribution—but one can also consider solutions that obtain as long as the energy-momentum tensor satisfies certain conditions. (Without such restrictions, Einstein's field equations can be taken to *define* the energy-momentum tensor for a given $g_{ab}$.) The *dominant energy condition* (DEC) holds if the energy-momentum flow measured by any observer at any point is a timelike or null vector; intuitively, this requires that energy-momentum propagates on or within the light cone. Technically, the DEC requires that for all pairs of future-directed, unit timelike vectors $\xi^a$, $\zeta^a$, $T_{ab}\xi^a\zeta^b \geq 0$. If we require only that the inequality holds when $\zeta^a = \xi^a$, we have the *weak energy condition* (WEC); intuitively, this requires that there are no negative energy densities according to any observer. Finally, the *strong energy condition* (SEC) requires that $T_{ab}\xi^a\xi^b \geq \frac{1}{2}\mathrm{Tr}(T_{ab})$ for every unit timelike $\xi^a$. (This terminology is misleading, in that the SEC does *not* entail the WEC.)

What is the status of these conditions? They were once taken as defining properties for "reasonable" fields, and they effectively guarantee that gravity is an "attractive force"—in the sense that geodesics converge in regions with a non-zero energy-momentum tensor. In contrast, geodesics *diverge* in regions with non-zero EC-violating fields, other things being equal. Due to this stark contrast, the ECs clearly play an essential role in various results in classical GR. Recently, physicists have seriously considered EC-violating fields in a variety of situations that seem to call for such repulsive behavior, such as inflationary cosmology and in modeling "dark energy" thought to drive the observed accelerated expansion of the universe. One can formulate classical theories with a scalar field which violate one or more of the ECs but are arguably "physically reasonable."[62] In addition, we will see in the next section that these inequalities are fundamentally at odds with the kinematics of quantum field theory.

The second qualification in the formulation above allows that there may be counterexamples based on highly "special" initial data sets (e.g. with a high degree of symmetry) that lead to a naked singularity. The refined bet raises the bar: any counterexample to cosmic censorship must work for an open set of initial data, which requires that "nearby" initial data would also lead to naked singularities. Specifying what qualifies as "special" or "generic" initial conditions requires introducing a measure on the space of initial data. We concur with Robert Wald's (1998) assessment that choosing an appropriate measure or topology demands greater insight into the dynamics of GR than we currently possess, so any proposed definitions of "special" initial data are provisional.

A successful proof of strong cosmic censorship would immediately rule out time machines as defined above, because it would establish that "suitable" initial data lead to an inextendible spacetime—with no Cauchy horizon, and no possibility of CTCs developing beyond it. Weak cosmic censorship, on the other hand, would only establish that asymptotic observers would be shielded from causal pathologies, including CTCs, by an event horizon. There are related censorship theorems whose proof may shed some light on these conjectures. Hawking's (1992) "chronology protection conjecture" is more specific than either version of cosmic censorship, as it states that the laws of physics do not allow CTCs to be created. But before sketching Hawking's argument, we will briefly discuss one result that has been proven: the topological censorship theorem.

Why is it the case that physical spacetime appears to have a surprisingly simple topology? Nothing within GR rules out attributing enchantingly baroque topological structures to spacetime. A topological space is simply connected if any closed loop can

---

[62]  Barceló and Visser (2002) argue that all of the energy conditions are "dead" or "moribund," based in part on theories with scalar fields. Mattingly (2001), perhaps the only paper focused on energy conditions in the philosophy of physics literature, endorses a similar position. We will discuss the status of the ECs in more detail in the next section.

be smoothly contracted to a point. A torus is an example of a multiply connected space; the loops going around the central hole and those looping around the ring itself cannot be contracted to a point. The basic requirements GR places on an underlying manifold in order to qualify as a spacetime do not rule out multiply connected topologies, and even requiring that a given three-manifold can serve as a "reasonable" initial data surface does not place constraints on the topology. Donald Witt (1986) showed that it is possible to specify vacuum initial data for *arbitrary* closed or asymptotically flat three-manifolds. Given this tolerance for varied topology, one may wonder why the GR models describing observed, macroscopic regions are simply connected.

The topological censorship theorem shows that in GR any "naked topology" is hidden in much the same fashion as naked singularities would be hidden according to weak cosmic censorship. Multiply connected topologies are hidden in that any causal curve connecting past and future null infinity in an asymptotically flat spacetime ($\mathscr{I}^-$ and $\mathscr{I}^+$) cannot pass through a non-simply connected region (Friedman, Schleich, and Witt 1993; cf. Galloway 1995). (Suppose that there is a curve $\gamma$ from $\mathscr{I}^-$ to $\mathscr{I}^+$ that "threads the topology," in the sense that it cannot be deformed into (i.e. is not homotopic to) a curve lying entirely in the asymptotic region, which is simply connected. The proof proceeds by showing that any such curve would have to thread a strongly outer trapped surface, but the properties of such a surface imply that $\gamma$ cannot reach $\mathscr{I}^+$.)[63] The theorem was inspired by a result of Dennis Gannon (1975), which shows (using the same methods as the Hawking-Penrose singularity theorems) that incomplete geodesics occur to the future of a non-simply connected Cauchy surface in an asymptotically flat spacetime. Gannon's proof implies that the incomplete geodesics arise as a result of the topology of the Cauchy surface, leading to the natural question whether event horizons develop as well—and the topological censorship theorem establishes that they do. Any observer lucky enough to enter a region of spacetime with multiply connected topology would be trapped behind an event horizon just as surely as an astronomer on the unfortunate mission of exploring the interior of a black hole.

Topological censorship is closely connected to the discussion of time machines due to its implications for traversable wormholes. The discovery by Thorne and co-workers that a traversable wormhole could be used as a time machine provided that

---

[63] More precisely (following Friedman, Schleich, and Witt 1993), the theorem states that given an asymptotically flat, globally hyperbolic spacetime satisfying the averaged null energy condition, every causal curve from $\mathscr{I}^-$ to $\mathscr{I}^+$ is homotopic to $\gamma_0$, a causal curve lying in the simply-connected region $U$ of $\mathscr{I}$. The proof of the theorem relies primarily upon one lemma, namely that none of the generators of $\dot{J}^+(\tau)$ for an outer-trapped surface $\tau$ intersect $\mathscr{I}^+$. If we suppose that one of the generators of $\dot{J}^+(\tau)$, say $\xi$, intersects $\mathscr{I}^+$, which implies that $\tau$ intersects $J^-(\mathscr{I})$, then $\xi$ would stretch from $\tau$ to $\mathscr{I}^+$ with infinite affine parameter length. $\xi$ may only remain a generator of the boundary if it has no conjugate points (i.e. a point where it intersects an infinitesimally neighboring member of a congruence of null geodesics). However, the energy condition guarantees that the matter-energy distribution in the spacetime will serve to focus null geodesics. Given that the expansion is initially non-positive (which is the case because $\tau$ is an outer-trapped surface), this focusing must lead to a conjugate point within finite affine parameter length. Thus, we have a contradiction, and the original hypothesis that $\xi$ intersects $\mathscr{I}^+$ is false. This implies that $\gamma$ cannot both be homotopically inequivalent to $\gamma_0$, as that would require threading an outer-trapped surface $\tau$, and connect $\mathscr{I}^-$ to $\mathscr{I}^+$.

the two mouths of the wormhole are in relative motion inspired much of the current literature on time machines.[64] Topological censorship might appear to rule out the use of traversable wormholes as time machines, but the theorem applies only granted two strong assumptions. The first assumption is that the averaged null energy condition holds. This is weaker than the pointwise energy conditions stated above: it allows that at some points $T_{ab}\xi^a\xi^b$ is negative (where $\xi^a$ is a null vector), but requires that the average value of $T_{ab}\xi^a\xi^b$ along a null geodesic (where $\xi^a$ are the null tangent vectors) is non-negative. The importance of this assumption indicates that maintaining a wormhole requires a form of "exotic" matter for which the averaged null energy condition fails to hold. Second, the theorem requires a strong causality condition—global hyperbolicity. Thus the theorem might be best thought of as elucidating the consequences of global hyperbolicity for the topology of the asymptotic region and for event horizons.[65] Obviously this second assumption undercuts the usefulness of this result as a no-go theorem for time machines, as the assumption rules out CTCs by *fiat*.

Hawking (1992) aimed to establish such a no-go theorem for time-machine spacetimes that does not depend on imposing a causality condition. In general terms one would like to prove a theorem of the following form:

**Conjecture 1 (Classical Chronology Protection)** Given initial data, satisfying "_____," specified on a surface $\Sigma$, there exists a solution of Einstein's field equations $\langle \mathcal{M}, g_{ab} \rangle$ (unique up to diffeomorphism), with properties "_____," that does not contain CTCs.

The importance of the result depends on exactly what goes into the two blanks: the result would be more decisive to the extent that there are unambiguous, well-motivated ways to fill in both blanks, such that the precisely formulated claim is amenable to proof. Filling in the blanks requires facing up to the same challenges we saw above in the discussion of strong and weak cosmic censorship: what is the status of the energy conditions one might impose in order to give Einstein's field equations some bite? How should we formulate the requirement that the resulting solution is not "special" within the space of solutions? It will thus come as no surprise that debate regarding results that have been established focuses on these questions.

Rather than attacking the general conjecture directly, one might instead formulate a necessary condition for a time-machine spacetime and then show that this condition is incompatible with some other requirement (such as the energy conditions). Hawking (1992) argued that a suitable necessary condition for a time-machine spacetime is the existence of a *compactly generated Cauchy horizon*, and then showed that the existence of such a horizon entails that the null energy condition must be violated.

---

[64] These ideas were introduced in Morris and Thorne (1988) and Morris, Thorne and Yurtsever (1988); see also Thorne's engaging account of how this line of research unfolded in Thorne (1994, Ch. 14).

[65] Galloway (1995) shows a weaker condition is all that is in fact required. Note that the main use of topological censorship in the physics literature is proving results regarding the topology of black holes; see Friedman and Higuchi (2006) for further discussion and references.

A Cauchy horizon is *compactly generated* if the null geodesics that are generators of the surface enter and remain within a compact set. (This condition is meant to rule out influences "coming from infinity" and emerging from singularities as having some impact on the creation of CTCs in the region $V$.) Hawking's argument then proceeds roughly as follows. The generators of the Cauchy horizon are null geodesics segments, which do not have past endpoints.[66] However, in order to enter and remain within a compact set (towards the past) the generators must converge. Imposing the null energy condition along with the assumption that the geodesics encounter some non-zero energy density (sometimes called the "generic condition") leads to a contradiction, because the presence of matter satisfying the energy condition implies, via the Raychaudhuri equation, that the generators must converge and have past endpoints. Thus the existence of a compactly generated Cauchy horizon requires violation of the null energy condition or the generic condition on the Cauchy horizon.

One response to Hawking's result is to question whether the presence of a compactly generated Cauchy horizon is an appropriate necessary condition (see Earman et al. (2009) for further discussion). We have argued above in favor of the Potency Condition (Condition 1) as an appropriate definition of a time-machine spacetime, and it is not the case that the Potency Condition fails in cases where the Cauchy horizon fails to be compactly generated (cf. Ori 2007). Even though our preferred definition of time machine is thus broader than Hawking's, his results might apply more broadly as well, to spacetimes with a Cauchy horizon. It is plausible that the existence of a Cauchy horizon, compactly generated or not, will be accompanied by violations of energy conditions in classical GR (cf. the earlier results of Tipler (1976, 1977)), although we do not know of any theorems that establish the general claim. There are also some results indicating that Cauchy horizons may be "special" in the sense of being measure zero, given a reasonable measure assigned to the space of solutions of Einstein's field equations. For example, Vincent Moncrief and James Isenberg (1983) prove that in a particular case (namely, granted that the Cauchy horizon is analytic and ruled by closed null geodesics) there are symmetries in the neighborhood of the Cauchy horizon. A more general result along these lines would support the strong cosmic censorship by showing that the existence of Cauchy horizons does not hold generically, for open sets of initial data rather than just for specific cases.

But the more intriguing question is whether we should trust the physical description of a Cauchy horizon offered by classical GR. Hawking (1992) hoped to formulate a quantum chronology protection conjecture based on a combination of ideas from quantum theory and classical GR. Quantum theory provides encouragement to time-travel fans because quantum fields do not satisfy the point-wise energy conditions.

---

[66] A *past endpoint* of a curve $\gamma(s)$ is a point $p$ such that for every neighborhood $\mathcal{O}$ of $p$ there exists $s'$ such that $\gamma(s) \in \mathcal{O}$ for all $s < s'$ (where the parameter $s$ increases with time along the curve). Note that although there is at most one past endpoint of a curve, the curve may "continually approach" $p$ without there being a value of $s$ such that $\gamma(s) = p$. The proof that the generators have no past endpoints uses the same techniques as proofs of the properties of other achronal surfaces such as $\dot{I}^+(S)$; the existence of a past endpoint in $H^+(\Sigma)$ is incompatible with the properties of such a surface.

This opens the possibility of treating results implying the violation of energy conditions as simply reflecting the fact that quantum effects will become important at the Cauchy horizon. In any case, Hawking sought a stronger no-go theorem based on semi-classical quantum gravity. This is a hybrid theory which incorporates the effects of quantum fields as sources within classical GR, without attempting to quantize spacetime geometry itself. The goal is to calculate the so-called "backreaction" of quantum fields as a perturbation to a classical spacetime, by putting the quantity $\langle \xi | T_{ab} | \xi \rangle$ (the expectation value of the renormalized energy-momentum tensor for the quantum state $| \xi \rangle$) into Einstein's field equations. Hawking (1992) argued that, *contra* Kim and Thorne's (1991) earlier results, the divergence of this quantity near the Cauchy horizon enforces chronology protection: the backreaction effects effectively prevent the formation of CTCs. The ensuing debate led to a theorem by Bernard Kay, Marek Radzikowski, and Wald (1997) which shows that the quantity $\langle \xi | T_{ab} | \xi \rangle$ is not well defined at all points of a compactly generated Cauchy horizon.[67] This result may be taken to support Hawking's conjecture, in that it clarifies the pathologies associated with compactly generated Cauchy horizons. However, one might instead read the result as showing that Cauchy horizons lie outside the domain of applicability of semi-classical quantum gravity. Even if we might have hoped to prove the chronology protection conjecture using a hybrid theory such as semi-classical quantum gravity, it seems that the full-fledged quantum gravity cannot be avoided. This is not so much a shortcoming of current results as a reason for interest in this topic: determining whether chronology protection holds, and if so *why* it holds, may provide some insight into a theory of quantum gravity.

# 7. ENERGY CONDITIONS IN QFT

At this stage we will turn to the status of time travel and time machines in theories that extend GR. The discussion is necessarily speculative and preliminary, given that the successor to GR has yet to be formulated, despite a great deal of effort. But we can at least pose the questions that we might expect the successor theory to answer, so that we can try to glean hints for an answer from the various research programs currently being pursued. Roughly put, does the successor to GR draw the bounds of physical possibility such that CTCs and similar causal pathologies are included or excluded? Given that the success of GR depends on abandoning non-dynamical

[67] They demonstrate, roughly speaking, that for a compactly generated Cauchy horizon there is a non-empty set of base points, which are past terminal accumulation points for some null geodesic generator $\gamma$—intuitively, $\gamma$ continually re-enters any given neighborhood of the point $p$. In any neighborhood of such points, there are points that can be connected by a null curve "globally" (within the full spacetime) even though they cannot be connected by a null curve "locally" (i.e. within the neighborhood). This conflict between local and global senses of null related undermines the standard prescription for defining $\langle \xi | T_{ab} | \xi \rangle$. See Earman et al. (2009) for further discussion, and Visser (2003), Friedman and Higuchi (2006) for more detailed reviews and further references.

global constraints on spacetime structure, it would be quite striking if a successor theory reinstated global constraints. However, the need for such constraints may come from the matter sector of the theory not included in classical GR. In this section we first consider recent results regarding energy conditions in QFT, which indicate that quantum fields satisfy "non-local" energy conditions even though the classical energy conditions do not hold.

In GR the term appearing on the left-hand side of the field equations (the Einstein tensor $G_{ab}$, constructed out of the metric and its first and second derivatives) characterizes spacetime geometry, whereas the energy-momentum tensor on the right-hand side ($T_{ab}$) contains information about the source of the gravitational field. Einstein frequently expressed his dissatisfaction with the need to put the energy-momentum tensor into the field equations by hand, by choosing a particular matter model. Einstein (1936, 335) described the field equations as a building with two wings: the left-hand side built of fine marble, and the right-hand side built of low-grade wood. Presumably Einstein's unified field theory project, if successful, would have produced a building constructed entirely of marble, in which fields act as their own sources and there is no need for independent matter models. But one need not share Einstein's goal of unification to have reason to avoid relying too heavily on low-grade wood, in the sense of proving results that hold only for particular energy-momentum tensors. In order to study dynamical evolution according to Einstein's field equations at a more general level, it is natural to consider properties shared by energy-momentum tensors for different types of matter. The energy conditions allow a more general approach and they are crucial assumptions in major results in classical GR such as the singularity theorems and positive mass theorems.[68] However, the point-wise energy conditions described above *all* fail for quantum fields. What does this failure imply regarding results such as the singularity theorems and the status of spacetimes with exotic causal structure?

As we mentioned above, the failure of point-wise energy conditions in QFT provides some encouragement, in the sense that quantum fields may provide the "exotic matter" needed to violate the energy conditions and, for example, maintain a traversable wormhole long enough to convert it into a time machine. The point-wise energy conditions fail in QFT because the energy density necessarily admits arbitrarily negative expectation values at a point (as shown by Epstein, Glaser, and Jaffe 1965). These negative energy densities occur even though the overall energy is positive, and they arise due to quantum coherence effects. Exploiting quantum fields as a kind of "exotic matter" requires understanding the failure of the energy conditions in more detail. For example, does QFT allow one to have quantum states with large negative energy densities, not just at a single point but over an extended region of spacetime? And what would be required in order to create macroscopic wormholes or other exotic structures, as opposed to "microscopic" (Planck scale) exotic structures? Recent work

---

[68] The positive mass theorems establish that the total energy associated with an isolated system is positive (see, e.g., Wald 1984, Ch. 11).

has demonstrated that QFT *does* place constraints on negative energy densities, in the form of "non-local" energy conditions (specifying limits on energy densities over spacetime regions rather than points). The precise nature of these constraints and their implications for time machines are still being debated, but the current results provide some evidence that the energy conditions enforced by QFT will be sufficient to rule out the exotic structures incompatible with the point-wise energy conditions in classical GR (within the domain of applicability of semi-classical quantum gravity).[69]

Lawrence Ford (1978) originally proposed non-local energy conditions (which he calls "quantum inequalities") based on thermodynamical considerations: he argued that negative energy densities without further constraints would lead to a violation of the second law of thermodynamics. If one could manipulate the quantum fields appropriately, it would be possible in principle to use negative energy fluxes to lower both the temperature and entropy of a hot body. Ford argued that if the magnitude of the flux is small enough (in particular, $\Delta E \geq -\hbar/\Delta t$), given the timescale during which it was transferred to the body $\Delta t$, the effect of the flux on the body's energy would be smaller than the uncertainty in the body's energy. Thus avoiding a conflict with thermodynamics requires placing constraints on the negative energies and fluxes allowed by QFT. Ford and Thomas Roman subsequently derived a number of quantum inequalities for different cases. These results generally take the form of showing that there is a (state-dependent) lower bound on the negative energy density "smeared" over a spacetime region, such that the bound on the energy varies inversely with the size of the region. As far as we know, the results obtained so far are still a patchwork quilt covering a variety of different cases—different choices of fields, flat vs curved spacetimes, etc. (see, e.g. Fewster 2005). But they are very suggestive that, while QFT allows for negative energy densities, the resulting violations of the point-wise energy conditions will not be sufficient to undermine the results of classical GR.

# 8. FROM CLASSICAL TO QUANTUM GRAVITY

Turning to full quantum theories of gravity now, we would like to take a brief look at three rather different approaches: causal sets, loop quantum gravity, and string theory. Our discussion is guided by asking how each approach treats the physical possibility of CTCs and other causal pathologies. If a *successful* theory of quantum gravity ruled out CTCs *ab initio* it would clearly show that GR erred on the side of permissiveness regarding global causal structure, and that the classical chronology-violating spacetimes will not be obtained as limits of QG solutions. It is difficult to distinguish an in-principle causal restriction from a practical restriction to globally hyperbolic for more pragmatic reasons. We have no objection to imposing global hyperbolicity, or a kindred condition, as a mathematical convenience, as long

---

[69] See, e.g. Flanagan and Wald (1996), Roman (2004), Ford (2005), Fewster (2005), and Friedman and Higuchi (2006).

as it is acknowledged that some further motivation is needed. Of course, the success of a quantum theory of gravity purged of causality-violating spacetimes may itself provide after-the-fact justification for such a restriction. It may be, however, that global hyperbolicity can be derived from the resources of the theory or from well-justified conditions on what is physically reasonable or even possible. If this second case were to materialize, it would constitute an important achievement, giving us principled reasons to reject general-relativistic spacetimes with CTCs as unphysical artifacts of the mathematical formalism of the theory. Thirdly, it may turn out that CTCs are prevalent in the space of solutions to a successor theory, or essential to physical applications and understanding the content of the theory, indicating that this intriguing aspect of GR will stay with us.

Causal sets is an iconoclastic approach to formulating a quantum theory of gravity that does not rely on known physics as a vantage point, instead trying to arrive at such a theory *ab initio* (Bombelli et al. (1987); Bombelli et al. (2003)). The causal sets approach postulates a fundamentally discrete spacetime structure that satisfies a few simple conditions and tries to establish that in the classical limit, the continuous spacetimes of GR can be recovered.[70] More particularly, the approach demands that the elements of the fundamental spacetime exhibit the structure of a *causal set*. Causal sets $\mathcal{C}$ are endowed with a binary relation $\prec$ such that for all $a, b, c \in \mathcal{C}$, (i) $a \prec b$ and $b \prec c$ imply $a \prec c$ (transitivity), (ii) $a \nprec a$ (acyclicity), and (iii) all *past sets* $\mathcal{P}(a) := \{b : b \preceq a\}$ are finite. Condition (ii) amounts to ruling out CTCs by stipulation, at least at the fundamental level. Although it is not yet clear how the theory relates these discrete structures to the continuous spacetimes of GR, since it fundamentally encodes the causal structure in the manner specified above, it cannot give rise to continuous spacetimes containing CTCs. Malament's (1977) theorem, mentioned in §3, establishes that if $\langle \mathcal{M}, g_{ab} \rangle$ and $\langle \mathcal{M}', g'_{ab} \rangle$ are both past and future distinguishing spacetimes, and if there exists a bijection $f$ between $\mathcal{M}$ and $\mathcal{M}'$ such that both $f$ and $f^{-1}$ preserve the causal precedence relations, then $f$ must be homeomorphism, that is, a topological isomorphism between the manifolds. This implies, unsurprisingly, that an approach encoding only the causal structure cannot allow closed causal curves. This means that such an approach does not command the resources to recover the metric structure of classical GR in its full generality. The spacetimes that can be captured in the continuum limit by a causal-set approach thus represent a proper subset of those admitted by the Einstein equations, excluding those with CTCs. In fact, the causal sets approach is wedded to a commitment to the first way of delineating physically reasonable spacetimes as listed in §6, that is, the one based on causality conditions. And we maintain, as above, that there are good reasons to prefer that causality conditions such as global hyperbolicity be deduced from independently motivated assumptions, rather than stipulated by hand.

---

[70] The causal sets approach is still regarded as a classical theory so far, as it fails to provide a proper quantum dynamics. To our knowledge, the classical probabilities involved in the dynamical evolution according to the causal sets theory have so far not been replaced by truly quantum dynamics, including e.g. transition amplitudes.

To be sure, if a particular approach to quantum gravity turns out to offer a successful, or even only a viable, quantum theory of gravity, then such success or viability would trump our objections to imposing causality conditions a priori. Unfortunately, it looks like the causal sets are far from delivering this. More promising, arguably, is another approach to quantum gravity: loop quantum gravity (Rovelli 2004, Thiemann 2007). Loop quantum gravity (LQG) attempts a canonical quantization of a Hamiltonian formulation of GR.[71] As a research program with the ambition of delivering a full quantum theory of gravity, and only of gravity, LQG has not been brought to a completion yet. Most importantly, the dynamics of the theory and the relationship to classical spacetimes theories remain ill-understood. Nevertheless, LQG is considered by many to be a very promising research program and is certainly the frontrunner of approaches starting out from classical GR.

In Earman et al. (2009), we stated, by way of conclusion, that LQG, like causal sets, simply ignores the possibility of CTCs as the canonical quantization procedure requires the spacetimes subjected to it to be globally hyperbolic, except in that as long as the classical limits of LQG states is so poorly understood, we cannot exclude that CTCs might emerge in this limit. This is a real possibility since in some cases, for example, such as classically singular spacetimes, the classical spacetime structure does not even approximate the well-defined corresponding quantum state. We would like to add a further reason for hesitation. The above remarks tacitly assume that LQG aspires to describe the global structure of quantum spacetime. This need not be so: one might just as well think of the theory as offering descriptions of much more local features of quantum spacetime, such as the spatial volume of a finite chunk of spacetime in a laboratory.[72] Of course, all these chunks of spacetime will be assumed to be globally hyperbolic. However, a spacetime patched together from globally hyperbolic spacetime need not be globally hyperbolic itself.[73] If conceived in this way, therefore, LQG might well permit time travel. It should be noted, however, that this may mean that LQG cannot be a fundamental quantum theory of spacetime as it doesn't account for the global structure of spacetime. Unless one thinks that the global structure of spacetime emerges from, or is supervenient on, the fundamental structure of patched together chunks of quantum spacetime, and barring the possibility of CTCs emerging in the classical limit of a loop quantum gravitational state, either LQG cannot be a fundamental quantum theory of spacetime or it rules out time travel.

The third, and by far most researched, approach to quantum gravity is string theory (Polchinski 1998). String theory takes as its vantage point, both historically and systematically, the standard model based on conventional QFT. It exists at two separate

---

[71] For GR as a Hamiltonian system with constraints, cf. Thiemann (2007, Ch. 1) and Wüthrich (2006, Ch. 4).

[72] We are indebted to Carlo Rovelli for arguing, in private conversations, for the validity of this approach to one of us (C.W.).

[73] For a simple example, just think of a rolled-up slice of Minkowski spacetime, which contains CTCs and is thus not globally hyperbolic, as a carpet glued together from globally hyperbolic, diamond-shaped tiles.

levels. At the perturbative level, on the one hand, string theory consists of a set of well-developed mathematical techniques which define the string perturbation expansion over a given background spacetime. On the other hand, attempts at formulating the elusive non-perturbative theory, supposed to be capable of generating the perturbation expansion, have not succeeded so far. Such a theory, conventionally named M-theory, for "membrane," "matrix," or "mystery" theory, currently consists of incipient formulations using non-perturbative compactifications of higher dimensional theories based on so-called duality symmetries, that is, symmetries relating strong coupling limits in one string theory to a weak coupling limit in another (dual) string theory.

To the best of our knowledge, there are so far no results in non-perturbative M-theory pertaining directly to the possibility of time travel. There are, however, a series of pertinent findings in supersymmetric gravity, a close relative of string theory. These results show that CTCs arise naturally in certain solutions of this theory. If supersymmetric gravity turned out to permit time travel in the sense of CTCs, then string theory will be hard pressed to eschew its possibility.[74] As far as we are aware, the story begins ten years ago when Gary Gibbons and Herdeiro (1999) asked whether supersymmetry allowed CTCs. The straightforward answer is that it does, at least in that there are supersymmetric solutions of flat space with periodically identified time coordinate analogous to the rolled-up Minkowski spacetime in GR depicted in Figure 20.3. Solutions of this type, however, are not topologically simply-connected and the CTCs can thus be avoided by passing to a covering spacetime. In many cases of supersymmetric solutions with CTCs, this move is not possible since the relevant supersymmetric spacetimes are topologically trivial. It turns out that there are at least two important types of supersymmetric solutions containing CTCs: a supersymmetric cousin of Gödel spacetime and the so-called BMPV black hole spacetime. Let us look at these in turn.

Jerome Gauntlett et al. (2003) have shown that there exists a solution of five-dimensional supersymmetric gravity that is very similar to the Gödel spacetime of GR in that it also describes a topologically trivial, rotating, and homogeneous—and thus not asymptotically flat—universe. Almost by return of mail, however, Petr Hořava and collaborators have argued that holography acts as a form of chronology protection in the case of this Gödel-like spacetime, in that the CTCs are either hidden behind "holographic screens," and thus inaccessible for timelike observers, or that they are broken up into causally non-circular pieces (Boyda et al. 2003).

The second important supersymmetric spacetime containing CTCs is the so-called BMPV black hole solution.[75] BMPV black holes are the supersymmetric counterparts of the Kerr-Newman black holes of GR: they are charged, rotating black holes in simply connected, asymptotically flat spacetime. And similarly to the Kerr-Newman case in

[74] Many of the following results have been gained in five-dimensional supersymmetric gravity, rather than its higher-dimensional relatives. Five-dimensional supergravity is an approximation to higher-dimensional string theory. It should be noted that all solutions in the five-dimensional case can easily be amended to be solutions for ten- and eleven-dimensional supergravity (Gauntlett et al. 2003, 4590).

[75] After the initials of Breckenridge et al. (1997).

GR, as Gibbons and Herdeiro (1999) have shown, the solution can be maximally analytically extended to contain a region with CTCs.[76] Below the critical value of angular momentum, we find a black hole with an event horizon, and CTCs in a region screened off by this horizon from asymptotic observers (Gauntlett et al. 2003, 4589). Not only can asymptotic observers not see the CTCs in this case, but they are inaccessible, in that they are hidden behind a "velocity-of-light surface," that is, a surface which can only be passed by accelerating beyond the speed of light. If the angular momentum is above the critical value, however, the black hole is shielded in the sense that geodesics from the asymptotic region cannot pass into the black hole (Gibbons and Herdeiro 1999). The solution is thus geodesically complete. In this case, we find "naked" CTCs outside the event horizon.[77] As Gibbons and Herdeiro (1999) show, no cosmic censorship seems to be able to rule out this case: This hyper-critical solution represents a geodesically complete, simply connected, asymptotically flat, non-singular, time-orientable, supersymmetric spacetime with finite mass that satisfies the dominant energy condition. Gibbons and Herdeiro note in their analysis, however, that this hyper-critical solution describes a situation where the CTCs have existed "forever," that is, it seems not amenable to an implementation of a time machine in our sense.[78]

Although this may rule out the practicability of time travel in BMPV spacetimes, Gauntlett et al. (2003) have more good news for the aspiring time traveler. In their classification of all supersymmetric solutions of minimal supergravity in five dimensions, they find that CTCs generically appear in physically important classes of solutions. In fact, they complain that it is difficult to find any new solutions of five-dimensional supersymmetric gravity that do *not* contain either CTCs or singularities.

This brief survey of one line of research in string theory is of course a slender basis upon which to make general claims regarding the fate of causal pathologies in the successor theory to GR. However, they do suggest that CTCs arise naturally in string theory. These results are provisional in that we do not know whether they translate to the full, non-perturbative M-theory, or whether non-perturbative string theory is a viable theory in its own right for that matter. But the possibility of time travel and perhaps of time machines seems likely to stay with the foundations of physics for some time to come.

---

[76] Strictly speaking, Gibbons and Herdeiro show this for the *extremal* case, i.e. black holes whose angular momentum is equal to its mass (in natural units). The result may generalize to the non-extremal case, but the hope that it does so is based on only very preliminary results, and the hope is not universally shared.

[77] It has been suggested that the emergence of "naked" CTCs may be the result of the breakdown of unitarity (Herdeiro 2000).

[78] The reason for this is that passing from the "under-rotating" case to the "over-rotating" one seems to require an infinite amount of energy. As Dyson (2004) has shown in her analysis of BMPV black holes made out of gravitational waves and D-branes, i.e. hypersurfaces in ten-dimensional spacetime, speeding up the rotation of an "under-rotating" BMPV black hole in order to produce naked CTCs leads to the formation of a shell of gravitons with the D-branes enclosed inside the black hole. This mechanism, which is akin to the "enhancon mechanism" that string theorists use to block a class of naked singularities, precludes the system from speeding up beyond the critical value.

# 9. CONCLUSIONS

In conclusion, let us return to the question posed in the introduction: what does the existence of solutions to Einstein's field equations with exotic causal structure imply regarding the nature of space and time according to GR; or, more generally, does physics permit such exotic causal structures, and if so, what does this permission mean for the nature of space and time? The possibility of exotic structures arises as a byproduct of GR's near elimination of global constraints on topology and geometry. While the resulting freedom opens up fascinating possibilities such as those described above, we should emphasize that the empirical success of GR does not appear to depend on the existence of these solutions. As a result the freedom looks excessive. However, it remains unclear whether this excessive freedom can be traced to an incompleteness or inaccuracy of GR that will be corrected in a successor theory. The vitality of the physics literature regarding time machines and time travel indicates the importance of this issue as well as its difficulty.

Our first focus has been on the implications of time travel, defined in terms of the existence of CTCs. Many philosophers have attempted to dismiss this question as illegitimate, on the grounds that a variety of paradoxes establish the logical impossibility, metaphysical impossibility, or improbability of time travel in this sense. We found these arguments wanting, although they do usefully illustrate the importance of consistency constraints in spacetimes with CTCs. It may come as a shock to discover that the consistent time-travel scenarios are not just the stuff of fiction: there are several chronology-violating spacetimes that exhibit the local-to-global property described in §3 for appropriate choices of fields. However shocking the existence of these solutions may be, we assert that there is no footing to reject them due to alleged paradoxes, and no basis for imposing a causality condition insuring "tame" causal structure as an a priori constraint.

Setting aside objections based on the paradoxes, attempting to answer our question leads into a tangle of interconnected issues in philosophy of science and the foundations of GR. We hope to have at least clearly identified some of these issues and illustrated how their resolution contributes to an answer. First, consider cosmological models such as Gödel's that are not viable models for the structure of the observed universe. Assessing the importance of these models turns on difficult questions of modality applied to cosmology. Even if we grant that GR provides the best guide to what is physically possible in cosmology, the existence of models like Gödel's does not directly undermine the use of special structures such as the preferred foliation in the FLRW models without a questionable modal argument, or claim that such models reveal something significant about the laws of GR. Thus, if we were only considering cosmological models with exotic causal structure, it would be difficult to answer Maudlin's challenge. Maudlin claims that metaphysicians can safely dismiss exotic spacetimes because dynamical evolution according to Einstein's field equations does not force CTCs to arise from possible initial data. But this assertion presumes a resolution of a second open issue, the cosmic censorship conjecture or (some form of) the weaker chronology protection conjecture. Given a proof of the cosmic censorship conjecture, one could clearly demarcate situations in which Einstein's field equations coupled to source equations satisfying constraints such as the energy conditions, generically lead to globally hyperbolic spacetimes from situations in which dynamical evolution leads to Cauchy horizons, and the possibility of extensions beyond them containing CTCs. There are still significant obstacles to a proof of cosmic censorship due to our lack of understanding of the space of solutions to GR. Similarly, a proof of a sufficiently powerful chronology protection conjecture imposing some principled conditions on a spacetime's properties would underwrite Maudlin's claims. Alas, this second issue remains open to date, not least because it is far from obvious how the blanks in Conjecture 1 concerning suitable initial data and physically reasonable spacetimes ought to be filled in.

A third issue concerns the impact of incorporating quantum effects. Does the space of solutions of semi-classical quantum gravity, or even full quantum gravity, include time-machine solutions or solutions with CTCs? Thus, our investigation went beyond a mere analysis of the foundations of GR, in at least two respects. First, we have turned to semi-classical quantum gravity and listed how the quantum can be more permissive in tolerating the violation of energy conditions and thus be more lax about the suitability of the matter sector. Although no one really takes semi-classical theories seriously as competitors for final theories of quantum gravity, important lessons of how spacetime and quantum matter interact may be gleaned from them. Second, in a brief survey of three approaches to full quantum gravity, causal set theory, loop quantum gravity, and string theory, we have found that string theory in particular seems to nourish the hopes of aspiring time travelers, while one shouldn't be too hasty in ruling time travel out in the case of loop quantum gravity. These results are very preliminary and much remains to be seen, not the least of which is whether any of the mentioned theories can offer a full quantum theory of gravity. But we hope that the reader walks away from this article with a firm sense that these foundational analyses in GR, semi-classical, and full quantum gravity constitute important attempts at both understanding the classical theory, as well as illuminating the path towards a quantum theory of gravity.

# ACKNOWLEDGEMENTS

We are indebted to Craig Callender for his patience and comments on an earlier draft. We also wish to thank John Earman, John Manchak, and the Southern California Reading Group in the Philosophy of Physics, as well as audiences at Utrecht and Paris for valuable feedback. C. W. gratefully acknowledges support for this project by the Swiss National Science Foundation (Project "Properties and Relations", grant 100011–113688).

## REFERENCES

Andréka, H., Németi, I., and Wüthrich, C. (2008), 'A Twist in the Geometry of Rotating Black Holes: Seeking the Cause of Acausality', *General Relativity and Gravitation* 40: 1809–1823.

Arntzenius, F. (2006), 'Time Travel: Double your Fun', *Philosophy Compass* 1: 599–616.

——and Maudlin, T. (2005), 'Time Travel and Modern Physics', in E. Zalta (ed.), *The Stanford Encyclopedia of Philosophy*. http://plato.stanford.edu/entries/time-travel-phys/.

Barceló, C., and Visser, M. (2002), 'Twilight for the Energy Conditions?', *International Journal of Modern Physics* D11: 1553–1560.

Belot, G. (2005), 'Dust, Time, and Symmetry', *British Journal for the Philosophy of Science* 56: 255–291.

Black, M. (1956), 'Why Cannot an Effect Precede its Cause?', *Analysis* 16: 49–58.

Bombelli, L., Lee, J., Meyer, D., and Sorkin, R. D. (1987), 'Space-Time and a Causal Set', *Physical Review Letters* 59: 521–524.

Bonnor, W. B. (2001), 'The Interactions Between Two Classical Spinning Particles', *Classical and Quantum Gravity* 18: 1381–1388.

Boyda, E. K., Ganguli, S., Hořava, P., and Varadarajan, U. (2003), 'Holographic Protection of Chronology in Universes of the Gödel Type', *Physical Review* D67: 106003.

Breckenridge, J. C., Myers, R. C., Peet, A. W., and Vafa, C. (1997), 'D-branes and Spinning Black Holes', *Physics Letters* B391: 93–98.

Callender, C., and Weingard, R. (2000), 'Topology Change and the Unity of Space', *Studies in History and Philosophy of Modern Physics* 31: 227–246.

Cho, H. T., and Kantowski, R. (1994), 'Measure on a Subspace of FRW Solutions and "the Flatness Problem" of Standard Cosmology', *Physical Review* D50: 6144–6149.

Coule, D. H. (1995), 'Canonical Measure and the Flatness of a FRW Universe', *Classical and Quantum Gravity* 12: 455–469.

Deutsch, D. (1991), 'Quantum Mechanics Near Closed Timelike Lines', *Physical Review* D44: 3197–3217.

Dorato, M. (2002), 'On Becoming, Cosmic Time, and Rotating Universes', in C. Callender (ed.), *Time, Reality, and Existence* (Cambridge University Press), 253–276.

Douglas, R. (1997), 'Stochastically Branching Spacetime Topology', in S. Savitt (ed.), *Time's Arrow Today* (Cambridge University Press), 173–190.

Dummett, M. (1964), 'Bringing About the Past', *Philosophical Review* 73: 338–359.

Dyson, L. (2004), 'Chronology Protection in String Theory', *Journal of High Energy Physics* 3: 024.

Earman, J. (1986), *A Primer on Determinism* (Kluwer Academic).

——(1995), *Bangs, Crunches, Whimpers, and Shrieks: Singularities and Acausalities in Relativistic Spacetimes* (Oxford University Press).

——(2008), 'Pruning Some Branches from "Branching Spacetimes"', in D. Dieks (ed.), *The Ontology of Spacetime II* (Elsevier), 187–205.

——, Smeenk, C., and Wüthrich, C. (2009), 'Do the Laws of Physics Forbid the Operation of Time Machines?', *Synthese* 169: 91–124.

—— and Wüthrich, C. (2004), 'Time Machines', in E. Zalta (ed.), *The Stanford Encyclopedia of Philosophy*. http://plato.stanford.edu/entries/time-machine/.

Einstein, A. (1936), 'Physik und Realität', *Journal of the Franklin Institute* 221: 313–337. Translated by S. Bargmann as 'Physics and Reality', in Einstein, A., *Ideas and Opinions* (Crown Publishers), 290–323.

Ellis, G. F. R. (1996), 'Contributions of K. Gödel to Relativity and Cosmology', in P. Hájek (ed.), *Gödel '96: Logical Foundations of Mathematics, Computer Science and Physics—Kurt Gödel's Legacy*, (Berlin: Springer-Verlag), 34–49.

Epstein, H., Glaser, V., and Jaffe, A. (1965), 'Nonpositivity of the Energy Density in Quantized Field Theories', *Nuovo Cimento* 36: 1016–1022.

Fewster, C. J. (2005), 'Energy Inequalities in Quantum Field Theory', in J. C. Zambrini (ed.), *XIVth International Congress on Mathematical Physics* (World Scientific). Extended version available at http://arxiv.org/abs/math-ph/0501073.

Flanagan, E., and Wald, R. (1996), 'Does Backreaction Enforce the Averaged Null Energy Condition in Semiclassical Gravity?', *Physical Review* D54: 6233–6283.

Ford, L. H. (1978), 'Quantum Coherence Effects and the Second Law of Thermodynamics', *Proceedings of the Royal Society London* A364: 227–236.

—— (2005), 'Spacetime in Semiclassical Gravity', in A. Ashtekar (ed.), *100 Years of Relativity: Space-Time Structure: Einstein and Beyond* (World Scientific), 293–310.

Friedman, J. L. (2004), 'The Cauchy Problem on Spacetimes that are not Globally Hyperbolic', in P. T. Chrusciel and H. Friedrich (eds.), *The Einstein Equations and the Large Scale Behavior of Gravitational Fields: 50 Years of the Cauchy Problem in General Relativity* (Birkhäuser), 331–346.

—— and Higuchi, A. (2006), 'Topological Censorship and Chronology Protection', *Annalen der Physik* 15: 109–128.

——, Morris, M. S., Novikov, I. D., Echeverria, F., Klinkhammer, G., Thorne, K. S., and Yurtsever, U. (1990), 'Cauchy Problem in Spacetimes with Closed Timelike Curves', *Physical Review* D42: 1915–1930.

——, Schleich, K., and Witt, D. M. (1993), 'Topological Censorship', *Physical Review Letters* 71: 1486–1489.

Galloway, G. J. (1995), 'On the Topology of the Domain of Outer Communication', *Classical and Quantum Gravity* 12: L99–L101.

Gannon, D. (1975), 'Singularities in Nonsimply Connected Space-times', *Journal of Mathematical Physics* 16: 2364–2367.

Gauntlett, J. P., Gutowski, J. B., Hull, C. M., Pakis, S., and Reall, H. S. (2003), 'All Supersymmetric Solutions of Minimal Supergravity in Five Dimensions', *Classical and Quantum Gravity* 20: 4587–4634.

Geroch, R. (1967), 'Topology in General Relativity', *Journal of Mathematical Physics* 8: 782–786.

—— (1977), 'Prediction in General Relativity', in J. Earman, C. Glymour, and J. Stachel (eds.), *Foundation of Spacetime Theories: Minnesota Studies in the Philosophy of Science VIII* (University of Minnesota Press), 81–93.

—— and Horowitz, G. (1979), 'Global Structure of Spacetimes', in S. W. Hawking and W. Israel (eds.), *General Relativity: An Einstein Centenary Survey* (Cambridge University Press), 212–293.

Gibbons, G. W., and Herdeiro, C. A. R. (1999), 'Supersymmetric Rotating Black Holes and Causality Violation', *Classical and Quantum Gravity* 16: 3619–3652.

Gödel, K. (1949a), 'An Example of a New Type of Cosmological Solutions of Einstein's Field Equations of Gravitation', *Review of Modern Physics* 21: 447–450.

—— (1949b), 'A Remark About the Relationship Between Relativity Theory and Idealistic Philosophy', in P. A. Schilpp (ed.), *Albert Einstein: Philosopher-Scientist* (Open Court), 557–562.

—— (1952), 'Rotating Universes in General Relativity Theory', in L. M. Graves et al. (eds.), *Proceeding of the International Congress of Mathematicians* (American Mathematical Society), 175–181.

Hawking, S. W. (1992), 'Chronology Protection Conjecture', *Physical Review* D46: 603–611.

—— and Ellis, G. F. R., (1973), *The Large Scale Structure of Space-time* (Cambridge University Press).

—— and Page, D. N. (1988), 'How Probable is Inflation?', *Nuclear Physics* B298: 789–809.

—— and Penrose, R. (1996), *The Nature of Space and Time* (Princeton University Press).

Herdeiro, C. A. (2000), 'Special Properties of Five Dimensional BPS Rotating Black Holes' *Nuclear Physics* B582: 363–392.

Horwich, P. (1987), *Asymmetries in Time: Problems in the Philosophy of Science* (MIT Press).

Kay, B. S., Radzikowski, M. J., and Wald, R. M. (1997), 'Quantum Field Theory on Spacetimes with Compactly Generated Cauchy Horizons', *Communications in Mathematical Physics* 183: 533–556.

Kim, S. W., and Thorne, K. S. (1991), 'Do Vacuum Fluctuations Prevent the Creation of Closed Timelike Curves?', *Physical Review* D43: 3929–3947.

Kogut, A., Hinshaw, G., and Banday, A. J. (1997), 'Limits to Global Rotation and Shear from the COBE DMR Four-Year Sky Maps', *Physical Review* D55: 1901–1905.

Krasnikov, S. (2002), 'No Time Machines in Classical General Relativity', *Classical and Quantum Gravity* 19: 4109–4129.

Lewis, D. (1976), 'The Paradoxes of Time Travel', *American Philosophical Quarterly* 13: 145–152. Reprinted in his *Philosophical Papers*, Volume II (Oxford University Press), 67–80.

Malament, D. (1977), 'The Class of Continuous Timelike Curves Determines the Topology of Spacetime', *Journal of Mathematical Physics* 18: 1399–1404.

—— (1985a), 'Minimal Acceleration Requirements for "Time Travel" in Gödel Spacetime', *Journal of Mathematical Physics* 26: 774–777.

—— (1985b), ' "Time travel" in the Gödel universe', in P. D. Asquith and P. Kitcher (eds.), *PSA 1984, Vol. 2* (Philosophy of Science Association), 91–100.

—— (1995), 'Introductory Note to *1949b', in S. Feferman et al. (eds.), *Kurt Gödel: Collected Works, Volume III* (Oxford University Press), 261–269.

—— (2002), 'A No-Go Theorem about Rotation in Relativity Theory', in D. Malament, (ed.), *Reading Natural Philosophy (Essays Dedicated to Howard Stein on His 70th Birthday)*, (Open Court Press), 267–293.

Manchak, J. B. (2009a), 'Is Spacetime Hole–Free?', *General Relativity and Gravitation* 41: 1639–1643.

—— (2009b), 'On the Existence of "Time Machines" in General Relativity', *Philosophy of Science* 76: 1020–1026.

Mattingly, J. (2001), 'Singularities and Scalar Fields: Matter Fields and General Relativity', *Philosophy of Science* 68: S395–S406.

Maudlin, T. (2007), *The Metaphysics Within Physics* (Oxford University Press).

Meiland, J. W. (1974), 'A Two-Dimensional Passage Model of Time for Time Travel', *Philosophical Studies* 26: 153–173.

Mellor, D. H. (1981), *Real Time* (Cambridge University Press).

Moncrief, V., and Isenberg, J. (1983), 'Symmetries of Cosmological Cauchy Horizons', *Communications in Mathematical Physics* 89: 387–413.

Monton, B. (2009), 'Time Travel without Causal Loops', *Philosophical Quarterly* 59: 54–67.

Morris, M. S., and Thorne, K. S. (1988), 'Wormholes in Spacetime and Their Use for Interstellar Travel: A Tool for Teaching General Relativity', *American Journal of Physics* 56: 395–412.

———— and Yurtsever, U. (1988), 'Wormholes, Time Machines, and the Weak Energy Condition', *Physical Review Letters* 61: 1446–1449.

Ori, A. (2007), 'Formation of Closed Timelike Curves in a Composite Vacuum/Dust Asymptotically Flat Spacetime', *Physical Review* D76: 044002.

Penrose, R. (1969), 'Gravitational Collapse: The Role of General Relativity', *Rivista del Nuovo Cimento* 1: 252–276 (Numero speciale).

——(1979), 'Singularities and Time-Asymmetry', in S. W. Hawking and W. Israel (eds.), *General Relativity: An Einstein Centenary Survey*, 581–638.

Polchinski, J. (1998), *String Theory* (Cambridge University Press).

Politzer, H. D. (1992), 'Simple Quantum Systems in Spacetimes with Closed Timelike Curves', *Physical Review* D46: 4470–4476.

Price, H. (1984), 'The Philosophy and Physics of Affecting the Past', *Synthese* 16: 299–323.

Rendall, A. (2008), *Partial Differential Equations in General Relativity* (Oxford University Press).

Rovelli, C. (2004), *Quantum Gravity* (Cambridge University Press).

Savitt, S. (1994), 'The Replacement of Time', *Australasian Journal of Philosophy* 72: 463–474.

Smith, N. J. J. (1997), 'Bananas Enough for Time Travel?', *British Journal for the Philosophy of Science* 48: 363–389.

Stein, H. (1970), 'On the Paradoxical Time-Structures of Gödel', *Philosophy of Science* 37: 589–601.

——(1995), 'Introductory Note to *1946/9', in S. Feferman et al. (eds.), *Kurt Gödel: Collected Works, Volume III* (Oxford University Press), 202–229.

Thiemann, T. (2007), *Modern Canonical Quantum General Relativity* (Cambridge University Press).

Thorne, K. S. (1994), *Black Holes and Time Warps: Einstein's Outrageous Legacy* (W.W. Norton and Company).

——(2002), 'Space-Time Warps and the Quantum World: Speculations about the Future' in S. W. Hawking et al. (eds.), *The Future of Spacetime* (W.W. Norton), 109–152.

Tipler, F. J. (1974), 'Rotating Cylinders and the Possibility of Global Causality Violation', *Physical Review* D9: 2203–2206.

——(1976), 'Causality Violation in Asymptotically Flat Space-Times', *Physical Review Letters* 37: 879–882.

——(1977), 'Singularities and Causality Violation', *Annals of Physics* 108: 1–36.

van Stockum, W. J. (1937), 'The Gravitational Field of a Distribution of Particles Rotating About an Axis of Symmetry', *Proceedings of the Royal Society of Edinburgh* 57: 135–154.

Visser, M. (1996), *Lorentzian Wormholes: From Einstein to Hawking* (American Institute of Physics).

Visser, M. (2003), 'The Quantum Physics of Chronology Protection', in G. W. Gibbons, E. P. S. Shellard, and S. J. Rankin (eds.), *The Future of Theoretical Physics and Cosmology: Celebrating Stephen Hawking's 60th Birthday* (Cambridge University Press), 161–176.

Wald, R. M. (1984), *General Relativity* (The University of Chicago Press).

——(1998), 'Gravitational Collapse and Cosmic Censorship', in B. R. Iyer and B. Bhawal (eds.), *Black Holes, Gravitational Radiation and the Universe: Essays in Honor of C. V. Vishveshwara* (Kluwer Academic), 69–85.

Weyl, H. (1921), *Space-Time-Matter*, translated by S. Brose (Methuen).

Witt, D. M. (1986), 'Vacuum Space-Times that Admit no Maximal Slice', *Physical Review Letters* 57: 1386–1389.

Wüthrich, C. (2006), *Approaching the Planck Scale From a Generally Relativistic Point of View: A Philosophical Appraisal of Loop Quantum Gravity*, PhD thesis, University of Pittsburgh. http://philosophy.ucsd.edu/faculty/wuthrich/pub/WuthrichChristian PhD2006Final.pdf

——(2007), *Zeitreisen und Zeitmaschinen*, in T. Müller (ed.), *Philosophie der Zeit: Neue analytische Ansätze* (Vittorio Klostermann), 191–219.

Yourgrau, P. (1991), *The Disappearance of Time: Kurt Gödel and the Idealistic Tradition in Philosophy* (Cambridge University Press).

# PART V

## TIME IN A QUANTUM WORLD

# CHAPTER 21

# THE CPT THEOREM

FRANK ARNTZENIUS

## 1. INTRODUCTION

THE CPT theorem says that any (restricted) Lorentz invariant quantum field theory must also be invariant under the combined operation of charge conjugation C, parity P, and time reversal T, even though none of those individual invariances need hold. The CPT theorem is prima facie a perplexing theorem. Why would the combination of two space-time transformations plus a charge reversing transformation have to be a symmetry of any relativistic quantum field theory? What does charge conjugation have to do with space-time symmetries? Is there an analogous theorem for classical relativistic field theories? What, if anything does the CPT theorem tell us about space-time structure? I will try to answer these questions. The basic idea of my answer is that what standardly is called the CPT transformation really amounts to a PT transformation, that is, a pure space-time transformation.

In section 2, I will briefly clarify the notion PT invariance, and briefly explain why one might be interested in such an invariance. In section 3, I examine the PT invariance of classical tensor field theories, and suggest that this transformation includes charge conjugation. I then discuss the same with respect to quantum tensor field theories. After that I briefly discuss classical and quantum spinor field theories. I end with some tentative conclusions.

## 2. HOW DO QUANTITIES TRANSFORM UNDER PT?

Suppose we describe a world (or part of a world) using some set of coordinates $\{x, y, z, t\}$. A passive PT transformation is what happens to this description when we describe the same world but instead use coordinates $\{x', y', z', t'\}$ where $x' = -x$, $y' = -y$, $z' = -z$, $t' = -t$. An active PT transformation is the following: keep using the same coordinates, but change the world in such a way that the description of the world in these coordinates changes exactly as it does in the corresponding passive PT transformation.[1] Suppose now that we have a theory which is stated in terms of coordinate dependent descriptions of the world, that is, a theory which says that only certain coordinate dependent descriptions describe physically possible worlds, that is, are solutions. Such a theory is said to be PT invariant iff PT turns solutions into solutions and non-solutions into non–solutions.[2] Why might one be interested in PT invariance or non-invariance of theories?

Failure of PT invariance tells us something about the structure of space-time: it tells us that space-time either has objective spatial handedness or has an objective temporal orientation, or both. Why? Well, suppose we start with a coordinate dependent description of a world which some theory allows. And suppose that after we do a passive coordinate transformation the theory says that the new (coordinate dependent) description of this world is no longer allowed. This seems odd: it's the same world after all, just described using one set of coordinates rather than another. How could the one be allowed by our theory and the other not? Indeed, this does not make much sense unless one supposes that the theory, as stated in coordinate dependent form, was true in the original coordinates but not in the new coordinates. And that means that, according to the theory, there is some objective difference between the $\{x, y, z, t\}$ coordinates and the $\{x', y', z', t'\}$ coordinates, that is, the $\{x, y, z, t\}$ coordinates do not stand in the same relation to the objective structure of the world as the $\{x', y', z', t'\}$ coordinates.

In the case of PT this could be because space-time has an objective temporal orientation, or an objective spatial handedness, or both. Note that failure of PT invariance does not indicate that space-time has a space-time handedness structure, for space-time handedness is invariant under PT.[3]

It is in fact more interesting to consider what one should infer if a theory is invariant under PT, but fails to be invariant both under P and under T, since it appears that our world is such.[4] What this, prima facie, would suggest is that space-time has a space-time handedness structure, while having neither a spatial handedness nor a

---

[1] Note that there are therefore many distinct PT transformations, one corresponding to each distinct inertial coordinate system. I will assume a flat space-time throughout this chapter, and am not here going to address how to talk about PT in General Relativity. For more on that issue see Malament 2004 and Arntzenius and Greaves 2007.

[2] If the theory is probabilistic, then the theory is PT invariant if its probabilities are invariant under PT.

[3] Space-time handedness is the 4-dimensional analogue of spatial handedness: when one mirrors a single coordinate of a 4-tuple of coordinates, one flips the space-time handedness of the coordinate system. So, if one mirrors all 4 coordinates, as one does in PT, one ends up in a coordinate system of the same handedness as the original coordinate system.

[4] That is to say, the world is PT-invariant on my understanding of what a PT-transformation amounts to, which in standard terminology means that the world is CPT-invariant.

temporal orientation, since space-time handedness is the minimal natural structure which explains the symmetry properties in question.

So that is why we should be interested, but how do we go about investigating PT invariance; in particular, how do we know how the quantities occurring in our theories transform under PT? Well, though we often give our theories in coordinate dependent form, nature itself, of course, is coordinate independent. We can use ($n$-tuples of) numbers to denote locations in space-time, and we can use ($m$-tuples of) numbers to indicate the magnitudes and directions of various quantities in space-time, but nature itself does not come equipped with numbers. How our numerical representation of a quantity should transform under a change of coordinates depends on the structure of the quantity, and on the way in which we have designed our coordinate representation. Let me illustrate this for a very simple case.

Suppose a vector $V$ at a point $p$ in a three-dimensional Euclidean space is a coordinate independent quantity which has a magnitude and picks out a direction. We can put Cartesian coordinates $\{x, y, z\}$ on the Euclidean space, and then use these coordinates to numerically indicate the direction and magnitude of $V$. To be precise: the coordinates on the Euclidean space naturally induce corresponding coordinates on the space of tangent vectors at $p$. It follows from the vector nature of $V$ that when we, say, switch to coordinates $\{x', y', z'\}$ where $x' = -x$, $y' = -y$, $z' = -z$, then the three numbers representing $V$ will each flip their sign. My point here is simply that one is not free to choose how one's numerical representation of quantities transforms under certain transformations. In particular, one cannot make a theory invariant under some transformation simply by judiciously choosing how the quantities occurring in the theory transform. For how a quantity transforms is determined by the coordinate independent nature of the quantity in question (together with the way in which we manufacture coordinate representations of it).

# 3. PT IN CLASSICAL TENSOR FIELD THEORIES

Let's start with a simple case, namely, the classical real Klein Gordon field. The classical real Klein Gordon field is a real scalar field whose field values are invariant under the restricted Lorentz transformations.[5] (The restricted Lorentz transformations are the ones that are continuously connected to the identityy. They include spatial rotations and Lorentz boosts. They include neither P nor T nor PT.) We can then impose a law of evolution on our Klein Gordon field, namely the Klein Gordon equation: $(\partial^\mu \partial_\mu + m^2)\varphi = 0$, and check whether this law of evolution is invariant under the restricted Lorentz transformations. It is.

---

[5] For the sake of simplicity I am assuming that there are objective, path independent, facts as to whether the values of the real Klein Gordon field at two different locations in space-time are the same. That is to say, I am here ignoring the idea that the Klein Gordon field configurations correspond to sections on a fibre bundle with a connection on it.

How about PT invariance? Well, we first need to know how the Klein Gordon Field transforms under PT. The standard assumption is that the Klein Gordon field is invariant under PT: under PT the field $\varphi$(t,r) transforms into $\varphi$(−t, −r). It immediately follows that the Klein Gordon equation is invariant under PT. I will later return to the issue as to whether this is the only possible way in which a scalar field could transform under PT. In the meantime let us turn to another field: the classical electromagnetic field.

Maxwell's equations for the free electromagnetic field, written in terms of the four-potential $A^\mu$ : $\Box A^\mu - \partial^\mu(\partial_\nu A^\nu) = 0$. In order to see whether this equation is invariant under PT, we need to know how the four-potential $A^\mu$ transforms when we move it from location (t,r) to location (−t, −r), or, equivalently, how it transforms when we switch from co-ordinates (t,r) to co-ordinates (t′, r′) where t′ = −t and r′ = −r. This depends on what kind of co-ordinate independent quantity we take the four-potential to be. Let's make the simplest assumption, namely that it is the same kind of quantity as $\partial^\mu \varphi$ (t,r), namely a tangent vector in a four-dimensional space-time.[6] Now, we know how $\partial^\mu \varphi$ (t,r) transforms under active PT. For the scalar field $\varphi$ just gets moved to its new location, and this means that its derivatives flip sign. So $\partial^\mu \varphi$(t, r) transforms to $-\partial^\mu \varphi$(−t, −r) . Assuming that $A^\mu$(t, r) is the same kind of quantity, it follows that $A^\mu$(t, r) transforms to $-A^\mu$(−t, −r) under active PT.

It is important to note that this is *not* the standard view of how $A^\mu$(t, r) transforms under PT. On the standard view $A^\mu$(t, r) transforms to $A^\mu$(−t, −r) under PT. (For a more extensive discussion of why the standard view says this, and why I find it an unattractive view, see Arntzenius and Greaves 2007.) Since this assumption is controversial, and since it will turn out to be crucial in what follows, let me give two additional justifications for the claim that $A^\mu$(t, r) transforms to $-A^\mu$(−t, −r) under PT.

The ordinary, 'real', Lorentz transformations form a group of transformations that splits into four disconnected components. (The full group of Lorentz transformations is the group of transformations that leaves the Minkowski metric invariant.) Here is why. Parity (mirroring of all three spatial axes) is a Lorentz transformation. But in the space of all possible Lorentz transformations there is no continuous path that starts out at the Identity and ends up at Parity. (The pure spatial rotations are all continuously connected to the Identity, and so are the pure Lorentz boosts, but one cannot reach Parity by pure boosts or pure rotations or combinations of the two.) So the real Lorentz group splits up into at least two disconnected components: the Lorentz transformations that one can reach via a continuous path from the Identity (the 'restricted' Lorentz transformations), and the Lorentz transformations that one can reach via a continuous path from Parity. And there is another split, namely the split between the Lorentz transformations that include Time Reversal and the ones that do not. So the Lorentz group has at least four disconnected components. In fact it has exactly four disconnected components.

[6]  Again, for simplicity, I am ignoring the idea that $A^\mu$ is a connection on a fibre bundle.

The Lorentz transformations of four-vectors can be represented as 4x4 matrices L with real entries acting on 'columns' of four real numbers representing the four-vectors, where these matrices have the property that they preserve the Minkowski inner product between the 'columns'. The demand that each L preserves the Minkowski inner product amounts to the demand that $L^{TR}GL = G$, where $L^{TR}$ is the transpose of L, and G is the matrix whose diagonal entries equal $1, -1, -1, -1$, and whose off-diagonal entries equal 0. Now, while it is natural to suppose that the matrix $L^{PT}$ representing PT (in some inertial frame of reference) should be the matrix whose diagonal entries each equal -1 and whose off-diagonal entries are 0, this is not the only possible representation of the real Lorentz group. Another possible one, for example, is one in which PT is represented as the identity matrix (though this is not a 'faithful' representation).[7] However, let us now turn to the complex Lorentz group.

One can introduce the complex Lorentz group abstractly as the so-called 'complexification' of the real Lorentz group. But it is easier to think of the complex Lorentz group in terms of its representation by means of 4x4 matrices with complex entries which preserve the Minkowski inner product, that is, the set of 4x4 complex matrices L such that $L^{TR}GL = G$. What is important for our purposes is that within the complex Lorentz group PT is connected to the Identity. Here is why. The following is a one parameter subset of the complex Lorentz matrices. where the parameter is $t$:

$$\begin{pmatrix} \cosh it & 0 & 0 & \sinh it \\ 0 & \cos t & -\sin t & 0 \\ 0 & \sin t & \cos t & 0 \\ \sinh it & 0 & 0 & \cosh it \end{pmatrix}$$

For t=0 we find that $L_t$ is the Identity. Continuously increasing t to $t = \pi$ we arrive at minus the Identity, which is the representation of PT. So $A^\mu(t, r)$ transforms to $-A^\mu(-t, -r)$ if it transforms as an element of the standard (four-dimensional) representation of the complex Lorentz group.[8]

Note that this argument similarly establishes that a scalar field (which is invariant under the restricted Lorentz transformations) must be invariant under PT. Note also that this argument does not tell us how $A^\mu(t,r)$ transforms under P or T separately; for in the complex Lorentz group P and T are not connected to the Identity. There can be so-called 'pseudo-scalar' fields, 'pseudo-vector' fields, and 'pseudo-tensor' fields, which flip sign under P and under T.

Another argument for the claim that $A^\mu(t,r)$ must transform to $-A^\mu(-t, -r)$ under PT, and that a scalar field must be invariant under PT, can be given if one makes the assumption that the only types of quantities that can occur in our theories must be (restricted) Lorentz invariant tensor quantities. Let me start on this argument by

[7] A 'faithful' representation of a group (by matrices) is one whereby each distinct element of the group gets represented by a distinct matrix.
[8] I am not sure whether it must do so in every non-trivial 4-dimensional representation of the complex Lorentz group.

indicating how one can manufacture 'pseudo-tensors', tensors whose sign flips under P and under T, using tensors that are invariant under the restricted Lorentz transformations. Consider the totally anti-symmetric Levi-Civita tensor $\varepsilon_{ABCD}$. It is invariant under restricted Lorentz transformations. It can be taken to represent an objective space-time orientation (space-time handedness). One can also use it to manufacture a pseudo-scalar field from four distinct vector-fields. Suppose that vector fields $V^A, W^B$, $X^C$, and $Y^D$ each flip their time component under time reversal T. Let's take it that space-time orientation is a geometric, invariant, object represented by $\varepsilon_{ABCD}$. Then the pseudo-scalar $\Sigma_{ABCD}\varepsilon_{ABCD}V^A W^B X^C Y^D$ will transform to $-\Sigma_{ABCD}\varepsilon_{ABCD}V^A W^B X^C Y^D$ under T. So in this manner one can manufacture pseudo-scalars whose signs flip under P and T. However, one cannot in this manner manufacture pseudo-scalars whose signs flip under PT.[9] So $A^\mu(t,r)$ transforms to $-A^\mu(-t, -r)$ under PT. It follows that Maxwell's equations for the free electromagnetic field are invariant under PT.

Let's do one more example: a complex Klein Gordon field $\varphi$ interacting with the electromagnetic field. For ease of presentation let me represent the electromagnetic field using both the four-potential $A^\mu$ and the Maxwell-Faraday tensor $F^{\mu\nu}$. The following Lagrangian then gives a dynamics for the interacting fields:

$$L = \partial^\mu\varphi^*\partial_\mu\varphi - m^2\varphi^*\varphi - 1/4(F^{\mu\nu}F_{\mu\nu}) + e^2 A_\mu A^\mu \varphi^*\varphi - ie(\varphi^*\partial^\mu\varphi - \varphi\partial^\mu\varphi^*)A_\mu,$$

Is this theory invariant under PT? Well, suppose that under PT the four-potential $A^\mu(t,r)$ transforms to $-A^\mu(-t, -r)$. The Maxwell-Faraday tensor and the four-potential are related via the following equation: $F_{\mu\nu} = \partial_\mu A_\nu - \partial_\nu A_\mu$. It follows that the Maxwell-Faraday tensor is invariant under PT. The complex scalar field is invariant under PT, so $\partial^\mu\varphi$ flips sign under PT. All of this taken together implies that each of the five terms of the Lagrangian is separately invariant under PT. So the Lagrangian is invariant under PT. So our dynamics is invariant under PT.

It should be obvious from this last example that the PT-invariance of my sample theories is not a coincidence. If one's theory derives its dynamics from a local Lagrangian, where this Lagrangian is a Lorentz-scalar built (by contractions and summations) from tensors, each of which transform as indicated under PT, then this Lagrangian, and hence one's dynamics, will be invariant under PT.

In fact, J. S. Bell has proved a general classical PT theorem along these lines (see Bell 1955). Here is a statement of Bell's PT theorem. Let us suppose that the equations of a theory can be stated in the following form: $F^i(\varphi, A^\mu, \ldots, \partial^\nu\varphi, \partial^\lambda A^\mu, \ldots\ldots) = 0$, where $\varphi$, $A^\mu, \ldots$ are fields which transform as tensors under the restricted real Lorentz transformations (the ones that are connected to the identity), that the $\partial^\nu\varphi$, $\partial^\lambda A^\mu, \ldots.$ are finite order derivatives of the tensor fields, and the $F^i$ are finite polynomials in these terms. Then the equations are invariant under PT, when the representation of PT in some frame is

---

[9] For a proof, see Greaves (manuscript). Basic idea: any tensor representing a temporal orientation (just as $\varepsilon_{ABCD}$ represents space-time handedness) would have to have an odd number of space-time indices, i.e. would have to have odd rank. But one can show that there are no tensor fields of odd rank that are invariant under the restricted Lorentz transformations.

$$L_{PT} = \begin{pmatrix} -1 & 0 & 0 & 0 \\ 0 & -1 & 0 & 0 \\ 0 & 0 & -1 & 0 \\ 0 & 0 & 0 & -1 \end{pmatrix}$$

At this point one might very well ask: but what does all of this have to do with charge conjugation? Well, notice that under the above PT transformation all four-vectors flip. In particular therefore, any charge-current four vector, such as $ie(\varphi^* \partial^\mu \varphi - \varphi \partial^\mu \varphi^*)$ in the case of the complex Klein Gordon field, will 'flip over' under PT. So any charge density, which is the first component of a charge-current four-vector, will flip sign under PT. So the PT transformation, when properly conceived, has as a consequence that the PT-transformed fields behave as if they have opposite charge. Indeed it is my contention that what standardly is called the CPT transformation should really have been called the PT transformation. This also provides an answer as to how it can be that what is allegedly a combination of two geometrical transformations (PT) and a non-geometrical transformation (C) has to be a symmetry of any quantum field theory. The answer I am suggesting is that what is standardly called the CPT transformation really is a geometric transformation, namely the PT transformation, and that invariance under PT, and lack of invariance under each of P and T, corresponds to the fact that our space-time has space-time handedness structure, but neither a spatial handedness nor a temporal orientation.

Note also that one can include classical particles (rather than just fields) in our considerations, by making the assumption that the four-velocities $V^\mu$ of particles flip over under PT. We can then, for example, consider an interaction between a charged particle and the electromagnetic field which is governed by the following Lagrangian:

$$L = -1/4(\partial_\mu A_\nu - \partial_\nu A_\mu)(\partial^\mu A^\nu - \partial^\nu A^\mu) - qV_\mu A^\mu$$

It is clear that this Lagrangian is invariant under our PT transformation. Moreover, note that if we switch the sign of q and switch the sign of the four-velocity $V^\mu$, while keeping $A^\mu$ invariant, then the Lagrangian is invariant. That is to say: flipping over the four-velocity of a particle is equivalent to flipping the sign of the charge q. So, again, PT, when properly conceived, includes charge conjugation. (For a more extensive discussion of the classical charged particle case, see Arntzenius and Greaves 2007.)

## 4. PT IN QUANTUM TENSOR FIELD THEORIES

The basic idea of this section is very simple. In quantum field theory particle states correspond to 'positive frequency' solutions of the corresponding classical field theory, while anti-particle states correspond to 'negative frequency' solutions. Since PT turns positive frequency solutions into negative frequency solutions, PT in quantum field theory turns particles into anti-particles. Now for some details.

Let's start with the classical free complex Klein Gordon equation: $(\partial^\mu \partial_\mu + m^2)\varphi = 0$. This equation has plane wave solutions: $\varphi_k = \exp(ik_\mu x^\mu)$, where $k_\mu k^\mu = m^2$. Any solution $\varphi(x)$ is a unique superposition of these plane wave solutions: $\varphi(x) = \int f(k)\exp(ik_\mu x^\mu)dk$. Relative to a given direction of time, we will call plane wave $\varphi_k$ with positive $k^0$ a 'positive frequency' plane wave, and a plane wave $\varphi_k$ with negative $k^0$ a 'negative frequency' plane wave. More generally any solution that is a superposition only of positive frequency plane waves will be called a positive frequency solution, and any solution that is a superposition only of negative frequency plane waves will be called a negative frequency solution. The particle Hilbert space $\mathcal{H}$ of the quantized Klein Gordon field, and the operators on it, can be constructed from the positive frequency solutions to the classical Klein Gordon equation, and the anti-particle Hilbert space $\underline{\mathcal{H}}$ can be constructed from the negative frequency solutions to the classical Klein Gordon equation. Here is how. (In this section I am largely following Geroch 1971. The errors are all mine.)

Start by assuming that there is a 1–1 correspondence between single particle Hilbert space states $|\varphi>$ and positive frequency solutions $\varphi(x)$ to the classical Klein-Gordon equation. The superposition of two Hilbert space states is the Hilbert space state corresponding to the addition of the two corresponding positive frequency solutions, that is $|\varphi_1 > +|\varphi_2 >$ corresponds to positive frequency solution $\varphi_1(x) + \varphi_2(x)$. Scalar multiplication of a particle Hilbert space state corresponds to scalar multiplication of the corresponding positive frequency solution: $c|\varphi >$ corresponds to $c\varphi(x)$. The inner product between two particle Hilbert space states is defined as $(|\varphi_1 >, |\varphi_2 >) = \int f_1(k)f_2^*(k)dk$. This defines the particle Hilbert space.

The single anti-particle Hilbert space is constructed in analogous fashion, now using the negative frequency solutions, *except* that multiplication of an anti-particle state by a complex number c corresponds to multiplication of the corresponding classical solution with the complex conjugate number of that number: $c^*$. That is, if $|\varphi >$ corresponds to $\varphi(x)$ then $c|\varphi >$ corresponds to $c^*\varphi(x)$. The inner product between two anti-particle Hilbert space states is defined as $(|\varphi_1 >, |\varphi_2 >) = \int f_1^*(k)f_2(k)dk$. This defines the anti-particle Hilbert space.

Since it is important for what follows let me explain why multiplication by a complex number c of the anti-particle Hilbert space states corresponds to multiplication by $c^*$ of the corresponding negative frequency solution. Associated with a constant time-like vector field $r^a$ on Minkowski space-time is an energy operator E, where E acts on solutions of the complex Klein Gordon equation in the following way: $E\varphi(x) = -ir^a \nabla_a \varphi(x)$. Here multiplication by i means multiplication of the Hilbert space state $|\varphi >$, not (pointwise) multiplication of the (complex) classical field $\varphi(x)$. Now, suppose that multiplication of a Hilbert space state did correspond to (pointwise) multiplication of the corresponding solution $\varphi(x)$, both for negative frequency solutions and positive frequency solutions. Then the expectation value of E for $\varphi(x)$ would be: $\int r^a k_a f(k)f^*(k)dk$. This is positive for any k which points in the same direction of time as r and negative for any k which points in the opposite direction of time from r. That is to say, anti-particles and particles would then have opposite signs of energy. This would be a disaster. In the first place, it is experimentally known that particles and anti-particles have the same sign of energy: they can annihilate and thereby produce energy (particles with non-zero energy), rather than that their total energy is 0. Moreover, in an interacting theory, one would beget radical instability: decays into deeper and deeper negative energy states would be allowed, which would release unlimited amounts of positive energy. This is no good: we need the energies of particles and anti-particles to have the same sign (though it is perfectly all right if what sign that is, is a matter of convention.) So we choose the Hilbert space structure of the anti-particle state-space such that $c|\varphi >$ corresponds to $c^*\varphi(x)$ for negative frequency solutions $\varphi(x)$. For then the expectation value of E equals $\int_{M+} r^a k_a f(k) f^*(k) dk - \int_{M-} r^a k_a f(k) f^*(k) dk$, where M+ is the positive mass shell, and M− is the negative mass shell, that is, the first integration is over momenta that point in the same direction of time as r, and the second integration is over momenta that point in the opposite direction of time as r. This has as a consequence that the energies of particles and anti-particles have the same sign.

Let me clarify and emphasize one more point. At the beginning of this section I arbitrarily picked some direction of time, and called plane wave solutions 'positive frequency solutions' if their wave-vector pointed in that direction of time. Later on I associated particles with positive frequency solutions and anti-particles with negative frequency solutions. And then I said that multiplication of a Hilbert space particle state by a complex number c corresponded to multiplication of the corresponding positive frequency solution by c, while in the anti-particle case it corresponded to multiplication of the corresponding negative frequency solution by c*. None of this implies that space-time has a temporal orientation. All I have made of this is that there is a fact of the matter as to whether two time-like vectors point in the same direction of time, or in opposite directions of time. Which direction of time gets called the future, which the past, which solutions get dubbed positive frequency, which negative frequency, which Hilbert space state multiplication corresponds to multiplication of the corresponding solution by c and which corresponds to multiplication by c*: all of that can be taken to be a matter of convention. But that the two directions, the two frequency types, and the two particle types are not identical: that is not a matter convention. Let me now continue with the construction of the full particle and anti-particle Hilbert spaces. (So far we only have the single particle and single anti-particle Hilbert space.)

Given the single particle Hilbert space $\mathcal{H}$ we can build the corresponding Fock space $\mathcal{F} = C \oplus \mathcal{H}^\alpha \oplus (\mathcal{H}^{(\alpha} \oplus \mathcal{H}^{\beta)}) \oplus \cdots\cdots$. Here C is the space of complex numbers and the $\mathcal{H}^\alpha$, $\mathcal{H}^\beta$ etc. are simply copies of $\mathcal{H}$. The brackets around the indices indicates the restriction to symmetrical states in the tensor product Hilbert spaces. A typical element of this Fock space is $|\psi >= (\chi, |\chi >^\alpha, |\chi >^{\alpha\beta}, \ldots)$. Here $\chi$ is a complex number, representing the amplitude of the vacuum state, $|\chi >^\alpha$ is an element of $\mathcal{H}^\alpha$, that is, a one-particle state, $|\chi >^{\alpha\beta}$ is an element of $\mathcal{H}^{(\alpha} \oplus \mathcal{H}^{\beta)}$, that is a two-particle state, etc… Similarly, given the single anti-particle Hilbert space $\underline{\mathcal{H}}$ we can build the corresponding Fock space $\underline{\mathcal{F}} = C \oplus \mathcal{H}^\alpha (\mathcal{H}^{(\alpha} \mathcal{H}^{\beta)} \oplus \cdots\cdots$. (Note the underlining of the symbols associated with the anti-particle Hilbert spaces.)

We can then define the particle momentum creation operators $C_k$ and $C_{-k}$ which, when operating on the vacuum, create a particle in momentum eigenstate $|k>$, $|-k>$ respectively, that is, create a particle, and the corresponding anti-particle, respectively. Similarly one can define particle momentum annihilation operators $A_k$, and $A_{-k}$ which when acting on momentum eigenstate $|k>, |-k>$ respectively, produce the vacuum. Then one can define a Klein Gordon field operator $\varphi(x) = \int A_k \exp(-ik_\mu x^\mu) + C_{-k} \exp(ik_\mu x^\mu))dk$, and its adjoint $\varphi^+(x) = \int C_k \exp(ik_\mu x^\mu) + A_{-k} \exp(-ik_\mu x^\mu))dk$. One can then show that these field operators, $\varphi(x)$ and $\varphi^+(x)$, satisfy the same equation that the corresponding classical fields, $\varphi(x)$ and $\varphi^*(x)$, satisfy, namely the complex Klein Gordon equation. (One has to be a bit careful. Strictly speaking the way I defined the field operators makes no sense: I should have smeared them out with test functions.) Now let us turn to PT. How do quantum states $|\varphi>$ transform under PT? Let's start with the single particle states. Well, under PT the corresponding classical solution $\varphi(x)$ should transform to $\varphi(PTx)$. If we choose our coordinate system x so that PT consists of mirroring in the origin of that coordinate system, then a classical solution $\varphi(x)$ transforms to $\varphi(-x)$ under this PT transformation. Given that multiplication in the anti-particle Hilbert space by c corresponds to multiplication of the corresponding solution by $c^*$, this is an anti-linear transformation on the Hilbert space.

How do the field operators transform? Well, a particle state $|k>$ transforms to the corresponding anti-particle state $|-k>$ (give a suitable choice of phases for the momentum states), and vice versa. So $C_k$ transforms to $C_{-k}$ and vice versa, and $A_k$ transforms to $A_{-k}$ and vice versa. This, together with the fact that the transformation is anti-unitary means that the field operators $\varphi(x) = \int A_k \exp(-ik_\mu x^\mu) + C_{-k} \exp(ik_\mu x^\mu))dk$ transform to $\int A_{-k} \exp(ik_\mu x^\mu) + C_k \exp(-ik_\mu x^\mu))dk = \varphi^+(-x)$. So $\varphi(x)$ transforms to $\varphi^+(-x)$ under PT. Similarly $\varphi^+(x)$ transforms to $\varphi(-x)$ under PT. But this is exactly how the standard view has it that the Klein-Gordon field operators transform under CPT! So I have argued that what standardly is called a CPT transformation in quantum field theory should really have been called a PT transformation. So standard proofs of the CPT theorem amount to proofs that relativistic quantum field theories must be invariant under what I have argued to be the PT transformation. (I should point out that such proofs can not be exactly the same as in the classical case. In the first place, as we have just seen, quantum fields get Hermitian conjugated under PT. Moreover, the fields are now operator fields, so we cannot re-order them as we please. Luckily, and somewhat mysteriously, these two effects manage to cancel each other, so that any quantum tensor Lagrangian must be invariant under the PT transformation.[10])

---

[10]  Here is a very brief sketch of how such a proof goes. Other than the Hermitian conjugation and order worries, the PT transformation is the same as the classical one. Assuming that the Lagrangian is Hermitian, this means that corresponding to any non-Hermitian term in the Lagrangian there is the Hermitian conjugate term. The ordering problem is solved by specifying that the Lagrangian must be normal ordered (all creation operators in front of annihilation operators). Now consider an interaction

# 5. PT IN SPINOR FIELD THEORIES

Let me start by considering classical non-integer spin fields. I will here restrict attention to Dirac spinor fields. Does including Dirac spinor fields affect the argument for the PT invariance of classical local Lagrangian theories? In order to find that out we need to know how Dirac spinor fields transform under PT. In order to find that out we need to quickly review some spinor theory. (For more detail see, e.g. Wald 1984, Chapter 13, or Bogolubov, Logunov, and Todorov 1975, Chapter 7.)

Let W be a two-dimensional vector space over the complex numbers. The elements $\xi^A$ of W are 'two-component spinors', or spinors for short. Given a choice of two basis vectors we can represent each spinor by a 'column' consisting of two complex numbers. The dual space $W^D$ is the space of all linear maps from elements of W to complex numbers. I will call the elements $\eta_A$ of $W^D$ 'dual spinors'. We will assume that W comes equipped with an inner product: a bilinear map $\varepsilon_{AB}$ from pairs of spinors to the complex numbers. ($\varepsilon_{AB}$ thus maps elements of W to elements of $W^D$ and vice versa). Next let us define the group of all SL(2,C) transformations on W to be the linear maps L from W into itself which have unit determinant. (The determinant of L is defined as $\det(L) = \varepsilon_{AB}\varepsilon^{CD}L^A_C L^B_D$.) Given a choice of basis vectors, one can represent each such transformation by a 2x2 complex matrix with unit determinant.

One can then show that the group of SL(2,C) transformations has the same structure as the group of 'restricted' Lorentz transformations. (The 'restricted' Lorentz transformations are the ones that are connected to the identity.) To be precise: one can show that there exists a two-to-one mapping h from the elements L of SL(2,C) to the elements $\Lambda$ of the restricted Lorentz group such that if $h(L_1) = \Lambda_1$ and $h(L_2) = \Lambda_2$ then $h(L_2 \circ L_1) = \Lambda_2 \circ \Lambda_1$. It will be helpful to have an explicit representation. Here is how one can do that. Pick a Lorentzian frame of reference. Given such a frame of reference each 4-vector V has four components $V_0, V_1, V_2, V_3$. We can then use these components to define the following associated Hermitian matrix:

term such as the following:

$$W^\mu =: \varphi_a^+(x)\partial^\mu \varphi_b(x) : + : \partial^\mu \varphi_b^+(x)\varphi_a(x) :$$

(Here : .... : denotes normal ordering.) Under PT this will transform to:

$$- : \varphi_a(-x)\partial^\mu \varphi_b^+(-x) : - : \partial^\mu \varphi_b(-x)\varphi_a^+(-x) :$$

Given the commutation relations between scalar fields we can re-order this as:

$$- : \partial^\mu \varphi_b^+(-x)\varphi_a(-x) : - : \varphi_a^+(-x)\partial^\mu \varphi_b(-x) :$$

And that is just equal to $-W^\mu(-x)$. More generally, the commutation relations for integer spin fields combine with the Hermitian conjugation and normal ordering so as to always produce invariance under PT for relativistic tensor quantum field theories.

$$V = \begin{pmatrix} V_0 + V_3 & V_1 - iV_2 \\ V_1 + iV_2 & V_0 - V_3 \end{pmatrix}$$

Now, we can use any SL(2,C) matrix $L$ to transform $M$ to a new Hermitian matrix $M' = LML^+$, where $L^+$ is the complex conjugate transpose of $L$. The four-vector V' associated with the transformed Hermitian matrix $M'$ will then be a Lorentz transformation of the original four-vector V. Thus we have associated a unique Lorentz transformation with each SL(2,C) matrix. Note that any SL(2,C) matrix L will induce the same transformation as $-L$. Indeed, it turns out that for every possible restricted Lorentz transformation there exist exactly two associated SL(2,C) matrices. Note, however, that there is no SL(2,C) matrix associated with PT. For PT should map $V_0$, $V_1$, $V_2$, $V_3$ into $-V_0$, $-V_1$, $-V_2$, $-V_3$, but there is no SL(2,C) matrix $L$ such that $LML^+ = -M$.

In order to have a faithful representation of the full Lorentz group, we need to consider four-component spinors. (The full Lorentz group is the group of all transformations that preserves the Minkowski metric. It includes P, T and PT.) Let me start by defining the complex conjugate dual spinor space $W^{*D}$ to be the space of all anti-linear maps from W to the complex numbers. Next let me define the complex conjugate spinor space $W^*$ to be the space of all linear maps from $W^{*D}$ to the complex numbers. I will denote elements of $W^*$ as $\chi^{A'}$, that is, with a raised primed index, and call them conjugate spinors. I will denote elements of $W^{*D}$ as $\mu_{A'}$, that is, with a lowered primed index, and call them conjugate dual spinors. There is a natural anti-linear 1–1 correspondence between spinors $\xi^A$ and conjugate spinors $\xi^{A'}$ generated by the demand that $\mu_{A'}(\xi^A) = \xi^{A'}(\mu_{A'})$ for all $\mu_{A'}$. It follows that if under a restricted Lorentz transformation $\xi^A$ transforms to $L\xi^A$ then $\xi^{A'}$ transforms to $L^*\xi^{A'}$. ($L^*$ is the SL(2,C) matrix whose entries are the complex conjugates of L's entries.)

Dirac spinors are ordered pairs of spinors and complex conjugate dual spinors. Since we know how spinors and complex conjugate spinors transform under restricted Lorentz transformations, we know how Dirac spinors transform under restricted Lorentz transformations. (We can use $\varepsilon_{AB}$ to transform conjugate spinors into conjugate dual spinors and vice versa.) But how about PT? In order to see how to do that let me introduce a relation between Dirac spinors and complex four-vectors V. Given a basis in spinor space, and the associated basis in conjugate dual spinor space, one can define a map from pairs of spinors and conjugate dual spinors to 2x2 complex matrices: $M_\alpha^\beta = \xi_\alpha \mu^{\beta'}$. (Here $M_\alpha^\beta$ denotes the components of matrix M, $\xi_\alpha$ denotes the components of spinor $\xi^A$, and $\mu^{\beta'}$ denotes the components of the complex conjugate dual spinor $\mu_{B'}$.) There is also a natural correspondence between such matrices $M_\alpha^\beta$ and the four components of a complex four-vector V (in some frame of reference), given by the following prescription: $2V_0 = M_1^1 + M_2^2$, $2V_1 = M_1^2 + M_2^1$, $2V_2 = -i(M_1^2 - M_2^1)$, $2V_3 = M_1^1 - M_2^2$. Now, suppose we take an SL(2,C) matrix L and transform spinor $\xi^A$ to $L\xi^A$, and we transform $\mu_{A'}$ to $\varepsilon L^*(\varepsilon)^{-1}\mu_{A'}$. (Here * denotes complex conjugation of the matrix entries of L.) One can show that then the complex four-vector V that is associated with the Dirac spinor $< \xi^A, \mu_{A'} >$ will transform according to the restricted Lorentz transformation that corresponds to L. Moreover, one can generate all the complex Lorentz transformations on V by transforming $\xi^A$ and $\mu_{A'}$ with two independent SL(2,C) matrices, that is, by transforming $\xi^A$ to $L_1\xi^A$ and $\mu_{A'}$ to $\varepsilon L_2^*(\varepsilon)^{-1}\mu_{A'}$. Finally, by looking at the complex Lorentz transformations that correspond to the rotations in complex space, one finds that there are exactly two pairs of SL(2,C) matrices that correspond to PT, namely the pair (Identity, -Identity) and the pair (-Identity, Identity). Obviously, both of these pairs transform the associated matrix M to -M, and hence transform the associated four-vector V to $-V$. So now we know, up to a sign, how Dirac spinors transform under PT. And this is the standard view as to how Dirac spinors transform under CPT!

How about PT invariance? Let's consider an example: the Dirac equation for the free Dirac spinor field. The Dirac equation, written in terms of spinors and complex conjugate dual spinors corresponds to the following pair of equations (see e.g. Peskin and Schroeder 1995. page 44):

(1)    $-m\xi^A + i(\partial_0 + \sigma \bullet \nabla)\mu_{A'} = 0$

(2)    $i(\partial_0 + \sigma \bullet \nabla)\xi^A - m\mu_{A'} = 0$

(Here $\sigma$ is the vector consisting of the three Pauli matrices, $\Delta$ is the spatial derivative operator, and $\bullet$ denotes the spatial inner product.) These equations are invariant under PT, since all the derivatives change sign under PT, and either $\mu_{A'}$ or $\xi^A$ flips sign under PT. Good!

Can we more generally argue that any local Lagrangian that contains ordinary tensor fields as well as tensor fields constructed from Dirac spinor fields must be invariant under PT? No, we cannot. The problem is that one can construct four-vectors from Dirac spinors that are *invariant* under PT, rather than that they flip over under PT. (That is to say, one can construct 'PT-pseudo-vectors' from spinors, which are types of quantities which Bell excludes *by fiat*.) For instance, consider the 'probability current' $j^\mu = \psi^+\gamma_0\gamma^\mu\psi$, where $\psi$ denotes a Dirac four-spinor, $\psi^+$ its complex conjugate transpose and the $\gamma^\mu$ denote Dirac's 'gamma matrices'. This probability current transforms like a four-vector under all restricted Lorentz transformations, but is invariant under PT. (Note e.g. that $j^0$ just equals $\psi^+\psi$ which is positive definite.) So, by using some four-vectors that are invariant under PT and some that are not, one can construct Lagrangians that are not invariant under PT.

Miraculously, this problem goes away when one goes to spinor *quantum* field theories: the anti-commutation relations between spinor quantum fields produce an extra sign flip under PT which makes it impossible to construct PT pseudo-tensors from quantum spinor fields. As yet I find it completely mysterious as to why this happens. But the fact remains that quantum spinor field theories must be invariant under what I have called the PT-transformation.

# 6. TENTATIVE CONCLUSIONS

Whether a particle has positive or negative charge is determined by the temporal direction in which the four-momentum of a particle points. What is standardly called the CPT-theorem should be called the PT-theorem. It holds for classical and quantum tensor field theories, fails for classical spinor field theories, but holds for quantum spinor fields. The fact that it holds for quantum field theories suggests that space-time has neither a temporal orientation nor a spatial handedness.

## REFERENCES

Arntzenius, F. and Greaves, H. (2007), 'Time Reversal in Classical Electromagnetism', http://philsci-archive.pitt.edu/archive/00003280/

Bell, J. S. (1955), 'Time Reversal in Field Theory', *Proceedings of the Royal Society of London. Series A, Mathematical and Physical Sciences*, Vol 231, No 1187, 479–495.

Bogolubov, N., Logunov, A. and Todorov, I. (1975), *Introduction to Axiomatic Quantum Field Theory*, (London: W. A. Benjamin).

Geroch, R. (1971), 'Special Topics in Particle Physics' (unpublished lecture notes, University of Texas at Austin).

Greaves, H. (manuscript), 'Towards a Geometrical Understanding of the CPT Theorem'.

Malament, D. (2004), 'On the Time Reversal Invariance of Classical Electromagnetic Theory', http://philsci-archive.pitt.edu/archive/00001406/

Peskin, M. and Schroeder, D. (1995), *An Introduction to Quantum Field Theory*, (Reading, Mass: Perseus Books).

Wald, R. (1984), *General Relativity*. (Chicago, Chicago University Press).

CHAPTER 22

..........................................................................................

# TIME IN QUANTUM
# MECHANICS

..........................................................................................

JAN HILGEVOORD AND DAVID ATKINSON

# 1. THE PROBLEM OF TIME IN QUANTUM
## MECHANICS

..........................................................................................

MANY physicists believe that time constitutes a serious problem in quantum mechanics. The difficulty was epitomized in Wolfgang Pauli's Handbook article of 1933 (Pauli 1933: 140):

> We conclude therefore that the introduction of an operator $t$ must be renounced as a matter of principle, and that time $t$ must necessarily be considered as an ordinary number ('c-number') in wave mechanics. (Translation: JH and DA).

Pauli's article signalled the conclusion of a period of rapid development of the new quantum theory that had been inaugurated by Heisenberg's revolutionary paper of 1925 (Heisenberg 1925). An essential feature of quantum theory is that the dynamical variables of classical mechanics are represented by self-adjoint operators on a Hilbert space. The basic example is that of the position and momentum variables $q$ and $p$ of a particle, which are replaced by self-adjoint operators satisfying the commutation relation

$$p\,q - q\,p = -i\hbar,\qquad(1)$$

where $\hbar = \frac{h}{2\pi}$, and $h$ is Planck's constant. If one assumes $q$ and $p$ to have continuous eigenvalues running from $-\infty$ to $+\infty$ on the real axis, the well-known unique solution of (1) is $q = q$, $p = -i\hbar\,d/dq$, on a suitable space of functions (up to unitary equivalence, and modulo irreducibility).

In a second famous article (Heisenberg 1927) Heisenberg sought to clarify the physical meaning of relation (1) by considering experiments in which the position and momentum of a particle could be measured. He concluded that these quantities could not both be measured with arbitrary precision in the same experiment. This was expressed by the uncertainty relation

$$\delta p \, \delta q \sim \hbar \tag{2}$$

where $\delta p$ and $\delta q$ are the precisions, or uncertainties, with which the values of $p$ and $q$ are known. Actually, by introducing appropriate definitions of the quantities $\delta p$ and $\delta q$, relation (2) can be shown to be a direct consequence of relation (1). For example, in terms of the standard deviation as a measure of uncertainty, the inequality

$$\Delta p \Delta q \geq \tfrac{1}{2}\hbar \tag{3}$$

can be derived. In the same article two more commutation relations are presented:

$$Et - tE = -i\hbar \quad \text{or} \quad Jw - wJ = -i\hbar. \tag{4}$$

The quantities $J$ and $w$ are the conjugate variables that typically appear in the description of periodic systems. Classically $J$ and $w$, like $p$ and $q$, form a pair of conjugate variables, and so the second of the relations (4) is analogous to (1). The meaning of the first equation (4) is however much less clear. In some of the passages following equations (4), Heisenberg identifies $E$ with $J$ and $t$ with $w$, calling $t$ a 'phase', that is, $t$ is considered to be an internal variable of the system. This interpretation reflects the 'or' between the equations (4). But in other passages, and notably in the examples leading up to the uncertainty relation $\delta E \, \delta t \sim \hbar$, $t$ is treated as a classical time parameter, clearly contradicting (4).

Why did Heisenberg present the 'same' formula (4) in two different guises? In classical mechanics the time parameter is sometimes turned into an internal dynamical variable conjugate to (minus) the Hamiltonian of the system. Heisenberg may have had this in mind in connection with the first equation (4), although the minus sign in that equation would not then be correct. The notation also suggests a connection to Eq.(1). In relativity theory the momentum $p$ and energy $E$ of a particle are the components of a four-vector, and it is quite common to consider the position $q$ of the particle and the time parameter $t$ to form a four-vector also. The first equation (4) might then be seen as a natural complement of Eq.(1), as dictated by relativity theory. We shall come back to this matter in the next section.

At about the same time it had become clear that Eqs.(4) can be mathematically problematic. In many cases (but not all) the range of the eigenvalues of $J$ is the positive real axis, while the eigenvalues of $w$ are angles in the interval $[0, 2\pi]$. It can be shown that self-adjoint operators whose eigenvalues satisfy these conditions cannot also satisfy the second relation (4). Also, it is not possible for the first equation (4) to be correct if $E$ is bounded from below, and if the eigenvalue spectrum of $t$ is the whole real axis (the latter condition being a necessary requirement if $t$ is to be interpreted as the time parameter). Since in many cases the energy operator is known to be a well-behaved self-adjoint operator that is indeed bounded from below, it seems to follow that an acceptable time operator does not exist in quantum mechanics. Whence Pauli's verdict mentioned at the beginning; and as a consequence the relations (4) have passed into oblivion. For a more complete account of this early period (see Hilgevoord 2005).

The asymmetry between space and time that seems to be implied by Pauli's statement, apparently contradicting the principles of relativity, has bothered physicists for a long time. Many proposals for circumventing the difficulty have been put forward, in particular a generalization of the axiom that observables must correspond to self-adjoint operators on Hilbert space. One can weaken the axiom to the postulate that "each observable is associated to a Positive Operator Valued Measure (POVM)" (Egusquiza, Gonzalo Muga and Baute, 2002: 283; see also Busch 1989). POVMs are interesting in their own right, having many practical applications, but we shall not discuss them here, since we believe their use as a way of nullifying Pauli's objection to be fundamentally misdirected. Again, much attention has been given to finding an analogue of the uncertainty relation (2) in the case of energy and time. If time is not an operator, a relation of this type cannot exist. Nevertheless there do exist 'uncertainty' relations between energy and time of a different kind, for example the relation between the energy spread and the lifetime of a quantum state. The existence of such relations, in which $t$ is an ordinary number, might suggest a certain similarity with momentum and position, but by the same token there appears to be a fundamental difference between position and time in quantum mechanics.

Notwithstanding all these considerations, we shall show in the next section that quantum mechanics does not involve a special problem for time, and that there is no fundamental asymmetry between space and time in quantum mechanics over and above the asymmetry that already exists in classical physics. In section 3 we study time operators in detail, and in section 4 various uncertainty relations involving time are discussed.

## 2. THE PROBLEM DISSOLVED

To see that time poses no problem for quantum mechanics, one must distinguish between two ways that it can appear in physics. First, time may figure as a general parameter of the theory, and second it may appear as a dynamical internal variable of some particular physical system described by the theory.

In its first guise, time $t$ is on a par with the spatial coordinates $x$, $y$, $z$. This is explicit in relativity theory, where $t$ is added as a fourth coordinate to the space coordinates to form a relativistic four-vector $(x, y, z, ct)$. Like the space coordinates, the time coordinate is independent of the physical systems the theory describes. In quantum mechanics the space and time coordinates remain c-numbers; neither $x$, $y$, $z$, nor $t$ become operators. The space-time coordinates appear as parameters in the definition of well-known space-time symmetries, for example rotation invariance, spatial and temporal translation invariance, and Lorentz-invariance.

In its second guise, time is a dynamical variable belonging to a particular physical system, and more than one such time variable may be present in the system. Like the other dynamical variables, dynamical time variables become operators in quantum mechanics. Examples of time operators are discussed in the following section, where we will see that time operators, although they are less common than position operators, are not fundamentally problematic in quantum theory.

The work of Heisenberg that we discussed in the previous section clearly betrays a confusion between the two ways in which time occurs in physics. On the one hand, $t$ in the first of Eqs.(4) is considered to be the analogue of $w$, an internal dynamical variable of the system, but on the other hand $t$ is looked upon as the unique external time parameter. Also, Pauli seems to understand by $t$ the general time parameter, the c-number character of which we have seen to be in fact unproblematic. The apparent problem of time arises when this time parameter is put on a par with dynamical position variables rather than with the coordinates of space. The confusion has proved to be quite persistent in the quantum mechanics literature, and in the remainder of this section we will try to see how this has come about.

Ironically, the origin of the problem is to be found in space rather than in time, and its roots lie in classical mechanics. Much of fundamental physics deals with 'point' particles. A point particle is a material object that can have a position, a momentum, a mass, an energy, a charge, etc. At any moment the particle is located at a point of space. Evidently a point particle and a point of space are very different things. Nevertheless they are not always clearly distinguished. Quite often the coordinates of space and the position variables of a point particle are denoted by the same symbols $x$, $y$, $z$ (e.g. when one writes $\psi(x, y, z, t)$ for the wave function of a particle). To avoid this confusion we shall denote the dynamical position variables of a particle by $q = (q_x, q_y, q_z)$, reserving the symbols $x$, $y$, $z$ for the coordinates of a point of space.

The same confusion has led to the erroneous view that the position $q$ of a particle and the time coordinate $t$ form a relativistic four-vector. Though this may be true numerically, since the numbers $q_x, q_y, q_z$ can coincide with the numbers $x, y, z$, the variable $q$ and the coordinate $t$ are conceptually quite distinct. This becomes evident when there are several particles, which must share one and the same $t$. Conflating position variables of particles and space coordinates is responsible for the view that space is quantized whereas time is not, creating the false impression of an asymmetry between the treatment of space and time in quantum mechanics. We have seen that such an asymmetry does not exist, since neither the space coordinates $x, y, z$, nor the time coordinate $t$ are quantized. If $t$ is not the relativistic partner of $q$, what is the true partner of the latter? The answer is simply that such a partner does not exist; the position variable of a point particle is a non-covariant concept. It is an interesting fact that, whereas in classical physics the non-covariance of $q$ may easily remain hidden because the pair $q, t$ behaves as a four-vector, in quantum mechanics the non-covariance of $q$ is very clear. Newton and Wigner were the first to show that an operator that represents the position of a particle must necessarily be non-covariant (Newton and Wigner 1949; Schweber 1962: 60–62).

It is sometimes said that a worldline $x^\mu(\tau)$ does provide a covariant description of a point particle, where $x^\mu = (x, y, z, ct)$ and $\tau$ is the proper time along the curve. However, just as $x^\mu$ is only a *point* of spacetime, conceptually unrelated to a particle, $x^\mu(\tau)$ is just a *curve* in spacetime. And, just as the position $q$ of a point particle at time $t$ may coincide with the spatial components of the point $x^\mu$, so the orbit $q(t)$ can coincide with the curve $x^\mu(\tau)$ (though only if the tangent vector to the curve is timelike at every point). But the dynamical position variable $q$ of the particle remains without a relativistic partner. Likewise, the dynamical time operators, to be discussed in the next section, are non-covariant quantities. We conclude that time gives rise to no special problem in quantum mechanics. For a fuller discussion see Hilgevoord (2005).

# 3. TIME OPERATORS

An ideal clock is a device or system which produces a reading that mimics coordinate time in much the same way that the position of a point particle mimics coordinate space. Let us consider this analogy in detail for a system of $n$ free particles. The canonical commutation relations are

$$[\boldsymbol{q}_{x,i}, \boldsymbol{p}_{x,j}] = i\, \boldsymbol{I}\, \delta_{ij} \qquad\qquad [\boldsymbol{q}_{x,i}, \boldsymbol{q}_{x,j}] = 0 = [\boldsymbol{p}_{x,i}, \boldsymbol{p}_{x,j}], \qquad (5)$$

for $i, j = 1, 2, \ldots, n$, where $\boldsymbol{q}_{x,i}$, $\boldsymbol{p}_{x,i}$ are the linear operators representing the $x$-components of the positions and momenta, respectively, of the $n$ particles. Similar relations apply of course for the $y$ and $z$ components. Units are such that $\hbar = 1$, and $\boldsymbol{I}$ is a unit operator on a suitable subspace of Hilbert space. In this section and the following one we shall consistently use the convention that operators are designated by bold letters, ordinary or c-numbers by normal fonts.

A clock reading is represented by an eigenvalue of a clock variable operator, which may be either any real number between $-\infty$ and $\infty$ (we then speak of the linear clock), or it may be limited to the interval $[0, 2\pi]$ (this is the cyclic or periodic clock). There are interesting differences between the mathematics of the linear and of the cyclic clocks, and we will concentrate first exclusively on the linear clock, since the mathematics parallels that of the point particles, returning to the cyclic clock in subsection 3.2.

## 3.1 The linear clock

Suppose that the clock variables associated with $n$ clocks are represented by the linear operators $\boldsymbol{\tau}_i$, and let the canonically conjugate variables be called $\boldsymbol{\eta}_i$. The canonical commutation relations are

$$[\boldsymbol{\tau}_i, \boldsymbol{\eta}_j] = i\, \boldsymbol{I}\, \delta_{ij} \qquad\qquad [\boldsymbol{\tau}_i, \boldsymbol{\tau}_j] = 0 = [\boldsymbol{\eta}_i, \boldsymbol{\eta}_j]. \qquad (6)$$

Evidently these commutation relations are analogous to Eq.(5).

The generator of a translation in space is the total momentum operator $\vec{P}$, whereas the generator of a translation in time is the Hamiltonian $H$. Consider first an infinitesimal translation, $dx$, in the $x$-direction. This is generated by $P_x$, so any canonical coordinate, $\Omega$, changes under the transformation to $\Omega + d\Omega$, where

$$d\Omega = [\Omega, P_x]dx/i.$$

The $y$ and $z$ components of $\vec{q}_i$, and all the Cartesian components of $\vec{p}_i$, should remain unchanged under this transformation. For the system of $n$ point particles, only the position operators $q_{x,i}$ should be changed under the translation generated by $P_x$. Evidently the $x$-component of

$$\vec{P} = \sum_{i=1}^{n} \vec{p}_i \tag{7}$$

generates precisely the required transformation. Thus

$$\frac{\partial q_{x,i}}{\partial x} = [q_{x,i}, P_x]/i = I,$$

and so the operators $q_{x,i}$ are linear functions of coordinate space, indeed

$$q_{x,i} = q^0_{x,i} + I x,$$

where the $q^0_{x,i}$ are (operator) constants of integration (i.e. they are independent of $x$). The eigenvalues of the shifted operators

$$Q_{x,i} = q_{x,i} - q^0_{x,i}$$

are all equal to $x$, and in this sense the particle positions mimic coordinate space. Indeed, it is from this fact that the pernicious error arises of *identifying* the shifted operators with coordinate space, $x$. These operators are equal to $I x$, not $x$, and the parallel distinction in the case of the clock is of cardinal importance in resolving much confusion about time in quantum mechanics.

An infinitesimal time-translation, $dt$, is generated by $H$, changing any canonical coordinate $\Omega$ to $\Omega + d\Omega$, where

$$d\Omega = [\Omega, H]dt/i.$$

For a system of $n$ ideal clocks, the clock variables $\tau_i$ should all be augmented under this transformation by $dt$,

$$\tau_i \rightarrow \tau_i + I dt, \tag{8}$$

the variables $\eta_i$ being unchanged. In this way the ideal clocks are close analogues of the point masses: the position operators of the point masses are boosted in space by the total momentum operator, while the ideal clock variables are boosted in time by the Hamiltonian. The analogue of Eq.(7) is

$$H = \sum_{i=1}^{n} \eta_i. \tag{9}$$

This ensures that the clock variables transform as in Eq.(8), and that the $\eta_i$ remain unchanged. The clock variables in fact satisfy

$$\frac{d\tau_i}{dt} = [\tau_i, H]/i = I,$$

and so they are linear functions of time,

$$\tau_i = \tau_i^0 + I\,t.$$

It is crucial to distinguish here coordinate time $t$ from the clock readings that are the eigenvalues of the clock variables $\tau_i$. In an eigenbasis $|\tau\rangle = |\tau_1, \ldots, \tau_n\rangle$ of these operators,

$$\tau_i |\tau\rangle = (\tau_i^0 + t)|\tau\rangle.$$

These $n$ eigenvalues serve as indicators of coordinate time; but they are no more conceptually identical to it than the positions of $n$ point particles are identical to coordinate space. The clock variables can be reset by defining

$$T_i = \tau_i - \tau_i^0,$$

so that $T_i|\tau\rangle = t|\tau\rangle$. The eigenvalues of the reset clock variables, $T_i$, are all equal to coordinate time $t$, much as the eigenvalues of the shifted position operators representing the particle positions were all equal to $x$. The temptation to identify the reset clock variables with coordinate time is great, but it must be resisted: these variables are equal to $I\,t$, not to $t$.

It should be noted that the Hamiltonian (9) is not bounded from below, indeed its eigenvalues extend over the entire real line (cf. subsection 3.3). We use a Fourier integral to express the eigenvector $|\tau\rangle$ as

$$|\tau\rangle = \frac{1}{(2\pi)^{\frac{n}{2}}} \int_{-\infty}^{\infty} \cdots \int_{-\infty}^{\infty} d\eta_1 \ldots d\eta_n \exp\left(-i \sum_{i=1}^{n} \eta_i \tau_i\right)|\eta\rangle, \tag{10}$$

where $|\eta\rangle = |\eta_1, \ldots, \eta_n\rangle$ is an eigenvector of the variables $\eta_i$. The inverse is

$$|\eta\rangle = \frac{1}{(2\pi)^{\frac{n}{2}}} \int_{-\infty}^{\infty} \cdots \int_{-\infty}^{\infty} d\tau_1 \ldots d\tau_n \exp\left(i \sum_{i=1}^{n} \eta_i \tau_i\right)|\tau\rangle. \tag{11}$$

In the analogue of configuration space, a state vector $|\psi\rangle$ is assigned a wave function $\psi(\tau)$. This takes the form

$$\psi(\tau) \equiv \langle\tau|\psi\rangle = \frac{1}{(2\pi)^{\frac{n}{2}}} \int_{-\infty}^{\infty} \cdots \int_{-\infty}^{\infty} d\eta_1 \ldots d\eta_n \exp\left(i \sum_{i=1}^{n} \eta_i \tau_i\right)\langle\eta|\psi\rangle,$$

and, in this space, $\tau_i$ is represented by the number $\tau_i$, while $\eta_i$ is represented by the differential operator $-i\partial/\partial\tau_i$:

$$\langle\tau|\eta_i|\psi\rangle = -i\frac{\partial}{\partial\tau_i}\psi(\tau) = \frac{1}{(2\pi)^{\frac{n}{2}}}\int_{-\infty}^{\infty}\cdots\int_{-\infty}^{\infty}d\eta_1\ldots d\eta_n\exp\left(i\sum_{i=1}^{n}\eta_i\tau_i\right)\eta_i\langle\eta|\psi\rangle.$$

Note that the Hamiltonian itself has the following representation as a differential operator on the space spanned by the eigenvectors of the clock variables:

$$H \sim -i\sum_{i=1}^{n}\frac{\partial}{\partial\tau_i}.$$

This is perfectly well defined, whereas the occasionally suggested equivalence $H = -id/dt$ is nonsense.

It follows from Eqs.(9)-(10) that the unitary time evolution operator, $U(t) = \exp(-itH)$, induces the transformation

$$U(t)|\tau_1, \ldots, \tau_n\rangle = |\tau_1 + t, \ldots, \tau_n + t\rangle, \tag{12}$$

that is, an eigenstate of the time operators simply evolves into an eigenstate of the same operators at a later time.

An intuitively appealing realization of the clock algebra (6),(9) is afforded by the following canonical transformation:

$$\hat{q}_i = \frac{g\,\tau_i^2}{2} - \frac{\eta_i}{m_i g} \qquad\qquad \hat{p}_i = m_i g\,\tau_i.$$

These new canonical coordinates satisfy the standard commutation relations, and the Hamiltonian (9) takes on the form

$$H = \sum_{i=1}^{n}\left[\frac{\hat{p}_i^2}{2m_i} - m_i g\,\hat{q}_i\right], \tag{13}$$

which we recognize as that of $n$ freely falling point masses $m_i$ in a uniform gravitational field, $g$ being the acceleration due to gravity. In this realization, the clock times are provided by the canonical momenta $\hat{p}_i$. This Hamiltonian is not bounded from below, of course, being simply (9) expressed in different variables. It is true that ideal clocks are 'unphysical' in a sense, but then so are point particles. Both are consistent with the formalism of quantum mechanics. Together they illustrate the similarity between the quantum mechanical treatment of indicators of position in space and indicators of position in time.

## 3.2 The cyclic clock

An alternative realization of the commutation relations (6) is afforded by the cyclic clock, to which we alluded above. The eigenvalues of these clock variables, $\tau_i$, are readings, $\tau_i$, that are limited to $[0, 2\pi]$. Such clock variables resemble angle rather than position variables, and their conjugate variables, $\eta_i$, resemble angular momenta rather than linear momenta.

For notational simplicity we restrict our attention to just one cyclic clock, but any number could be treated. The eigenvectors can be expanded in a Fourier series:

$$|\tau\rangle = \frac{1}{\sqrt{2\pi}} \sum_{m=-\infty}^{\infty} e^{-im\tau} |\eta_m\rangle, \tag{14}$$

where $|\eta_m\rangle$ designates a discrete set of vectors, and $|\tau\rangle$ refers to a continuous set of vectors in the same space, the two sets being related by the discrete Fourier transformation (14). The inverse of Eq.(14) is the finite Fourier integral

$$|\eta_m\rangle = \frac{1}{\sqrt{2\pi}} \int_0^{2\pi} d\tau \, e^{im\tau} |\tau\rangle. \tag{15}$$

The eigenvector $|\eta_m\rangle$ of $\eta$ evidently belongs to the eigenvalue $\eta_m \equiv m$, where $m = 0, \pm1, \pm2, \ldots..$

In the analogue of configuration space, a state vector $|\psi\rangle$ is assigned a wave function $\psi(\tau)$:

$$\psi(\tau) \equiv \langle\tau|\psi\rangle = \frac{1}{\sqrt{2\pi}} \sum_{m=-\infty}^{\infty} e^{im\tau} \langle\eta_m|\psi\rangle. \tag{16}$$

In this space, $\tau$ is represented by the number $\tau$, and $\eta$ by the differential operator $-i\frac{d}{d\tau}$:

$$\langle\tau|\eta|\psi\rangle = -i\frac{d}{d\tau}\psi(\tau) = \frac{1}{\sqrt{2\pi}} \sum_{m=-\infty}^{\infty} m \, e^{im\tau} \langle\eta_m|\psi\rangle.$$

The Hamiltonian of the cyclic clock is $\eta$, and the unitary time evolution operator is accordingly $U(t) = \exp(-itH) = \exp(-it\eta)$. From Eq.(14), and recalling that $\eta_m = m$, we find

$$U(t)|\tau\rangle = |\tau + t\rangle \qquad (\text{mod } 2\pi).$$

An operator on a Hilbert space of vectors is fully defined only if one gives its domain, which is a subset of vectors in the space that are mapped by the operator on to vectors in the space. It turns out that a careful definition of the relevant domains is of crucial importance in resolving certain conceptual difficulties with respect to the cyclic clock (and for the correct treatment of angular momentum), and we will now give the necessary attention to these matters. In particular, it will prove important to specify the domain of the unit operator appearing on the right of Eq.(6).

It may be helpful first to explain some basic properties of unbounded linear operators on Hilbert space. We shall then briefly consider their relevance to the linear clock variables of the previous subsection, and then in more detail to those of the cyclic clock, where their application is indispensable. An operator $\Omega$ is defined by specifying a subset of the space called its *domain,* $D(\Omega)$, and a linear mapping of a vector in that domain to another vector in the space. The domain will not be the whole space if $\Omega$ is an unbounded operator; but if, for any $|\phi\rangle$ in the whole space, one can nevertheless find a $|\psi\rangle$ in $D(\Omega)$ that is arbitrarily close to $|\phi\rangle$ in the sense of the norm, then the domain is said to be *dense* in the space.

If, for every $|\psi_1\rangle$ and $|\psi_2\rangle$ in the domain, which is presumed dense, it is the case that

$$\langle\psi_1|\Omega\psi_2\rangle = \langle\Omega\psi_1|\psi_2\rangle,$$

then $\Omega$ is said to be Hermitian (or symmetric). This property is however not sufficient to guarantee the *self-adjointness* of $\Omega$. For a fixed $\psi_2$ in $D(\Omega)$, consider the subset of vectors in the Hilbert space comprising all $\psi_1$ such that there is a $\phi$ in the space for which

$$\langle\psi_1|\Omega\psi_2\rangle = \langle\phi|\psi_2\rangle.$$

The set of all these $\psi_1$ is called the domain of the operator adjoint to $\Omega$. If $\Omega$ is Hermitian, *and* this adjoint domain is equal to $D(\Omega)$, then $\Omega$ is said to be self-adjoint. It is a basic assumption of quantum mechanics that any physical observable is represented by a self-adjoint operator on Hilbert space, for then a complete orthogonal set of eigenvectors exists (not admittedly always in the Hilbert space itself, but always in the extended space of the associated rigged Hilbert space, see Atkinson and Johnson 2002: 3). It should be noted that self-adjointness is a stronger constraint than mere Hermiticity, a fact that is ignored in elementary introductions to quantum mechanics.

For the linear clock, the algebra of the operators $\tau$ and $\eta$ is isomorphic to that of the position and momentum operators of point particles, as we saw, and the following statements can be demonstrated by adapting standard proofs, for example those given in Yosida (1970: 198). In the space spanned by the eigenvectors of $\tau$, the operators $\tau$ and $\eta$ are represented by $\tau$ and $-i\frac{d}{d\tau}$ respectively. Consider these as unbounded operators on the Hilbert space $L^2(-\infty, \infty)$, which is defined to be the set of all square-integrable functions, that is, all complex functions $\psi(\tau)$ for which $\int_{-\infty}^{\infty} d\tau\,|\psi(\tau)|^2 < \infty$, equipped with the inner product

$$\langle\phi|\psi\rangle = \int_{-\infty}^{\infty} d\tau\,\phi^*(\tau)\psi(\tau).$$

From Yosida's results we read off

1. $\tau$ is self-adjoint on the (dense) domain defined by the requirements that both $\psi(\tau)$ and $\tau\psi(\tau)$ lie in $L^2(-\infty, \infty)$.
2. $\eta$ is self-adjoint on the (dense) domain defined by the requirements that both $\psi(\tau)$ and $\psi'(\tau)$ lie in $L^2(-\infty, \infty)$, and moreover that $\psi(\tau)$ be absolutely continuous, i.e. that $\psi(\tau)$ be expressible in the form

$$\psi(c) + \int_c^\tau dx\,g(\sigma)$$

where $g(\sigma)$ is a locally integrable function.

Returning to the cyclic clock, one defines the Hilbert space $L^2(0, 2\pi)$ as the space of all complex functions $\psi(\tau)$ for which $\int_0^{2\pi} d\tau \, |\psi(\tau)|^2 < \infty$, equipped with an inner product like that in $L^2(-\infty, \infty)$, except that the integration domain is limited to $(0, 2\pi)$. Whereas $\tau$ is self-adjoint on $L^2(0, 2\pi)$, $\eta$ is not self-adjoint on the whole of the space. The most useful self-adjoint extension is defined by the requirements that both $\psi(\tau)$ and $\psi'(\tau)$ lie in $L^2(0, 2\pi)$, that $\psi(\tau)$ be absolutely continuous, and that $\psi(0) = \psi(2\pi)$. This periodicity is in accord with the Fourier series representation Eq.(16).

Consider, however, the commutation relation,

$$[\tau, \eta] = i I. \tag{17}$$

On what subspace of $L^2(0, 2\pi)$ is $I$ the unit operator? The difficulty is that, whereas $\psi(\tau)$ must satisfy $\psi(0) = \psi(2\pi)$ to qualify for inclusion in a domain on which $\eta$ is self-adjoint, the function $\phi(\tau) = \tau\psi(\tau)$ satisfies this condition only if $\psi(0) = \psi(2\pi) = 0$. Hence the domain on which the commutator (17) is valid is specified by the requirements that both $\psi(\tau)$ and $\psi'(\tau)$ lie in $L^2(0, 2\pi)$, that $\psi(\tau)$ be absolutely continuous, and that $\psi(0) = \psi(2\pi) = 0$. In Eq.(17), $I$ is to be understood as the unit operator on this subspace alone. While there is no objection to this limitation, the restricted subspace being also dense in $L^2(0, 2\pi)$, there are consequences for some of the uncertainty relations, as we will see in Section 4.

It may be noted that the above discussion of the cyclic clock variable and its conjugate is equally applicable to the angle, $\theta$, and the orbital angular momentum, $L$, for the motion of a point particle in a central force field. The algebra of the operators $\tau$ and $\eta$ for the cyclic clock is isomorphic to that of $\theta$ and $L$.

In the physics literature it is sometimes asserted that if two operators satisfy the canonical commutation relation, then both must have continuous eigenvalues on the whole real axis. In fact, Pauli's negative conclusion about the existence of a time operator mentioned at the beginning of this article was based on this belief. For a proof, Pauli referred to the first edition of Dirac's famous book on quantum mechanics (Dirac 1930: 56). Our adaptation of his proof goes as follows: by repeated application of relation (17), one obtains the equality

$$\exp(ic\tau)\, \eta \exp(-ic\tau) = \eta - cI, \tag{18}$$

where $c$ is a real number. Let $|\eta\rangle$ be some eigenvector of $\eta$ belonging to the eigenvalue $\eta$. So

$$\eta \exp(-ic\tau)|\eta\rangle = \exp(-ic\tau)(\eta - c)|\eta\rangle = (\eta - c)\exp(-ic\tau)|\eta\rangle, \tag{19}$$

thus $\eta - c$ is an eigenvalue of $\eta$, and since $c$ may have any real value, so may the eigenvalues of $\eta$. Note that the eigenvector $\exp(-ic\tau)|\eta\rangle$ is guaranteed not to be the nullvector, since its norm is the same as that of $|\eta\rangle$, and that is surely nonvanishing.[1] On interchanging the roles of $\eta$ and $\tau$ we find a similar result for the eigenvalues of $\tau$.

---

[1] In the second edition (1935: 94) of the book of Dirac just cited, the author adds that *complex* values of $c$ are not allowed for physical reasons, since the putative eigenvector would blow up exponentially at infinity. The mathematical version of this objection is that such a complex value leads to a vector that is not contained in the extended Hilbert space.

Now we have seen that, in the case of the cyclic clock, $\eta$ has discrete eigenvalues. In view of Dirac's result, how can this be? To understand that, we need to consider more carefully the conditions under which the above derivation is valid. In fact, in order to use the commutator (17) repeatedly, so as to obtain Eq.(18), we must find a domain on which the products $\tau\eta$ and $\eta\tau$ are well-defined, and on which both operators are self-adjoint. For the linear clock, this common domain is dense in the whole of the Hilbert space $L^2(-\infty, \infty)$, so there is no difficulty, but for the cyclic clock the matter is quite different.

As we have noted, for the cyclic clock $\eta$ is self-adjoint on a subspace of $L^2(0, 2\pi)$ for which $\psi(0) = \psi(2\pi)$, but the problem is that Eq.(17) does not hold on the whole of this space. In fact it can be shown that, if $|\psi\rangle$ and $|\phi\rangle$ are any two vectors in the domain of $\eta$,

$$\langle\psi|[\tau, \eta] - i\boldsymbol{I}|\phi\rangle = -2\pi i\psi^*(2\pi)\phi(2\pi), \qquad (20)$$

which means that Eq.(17) is simply not true on this space. It is only true on the subspace of the domain specified by $\psi(0) = \psi(2\pi) = 0$, but the difficulty is that $\eta$ is not self-adjoint on this restricted domain, and Eq.(18) does not hold, so the argument of Dirac does not go through. There is no space on which $\tau$ and $\eta$ are self-adjoint, and on which Eq.(19) is valid. For further details, (see Kraus 1965 and Uffink 1990).

## 3.3 Discussion

We have seen that the formalism of quantum mechanics allows for the existence of ideal clocks, and that the energy of such systems is unbounded. While the energy eigenvalues extend from $-\infty$ to $\infty$, it is particularly the lack of a *lower* bound that is sometimes thought to be a serious defect, since it is feared that such a system could act as an infinite source of energy. However, as long as the system is isolated or coupled to a system that cannot take up infinite amounts of energy, nothing untoward will happen. In a sense, an ideal clock is better behaved than a point particle, for an eigenstate of such a particle's position spreads with infinite velocity. By contrast, the eigenstates of time operators do not spread at all, but rather transform into eigenstates belonging to a different eigenvalue, cf. Eq.(12). We conclude that the existence of ideal clocks is perfectly consistent with the formalism of quantum mechanics.

# 4. UNCERTAINTY RELATIONS

The commutation relation (1) and the inequality (3) are generally considered to embody the essential content of elementary quantum mechanics. They express two fundamental aspects of the theory: non-commutativity and irreducible uncertainty. In Section 3 we have seen that energy and time operators also exist satisfying the relation

$[\tau, \eta] = iI$ on suitably defined domains. Concerning (3) we remark that although this inequality is traditionally considered to be the mathematical expression of the quantum mechanical uncertainty principle, it is actually quite unsatisfactory in this respect. The standard deviation is not the most obvious, and certainly not the most adequate measure of uncertainty in quantum mechanics. For many perfectly normal quantum states the standard deviation diverges, and even when the wave-function approximates a $\delta$-function, the standard deviation may remain arbitrarily large. A consequence of this fact is that inequality (3) permits probability distributions of $p$ and $q$ to be simultaneously arbitrarily narrow, contrary to what might be expected from an uncertainty relation (Uffink and Hilgevoord 1985, Hilgevoord 2002). A more adequate measure of the spread of a probability distribution is the length $W_\alpha$ of the smallest interval on which a sizeable fraction $\alpha$ of the distribution is situated. An inequality of type (2) also holds for this measure:

$$W_\alpha(p)W_\alpha(q) \geq c_\alpha \hbar, \qquad \text{if} \qquad \alpha \geq \tfrac{1}{2}, \tag{21}$$

where $c_\alpha$ is of order 1 (Landau and Pollak 1961, Uffink 1990). This relation expresses the intuitive content of the uncertainty principle in a much more satisfactory manner than does Eq.(3).

As to the time and energy variables, since the linear clock is mathematically isomorphic to the case of $p$ and $q$, we simply have

$$\Delta\eta\Delta\tau \geq \tfrac{1}{2}\hbar \tag{22}$$

and

$$W_\alpha(\eta)W_\alpha(\tau) \geq c_\alpha \hbar, \qquad \text{if} \qquad \alpha \geq \tfrac{1}{2}. \tag{23}$$

In the case of the cyclic clock the eigenvalues of $\eta$ are discrete and proper eigenstates exist. For these states $\Delta\eta = 0$, and this contradicts Eq.(22)! The problem is in fact well known, being analogous to the much-discussed and mathematically identical case of angle and angular momentum. It turns out, however, that inequalities using more appropriate measures of uncertainty of the type (23) exist in this case too, as might be expected, since they are simply consequences of the Fourier transformation that relates the conjugate variables of the cyclic clock (Eqs. (14)–(15)). We refer to Uffink (1990) for a full discussion.

In Section 1 we mentioned that the apparent lack of a counterpart to inequality (3) for a time *operator* stimulated people to look for uncertainty relations containing time as a *parameter*. Many instances of such relations in quantum mechanics are known, for example the familiar relation between the lifetime and energy-spread of a decaying state. An extensive discussion of such energy-time uncertainty relations is given in Busch (1989, 2000). Here we will briefly discuss a powerful and general class of relations which brings out the full symmetry between the four coordinates $t, x, y, z$. For a detailed discussion, see Hilgevoord (1998). Basic to these relations is a notion of uncertainty that differs from the one in Eq.(3) and Eq.(21). It is an uncertainty concerning the *state* in which a quantum system finds itself (Hilgevoord and Uffink 1991).

If a system is in quantum state $|\Psi\rangle$, the probability of finding it in a different state $|\Phi\rangle$ is generally non-zero; in fact this probability is given by the number $|\langle\Phi|\Psi\rangle|^2$. The closer $|\langle\Phi|\Psi\rangle|^2$ is to 1 the harder it will be to distinguish between the two states by measurements. Accordingly, it is useful to define a number $\rho$ as follows: $|\langle\Phi|\Psi\rangle| = 1 - \rho$, with $0 \leq \rho \leq 1$, and call it the *reliability* with which the states $|\Psi\rangle$ and $|\Phi\rangle$ can be distinguished. If the states coincide, this reliability is 0, whereas it has its maximal value 1 when the states are orthogonal.

Let us apply these ideas to states that are translated with respect to each other in time and in space, respectively. For simplicity we consider only one space coordinate $x$, and we once more put $\hbar = 1$. The unitary operators $U_t(\tau)$ and $U_x(\xi)$ of translations in time and space are given by

$$U_t(\tau) = \exp(-i\,H\,\tau) \qquad \text{and} \qquad U_x(\xi) = \exp(i\,P_x\,\xi), \tag{24}$$

where $H$ is the operator of the total energy and $P_x$ the $x$-component of the operator of the total momentum of the system, as in Section 3. Let the system be initially in state $|\Psi\rangle$. Then we define $\tau_\rho$ to be the smallest time for which the following equality is valid:

$$|\langle\Psi|U_t(\tau_\rho)|\Psi\rangle| = 1 - \rho. \tag{25}$$

Similarly, $\xi_\rho$ is the smallest distance for which

$$|\langle\Psi|U_x(\xi_\rho)|\Psi\rangle| = 1 - \rho. \tag{26}$$

That is, the state $|\Psi\rangle$ must be translated over at least an interval $\tau_\rho$ in time to become distinguishable from the original state with reliability $\rho$, or it must be translated over at least a distance $\xi_\rho$ in space to become distinguishable with reliability $\rho$.

We shall call $\tau_\rho$ and $\xi_\rho$ the temporal and spatial translation widths of the state $|\Psi\rangle$. Both quantities have well-known physical meanings. If $(1 - \rho)^2 = \frac{1}{2}$, $\tau_\rho$ is the so-called half-life of the state; and relation (26) is closely related to Rayleigh's criterion for the distinguishability of two spatially translated states, $1/\xi_\rho$ being a generalization of the notion of the resolving power of an optical instrument.

These translation widths are subject to very general uncertainty relations, which connect them to the width of the energy and momentum spectrum of the state $|\Psi\rangle$, respectively. Let $|E\rangle$ and $|P_x\rangle$ denote complete sets of eigenstates of $H$ and $P_x$, where a possible degeneracy is ignored for simplicity. Using Eq.(24), we have

$$\langle\Psi|U_t(\tau_\rho)|\Psi\rangle = \int |\langle E|\Psi\rangle|^2\, e^{-i\tau E}\, dE \quad \text{and} \quad \langle\Psi|U_x(\xi_\rho)|\Psi\rangle = \int |\langle P_x|\Psi\rangle|^2\, e^{i\xi P_x}\, dP_x, \tag{27}$$

where the integrals may include summation over discrete eigenvalues. Define the overall widths $W_a(E)$ and $W_a(P_x)$ of the energy and momentum distributions as the smallest intervals such that $\int_{W_a(E)} |\langle E|\Psi\rangle|^2 dE = a$ and similarly $\int_{W_a(P_x)} |\langle P_x|\Psi\rangle|^2 dP_x = a$. It can then be shown (Uffink and Hilgevoord 1985, Uffink 1993) that

$$\tau_\rho W_a(E) \geq C(a, \rho)\hbar \quad \text{and} \quad \xi_\rho W_a(P_x) \geq C(a, \rho)\hbar \tag{28}$$

for $\rho \geq 2(1 - a)$. For sensible values of the parameters, say $a = 0.9$ or $0.8$, and $0.5 \leq \rho \leq 1$, the constant $C(a, \rho)$ is of order 1. Inequalities (28) hold for all states $|\Psi\rangle$, and the only assumptions needed for their validity are the existence of the translation operators Eq.(24) and the completeness of the energy and momentum eigenstates. Furthermore, since in relativity theory $t$ and $\vec{x}$ on the one hand, and $E$ and $\vec{P}$ on the other, are united to form a Four-vector, the inequalities (28) lead to a relativistically covariant set of uncertainty relations.

Let us expand a little on the physical meaning of inequalities (28). The first of these is a useful general expression of the relation between the lifetime and the energy spread of a state. Usually such an inequality is obtained in the approximation in which the decay is exponential, but here it is completely general. Though these inequalities deal with coordinate time and coordinate space, they also have relevance to time operators and position operators. This may be seen as follows. Taking position $q$ as an example, let us see how the second inequality (28) relates to inequality (21). If the main part of the probability distribution $|\langle q|\Psi\rangle|^2$ of $q$ in state $|\Psi\rangle$ is concentrated on an interval of length $a$, the width $W_a(q)$ is of order $a$. In this case the translation width $\xi_\rho$ will be of the same order $a$. If, on the other hand, the distribution consists of a number of narrow peaks of width $b$, as in an interference pattern, the translation width $\xi_\rho$ will be of order $b$, whereas $W_a(q)$ remains of the order of the overall width $a$ of the distribution. Thus, inequalities (28) are stronger than (21). Next consider the case where a number of position operators $q_i$ are present in the system. Suppose the spread $W_a(P)$ of the total momentum spectrum is small. From Eq.(28) it follows then that the translation width $\xi_\rho$ must be large. This implies that the spread in all position variables of the system must be large. Conversely, if the spread in only one position variable is small, then $\xi_\rho$ is small and Eq.(28) implies that the spread in the total momentum must be large. Thus, it is the total momentum that determines whether or not the position variables of a system can be sharply determined. Similar conclusions follow for the relation between the spread $W_a(E)$ of the total energy and the spread of all time operators. It is the total energy that decrees whether or not the time variables of a system can be sharply determined.

## REFERENCES

Atkinson, D., and Johnson, P. W. (2002), *Quantum Field Theory: A Self-Contained Course* (Rinton Press, Princeton).

Busch, P. (1989), 'On the Energy-Time Uncertainty Relations, I and II', in *Foundations of Physics* 20: 1–43.

—— (2002), 'The Time-Energy Uncertainty Relation', in J. G. Muga, R. Sala Mayato and I. L. Egusquiza (eds.), *Time in Quantum Mechanics* (Springer, Berlin), 69–98.

Dirac, P. A. M. (1930), *The Principles of Quantum Mechanics* (Clarendon Press, Oxford).

Egusquiza, I. L., Gonzalo Muga, J. and Baute, A. (2002), '"Standard" Quantum Mechanical Approach to Times of Arrival', in J. G. Muga, R. Sala Mayato and I. L. Egusquiza (eds.), *Time in Quantum Mechanics* (Springer, Berlin), 279–304.

Hilgevoord, J. and Uffink, J. (1990), 'A New View on the Uncertainty Principle', in A. I. Miller (ed.), *62 Years of Uncertainty, Historical and Physical Inquiries into the Foundations of Quantum Mechanics* (Plenum, New York), 121–139.

———— (1991), 'Uncertainty in Prediction and in Inference', in *Foundations of Physics* 21: 323–341.

———— (1996), 'The Uncertainty Principle for Energy and Time', in *American Journal of Physics* 64: 1451–1456.

———— (1998), 'The Uncertainty Principle for Energy and Time II', in *American Journal of Physics* 66: 396–402.

———— (2002a), 'The Standard Deviation is Not an Adequate Measure of Quantum Uncertainty', in *American Journal of Physics* 70: 983.

———— (2002b), 'Time in Quantum Mechanics', in *American Journal of Physics* 70/3: 301–306.

———— (2005), 'Time in Quantum Mechanics: A Story of Confusion', in *Studies in History and Philosophy of Modern Physics* 36: 29–60.

Kraus, K. (1965), 'Remark on the Uncertainty Between Angle and Angular Momentum', in *Zeitschrift für Physik* 188: 374–377.

Newton, T. D. and Wigner, E. P. (1949), 'Localized States for Elementary Systems', in *Rev. Mod. Phys.* 21: 400–406.

Pauli, W. (1933), 'Die allgemeine Prinzipien der Wellenmechanik', in *Handbuch der Physik*, 2. Auflage, Band 24, 1. Teil (Springer-Verlag, Berlin), 83–272. An English translation of the first nine chapters exists under the title: *General Principles of Quantum Mechanics* (Springer-Verlag, Berlin, 1980).

Pollak, H. O. and Landau, H. J. (1961), 'Prolate Spheroidal Wave Functions, Fourier Analysis and Uncertainty, 11'; *Bell System Technical Journal*, 40:1 (January 1961) 65–84.

Schweber, S. S. (1962), *An Introduction to Relativistic Quantum Field Theory* (Harper and Row, New York).

Uffink, J. and Hilgevoord, J. (1985), 'Uncertainty Principle and Uncertainty Relations', in *Foundations of Physics* , 15: 925–944.

Uffink, J. (1990), 'Measures of Uncertainty and the Uncertainty Principle', (PhD thesis, Utrecht University).

———— (1993), 'The Rate of Evolution of a Quantum State', in *American Journal of Physics* 61: 935–936.

Yosida, K. (1970), *Functional Analysis* (Springer, Berlin).

# CHAPTER 23

# TIME IN QUANTUM GRAVITY

## CLAUS KIEFER

# 1. INTRODUCTION

TIME is a fundamental concept in all physical theories. Because it enters the dynamical laws, changing these laws is inextricably linked with a change in the notion of time.

Modern science started with the advent of Newtonian mechanics. In a certain sense, Newton 'invented' time in order to formulate his laws. Like space, time in Newtonian physics is absolute, that is, it is externally given and unaffected by any material agency. This reflects the ontological idea that the world evolves *in* time in an objective sense. It is possible to introduce a Newtonian picture of four-dimensional spacetime, which can be foliated into three-dimensional hypersurfaces of Euclidean geometry in an absolute sense. There is an absolute notion of simultaneity: It is objectively clear whether two points in spacetime are simultaneous or not. As pointed out by Ludwig Lange in 1885, a central property of Newtonian spacetime is its affine structure: straight timelike worldlines are distinguished because they describe the free (inertial) motion of objects, cf. Ehlers (1973).

The notions of absolute space and time were criticized early on, by Leibniz, Huygens, and Berkeley, and in the nineteenth century by Ernst Mach, with the argument that they are unobservable. These authors favoured a picture of the world in which all motion is relative. They were, however, not able to construct a viable alternative dynamics. Models for relative motion in mechanics were only developed in the twentieth century (Barbour 1986).

Albert Einstein, in 1905, recognized that the notion of an absolute simultaneity is empirically unfounded. This led him to his special theory of relativity. While its causal structure has changed, spacetime is still absolute in the sense of being non-dynamical: it acts on matter and fields, but is itself not acted upon; it is only an arena for the dynamics.

Spacetime becomes dynamical only with the advent of general relativity in 1915. Its geometry is a manifestation of the gravitational field, and gravity is dynamical. There is now a complicated non-linear interaction between gravity and non-gravitational fields, as encoded in the Einstein field equations. The running of a clock depends on its position in spacetime, and the clock acts back on spacetime due to its mass. This back action is very natural because 'it is contrary to the scientific mode of understanding to postulate a thing that acts, but which cannot be acted upon' (Einstein 1922).

On the other hand, quantum mechanics retains the Newtonian concept of absolute time. The time parameter $t$ in the Schrödinger equation is non-dynamical and not transformed into an operator. The wave function evolves *in* time and gives probability amplitudes and transition rates for physical entities with respect to external time; the total probability is conserved in time.

The situation does not change much in special-relativistic quantum field theory. Quantum fields evolve *on* the externally given, rigid Minkowski spacetime of special relativity. The Standard Model of strong and electroweak interactions is formulated on this background.

This is no longer possible for gravity. Since it entails dynamical degrees of freedom for spacetime, the latter can no longer serve as a background. One thus has to look for a *background-independent* quantum theory. If one sticks to the universality of quantum theory—and there is no evidence to the contrary—gravity and spacetime must be quantized, too, leading to a theory of quantum gravity. It is obvious that this will have drastic consequences for the concept of time. The clash between the external time in quantum theory and the dynamical time in general relativity is one aspect of the *problem of time* in quantum gravity, on which I shall elaborate in section 3.

In the following, I shall first motivate why a theory of quantum gravity is needed and give a brief overview of the existing approaches (section 2). I shall then discuss the problem of time in more detail (section 3). Sections 4 and 5 are devoted to two methods towards its solution: the choice of time before and after quantization. Section 6 will deal with the recovery of the old, 'external' time of quantum theory as an approximate notion in an appropriate semiclassical limit. In section 7, I shall briefly discuss the concept of time in loop quantum gravity and string theory, and section 8 is devoted to the arrow of time in quantum gravity. I shall end with a brief Conclusion in section 9.

I want to end this introduction with a quote by Einstein. In his foreword to the book *Concepts of Space* by Max Jammer, he writes

> It was a hard struggle to gain the concept of independent and absolute space which is indispensible for the theoretical development. And it has not been a smaller effort to overcome this concept later on, a process which probably has not yet come to an end.[1]

---

[1] Es hat schweren Ringens bedurft, um zu dem für die theoretische Entwicklung unentbehrlichen Begriff des selbständigen und absoluten Raumes zu gelangen. Und es hat nicht geringerer Anstrengung bedurft, um diesen Begriff nachträglich wieder zu überwinden—ein Prozeß, der wahrscheinlich noch keineswegs beendet ist.

Although he writes here only about space, the same holds for time. The 'process which probably has not yet come to an end' can perhaps be interpreted as referring to a future fundamental theory such as quantum gravity will be.

The literature on this subject is enormous. I have therefore cited only a few original articles. More details and a guide to the literature can be found, for example, in Butterfield and Isham (1999), Isham (1993), Kiefer (2007), Kuchař (1992), and Rovelli (2004).

## 2. WHY QUANTUM GRAVITY?

What are the main motivations for developing a quantum theory of gravity? Since there are currently no experimental hints, the main reasons are conceptual.

Within general relativity, one can prove singularity theorems which show that the theory is incomplete: under very general conditions, singularities are unavoidable, cf. Hawking and Penrose (1996). Singularities are borders of spacetime beyond which geodesics cannot be extended; the proper time of any observer comes to an end. Such singularities can be rather mild, that is, of a topological nature, but they can also have diverging curvatures and energy densities. In fact, the latter situation seems to be realized in two important physical cases: the Big Bang and black holes. The presence of the cosmic microwave background (CMB) radiation indicates that a Big Bang has happened in the past. Curvature singularities also seem to lurk inside the event horizon of black holes. One thus needs a more comprehensive theory to understand these situations, and the general expectation is that a *quantum* theory of gravity is needed, in analogy to quantum mechanics in which the classical instability of atoms has disappeared. The origin of our universe cannot be described without such a theory, so cosmology remains incomplete without quantum gravity.

Because of its geometric nature, gravity interacts universally with all forms of energy. Since it thus interacts with the quantum fields of the Standard Model, it would seem natural that gravity itself is described by a quantum theory. It is hard to understand how one could construct a unified theory of all interactions in a hybrid classical–quantum manner. In fact, all attempts to do so have failed up to now.

Since gravity is a manifestation of spacetime geometry, quantum gravity should make definite statements about the microscopic behaviour of spacetime. For this reason it has been speculated long ago that the inclusion of gravity can avoid the divergences that plague ordinary quantum field theories. These divergences arise from the highest momenta and thus from the smallest scales. This speculation is well motivated, and can be traced back to the background independence discussed in the Introduction. If the usual divergences have really to do with the smallest scales of the background spacetime, they should disappear together with the background.

A direct experimental test of quantum gravity is hard to perform. This is connected with the smallness of the corresponding length and time scales and the largeness of the mass (energy) scale: combining the gravitational constant, $G$, the speed of light, $c$, and the quantum of action, $\hbar$, into units of length, time, and mass, respectively, one arrives at the famous Planck units,

$$l_P = \sqrt{\frac{\hbar G}{c^3}} \approx 1.62 \times 10^{-33} \text{ cm},$$ (1)

$$t_P = \sqrt{\frac{\hbar G}{c^5}} \approx 5.40 \times 10^{-44} \text{ s},$$ (2)

$$m_P = \sqrt{\frac{\hbar c}{G}} \approx 2.17 \times 10^{-5} \text{ g} \approx 1.22 \times 10^{19} \text{ GeV}/c^2.$$ (3)

To probe, for example, the Planck length with contemporary accelerators, one would have to build a machine of the size of our Milky Way! Direct observations should thus be possible mainly in astrophysics—probing the early universe and the structure of black holes.

Concerning now the attempt to construct a full quantum theory of gravity, the question arises: what are the main approaches? In brief, one can distinguish between

- *Quantum general relativity*: The most straightforward attempt, both conceptually and historically, is the application of 'quantization rules' to classical general relativity. This approach can be divided further into
  - *Covariant approaches*: These are approaches that employ four-dimensional covariance at some stage of the formalism. Examples include perturbation theory, effective field theories, renormalization-group approaches, and path integral methods.
  - *Canonical approaches*: Here one makes use of a Hamiltonian formalism and identifies appropriate canonical variables and conjugate momenta. Examples include quantum geometrodynamics and loop quantum gravity.
- *String theory (M-theory)*: This is the main approach to construct a unifying quantum framework of all interactions. The quantum aspect of the gravitational field only emerges in a certain limit in which the different interactions can be distinguished.
- Other fundamental approaches, such as a direct quantization of topology, or the theory of causal sets.

All these approaches, in their non-perturbative version, are either already formulated in a background-independent way, or have at least the ambition to aim at such a formulation. They thus all face the problem of time, to which I shall now turn in more detail.

## 3. THE PROBLEM OF TIME

As we have seen, time in quantum theory is a non-dynamical quantity. The parameter *t* in the Schrödinger equation,

$$i\hbar \frac{\partial \psi}{\partial t} = \hat{H}\psi, \tag{4}$$

is identical to Newton's absolute time. On the other hand, time in general relativity is dynamical because it is part of spacetime as described by Einstein's equations,

$$R_{\mu\nu} - \frac{1}{2} g_{\mu\nu} R + \Lambda g_{\mu\nu} = \frac{8\pi G}{c^4} T_{\mu\nu}. \tag{5}$$

Spacetime and matter are inextricably linked by these non-linear dynamical equations, and it is conceptually impossible to introduce a non-dynamical background into this framework.

What, then, happens with the concept of time if gravity is quantized? Although the problem of time is present in all approaches, I shall restrict myself to quantum geometrodynamics, because there the discussion can be presented in a most transparent way. Quantum geometrodynamics is the oldest version of canonical quantum gravity; its fundamental variable is the *three-dimensional metric*. Some remarks on loop quantum gravity and string theory are made in section 7 below.

Quantum geometrodynamics is one of the most conservative approaches. Quantization is here performed in a spirit similar to Erwin Schrödinger's original heuristic approach to quantum mechanics. It leads to a wave equation which correctly produces the Einstein equations (5) in the semiclassical limit. To the very least, quantum geometrodynamics should provide a good approximation to any full quantum theory of gravity for not-too-small length scales.

In order to understand the concept of time, let us inspect in more detail the canonical formalism. Starting point is the '3+1 decomposition', that is, the split of spacetime into a foliation of three-dimensional spacelike hypersurfaces. This is schematically shown in Figure 23.1. A necessary requirement for this to work is that the spacetime

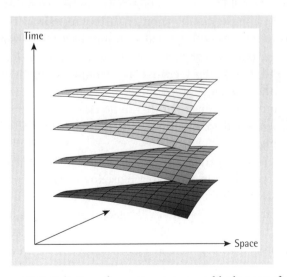

FIGURE 23.1 Foliation of spacetime into spacelike hypersurfaces

manifold, $\mathcal{M}$, be globally hyperbolic, that is, isomorphic to the cartesian product $\mathbb{R} \times \Sigma$, where $\Sigma$ denotes a three-dimensional manifold.

It is illustrative to compare the situation with mechanics. A particle trajectory is the succession of spatial points in time. Similarly, spacetime can be understood as a generalized trajectory in the sense that it is a succession of three-dimensional spaces. In quantum mechanics, trajectories are no longer part of the formalism; instead one has wave functions that only depend on the spatial configurations. In the same manner, upon quantization, only a wave functional depending on the *three*-dimensional spaces remains. There is a difference, though. Whereas quantum mechanics still contains the absolute time $t$, no such time is available in canonical quantum gravity because there is no absolute time in general relativity. One should thus expect that the equations of quantum gravity are fundamentally timeless.

This is indeed what happens. Already the classical theory contains *constraints*, that is, equations which constrain the possibilities to choose initial data on a space-like hypersurface. The presence of such constraints is intimately connected with the presence of the freedom to choose the coordinates in general relativity. In the picture of Figure 23.1, this encapsulates the freedom to choose both the foliation as well as the three spatial coordinates on each space. In fact, the first four of the ten Einstein equations (5) are constraints. They are called *Hamiltonian constraint* (the one responsible for the freedom to choose the foliation in an arbitrary way) and *momentum* or *diffeomorphism constraints* (the three that are responsible for the choice of spatial coordinates). An analogy to these constraints in electrodynamics is Gauss law, $\nabla \mathbf{E} = (4\pi/c)\rho$, whose presence can be traced back to the gauge invariance. Since Gauss' law is devoid of time, it is a constraint on the initial data of the theory. Quite generally, constraints generate redundancy transformations such as gauge or coordinate transformations.

It turns out that the full Hamiltonian of gravity can be written, apart from possible boundary terms, as a linear combination of these constraints. The Hamiltonian is thus itself a constraint. This is connected with the absence of absolute time. Unlike theories with a background, the gravitational Hamiltonian can no longer generate time translations. Instead it generates a change of foliation and spatial coordinate transformations; since these changes are redundancies without observable consequence, the Hamiltonian must be a constraint. It will be clear from these remarks that *any* background-independent theory, not only general relativity, leads to a Hamiltonian which is a constraint.

In a certain sense, the constraints in general relativity already contain all the information of the theory. A theorem states that Einstein's equations are the unique propagation laws consistent with the constraints. To demand the validity of the constraints on each hypersurface necessarily entails the fact that all Einstein's equations must hold on the foliated spacetime. The analogy in electrodynamics states that Maxwell's equations are the unique propagation laws consistent with the Gauss constraint. Constraints and dynamical equations are inextricably mixed.

After quantization, the 'trajectory', that is, the spacetime has vanished. Only the three-dimensional space $\Sigma$ remains. Consequently, only the constraints are left: the remaining six Einstein equations have disappeared. All the information about the quantum theory is therefore contained in the quantum version of the constraints. They thus constitute the central equations of quantum gravity.

## 4. TIME BEFORE QUANTIZATION

The constraints of the classical theory are constraints on the initial data, that is, on the generalized positions and their momenta. As already mentioned, the generalized positions are the components of the three-dimensional metric, $h_{ab}(\mathbf{x})$. The conjugate momenta, $p^{cd}(\mathbf{x})$, are a linear combination of the components of the extrinsic curvature. (The extrinsic curvature is a measure of the embedding of space into spacetime.) How should one transform these constraints into the quantum theory?

A first attempt would be to solve the constraints already on the classical level. What does this mean? It is well known that classical mechanics can be put into a parametrized form, where time $t$ is elevated to a dynamical variable; $t$ as well as all the $q^i$ are then supposed to depend on an arbitrary parameter $\tau$ (cf. Lanczos 1986). It turns out that the resulting formalism is invariant under arbitrary reparametrizations of $\tau$; it is called 'time-reparametrization invariant'. As a consequence of this invariance, a constraint appears. It is of the form

$$p_t + H = 0, \tag{6}$$

where $p_t$ is the momentum conjugate to $t$ (which in this formalism is dynamical), and $H$ is the usual Hamiltonian. Following a general prescription introduced by Dirac, the constraint (6) can be transformed into the quantum theory by turning all variables into operators and interpreting the resulting constraint as a restriction on physically allowed wave functions. Substituting $p_t$ by $(\hbar/i)\partial/\partial t$ and $H$ by $\hat{H}$, one then arrives at the Schrödinger equation (4).

The question now is whether this procedure can also be performed in the much more complicated case of the constraints of general relativity. If it could, one would be able to identify a function of the canonical variables $h_{ab}(\mathbf{x})$ and $p^{cd}(\mathbf{x})$ that could play the role of an appropriate time variable already at the classical level. Note that this would *not* be a time coordinate on spacetime, but a function of the canonical variables which are defined solely on space.

After such an identification, the constraints would then be in a form similar to (6). One could then quantize the constraints in the same way and arrive at a functional version of the Schrödinger equation (4). This would have great advantages: one could extrapolate all the interpretational structure, such as the usual inner product and its conservation in time (unitarity) from quantum mechanics to quantum gravity.

There are, however, a lot of problems with such an attempt. Firstly, a transformation of the constraints into a form similar to (6) is not possible globally, that is, on the whole phase space. Secondly, even if attention is restricted to a local identification of time, there is the 'multiple-choice problem': there are many choices for such a time, and the corresponding quantum theories are generically not unitarily equivalent. Thirdly, the ensuing reduced Hamiltonian depends on this time and has in general a very complicated structure (containing square roots, etc.). In addition, there is the practical task of actually transforming the constraints into a form similar to (6), a task that has been accomplished only in a few relatively simple cases (such as cylindrical gravitational waves and spherically-symmetric black holes). Most authors do not, therefore, prefer this approach but try to identify an appropriate time, if any, after quantization.

## 5. TIME AFTER QUANTIZATION

In this alternative approach one takes the contraints as they appear and tries to transform them directly into quantum constraint equations. As we have seen, the constraints combine into the Hamiltonian of general relativity, which is then again a constraint, $H = 0$. (We shall neglect boundary terms.) The application of Dirac's prescription then leads to the equation

$$\hat{H}\Psi = 0, \tag{7}$$

where $\Psi$ is the quantum gravitational wave functional, which depends on the three-dimensional metric $h_{ab}(\mathbf{x})$. There remain, of course, the usual problems of factor ordering and regularization.

Strictly speaking, (7) are infinitely many equations, one equation at each space point. If non-gravitational fields are present, they will be included into the constraint, and (7) then refers to the full Hamiltonian of gravity and matter. In honour of the work by Bryce DeWitt (1967) and John Wheeler (1968), Equation (7) is called the *Wheeler–DeWitt equation*. It is the central equation of quantum geometrodynamics.

The Wheeler–DeWitt equation is fundamentally timeless (in the sense that a time parameter is absent); all components of the three-dimensional metric are on equal footing. Its solutions are thus static waves. A closer inspection of its kinetic term exhibits, however, a particular and important feature: the kinetic term is of an indefinite nature. More precisely, the Wheeler–DeWitt equation is locally (that is, at each space point) of a hyperbolic structure. Unlike the Schrödinger equation, it has the form of a wave equation similarly, for example, to the Klein–Gordon equation. A wave equation is characterized by the fact that one of the variables comes with the opposite sign in the kinetic term; this variable is usually related to time. The structure of the Wheeler–DeWitt equation thus suggests the presence of an intrinsic timelike variable, in short: *intrinsic time*.

The intrinsic time is constructed from part of the three-dimensional metric $h_{ab}(\mathbf{x})$. More precisely, it is the size (instead of the shape) of the three-dimensional geometry. This becomes evident in models of quantum cosmology, cf. Kiefer and Sandhöfer (2008). There, only very few variables are quantized, such as the scale factor ('radius') of the universe and a homogeneous scalar field (representing matter). Consider, for example, a simple model of a closed Friedmann-Lemaître universe with scale factor $a$, containing a massive scalar field $\phi$. The Wheeler–DeWitt equation then reads (after a suitable redefinition of variables)

$$\left( \frac{\hbar^2}{m_p^2} \frac{\partial^2}{\partial a^2} - \hbar^2 \frac{\partial^2}{\partial \phi^2} - m_p^2 e^{4\alpha} + e^{6\alpha} m^2 \phi^2 \right) \psi(\alpha, \phi) = 0 , \tag{8}$$

where $\alpha \equiv \ln a$; the variable $\alpha$ has the advantage that its range is from $-\infty$ to $+\infty$ instead of 0 to $\infty$, which holds for $a$. One recognizes that the sign of the kinetic $\alpha$-term has the 'wrong' sign compared to the standard matter kinetic term; it is thus the size of the universe which can serve as the intrinsic time. (Additional cosmological and matter variables will all come with the same sign as the $\phi$-term.)

This new concept of time has far-reaching consequences: the classical and the quantum model exhibit two drastically different concepts of determinism, see Figure 23.2.

Let us consider the case of a classically recollapsing universe. In the classical case (left) we have a trajectory in configuration space: although it can be parametrized in many ways, the important point is that it *can* be parametrized by some time parameter. Therefore, upon solving the classical equations of motion, the recollapsing part of the trajectory is the deterministic successor of the expanding part: the model universe expands, reaches a maximum point, and recollapses.

Not so for the quantum model. There is no classical trajectory and no classical time parameter, and one must take the wave equation as it stands. The wave function only distinguishes small $a$ from large $a$, not earlier $t$ from later $t$. There is thus no intrinsic difference between Big Bang and Big Crunch. If one wants to construct a wave packet following the classical trajectory as a narrow tube, one has to impose the presence of two packets as an initial condition at small $a$; if one chose only one packet, one would obtain a wave function which is spread out over configuration space and which does not resemble anything close to a narrow wave packet. (Equation (8) with $m = 0$ directly corresponds to the situation of Figure 23.2.)

Wave packets are of crucial importance when studying the validity of the semiclassical approximation. In quantum cosmology, this issue has to be discussed from the viewpoint of the Wheeler–DeWitt equation (8). If the classical model describes a recollapsing universe, one has to impose in the quantum theory onto the wave function the restriction that it go to zero for $a \to \infty$; with respect to intrinsic time, this corresponds to a 'final condition'. Calculations show that it is then *not* possible to have narrow wave packets all along the classical trajectory: the packet disperses, and the references therein. This is again a consequence of the novel concept of time in quantum gravity.

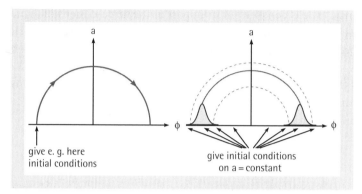

**FIGURE 23.2** The classical and the quantum theory of gravity exhibit drastically different notions of determinism

From Kiefer (2007)

But how do classical properties arise if wave packets necessarily disperse? The answer to this question is *decoherence*—the irreversible emergence of classical behaviour through the unavoidable interaction with an ubiquitous environment (Joos et al. 2003). 'Environment' is a general name for uncontrollable or irrelevant variables. In quantum cosmology, such degrees of freedom can be small density fluctuations or weak gravitational waves. They can act as an 'environment' which becomes quantum entangled with $a$ and $\phi$, causing their classical appearance. This classical appearance holds for most of the evolution of the universe. Possible exceptions are the Planck regime (small $a$) and the turning point of a classically recollapsing quantum universe (section 8).

## 6. Semiclassical time

The Wheeler–DeWitt equation (7) is timeless in the sense that classical spacetime is absent and only space is present. While this holds at the most fundamental level, spacetime and with it the familiar time parameter $t$ must emerge at an appropriate level of approximation. In particular, the functional (field-theoretic) version of the Schrödinger equation (4) must reappear as an effective equation. Together with the $t$, the imaginary unit i must show up: whereas the Wheeler–DeWitt equation is a real equation, allowing for real solutions, the Schrödinger equation is complex, a feature that is of the highest importance for the interpretational structure of quantum mechanics. How can the emergence of both $t$ and i be understood (Kiefer 2007)?

It can be seen from (8) that the Planck mass $m_P$ appears explicitly in the gravitational part of the kinetic term. (This is also true for the full Wheeler–DeWitt equation (7)). Assuming, then, that one is in a regime where the relevant masses and energies are much smaller than the Planck mass, one can make a formal expansion in inverse powers of the Planck mass. This is similar to what is called Born–Oppenheimer expansion in molecular physics. In this way, one arrives at the following approximate solution of (7):

$$\Psi \approx \exp(iS_0[h]/\hbar)\,\psi[h,\phi], \tag{9}$$

where $h$ is an abbreviation for the three-dimensional metric, and $\phi$ stands for non-gravitational fields. In short, the approximation scheme leads to the following results:

- $S_0$ obeys the Hamilton–Jacobi equation for the gravitational field and thereby defines a classical spacetime which is a solution to Einstein's equations. (This order is akin to the recovery of geometrical optics from wave optics via the eikonal equation.)
- $\psi$ obeys an approximate (functional) Schrödinger equation,

$$i\hbar \underbrace{\nabla S_0 \, \nabla \psi}_{\frac{\partial \psi}{\partial t}} \approx H_{\mathrm{m}}\,\psi, \tag{10}$$

where $H_{\mathrm{m}}$ denotes the Hamiltonian for the non-gravitational fields $\phi$. Note that the expression on the left-hand side of (10) is a shorthand notation for an integral over space, in which $\nabla$ stands for functional derivatives with respect to the three-metric. In the case of (8) one has, for example,

$$\frac{\partial}{\partial t} \propto \frac{\partial S_0}{\partial a}\frac{\partial}{\partial a}.$$

Semiclassical time $t$ is thus defined in this limit from the dynamical variables. This bears a resemblance to the notion of ephemeris time in astronomy.

- The next order of the Born–Oppenheimer scheme yields quantum gravitational correction terms proportional to the inverse Planck mass squared, $1/m_{\mathrm{p}}^2$. The presence of such terms may in principle lead to observable effects, for example, in the anisotropy spectrum of the cosmic microwave background radiation.

The Born–Oppenheimer expansion scheme distinguishes a state of the form (9) from its complex conjugate. In fact, in a generic situation both states will decohere from each other, that is, they will become dynamically independent from each other (Joos et al. 2003). This is a type of symmetry breaking in analogy to the occurrence of parity violating states in chiral molecules. It is through this mechanism that the $t$ and the $i$ in the Schrödinger equation emerge.

The recovery of the Schrödinger equation (10) raises an interesting issue. It is well known that the notion of Hilbert space is connected with the conservation of probability (unitarity) and thus with the presence of an external time (with respect to which the probability is conserved). The question then arises whether the concept of a Hilbert space is still required in the *full* theory where no external time is present.

It could then be that this concept makes sense only on the semiclassical level where (10) holds. This question is not yet settled.

This idea of recovering time is also of interest in ordinary quantum mechanics. One can adopt there the idea that the fundamental equation is the stationary Schrödinger equation, not the time-dependent one. Mott (1931), for example, considered a time-independent Schrödinger equation for a total system consisting of an alpha-particle and an atom. If the state of the alpha-particle can be described by a plane wave (corresponding in this case to high velocities), one can make an ansatz similar to (9) and derive a time-dependent Schrödinger equation for the atom alone, for which time is defined by the state of the alpha-particle.

# 7. TIME IN LOOP QUANTUM GRAVITY AND STRING THEORY

So far, we have restricted our discussion to quantum geometrodynamics. I have argued that the situation encountered there should be typical for all theories of quantum gravity, at least for those which start from a classical theory devoid of an absolute background. Let us see how the situation is in two highly discussed approaches: loop quantum gravity and string theory.

Loop quantum gravity is a particular version of canonical quantum gravity (Rovelli 2004). Instead of three-metric and extrinsic curvature, the canonical variables are now holonomies along loops and fluxes of generalized electric fields through the loops, concepts well known from gauge theories. In fact, one of the main merits of loop quantum gravity is the use of concepts akin to Yang–Mills theories. It is thus hoped that this will be helpful in the search for a unified theory. It seems that loop quantum gravity is mathematically better behaved than quantum geometrodynamics.

Since loop quantum gravity is a canonical theory, one arrives again at constraint equations of the form (7). The detailed structure of this equation is different, but its interpretation with regard to the problem of time is similar: there is no time parameter at the fundamental level. An important new feature is the emergence of a *discrete* structure of space: one can rigorously define geometric operators (such as an area operator) on the kinematical level (that is, before solving all constraints), which has a discrete spectrum: there thus exists a smallest quantum of area.

One would expect that this discreteness in space leaves its imprints on the time that emerges on a semiclassical level (section 6). Unfortunately, the semiclassical approximation has not yet been completed on the level of full loop quantum gravity. Some insights can, however, be gained from loop quantum cosmology (Bojowald 2005), that is, the application of loop quantum gravity to cosmology. The Wheeler–DeWitt equation (8) is there replaced by a *difference equation*: only discrete values of the scale factor $a$ are allowed. For large-enough values of $a$, that is, for values sufficiently above the Planck scale, this difference equation becomes indistinguishable from the Wheeler–DeWitt equation. The semiclassical approximation with its recovery of both the time parameter $t$ and the imaginary unit thus proceeds as in section 6. But if viewed from the fundamental perspective of loop quantum cosmology, this semiclassical time parameter inherits a small discrete structure recognizable only at the scale of the Planck time.

String theory, on the other hand, is conceptually different from all of the above approaches. It does not involve a direct quantization of the classical theory of relativity, but instead aims at directly constructing a fundamental quantum theory of all interactions, see, for example, Zwiebach (2004). Quantum general relativity should thus only emerge in an appropriate limit in which the various interactions such as gravity are distinguishable.

String theory starts from the formulation of a string *on* a given background spacetime. In more recent years, one has learnt that also higher-dimensional objects ('branes') have to be taken into account; they, too, propagate on the background. Eventually, spacetime itself should be constructed out of strings and branes. It seems, however, that string theory has not yet rescued itself from a background-independent formulation. But an important step towards such a formulation could be provided by the so-called AdS/CFT-correspondence, cf. Horowitz (2005). This correspondence states that non-perturbative string theory in a background spacetime which is asymptotically anti-de Sitter (AdS) is equivalent to a conformal field theory (CFT) defined in a flat spacetime of one less dimension. In a sense, a theory containing gravity (the AdS sector) is equivalent to a theory without gravity in one less dimension (the CFT sector), cf. Maldacena (2007). In addition to the problem of time, string theory thus seems to lead to a 'problem of space'.

The AdS/CFT-correspondence can be interpreted as a non-perturbative and partly background-independent definition of string theory, since the CFT is defined non-perturbatively, and the background metric enters only through boundary conditions at infinity. Full background independence in the sense of canonical quantum gravity has, however, not yet been achieved. The problem of time in string theory thus still awaits its solution.

# 8. THE DIRECTION OF TIME

Although most of the fundamental physical laws are invariant under time reversal, there are several classes of phenomena in Nature that exhibit an arrow of time, see, for example, Zeh (2007). Because most subsystems in the universe cannot be considered as isolated, these various arrows of time all point in the same direction. The question then arises whether there exists a master arrow of time underlying all these arrows. The tentative answer is yes. Already Ludwig Boltzmann has speculated about a possible foundation of the Second Law of thermodynamics from cosmology:

it is the huge temperature gradient between the hot stars and the cold space which provides the entropy capacity which is necessary for the entropy to increase (instead of being already at its maximum).

This is also the modern point of view, although the arguments are somewhat different. The global expansion of the universe, together with the gravitational collapse of substructures (leading to stars etc.), defines a gravitational arrow that seems to monitor the other arrows. If there is a special initial condition of low entropy in the early universe, statistical arguments can be invoked to demonstrate that the entropy of the universe will increase with increasing size. But where does such an initial condition come from, and how can the entropy of the universe be calculated?

There are several subtle issues connected with these questions. First, one does not yet know a general expression for the entropy of the gravitational field, except for the black-hole entropy, which is given by the Bekenstein–Hawking formula. As for the general case, Roger Penrose has suggested the use of the Weyl tensor as a measure of gravitational entropy. He has also estimated from the Bekenstein–Hawking formula how unlikely our universe in fact is (Penrose 1981). Second, because the very early universe is involved, the problem has to be treated within quantum gravity. But as we have seen, there is no external time in quantum gravity – so what does the notion 'arrow of time' mean in a timeless theory?

The following discussion will again be based on quantum geometrodynamics, that is, on the Wheeler–DeWitt equation. It should be possible to implement an analogous reasoning in loop quantum gravity and string theory.

An important observation is that the Wheeler–DeWitt equation exhibits a fundamental asymmetry with respect to the intrinsic time introduced above. Very schematically, one can write this equation as (with a convenient choice of units)

$$\hat{H}\,\Psi = \left( \frac{\hbar^2}{m_{\mathrm{P}}^2} \frac{\partial^2}{\partial a^2} + \sum_i \left[ -\hbar^2 \frac{\partial^2}{\partial x_i^2} + \underbrace{V_i(a, x_i)}_{\to 0 \text{ for } a \to -\infty} \right] \right) \Psi = 0, \qquad (11)$$

where again $a = \ln a$, and the $\{x_i\}$ denote inhomogeneous degrees of freedom describing perturbations of the Friedmann universe; they can describe weak gravitational waves or density perturbations. The important property of the equation is that the potential becomes small for $a \to -\infty$ (where the classical singularities would occur), but complicated for increasing $a$; the Wheeler–DeWitt equation thus possesses an asymmetry with respect to 'intrinsic time' $a$. One can in particular impose the simple boundary condition

$$\Psi \xrightarrow{a \to -\infty} \psi_0(a) \prod \psi_i(x_i), \qquad (12)$$

which would mean that the degrees of freedom are initially *not* entangled. Defining an entropy as the entanglement entropy between relevant degrees of freedom (such as $a$) and irrelevant degrees of freedom (such as most of the $\{x_i\}$), this entropy vanishes initially but increases with increasing $a$ because entanglement increases due to the presence of the potential. In the semiclassical limit where $t$ is constructed from $a$

(and other degrees of freedom), cf. (10), entropy increases with increasing $t$. This then *defines* the direction of time and would be the origin of the observed irreversibility in the world. The expansion of the universe would then be a tautology. Due to the increasing entanglement, the universe rapidly assumes classical properties for the relevant degrees of freedom due to decoherence.

This process has interesting consequences for a classically recollapsing universe (Kiefer and Zeh 1995). Since Big Bang and Big Crunch correspond to the same region in configuration space ($a \rightarrow -\infty$), an initial condition for $a \rightarrow -\infty$ would encompass both regions, cf. Figure 23.2. This would mean that the above initial condition would always correlate increasing size of the universe with increasing entropy: the arrow of time would formally reverse at the classical turning point. As it turns out, however, a reversal cannot be observed because the universe would enter a quantum phase. Further consequences concern black holes within such a universe because no horizon and no singularity would ever form. It is, of course, not clear whether this situation is actually realized in Nature. This scenario shows, however, that drastic consequences for our understanding of the universe can arise from the direct combination of two well-established theories: general relativity and quantum theory.

## 9. CONCLUSION

Quantum gravity places the concept of time on a new level. In the absence of experimental hints, mathematical and conceptual issues must be chosen as the guides in the search for such a theory. The situation can be compared with Einstein's investigation into the meaning of time in 1905, which led him to develop his special theory of relativity.

We have seen that general relativity does not contain a non-dynamical background spacetime. Upon quantization, spacetime disappears in the same way as a particle trajectory has disappeared in quantum mechanics; only space remains. There is thus no time parameter in quantum gravity. I have discussed this explicitly within the framework of quantum geometrodynamics, but the situation should be similar in all reasonable approaches. String theory, for example, leads to general relativity in an appropriate limit, and its quantization should thus lead to the absence of a time parameter, too.

In quantum geometrodynamics, a sensible notion of intrinsic time can be introduced. In simple models, it can be constructed from the size of the universe—the universe thus provides its own clock. The standard time parameter appears on a semiclassical level. In loop quantum gravity, it should exhibit a discrete structure recognizable for times of the order of the Planck time.

The origin of the arrow of time can in principle be understood from quantum gravity. Intrinsic time enters asymmetrically into the Wheeler–DeWitt equation and thus allows for a natural initial condition that leads to an entropy correlated with the size of the universe. In the semiclassical limit, where a time parameter $t$ appears, this entails the Second Law of thermodynamics. Both the familiar time and its arrow can thus be understood from quantum gravity, which itself is fundamentally timeless.

## REFERENCES

Barbour, J. B. (1986), 'Leibnizian Time, Machian Dynamics, and Quantum Gravity', in *Quantum Concepts in Space and Time* (ed. R. Penrose and C. J. Isham), 236–46. (Oxford University Press, Oxford).

Bojowald, M. (2005), 'Loop Quantum Cosmology', *Living Reviews of Relativity*, **8**, 11. URL (cited on June 15, 2008): http://www.livingreviews.org/lrr-2005-11.

Butterfield, J. and Isham, C. J. (1999), 'On the Emergence of Time in Quantum Gravity', in *The Arguments of Time* (ed. J. Butterfield), 111–68, (Oxford University Press, Oxford).

DeWitt, B. S. (1967), 'Quantum Theory of Gravity. I: The Canonical Theory', *Physical Review*, **160**, 1113–48.

Ehlers, J. (1973), 'The Nature and Structure of Spacetime', in *The Physicist's Conception of Nature* (ed. J. Mehra), 71–91. (D. Reidel, Dordrecht).

Einstein, A. (1922), *The Meaning of Relativity* (Princeton University Press, Princeton).

Hawking, S. W. and Penrose, R. (1996), *The Nature of Space and Time*, (Princeton University Press, Princeton).

Horowitz, G. T. (2005), 'Spacetime in String Theory', *New Journal of Physics*, 7, 201 [13 pages].

Isham, C. J. (1993), 'Canonical Quantum Gravity and the Problem of Time', in *Integrable Systems, Quantum Groups, and Quantum Field Theory* (ed. L. A. Ibort and M. A. Rodríguez), 157–287. (Kluwer, Dordrecht).

Joos, E., Zeh, H. D., Kiefer, C., Giulini, D., Kupsch, J., and Stamatescu, I.-O. (2003), *Decoherence and the Appearance of a Classical World in Quantum Theory*, 2nd edn. (Springer, Berlin).

Kiefer, C. (2007), *Quantum Gravity*, 2nd edn. (Oxford University Press, Oxford).

—— and Sandhöfer, B. (2008), 'Quantum Cosmology'. Available online as arXiv: 0804.0672. To appear in *Beyond the Big Bang* (ed. R. Vaas) (Springer, Berlin).

——, and Zeh, H. D. (1995), 'Arrow of Time in a Recollapsing Quantum Universe', *Physical Review D*, **51**, 4145–53.

Kuchař, K. V. (1992), 'Time and Interpretations of Quantum Gravity', in *Proc. 4th Canadian Conf. General Relativity Relativistic Astrophysics* (ed. G. Kunstatter, D. Vincent, and J. Williams), 211–314. World Scientific, Singapore.

Lanczos, C. (1986), *The Variational Principles of Mechanics*, 4th edn. (Dover Publications, New York).

Maldacena, J. (2007), 'The Illusion of Gravity', *Scientific American Reports*, April 2007, 75–81.

Mott, N. F. (1931), 'On the Theory of Excitation by Collision with Heavy Particles'. *Proceedings of the Cambridge Philosophical Society*, **27**, 553–560.

Penrose, R. (1981), 'Time-Asymmetry and Quantum Gravity', in *Quantum Gravity*, Vol. 2 (ed. C. J. Isham, R. Penrose, and D. W. Sciama), 242–72. (Clarendon Press, Oxford).

Rovelli, C. (2004), *Quantum Gravity*. (Cambridge University Press, Cambridge).

Wheeler, J. A. (1968), 'Superspace and the Nature of Quantum Geometrodynamics', in *Battelle Rencontres* (ed. C. M. DeWitt and J. A. Wheeler), 242–307, (Benjamin, New York).

Zeh, H. D. (2007), *The Physical Basis of the Direction of Time*, 5th edn. (Springer, Berlin).

Zwiebach, B. (2004), *A First Course in String Theory*, (Cambridge University Press, Cambridge).

# INDEX

electrodynamics:
  Wheeler-Feynman
      action-at-a-distance 494, 496–7,
      517, 520
  classical 487, 490–1, 495–7, 517–8, 520,
      522
  quantum 487, 497, 519–20, 522–4
endurance 17–20, 23–6, 32, 34
endurantism 14–5, 30, *see also* endurance
energy 193, 210–11, 218, 327, 578, 600–1,
      605, 611, 640–1, 648–50, 658–61, 665
  condition 607, 609, 612, 617, 619, 625
    dominant 612, 623
    null 615–16
    point-wise 616, 618–9
    in QFT 618
    strong 612
    weak 612
  dark 509, 613
  densities 616, 619, 665
    negative 612, 618, 619
  electromagnetic 512
  kinetic 319, 322, 496
  -momentum 496, 504, 518, 612, *see also*
      tensor, energy-momentum
  principle 496
entanglement 676–7
entities:
  abstract 124, 133, 134, 136
  concrete 136
  ersatz 130, 134
  event-like 174
  evidence-transcendent 139
  exactly co-located 27
  external 258
  fictional 130
  instantaneous 26
  macroscopic 261
  non-present 125–6, 134
  non-propositional 128
  sub-personal 364
  substantival 196
  theoretical 200
entropy 284, 314–15, 318–26, 329, 330–5,
      339–40, 489, 619, 676–7
  gradient 306–9, 489
  gravitational 676

initial low 295–6, 322, 324–9, 339–41, 489,
      502, 676
  method 285
  thermodynamic 321, 327
epistemicism-presentism 139, 141
ergodicity 331–2
ergodic theory 331–2
eternalism 16–17, 70, 131, 134–5, 141, 171,
      176, 383
Euclidean space 177, 182, 184–5, 537, 540,
      544, 635
exdurance 17–20, 23, 26, 29, 32, 34
exdurantism 14–5, *see also* exdurance

fatalism 41, 44, 46, 56, 60, 62, 63
field:
  advanced 486, 492, 493, 502, 503, 518
  charge-current 487
  half-retarded-half advanced 498
  Liénard-Wiechert 493
  retarded 486, 490, 492, 493, 498, 502,
      503, 518
  vector 488–9, 590, 638, 640
fixity 42, 61
  of interests 369
  of our own perspective 307
  of the past 81–2, 88, 247–8, 481
flow
  energy-momentum 612
  experiential 408
  kinetic 426
  in time 602
  phenomenal 407
  of space 602
  of time 56–7, 74, 88, 146, 248, 276, 283,
      300, 303–6, 481, 529
    objective 281, 301
  time's, *see* flow, of time
foreknowledge 62, 63
four-dimensionalism 14, 16 n.10, *see also*
      perdurantism
freedom 57–9, 430, 624, 668
  degree of 321, 431, 664, 672, 676–7
  gauge 491, 601
  lack of 57
  phenomenology of 63
  real 53